EARTH

EARTH

THE DEFINITIVE VISUAL GUIDE

Editors-in-Chief
James F. Luhr and Jeffrey E. Post

REVISED EDITION

Senior Editor Miezan Van Zyl
Project Editor Glenda R Fernandes
Senior Picture Researcher Sumedha Chopra
Assistant Picture Researcher Nunhoih Guite
Senior Jackets Coordinator Priyanka Sharma Saddi
Production Manager Pankaj Sharma
Pre-Production Manager Balwant Singh
Production Editor Jaypal Chauhan
Managing Editors Angeles Gavira Guerrero, Rohan Sinha
Associate Publishing Director Liz Wheeler
Publishing Director Jonathan Metcalf
Design Director Philip Ormerod

Senior Art Editors Ina Stradins, Mahua Sharma
Project Art Editor Rupanki Arora Kaushik
Art Editor Sanjay Chauhan
Assistant Art Editor Tanya Varkey
Jacket Designer Juhi Sheth
Jacket Design Development Manager Sophia MTT
DTP Designers Nand Kishor Acharya, Deepak Mittal
DTP Coordinator Vishal Bhatia
Production Controller Meskerem Berhane
Deputy Managing Editor Vaibhav Rastogi
Managing Art Editor Michael Duffy
Creative Head Malavika Talukder
Art Director Karen Self

Smithsonian Project Co-ordinators
Avery Naughton, Paige Towler

PREVIOUS EDITIONS

Senior Editor Peter Frances
Project Editors Sophie Bevan, Kim Dennis-Bryan, Sean O'connor, Gill Pitts, David Summers, Miezan van Zyl
Editors Dharini Ganesh, Georgina Garner, Ben Hoare, Giles Sparrow, Nikky Twyman
Indexer John Noble

Senior Art Editor Caroline Buckingham
Art Editors Clare Joyce, David Ball, Kirsten Cashman, Kenny Grant, Thomas Keenes, Helen Taylor, Duncan Turner
Designers Paul Drislane, Lee Riches, Vanessa Thompson
DTP Designers Rajen Shah, Simon Longstaff, Martin Nilsson

Pre-Production Producer Adam Stoneham
Picture Researchers Louise Thomas, Aditya Katyal
Illustrators Tom Coulson (Encompass Graphics Limited), Arran Lewis (ecorche), Peter Bull (Peter Bull Art Studio) John Plumer, Planetary Visions Limited
Cartographers David Roberts, Iorwerth Watkins
Production Controllers Elizabeth Cherry, Sarah Dodd
Producer Mary Slater
Jacket Designer Natasha Rees
Jacket Editor Manisha Majithia
Jacket Design Development Manager Sophia MTT
Managing Art Editor Michelle Baxter
Managing Editor Angeles Gavira Guerrero
Publisher Sarah Larter
Art Directors Philip Ormerod, Bryn Walls
Associate Publishing Director Liz Wheeler
Publishing Director Jonathan Metcalf

COOLING BROWN
(Rocks, Minerals, Environmental Issues)
Art Editor Philip Lord
Project Editor Joanna Chisholm
Creative Director Arthur Brown
Managing Editor Amanda Lebentz

Smithsonian Project Co-ordinator
Kealy Wilson

This edition published in 2024
First published in Great Britain in 2003
Published as *Illustrated Encyclopedia of Earth* in 2011 by
Dorling Kindersley Limited
DK, One Embassy Gardens, 8 Viaduct Gardens,
London, SW11 7BW

The authorised representative in the EEA is
Dorling Kindersley Verlag GmbH. Arnulfstr. 124,
80636 Munich, Germany

Copyright © 2003, 2011, 2013, 2024 Dorling Kindersley Limited
A Penguin Random House Company
Satellite Imagemap copyright © 2003 Planetary Visions Limited
10 9 8 7 6 5 4 3 2
003-325006-May/2024

A CIP catalogue record for this
book is available from the British Library.
ISBN: 978-0-2415-1561-7

Colour reproduction by Colourscan, Singapore
Printed and bound in UAE

www.dk.com

MIX
Paper | Supporting
responsible forestry
FSC™ C018179

This book was made with Forest
Stewardship Council™ certified
paper – one small step in DK's
commitment to a sustainable future.
Learn more at
www.dk.com/uk/information/sustainability

CONTENTS

ABOUT THIS BOOK 6

PLANET EARTH

THE HISTORY OF THE EARTH   **10**
THE EARTH'S PAST   12

THE ANATOMY OF THE EARTH   **32**
THE EARTH'S STRUCTURE   34
THE CORE   36
THE MANTLE   38
THE CRUST   40
MINERALS   42
ROCKS   62

THE CHANGING EARTH   **84**
TECTONIC PLATES   86
PLATE BOUNDARIES   88
WEATHERING   90
EROSION   92
DEPOSITION   94
MASS MOVEMENT   96
METEORITE IMPACTS   100
WATER   106
LIFE   112

LAND

MOUNTAINS AND VOLCANOES	**120**
FAULT SYSTEMS	122
MOUNTAINS	136
VOLCANOES	154
IGNEOUS INTRUSIONS	178
HOT SPRINGS AND GEYSERS	185
RIVERS AND LAKES	**196**
RIVERS	198
LAKES	224
UNDERGROUND RIVERS AND CAVES	240
GLACIERS	**250**
DESERTS	**276**
FORESTS	**298**
WETLANDS	**318**
GRASSLANDS AND TUNDRA	**328**
GRASSLANDS	330
TUNDRA	338
AGRICULTURAL AREAS	**340**
URBAN AREAS	**352**
INDUSTRIAL AREAS	**372**

OCEAN

OCEANS AND SEAS	**382**
OCEAN WATER	384
OCEAN TECTONICS	386
SHELVES AND PLAINS	392
CIRCULATION AND CURRENTS	394
REEFS	396
POLAR OCEANS	398
OCEANS OF THE WORLD	400
COASTS	**430**
TIDES AND WAVES	432
COASTS AND SEA-LEVEL	434
EROSIONAL AND DEPOSITIONAL COASTLINES	436

ATMOSPHERE

CLIMATE	**442**
STRUCTURE OF THE ATMOSPHERE	444
ENERGY IN THE ATMOSPHERE	446
ATMOSPHERIC CIRCULATION	450
CLIMATE CHANGE	454
CLIMATE REGIONS	460
WEATHER	**464**
AIR MASSES AND WEATHER SYSTEMS	466
PRECIPITATION AND CLOUDS	472
WIND	482

TECTONIC EARTH

THE EARTH'S PLATES	**488**
NORTH AMERICAN	490
SOUTH AMERICAN	492
EURASIAN	494
AFRICAN	496
AUSTRALIAN	498
PACIFIC	500
ANTARCTIC	502
GLOSSARY	**504**
INDEX	**510**
ACKNOWLEDGMENTS	**526**

ABOUT THIS BOOK

Earth is divided into five main sections. The first section, PLANET EARTH, is an introduction to the Earth as a whole. The next three sections are about the planet's main environments – the LAND, OCEAN, and ATMOSPHERE. Within these sections, the Earth's features are divided into categories, such as Rivers and Lakes. Introductory pages describe typical features found in each category and explain how they are formed. On the succeeding pages, major features in that category are profiled individually. Thematic panels (see right) appear through-out. The last section, TECTONIC EARTH, is a three-dimensional atlas.

Planet Earth

This overview of our planet is divided into three parts. The History of the Earth is a narrative account of how our planet and the Universe as a whole came to be the way they are now. The Anatomy of the Earth looks at the Earth's structure and materials, including visual profiles of major rocks and minerals, while The Changing Earth explains those processes that operate on a global scale.

text explains key concepts to support later sections

three-dimensional artworks reveal structural detail

OVERVIEW OF A PLANET
These pages include a history of the Earth (above) and an account of its anatomy (right).

Tectonic Earth

The Earth's rocky outer crust is divided into large sections called tectonic plates. It is the movement of these plates, more than any other process, that accounts for the changing shape of the planet's surface. This section contains profiles of the Earth's seven major plates, plus the adjoining minor ones. Supporting maps show human population density and political boundaries.

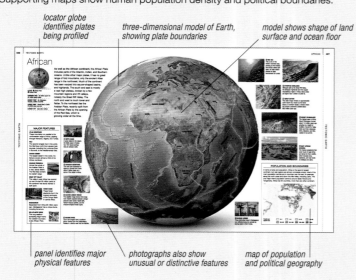

locator globe identifies plates being profiled

three-dimensional model of Earth, showing plate boundaries

model shows shape of land surface and ocean floor

panel identifies major physical features

photographs also show unusual or distinctive features

map of population and political geography

Land, Ocean, and Atmosphere

These three sections focus on specific features – on the Earth's landmasses, in the five great oceans, and in the layer of gases separating the Earth from space. Within each section, features are grouped by type into smaller sections, and then explained in general and individually profiled. Most profiles describe actual features (such as particular volcanoes), while others characterize general kinds of phenomena (such as types of clouds). Profiles of actual features are presented according to the continent on which they are found, in the following order: the Arctic, North America, Central America, South America, Europe, Africa, north and west Asia, south and east Asia, Australasia, and Antarctica. Within continents, features are arranged approximately in north–south order.

◄ MAIN SECTIONS
The three main domains of the Earth's exterior are described in separate sections.

GROUP INTRODUCTIONS ►
The main sections are subdivided so that particular types of feature are grouped together. Some sections are divided again to create a further level of detail.

colour-coded panel contains references to other relevant sections

EXPLANATORY PAGES ►
Throughout the book, these pages contain explanations of forces and processes, as well as descriptions of typical features. Most of these pages are followed by profiles of actual features.

compass direction indicates position within continent (locations within Australasia are identified by name of country)

name of continent on which feature is found

NORTH AMERICA *north*

name of feature being profiled

North American boreal forest

location of feature identified on world map by red dot (or red rectangle for larger features)

LOCATION Extending from central Alaska, USA, in the west to central Labrador, Canada, in the east

description of detailed location, including name of country (or countries) in which feature is found

detailed map shows feature in regional context

Anchorage

ATLANTIC OCEAN

Hudson Bay

purple shading shows extent of larger features

PACIFIC OCEAN

Vancouver

Montréal

table of summary information (categories vary between sections)

TYPE Boreal forest

AREA 6 million square km (2.3 million square miles)

scale bar divided into blocks, each representing 100km (60 miles)

LOCATION AND DATA ▲

Thematic Panels

GLACIER LABORATORY

Tunnels have been drilled in the bedrock under the Engabreen Glacier as part of a hydroelectric power scheme. At the side of these tunnels, a laboratory has been excavated under the glacier, where the ice is about 200m (660ft) thick. This has allowed researchers to install instruments that continuously record the stress on the bedrock and the speed, temperature, and pressures within the ice as it slips across the rock. The results are helping scientists develop mathematical models for predicting glacier movement.

◀ SCIENCE
These features reveal how scientists have learned about different aspects of the workings of our planet.

JOSEPH BANKS

Englishman Joseph Banks (1743–1820) was one of the great botanical explorers and collectors. Aged 25, he joined James Cook's expedition to the South Pacific on the *Endeavour*. In 1770, they landed in eastern Australia, where Banks amassed a vast collection of botanical specimens then unknown to Europeans. Appropriately, Captain Cook named their landing place Botany Bay.

▲ BIOGRAPHY
Notable earth scientists, explorers, and others are profiled in this type of feature.

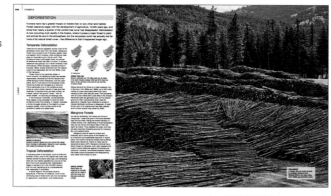

RAINFOREST FIRES

Fires are a major threat, even to moist rainforest, and Indonesia has suffered badly, both from natural fires and from those started deliberately as an aid to logging operations or for clearing trees to make way for alternative land-use. Once a forest fire has taken hold, particularly in dry conditions, such as those of the droughts in 1997 and 2003, the proximity of the trees allows it to spread rapidly, destroying not only the trees themselves but virtually all the associated wildlife.

◀ ▲ ENVIRONMENT
The ways in which humans are changing the Earth – and how the Earth, in turn, affects humans – are described in double-page features (above) and smaller panels (left) throughout the book. Each double-page article takes a global view of a particular subject.

thematic panels (see above) on explanatory pages describe general trends or issues; panels in feature profiles cover localized effects

in most sections, a world map shows global distribution of features being described

introductory panel defines terms and measurements given in the feature profiles that follow

▼ FEATURE PROFILES
All profiles contain a text description and a locator or distribution map. Most profiles are also illustrated with a colour photograph.

profiles are arranged on page in geographical order

CONTRIBUTORS AND CONSULTANTS

Earth is the product of a collaboration between Dorling Kindersley, several contributors (listed below), and staff members of the Smithsonian (listed at the bottom of this panel).

CONTRIBUTORS

Michael Allaby Atmosphere
David Burnie Agricultural Areas, Environment features
Robert Dinwiddie The History of the Earth, Glaciers, Ocean
John Farndon The Changing Earth, Fault Systems, Urban Areas, Industrial Areas, Tectonic Atlas
Douglas Palmer The History of the Earth, The Anatomy of the Earth, The Changing Earth, Mountains, Volcanoes, Hot Springs and Geysers
Clint Twist Rivers, Lakes, Underground Rivers and Caves, Urban Areas, Industrial Areas, Atmosphere
Martin Walters The Changing Earth, Deserts, Forests, Wetlands, Grasslands and Tundra
Tony Waltham Igneous Intrusions, Underground Rivers and Caves
Richard Beatty Glossary

THIRD EDITION
This edition was prepared with the help of the following people:

Robert Dinwiddie
John Farndon
David Holmes
Douglas Palmer
Dorrik Stow

SMITHSONIAN CONSULTANTS

ORIGINAL EDITION
James F. Luhr Editor-in-Chief
Bruce B. Collette Ocean
Douglas H. Erwin The History of the Earth
Richard S. Fiske Igneous Intrusions, Hot Springs and Geysers
Bevan M. French Meteorite Impacts
Andrew K. Johnston Urban Areas, Industrial Areas
Ian G. MacIntyre Ocean
Timothy McCoy The History of the Earth, Rocks, Meteorite Impacts
J. Patrick Megonigal Water, Rivers and Lakes, Wetlands, Atmosphere
William G. Melson Ocean
Jeffrey E. Post Minerals
Richard Potts The History of the Earth
Stanwyn G. Shetler Life, Deserts, Forests, Grasslands and Tundra, Agricultural Areas
Lee Siebert Volcanoes, Mass Movement
Timothy Rose Underground Rivers and Caves
Tom Simkin Tectonic Plates, Plate Boundaries, Volcanoes
Sorena S. Sorensen Rocks, Mountains, Glaciers
Edward P. Vicenzi Mountains
Scott L. Wing The History of the Earth

SECOND EDITION
This edition was prepared with the help of the following staff of the National Museum of National History:
Dr. Jeffrey E. Post, Curator-in-Charge, National Gem Collection, **Editor-in-Chief**
Dr. Benjamin J. Andrews, Research Geologist, Department of Mineral Sciences
Dr Doug Erwin, Curator for Paleozoic Invertebrates, Department of Paleobiology
Brian T. Huber, Chair and Curator of Foraminifera, Department of Paleobiology
Gene Hunt, Curator, Department of Paleobiology
Dr. Briana Pobiner, Research Scientist and Educator, Human Origins Program
Nicholas D. Pyenson, Ph.D., Curator of Fossil Marine Mammals, Department of Paleobiology
Hans-Dieter Sues, Curator of Vertebrate Paleontology

PLANET EARTH

ANCIENT LIFE-FORMS
Some of the first organisms to inhabit
the Earth can still be seen. Over 3 billion
years ago, bacteria in shallow seas began
to form mounds called stromatolites.
Such structures are still forming at
Shark Bay, Australia.

THE HISTORY OF THE EARTH

Over the last 200 years scientists have carefully studied the Earth's rocks and recovered the remains of past animals and plants in order to piece together the history of our planet. The Earth came into existence about 4,560 million (4.56 billion) years ago, as a hot, rocky body in our Solar System. Primitive life first appeared in the oceans about 3,700 million years ago and has been spreading and diversifying ever since, but its evolutionary path has been far from smooth. The Earth's environments have always been subject to change, as volcanic activity, meteorite impacts, and climate changes have produced often life-threatening and sometimes catastrophic effects. The development of our planet has turned out to be remarkably eventful – and, from the evidence of history, it will continue to be so.

The Earth's Past

Rocks	62–83
Tectonic plates	86–87
Life	112–15
Tectonic Earth	486–503

Geologists use information from many sources to help them reconstruct the Earth's long history within the even longer evolution of the Universe. Valuable information about the Earth's early history can be gathered from extra-terrestrial sources, such as meteorites. Analysis of rocks, minerals, and fossils found at the surface tells us directly about the crust and provides some clues about the interior. However, new techniques and analyses are revealing a more detailed view of the Earth's interior, its evolution, and its interaction with the crust. They also provide information about the interaction of these geological processes and events with the atmosphere and the evolution of life.

The Rock Record

Natural processes of deposition lay down successive layers of rock and mineral material as sedimentary strata (or layers), with relatively young strata lying on top of older ones. Geologists have matched (or correlated) strata from around the world using distinctive fossils and rock types to produce what is known as a stratigraphical column, stretching over the Earth's entire history. However, the process is complicated by many breaks in the record through plate movement and lack of deposition.

ROCK LAYERS
The angular relationship of strata at Siccar Point in Scotland represents a break in the record.

Geological Time

The Universe formed about 13–14,000 million years ago and, according to radiometric dating (see panel, opposite), the history of the Earth spans the last 4,560 million years. To make sense of the immensity of the Earth's history, it has been subdivided by geologists into a hierarchical scheme. The largest divisions are eons, many hundreds of millions of years long, including the Phanerozoic Eon, which extends from 539 million years ago to the present day. Within the Phanerozoic, there are three eras, based on the history of life. Eras are divided into smaller segments called periods, which in turn are divided into epochs. Decisions about where boundaries between the main periods of time should be are ongoing.

Evidence from fossils plays an important part in positioning these boundaries. However, international correlation between rock strata can be very difficult because the original sediments may have been deposited in different environments and climatic conditions and therefore contain different fossils.

AFRICAN SHIELD
Diamond-rich sands of the Namibian diamond field cover the Precambrian shield of southern Africa, one of the world's most ancient landmasses.

THE DIVISION OF GEOLOGIC TIME
Just 250 years ago, the Earth was generally thought to be a few thousand years old. It was not until the 1950s that a near-accurate age of 4,550 million years was determined. It took another decade to establish that the Universe is 13,000–14,000 million years old.

GRAND CANYON
Rocks revealed in the Grand Canyon, USA, were laid down over a period of almost 1,500 million years. The oldest are 1,700 million years old.

The Big Bang   Milky Way forms

first galaxies (200–400 mya after Big Bang)   13,000 million years ago (mya)   12,000 mya

9,000 mya   10,000 mya   11,000 mya

8,000 mya   7,000 mya

5,000 mya   6,000 mya

4,500 mya   4,000 mya   3,700 mya   3,000 mya
formation of the Earth   first evidence of life   first landmasses

Fossils

Fossils are the remains of past life that have been entombed by geological processes in rock strata. They range from pollen grains to the skeletons of giant dinosaurs or whales. Life can be preserved in many ways, and fossils vary from vague traces of past activity (such as burrows or footprints) or body chemistry (biomolecules) to encapsulated bodies (such as insects caught in amber). Most processes of preservation involve considerable loss of information about organisms. Nevertheless, the fossil record shows that life began in the seas approximately 3,700 million years ago and that from about 575 million years ago life diversified as it colonized, in succession, fresh water, land, and air. However, the record is highly biased towards marine organisms with hard parts, such as shells. By studying the processes of burial and fossilization, scientists have been able to search out rarer examples of fossils in which soft parts have been preserved. For example, the beautifully preserved fossils from the Cambrian Period found in the Burgess Shale, western Canada (see right), give better insights into the total diversity of past life and its biology.

PLANT FOSSIL
Rosettes of late Carboniferous horsetail (calamite) leaves are preserved here in rock.

BURGESS SHALE
High in the Canadian Rocky Mountains, the World Heritage Site of the Burgess Shale has yielded a remarkable diversity of late-Cambrian marine fossils.

PETRIFIED TREE STUMPS
Silica replacement of the woody tissues of these Triassic trees from the Chinle formation in Arizona, USA, has made them resistant to weathering and erosion.

HOW FOSSILS FORM

Fossilization is a lengthy and complex process. It requires the rapid covering of an organism's remains by sediment, followed by the deep burial, compression, and chemical changes involved in rock formation. Later uplift and erosion – sometimes combined with excavation – re-expose the fossil at the surface.

1. ANIMAL DIES

accumulating sediment | decomposing body

2. SKELETON BURIED

rapidly buried remains

3. ROCK FORMS
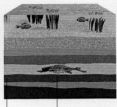
compressed rock | skeleton replaced by minerals

4. FOSSIL REVEALED

erosion of rock and excavation
evidence of hard body parts

RADIOMETRIC DATING

Until the advent of radiometric dating at the beginning of the 20th century, there was no reliable method of dating minerals, rocks, and fossils. The discovery of radioisotopes of some elements (such as uranium, carbon, and potassium) and their decay products that occur in minerals, rocks, and some ancient organic material, together with the measurement of their rates of decay, make it possible to calculate the age of these materials. The limits of radiometric dating depend on the element being used – carbon dating (used for most fossils) cannot provide dates for material that is more than 50,000 years old.

KEY TO TIMELINE OF EARTH HISTORY

The pages that follow contain a visual summary of the history of the Earth, beginning with the formation of the Universe. Information is summarized in the following bands.

EON
ERA
PERIOD
EPOCH

Divisions of geological time. (Decisions on boundaries are ongoing.)

Climate — Average temperatures and composition of the atmosphere. The intensity of the colour in this band indicates temperature trends

Life — Events and trends in the evolution of flora and fauna

Geologic Events — Includes maps indicating continental movements

SYMBOL KEY
◆ Event occurs at this time
Trend occurs over this period

Event occurs during this period
◆------------|------------◆

MAP KEY
N. America
Siberia/N. Asia
Baltica/N. Europe
Gondwana
Ice-caps

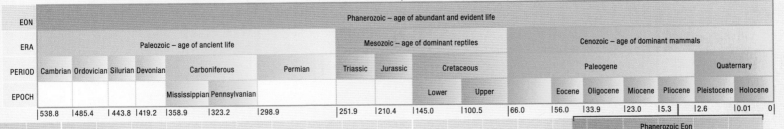

first birds | first primates | early human ancestors | Homo sapiens

EON	Phanerozoic – age of abundant and evident life															
ERA	Paleozoic – age of ancient life					Mesozoic – age of dominant reptiles				Cenozoic – age of dominant mammals						
PERIOD	Cambrian	Ordovician	Silurian	Devonian	Carboniferous	Permian	Triassic	Jurassic	Cretaceous		Paleogene					Quaternary	
EPOCH					Mississippian	Pennsylvanian				Lower	Upper		Eocene	Oligocene	Miocene	Pliocene	Pleistocene	Holocene

538.8 | 485.4 | 443.8 | 419.2 | 358.9 | 323.2 | 298.9 | 251.9 | 210.4 | 145.0 | 100.5 | 66.0 | 56.0 | 33.9 | 23.0 | 5.3 | 2.6 | 0.01 | 0

Phanerozoic Eon

2,000 mya
first multi-celled organisms

1,000 mya

541.0 mya
first land plants

first land vertebrates | first dinosaurs

Present day

PLANET EARTH

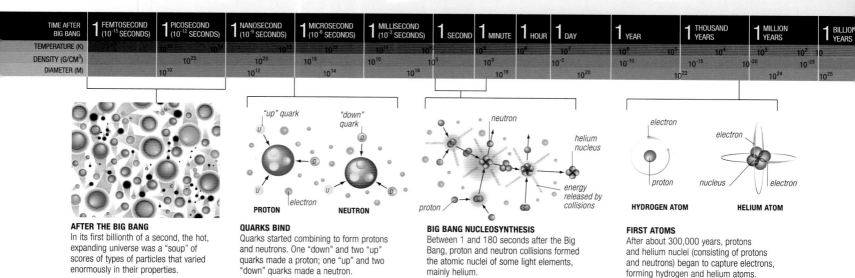

TIME AFTER BIG BANG	1 FEMTOSECOND (10^{-15} SECONDS)	1 PICOSECOND (10^{-12} SECONDS)	1 NANOSECOND (10^{-9} SECONDS)	1 MICROSECOND (10^{-6} SECONDS)	1 MILLISECOND (10^{-3} SECONDS)	1 SECOND	1 MINUTE	1 HOUR	1 DAY	1 YEAR	1 THOUSAND YEARS	1 MILLION YEARS	1 BILLION YEARS
TEMPERATURE (K)	10^{15} 10^{14}		10^{13}	10^{12}	10^{11} 10^{10}	10^{9}	10^{8}	10^{7}		10^{6} 10^{5}	10^{4}	10^{3} 10^{2} 10	
DENSITY (G/CM³)	10^{25}		10^{20}	10^{15}	10^{10}	10^{5}	10^{0}	10^{-5}		10^{-10}	10^{-15}	10^{-20} 10^{-25}	
DIAMETER (M)	10^{10}		10^{12}	10^{14}	10^{16}		10^{18}	10^{20}			10^{22}	10^{24}	10^{25}

AFTER THE BIG BANG
In its first billionth of a second, the hot, expanding universe was a "soup" of scores of types of particles that varied enormously in their properties.

QUARKS BIND
Quarks started combining to form protons and neutrons. One "down" and two "up" quarks made a proton; one "up" and two "down" quarks made a neutron.

"up" quark U
"down" quark D
D electron U D
PROTON **NEUTRON**

BIG BANG NUCLEOSYNTHESIS
Between 1 and 180 seconds after the Big Bang, proton and neutron collisions formed the atomic nuclei of some light elements, mainly helium.

neutron
helium nucleus
energy released by collisions
proton

FIRST ATOMS
After about 300,000 years, protons and helium nuclei (consisting of protons and neutrons) began to capture electrons, forming hydrogen and helium atoms.

electron
proton
HYDROGEN ATOM

electron
nucleus
electron
HELIUM ATOM

Before the Earth

To understand the origins of the Earth, it is necessary to go back to the origin of the Universe itself, some 13,000–14,000 million years ago in the event described by the Big Bang theory, when matter, time, energy, and space all came into existence. The early Universe was small, hot, and dense. Ever since, it has been expanding and cooling. Within a nanosecond (a billionth of a second), it was hundreds of millions of kilometres in diameter with a temperature of tens of trillions Kelvin (degrees above -273°C or absolute zero).

BUILDING BLOCKS

At this stage, the Universe was a seething "soup" of particles created out of energy, together with vast numbers of photons (little packets of radiant energy). Among the most numerous particles were electrons

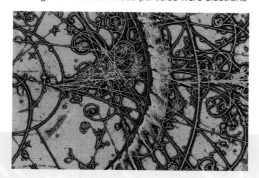

PARTICLE TRACKS
Today, scientists try to simulate what happened in the Big Bang by using particle accelerators to smash subatomic particles together. The resulting tracks can then be studied.

and quarks, but there were also many other kinds that no longer exist today. Although it contained plenty of electrons, the early Universe held neither of the other main building blocks of atoms: protons and neutrons. However, after just one microsecond (a millionth of a second), it had cooled sufficiently for vast quantities of protons and neutrons to form as two different types of quark combined.

Most of the protons were destined to become the nuclei of hydrogen atoms, but from about one second after the Big Bang, collisions between protons and neutrons started to form the nuclei of some other light elements – helium and tiny amounts of lithium and beryllium. This process, termed Big Bang nucleosynthesis, was completed in three minutes and formed the nuclei of 98 per cent of the helium atoms present in the Universe today. It also mopped up all the neutrons.

THE FIRST ATOMS

For the next few hundred thousand years, the Universe continued to expand and cool, but it was still too energetic for atoms to form. If electrons and

atomic nuclei came together momentarily, they were quickly split apart by photons, which were themselves trapped in a process of continual collision with the particles. Eventually, after 300,000 years, when the temperature had dropped to about 3,000 Kelvin, the protons and other atomic nuclei started capturing electrons permanently, forming the first atoms, which were primarily hydrogen and helium. At the same time, the photons were released and streamed freely in all directions. At this stage, the Universe became transparent, as the earlier "fog" of particles and energy cleared.

BACKGROUND EVIDENCE

A crucial piece of evidence supporting the Big Bang theory is the existence of Cosmic Microwave Background Radiation (CMBR). This is a faint heat radiation that emanates uniformly from all points in the sky. The only plausible explanation is that this originated in the hot fireball conditions of the early Universe, as described by the Big Bang theory.

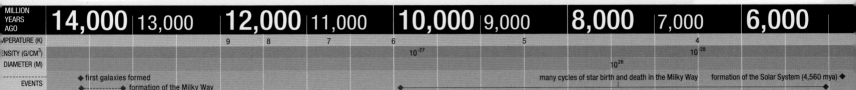

MILLION YEARS AGO	14,000	13,000	12,000	11,000	10,000	9,000	8,000	7,000	6,000
TEMPERATURE (K)			9	8	7	6	5	4	
DENSITY (G/CM³)					10^{-27}		10^{26}	10^{28}	
DIAMETER (M)									
EVENTS	◆ first galaxies formed	◆ formation of the Milky Way					many cycles of star birth and death in the Milky Way		formation of the Solar System (4,560 mya) ◆

GALAXIES FORM

Gravity now began to cause gas atoms to come together. Over hundreds of millions of years, swirling clouds of hydrogen and helium gas formed and started to extend into long thin strands. About 13,600 million years ago (around 200–400 million years after the Big Bang), the strands began to clump together to form the first galaxies. Further concentration of matter within galaxies led to the creation of the first stars. The production of energy (including heat and light) in stars began when hydrogen nuclei in their centres started fusing to form helium.

NEW ELEMENTS

When the first galaxies and stars formed, there were still just four chemical elements in the Universe: hydrogen, helium, lithium, and beryllium. The formation of the first stars was highly significant, because it is within these that heavier chemical elements are created from lighter ones, through various processes of fusion. Many of the most common chemical elements on Earth, such as oxygen, silicon, and iron, were made in this way. The very heaviest elements, such as lead, cannot be created in ordinary stars but form only in supernovae – the massive explosions of giant stars in their final death throes. These explosions also distributed new elements throughout galaxies, where they were incorporated into new stars and planets.

REMNANTS OF A SUPERNOVA
Supernovae are huge star explosions that distribute new elements through galaxies. This is the remnant of Cassiopeia A, which exploded about 10,000 years ago.

BIRTH OF THE SOLAR SYSTEM

The exact age of our own spiral galaxy, the Milky Way, is not known, but it was probably formed by 13,200–13,600 million years ago. About 4,560 million years ago, within our galaxy, a clump of gas and dust, known as a nebula, began to condense into what became the Solar System.

Within this nebula, matter coalesced into a denser central region (the proto-Sun) and more diffuse outer regions. Eventually the nebula shrank into a spinning, disk-like object, called the protoplanetary disk.

Within the disk, dust and ice collided randomly to form ever larger particles. In the centre of the disk, as matter was drawn in by gravity, temperatures rose to the point where hydrogen fused to form helium, and a fully-fledged star – our Sun – was born.

In other parts of the disk, the predominant matter was solid particles. Nearest the Sun, rocky materials became the main component. In the colder outer regions of the disk, ice particles were more common. As particles throughout the disk became larger, gravity drew them into collisions. This process, known as accretion, caused larger and larger bodies to merge, eventually forming planetesimals. These were the size of boulders, or larger, and composed of rock or rock and ice.

As well as heat and light, the new Sun also emitted a stream of energetic particles known as the solar wind. This "wind" blew volatile gases from the inner areas into the outer areas of the disk.

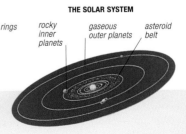

NEBULA
The Solar System condensed out of huge clouds of gas and dust, similar to these of the Lagoon Nebula, a star-forming region in the Milky Way.

Most of the remaining planetesimals collided to form the four rocky inner planets. They also formed the cores of the outer planets, around which the volatile gases collected.

Not all of the planetesimals merged to form planets. Some "leftovers" remain in the Solar System as two types of bodies: asteroids and comets. Most of the asteroids orbit in a belt between Mars and Jupiter, but some follow Jupiter's orbit, and some others have paths that cross the Earth's orbit. Comets formed from icy planetesimals in the outer edge of the disk.

FORMATION OF THE SOLAR SYSTEM

The Solar System formed in several stages. First a huge nebula (cloud) of gas and dust condensed into a spinning, disk-like object. The centre of the disk condensed further to form the Sun, while particles in the outer parts of the disk collided and accreted to form the planets.

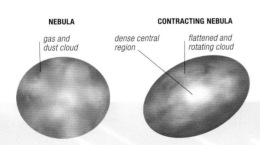

NEBULA
gas and dust cloud

CONTRACTING NEBULA
dense central region
flattened and rotating cloud

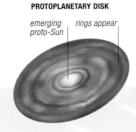

PROTOPLANETARY DISK
emerging proto-Sun
rings appear

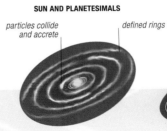

SUN AND PLANETESIMALS
particles collide and accrete
defined rings

THE SOLAR SYSTEM
rocky inner planets
gaseous outer planets
asteroid belt

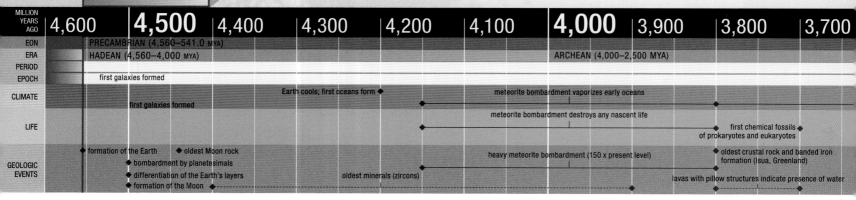

PRECAMBRIAN

MILLION YEARS AGO	4,600	4,500	4,400	4,300	4,200	4,100	4,000	3,900	3,800	3,700
EON	PRECAMBRIAN (4,560–541.0 MYA)									
ERA	HADEAN (4,560–4,000 MYA)						ARCHEAN (4,000–2,500 MYA)			
PERIOD										
EPOCH	first galaxies formed									
CLIMATE	first galaxies formed			Earth cools; first oceans form ◆			meteorite bombardment vaporizes early oceans			
LIFE						meteorite bombardment destroys any nascent life			first chemical fossils ◆ of prokaryotes and eukaryotes	
GEOLOGIC EVENTS	◆ formation of the Earth ◆ oldest Moon rock	◆ bombardment by planetesimals	◆ differentiation of the Earth's layers	◆ formation of the Moon		oldest minerals (zircons)	heavy meteorite bombardment (150 x present level)		◆ oldest crustal rock and banded iron formation (Isua, Greenland) lavas with pillow structures indicate presence of water	

Precambrian Earth

At one time, the Precambrian Eon was seen as a geological *terra incognita*, of unknown age and without fossil remains. Not until 1956, when the age of the Earth was accurately measured, did the immensity of the Precambrian's 4,000 million years become glaringly obvious. Since then, fossils have been found, rocks dated, and the formation of the Solar System unravelled. The Earth's early history is now known to be a time of cataclysmic change. Early accretion and differentiation of the core and mantle, along with bombardment from space, culminated in the Moon's formation. Continued

FORMATION OF THE MOON
More than 4,500 million years ago, a planetary body the size of Mars collided with the Earth, tearing away a large volume of rock. The debris was held by the Earth's gravity and cooled and coalesced to form the satellite we call the Moon.

ZIRCON CRYSTALS
Crystals from rocks in Western Australia have been dated at 4,400–3,900 million years old.

layering was followed by the formation of a primitive atmosphere, oceans, crust, and possibly life. However, renewed meteorite bombardment, which so scarred the Moon, also devastated the Earth, causing melting of its rocks.

Not until about 3,800 million years ago, were oceans and a primitive atmosphere regenerated and the first chemical evidence for photosynthesis by marine micro-organisms preserved. However, these life-forms had to tolerate a lack of oxygen and high ultraviolet radiation because there was no ozone layer to screen it. It took time for the atmosphere and oceans to develop to their present condition. Surface

THE EARTH'S LAYERS FORM
Early accretion of cosmic material formed a larger, growing body, in which melts formed and migrated. Heavy elements concentrated in the core and lighter ones in the overlying mantle.

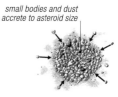
small bodies and dust accrete to asteroid size

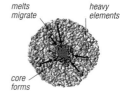
melts migrate — heavy elements — core forms

lighter elements form mantle — molten core

temperatures fell slowly and oxygen levels rose slowly as increasing numbers of photosynthesizing microbes produced oxygen and a protective ozone layer, aided by the emission of water vapour that split to release oxygen and ozone into the upper atmosphere.

The Precambrian Eon is divided into the Hadean, Archean, and Proterozoic eras, during which dynamic processes originating from within the Earth generated new ocean-floor rocks and destroyed them elsewhere, while continents grew and were moved through the processes of plate tectonics (see pp.86–89). Meanwhile, an increasing diversity of marine micro-organisms evolved slowly. Major events continued to perturb evolving ecosystems, with large-scale volcanism, impact events, and climate change culminating in runaway glaciations. Although often catastrophic, these upheavals may have stimulated evolution.

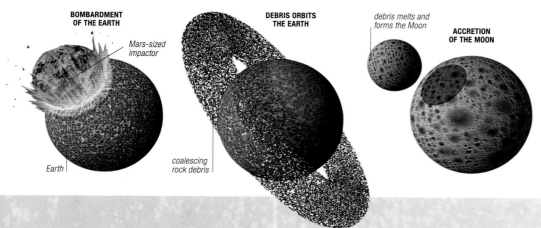

BOMBARDMENT OF THE EARTH
Mars-sized impactor
Earth

DEBRIS ORBITS THE EARTH
coalescing rock debris

debris melts and forms the Moon

ACCRETION OF THE MOON

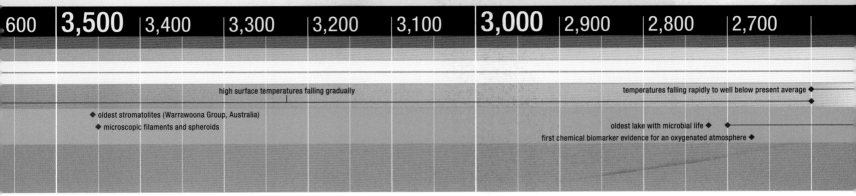

high surface temperatures falling gradually

temperatures falling rapidly to well below present average ◆

◆ oldest stromatolites (Warrawoona Group, Australia)

◆ microscopic filaments and spheroids

oldest lake with microbial life ◆

first chemical biomarker evidence for an oxygenated atmosphere ◆

METEORITE BOMBARDMENT

Dating of the Moon's rocks and the appearance of its heavily cratered surface provide evidence that for some 600 million years after their formation, both the Moon and Earth were subject to intense bombardment from space. The Moon has impact craters, such as the Aitken Basin, up to 2,500km (1,500 miles) across, and because of its greater size and gravity, the Earth underwent greater bombardment than the Moon. The results were melting of rock and destruction of any nascent life and atmosphere.

A LIFE-SUSTAINING ATMOSPHERE
The Earth's first atmosphere was blasted away by impacts and the solar wind. Volcanism built up a secondary atmosphere by releasing nitrogen, carbon dioxide, and water vapour. The latter was split by ultraviolet light into hydrogen, oxygen, and ozone. The lightest gas, hydrogen, was released into space.

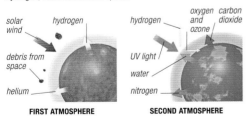

solar wind

hydrogen

debris from space

helium

FIRST ATMOSPHERE

hydrogen

UV light

water

nitrogen

oxygen and ozone

carbon dioxide

SECOND ATMOSPHERE

OCEANS AND CONTINENTS

The Hadean Era, as its name suggests, was hellish: the early Earth's turmoil destroyed any primary atmosphere, oceans, and life. The Archean times that followed saw the water vapour released from volcanoes collecting to form oceans, where salts dissolved to increase salinity gradually. Low-density silicate minerals accumulated to form the Earth's outer crust. But continuing volcanic eruptions resulted in the constant formation of new surface rocks. As the Earth was no longer expanding through accretion, the cooler and denser rocks, which had been formed previously, sank into the interior to accommodate the new crustal rocks. The Earth's crust was fragmented into a number of plates, with both convergent and divergent margins (see pp.88–89). Low-density rocks accreted into continents and higher density rocks formed the ocean floor.

EARLY LIFE

Fossil evidence for early life is rare in ancient sedimentary rocks, which were altered by heat and pressure. The oldest fossils are the degraded remains of organic carbon isotopes from 3,700-million-year-old seabed muds found in Western Greenland. They derived from photosynthesizing prokaryote micro-organisms that used light energy from the Sun. Within 250 million years, such marine microbes were interacting with

PROKARYOTES
These single-celled organisms have a simple cell that lacks structures (such as a nucleus) found in advanced cells.

seabed sedimentary processes and producing distinctive structures known as stromatolites. The oldest stromatolites date from around 3,480 million years ago and they become increasingly common, but the most ancient do not preserve fossils of the microbes that made them.

Although the first carbonaceous microfossils date from 3,500 million years ago, the first well-defined microfossils from China, known as acritachs, date from 1,800 million years ago and probably formed microbial mats in sunlit nearshore waters. There is some chemical biomarker evidence for the production of oxygen by photosynthesis by 2,700 million years ago. There is geochemical evidence (from biomarkers) for oxygen in the atmosphere by 2,400 million years ago.

hydrogen atom

carbon atom

METHANE (CH$_4$)
The simplest organic compound is one carbon atom combined with four hydrogen atoms.

CLAIR PATTERSON

After working on the first atomic bomb and studying radioactivity, American physicist Clair Patterson became interested in calculating the age of the Earth. In 1956, he compared measurements from meteorites and Earth minerals to obtain the age of 4,550 million years. This was the first accurate estimate.

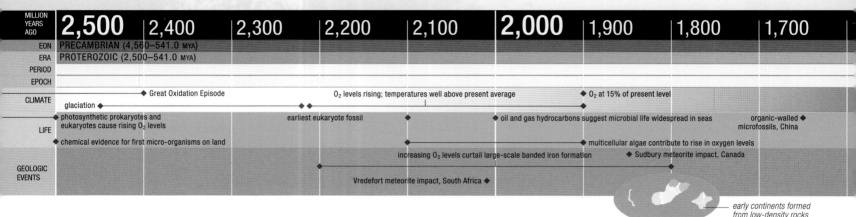

MILLION YEARS AGO	**2,500**	2,400	2,300	2,200	2,100	**2,000**	1,900	1,800	1,700

EON PRECAMBRIAN (4,560–541.0 MYA)

ERA PROTEROZOIC (2,500–541.0 MYA)

PERIOD

EPOCH

CLIMATE

◆ Great Oxidation Episode   O₂ levels rising; temperatures well above present average   ◆ O₂ at 15% of present level

glaciation ◆

LIFE

◆ photosynthetic prokaryotes and eukaryotes cause rising O₂ levels   earliest eukaryote fossil   ◆ oil and gas hydrocarbons suggest microbial life widespread in seas   organic-walled ◆ microfossils, China

◆ chemical evidence for first micro-organisms on land   ◆ multicellular algae contribute to rise in oxygen levels

GEOLOGIC EVENTS

increasing O₂ levels curtail large-scale banded iron formation   ◆ Sudbury meteorite impact, Canada

Vredefort meteorite impact, South Africa ◆

early continents formed from low-density rocks

OXYGENATING THE ATMOSPHERE

About 2,700 million years ago, primitive photosynthesizing micro-organisms began releasing oxygen into the early atmosphere. This oxygen was initially used up oxidizing iron in the oceans. Between 2,430 and 2,200 million years ago, during the Great Oxidation Episode, oxygen became a major component of the atmosphere and oceans, with oxygen levels rising to between 1 and 10 per cent of present levels. By 1,900 million years ago, oxygen levels were about 15 per cent of present levels. Micro-organisms preferring the anoxic (low-oxygen) conditions were forced to adapt by retreating within sediments and below ground.

An important development for evolving life was the formation of the protective ozone layer above the atmosphere to filter out incoming ultraviolet rays that

CHEMICAL FOSSILS

Even when organisms die and their tissues decay, the organic chemicals (hydrocarbons) of which they are made may survive in sediments, albeit in a degraded form. These are known as chemical fossils. Commonly, these hydrocarbons form mobile oil and gas, but some residues can persist within the rock record. Chemical analysis of these residues can differentiate the organic molecules present, their structural complexity and, broadly, what kind of organisms they are derived from.

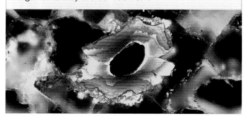

MODERN STROMATOLITES
Known as stromatolites, these mounds of layered sediment with surface films of bacteria are identical to those found in Precambrian sedimentary rocks.

BANDED IRON
These layered iron deposits were formed by oxidation reactions in the oceans before the formation of the atmosphere.

are particularly harmful to DNA. Above a 2-per-cent oxygen level, an ozone layer will begin to form, so it should have been well established by 1,900 million years ago. An oxygen-rich atmosphere, protected by an ozone shield, permitted new life-forms to evolve and thrive. Oxygen-tolerant photosynthesizers inherited the Earth.

MULTICELLULAR LIFE

Coiled filamentous microfossils from Michigan, USA, called *Grypania,* are 2,100 million years old and may provide the first direct fossil evidence for eukaryotes. However, the oldest convincing direct fossil evidence comes from some 1,650 million-year-old organic-walled microfossils from China. The first known multicells are 1,200 million-year-old fossils of a red alga called *Bangiomorpha* from arctic Canada, whose many cells show some specialization, including structures indicating that they reproduced sexually.

By 1,000 million years ago, towards the end of the Proterozoic Era and following a prolonged evolutionary gestation, the "big bang" of eukaryotic evolution had started. The evolution of multicellular organisms with sexual reproduction paved the way for larger and more diverse organisms. It used to be thought that it was the development of sexual reproduction, with its exchange of genetic material, that was the key to this diversification. However, it is now known that this innovation alone did not promote eukaryotic diversification of life, as even bacteria exchange

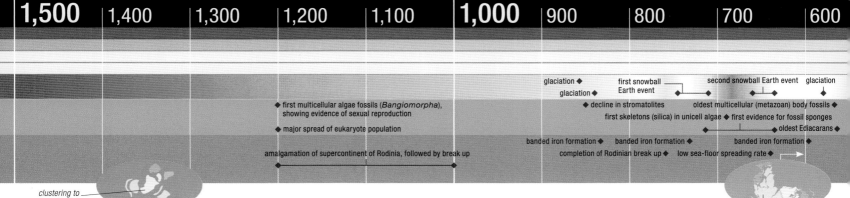

glaciation ◆ | first snowball Earth event | second snowball Earth event | glaciation
glaciation ◆ | |
◆ first multicellular algae fossils (*Bangiomorpha*), showing evidence of sexual reproduction | ◆ decline in stromatolites | oldest multicellular (metazoan) body fossils ◆
◆ major spread of eukaryote population | first skeletons (silica) in unicell algae ◆ first evidence for fossil sponges
| | ◆ oldest Ediacarans
banded iron formation ◆ | banded iron formation ◆ | banded iron formation ◆
amalgamation of supercontinent of Rodinia, followed by break up | completion of Rodinian break up ◆ | low sea-floor spreading rate ◆

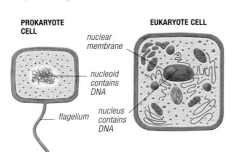

clustering to form Rodinian supercontinent

formation of Gondwanan supercontinent

genetic material. More important perhaps is the ability of multicellular organisms to increase in size beyond the microscopic, with specialization of certain cells for certain tasks within the organism. By 720 million years ago, the first fossil sponges from Australia and the traces they left behind in Oman, provide evidence for the increasing complexity of marine animals. They are followed by Ediacaran fossils from 579 million years ago and the first shelled animals, such as *Cloudina*, from 548 million years ago.

COMPLEX CELLS EVOLVE
Unlike the more primitive prokaryotes, eukaryotes have their nuclear material enclosed in a membrane. Eukaryote cells are sometimes part of a larger, multicellular organism, while prokaryotes are exclusively single-celled organisms and often incorporate a flagellum for mobility in a watery environment.

PROKARYOTE CELL

EUKARYOTE CELL

nuclear membrane

nucleoid contains DNA

nucleus contains DNA

flagellum

GRYPANIA
Coiled ribbons, at least 20cm (8in) long, called *Grypania* may be the oldest multicellular eukaryote algae.

SNOWBALL EARTH
There is evidence that the Earth suffered extensive glaciation in Late Proterozoic times, with ice-caps that extended from the poles to the tropics. Depending on whether most of the Earth's surface was covered, or just a large part of it, they have been interpreted snowball or slushball events. The occurrence of glacial deposits known as diamictites close to sea-level in the tropics provides evidence for low latitude glaciation. Measurements of magnetically orientated minerals within Precambrian glacial sediments indicate that 750 and 635 million years ago there were were two snowball events (with glacial retreat in between), and a glacial event between 2,450 and 2,220 million years ago, whose extent is not yet well established. Important support for the theory comes from the presence of carbonate sediments, which are typical of low latitudes, immediately above glacially related ones. The appearance of sedimentary iron formations in the oceans suggests that conditions were anoxic, which is consistent with glacial events. The cause of such

SNOWBALL EARTH
Geological evidence shows that low-latitude continents were glaciated; the extent of sea-ice cover is less clear.

EDIACARAN FOSSIL
Spriggina, an Ediacaran fossil with a head-like structure, is named after Australian geologist Reg Sprigg, who found the Ediacarans in Australia.

events is not fully understood but is thought to be related to the clustering of the continents within the tropics, raising the amount of light reflected from the planet and cooling global climates.

SOFT-BODIED ANOMALIES
The Ediacarans are among the most intriguing Precambrian fossils. A group of sea-dwelling organisms that lived 580–543 million years ago, they include some of the first known animals. Their bodies were up to 2m (6ft) long, which were entirely soft without any hard skeletal parts (see below) but often ribbed or quilted. Their relationship to living animal groups is unclear. They had a variety of body shapes, including a flat disc or a frond, while some had attachment discs and curiously quilted or serially divided surfaces. About 200 species are known, all entirely soft-bodied and preserved as sediment moulds and infills, suggesting that some lived within the sediment and that their body tissues were tougher than those of jellyfish, which are rarely preserved as fossils. Their first known appearance follows the last glacial event and their evolution may have been stimulated by its meltdown.

MILLION YEARS AGO	550	540	530	520	510	500	490	480	4
EON	PRECAMBRIAN	PHANEROZOIC (541.0 MYA–PRESENT DAY)							
ERA	PROTEROZOIC	PALEOZOIC (541.0–252.2 MYA)							
PERIOD		CAMBRIAN (541.0–485.4 MYA)					ORDOVICIAN (485.4–443.4 MYA)		
EPOCH									
CLIMATE		◆ rising O₂ levels linked to increasing biodiversification; temperature high			temperature falling; carbon dioxide (CO₂) to high of over 16 x present level		◆ CO₂ at less than 16 x present level; temperature low, then fluctuating above present average		
LIFE	maximum Ediacaran diversity ◆	◆ increasing trace fossils; diverse small shelly fossils ◆ decline in Ediacarans		◆ Cambrian explosion: unarmoured vertebrates, graptolites, arthropods including trilobites, and the first stem vertebrates (yunnanozoans) and animals with a notochord (*Haikouella*)		◆ Burgess Shale; conodonts (eel-like marine chordates) and jawless fish		first complete armoured jawless fish	
			◆ first metazoan skeletal reefs: archaeocyathan sponges, cyanobacteria				first terrestrial bryophyte (moss-type) fossils ◆		
GEOLOGIC EVENTS	◆ brief assembly of Pannotia supercontinent	◆ Gondwanan continental assembly stretches from pole to pole	◆ North America separated from Gondwana by the Iapetus ocean						
				◆ maximum sea-floor spreading rate					

rotation of Gondwana towards south pole

The Paleozoic Era

Paleozoic means "ancient life", and this era saw the appearance of abundant shelly fossils and the development of land plants. It ended with life's biggest ever extinction event, at the end of the Permian Period, when 90 per cent of all life on the Earth was wiped out.

PLATE TECTONICS

The Earth's continents are constantly reconfigured by movements of crustal plates, with new oceans opening and the subduction of the old ocean floor (see p.89). During the Paleozoic, amalgamation of the southern continents formed the Gondwanan supercontinent, which drifted north over time to form the even larger supercontinent of Pangea. By the end of the Paleozoic, Pangea stretched from pole to pole.

THE CAMBRIAN EXPLOSION

In the mid-19th century, fossil evidence seemed to indicate that life began in the earliest Cambrian with the appearance of a variety of sea-dwelling organisms, such as sponges and trilobites, with mineralized shells and skeletons. We now know that life is much more ancient and that shelled animals first appear in late Proterozoic rocks. However, it still seems that a large variety of fossil shells appear suddenly at the beginning of the Cambrian. Many of these fossils are very small and represent several different kinds of molluscs and armoured arthropods (creatures with paired and jointed limbs) – it is as if there was an explosion in the diversity of life.

Some scientists have argued from early molecular clock estimates that there was no great explosion of life. Rather, it was the rapid acquisition of readily fossilizable hard part, such as shells and skeletons that produces the appearance of an explosion. However, the overwhelming empirical evidence is that while animal lineages did begin to diversify during the late Neoproterozoic, as suggested by recent molecular clock estimates, the Cambrian explosion of fossils does accurately reflect the acquisition of new body plans and larger body size. Within Cambrian times there is fossil evidence for the appearance of most of the major groups of marine invertebrates (those animals that lack a backbone), including a number of groups that subsequently became extinct, such as the trilobites, conodonts, and graptolites. Fossils from Cambrian strata in China include the yunnanozoans (the first "stem" vertebrates) and the first more advanced fish-like animals with traces of a backbone (a stiffening rod called a notochord). The vast majority of this life

CHARLES WALCOTT

The American paleontologist Charles Walcott (1850–1927) quarried some 70,000 mid-Cambrian fossils from the Burgess Shale in the Canadian Rocky Mountains, which is now a World Heritage Site. The soft-part preservation of its fossils provides insights into marine life of the time. Walcott became director of the US Geological Survey, then secretary of the Smithsonian Institution in Washington DC, USA.

MOLECULAR CLOCKS

The molecular clock uses genetic difference between living groups and assumptions of constant rates of evolution to estimate the origin and divergence of groups of organisms. However, it needs to be calibrated by appropriate fossils of known age. For instance, according to the molecular clock, sharks evolved 528 million years ago, yet fossils first occur in 385 million-year-old strata. The discrepancy may reflect a lack of fossils, or a misunderstanding about rates of evolution.

SILURIAN TRILOBITE
Marine arthropods called trilobites, with mineralized exoskeletons, evolved in Cambrian times.

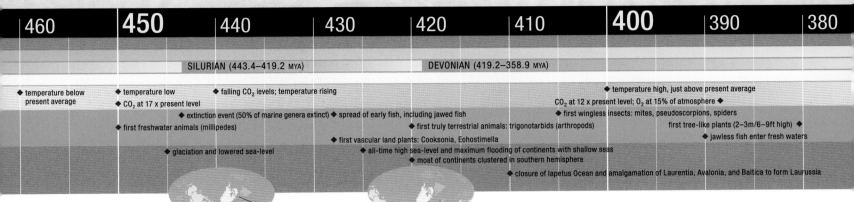

SILURIAN (443.4–419.2 MYA)   DEVONIAN (419.2–358.9 MYA)

◆ temperature below present average

◆ temperature low
◆ CO_2 at 17 x present level

◆ falling CO_2 levels; temperature rising

◆ temperature high, just above present average
CO_2 at 12 x present level; O_2 at 15% of atmosphere ◆

◆ extinction event (50% of marine genera extinct) ◆ spread of early fish, including jawed fish

◆ first wingless insects: mites, pseudoscorpions, spiders

◆ first freshwater animals (millipedes)

◆ first truly terrestrial animals: trigonotarbids (arthropods)

first tree-like plants (2–3m/6–9ft high) ◆

◆ jawless fish enter fresh waters

◆ first vascular land plants: Cooksonia, Eohostimella

◆ glaciation and lowered sea-level

◆ all-time high sea-level and maximum flooding of continents with shallow seas

◆ most of continents clustered in southern hemisphere

◆ closure of Iapetus Ocean and amalgamation of Laurentia, Avalonia, and Baltica to form Laurussia

Laurentia and Baltica converge   *Iapetus Ocean closing*   *Avalonia moves north*

was confined to the oceans and seas. However, recent discoveries indicate that shallow water, fresh water, and the land were invaded back in the early Proterozoic by microbial life.

MARINE LIFE

Because the fossil record mostly preserves mineralized hard parts such as shells, skeletons, and teeth, it tends to be highly biased. The fossil record from the Cambrian preserves the first appearances of hardparts belonging to a number of groups of organisms, which must have originated in late Proterozoic times. Initially, the most common fossils were the remains of seabed sponges; various arthropods, such as the extinct trilobites; molluscs, especially cephalopods; and brachiopods, which look like molluscan clams but are anatomically different. Additionally, there were numerous soft-bodied groups, especially worms, such as the priapulids. Other important groups soon evolved, such as the echinoderms and the vertebrates along with the corals, whose reef growth provided new environments for life ranging from calcareous algae and bryozoans (moss-animals) to crinoids (sealilies) and fish.

ICE AGE

The discovery of ice-scratched rock surfaces, ice-rafted boulders, and other glacial deposits in the late Ordovician strata of North Africa might seem hard to explain in relation to today's climate. However, measurement of iron-rich grains orientated by the Earth's Ordovician magnetic field shows that these rocks were clustered near the South Pole at the time of their formation and have since been moved by the processes of plate tectonics. Africa and South

DEVONIAN JAWLESS FISH
Strange-looking fish without jaws or teeth, such as this Devonian cephalaspid, were among the first animals with backbones.

America, both part of the supercontinent of Gondwana, were glaciated by the growth of a polar ice-cap. The resulting drastic climate and sea-level change caused the end-Ordovician mass extinction, which wiped out about 50 per cent of the genera of marine organisms.

COOKSONIA
The tiny forked, leafless stems of *Cooksonia* from Silurian strata are among the earliest upright-growing land plants.

LIFE ON LAND

Evidence for macroscopic life on land begins with rare fossil footprints of freshwater arthropods and spores of primitive moss-like plants, both in Ordovician strata. The Silurian saw upright-growing (vascular) plants, such as *Cooksonia*, still dependent on watery environments for reproduction, and tiny arthropods that fed on their decaying remains.

GREENING OF THE LAND

The earliest fossil record of land-plant evolution comprises mainly spores and pollen. Plant fossils are increasingly common from the Devonian onwards, reflecting the growth of low-lying, wet woodlands.

extensive forests, which formed coal deposits

ORDOVICIAN PERIOD
first land plants – moss-like bryophytes

SILURIAN PERIOD
tiny, upright-growing land plants

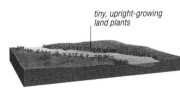

DEVONIAN PERIOD
first tree-sized land plants

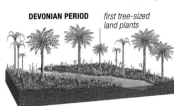

CARBONIFEROUS PERIOD

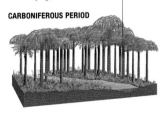

MILLION YEARS AGO	370	360	350	340	330	320	310	300	2
EON	PHANEROZOIC (541.0 MYA–PRESENT DAY)								
ERA	PALEOZOIC (541.0–252.2 MYA)								
PERIOD	DEVONIAN (419.2–358.9 MYA)	CARBONIFEROUS (358.9–298.9 MYA)						PERMIAN (298.9–252.2	
EPOCH		MISSISSIPPIAN (354.0–323.0 MYA)			PENNSYLVANIAN (323.0–290.0 MYA)				

CLIMATE

◆ CO_2 at 10 x present level; O_2 rising; temperature falling to present average ◆ temperature high (20°C/68°F average) ◆ temperature falling to below present average ◆ temperature low (10°C/50°F average); O_2 at all-time high

LIFE

◆ oldest tetrapod ◆ first forests; first ferns
◆ oldest terrestrial tetrapod
◆ extinction event ◆ *Tulerpeton* (closest relative of the amniotes)

◆ first temnospondyl amphibians and anthracosaur amniotes   first beetles ◆ first land snails
◆ first extensive tropical coal-measure forests ◆ earliest certain reptile amniotes
◆ first cockroaches ◆ first pelycosaurs (mammal-like reptiles)

GEOLOGIC EVENTS

◆ low sea-level (associated with glaciation)

Laurentia straddles equator (coinciding with development of coal-measure forests) ◆

◆ Gondwana rotates clockwise; low sea-floor spreading rate   maximum deposition of coal-bearing s

Gondwana converges with Laurentia and Baltica

Gondwana moves north

Diversifying land plants in early Devonian times developed into low-lying wetland forests in whose waters the earliest tetrapod-like fish evolved paired fins. By mid-Devonian times, tetrapod trackways show that land-going tetrapods had evolved but they still had to breed in water and their bony remains are not found until Late Devonian times.

CLIMATE CHANGE

Today, the sequence of rock strata preserved within individual continents, such as North America, shows a Paleozoic history of drastic climate change, from the

FOSSILIZED TREE STUMP
Fossil trees from Antarctica indicate times when the climate of the continent was much warmer than it is today.

cool marine waters of Ordovician times through Devonian deserts to Carboniferous tropical seas (during the Mississippian Epoch), tropical rain forests (during the Pennsylvanian Epoch), and back to deserts again in the Permian Period. However, this sequence of changes is not, as might be expected, due to dramatic changes in global climates but mostly to the movement of continents through the Earth's climate zones due to plate tectonics. From Ordovician to Permian times, North America (on the Laurentian Plate) moved from the southern hemisphere across the equator into the northern hemisphere.

PERMIAN SANDSTONE
Thick piles of sandstone strata were deposited within the vast arid deserts of the Pangean supercontinent. Structures in the beds show that these were once part of sand dunes.

LAND VERTEBRATES

Early land tetrapods were salamander-like animals about 1m (3ft) in length. Little is known of their biology except that they had to return to water to breed. By the mid-Carboniferous, both amphibian and more reptile-like groups had evolved. The amphibians were 2–3m (6–10ft) long and resembled crocodilians. However, the critical reptilian feature – the ability to lay a shelled (amniote) egg on dry land – is difficult to recognize in the fossils.

From the late Carboniferous into the Permian Period, however, true reptiles were diversifying. The Pangean supercontinent was

CARNIVOROUS AMPHIBIAN
These early Triassic fossil remains of several meat-eating amphibians were found in South Africa, clustered around those of a plant-eating animal called *Lystrosaurus*.

occupied by many short-lived reptile groups, some of which were to eventually evolve into dinosaurs, others into various marine reptiles and turtles, and others into mammals.

EXTINCTION

At the end of the Permian, an estimated 90 per cent of all living organisms (60 per cent of genera), died out in the biggest mass extinction known. Groups that died out include the Paleozoic corals, trilobites, and water scorpions, while the echinoderm crinoids (sea-lilies), brachiopods, and clams suffered serious losses, as did 60–70 per cent of life on land.

Geologists have searched for a single cause, especially a major meteorite impact like that which brought about the end-Cretaceous extinction event (see pp.24–25), but without success. There is evidence for widespread anoxia in the oceans, which may have disrupted food chains, causing many marine organisms to die out. The source of this anoxia was thought to have been a sudden release of methane from gas hydrates buried in ocean-bed deposits. However, since sea-level was rising during the extinction, this is unlikely. Instead, it is much more likely that the extinction coincided with a vast outpouring of lavas and volcanic gasses in Siberia (known as the Siberian Traps), which may well have altered global climates.

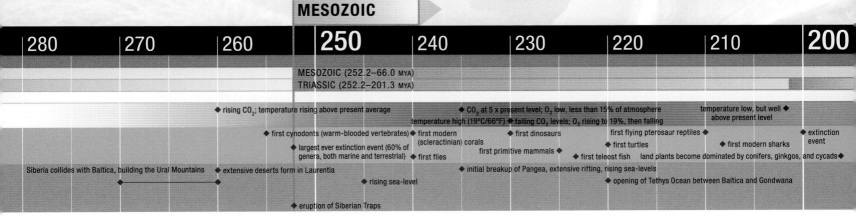

MESOZOIC (252.2–66.0 MYA)
TRIASSIC (252.2–201.3 MYA)

◆ rising CO_2; temperature rising above present average   ◆ CO_2 at 5 x present level; O_2 low, less than 15% of atmosphere   temperature low, but well ◆
above present level

temperature high (19°C/66°F) ◆ ◆ falling CO_2 levels; O_2 rising to 19%, then falling

◆ first cynodonts (warm-blooded vertebrates) ◆ first modern   ◆ first dinosaurs   first flying pterosaur reptiles ◆   ◆ extinction
(scleractinian) corals   ◆ first turtles   event

◆ largest ever extinction event (60% of   first primitive mammals ◆   ◆ first modern sharks
genera, both marine and terrestrial) ◆ first flies   ◆ first teleost fish   land plants become dominated by conifers, ginkgos, and cycads ◆

Siberia collides with Baltica, building the Ural Mountains   ◆ extensive deserts form in Laurentia   ◆ initial breakup of Pangea, extensive rifting, rising sea-levels

◆ rising sea-level   ◆ opening of Tethys Ocean between Baltica and Gondwana

◆ eruption of Siberian Traps

The Mesozoic Era

Spanning the 187 million years of the Triassic, Jurassic, and Cretaceous periods, the Mesozoic (or "middle life") Era lies between two major extinction events, the second one known as the end-Cretaceous extinction. It is often characterized the Age of the Reptiles, because many major reptile groups, such as the land-lving dinosaurs and the marine ichthyosaurs and turtles all diversified.

MARINE LIFE

The end-Permian extinction and collapse of the Paleozoic reefs caused a major turnover in marine life. Eventually, modern reef corals evolved and marine diversity and food chains were reestablished. The more adaptable molluscan clams and snails gradually took over from brachiopods. Both bony and cartilaginous fish diversified, with some sharks becoming top predators. But the waters were also occupied by new marine reptiles, along with the now extinct plesiosaurs, dolphin-like ichthyosaurs, and mosasaurs.

MARINE TURTLES
The first turtles evolved during the Triassic, although it is uncertain whether they did so on land or in the sea.

SEA REPTILES
New predators occupied the seas, such as the fast-swimming ichthyosaurs. Shaped like dolphins with toothed, beak-like jaws, these reptiles lived from the Triassic to the Cretaceous.

PLANT LIFE

During the late Permian and early Triassic, the dominant Paleozoic plants – such as ferns, clubmosses, and horsetails – declined. Newly diversifying groups included conifers, which reproduce by means of seeds borne in cones. They became particularly important in Jurassic and Cretaceous times, along with cycads and cycad-like bennettitaleans. Most of the cycads also had seed-bearing cones and were distributed worldwide, even in polar regions when they were free of ice. Flowering plants (angiosperms) evolved through the Cretaceous. They combined a number of features found in other plants of the

SURVIVING CYCADS
The few living cycads are survivors of a far larger group of seed-bearing plants that were key members of Jurassic and Cretaceous forests.

Mesozoic, such as flower-type reproductive structures (also found in the bennettitaleans), with the one unique feature that distinguishes flowering plants – unfertilized seeds enclosed in a carpel.

DINOSAURS

For 150 million years, from the late Triassic to the end of the Cretaceous, terrestrial life was dominated by reptiles. One very varied group, known as the dinosaurs, with roughly 700 valid genera, diversified from crow-sized bipedal forms to the largest land-living animals ever known – the four-limbed sauropods at up to 34m (110ft) long.

Dinosaurs were first recognized as a distinct group of reptiles in 1842, their distinguishing characteristic being that they walked with their legs tucked under their bodies, unlike living reptiles, which have splayed legs. Most very large dinosaurs were plant-eaters with long necks. However, there were also very large carnivorous predators such as the Jurassic allosaurs, which were up to 12m (40ft) long.

PLACERIAS
This late Triassic 3-m- (10-ft-) long giant from North America was one of last surviving therapsids.

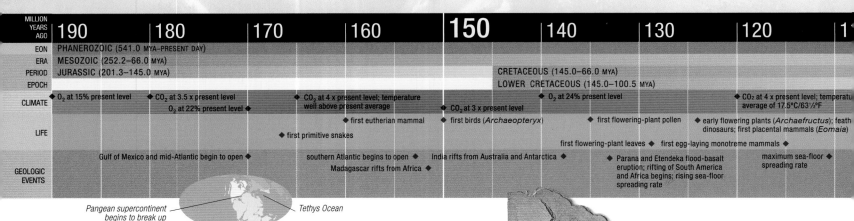

MILLION YEARS AGO	190	180	170	160	150	140	130	120	1
EON	PHANEROZOIC (541.0 MYA–PRESENT DAY)								
ERA	MESOZOIC (252.2–66.0 MYA)								
PERIOD	JURASSIC (201.3–145.0 MYA)					CRETACEOUS (145.0–66.0 MYA)			
EPOCH						LOWER CRETACEOUS (145.0–100.5 MYA)			
CLIMATE	◆ O₂ at 15% present level	◆ CO₂ at 3.5 x present level O₂ at 22% present level ◆		◆ CO₂ at 4 x present level; temperature well above present average	◆ CO₂ at 3 x present level	◆ O₂ at 24% present level		◆ CO₂ at 4 x present level; temperatu average of 17.5°C/63½°F	
LIFE				◆ first eutherian mammal ◆ first primitive snakes	◆ first birds (Archaeopteryx)	◆ first flowering-plant pollen first flowering-plant leaves ◆	◆ early flowering plants (Archaefructus); feath dinosaurs; first placental mammals (Eomaia) ◆ first egg-laying monotreme mammals ◆		
GEOLOGIC EVENTS	Gulf of Mexico and mid-Atlantic begin to open ◆		southern Atlantic begins to open ◆ Madagascar rifts from Africa ◆	India rifts from Australia and Antarctica ◆		◆ Parana and Etendeka flood-basalt eruption; rifting of South America and Africa begins; rising sea-floor spreading rate	maximum sea-floor ◆ spreading rate		

Pangean supercontinent begins to break up — Tethys Ocean

The dinosaurs filled most available habitats of the Mesozoic. There are two distinct types of dinosaur: one with a bird-like pelvis and the other reptile-like. The latter type included a group called the maniraptoran dinosaurs, a number of which were feathered, and one Cretaceous group of these small predators was probably the ancestors of the birds.

DINOSAUR FOOTPRINTS
These three-toed footprints of dinosaurs, initially mistaken for those of giant birds, are now used to estimate how fast dinosaurs could run.

MAMMALS AND BIRDS
Mammals (about 6,400 living species) and birds (about 10,900 living species) are the two groups of warm-blooded vertebrates that came to dominate Cenozoic landscapes. Although reptiles were still abundant (over 10,000 living species), they were mostly smaller than mammals. Fossils show that the reptile origins of mammals and birds lie deep within the Mesozoic, the former in the late Triassic and the latter in the Jurassic. According to the fossil record, a significant spread of both groups did not occur until after the end-Cretaceous extinction event, which wiped out most of the evolving bird groups and some of the early mammals. However, so many different bird and mammal groups appear immediately after the Mesozoic Era in the Paleocene Period that there were probably more late-Cretaceous ancestral groups than the fossil record preserves.

Primitive Mesozoic mammal groups ranged in size from that of a shrew to badger-size, and most died out. However, the marsupials and egg-laying monotremes have survived, especially in Australasia and the Americas.

ARCHAEOPTERYX
The first fossil bird was first found in 1861 complete with impressions of flight feathers.

MEGAZOSTRODON
Morganucodontids, such as the insect-eating Megazostrodon, were among the first primitive shrew-like mammals.

CLIMATE CHANGE
High global temperatures in the Mesozoic, averaging 19°C (66°F), generally prevented the growth of polar ice-caps. However, there is evidence of glaciation in Antarctica and downward fluctuation carbon-dioxide levels in the Late Cretaceous. During the end-Cretaceous extinction event 66 million years ago, climates were disturbed on a global scale.

The details are still debated, but fossil charcoal indicates there was a short period of global wildfire that devastated life and perhaps longer-term warming, enhanced by greenhouse gasses derived from a massive outpouring of lava known as the Deccan eruptions, which formed the upland region of India known as the Deccan Plateau.

EXTINCTION
More than 75 per cent of all species died out at the end of the Cretaceous, in an event known as the K-PG extinction event. Victims included the dinosaurs (apart from the birds), the remaining marine and flying reptiles, and ammonites (shelled, squid-like animals). The extinction coincided with the impact of a large meteorite in the Caribbean Sea, off the Yucatan Peninsula at Chicxulub, Mexico, and the Deccan lava eruptions. Both had a significant effect on the global climate, but the several effects of the meteorite impact probably caused the extinctions.

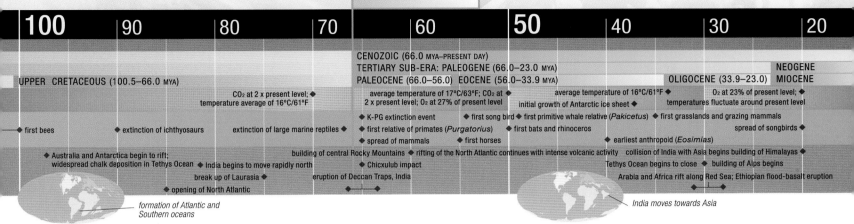

100	90	80	70	60	50	40	30	20

CENOZOIC (66.0 MYA–PRESENT DAY)
TERTIARY SUB-ERA: PALEOGENE (66.0–23.0 MYA)
NEOGENE

UPPER CRETACEOUS (100.5–66.0 MYA)
PALEOCENE (66.0–56.0) EOCENE (56.0–33.9 MYA)
OLIGOCENE (33.9–23.0) MIOCENE

CO₂ at 2 x present level; ◆
temperature average of 16°C/61°F

average temperature of 17°C/63°F; CO₂ at ◆
2 x present level; O₂ at 27% of present level

average temperature of 16°C/61°F ◆
initial growth of Antarctic ice sheet ◆

O₂ at 23% of present level; ◆
temperatures fluctuate around present level

◆ first bees

◆ extinction of ichthyosaurs

extinction of large marine reptiles ◆

◆ K-PG extinction event

◆ first song bird ◆ first primitive whale relative (Pakicetus) ◆ first grasslands and grazing mammals

◆ first relative of primates (Purgatorius) ◆ first bats and rhinoceros

spread of songbirds ◆

◆ spread of mammals ◆ first horses

◆ earliest anthropoid (Eosimias)

◆ Australia and Antarctica begin to rift;
widespread chalk deposition in Tethys Ocean ◆ India begins to move rapidly north

building of central Rocky Mountains ◆ rifting of the North Atlantic continues with intense volcanic activity   collision of India with Asia begins building of Himalayas ◆

Tethys Ocean begins to close ◆ building of Alps begins

◆ Chicxulub impact

break up of Laurasia ◆

eruption of Deccan Traps, India

Arabia and Africa rift along Red Sea; Ethiopian flood-basalt eruption

◆ opening of North Atlantic

formation of Atlantic and
Southern oceans

India moves towards Asia

LUIS AND WALTER ALVAREZ

Walter Alvarez (b.1940), an American geologist, found high concentrations of the rare element iridium in a marine clay dated to the Cretaceous–Cenozoic boundary. From this finding, in 1980 he developed the hypothesis with his father Luis (1911–88), a Nobel physicist, that the iridium concentration is best explained by an extra-terrestrial body, such as a very large meteorite, impacting upon the Earth. They further speculated that a global catastrophe accompanying the impact caused the demise of the dinosaurs.

The Cenozoic Era

The Cenozoic Era extends from 66 million years ago to the present day (Cenozoic means "recent life"). The era is characterized by a diversification of modern bony fish, flowering plants, pollinating insects, birds, and mammals, including our primate relatives.

RIFTING AND VOLCANISM

Cenozoic movements of the Earth's plates had a considerable influence on the evolution of life. Mountain building and large-scale uplift down the western flank of the Americas transformed regional climates, while the formation of the Himalayas (see pp.168–69) and the Tibetan Plateau generated the Southeast Asian monsoon (see p.463). Rifting and volcanism opened up the North Atlantic (see pp.402–403) and created the Great Rift Valley of east Africa (see pp.148–49), where many of our primate ancestors flourished.

Cenozoic volcanism, associated with rising plumes of heat from within the Earth, led to doming and rifting at the surface (see p.156). The early Cenozoic saw a region 2,000km (1,250 miles) wide,

PRESERVED LAVAS AT GIANT'S CAUSEWAY
As early Cenozoic lavas erupted in the rift between Norway, Britain, and Greenland, the North Atlantic widened. Evidence of this outpouring can still be seen in Ireland (shown here).

from Greenland to Norway, domed by up to 2km (1¼ miles), leading to rifting and the outpouring of lava, with the Atlantic Ocean flooding north through the rift, a process continuing today in Iceland. Similar doming in Africa led to rifting and the outpouring of the Ethiopian flood basaltic lava (31–28 million years ago) and outpouring of the Columbia River lava in North America (15 million years ago).

CHICXULUB IMPACT
Sixty-six million years ago, a meteorite crashed into shallow seas, which are now part of the Yucatan Peninsula, Mexico (see p.123). The impact blasted huge volumes of carbonate rock through the atmosphere, blocking out sunlight and shutting down photosynthesis.

EXTRA-TERRESTRIAL IMPACTOR

10-km- (6-mile-) wide impactor

EXPLOSION ON IMPACT

front of impactor collapses

back of impactor continues forward

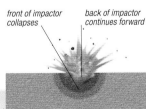

CRATER FORMATION

rocks blast into atmosphere

crater 100km (60 miles) wide and 12km (7½ miles) deep

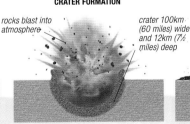

CRATER COLLAPSE

steep sides fall in

crater up to 240km (150 miles) wide

EON	PHANEROZOIC (541.0 MYA–PRESENT DAY)					
ERA	CENOZOIC (66.0 MYA–PRESENT DAY)					
PERIOD	TERTIARY SUB-ERA: NEOGENE (23.0–2.6 MYA)					
EPOCH	MIOCENE (23.0–5.3 MYA)					
CLIMATE						temperature high within ◆ general ice-house state
LIFE	◆ spread of primitive apes in Central Africa	spread of grazing *Hipparion* horses in North America and dispersal through Eurasia to Africa	molecular divergence of ◆ Orang-Utan from other great apes		African forests ◆ spread of whales diminish and dolphins	spread of snakes, ◆ frogs, rats, and mice
					expansion of open grasslands into middle and high latitudes continues throughout Miocene ◆	
GEOLOGIC EVENTS		◆ Australia moves north	uplift of Tibet ◆		◆ Columbia River flood-basalt eruption	
		◆ rifting and volcanism in east Africa				

MARINE LIFE

Following the extinction of the large predatory marine reptiles, early Cenozoic oceans mostly lacked predators of similar size, except for sharks and some bony fish. By the Early Eocene, the appearance of the earliest whales and sea cows heralded the start of several different mammalian groups that returned to the seas. These marine mammals diversified into large predators that inhabited the open sea from 35 million years ago to the present day.

LIFE ON LAND

In early Cenozoic times, mammals took over the habitats that had previously been occupied by terrestrial dinosaurs. However, there were major differences between the emerging mammals, with Australia and South America inheriting the pouched marsupials and egg-laying monotreme mammals, while the rest of the world was soon dominated by placental mammals. The latter give birth to bigger, more advanced young, nurtured for longer within the mother's body via a placenta.

Initially marsupials thrived, evolving both plant-eaters and carnivorous predators, which diversified into forms that were subsequently mimicked by the placentalrodents, hippopotamuses, horses, dogs, and big cats. However, as the more advanced placentals evolved and diversified, they tended to diplace the marsupials. Formation of a land bridge between North and South America allowed an interchange of placentals and marsupials between the continents. Today, the surviving marsupials of the Americas are the opossums, which are still represented by 63 species. But even more marsupials held out for longer on the isolated continent of Australia.

The earliest Cenozoic placental mammals were mostly shrew-like insect-eaters, but soon bigger (sheep-sized) browsing and rooting plant-eaters evolved. By the late Paleocene, there were rhinoceros-sized browsers and some dog-sized carnivores. The number of placental families more than quadrupled from the Cretaceous to the Eocene. The Earth's habitats were soon filled by placental mammals, from otters, seals, and whales in the seas, through the vast range of land mammals – from shrews to giant plant-eaters, such as the extinct 8-m- (26-ft-) long giant rhinoceros *Indricotherium*, and numerous carnivores – to bats in the air. Environmental and climatic changes also affected mammal evolution, especially decreasing forest cover and increasing grasslands, with the co-evolution of pollinating insects, song-birds, and grazing mammals. These changes are also thought to have affected the evolution of hominoids (higher apes and humans).

MOUNTAIN BUILDING

Cenozoic times saw the formation of four major mountain belts around the world: the Andes (see pp.158–59), the Rocky Mountains (see p.157), the European Alps (see pp.162–63), and the Himalayas

FLOWERING TREES
By earliest Cenozoic times flowering plants had evolved into a diverse range, including the magnolia (above) and laurels that formed the woodlands.

TARSIER
Today's tiny insect-eating tarsiers are thought to be descended from a primitive Cenozoic primate group.

MAMMALS OF LAKE MESSEL
Fifty million years ago, the oil-rich muds of a German lake bed preserved an astonishing diversity of its inhabitants, such as this *Ailuravus*, a rodent slightly larger than a squirrel.

AMBER LIFE
Since Cretaceous times, amber from some resin-producing trees has trapped and preserved small animals, especially insects.

primates disperse from Africa to Europe (eg *Dryopithecus*) and Asia (eg *Sivapithecus*)

◆ expansion of apes from Africa into Asia
mammals diversify, especially rodents, grazers, and large carnivores ◆

◆ extensive savannas in low to intermediate latitudes
insects, especially ants and termites, diversity ◆

◆ dispersal of great apes from Asia to Africa

◆ spread of pigs and camels

◆ South America moves slowly north

◆ Red Sea ocean floor spreads, pushing Arabia northwest, away from Africa

Africa's northward move halted by Europe ◆

(see pp.168–69). The Andes are a classic example of mountain building, resulting from the convergence of an oceanic (Pacific) plate with a continental plate (South America), combined with the subduction of the oceanic plate (see p.109) and frequent earthquakes and volcanoes. By contrast, the Alpine–Himalayan belt resulted from the convergence of the African and Indian plates with Europe and Asia respectively. Africa's northward movement subducted a body of water known as the Tethys Ocean; the intervening rocks were compressed, thickened, and elevated with intense folding and faulting, leading to the formation of the Alps. The breakaway of India from Africa and Antarctica, which began in the Mesozoic, eventually caused the subduction of the eastern Tethys and its convergence with Asia, beginning 20 million years ago. Again, the intervening rocks were compressed, thickened, and elevated with intense folding and faulting, resulting in the formation of the Himalayas. The continuing northward drive of India is thought to

HIMALAYAS
The spectacular mountain belt of the Himalayas, with the high plateau of Tibet to the north and low sediment-filled floodplains to the south, is the result of the convergence of the Indian and Asian plates.

have pushed the deeper part of the Indian plate beneath Tibet, thickening and elevating it without folding. The formation of this high and elongated physical barrier caused significant change in the regional climate, with the development of the Southeast Asian monsoon.

THE ORIGIN OF HUMANS
Humans, apes, lemurs, and monkeys are all mammals and were first grouped together as primates in 1758 by the Swedish naturalist

Carolus Linnaeus (1707–78). The fossil record shows that the earliest primate-like mammals originated as small insect-eating animals, such as the shrew-like *Purgatorius* in the early Paleocene. In the early Cenozoic, a group of tree-climbing, squirrel-like animals (the plesiadapiforms) had evolved in North America and Europe. They were followed by tarsier- and lemur-like primates, which spread into Africa and Asia along with two groups of higher primates, the New World monkeys and the Old World monkeys and apes. It was from the latter group that hominoids evolved in Africa. These included the 25- to 23-million-year-old (early Miocene) tail-less and monkey-like *Proconsul,* which could climb trees and walk on all fours. In mid-Miocene times, the African apes expanded into Asia, from where the great apes evolved and returned to Africa in Late Miocene times.

SPREADING GRASSLANDS
By the Miocene, cooling climates and greater aridity caused forests to break up and grasslands to expand, along with fleet-footed grazing mammals and their predators, such as big cats.

BIPEDAL BEGINNINGS
It seems likely that our ape–human ancestor was a knuckle-walker, like today's apes. However, fossil hominids from 6–7 million years ago, appear already to have walked on two feet.

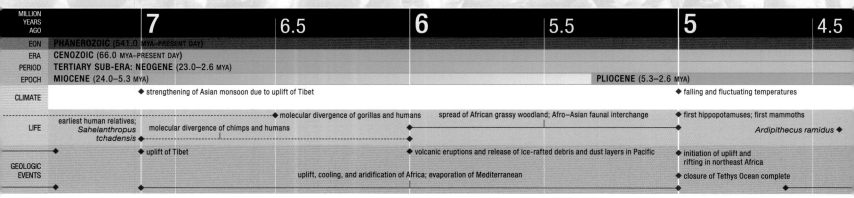

MILLION YEARS AGO	**7**		6.5	**6**		5.5	**5**		4.5
EON	PHANEROZOIC (541.0 MYA–PRESENT DAY)								
ERA	CENOZOIC (66.0 MYA–PRESENT DAY)								
PERIOD	TERTIARY SUB-ERA: NEOGENE (23.0–2.6 MYA)								
EPOCH	MIOCENE (24.0–5.3 MYA)						PLIOCENE (5.3–2.6 MYA)		

CLIMATE: ◆ strengthening of Asian monsoon due to uplift of Tibet ◆ falling and fluctuating temperatures

LIFE: earliest human relatives; *Sahelanthropus tchadensis* · molecular divergence of chimps and humans · ◆ molecular divergence of gorillas and humans · spread of African grassy woodland; Afro–Asian faunal interchange · ◆ first hippopotamuses; first mammoths · *Ardipithecus ramidus* ◆

GEOLOGIC EVENTS: ◆ uplift of Tibet · ◆ volcanic eruptions and release of ice-rafted debris and dust layers in Pacific · ◆ initiation of uplift and rifting in northeast Africa · uplift, cooling, and aridification of Africa; evaporation of Mediterranean · ◆ closure of Tethys Ocean complete

---

It was the great apes of Africa that evolved into gorillas, chimpanzees, and humans. According to the molecular clock (see p.30), gorillas diverged first, between 8–19 million years ago, while the chimpanzees and humans share a common ancestor who lived 7–8 million years ago.

THE SPREAD OF HOMINIDS

The fossil record of our extinct human relatives is sparse but increasing beyond the fossil hotspots of the Great Rift Valley of east Africa and caves in South Africa, Europe, and China with new sites in Morocco, Siberia, and Indonesia. This preservational bias has distorted our view of human evolution but important finds have been made elsewhere – such as a skull of the 6–7 million year old *Sahelanthropus* found in Chad. It combines features of a chimp-sized, small-brained ape with human features, suggesting that it may have been able to walk upright. Fossil remains of *Orrorin* (6.2–5.8 million years old),

SKULL RECORD
By examining the skulls, it is clear that brain size increases significantly from Australopithecus to the large-brained *Homo* species: *H. neanderthalensis* and *H. sapiens*.

AUSTRALOPITHECUS

HOMO NEANDERTHALENSIS

HOMO SAPIENS

JANE GOODALL

British primatologist Jane Goodall (b.1934) is the world's leading expert in chimpanzees. In 1957 she travelled to Kenya, where she met paleoanthropologist Louis Leakey (see p.128). With Leakey's support, Goodall set up a research base in Gombe, Tanzania, in 1960. Her observations of chimpanzee behaviour and their use of tools changed the understanding of not just chimpanzees, also led to critical steps in understanding the close relationship between *Homo sapiens* and chimpanzees.

Ardipithecus (4.4 million years old), and *Australopithecus afarensis* (3.85–2.95 million years old) support the early evolution of upright walking among hominids and also an African base for early human evolution. From 4 million years ago, evolution of the small, ape-like australopithecines and their relatives gave rise to the first more human-like *Homo habilis* by 2.4 million years ago.

LAETOLI FOOTPRINTS
Some 3.6 million years ago in Laetoli, east Africa, two upright-walking human-related adults and a juvenile walked on wet volcanic ash, leaving the oldest known footprints.

HUMAN ATTRIBUTES

The attributes that distinguish human species (members of the genus *Homo*) from other extinct human relatives have become less definite in recent decades. Traits that were once attributed to human species are now known to be more ancient. Upright walking originated some 6 million years ago. The use and manufacture of basic tools began at least 3.3 million years ago and of more sophisticated tools by 2.9 million years ago, well before the earliest *Homo* species. The latter tools are associated with remains of the muscular-jawed australopithecine *Paranthropus*. However, the earliest known human species, *Homo habilis* can be distinguished by its smaller cheek teeth and enlarged incisors, which indicate a change from the australopithecine diet including tough plant materials to a more varied and nutritious one that included animal tissues, perhaps scavenged from carcasses.

By 1.9 million years ago, at least one other tool-making species of *Homo* had evolved in Africa. *Homo erectus* had an even more human anatomy. It was taller (145–185 cm/4ft 9in–6ft 1in), more slender, more mobile, and had a larger brain (600–910 cubic cm/ 37–56 cubic in). These were the first

DEVELOPING SKILLS
The fossil record tends to preserve only stone and bone tools, but some of the most primitive tools were probably digging sticks.

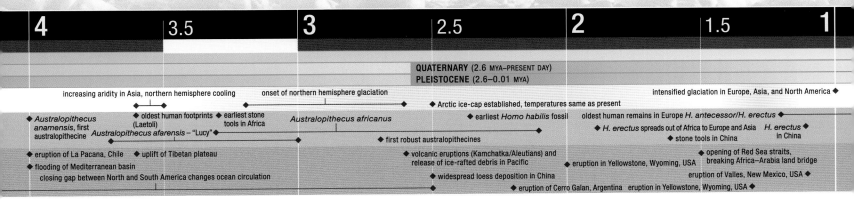

QUATERNARY (2.6 MYA–PRESENT DAY)
PLEISTOCENE (2.6–0.01 MYA)

increasing aridity in Asia, northern hemisphere cooling   onset of northern hemisphere glaciation   intensified glaciation in Europe, Asia, and North America ◆

◆ Arctic ice-cap established, temperatures same as present

◆ *Australopithecus anamensis*, first australopithecine   ◆ oldest human footprints (Laetoli)   ◆ earliest stone tools in Africa   *Australopithecus africanus*   ◆ earliest *Homo habilis* fossil   oldest human remains in Europe *H. antecessor/H. erectus* ◆

Australopithecus afarensis – "Lucy" ◆   ◆ *H. erectus* spreads out of Africa to Europe and Asia   *H. erectus* ◆ in China

◆ first robust australopithecines   ◆ stone tools in China

◆ eruption of La Pacana, Chile   ◆ uplift of Tibetan plateau   ◆ volcanic eruptions (Kamchatka/Aleutians) and release of ice-rafted debris in Pacific   ◆ opening of Red Sea straits, breaking Africa–Arabia land bridge

◆ flooding of Mediterranean basin   ◆ eruption in Yellowstone, Wyoming, USA   eruption of Valles, New Mexico, USA ◆

closing gap between North and South America changes ocean circulation   ◆ widespread loess deposition in China

◆ eruption of Cerro Galan, Argentina   eruption in Yellowstone, Wyoming, USA ◆

CAVE ART
Representations of animals, some extinct and others no longer living in the same region, point to the antiquity of much rock art, which is otherwise very difficult to date.

human relatives to spread beyond Africa, reaching Georgia in western Asia by 1.8 million years ago and Southeast Asia (Java) by at least 1.6 million years ago.

In Europe, the earliest human relatives date from around 1.2 million years ago. Found in Spain, *Homo antecessor* has evolutionary links to the older *Homo erectus* and the younger *Homo heidelbergensis*. The latter dates from around 700,000 years ago and was widespread from southern Africa, Eastern Asia, and

Europe. It was well built, with males averaging 1.57m (5ft 2 in) in height and females 1.75m (5ft 9in). *Homo heidelbergensis* was probably ancestral to the extinct Eurasian species *Homo neanderthalensis*, around 300,000 years ago, and the African *Homo sapiens* from around 200,000 years ago. *Homo heidelbergensis* is also ancestral to another species from western Asia called the Denisovans, from 200,000–30,000 years ago, which is known only from its DNA and a few fossil bones. All three species may have met up in western Asia, but only *Homo sapiens* survived the rigours of the Quaternary Ice Ages.

The evolution of sophisticated tools and culture allowed *Homo sapiens* to expand globally with a population of more than 8 billion. Recent discoveries include *Homo naledi* from South Africa (living between 335,000–235,000 years ago) and *Homo*

HUMAN EVOLUTION
Recent discoveries, mostly in Africa, show that human ancestry extends further back than previously thought. The result is a shrub-like evolutionary tree with about 20 known species of extinct human relatives, arranged in seven genera.

floresiensis, from the Indonesian island of Flores (living between 100,000–50,000 years ago), whose ancestry is still unclear.

The Ice Ages

The recent geological past has seen the Earth plunge into an ice-house phase of alternating cold and warm climates known as Ice Ages. Global climate cooled after the mid-Miocene, around 16 million years ago. From around 3 million years ago, the acceleration of cooling led to the onset of glaciation in the northern hemisphere and the formation of the northern hemisphere ice caps around 3 million years ago.

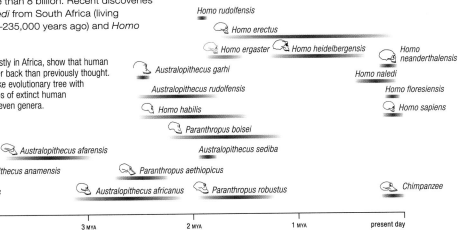

Homo rudolfensis

Homo erectus

Homo ergaster   *Homo heidelbergensis*   *Homo neanderthalensis*

Australopithecus garhi   *Homo naledi*

Australopithecus rudolfensis   *Homo floresiensis*

Homo habilis   *Homo sapiens*

Paranthropus boisei

Australopithecus sediba

Australopithecus afarensis

Australopithecus anamensis   *Paranthropus aethiopicus*

Ardipithecus ramidus

Orrorin tugenensis

Sahelanthropus tchadensis   *Ardipithecus kadabba*   *Australopithecus africanus*   *Paranthropus robustus*   Chimpanzee

| 7 million years ago | 6 MYA | 5 MYA | 4 MYA | 3 MYA | 2 MYA | 1 MYA | present day |

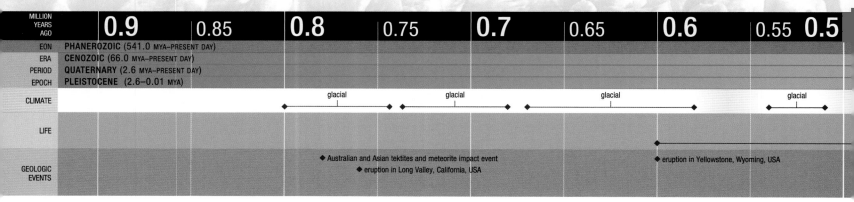

MILLION YEARS AGO	0.9	0.85	0.8	0.75	0.7	0.65	0.6	0.55	0.5
EON	PHANEROZOIC (541.0 MYA–PRESENT DAY)								
ERA	CENOZOIC (66.0 MYA–PRESENT DAY)								
PERIOD	QUATERNARY (2.6 MYA–PRESENT DAY)								
EPOCH	PLEISTOCENE (2.6–0.01 MYA)								
CLIMATE			glacial		glacial		glacial		glacial
LIFE									
GEOLOGIC EVENTS			◆ Australian and Asian tektites and meteorite impact event ◆ eruption in Long Valley, California, USA				◆ eruption in Yellowstone, Wyoming, USA		

The causes of global climate cooling are complex (see pp.450–53) but include the interplay of astronomical cycles related to fluctuations in the Earth's orbit around the Sun, which produces a 100,000-year cycle, and variation in the tilt of the Earth's axis of rotation, which produces a 41,000-year cycle. Other factors include changes to the circulation of ocean currents brought about by plate movement. The Quaternary Ice Ages began around 2.6 million years ago and have been marked by 50 cycles of alternating colder glacial and warmer interglacial climates.

SEA-LEVEL AND TEMPERATURE
Changes in global average temperatures over the latter part of the Quaternary ice ages, with their fluctuating cold and warm spells, are closely tracked by related sea-level changes.

Temperature

Sea-level

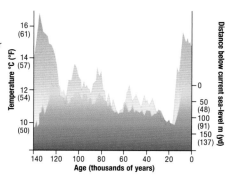

POST-GLACIAL TEMPERATURE
Global temperatures over the last 10,000 years, since the end of the last glacial advance, have been unusually stable, but may not remain so for long.

FORAMINIFERANS

The shell composition of tiny, sea-dwelling single-celled animals called foraminiferans records changes in the chemistry of ocean waters. By recovering such fossil shells from ocean-floor sediments and analysing their chemistry over successive generations, the changes in ocean chemistry can be reconstructed. Oxygen-isotope ratios in the calcium carbonate of the shells are measured, and from this information past fluctuations in ocean temperature, ice volume, and hence climate change can be recovered.

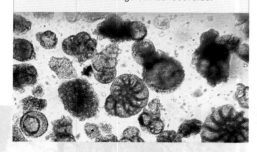

ADVANCING GLACIERS
Extensive ice-sheets and glaciers first developed in the northern hemisphere about 2.6 million years ago. An ice-cap formed over the Arctic Ocean, which eventually extended as far south as today's New York in North America, close to London, England, and St. Petersburg in Russia. Beyond the ice, permafrost extended as far south as the Black and Mediterranean seas. Mountain glaciers developed globally, even in the tropics. Landscapes were dramatically altered with upland glacial erosion and lowland deposition of glacial debris.

From such extremes of cold, the numerous ice-age cycles saw interglacial climates, which were occasionally warmer than those of today. For

instance, 125,000 years ago elephants and hippos flourished in England, where previously there had been mammoths and giant deer. The climate cycles were accompanied by changes in sea-level, allowing migration of animals and peoples and then isolating them on offshore islands.

LIFE ADAPTS TO CHANGE
Climate chage, such as that of the ice ages, has a considerable impact on life. In the face of climate change, some species evolve in response to the

PURPLE SAXIFRAGE
Some flowering plants, such as the Arctic bog cotton and purple saxifrage, are well adapted to tundra conditions.

QUATERNARY ICE-CAPS
Polar ice-caps grew and retreated several times during the Quaternary, reaching the maximum extent shown here.

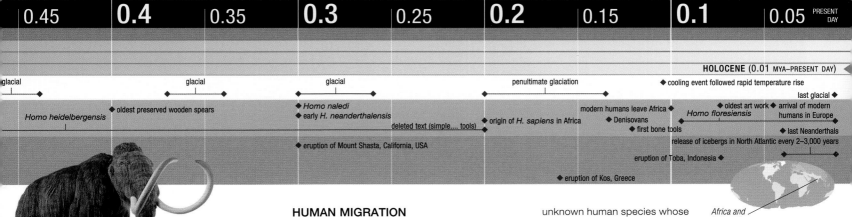

HOLOCENE (0.01 MYA–PRESENT DAY)

glacial · glacial · glacial · penultimate glaciation · cooling event followed rapid temperature rise

last glacial ◆

Homo heidelbergensis · ◆ oldest preserved wooden spears · ◆ *Homo naledi* · ◆ early *H. neanderthalensis* · modern humans leave Africa ◆ · *Homo floresiensis* · ◆ oldest art work ◆ arrival of modern humans in Europe

deleted text (simple.... tools) · ◆ origin of *H. sapiens* in Africa · ◆ Denisovans

◆ first bone tools

◆ last Neanderthals

◆ eruption of Mount Shasta, California, USA

release of icebergs in North Atlantic every 2–3,000 years

eruption of Toba, Indonesia ◆

◆ eruption of Kos, Greece

Africa and India still moving north

MAMMOTH
Well adapted to the cold, the woolly mammoth had insulating hair, a thick layer of fat, and tusks to clear snow from vegetation.

new conditions, others shift their geographic ranges to track suitable conditions, and still others go extinct. As climate cooled, some animals adapted to conserve heat better through improved insulation (provided by fatty tissues, hair, feathers, or clothing) and more compact bodies in which surface area was lower in relation to volume. The mobility of many animals allows them to migrate towards their preferred climates but rooted plants have more of a problem. They have to rely upon reproduction and dispersal for survival of successive generations. Cold-adapted plants have to be able to withstand freezing air and soils with short growing seasons. As a result they tend to have small leaves, shallow roots, and ground-hugging form.

CARIBOU MIGRATION
Reindeer, also known as Caribou, are cold-adapted herd animals that survive in polar regions by migrating over long distances between winter and summer feeding grounds.

HUMAN MIGRATION
Both archaeological and biomolecular evidence show that the first migration of modern humans *(Homo sapiens)* from Africa was initiated about 100,000 years ago. The migration reached the western shores of the Mediterranean Sea by around 90,000 years ago, but then went no further. Around 60,000 years ago, another wave of migration left the African continent, reaching China 68,000 years ago, Australia 50,000 years ago, and then Eurasia 40,000 years ago. With the lowered sea-level at the time, the existence of the Bering landbridge from Siberia to North America allowed humans to reach South America by 11,500 years ago.

Newly discovered fossil remains from the remote Denisova Cave in Siberia, Russia, and biomolecular data have revealed previously

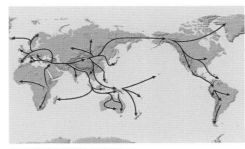

MIGRATION OF HOMO SAPIENS
Through the collection of fossil records, the pathways by which modern humans *(Homo sapiens)* are thought to have achieved global distribution have been reconstructed. The journey is believed to have begun in east Africa about 120,000 years ago.

unknown human species whose existence indicates that human migration was a lot more complex than once thought, especially within Eurasia.

RECENT EXTINCTIONS
Over the last 50,000 years, there has been continuing global extinction of the megafauna – animals over 45kg (100lb) in weight, such as the mammoth. Other victims were the Neanderthals and Denisovans whose demise followed the arrival of modern humans about 40,000 years ago. However, comparison of Neanderthal, Denisovan, and modern human DNA suggests that they interbred before dying out. Genes were shared between western Eurasians and Neanderthals (some 1–4 per cent), some Asians and the Denisovans (some 5 per cent), and between the Denisovans and some Neanderthals (some 17 per cent). The exact cause of the megafauna extinction, whether climatic and environmental change, human hunting, or a combination of the two, is still debated. However, recent accurate dating, especially in Australasia, which suffered little climate change, shows that extinctions closely follow the arrival of modern human hunters. The exception is Africa, where modern humans co-existed with large mammals for much longer, but now rising populations are threatening even Africa's big game.

Of course, human influence on the Earth goes far beyond that of hunter. The encroachment on almost all natural habitats, from the rainforests to the polar wastes, along with the effects of human-enhanced global warming, pose a far greater threat to the planet's flora and fauna.

RAW MATERIAL
Lava pouring across the Pacific island of
Hawaii is a vivid demonstration of Earth's
hot interior, and the processes that
formed the diverse minerals and rocks
at the surface.

THE ANATOMY OF THE EARTH

Planet Earth is a complex structure that is made up of three primary layers: the core, the mantle, and the crust. The density and temperature of the materials that make up the Earth increase approaching the metallic iron core, which has a solid interior and a molten exterior. The brittle outer layer of the planet, however, is made of many different kinds of rock, some up to 4 billion years old. Rocks consist of combinations of minerals, and are formed by various geological processes. These geological processes include internal forces, such as the movement of tectonic plates and volcanic eruptions, layering and compression of sediment, and interaction with living organisms and the atmosphere.

PLANET EARTH

The Earth's Structure

10–31	The history of the Earth
	The core 36–37
	The mantle 38–39
	The crust 40–41
	Minerals 42–61
	Rocks 62–83
	Tectonic plates 86–87
	Plate boundaries 88–89

Everyday experience on the Earth might suggest that our planet is remarkably varied – its materials range from water and ice to atmospheric gases and a host of rocks and minerals. However, the thin surface biosphere we occupy is not typical of the Earth as a whole. Overall, the planet is much less varied – within a few tens of kilometres below the surface, it consists only of rocks, minerals, and metallic compounds. Measurements of earthquake waves passing through the Earth allow us to probe its structure, since different layers propagate the waves at different speeds. They reveal the existence of a hot, dense core, surrounded by a mantle and a thin, rocky outer crust. The crust and a thin layer of mantle together form the jigsaw of tectonic plates that make up the Earth's outer shell.

INSIDE THE EARTH
The Earth has a layered internal structure, with each layer increasing in density and temperature. It is heated internally mainly by radioactive decay and by heat left over from its formation. Heat flows from the top of the core through the mantle, until it eventually reaches the cooler crust and escapes.

North-American Plate
Under Yellowstone Park, on the North American plate, lies a huge reservoir of magma (hot, melted rock) possibly originating from the mantle.

Shape and Form

The overall shape of the Earth, called the geoid, is determined by the effects of gravity and rotation on the materials that make up our planet. Gravity will pull any sufficiently massive object into a sphere – only smaller Solar System objects have non-spherical shapes. The Earth is an almost perfect sphere, but its rapid rotation every 24 hours – equivalent to the surface at the Equator moving at more than 1,600kph (1,000mph) – reduces the effect of gravity around the Equator, and means that equatorial regions bulge outwards by about 21km (13 miles) compared to the poles. The Earth's surface topography varies by about 20km (12½ miles) from the highest mountains to the deepest ocean trenches, and variations in surface elevation reflect two fundamentally different types of surface crust: continental crust with average elevation of less than 1km (⅗ mile) above sea-level; and oceanic crust with average depth about 4.5km (2⅔ miles) below sea-level. Gravity, coupled with processes such as tectonics (see pp.86–87) and erosion (see pp.92–93), make larger variations in elevation unsustainable over long periods.

Cascadia subduction zone
This marks where the small Juan de Fuca tectonic plate dips below the North American plate.

Gorda Ridge
This spreading ridge is formed by a shallow magma upwelling.

Hawaiian Island chain
These islands have formed during the past 30 million years from movement of the Pacific plate over a "hotspot" currently under the Big Island of Hawaii.

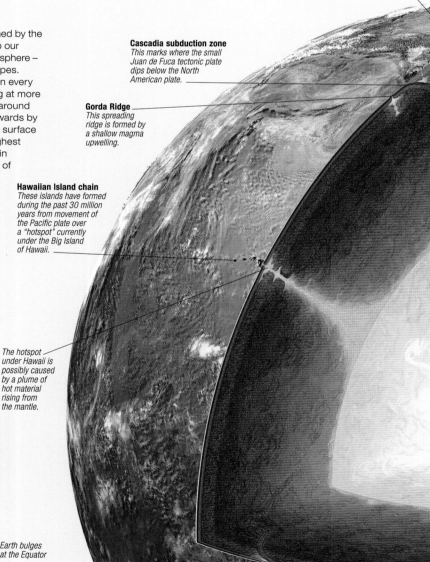

The hotspot under Hawaii is possibly caused by a plume of hot material rising from the mantle.

GREAT RIFT VALLEY
Tectonic processes have pushed up these mountains in east Africa, but erosion over millions of years will reduce them to insignificant hills.

EARTH IN SPACE
Seen from space, Earth is a blue planet. Two-thirds of the surface is covered by sea, and much of the land is often obscured by cloud.

direction of rotation (1,600kph/1,000mph at Equator)

Earth bulges at the Equator

CHANGING SHAPE
The Earth's high rotation rate slightly distorts its structure so that it bulges at the Equator and is flattened at the poles.

flattened pole

Tonga Trench
Up to 10,880m (35,702ft) deep, this marks where the Pacific plate dips below the small neighbouring Tonga plate.

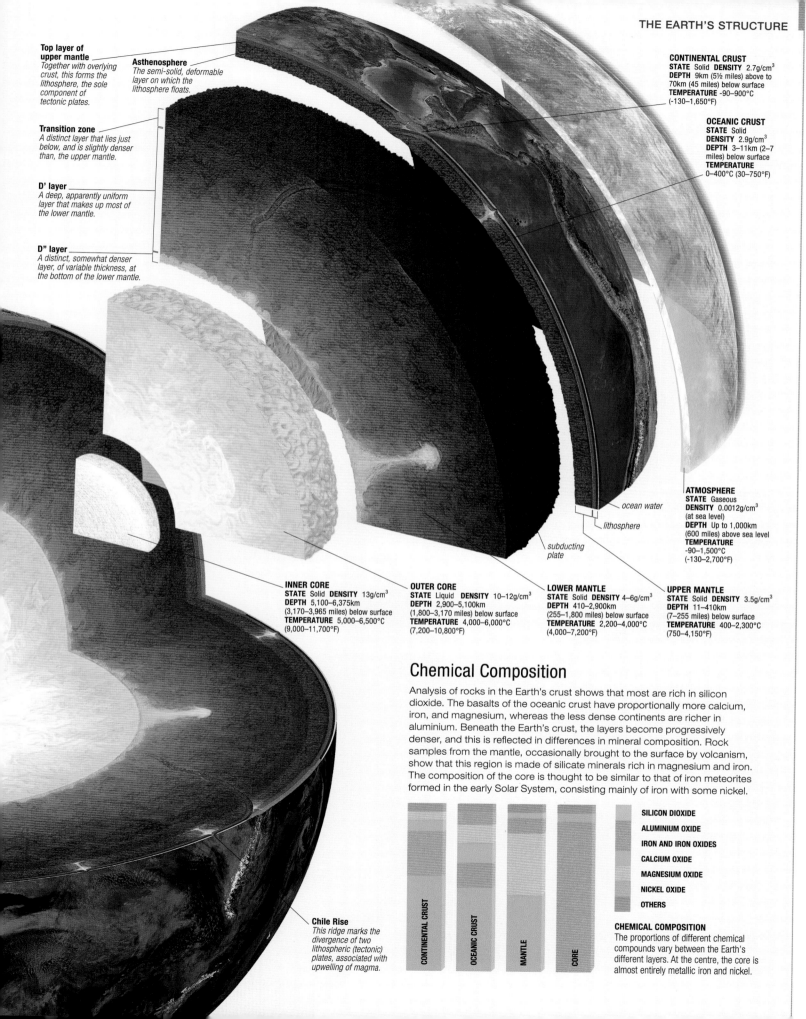

Top layer of upper mantle
Together with overlying crust, this forms the lithosphere, the sole component of tectonic plates.

Asthenosphere
The semi-solid, deformable layer on which the lithosphere floats.

Transition zone
A distinct layer that lies just below, and is slightly denser than, the upper mantle.

D' layer
A deep, apparently uniform layer that makes up most of the lower mantle.

D" layer
A distinct, somewhat denser layer, of variable thickness, at the bottom of the lower mantle.

CONTINENTAL CRUST
STATE Solid **DENSITY** 2.7g/cm³
DEPTH 9km (5½ miles) above to 70km (45 miles) below surface
TEMPERATURE -90–900°C (-130–1,650°F)

OCEANIC CRUST
STATE Solid
DENSITY 2.9g/cm³
DEPTH 3–11km (2–7 miles) below surface
TEMPERATURE 0–400°C (30–750°F)

ATMOSPHERE
STATE Gaseous
DENSITY 0.0012g/cm³ (at sea level)
DEPTH Up to 1,000km (600 miles) above sea level
TEMPERATURE -90–1,500°C (-130–2,700°F)

ocean water

lithosphere

subducting plate

INNER CORE
STATE Solid **DENSITY** 13g/cm³
DEPTH 5,100–6,375km (3,170–3,965 miles) below surface
TEMPERATURE 5,000–6,500°C (9,000–11,700°F)

OUTER CORE
STATE Liquid **DENSITY** 10–12g/cm³
DEPTH 2,900–5,100km (1,800–3,170 miles) below surface
TEMPERATURE 4,000–6,000°C (7,200–10,800°F)

LOWER MANTLE
STATE Solid **DENSITY** 4–6g/cm³
DEPTH 410–2,900km (255–1,800 miles) below surface
TEMPERATURE 2,200–4,000°C (4,000–7,200°F)

UPPER MANTLE
STATE Solid **DENSITY** 3.5g/cm³
DEPTH 11–410km (7–255 miles) below surface
TEMPERATURE 400–2,300°C (750–4,150°F)

Chemical Composition

Analysis of rocks in the Earth's crust shows that most are rich in silicon dioxide. The basalts of the oceanic crust have proportionally more calcium, iron, and magnesium, whereas the less dense continents are richer in aluminium. Beneath the Earth's crust, the layers become progressively denser, and this is reflected in differences in mineral composition. Rock samples from the mantle, occasionally brought to the surface by volcanism, show that this region is made of silicate minerals rich in magnesium and iron. The composition of the core is thought to be similar to that of iron meteorites formed in the early Solar System, consisting mainly of iron with some nickel.

Chile Rise
This ridge marks the divergence of two lithospheric (tectonic) plates, associated with upwelling of magma.

CONTINENTAL CRUST

OCEANIC CRUST

MANTLE

CORE

SILICON DIOXIDE

ALUMINIUM OXIDE

IRON AND IRON OXIDES

CALCIUM OXIDE

MAGNESIUM OXIDE

NICKEL OXIDE

OTHERS

CHEMICAL COMPOSITION
The proportions of different chemical compounds vary between the Earth's different layers. At the centre, the core is almost entirely metallic iron and nickel.

10–31 The history of the Earth

The mantle 38–39

The crust 40–41

Minerals 42–61

Rocks 62–83

Tectonic plates 86–87

PLANET EARTH

The Core

The Earth's core is far too deep to be studied by direct sampling, so our understanding of it depends on indirect investigations. Comparing the Earth's overall mass with the density of rocks close to its surface shows that parts of the Earth's interior must be much denser than its crustal rocks. Confirmation of a distinct core comes mainly from studies of earthquake waves travelling through the planet. These not only reveal the size of the core, they also show it has solid inner and liquid outer parts. Density calculations and the examination of meteorites formed at the same time and from the same materials as the Earth itself, have helped to establish that the core must consist mainly of iron, together with some nickel and small amounts of other elements. The nature of the Earth's magnetic field has also provided evidence of the core's structure and composition.

Evidence for the Core

An earthquake produces shock waves, some of which travel through the Earth's interior. Seismometers can be used to detect these planet-penetrating waves, where they arrive back at the surface. There are two different types of waves: P-waves and S-waves. P-waves passing straight through the Earth are significantly slowed down within a central region, indicating that this region must have a fundamentally different composition from the rest of the planet. There are "shadow zones" on Earth's surface where no P-waves are detectable following an earthquake, an effect of the core refracting P-waves. For every earthquake, there is also a large shadow zone for S-waves on the opposite side of the Earth. Since S-waves cannot pass through liquids, this implies that at least part of the core is liquid.

The Inner Core

The inner core was discovered in the 1930s, when Danish seismologist Inge Lehmann detected some weak P-waves (which she called P'-waves) in the P-wave shadow zone. She suggested that these must have reflected off a boundary between inner and outer regions within the core. It is now known that the inner core is a solid ball, consisting mainly of a crystalline iron-nickel alloy. Over recent years, evidence has been found that it may rotate within the outer core at a slightly different rate from the Earth as a whole, and that it may be gradually growing, or is possibly growing on one side and melting on the other.

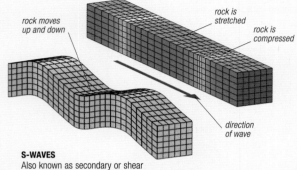

P-WAVES
Also known as primary or compressional waves, P-waves produce alternate stretching and compression in the material they pass through.

rock moves up and down

rock is stretched

rock is compressed

direction of wave

S-WAVES
Also known as secondary or shear waves, S-waves shake rocky material from side to side or up and down as they pass. Unlike P-waves, they cannot pass through liquids.

P-wave shadow zone

waves can be detected with seismometers where they arrive back at surface, which can be anything from a few seconds to more than 20 minutes after earthquake

S-wave

P-wave

waves refract (change direction) sharply when they encounter sudden changes in density

waves follow curved paths due to gradual changes in density with depth through the interior

S-waves are blocked where they encounter the liquid outer core

earthquake focus is point of origin of both P- and S-waves that travel through Earth's interior

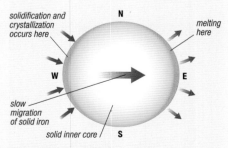

solidification and crystallization occurs here

N

melting here

W

E

slow migration of solid iron

solid inner core

S

CORE MELT AND FORMATION
Some evidence indicates that the Earth's inner core may be growing on one side, by solidification of outer core material onto it, and melting on the other.

SHOCK WAVE PATTERN
The pattern in which P- and S-type waves from an earthquake are detected around the surface of the world, and the time it takes them to reach various points, has proved that the Earth has a core that is partly liquid and partly solid.

Earth's Magnetic Field

An important effect of the Earth having an outer core of liquid iron (see below) is the generation of a substantial magnetic field. The lines of the field converge at two points on Earth's surface called the magnetic poles. The magnetic field extends far into space and creates a protective bubble that deflects high-speed streams of charged particles from the Sun, called the solar wind. Without this field, the solar wind would likely have stripped away Earth's atmosphere long ago and life on Earth would be very different. Studies of iron-rich materials in the Earth's crust have shown that at variable time intervals – from less than 100,000 to millions of years – the Earth's north and south magnetic poles have switched. These reversals are recorded in iron-containing minerals in sedimentary rocks and have proved important in deciphering Earth's history.

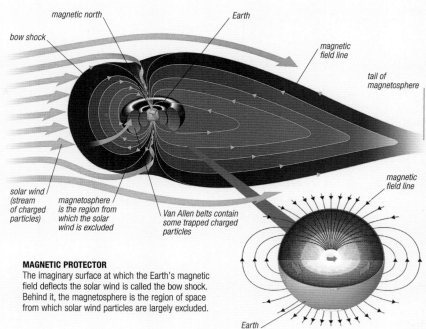

magnetic north

Earth

bow shock

magnetic field line

tail of magnetosphere

magnetic field line

solar wind (stream of charged particles)

magnetosphere is the region from which the solar wind is excluded

Van Allen belts contain some trapped charged particles

Earth

AURORAE
Electrically charged particles from the Sun, funnelled into the Earth's magnetic field above the poles, collide with atoms in the upper atmosphere to produce luminous streams of light known as aurorae.

MAGNETIC PROTECTOR
The imaginary surface at which the Earth's magnetic field deflects the solar wind is called the bow shock. Behind it, the magnetosphere is the region of space from which solar wind particles are largely excluded.

The Outer Core

The Earth's outer core is about 2,200km (1,370 miles) deep and makes up 95 per cent of the volume of the whole core. It is thought to consist mainly of iron with some nickel, and its temperature ranges from around 4,000°C (7,200°F) at the boundary with the mantle up to about 6,000°C (10,800°F) at its interface with the inner core. At these temperatures, the pressure in the outer core is insufficient to keep it solid; instead it is a low-viscosity liquid. An overall flow of heat occurs through the outer core. This heat is generated principally by solidification of molten metal onto the inner core (releasing latent heat). The heat flow sets up convection currents, which combined with the Earth's rotation sets up spiralling motions in the liquid metal, creating a magnetic field. The inner core may act to stabilize this magnetic field.

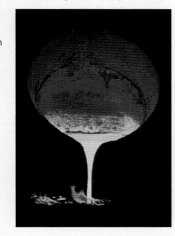

MOLTEN IRON
Metallic iron (seen here being poured in a foundry) has a high density and strongly magnetic character. It probably forms most of the Earth's core, along with a small percentage of nickel and some other lighter element, perhaps sulphur.

S-wave shadow zone, where no S-waves can be detected, provides evidence that the core is partly liquid

arrival of P-waves that have passed right through the planet is appreciably delayed because they have been slowed during their journey through the core

P-wave shadow zone, where no strong, direct, P-waves can be detected, provides evidence of a core that refracts P-waves

P'-wave (weak P-wave) within P-wave shadow zone provides evidence of an inner core that reflects some P-waves

Earth's rotational axis

spiralling convection currents influenced by rotation of the Earth

SPIRALLING CONVECTION
Acting like the spinning conductor in a bicycle dynamo, the liquid iron and nickel in the outer core continuously churns under the combined influence of the Earth's rotation and heat fluxes.

outer core

inner core

PLANET EARTH

PLANET EARTH

The Mantle

34–35	The Earth's structure
36–37	The core
The crust	40–41
Minerals	42–61
Rocks	62–83
How plates move	87
Volcanoes	154–77
Igneous intrusions	178–84

The Earth's mantle is known to be basically solid, but seismic studies indicate that it is not of uniform composition. Instead, it contains layers that increase in density with depth. Despite being solid, the rocks of the mantle appear to slowly deform, producing heat transfer by convection. Over the past 20 years, imaging of the mantle by means of earthquake waves, called seismic tomography, and other investigations have shown that the pattern of this convection is more complex than originally thought, and the exact nature of it is still being debated.

ANDRIJA MOHOROVICIC

The boundary between the crust and mantle was first detected and described by Croatian geophysicist Andrija Mohorovicic (1857–1936). By studying records of seismic waves generated by an earthquake near Zagreb in 1909, Mohorovicic noticed that some waves seemed to arrive earlier than others. He reasoned that they were affected by a density change within the Earth, around 30km (18 miles) deep. Now known as the Mohorovicic discontinuity, or Moho, its depth changes from 25–70km (16–45 miles) beneath the continents to 7–11km (4–7 miles) below the ocean floor. The discontinuity marks the compositional change between the crust and the underlying mantle.

The Upper Mantle and Transition Zone

From the base of the Earth's crust down to a depth of about 660km (410 miles) is a region now considered to have two main parts: the upper mantle down to about 410km (255 miles); and a denser region, the transition zone, forming the rest. Earthquake waves passing across the boundary at 410km suddenly speed up due to a change in density. Within the upper mantle are two distinct zones. The top layer is fused to the crust, forming the Earth's rigid outer shell (lithosphere), which is split into tectonic plates. Beneath this is a warmer, slightly more plastic layer, about 200km (120 miles) deep, called the asthenosphere. The predominant rock type in the upper mantle is peridotite, which consists mainly of the minerals olivine and pyroxene. In the transition zone, the increase in pressure alters the crystal structure or arrangement of atoms within olivine to produce denser versions of it.

LOCATION OF SLICE THROUGH THE EARTH

MANTLE XENOLITHS
A xenolith is a rock fragment of different origin from the igneous rock in which it is embedded. The dark xenolith here has come from the upper mantle.

PERIDOTITE
These igneous rocks in Newfoundland, Canada, are the remains of ancient ocean floor. They include coarse-grained, olivine-rich rocks called peridotites, thought to have similar composition to upper mantle rocks.

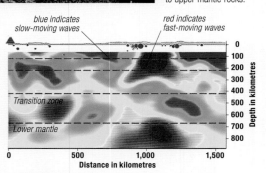

blue indicates slow-moving waves   *red indicates fast-moving waves*

Transition zone

Lower mantle

| Depth in kilometres |
| 0 |
| 100 |
| 200 |
| 300 |
| 400 |
| 500 |
| 600 |
| 700 |
| 800 |

0   500   1,000   1,500
Distance in kilometres

SEISMIC TOMOGRAPHY
This technique is used to generate images of the Earth's interior from measurements of seismic wave velocity, which varies with the temperature and composition of rocks.

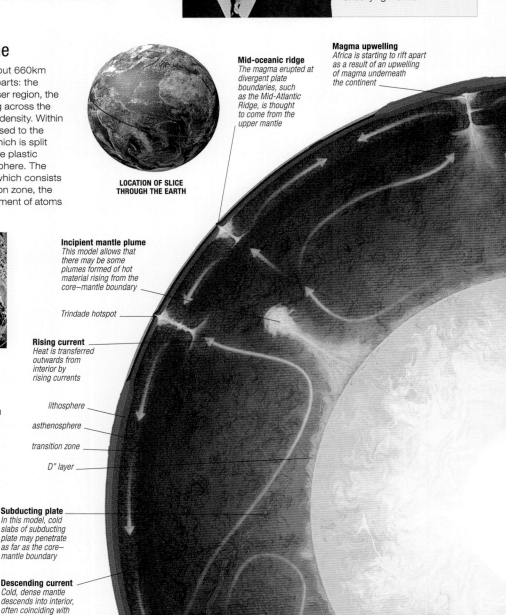

Magma upwelling
Africa is starting to rift apart as a result of an upwelling of magma underneath the continent

Mid-oceanic ridge
The magma erupted at divergent plate boundaries, such as the Mid-Atlantic Ridge, is thought to come from the upper mantle

Incipient mantle plume
This model allows that there may be some plumes formed of hot material rising from the core–mantle boundary

Trindade hotspot

Rising current
Heat is transferred outwards from interior by rising currents

lithosphere

asthenosphere

transition zone

D" layer

Subducting plate
In this model, cold slabs of subducting plate may penetrate as far as the core–mantle boundary

Descending current
Cold, dense mantle descends into interior, often coinciding with zones of plate subduction

South America

Colombian trench

The Lower Mantle

Earthquake waves reaching a depth of around 660km (410 miles) in the mantle significantly speed up, indicating a change in rock density. The region of Earth's interior that extends from this boundary down to the core–mantle boundary, at around 2,900km (1,800 miles), is known as the lower mantle. It is of mostly uniform composition, but at its base is a distinct sublayer, varying from about 50 to 200km (30 to 125 miles) deep, called the D'' layer. This appears to have a complex structure, involving local layering and regions of partial melt, possibly reflecting processes occurring at the core–mantle boundary. The D'' layer may be the site from which upwelling mantle plumes originate, as well as a repository for material from subducted plates. The most abundant minerals in the lower mantle contain magnesium, iron, or calcium in combination with silicate (silicon and oxygen). Of these, a substance called magnesium silicate perovskite is probably the most abundant mineral on Earth.

octahedral SiO$_6$ silicate units

encased magnesium atoms

magnesium atoms become arranged in layers

○ **MAGNESIUM ATOM**
● **OXYGEN ATOM**
● **SILICON ATOM**

MAGNESIUM SILICATE PEROVSKITE

POST PEROVSKITE

CRYSTAL COMPRESSION
High pressure in the D'' layer alters the crystal structure of magnesium silicate perovskite, turning it into denser post-perovskite. The atoms in post-perovskite are about 1 per cent more tightly packed.

WHOLE-MANTLE CONVECTION MODEL
In this "single layer" model of mantle convection, which is partly speculative, a series of convection cells gradually carry material from the bottom of the mantle to the top and vice versa.

Volcanic hotspot
Thought to have been intermittently active for 65 million years, this hotspot has most recently created the volcanic island of Réunion

Mantle plume
Some scientists think the Réunion hotspot is caused by a rising plume of hot material from deep in the lower mantle

Madagascar

FIRES WITHIN
Investigation of the chemistry of magma that reaches the Earth's surface in various volcanic settings, including hotspots and spreading ocean ridges, has provided some clues to the pattern of mantle convection.

Mohorovicic discontinuity

Sunda trench

Sumatra

Subduction zone
Volcanism in these regions results from magma produced by relatively shallow, upper-mantle processes

Borneo

Philippines

Philippines trench

Descending slabs
Lithosphere slabs do not penetrate as far as the core–mantle boundary

Mariana Islands

Mariana trench

Mantle Convection

At the base of the mantle, heat continuously flows out of the Earth's core. This rising heat causes the mantle to slowly circulate, its rocks shifting a few centimetres a year, carrying heat outwards by convection. At the top of the mantle, convection plays a part in driving the movement of tectonic plates. However, the overall pattern of convection in the mantle is not fully understood – a point of debate, for example, is whether it occurs in a single layer, as in the whole mantle convection model (see left) or in two layers, as in the deep layer model (below). In formulating different models, geophysicists strive to explain findings obtained from studies such as seismic tomography and chemical analysis of magma (presumed to come from the mantle) erupted in different settings at the Earth's surface. For example, any model has to explain why magmas erupted at volcanic hotspots in the middle of plates are different from those erupted at mid-ocean ridges. Some scientists think that the former are in many cases the result of streams of hot material, called mantle plumes, coming from the core–mantle boundary, whereas the latter are caused by shallow, upper mantle processes. Another point to be settled is whether slabs of plate subducted in zones such as the Sunda trench penetrate all the way down to the core–mantle boundary as part of mantle convection cells, or disintegrate at a shallower depth.

Deep layer
Convection of material here is largely separate from that in the rest of the mantle

Hotspot volcanism
Some deep layer material flows upwards as mantle plumes, creating hotspots

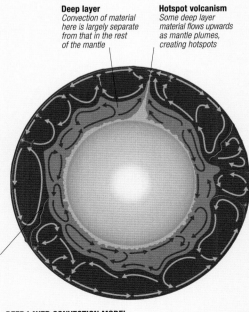

DEEP LAYER CONVECTION MODEL
In this "two layer" model, heat from the Earth's interior causes two distinct layers to slowly swell and shrink in complex patterns, without much mixing.

PLANET EARTH

The Crust

34–35	The Earth's structure
36–37	The core
38–39	The mantle
Minerals	42–61
Rocks	62–83
Tectonic plates	86–87
Plate boundaries	88–89
Volcanoes	154–77
Ocean tectonics	386–89

The thinnest and outermost layer of the Earth is the crust, with an average thickness of about 30km (18 miles) below the land surface and about 7.5km (4½ miles) below the seafloor. The crust and the uppermost layer of the mantle, to which the crust is fused, constitute Earth's lithosphere, which is split up into plates (see pp.96–97). In general, the rocks of the crust are less dense than mantle rocks, because they are richer in minerals containing relatively light elements. However, there are two distinctly different different types of crust – oceanic and continental. Variations in their composition, density, and thickness help to account for differences in their topography, relative age, and history of formation.

Oceanic Crust

The oceanic crust forms over two-thirds of the Earth's surface, but even the oldest parts of the ocean floor are no more than 200 million years old and covered only by a thin veneer of sediment. Composed of relatively dense basaltic lavas and related rocks, oceanic crust varies from 7–8km (4–5 miles) thick. Because of its density, it is less elevated than the lighter continental crustal rocks, with an average depth of about 3km (2 miles) below sea-level. Oceanic crust is continually formed from mantle material within the long rifts known as spreading ridges (see p.88). Here, the mantle is rising, heating, and expanding the overlying rocks to form an elevated submarine mountain chain, where oceanic plates on either side of the rift are carried apart like two opposing conveyor belts. Basaltic lavas erupt from fissures and cones, and cool to form new oceanic floor that is also punctuated by numerous individual volcanoes. These can grow large enough to rise above sea-level, and where the crust is moving above a semi-permanent mantle hotspot plume (see p.87), they may form chains of volcanic islands. Elsewhere, in areas called subduction zones, oceanic crust descends into the mantle at the same rate that it is created at the ridges (see p.89).

ANDESITE CONES, JAVA
When the energy released by subduction of oceanic crust partially melts mantle rocks, it can trigger the formation of curved chains of island volcanoes.

ICELANDIC RIDGE
Because of the density of their basaltic rocks, oceanic crust and mid-ocean ridges do not usually appear above sea-level. Iceland is an exception, raised above the waters by exceptional heat flow.

ATMOSPHERE
Above the crust lies the Earth's gaseous atmosphere. Its interaction with oceans and landmasses, and the protection it offers from damaging radiation, are vital for the survival of life.

OCEAN
The Earth is the only planet in the Solar System with persistent oceans of liquid water that cycle moisture through the atmosphere to the land and back to the oceans.

Basalt makes up much of the upper part of oceanic crust.

Sediments

Oceanic crust
This consists mainly of gabbro and basalt rock.

Mohorovicic discontinuity
The boundary between crust and mantle.

Uppermost mantle layer
This consists mainly of the coarse-grained igneous rock peridotite.

Asthenosphere
This deformable mantle layer lies just below the lithosphere.

OCEANIC LITHOSPHERE

OCEANIC LITHOSPHERE
This consists of a top layer of oceanic crust, about 7 to 8km (4 to 5 miles) thick, underlain by about 40 to 100km (25 to 60 miles) of the uppermost mantle layer. On top of the oceanic crust is a thin layer of sediments.

PLANET EARTH

Continental Crust

The continental crust accounts for only about one-third of the Earth's global surface, but it forms all the major landmasses and their fringing shallow seas. Its thickness ranges from 25 to 70km (16 to 45 miles), with the thickest portions underlying young mountain belts, and it has huge variations in composition – from sedimentary rocks such as sandstone, coal, and limestone (see pp.76–82) through metamorphic rocks such as marble and slate (see pp.71–75) to igneous rocks such as granite and gabbro (see pp.66–70). This variety of rock types has come about mainly because light continental crust is not recycled within the Earth to the same extent as denser oceanic crust. As a result, some continental rocks are up to 4 billion years old, and much of the material now above sea-level has been through repeated cycles of erosion, formation into sedimentary rocks, and metamorphosis. Tectonic forces have also subjected the continents to phases of fragmentation, amalgamation, and long-term movement, resulting in the formation of new oceans, mountain building, widespread volcanism, and occasional coalescence into single landmasses called supercontinents. Throughout all this, new layers of rock have constantly been added to the surfaces and margins of the continents, forming the strata that are the basis for much of our understanding of the geological history of the Earth and, through the fossil record, its life.

TIBETAN PLATEAU
Regions of high elevation such as this windswept plateau are underlain with continental crust that is far thicker than average.

ARAL BASIN
Low-lying areas of continental crust are also the thinnest. When surrounded by more elevated regions, they form natural drainage basins.

Isostasy

The continental crust is at its thickest beneath young mountain belts, where deep roots of crust stretch down to depths of about 70km (45 miles) – well into the underlying mantle. Crustal blocks can be thought of as floating on the mantle – a concept known as isostasy. Because the continental crust has a much lower density than the mantle, it is buoyant, in the same way that an iceberg is buoyant in seawater. But the greater the mass of rock above sea-level, the more buoyancy is required to support it, and the deeper its roots must go. By comparison, thinner, denser oceanic crust is less buoyant, and so reaches an equilibrium point well below sea-level which is why it is rarely seen as dry land.

SEA-LEVEL CHANGE
Old beaches and cliffs raised above present sea-level attest to past changes in the levels of both sea and land.

PLANET EARTH

FRESH WATER
Water precipitated from the atmosphere forms lakes and rivers as it collects and flows back towards the ocean, eroding the landscape through which it flows.

VEGETATION
Land vegetation has evolved over more than 450 million years and has played a significant role in altering global climates and forming deposits such as soils and coals (see p.115).

Metamorphic rocks such as hornfels

Intrusion of granite, an igneous rock

Layers of sedimentary rock, such as sandstone and limestone

Continental crust
This consists of a wide variety of rock types.

Mohorovicic discontinuity

CONTINENTAL LITHOSPHERE

Uppermost mantle
As in oceanic lithosphere, this layer consists mainly of peridotite.

Asthenosphere
A deformable mantle layer, the asthenosphere lies just below the lithosphere.

CONTINENTAL LITHOSPHERE
This consists of a top layer of about 25 to 80km (16 to 50 miles) of continental crust underlain by about 80km (50 miles) of uppermost mantle layer. As in oceanic lithosphere, a change in chemical composition occurs at the Mohorovicic discontinuity.

Minerals

34–35 The Earth's structure

40–41 The crust

Rocks 62–63

Water content and deposits 185

Underground rivers and caves 240–41

Mineral extraction 375–77

Minerals are the natural building blocks of the rock materials of the Earth and all the solid bodies of the Universe. Studying minerals helps us to understand the origin and evolution of the Earth and planets. Most minerals are solid crystalline substances composed of atoms, which are generally arranged in orderly, repeating patterns, giving the mineral its crystalline structure and shape. A few minerals, however, have no such orderly crystalline structure but are amorphous solids similar to glass. Although more than 5,000 minerals have been discovered, only about 30 (known as rock-forming minerals) are common at the Earth's surface.

GEODE INTERIOR
Natural rock cavities, or geodes, often become infilled with concentric or radial crystal growths when permeated by mineral-rich solutions. Here, an agate geode has quartz minerals growing in its centre.

Mineral Formation

Individual minerals are distinguished by unique combinations of chemical elements and the arrangement of their atoms. This is determined when a gas or liquid crystallizes into a solid. Generally, the resulting structure is a regular, repeating three-dimensional array, or lattice, of atoms. The lattice grows by the ordered addition of atoms and usually maintains the same shape and composition. Some minerals have well-developed crystal form, but most are aggregates with other textures and habits. Similarly sized mineral grains have a granular texture, while a mineral lacking crystal faces is called massive. Primary minerals are formed at the same time as the host rock; later changes from, for example, erosion or metamorphism, produce secondary minerals.

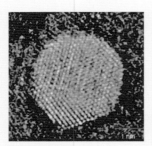

REGULAR MATRIX
Gold atoms grown on a carbon substrate naturally bond and pack together in an orderly cubic structure.

Crystal Structure and Shape

A crystal is a solid such as a mineral with an orderly, repeating atomic structure. With unrestricted growth, it forms a geometric shape with naturally flat planes, called faces, which are often arranged symmetrically. Faces are the external expression of a mineral's internal atomic regularity of structure, and they lie at specific angles one to another, depending on the crystal system to which the mineral belongs. Six main crystal systems – cubic, hexagonal/trigonal, monoclinic, orthorhombic, tetragonal, and triclinic – are defined by their geometric symmetry. Slight misalignment of the structure during growth often produces twinned crystals that are the mirror image of one another. Crystal shape is described using terms such as platy for thin, flat forms and prismatic for columnar crystals.

CRYSTAL SHAPES
Crystal systems are defined by the proportions of their axes and the angles between them (see open figures below). Crystals in any one system can assume various shapes (one of which is represented by a solid figure below).

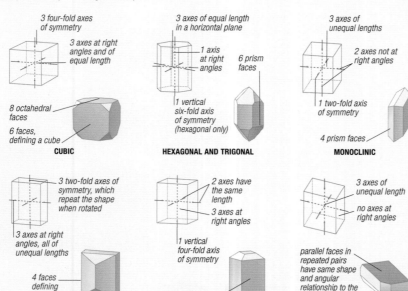

CUBIC
- 3 four-fold axes of symmetry
- 3 axes at right angles and of equal length
- 8 octahedral faces
- 6 faces, defining a cube

HEXAGONAL AND TRIGONAL
- 3 axes of equal length in a horizontal plane
- 1 axis at right angles
- 6 prism faces
- 1 vertical six-fold axis of symmetry (hexagonal only)

MONOCLINIC
- 3 axes of unequal lengths
- 2 axes not at right angles
- 1 two-fold axis of symmetry
- 4 prism faces

ORTHORHOMBIC
- 3 two-fold axes of symmetry, which repeat the shape when rotated
- 3 axes at right angles, all of unequal lengths
- 4 faces defining a prism

TETRAGONAL
- 2 axes have the same length
- 3 axes at right angles
- 1 vertical four-fold axis of symmetry
- 4 prism faces

TRICLINIC
- 3 axes of unequal length
- no axes at right angles
- parallel faces in repeated pairs have same shape and angular relationship to the crystal axis

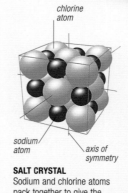

SALT CRYSTAL
- chlorine atom
- sodium atom
- axis of symmetry

Sodium and chlorine atoms pack together to give the mineral halite (salt) a cubic form. Its crystals growing as perfect cubes with six square faces at right angles to one another, as can be seen here.

SALT CRYSTAL

Mineral Identification Tests

For 2.6 million years, humans and their extinct relatives have exploited particular minerals, and rocks for a variety of purposes, from making stone tools to swords, ploughshares, space shuttles, and exquisite jewellery. To use minerals to their best advantage, it is first necessary to identify their individual characteristics. Over the millennia, many minerals have been discovered to possess unique combinations of physical properties, such as colour, hardness, and specific gravity, which allow them to be picked out even when surrounded by other minerals in a rock. Several of these properties, such as crystal shape, hardness, specific gravity, and cleavage, can be related to actual crystal structures, and are thus predictable. The specific gravity of an open-structured mineral such as graphite, for example, is low, while diamond, with the same composition but a more closely packed structure, has a higher specific gravity. Expert mineralogists can recognize hundreds of different minerals largely on the basis of their appearance and a few tests with a steel point and a hand lens. The identification of otherminerals, however, needs optical, chemical, X-ray, or other kinds of analysis.

cleavage plane
CALCITE

MUSCOVITE

CLEAVAGE
This describes a plane of weak bonding between atoms in a crystal, along which the crystal may split. Calcite has three cleavage planes, which meet to form rhomb-shaped cleavage fragments, while muscovite mica has well-developed and distinct cleavage in just one plane.

FRACTURE
Resulting from mineral breakage other than along cleavage planes, fracture frequently shows characteristic forms, such as opal's glass-like, conchoidal (rounded) fracture or kaolinite's irregular one.

OPAL
curved fracture plane

KAOLINITE

TALC
DIAMOND

HARDNESS
The strength of a mineral's chemical bonds determines its hardness. This is measured by its resistance to scratching on a 1–10 scale, which was invented by Friedrich Mohs in 1822; soft (talc) is 1 and hard (diamond) is 10.

SPECIFIC GRAVITY
This is determined by the structure and composition of a mineral. The weight of a mineral's atoms relate to its specific gravity. Gold has a higher specific gravity than halite because its atoms are heavier. Specific gravity is measured by comparing the weight of the mineral with an equal volume of water.

HALITE

GOLD

COLOUR
The composition of a mineral, its atomic structure, often the presence of certain trace elements, and selective light absorption combine to generate the colour of a mineral. The element copper in azurite produces its typical blue colour. Other minerals are uncoloured or white, because no light is absorbed, as in milky quartz.

AZURITE
MILKY QUARTZ

LUSTRE
The way in which light is absorbed or reflected off the surface of a mineral determines its lustre, which is independent of a mineral's colour. It may be diagnostic, as in smoky quartz, which shows a vitreous or glassy lustre, and galena, which has a metallic one.

glassy lustre
SMOKY QUARTZ

metallic lustre
GALENA

TRANSPARENCY
This is the extent to which light can pass through a mineral. It ranges from the water-clear transparency of the rock crystal form of quartz, to the complete opacity of many metal ore minerals, such as magnetite.

ROCK CRYSTAL
MAGNETITE

STREAK
The colour of a mineral when crushed to a powder provides its streak, which often differs from the mineral's surface colour, especially in certain ore minerals. For example, hematite varies in its colour, but its streak is always red.

red streak

HEMATITE

LIGHT STUDIES
As light passes through many minerals, it is altered by the optical properties of the structure and chemistry of the mineral. By using polarizing microscopes and extremely thin slices of a rock, which transmit light, mineralogists can distinguish the minerals present. Some minerals, however, are opaque and appear black – such as spinel (shown here).

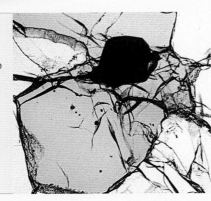

PLANET EARTH

Classifying Minerals

Many of the 4,000 or more known natural minerals have similarities in their properties. These have led to them being grouped according to whether they are chemical compounds, such as oxides or silicates, rather than minerals of particular elements, for example, copper or calcium. These groupings are reinforced by their occurrence in similar geological settings. Minerals are chemical compounds with a stability based on an electrical balance between two parts. One positively charged (cationic) part is balanced by a negatively charged (anionic) part, and this is reflected in their chemical formula. The positive part is usually a metal, such as iron (Fe), and is written first. After it comes the negative part, usually a non-metallic ion, such as oxygen (O) or sulphur (S), or a combination of negatively charged elements, such as sulphate (SO_4). Minerals are classified by their anionic groupings, and are commonly divided into the ten groups shown here. Each grouping contains many minerals: for example, the carbonates include over 200 minerals ranging from calcite to azurite.

GOLD

NATIVE ELEMENTS
A group of about 20 elements, of which 10 are geologically significant. Native elements include minerals such as gold (Au) and platinum (Pt). In each of these, the single element is free and uncombined. Metallic native elements with relatively low melting-points, such as copper, formed the historical basis for metal technology. The non-metals sulphur and carbon (especially as diamond) are also economically important.

SULPHIDES
A group of about 600 minerals, such as stibnite (Sb_2S_3) and galena (PbS), in which sulphur (S) combines with metallic and metal-like elements. Most sulphides are opaque, with distinctive colours and characteristic streaks.

STIBNITE

OXIDES AND HYDROXIDES
Two closely related groups of about 400 minerals, in which metallic elements combine with oxygen (O_2) or hydroxyl (OH). Oxides and hydroxides contain some important ore minerals, such as chromite ($FeCr_2O_4$) and goethite (FeO(OH)).

CHROMITE

HALIDES AND FLUORIDES
Two closely related groups of about 140 minerals, in which fluorine (F), chlorine (Cl), bromine (Br), or iodine (I) combines with elements, such as calcium (Ca) or sodium (Na). Fluorite (CaF_2) is an example.

FLUORITE

CARBONATES
A group of more than 200 minerals, in which a metal combines with strongly bonded carbonate (CO_3) units. Most do not contain water (H_2O), but a few do, such as azurite. Calcite ($CaCO_3$) is an important carbonate.

CALCITE

BORATES
A group of 125 minerals in which metal elements combine with borate (BO_3), for example borax ($Na_2B_4O_5(OH)_4.8H_2O$). Borate units form chains and sheets that are similar to those found in SiO_4 groups in silicates.

BORAX

SULPHATES
A group of about 300 minerals in which metallic elements combine with sulphate (SO_4), for example anhydrite ($CaSO_4$). Sulphate forms a fundamental structural unit, because sulphur has a strong bond with oxygen.

ANHYDRITE

MOLYBDATES AND TUNGSTATES
Two closely related groups of minerals, in which metallic elements combine with molybdate (MoO_4) or tungstate (WO_4), such as in wulfenite ($PbMoO_4$) and wolframite $(Fe,Mn)WO_4$.

WULFENITE

PHOSPHATES AND VANADATES
Two closely related groups of more than 400 minerals, in which elements combine with phosphate (PO_4) or vanadate (VO_4); examples include apatite $Ca_5(PO_4)_3$ (F,Cl,OH) and carnotite ($K_2(UO_2)2(VO_4)_2.3H_2O$).

APATITE

SILICATES
A group of about 500 minerals in which metallic elements combine with silicon and oxygen (SiO_4) to form structural units known as tetrahedra, for example olivine ($(Mg,Fe)_2SiO_4$). These can link singly, in sheets, in chains, or three-dimensionally.

OLIVINE

ABRAHAM GOTTLOB WERNER

Abraham Werner (1749–1817) was a German mineralogist and teacher, who attracted students from all over Europe, even the Americas, to the Freiberg Mining Academy. He produced one of the earliest classification systems for minerals, and was a leading proponent of the Neptunist theory, which claimed that all rocks were laid down sequentially in an ocean created by the Flood described in the Bible. Igneous and metamorphic rocks were deposited, then sedimentary rocks, and finally surface sediments.

Rock-Forming Minerals

By far the most abundant elements in the Earth's crust are oxygen and silicon, respectively 46.6 and 27.7 per cent by weight, then aluminium at 8.3 per cent and iron at 5.0 per cent. Together these elements account for 87.6 per cent by weight of the crust. Therefore, silicate minerals, such as olivine, pyroxene, amphibole, feldspar, and quartz, which combine silicon with oxygen, make up more than 90 per cent of the rocks of the Earth's surface. They are the main constituents of the commonest rocks, especially igneous ones, many metamorphic rocks, and clay-rich sedimentary ones. The non-silicate minerals contain different groups of chemicals, such as oxygen combined with carbon to form carbonates (as in limestone), which are important sedimentary rocks. Oxygen can also combine with metals to form oxides such as hematite. Most rocks are comprised of just a few essential minerals, but may also contain small quantities of various other – so-called accessory – minerals.

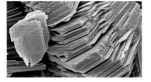

KAOLINITE
Many silicate clay minerals, such as kaolinite (above), form minute platy crystals. These produce a single, well-defined cleavage that is parallel to the sheet-like, platy surfaces.

RIVER MUD
This deposit of river mud (above) is largely made of silicate clay minerals derived from the weathering or alteration and erosion of other silicate minerals such as feldspars.

MAMMOTH MINERAL GROWTH
One of the most accessible processes of mineral formation is to be seen around hot springs, here at Yellowstone, USA. Evaporation of hot, mineral-enriched water leads to the formation of minerals, especially carbonates.

IRON DEPOSITS
The sedimentary iron ore deposits of the Hamersley Range in Australia contain about 40 billion tonnes of ore with a lucrative 55 per cent iron concentration. They date from the Precambrian Eon.

HYDROTHERMAL VEINS
Many ore deposits are derived from hot, mineral-enriched solutions, which invade the surrounding rocks through any available cracks, cavities, or pores, where they form hydrothermal veins.

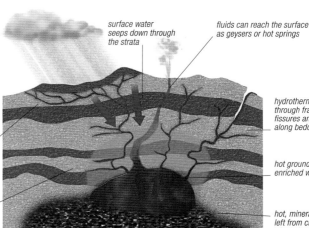

surface water seeps down through the strata

fluids can reach the surface as geysers or hot springs

weathering turns ore minerals at surface into different minerals

hydrothermal fluids rise through fractures or fissures and migrate along bedding planes

hot groundwater enriched with minerals

mineral deposits form in small cracks or permeable zones adjacent to veins

hot, mineral-rich liquid left from crystallization of granite pluton

Ore Deposits

Mineral deposits from which valuable metals can be extracted profitably are known as ores. Market conditions determine whether a naturally occurring concentration of a mineral can be regarded as an ore: the cost of extraction is justified only if it is outweighed by the deposit's market value. Platinum, for example, is so valuable that it need only have a concentration of more than 0.1 parts per million to be exploitable, whereas iron is less valuable and so the ore needs to contain a much higher iron content, perhaps 50 per cent, if extracting it from the ground is to be a viable proposition

SUB-SURFACE EXTRACTION
Valuable ore deposits may be found underground and tunnels are dug deep into the Earth to reach and extract the ore.

Gems

Minerals that have been fashioned for use as personal adornment are known as gems. The most ancient known decorations, some 30,000 years old, were made of bone, shell, and amber, but, since stone cutting and polishing technology has developed, most materials selected as gems today are natural, inorganic minerals and commonly crystalline. They are valued as such because they are beautiful, rare, and durable. Out of the 4,000 or more different minerals in the Earth, only 50 or so are typically used as gems. Such is the value of many gems that there has been a very successful technological drive to manufacture synthetic ones, generally using the same chemical compounds as the natural stones themselves.

fluorite

amethyst quartz

OCTAGONAL STEP CUT

OVAL STEP CUT

sphalerite

BRILLIANT CUT

TYPES OF CUT
Brilliant cuts are best for delicately coloured gems such as sphalerite, to give brightness and "fire", whereas step cuts show deeply coloured stones to best advantage.

CUTTING STONES
Gem cutting is designed to enhance a stone's colour and beauty. For each gem mineral, selected and angled flat faces (facets) are cut or ground and polished to help reflect light from the stone.

uncut Burmese ruby crystal

cushioned mix cut ruby

MINERAL PROFILES

The pages that follow contain profiles of a selection of minerals. Each profile begins with the following summary information:

COMPOSITION Chemical formula

CRYSTAL SYSTEM Cubic, tetragonal, orthorhombic, monoclinic, triclinic, or hexagonal/trigonal

HARDNESS Resistance to scratching, as measured on Mohs' scale, ranging from 1 to 10 and changing in steps of half a unit

SPECIFIC GRAVITY The ratio of the mineral's weight to the weight of an equivalent volume of water, expressed to one decimal place

COLOUR A brief description, including common variations

NATIVE ELEMENTS
Gold

COMPOSITION Au	CRYSTAL SYSTEM Cubic
HARDNESS 2½–3	SPECIFIC GRAVITY 19.3

COLOUR Golden-yellow; golden-yellow streak

The appeal of gold can be traced through several millennia of human history. This precious metal is malleable, soft, and has a relatively low melting-point (1,062°C/1,944°F), which means that it can easily be shaped into a variety of forms, from thin sheets to wire; it can be melted and cast, too. Although generally golden, its colour depends

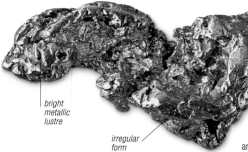

GOLD IN QUARTZ GROUNDMASS
Native gold grains, some as branching, dendritic crystals, are scattered through this vein quartz specimen from a hydrothermal vein.

tiny, branching crystals and grains of gold

vein quartz

bright metallic lustre

irregular form

on impurities: for example, red or pink gold contains traces of copper, and white gold is mixed with silver, platinum, zinc, or nickel, which also makes it harder. Gold is

NATIVE GOLD
Isolated nuggets of gold are occasionally found by miners panning placer deposits in streams, but rarely are they as big as this one.

relatively non-reactive, thereby retaining its lustre. Its abundance in the Earth's crust (less than 0.01 parts per million) is low compared with most other metals, and its deposits are scattered around the world. Even low-grade deposits can still be of economic value because of the high price this mineral commands. It is sometimes found associated with quartz in hydrothermal veins and with copper ores. Weathering and erosion of gold-bearing rocks and the waterborne transport of their debris result in gold and other heavy minerals with high specific gravity being set down as placer deposits. They can be of any geological age, and some are extensive. Most of the world's extracted gold is held as bullion for international financial settlements and increasing amounts as coinage and small bars for investment. Gold is also used to make jewellery and has some more utilitarian applications – it is used to make dental instruments, electronic components, and tarnish-free platings. About 30 per cent of world gold production is currently mined by China, Russia, and Australia.

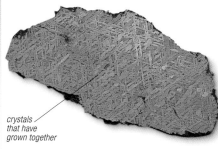

NATIVE ELEMENTS
Silver

COMPOSITION Ag	CRYSTAL SYSTEM Cubic
HARDNESS 2½–3	SPECIFIC GRAVITY 10.0–11.0

COLOUR White to grey, tarnishing to black; exhibits greyish white streak

wiry growth habit

greyish tarnish on weathered surfaces

Native silver, like gold, is a rare, valuable mineral that occurs as nuggets, grains, and wiry branching growths in hydrothermal veins. It is soft and easy to shape; consequently, over the ages, this metal has been used for decorative and utilitarian objects from crowns to cutlery and coins. When polished, it has a bright grey-white metallic colour, but it gradually tarnishes black on exposure to the atmosphere. Almost all silver mined today is separated from other minerals, especially large-scale sulphide ore deposits. Mexico and China are major producers.

NATIVE ELEMENTS
Copper

COMPOSITION Cu	CRYSTAL SYSTEM Cubic
HARDNESS 2½–3	SPECIFIC GRAVITY 8.9

COLOUR Copper-red, tarnished brown, may stain green; pinkish red streak

Copper is a relatively abundant metal in the Earth's crust and occurs mostly in sulphide ores such as chalcopyrite. Native copper is found mainly in basaltic lavas, and is formed by the reaction of hydrothermal solutions with iron oxide minerals. It typically occurs as masses or dendrites, and rarely as crystals. Native copper is found as a mineral only in small amounts, while compounds of copper and other elements are more widespread. Historically, copper has been very important in the development of metal technology. Its abundance and ease of working led to widespread use from the fifth to third millennia BCE, in the Copper Age. Later, copper was combined with tin to make bronze, a tougher and harder alloy; the Bronze Age lasted from 2200 to 800 BCE. Copper can

branching copper dendrites

goethite matrix

COPPER ON GOETHITE BASE
Here, dendritic native copper, with its typical branching form, has developed on a goethite mineral surface. The surface of the dendrites has been oxidized to a reddish brown colour.

easily be drawn into wire and is a very good conductor of heat and electricity, so is widely used in the electrical industry as well as to make pipes, boilers, and roofs (see panel, below).

COPPER ROOFS
Native copper is soft and malleable, and can be beaten or rolled into thin but tough and weatherproof sheets. These are readily formed into intricate curves and so are invaluable for cladding roofs with complex shapes, such as domes, spires, and cupolas. Although the copper on the roof surface typically weathers to a characteristic green colour, the metal will not deteriorate underneath, and the roof will remain in good condition for hundreds of years.

NATIVE ELEMENTS
Iron

COMPOSITION Fe	CRYSTAL SYSTEM Cubic
HARDNESS 4½	SPECIFIC GRAVITY 7.3–7.9

COLOUR Steel-grey to black; grey streak

The Earth's core is predominantly made of metallic iron with some nickel, and it is this iron composition in the core's outer liquid and inner solid parts that gives the Earth its magnetic field. Native iron is also found in some meteorites and as a mineral in the Earth's crust, although this is rare since iron readily combines with oxygen and water to form oxides and other minerals. It is occasionally found in some altered basalts where iron minerals have been reduced to native iron. The large amounts of iron that have been used since the Iron Age in the first millennium BCE are derived from smelting of iron minerals such as hematite.

crystals that have grown together

WEATHERED SURFACE
In iron meteorites, iron and nickel combine to form metal alloys with complex crystalline intergrowths. Transit through the atmosphere melts the outer layer to produce a shiny, pitted surface texture.

NATIVE ELEMENTS

Sulphur

COMPOSITION S   **CRYSTAL SYSTEM** Orthorhombic

HARDNESS 1½–2½   **SPECIFIC GRAVITY** 2.0–2.1

COLOUR Bright yellow to brownish yellow; white streak

Native sulphur typically forms bright yellow crystals, often with prism or pyramid shapes. It is also common as encrusting masses around volcanic vents and fumaroles and in some sedimentary rocks, especially evaporites. The crystals grow from sulphur-rich liquids and gases, such as sulphur dioxide, on the rock surfaces. Sulphur can be distinguished by its colour, low melting-point (112.8°C/235°F) – it ignites in a candle flame – and relatively low hardness. While native sulphur is unusual as a mineral, the element sulphur is common in many different minerals, especially the sulphide ore minerals and sulphates such as gypsum. The latter occurs in evaporite deposits, especially salt domes around the Gulf of Mexico, dating from the Cenozoic Era. Native sulphur is an extremely important source of the element, which is widely used in the chemical and

tabular crystal

resinous lustre

YELLOW COATING
These distinctive sulphur crystals have grown on a rock surface.

pharmaceutical industries: for example, in making sulphuric acid, dyes, insecticides, and fertilizers. About 3.5 million tonnes of native sulphur are obtained globally each year from mining and extracting, especially in the USA and Poland, and nearly 50 million tonnes are recovered each year from other sulphur ores such as pyrite. Sulphur, along with carbon dioxide, chlorine, and fluorine, is released in large quantities in volcanic eruptions – hence its occurrence at or near the craters of volcanoes.

MINERAL FORMATION
These sulphur deposits and other evaporite minerals develop from vents on Anak Krakatau volcano.

NATIVE ELEMENTS

Graphite

COMPOSITION C   **CRYSTAL SYSTEM** Hexagonal

HARDNESS 1-2   **SPECIFIC GRAVITY** 2.1–2.3

COLOUR Black; grey-black streak

The name graphite is derived from the Greek *grapho* ("to write"), because this soft, opaque, and dark-coloured mineral readily marks paper with an erasable grey deposit, a property used to make pencil "lead". Graphite is just one form of carbon found in the natural world – among others are diamonds, organic-derived charcoal, and coal. Graphite is uncommon, but occurs as scattered grains or flakes in some carbon-rich metamorphic rocks and as veins in pegmatites. Graphite's scaly crystals slide over one another to give a greasy feel, promoting its use as a lubricant in many industries.

metallic lustre

perfect cleavage

NATIVE ELEMENTS

Diamond

COMPOSITION C   **CRYSTAL SYSTEM** Cubic

HARDNESS 10   **SPECIFIC GRAVITY** 3.5

COLOUR Mainly colourless, but coloured varieties occur; white streak

Perhaps the most famous mineral of all, because of its remarkable hardness and its brilliance when cut as a gemstone, diamond's real value as jewellery has been appreciated only since modern techniques of cutting have fully revealed its "fire" and sparkle. This results from the internal reflection of light after the careful cutting of facets at certain angles to one another. Coloured diamonds occur naturally and are highly valued, especially if red,

octahedral diamond crystal

kimberlite rock

TRUE GEM
This naturally occurring, eight-sided diamond crystal is still embedded in the kimberlite rock in which it was formed by high pressures and temperatures deep within the Earth's interior.

green, or blue, although colouring is usually the result of artificial irradiation. In recent decades, the value of diamond as an abrasive, for cutting metal and other stones, and providing hard-wearing bearings, has outweighed its decorative value. More than 116 million carats (23 tonnes) are mined each year, mainly in Australia, Canada, the Democratic Republic of Congo, Botswana, South Africa, and Russia (see panel, above). Diamonds originate in pipe-like intrusions called kimberlites, which have arisen from great depths (150km/ 90 miles or more) in the Earth's crust. Within kimberlite pipes, diamonds are formed from carbon at high temperature

ROUND-SHAPED

PEAR-SHAPED

BRILLIANT CUT

DIFFERENT CUTS
As well as occurring in many different colours, diamonds can be cut, or faceted, to enhance their natural "fire".

MINING DIAMONDS

The majority of diamonds are now mined on a large industrial scale by just a few major corporations, such as Alrosa, using heavy machinery and automated processing. The value of diamonds, however, is so high that there are numerous smaller mining operations around the world, such as this one in Koidu, Sierra Leone. Often these operations are in disputed or war-torn territories in less-developed countries, and have lasting impacts on the surrounding communities and environments. Here, impoverished miners work in dangerous conditions in the hope of finding diamonds, which they sell cheaply to middlemen or warlords.

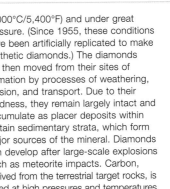

(3,000°C/5,400°F) and under great pressure. (Since 1955, these conditions have been artificially replicated to make synthetic diamonds.) The diamonds are then moved from their sites of formation by processes of weathering, erosion, and transport. Due to their hardness, they remain largely intact and accumulate as placer deposits within certain sedimentary strata, which form major sources of the mineral. Diamonds can develop after large-scale explosions such as meteorite impacts. Carbon, derived from the terrestrial target rocks, is fused at high pressures and temperatures, resulting in exceedingly small crystals, called micro-diamonds, found in deposits that have fallen from the air.

THE OPPENHEIMER DIAMOND
This 253.7-carat diamond crystal was found near Kimberley, South Africa, in 1964. It was later given to the Smithsonian Institution, USA.

MINERAL TREASURE
Human society needs Earth's rocks and minerals.
Here, gold is being mined in Alaska, USA. This kind
of open-cast mining is safer than deep mining
but scars landscapes and has the potential to
contaminate local groundwater.

Pyrite

COMPOSITION FeS$_2$	CRYSTAL SYSTEM Cubic
HARDNESS 6–6½	SPECIFIC GRAVITY 4.9–5.2

COLOUR Pale brass-yellow, darker if tarnished; greenish black streak

The name pyrite is derived from the Greek *pyros* ("fire"), because this mineral gives off sparks when struck sharply. It is commonly known as "fool's gold", as its golden yellow colour can lead to it being mistaken for gold, but it is much harder than gold and not malleable. It usually forms characteristic, well-shaped cubic crystals or twelve-faced crystals called pyritohedrons, with each face having five sides. Pyrite is a relatively common and widespread iron mineral, that occurs in a number of different geological settings. In igneous rocks, it is prevalent as an accessory mineral but it may also occur concentrated in

a variety of ore deposits. These range from massive, economically important sulphide ore bodies to smaller hydrothermal veins and replacement deposits. Pyrite is also widely distributed in sedimentary rocks, particularly fine-grained mud-rocks deposited in stagnant, low-oxygen conditions. Sulphur-reducing bacteria promote the formation of tiny pyrite crystals within the mud, and these are preserved as the sediments are consolidated into rock. When mud-rock is metamorphosed into slate, the pyrite may grow into larger crystals and sometimes replaces shells or bones of fossils.

octahedral pyrite crystal

quartz crystals

OCTAHEDRAL PYRITE
Perfectly formed pyrite crystals are here associated with clusters of small quartz crystals from a hydrothermal vein. The crystals grew from solutions enriched in iron sulphide and silica, respectively.

Sphalerite

COMPOSITION ZnS	CRYSTAL SYSTEM Cubic
HARDNESS 3½–4	SPECIFIC GRAVITY 3.9–4.1

COLOUR Yellow-brown to black; variable streak from white to pale yellow to brown

Also known as zincblende, sphalerite is the most common zinc mineral and thus the main source of this metal and rare elements such as cadmium. Its high zinc content (of about 67 per cent) was, however, discovered only in the 18th century. As a mineral, sphalerite is variable in colour and can be difficult to distinguish from minerals such as magnetite. Indeed, the name sphalerite is derived from the Greek *sphaleros* ("treacherous"), because of its lack of identifying features apart from a resinous lustre and well-developed cleavage. Sphalerite frequently occurs with galena and other sulphide minerals in hydrothermal veins, which are mined for their zinc, mainly in China, Australia, and Canada. Sphalerite also occurs in limestone by replacement of the carbonate minerals to form ore bodies with magnetite and pyrrhotite. Transparent brown or green crystals are sometimes found, and these can be faceted as gems but are too soft to be useful jewels.

Stibnite

COMPOSITION Sb$_2$S$_3$	
CRYSTAL SYSTEM Orthorhombic	
HARDNESS 2	SPECIFIC GRAVITY 4.5–4.6

COLOUR Lead-grey; grey streak

This mineral is the major source of the relatively rare metal antimony (only 0.2 parts per million in the Earth's crust), a toxic element used for hardening metal alloys for bearings, lead in batteries, and semiconductors. Stibnite occurs as elongate radiating crystals or massive forms. The latter can be confused with galena, but stibnite's crystal form is distinctive, as is its low melting-point – it liquifies in a match flame. It occurs associated with other sulphides in hydrothermal veins, hot-spring deposits, and within limestone. Most of the world's annual stibnite production comes from China and Tajikistan.

long stibnite crystals

quartz and barite matrix

Chalcopyrite

COMPOSITION CuFeS$_2$	CRYSTAL SYSTEM Tetragonal
HARDNESS 3½–4	SPECIFIC GRAVITY 4.1–4.3

COLOUR Golden brassy-yellow; greenish black streak

quartz

metallic lustre

As a copper-iron sulphide, chalcopyrite is the most important copper ore. It occurs as masses and sometimes crystals in many geological settings, such as hydrothermal veins and a number of different metamorphic and igneous rocks. Economically, the most important occurrences of chalcopyrite are in porphyry copper deposits, where veins of sulphide minerals are associated with large igneous intrusions. Pyrite and bornite are commonly present there, too. With its golden colour, chalcopyrite is similar to pyrite but is more yellow and has tetragonal, not cubic, crystals.

Bornite

COMPOSITION Cu$_5$FeS$_4$	CRYSTAL SYSTEM Cubic
HARDNESS 3	SPECIFIC GRAVITY 5.0–5.1

COLOUR Brownish bronze, tarnishing purple-blue and black; black streak

A relatively dense and brittle mineral with a characteristic iridescent tarnish, bornite contains about 63 per cent copper, 11 per cent iron, and 26 per cent sulphur. Only occasionally does it form crystals; normally it occurs in masses in pegmatites and other igneous rocks and also in hydrothermal veins with other sulphide minerals such as chalcopyrite. Bornite is relatively common and, because of its high metal content, it is an economically important copper mineral. The largest deposits are in Mexico and Butte, Montana, USA. It owes its common name of peacock ore to its characteristic tarnish, with iridescent colours of blue to violet to red.

blue-black tarnish

Cinnabar

COMPOSITION HgS	
CRYSTAL SYSTEM Trigonal	
HARDNESS 2–2½	
SPECIFIC GRAVITY 8.0–8.2	

COLOUR Blood-red; deep vermilion-red streak

The very high density of cinnabar distinguishes it from another bloodred mineral – realgar. Cinnabar is the most common and economically important mercury mineral and contains nearly 87 per cent mercury. Both massive and crystalline forms occur within fracture zones around active volcanoes and hot

springs. Mined since the 7th century BCE, cinnabar was a source of black and red pigments and quicksilver (mercury) for mirrors; it is also used in drugs, batteries, electrical instruments, and many chemicals.

massive form

small crystals

Realgar

COMPOSITION AsS	CRYSTAL SYSTEM Monoclinic
HARDNESS 1½–2	
SPECIFIC GRAVITY 3.5	

COLOUR Orange-red; orange-red streak

This naturally occurring but rare arsenic sulphide mineral forms grains and well-shaped crystals. The latter have a striking, orange-red colour with a resinous lustre. Realgar looks similar to cinnabar but is softer and less dense. It typically occurs as a secondary mineral along with yellow orpiment, another arsenic sulphide, in hydrothermal veins

and hot-spring deposits. In this context, realgar (70 per cent arsenic) is produced by the decomposition of other arsenic minerals such as arsenopyrite. Highly toxic, it was used in medieval medicine and glassmaking, and is now used in fireworks and pesticides.

rock matrix

prismatic realgar crystal

quartz

SULPHIDES

Galena

COMPOSITION PbS **CRYSTAL SYSTEM** Cubic

HARDNESS $2\frac{1}{2}$

SPECIFIC GRAVITY 7.4–7.6

COLOUR Lead-grey; lead-grey streak

The most common and economically important lead ore (it is about 87 per cent lead), galena forms crystals, masses, and granules. It occurs in a variety of geological settings, from metamorphic rocks to volcanic related, massive sulphide deposits (including sphalerite and copper ores), and large ore bodies in reef limestones and dolomites associated with hydrothermal fluids. Galena is easily distinguished by its grey colour, metallic sheen, high density, and perfect cubic cleavage – it readily breaks into small cubes. Galena was one of the first metal ores to be mined:

cubic crystal

metallic lustre

it was smelted by the Babylonians to make lead vases, and it was much sought after by the Romans (see panel, above). The word galena is Latin for lead ore. Lead derived from galena has been used for printing, in paint, batteries, and in plumbing and electrical wiring, but its toxicity has now greatly reduced its worth, especially for domestic uses. China, Australia, and the USA produce half of the world's annual lead output, which comes mostly from galena.

CRYSTALS OF GALENA
Well-formed crystals, such as these cubes of galena, have perfect cubic cleavage and can easily break into even smaller cubes.

LEAD AND THE ROMANS

Archaeological finds show that, more than two thousand years ago, the Romans made extensive use of lead for domestic purposes. They discovered that this malleable metal could be cast and beaten into complex shapes, allowing the manufacture of lead vessels and pipes for conducting and storing water around their buildings. Demand for the metal was one reason for the Roman invasion of the British Isles, as there were lead and copper ores in Cornwall. Roman smelting of lead and copper ores between 500 BCE and 300 CE caused a small but significant rise in atmospheric pollution, as recorded in Greenland ice cores, before it declined again.

DUAL PURPOSE
Under Emperor Hadrian, the Romans used lead ingots both as a unit of currency and in manufacturing.

massive galena

COMMON LEAD ORE
Galena, with its bright lustre, is frequently found in massive form (such as here) along with other sulphide minerals in hydrothermal veins.

OXIDES

Spinel

COMPOSITION $MgAl_2O_4$

CRYSTAL SYSTEM Cubic

HARDNESS $7\frac{1}{2}$–8 **SPECIFIC GRAVITY** 3.5–4.1

COLOUR Very variable or colourless; white-brown streak

quartz matrix

eight-faced spinel crystal

The composition of this magnesium oxide mineral varies greatly, with iron, chromium, zinc, and manganese atoms substituting for magnesium. Consequently, spinel forms a series of minerals with variable properties, especially colour. Some varieties are gem minerals, but large crystals are rare. Red spinel resembles ruby, and can be produced synthetically. Typically, spinel is finely granular, but it also occurs as octahedral crystals in gabbroic igneous rocks and as a result of thermal metamorphism of dolomites. These latter occurrences, along with river sediments derived from them, produce most gem spinel.

OXIDES

Hematite

COMPOSITION Fe_2O_3

CRYSTAL SYSTEM Trigonal

HARDNESS 5–6 **SPECIFIC GRAVITY** 4.9–5.3

COLOUR Reddish brown, steel-grey to black; red streak

This is probably one of the first minerals to have been exploited by humans. When ground, hematite produces a powder called red ochre, which has been used as a pigment and is found in association with burials dating back some 80,000 years. It is distinguished by its hardness and blood-red streak; indeed, its name is derived from the Greek *haima* ("blood"). Widely distributed, hematite is the most

hexagonal outline

metallic lustre

MIRROR IMAGE
Specular hematite, a crystalline form of hematite, has such a highly reflective crystal face that it was once used for mirrors.

important iron ore mineral. It is found in a variety of forms ranging from tabular crystals to bulbous masses with radiating and striated interiors. Hematite occurs in a number of different geological settings, including hydrothermal veins and as a rare component of igneous rocks. It is most common in ancient sedimentary rocks known as banded ironstones (which date from the early Precambrian Eon). More than half of the world's annual production of iron ore comes from Australia, Brazil, and China, mostly from hematite.

quartz crystals

mass of specular hematite crystals

MIXED CRYSTALS
Numerous tiny hematite crystals grow alongside prismatic quartz crystals, as here, on a surface of massive red hematite ore.

OXIDES

Magnetite

COMPOSITION Fe_3O_4

CRYSTAL SYSTEM Cubic

HARDNESS $5\frac{1}{2}$–$6\frac{1}{2}$ **SPECIFIC GRAVITY** 5.2

COLOUR Black; black streak

Like hematite, magnetite is an important iron ore. It has strong magnetic properties that were recognized by the Chinese in the 11th century BCE. Magnetite is also mentioned by the Roman naturalist Pliny the Elder, who suggested that the name was derived from Magnes, a shepherd whose iron-nailed shoes stuck to rocks containing that mineral. Magnetite occurs in massive, granular, and crystalline forms with crystals of eight triangular faces. Widely distributed, it is a common but mostly minor mineral in many igneous rocks, where it sometimes occurs in

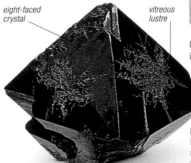

eight-faced crystal

vitreous lustre

MIXED CRYSTALS

PANNING STREAMS

Dense and hard minerals such as magnetite resist weathering and erosion. With patience, they can be washed from other placer minerals and sediment in a shallow pan – but this practice is barely viable. Magnetite is shown here being panned in Ivory Coast.

layers. Magnetite is also often found in metamorphic rocks, some mineral veins, and alluvial placer deposits (see panel, above).

DIRECTION FINDERS
For at least three millennia, humans have used naturally magnetic crystals of magnetite, such as this one, as compasses for finding the way.

Ilmenite

COMPOSITION $FeTiO_3$

CRYSTAL SYSTEM Trigonal

HARDNESS 5–6 **SPECIFIC GRAVITY** 4.5–5.0

COLOUR Black; brownish red to black streak

tabular ilmenite

Named after the Ilmen Mountains in Russia, ilmenite is an oxide of iron (48 per cent) and titanium (52 per cent) oxide, which occurs in massive form and as tabular crystals. Its appearance is similar to magnetite, and it is slightly magnetic. It is the world's main source of titanium, occurring as a minor constituent of igneous rocks such as gabbro and diorite, and occasionally in quartz veins and in some metamorphic rocks. Resistant to weathering, ilmenite can accumulate as placer deposits in some river sands, along with magnetite and rutile. Canada, Australia, and South Africa produce some two-thirds of the world's annual output of titanium concentrates, mainly from ilmenite.

Corundum

COMPOSITION Al_2O_3

CRYSTAL SYSTEM Trigonal

HARDNESS 9 **SPECIFIC GRAVITY** 3.9–4.1

Colour Variable, including brown, pink, red (ruby), blue (sapphire), and colourless forms

Several famous and very valuable gem forms, such as ruby (red) and sapphire (blue; see panel below), occur in corundum. When polished, some of these show a six-rayed star of light reflected from tiny, needle-shaped crystals of rutile within the corundum crystal, when viewed at a particular angle. Corundum's natural crystals

reflective crystals

STAR CABOCHON
The rounded shape of a cabochon (a stone cut without facets), such as this ruby, best reveals the star effect from light reflected by tiny rutile crystals.

have rough, barrel, or spindle shapes, and it also occurs with spinel and magnetite as a massive, black, granular

form known as emery, which is used as an industrial abrasive. Corundum is found in sodium-rich, granite-related igneous rocks and in their pegmatites, occasionally growing into large crystals. It also develops in a variety of metamorphic rocks, ranging from gneiss and schist to marble. Second only in hardness to diamond, corundum survives erosion to accumulate in alluvial sands and gravels as placer deposits, especially in Sri Lanka and Myanmar. Synthetic production of gem-quality corundum began in the late 19th century, and by 1902 the Frenchman Auguste Verneuil had produced the first synthetic ruby by fusing aluminium oxide and colouring material with the flame from a blowtorch. Different trace elements vary the colour of the synthetic gem: for example, chromium creates ruby red, and iron and titanium makes sapphire blue.

rough, columnar crystal

BRILLIANT CUT SAPPHIRE

SAPPHIRE CRYSTAL
Blue corundum looks undistinguished as a decorative jewel but careful cutting reveals the outstanding gem quality of sapphire.

CORUNDUM

SAPPHIRES

Since the Middle Ages, corundum gems of any colour except ruby red have been known as sapphires, but the name is now associated only with blue gems. These can vary from pale to very dark blue to almost black; some gems appear as different shades of blue according to the light. One colour description – heavenly blue – has inevitably become associated with tranquillity, purity, and peace.

Most of the early stones used to come from Sri Lanka and India, where they occurred in pegmatites and placer deposits. They are now found around the world, from the metallic-blue sapphires of Montana, USA, to the dark blue and nearly black stones from Nigeria.

TREASURED JEWELLERY
Blue sapphires are much sought after for their putative spiritual qualities.

Chromite

COMPOSITION $FeCr_2O_4$

CRYSTAL SYSTEM Cubic

HARDNESS 5½ **SPECIFIC GRAVITY** 4.1–5.1

COLOUR Black to brownish black; brown streak

nodular chromite crystal

Cassiterite

COMPOSITION SnO_2

CRYSTAL SYSTEM Tetragonal

HARDNESS 6–7 **SPECIFIC GRAVITY** 6.8–7.1

COLOUR Yellow-brown-black-red; white or grey streak

One of the earliest ores to be exploited for metal (it has been mined since 6000 BCE), cassiterite can be smelted to produce tin at about 1,000°C (1,800°F). This tin mineral is the world's main source of the metal (78 per cent tin), and it occurs in hydrothermal veins and pegmatites associated with granites.

This most important chromium mineral (chromium 73 per cent, iron 27 per cent) is typically massive or granular, but rare octahedral crystals do occur. Chromite is commonly found as a minor constituent of olivine-rich igneous rocks such as peridotite and serpentinite, but it may be concentrated into layers of sufficient commercial value to exploit. Because chromite is hard, dense, and resistant to erosion and weathering, it accumulates in alluvial sediments as a placer mineral. Its melting-point is 1,900°C (3,500°F). When alloyed with nickel, chromium is used for hardening steel; with iron, it is for making stainless-steel and chromium plating.

Placer deposits of alluvial cassiterite are also found. Cassiterite can be massive or granular or form columnar or pyramidal crystals. China, Indonesia, and Peru are major producers of this tin oxide mineral. Tin is used as a non-toxic, rust-proof coating for steel cans and, alloyed with lead, as a low-melting-point solder for electrical connections.

massive, rounded form
metallic lustre

Rutile

COMPOSITION TiO_2

CRYSTAL SYSTEM Tetragonal

HARDNESS 6–6 **SPECIFIC GRAVITY** 4.2–4.4

COLOUR Reddish brown to black; pale brown streak

prismatic crystal

This mineral was originally mistaken for tourmaline. It was only in 1795 that rutile was found to contain titanium (60 per cent), for which it is now the second most important source – ilmenite being the main one. Rutile occurs in massive and prismatic crystals, often striated and terminated with pyramidal faces. It grows, too, as fine needles within other minerals such as quartz and corundum. Found in various igneous and metamorphic rocks, rutile can also be produced by the breakdown of minerals such as sphene, and can occur as a placer deposit. Its Latin-derived name (rutilus, meaning "golden yellow") was suggested by Abraham Werner (see p.45).

Uraninite

COMPOSITION UO_2

CRYSTAL SYSTEM Cubic

HARDNESS 5–6 **SPECIFIC GRAVITY** 6.5–10.0

COLOUR Black to dark brown; grey to brownish black

The radioactive properties of this uranium oxide mineral (also known a pitchblende) were first discovered in 1896 by the French physicist Henri Becquerel, and the associated element radium was subsequently identified within it by the Polish chemist Marie Curie – both scientists using uraninite specimens from Jáchymov in Bohemia. Uraninite is usually massive and occurs in sulphide-bearing hydrothermal veins. Rare cubic crystals occur in granite pegmatites associated with minerals such as tourmaline and zircon. Sedimentary grains of uraninite are found concentrated in some placer deposits.

black pitchblende

HYDROXIDES
Geothite

COMPOSITION FeO(OH)

CRYSTAL SYSTEM Orthorhombic

HARDNESS 5–5½ **SPECIFIC GRAVITY** 3.3–4.3

COLOUR Yellowish to dark brown; yellow-brown streak

First identified as a mineral in its own right in 1806, goethite was named after the German poet J.W. Goethe, who was also a mineral collector. This hydrous iron oxide often occurs as striking, globular or stalactitic masses. Internally, it has a radiating, needle-shaped or fibrous structure. Larger crystals are rare but may be bladed or prismatic. Goethite is widespread but of little economic value; it is an oxidation product of iron minerals such as magnetite and pyrite. It is also a hydrothermal mineral and occurs in a form called bog iron ore in some sediments.

goethite crystals
quartz groundmass

HALIDES
Fluorite

COMPOSITION CaF$_2$

CRYSTAL SYSTEM Cubic

HARDNESS 4 **SPECIFIC GRAVITY** 3.2

COLOUR Variable, especially blue and green, also colourless; white streak

cubes that have grown together
cubic cleavage

Beautiful cubic crystals of fluorite in a wide range of colours are sometimes found in hydrothermal mineral veins, either alone or with metal ores, but they are too soft to be worked as gems. With its low melting-point, fluorite is used in smelting iron, and, since the times of ancient Greece, has been used as an ornamental stone. In 1824, the German mineralogist Friederich Mohs, who invented Mohs scale of hardness, discovered the marked fluorescence shown by some fluorite, especially in ultraviolet light.

HALIDES
Halite

COMPOSITION NaCl

CRYSTAL SYSTEM Cubic

HARDNESS 2½ **SPECIFIC GRAVITY** 2.1–2.2

COLOUR Variable, including place brown or reddish brown, and colourless forms; white streak

Of great economic importance as a source of salt, halite (also known as rock salt) is a common mineral in certain sedimentary rocks, where it occurs in massive, granular, or crystal form. Halite is mainly derived from the drying out of seawater in the geological past. The process frequently leaves vast stratified evaporite deposits in which halite is associated with other soluble minerals such as gypsum and anhydrite. Because of halite's low density and ability to recrystallize at low temperatures and pressures, deeply

stepped cubes

buried halite deposits deform into large dome structures known as diapirs. Although some rock salt is still mined (see panel, above), particularly in the USA, Canada, and Germany, most commercial salt is extracted directly from seawater by evaporation.

CONSISTENT SHAPE
Invariably, halite crystallizes as well-formed cubes, with stepped, hopper-shaped faces. It is typically colourless, except when chemical impurities introduce, for example, shades of red.

MINING SALT

Since ancient times, salt has been an essential mineral for domestic and industrial purposes. For countries far from the sea, salt mining is a necessity, and many famous salt mines have been excavated over the centuries. Some, such as those in Siberia, have achieved notoriety for the use of slave labour, while others have become famous for the beauty of their cavernous interiors, such as those of Wieliczka, in Poland, and Winsford, UK, which is shown here.

CARBONATES
Dolomite

COMPOSITION CaMg(CO$_3$)$_2$

CRYSTAL SYSTEM Trigonal

HARDNESS 3½–4 **SPECIFIC GRAVITY** 2.8–2.9

COLOUR White, grey to pale brown; white streak

Like its close chemical relative calcite, dolomite is an important rock-forming mineral that has formed in mountain-sized masses. The magnesium in the composition of this calcium and magnesium carbonate generates the small but important differences between

dolomite and calcite: for example, dolomite does not dissolve readily in acidic groundwater, and dolomitic carbonate rock does not form solution features such as caves. Most dolomite is a secondary product formed by magnesium-rich solutions interacting with calcium carbonate sediments, especially reef limestones. It can also occur as a primary mineral either alone in magnesium-enriched water or in hydrothermal veins associated with sulphide ores such as sphalerite and galena. Dolomite can be difficult to distinguish from calcite, but it frequently

trigonal crystal
UNUSUAL FORM
Transparent dolomite crystals are rare and, although sometimes attractively coloured pink and yellow, are too soft to be used as gemstones.

SADDLE-SHAPED CRYSTALS
Dolomite crystals commonly develop with curved surfaces that are composites of overlapping crystals whose orientation changes slightly as they grow.

curved crystal intergrowths
quartz

has a characteristic brown colour, and develops into curved surfaces made of composite intergrown, or twinned, crystals. It was not recognized as a separate mineral until the end of the 18th century, when the Swiss mineralogist H.B. de Saussure named it after the French geologist Déodat de Dolomieu, who had first noticed the distinct properties of dolomite rock. Dolomitic limestone makes good building stone and hardcore. It is also quarried for the manufacture of special cements and refractory linings.

DOLOMITE MOUNTAINS
Massifs such as the Italian Dolomites consist almost entirely of dolomite-rich sedimentary carbonate rocks. These are less prone to chemical weathering than limestone.

CARBONATES
Rhodochrosite

COMPOSITION MnCO$_3$

CRYSTAL SYSTEM Trigonal

HARDNESS 3½–4½ **SPECIFIC GRAVITY** 3.4–3.7

COLOUR Rose-red; also brown to grey; often has a brown or black outer crust; white streak

The name of this manganese carbonate mineral, which is often coloured a striking rose-pink, is derived from the Greek *rhodokhros* ("rose coloured"). Rhodochrosite generally occurs as crystals or in granular or massive form and sometimes as globular nodules with internal concentric colour bands. It is found in hydrothermal veins, with ores of copper, silver, and lead, and in some metamorphic rocks of sedimentary origin. It is a source of manganese, which is used for hardening steel. Argentinian rhodochrosite is made into jewellery.

concentric bands

PLANET EARTH

Calcite

COMPOSITION $CaCO_3$

CRYSTAL SYSTEM Trigonal

HARDNESS 3   **SPECIFIC GRAVITY** 2.7

COLOUR Variable, but generally white or colourless; exhibits white streak

One of the commonest minerals in the Earth's crust, calcite ranges from massive to granular and crystalline in form. Its crystals grow in many different shapes from tabular to prismatic, often with well-developed faces. Calcite is found in several types

nail-head crystal

galena matrix

NAIL-HEAD CALCITE
Calcite crystals grow in a variety of shapes, and these have been given common names, such as nail-head calcite, which develops tabular heads on more slender columns.

of rock, ranging from sedimentary (limestone) to metamorphic (marble) and igneous (carbonatite lavas). It also occurs in veins, with or without other minerals, and often replaces minerals such as calcium-rich feldspar and pyroxene in igneous rocks. Calcite crystals can grow at the pressures and temperatures found at the Earth's surface, and they readily dissolve in slightly acidic water. This means that calcite is easily leached from rocks and transported in solution to be redeposited elsewhere. At the Earth's surface, this results in the growth of limestone pavements, solution hollows, and cave systems (see pp.240–41). The dissolution of calcite can be reversed: calcium-carbonate-enriched waters deposit calcite in the form of stalactites, stalagmites, and flowstone. Most calcite is associated with sedimentary limestones, especially those of marine origin. Within these rocks, the calcite is deposited as tiny

POINTED CRYSTAL
High, pyramid-shaped calcite crystals are called dog-tooth spar, because of their resemblance to teeth.

line of cleavage

(see pp.240–41)

DOUBLE REFRACTION

Light passing through calcite is split into two separate rays because of the atomic structure of this mineral. The same property is shown by all translucent crystals other than those that belong to the cubic system. The rays have different vibration directions and velocities, with the slow ray bent (refracted) from its original path. The measurable difference between the two paths is known as the refractive index and is useful, especially to gemmologists, in helping distinguish between otherwise similar translucent crystals. In calcite the refractive index is very high (0.172), as can be seen by the apparent doubling of any object viewed through a cleavage fragment where the vibration directions show their maximum divergence.

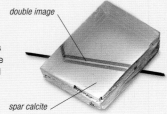

double image

spar calcite

calcium-carbonate crystals and indirectly from the skeletons and shells of sea creatures, especially where reefs are formed. Metamorphism of limestone at depth within the Earth's crust produces marble. Apart from its ubiquitous use as a building stone and and aggregate, calcite in the form of limestone has been used in great quantities to manufacture lime fertilizer and cement, and in the smelting of metal ores. Clear calcite crystals have also been used in microscope optical systems (see panel, above) due to their refractive properties.

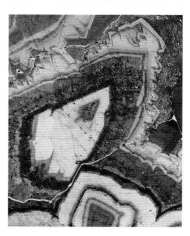

CALCITE GROWTH
When stained, calcite crystals reveal their growth patterns along with changes in their composition.

Malachite

COMPOSITION $Cu_2CO_3(OH)_2$

CRYSTAL SYSTEM Monoclinic

HARDNESS $3\frac{1}{2}$–4   **SPECIFIC GRAVITY** 3.9–4.0

COLOUR Pale to dark green; pale green streak

rounded form

This distinctive, green copper mineral is the common weathering product of copper ores. Malachite contains about 57 per cent copper and mostly occurs as banded encrustations with bulbous surfaces and an internal, fibrous, radiating habit. It is commonly found in the oxidation zone of other copper mineral deposits, where it is frequently associated with minerals such as goethite and calcite. The ancient Greeks mined malachite to make amulets, which reputedly warded off misfortune. It has since been used as a semi-precious stone.

Azurite

COMPOSITION $Cu_3(CO_3)_2(OH)_2$

CRYSTAL SYSTEM Monoclinic

HARDNESS $3\frac{1}{2}$–4   **SPECIFIC GRAVITY** 3.8–3.9

COLOUR Deep azure-blue; pale blue streak

Readily distinguished by its deep blue colour, which was widely used as a pigment in early illustrated manuscripts, azurite is compositionally similar to malachite. It develops not only as short, prismatic or tabular crystals but also in massive and earth-like forms. Its occurrence is also similar to malachite, with which it is often associated as a weathering product of copper ores. Azurite is also mined as a minor copper ore.

concentric layers

prismatic azurite crystal

Borax

COMPOSITION $Na_2B_4O_5(OH)_4 \cdot 8H_2O$

CRYSTAL SYSTEM Monoclinic

HARDNESS 2–$2\frac{1}{2}$ **SPECIFIC GRAVITY** 1.7

COLOUR White or colourless; white streak

white crystals

Borax is one of several relatively rare minerals to contain the element boron. An evaporite mineral, its short prismatic crystals are formed by the evaporation of water in salt lakes, where they then become part of the sediment. Borax can also be found in volcanic exhalations. This massive or crystalline mineral melts easily, is soluble in water, and can crumble to a powder if it dehydrates further. Borax is widely used in the glass, chemical, soap, and food industries.

Anhydrite

COMPOSITION $CaSO_4$

CRYSTAL SYSTEM Orthorhombic

HARDNESS 3–$3\frac{1}{2}$ **SPECIFIC GRAVITY** 2.9–3.0

COLOUR White or colourless; white streak

Usually in massive form, anhydrite crystals are rare and show three well-developed cleavages at right angles. The cleavage helps to distinguish this evaporite mineral from gypsum, with which it can be confused. Anhydrite's name refers to its composition and is derived from the Greek *anudros* ("without water"). This mineral can be deposited directly from heated seawater (in excess of 42°C/108°F), but is mostly found interbedded with gypsum and halite in evaporite deposits. When ground up, it is used as a soil conditioner and in cement.

three cleavages at right angles

SULPHATES

Gypsum

COMPOSITION $CaSO_4.2H_2O$

CRYSTAL SYSTEM Monoclinic

HARDNESS 2   **SPECIFIC GRAVITY** 2.3

COLOUR White or colourless; white streak

In Ancient Greece, gypsum was used as an ornamental stone because it was very easy to carve. The Romans then discovered that heating gypsum to 300°C (600°F) made a plaster that sets hard when mixed with water; such a plaster is still widely used in the building industry. Gypsum ranges from massive to granular (alabaster), fibrous (satin spar), and crystalline (selenite) in form. With low solubility, it is the first mineral to separate from evaporating seawater, followed

tabular crystal

SELENITE
A crystalline form of translucent gypsum, known as selenite, develops into diamond-shaped, tabular crystals in some sedimentary rocks.

by anhydrite and other similar minerals. Gypsum also occurs in mineral veins and volcanoes, where sulphuric acid reacts with limestone. In hot deserts, gypsum crystals full of sand grains develop after the evaporation of mineral-enriched waters from flash floods. Such crystals are known as desert roses.

MINERAL DUNES
When inland seas dry up, as here at White Sands, New Mexico, USA, they deposit gypsum as one of several evaporate minerals. Gypsum grains are then picked up by the wind and blown into dune formations.

TUNGSTATES

Wolframite

COMPOSITION $(Fe,Mn)WO_4$

CRYSTAL SYSTEM Monoclinic

HARDNESS 5–5½   **SPECIFIC GRAVITY** 7.0–7.5

COLOUR Grey to brownish black; brownish black streak

lined crystal face

This mineral is the most economically important tungsten ore. Originally known as wolfram, tungsten is symbolized as an element by the letter W, for wolframite. Tungsten typically forms tabular or prismatic crystals, yet can be massive or granular, too. It is mainly found in granite-related pegmatites and quartz veins but also occurs in some hydrothermal veins associated with other ore minerals such as sphalerite, cassiterite, and galena. Wolframite can accumulate in placer deposits as well. It used to be considered a waste material because it did not melt – the element tungsten was not recognized until 1781. Because of its exceptionally high melting-point (3,387°C/6,000°F), tungsten is used today for electric filaments and steels for high-speed tools.

MOLYBDATES

Wulfenite

COMPOSITION $PbMoO_4$

CRYSTAL SYSTEM Tetragonal

HARDNESS 3   **SPECIFIC GRAVITY** 6.5–7.0

COLOUR Shades of orange-yellow to red or brown; white streak

A striking lead-and-molybdate mineral, wulfenite can be found in massive and crystalline forms. Its crystals develop as beautiful square tablets of orange-yellow to reddish brown. This secondary mineral forms in the oxidized zone of ore deposits of lead and molybdenum, and may be associated with other lead

thin, platy crystal

rock matrix

minerals, such as anglesite and cerussite. When present in sufficient quantities, it is mined as a lead ore since its composition is around 61 per cent lead oxide. Wulfenite was named after F. X. Wülfen, the Austrian mineralogist who first described it in 1785.

PHOSPHATES

Apatite

COMPOSITION $CA_5(PO_4)_3(F,Cl,OH)$

CRYSTAL SYSTEM Hexagonal

HARDNESS 5   **SPECIFIC GRAVITY** 3.1–3.2

COLOUR Variable, including green, brown, and white; exhibits white streak

Difficult to distinguish from minerals such as tourmaline, apatite's name is Greek for "deceit", because of its misleading appearance, especially in crystalline form. It is also found in massive and granular states. Apatite occurs in hydrothermal veins, pegmatites, and metamorphosed limestone. In sedimentary rocks, it is formed from organic materials. It is used as a fertilizer and in the chemical industry.

calcite groundmass

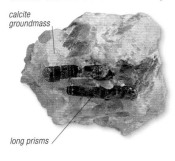

long prisms

PHOSPHATES

Turquoise

COMPOSITION $CuAl_6(PO_4)_4(OH)_8.5H_2O$

CRYSTAL SYSTEM Triclinic

HARDNESS 5–6   **SPECIFIC GRAVITY** 2.6–2.8

COLOUR Sky-blue to bluish green to greenish grey; greenish white streak

A phosphate mineral, often with a distinctive, pale blue-green colour, turquoise was one of the first gemstones to be mined. Normally massive or granular, it is rarely crystalline. Turquoise is secondary in origin, and occurs in highly altered veins within igneous rocks and metamorphosed

SYNTHETIC TURQUOISE

filled cracks

NATURAL AND SYNTHETIC
Turquoise is much in demand for jewellery, but it can be extremely difficult to tell the difference between synthetic (above) and natural (right) stones.

NATURAL TURQUOISE

turquoise layers

BLUE-STREAKED ROCK
Natural turquoise often occurs in thin veins and is highly porous and full of cracks, so it is unusual to obtain gem-quality pieces of large size.

sedimentary rocks rich in alumino-silicates. Sometimes it is highly porous and full of cracks. Its name is derived from Turkey, where during the Ottoman Empire it was traded as a precious stone, especially with Persia. Today, localities in Mexico, the USA, China, and Iran produce turquoise. Imitation turquoise was first created in France in 1972 using the Gilson process, yet it is not a true synthetic since its properties differ from the natural forms. Nevertheless, much turquoise sold today is synthetic.

VANADATES

Carnotite

COMPOSITION $K_2(UO_2)_2(VO_4)_2.3H_2O$

CRYSTAL SYSTEM Monoclinic

HARDNESS 2   **SPECIFIC GRAVITY** 4.0–5.0

COLOUR Lemon to greenish yellow; yellow streak

powdery carnite

Containing about 63 per cent uranium oxide, carnotite is highly radioactive and is mined for its uranium content. It is deposited in sedimentary rocks from vanadium- and uranium-enriched waters. Rarely crystalline, carnotite typically has a powdery or earthy texture. It is hard to distinguish from other uranium minerals except by using X-ray diffraction.

SILICATES

Olivine group

COMPOSITION $(Mg,Fe)_2SiO_4$

CRYSTAL SYSTEM Orthorhombic

HARDNESS 6½–7 **SPECIFIC GRAVITY** 3.2–4.4

COLOUR Clear yellowish green, also yellow–brown–black; white streak

GEM OLIVINE
Perfect crystals of green, transparent olivine, known as peridot, are rare and so especially prized as gems.

Commonly found in igneous rocks, olivine is the most abundant of all silicate minerals. It actually comprises a group of minerals whose composition ranges from magnesium-rich forsterite to iron-rich fayalite. Perfect crystals are rare, but there is a green gem form (see above). This was first brought to Europe by returning Crusaders from the Red Sea island of Zebirget, where it had been mined for 3,500 years. Olivine's primary origin is in silica-poor intrusive igneous rocks (such as gabbro and peridotite) and extrusive rocks (such as basalt). Dunite is a rare intrusive rock mainly made of olivine, and forsterite crystals up to 14cm (5½in) long have been found in dunite pegmatites in the Urals, Russia. The metamorphism of

magnesium-rich dolomitic limestone produces green forsterite marbles. Olivine was first recognized in 1772 in a meteorite. It is readily altered by hydrothermal solutions and prolonged weathering to create serpentinite minerals. It is used to make heat-resistant glass.

OLIVINE IN BASALT
Through a microscope, a transparent section of basalt rock, as here, reveals fissured yellow olivine crystals as a main constituent mineral.

characteristic olivine fissures

SANDS OF TIME
Weathering and erosion of Hawaii's basalt lavas have produced these dark green sands, which consist mostly of olivine.

UNEXPECTED PROPERTIES
Although a fairly hard mineral, olivine is susceptible to change and so is usually full of cracks.

SILICATES

Sillimanite

COMPOSITION Al_2SiO_5

CRYSTAL SYSTEM Orthorhombic

HARDNESS 6½–7½ **SPECIFIC GRAVITY** 3.2–3.3

COLOUR White-grey or colourless; white streak

A hard alumino-silicate mineral, sillimanite is similar in structure and mode of formation to kyanite and andalusite. Typically found in regional metamorphic rocks, sillimanite forms long, fibrous, prismatic crystals with slightly variable hardness, and in some pegmatites it can be of gem quality. It is second in density to kyanite, and is named after the late 18th-century American chemist Benjamin Silliman.

long, prismatic crystal

SILICATES

Garnet group

COMPOSITION $X_3Y_2Si_3O_4$ where X = Ca, Mg, Fe, or Mn; Y = Al ,Fe, Ti, or Cr

CRYSTAL SYSTEM CUBIC

HARDNESS 6–7½ **SPECIFIC GRAVITY** 3.5–4.3

COLOUR Variable, including red-brown, yellow-green, and black; white streak

Within this group of silicate minerals, some elements substitute for others in the atomic structure, resulting in great variation in colour. Crystals are common and may grow to several centimetres at relatively high temperatures. The kind of garnet found within various types of rock is linked to the chemistry of the host rock, especially in metamorphic and some igneous rocks: for example, magnesium-rich reddish pyrope is common in peridotites and associated

METAMORPHIC SCHIST
This thin section of garnet reveals typical spherical garnet crystals, despite rock compression.

serpentinites, while the rare green uvarovite garnet is found in some chrome-bearing serpentinites. Garnet minerals are particularly common in metamorphic rocks such as schist, gneiss, and eclogite. They also occur in placer deposits, because of their durability. Garnet has been used as a gemstone since Greek and Roman times but was especially popular in the 19th century, when Bohemian garnets – particularly pyropes – were traded extensively throughout Europe and North America. Since the 1960s, synthetic garnet has been produced mainly for industrial use – as abrasives and as bearings in fine instruments – and also as gemstones.

GROSSULAR CRYSTALS
Intergrown garnet crystals with well-developed, four-sided rhombic faces, and coloured orange by calcium, here form the variety known as grossular garnet.

rhombic face

PYROPE

ANDRADITE

SPESSARTINE

DIFFERENT ELEMENTS
The variety of environments in which garnet grows affects its composition and leads to a considerable range of gem colouring.

SILICATES

Zircon

COMPOSITION $ZrSiO_4$

CRYSTAL SYSTEM Tetragonal

HARDNESS 7½ **SPECIFIC GRAVITY** 4.6–4.7

COLOUR Variable, including brown, green, blue, and black; exhibits white streak

One of the oldest minerals on the Earth, zircon's hardness has enabled it to survive in many types of rocks (see

short, prismatic crystal

panel, below). In particular, it occurs as an accessory mineral in granites and alkali-rich intrusive igneous rocks such as syenite and certain pegmatites. Because of its resistance to weathering and erosion, zircon accumulates in alluvial placer deposits. It can also tolerate considerable metamorphism and is widely found in regional metamorphic terrains. For a long time zircon was confused with diamond, especially in gems from Sri Lanka. It is also the main source of the element zirconium, which is used in nuclear reactors and as an abrasive. Australia, South Africa, and China are major producers.

STEP-CUT ZIRCON GEM

RESILIENT CRYSTALS
Zircon's well-formed crystals, which are formed in igneous rocks, survive considerable metamorphism without deformation of their shape.

ZIRCON DATING

The resistance of zircon crystals to chemical and physical change makes them ideal for use in radiometrically dating very old rocks. Some radioactive elements, such as uranium, decompose over time, forming what are known as daughter isotopes. The rate of this decay has been established. This means that, by detecting single zircon crystals (or even zones within crystals) in old rocks and measuring the concentration of daughter isotopes in them against their original content, it is possible to calculate how much decay has taken place and thereby estimate the age of the rock.

SILICATES
Andalusite

COMPOSITION Al_2SiO_5

CRYSTAL SYSTEM Orthorhombic

Hardness $6\frac{1}{2}$–$7\frac{1}{2}$ **SPECIFIC GRAVITY** 3.1–3.2

COLOUR Pale pink–grey–brown–green; white streak

prismatic andalusite crystal

quartz groundmass

Sometimes found in massive form, andalusite generally forms prismatic crystals with a square cross-section. This alumino-silicate mineral usually occurs in thermally metamorphosed rocks formed under low pressure, such as hornfel. It is also found in some pegmatites, where gem-quality crystals may develop, and occasionally in placer deposits. It is used in the production of porcelain and other heat-resistant materials.

CHIASTOLITE
This variety of andalusite contains carbon inclusions in a dark cross-shape.

SILICATES
Kyanite

COMPOSITION Al_2SiO_5 **CRYSTAL SYSTEM** Triclinic

HARDNESS $5\frac{1}{2}$–7 **SPECIFIC GRAVITY** 3.5–3.7

COLOUR Blue, white, or grey; white streak

Flat, blade-shaped crystals are typical of kyanite. Like its polymorphs andalusite and sillimanite, kyanite is found in regionally metamorphosed rocks, but it has a denser structure. Consequently, it is found in schist and gneiss formed at high pressures and temperatures. Kyanite also occurs in pegmatites, sometimes as deep blue gem-quality stones – its name deriving from the Greek *kuanos* ("dark blue").

STEP CUT

FLAWED BEAUTY
Despite their closely packed structure, kyanite gems commonly have pressure cracks.

long, bladed crystals

ROCK MATRIX WITH CRYSTALS

SILICATES
Topaz

COMPOSITION $Al_2SiO_4(OH,F)_2$

CRYSTAL SYSTEM Orthorhombic

HARDNESS 8 **SPECIFIC GRAVITY** 3.5–3.6

COLOUR Variable yellow–brown–blue or colourless; white streak

Characteristically, topaz forms prismatic crystals of variable colour with vertically lined faces, but it can also occur in massive or granular forms. It is mostly found in pegmatites associated with granite, along with tourmaline, beryl, and apatite. Topaz also occurs in placer deposits and is well known as a gemstone. The discovery of large deposits in Brazil has provided spectacularly large crystals, including one weighing 271kg (596lb), and another that is

natural, prismatic crystal

flat-cut, faceted faces

COLOUR RANGE
Commonly coloured yellow to brown, topaz is occasionally found in other colours, such as blue.

GIANT TOPAZ
This 22,892-carat topaz gemstone was cut from a crystal that was found in Brazil. It is one of the world's largest gemstones.

the biggest cut blue topaz (21,005 carats). Another beautiful topaz, found in 1740, was set in the Portuguese crown in the mistaken belief that it was a diamond (the so-called Braganza "diamond" of 1,640 carats). The first-century Roman naturalist Pliny the Elder thought that the name topaz was derived by the Romans from the Red Sea island Topazos. The mineral found there was long mistaken for topaz and is now known to be olivine.

PEAR-SHAPED TOPAZ
Tiny internal flaws demonstrate that this topaz gem has been cut from a natural crystal.

SILICATES
Staurolite

COMPOSITION $(Fe,Mg,Zn)_2Al_9(Si,Al)_4O_{22}(OH)_2$

CRYSTAL SYSTEM Monoclinic

HARDNESS 7–$7\frac{1}{2}$ **SPECIFIC GRAVITY** 3.7–3.8

COLOUR Reddish brown to brown-black; white streak

This hard alumino-silicate mineral is associated with kyanite and mica in regionally metamorphosed phyllite, schist, and gneiss. The composition of staurolite is complex, with considerable substitution to give a number of varieties, such as Zambian lusakite, which has been mined as a blue pigment. Staurolite is often found in well-formed prismatic crystals, which are generally opaque and frequently twinned in the form of a distinctive cross (see panel, right) – the name staurolite being Greek for "cross stone". Crystals of staurolite

prismatic staurolite crystal

RELIGIOUS SYMBOLS
The frequent occurrence of staurolite as twinned cruciform crystals has led to their use as amulets for Christians over the centuries. Some of the best specimens were found in the mica schists of the Swiss Alps and were sold as *Baseler taufstein* ("Basel baptismal stone"). Because they were so readily available, staurolite amulets were not expensive and became widely worn.

are frequently larger than those that surround them, in which case they are referred to as porphyroblasts. Staurolite is also found in placer deposits.

DOMINANT ROLE
In places, staurolite is such a common mineral that its crystals comprise a significant proportion of the rock in which it is embedded.

SILICATES
Epidote group

COMPOSITION $X_2Y_3Si_3O_{12}OH$ where X = Ca; Y = Al, Fe, or Mg

CRYSTAL SYSTEM Orthorhombic and monoclinic

HARDNESS 5–7 **SPECIFIC GRAVITY** 3.2–4.5

COLOUR Variable, including green-black; grey streak

In this group of complex silicates, magnesium and iron (as in piedmontite) can be chemically substituted by aluminium and iron (as in epidote), depending on the chemical environment within which the mineral is developing. Epidote is mineralogically close to clinozoisite, and both minerals typically form columnar, prismatic, and needle-shaped crystals, but they may also be massive. While the natural forms are

vitreous lustre

lined, columnar crystal

FASHIONING EPIDOTE
Cut epidote gems such as this one are unusual, because their crystals are difficult to shape due to their perfect cleavage.

commonly black to green in colour, gem varieties may be yellow, dark green, or brown. This mineral is rarely cut as a gem, however, since there is a well-developed cleavage along which the crystals can easily break. The name epidote is Greek for "an addition", because this mineral was long mistaken for tourmaline until the French mineralogist René Haüy distinguished epidote as a separate mineral in 1801. Epidote is a relatively common calcium alumino-silicate in low- and medium-grade metamorphic rocks, especially those derived from igneous rocks such as basalt (amphibolite) and from limestone (marble). It also occurs in veins in igneous rocks. A new valuable blue variety of the epidote group is tanzanite, found at Arusha, Tanzania, in 1967 by an Indian tailor, Manuel d'Souza.

PROBLEMS OF IDENTITY
The elongate and lined prisms of epidote crystals look extremely similar to those of tourmaline, for which they can at times be mistaken.

PLANET EARTH

SILICATES

Beryl

COMPOSITION $Be_3Al_2Si_6O_{18}$

CRYSTAL SYSTEM Hexagonal

HARDNESS 7½–8 **SPECIFIC GRAVITY** 2.6–2.8

COLOUR Variable, including blue (aquamarine), green (emerald), yellow (heliodor), white, and colourless; white streak

Beryl is best known for its gem varieties, especially green emerald and pale blue-green aquamarine, both of which are highly valued. As well as occurring as a crystal, this hard beryllium mineral is also found in a massive form, which may be mistaken for quartz. Beryl crystals are usually marked by characteristic longitudinal lines and commonly develop in pegmatitic veins, especially in granite. The mineral also occurs in metamorphic schist and gneiss, and sedimentary placer deposits.

prismatic crystal

LARGE CRYSTALS
Beryl crystals, such as this fine gem-quality prismatic one grown from a metamorphic rock groundmass, can be extremely long and heavy.

BERYL MINING

Emeralds were mined by the ancient Egyptians as long ago as 1300 BCE, and for a thousand years the mines of Sikait and Zabara, in Egypt, were Europe's principal source of these gems. When the Spanish invaded the Americas, however, they discovered that Colombian emeralds were widely traded by native peoples – from Mexico to Chile. In 1573, the Spanish captured the Muzo beryl mine, in Colombia, which then supplanted the Egyptian source for Europe. Nevertheless, the value of beryl gems is so great that they are still mined wherever they occur, regardless of the quantity.

Beryl crystals can be very large, with specimens up to 6m (20ft) long and weighing 1.5 tonnes being found in Maine, USA; such crystals can be mistaken for apatite, but beryl is harder. The largest aquamarine found was a hexagonal prism discovered in Brazil in 1910; it weighed 110.5kg (243lb) and measured 48 x 40cm (19 x 16in). By comparison, emerald crystals rarely grow more than a few centimetres long and are commonly flawed with cracks and tiny crystals of other minerals, but this has not detracted from their value. Emeralds have in the past been particularly associated with Ottoman, Persian, and Mogul rulers, although the stones actually originated in Colombia, South America, and were sold by indigenous peoples and Spanish colonialists (see panel, above). Beryl is the main source of beryllium, which is one of the lightest metals known and an important ingredient of special metal alloys. More than 80 per cent of the world's annual production of beryl comes from the USA.

octagonal step cut

AQUAMARINE

pear-shaped fancy cut

EMERALD

POPULAR CUT STONES
Aquamarine and emerald are among the best known of several attractive and valuable gem varieties of the mineral beryl.

SILICATES

Tourmaline group

COMPOSITION $(Ca,Na,K)(Al,Fe,Li,Mg,Mn)_3(Al,Cr,F,V)_6$ $Al_6(BO_3)_3Si_6.O_{18}(OH,F)_4$

CRYSTAL SYSTEM Trigonal

HARDNESS 7 **SPECIFIC GRAVITY** 3.0–3.2

COLOUR Variable, including yellow, green, blue, pink, brown, and black; white streak

Colour variation may occur along the length of a single tourmaline crystal. Tourmaline crystals exhibit a piezoelectric effect, whereby an electric current applied to one end of a crystal has its charge reversed at the other end (for example, from positive to negative) – a property that has been used in electronics. Tourmaline is a hard, silicate gem mineral with long, lined, prismatic crystals. These have a triangular cross-section with curved faces. Crystals commonly occur in radiating or parallel clusters in granite pegmatites and metamorphic schist and gneiss, especially those that have been altered by boron bearing hydrothermal solutions.

lined columnar crystal

feldspar matrix

SILICATES

Pyroxene group

COMPOSITION $X_2Si_2O_6$ where X = Mg, Fe, Mn, Li, Ti, Al, Ca, or Na

CRYSTAL SYSTEM Orthorhombic and monoclinic

HARDNESS 5–6½ **SPECIFIC GRAVITY** 3.2–4.0

COLOUR Pale to dark brownish green or bronze; greenish white streak

Pyroxenes are some of the most widespread of all rock-forming silicate minerals. They can be broadly divided into clinopyroxenes, which have monoclinic crystals, and orthopyroxenes, which have

short, prismatic augite crystal

AUGITE PHENOCRYST
Crystals that grow larger than their surrounding ones are called phenocrysts. Here, a well-formed, short, stubby prismatic phenocryst of augite has developed within a finer-grained rock matrix.

orthorhombic crystals. Well-formed crystals are typically short, dark-coloured prisms, some of which occur as gems: for example, green enstatite, reddish brown hypersthene, and greenish diopside. Clinopyroxenes (such as aegirine and gabbro) contain combinations of calcium, sodium, aluminium, iron, or lithium, while orthopyroxenes (such as hypersthene and enstatite) have very little calcium but may contain magnesium and iron. Pyroxene minerals are commonly found as masses, grains, or better-defined crystals within quartz-poor igneous rocks such as basalt,

DOLERITE
The pyroxenes are the brown grains in this thin section of dolerite igneous rock.

GABBRO
The pale and golden-yellow grains in this thin section of gabbro are clinopyroxene; the black and white grains are feldspar.

gabbro, and pyroxenite. Calcium-rich clinopyroxene may also occur in some metamorphosed limestone, whereas orthopyroxene may be found in stony meteorites. Pyroxene can easily be confused with amphibole. Pyroxene crystals, however, have two sets of cleavages at 90° to one another, while amphibole has two sets at about 120°. The green-coloured pyroxene jadeite is usually massive with a very compact structure. These features give it great strength, which is why it has been used to make objects such as axe blades and carved ornaments.

DIFFERENT GEM TYPES
Olive-green diopside is a gem-quality clinopyroxene, while emerald-green enstatite is an orthopyroxene.

long, prismatic aegirine crystal

AEGIRINE CRYSTAL
Named after Aegir, the Scandinavian sea god, this very dark green clinopyroxene mineral was first discovered and described in Norway, in 1835.

DIOPSIDE   **ENSTATITE**

SILICATES

Amphibole group

COMPOSITION $X_2Y_5(OH)_2(Si,Al)_8O_{22}$ where X = Ca or Na; Y = Mg, Fe, or Al

CRYSTAL SYSTEM Monoclinic

HARDNESS 5–6 **SPECIFIC GRAVITY** 3.0–3.6

COLOUR Variable, including white–green–black; pale green streak

Amphiboles are a group of rock-forming silicate minerals that have a complex atomic structure in which the elements magnesium, iron, calcium, aluminium, and sodium substitute for one another. Amphibole resembles pyroxene except for its 120° cleavages and its hydroxyl (–OH) groups, which form an integral part

long, ribbed prism

vitreous lustre

CRYSTALLINE RIEBECKITE
The crystalline form, shown here, of this amphibole mineral is not a danger to health, unlike riebeckite in its fibrous form – asbestos.

of its composition. One of the most important amphibole minerals is hornblende, which contains calcium, sodium, magnesium, and iron. It is widespread in both igneous rocks and metamorphic rocks, such as amphibolite, while other amphiboles, such as tremolite, actinolite, and nephrite (jade), are found mainly in metamorphic rocks. Another amphibole, called riebeckite (also known as crocidolite or blue asbestos), is now notorious for its long-term toxicity and link with lung diseases such as asbestosis and mesothelioma cancer.

long, prismatic crystal

HORNBLENDE CRYSTALS
Long, prismatic crystals and green-black colouring are characteristic of hornblende crystal growth.

ANDESITE MAGNIFIED
Amphibole crystals appear brown in this thin section of the volcanic lava andesite.

SILICATES

Talc

COMPOSITION $Mg_3Si_4O_{10}(OH)_2$

CRYSTAL SYSTEM Monoclinic

HARDNESS 1 **SPECIFIC GRAVITY** 2.6–2.8

COLOUR White, grey-green; white streak

This soft silicate mineral (which on Mohs' scale has a hardness value of one, the lowest possible) has a soapy or greasy feel. Talc is formed by the metamorphism of magnesium-silicate minerals such as pyroxene, amphibole, or olivine. Talc usually occurs as foliated or granular masses and only rarely as crystals. Massive talc is known as steatite or soapstone and can easily be carved, for decorative purposes. The heat-resisting properties of talc are extensively exploited for the manufacture of fire-resistant materials, especially ceramics. China and the USA produce more than a third of the world's annual output of talc.

greasy lustre

GREAT ALTERATION
Olivine-rich rocks with magnesium silicate minerals metamorphose into talc. Typically massive, as here, it is cut through by lines of fracture and stress.

TALC QUARRIES
Uplift and erosion have brought talc rocks to the Earth's surface, where they can be extracted by open-cast quarrying, as here in the Shetland Isles, UK.

SILICATES

Mica group

COMPOSITION $XY_{2-3}(OH)_2(Si,Al)_{4-5}O_{10}$ where X = K, Na, or Ca; Y = Al, Mg, Fe, or Li

CRYSTAL SYSTEM Monoclinic

HARDNESS 2–4 **SPECIFIC GRAVITY** 2.7–3.3

COLOUR Variable, including white, pink, green, and brown-black; white streak

The micas are an important group of complex silicate minerals that are characterized by their distinctive atomic structure, which consists of silicon and oxygen tetrahedra linked together to

tabular crystal

BIOTITE
This dark brown-coloured but slightly translucent mica mineral has a perfect cleavage between flakes.

vitreous lustre

form sheets with relatively weak bonds between the sheets. Cleavage within crystals is thus well developed between successive sheets. Crystals are tabular in shape and often hexagonal in outline. They occur as scattered flakes or book-like aggregates of flakes. Mica is widespread throughout many igneous, metamorphic, and sedimentary rocks. Composition varies from white micas rich in potassium and aluminium (such as muscovite) to darker micas rich in magnesium and iron (such as biotite) and rarer varieties such as pink-coloured, lithium-rich micas (such as lepidolite). Muscovite is characteristic of quartz-rich igneous rocks such as granite and granitic pegmatites, as well as many metamorphic rocks, especially phyllite and schist. It also survives

TYPICAL MUSCOVITE
White-coloured muscovite crystals here exhibit their characteristic tabular form, with each crystal having a single, well-developed cleavage parallel to the largest (tabular) face.

pegmatite

tabular crystal

weathering and erosion well and is an important component of sedimentary rocks, from sandstone through siltstone to shale. Biotite is more typical of a

CRYSTALS OF LEPIDOLITE
Found in granite and pegmatites (as here), this mica mineral contains lithium and is usually distinguished by its lilac to pink colouring.

wider range of igneous rocks, from granite through to diorite along with their pegmatites, as well as a great variety of metamorphic rocks. Dark mica is less commonly found in sediments, because it is more readily altered and weathered into residual minerals. Some micas are extensively used in industry, because they are good thermal and electrical insulators. China and Madagascar produce more than half of the annual global output of mica.

MICA WINDOWS

In some pegmatites, mica crystals grow to great size – one found in the Urals, Russia, was 5 square metres (54 square feet) in area and 0.5m (1.6ft) thick. Like slate rocks, micas can be split along the well-developed cleavage plane into very thin sheets, which are transparent and can readily be cut. They have been widely used as substitutes for glass in industrial furnaces and domestic stoves, because mica has a very high melting-point and does not easily break. Wherever available and cheap, mica sheets have also been used as window panes in domestic buildings, especially in rural areas where glass was difficult to obtain or was too expensive.

SILICATES

Chlorite group

COMPOSITION $(Mg,Fe,Mn,Al)_{4-6}(Si,Al)_4O_{10}(OH,O)_2$

CRYSTAL SYSTEM Monoclinic

HARDNESS 2–3   **SPECIFIC GRAVITY** 2.6–3.3

COLOUR Green, also yellow-brown; white streak

This widespread group of minerals includes chamosite and clinochlore. They are similar to mica in composition, structure, and cleavage, except their cleavage flakes are brittle not elastic. Chlorite is commonly massive or fine grained; crystals do occur, often in scaly aggregates. It is found in igneous rocks as well as low-grade metamorphic rocks and sediments. The name chlorite is derived from the Greek *khloros* ("green"). Green chamosite develops in certain iron-rich sediments and has been mined as an iron ore.

scaly, crystal aggregates

SILICATES

Feldspar group

COMPOSITION $X(Al,Si)_4)_8$ where X = Na, K, Ca, or Ba

CRYSTAL SYSTEM Monoclinic and triclinic

HARDNESS 6–6½   **SPECIFIC GRAVITY** 2.5–2.8

COLOUR Variable, including white, pink, green, blue, brown, and colourless; white streak

One of the most important groups of rock-forming alumino-silicate minerals, feldspars are widely distributed throughout igneous, metamorphic, and sedimentary rocks. They are found in one of two types: potassic feldspars (such as orthoclase and microcline); and plagioclase feldspars (such as albite and labradorite), in which calcium and sodium are substituted for one another. The formation of feldspar minerals in igneous rocks depends on the chemical composition and temperature of the original melt

MICROCLINE

MOONSTONE ORTHOCLASE

POLISHED LABRADORITE

YELLOW ORTHOCLASE

DIFFERENT FELDSPARS
Varying enormously in appearance, feldspar may occur as opaque microcline crystals, gem-quality orthoclase, or reflective and decorative labradorite.

tabular, white crystal

ALBITE CRYSTALS
These white albite crystals exhibit the typical tabular form of many feldspar minerals, although a close look reveals repeated twinning.

(magma). Low-temperature potassic feldspars develop in silica-rich magmas that result in granites and rhyolite lavas; while plagioclase feldspars grow in more silica-poor, higher-temperature magmas that result in gabbros and basaltic lavas. Typically, feldspar forms tabular or prismatic crystals, many of which are twinned in highly distinctive patterns.

SILICATES

Zeolite group

COMPOSITION Complex and variable alumino-silicates

CRYSTAL SYSTEM Cubic, orthorhombic, and monoclinic

HARDNESS 3½–5½   **SPECIFIC GRAVITY** 2.0–2.5

COLOUR White, yellow, pink, brown, or colourless; white streak

radiating, scolecite crystals

There is considerable substitution of sodium and calcium, and occasionally barium and potassium, in the chemical composition of this complex group of alumino-silicates, which includes scolecite, analcime, and natrolite. Zeolite contains water that can be expelled on heating and then regained on exposure to moist air, without destroying the structure. Typically found in basalt and occasionally sedimentary rocks, it occurs as fibres, needles, prisms, or complex cubes. It has been used for water softening and extensively as industrial catalysts.

SILICATES

Serpentine group

COMPOSITION $(Mg,Fe,Ni)_3Si_2O_5(OH)_4$

CRYSTAL SYSTEM Monoclinic

HARDNESS 2½–4   **SPECIFIC GRAVITY** 2.5–2.6

COLOUR Pale to dark green, also brown, yellow, or white; white streak

One of the serpentine minerals – the fibrous chrysotile asbestos – has been identified as carcinogenic (see panel, right), but the common, non-fibrous forms of this widespread group of magnesium silicates are deemed safe. Such non-fibrous forms are either massive or scaly. Massive serpentine is used as an ornamental stone, and a nickel-bearing variety (garnierite) is mined for nickel in New Caledonia. Serpentine is essentially a secondary mineral, formed from magnesium-rich orthopyroxenes or olivine, and may be abundant enough to form masses called serpentinite bodies.

mass of fibres

SILICATES

Kaolinite group

COMPOSITION $Al_2Si_2O_5(OH)_4$

CRYSTAL SYSTEM Triclinic and monoclinic

HARDNESS 2–2½   **SPECIFIC GRAVITY** 2.6–2.7

COLOUR White or yellow; white streak

In detail, kaolinite is made up of microscopic, flaky-layered crystals with perfect basal cleavages, like mica, even though this clay mineral mostly looks like an earthy mass. This cleavage confers important properties on kaolinite and other related clay minerals such as illite and dickite: they are pliable and tend to act as lubricants. Kaolinite is formed through intense

dull, earthy form

weathering or hydrothermal action on alumino-silicate minerals such as feldspar. It has been used in the manufacture of porcelain since the 6th century in China, from where the name derives. It is also used as a filler in paper and paint. More than half of the annual output of kaolinite is produced by the USA and Uzbekistan.

SILICATES

Vermiculite

COMPOSITION $(Mg,Ca)_{0.7}(Mg,Fe,Al)_6(Al,Si)_8O_{22}(OH)_4 \cdot 8H_2O$

CRYSTAL SYSTEM Monoclinic

HARDNESS 1½   **SPECIFIC GRAVITY** 2.3

COLOUR Yellow-brown; white streak

The unusual properties of vermiculite have led to its widespread use in the manufacture of thermal and sound-insulation materials, especially as it is not carcinogenic like asbestos. This chemically complex alumino-silicate mineral is similar in structure and appearance to mica, forming flaky-

flat, tabular habit

pseudo-hexagonal outline

layered crystals with perfect basal cleavage. When heated, however, it expands greatly along the cleavage, to form worm-like masses – hence its name, which is derived from the Latin *vermiculus* ("little worm"). South Africa produces about half the annual vermiculite output.

ASBESTOS

Until recently, asbestos was regarded as a very useful fire- and chemical-resistant material. Mixed with cement, it was moulded into pipes and sheets, and added to paints, in the home and in numerous industries. Asbestos is made from a fibrous serpentinite (chrysotile) or a fibrous amphibole (crocidolite), but tragically, both these minerals are carcinogenic. Inhalation or contact can cause long-term damage that may even result in death many years later.

CHRYSOTILE ASBESTOS
This fibrous form of the serpentine group is a health hazard, because its numerous fibres can readily be separated from the solid mineral.

Quartz

COMPOSITION SiO$_2$	CRYSTAL SYSTEM Trigonal
HARDNESS 7	SPECIFIC GRAVITY 2.65

COLOUR Very variable, from white to black, or colourless; white streak

Due to its resistance to weathering and erosion, this common, rock-forming, silica mineral accounts for most of the grains in sediments found in deserts, river deposits, and coastal areas. When compressed or cemented, such sediment forms quartz sandstones, which are an important component

vitreous lustre

CITRINE CRYSTAL
This hexagonal prism with its pyramidal faces is the characteristic form of quartz, and is seen here in a yellow variety known as citrine.

of sedimentary rock. Quartz minerals are common in most sedimentary, metamorphic, and igneous rocks, except for those igneous rocks derived from silica-poor magmas (such as gabbro) and their metamorphosed equivalents (such as pyroxenite). They develop mostly in the continental crust, and range from well-developed crystals to surface encrustations and zoned cavity linings (such as in chalcedony and agate). Crystals are usually six-sided prisms, whose faces are often covered with horizontal lines and are topped with pyramidal faces. Giant crystals weighing several tonnes are known. In igneous rocks, quartz is very common in granite and rhyolite, and in pegmatites or hydrothermal veins related to them. Widespread throughout schist and gneiss, it may form pure quartz veins and be an important constituent of intrusive mineral veins.

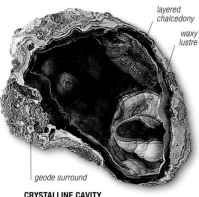
layered chalcedony
waxy lustre
geode surround

CRYSTALLINE CAVITY
Some rocks contain a cavity, or geode, which here is layered with a microcrystalline form of quartz known as chalcedony.

QUARTZ OUTCROP
These quartz-bearing rocks at Shining Rock Wilderness, USA, stand out against the rest of the landscape, because such rocks are very resistant to surface stresses, such as weathering and erosion.

QUARTZ ARTEFACTS

Differences in quartz-related minerals can be very great and have resulted in the formation of a large number of coloured varieties. Because they are relatively common, are hard, and can be cut and polished, a number have been adopted as precious stones by different cultures over the millennia. These include rock crystal, which is pure, colourless, and transparent quartz; purple (amethyst) and yellow (citrine) varieties; multi-coloured agates; and fine-grained or massive forms (such as blood-red jasper and black-and-white-striped onyx). Reflective bands of tiger's-eye or agate with internal mineral growths can also be found.

EGYPTIAN CHEST ORNAMENT
The quartz gems shown here are typical of those used in South America and Egypt for decoration.

ZONED CRYSTALS
The colour zones in this amethyst formed as the crystals grew, with traces of iron colouring the purple sectors.

quartz
amethyst

COLOURED BANDS
The layered structure of this sardonyx is emphasized by its coloration. Sardonyx forms in lava cavities and is a variety of chalcedony.

parallel bands

PETRIFIED FORESTS

Woody plant tissues are frequently fossilized by quartz and related silica minerals such as opal. Silica-enriched groundwater fills the porous tissue with mineral particles, which preserve the detailed cell structure of the original plant (such as these trees at Blue Mesa, Arizona). Such cell information is of use to paleobotanists when identifying a plant and the conditions under which it grew.

PLANET EARTH

Rocks

40–41	The crust
42–61	Minerals
Fossil fuels **80–81**	
Weathering, erosion, and deposition **90–95**	
Meteorite impacts **100–105**	
Volcanoes **154–77**	
Igneous intrusions **178–84**	

The Earth's crust is made of solid materials known as rocks, most of which are naturally occurring assemblages of minerals. The Earth's rocks are largely concealed beneath soil and vegetation, but in some places they are exposed at the surface, where they may form features such as rocky cliffs and mountains. The huge diversity of the Earth's rocks has developed over thousands of millions of years through the geological processes of igneous activity, changes in form known as metamorphism, and the formation of sediments and sedimentary rocks (including organic materials such as coal). Although most of the Earth's rocks are formed within the planet, meteorites derived from outer space are also found on its surface.

Minerals and Rocks

Rocks are made of minerals, and although they vary enormously across the Earth, rocks all form from three basic processes. Igneous rocks solidify from a hot, molten state. Some cool slowly underground and produce the relatively large crystals of plutonic rocks, such as granite. Others erupt at the surface to form volcanic rocks such as obsidian. Metamorphic rocks are transformed from older rocks of any type by heat and pressure within the Earth. Surface processes, such as erosion, transport, and deposition form sediments and, with burial and compaction, sedimentary rocks.

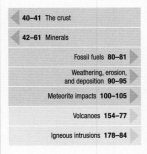

mica crystal   *glass shard*

MAGNIFIED IGNIMBRITE
This rock formed as magma cooled rapidly after a volcanic eruption. It consists of crystals, rock fragments, and glass shards welded together.

PLUTONIC IGNEOUS ROCK FORMATION
In a plutonic igneous rock (see p.79), the first minerals to grow form large crystals. Crystals that form later are smaller and less distinct.

large, first-formed crystals
remaining magma
crystalline matrix

quartz grain   *rock fragment*

MAGNIFIED SANDSTONE
Sedimentary sand grains of different shapes, sizes, and minerals are held together by a fine-grained matrix, or mineral groundmass, in this sandstone.

SEDIMENTARY ROCK AGGREGATIONS
When mineral grains are deposited by wind or water, they form loose sediments. Over time, these are compacted to form solid rock strata.

sediment grains
matrix (or cement)
substrate

water
settling grains

HOW ROCKS ARE MADE
Igneous, metamorphic, and sedimentary rocks form in a self-perpetuating cycle. Volcanic activity creates rocks at the Earth's surface. Rock is broken down by weathering and erosion, and fragments are carried away, settling in lakes, at coasts, and on seabeds. Burial transforms sediments into sedimentary rocks. Metamorphism induces further changes, and uplift and erosion expose them all.

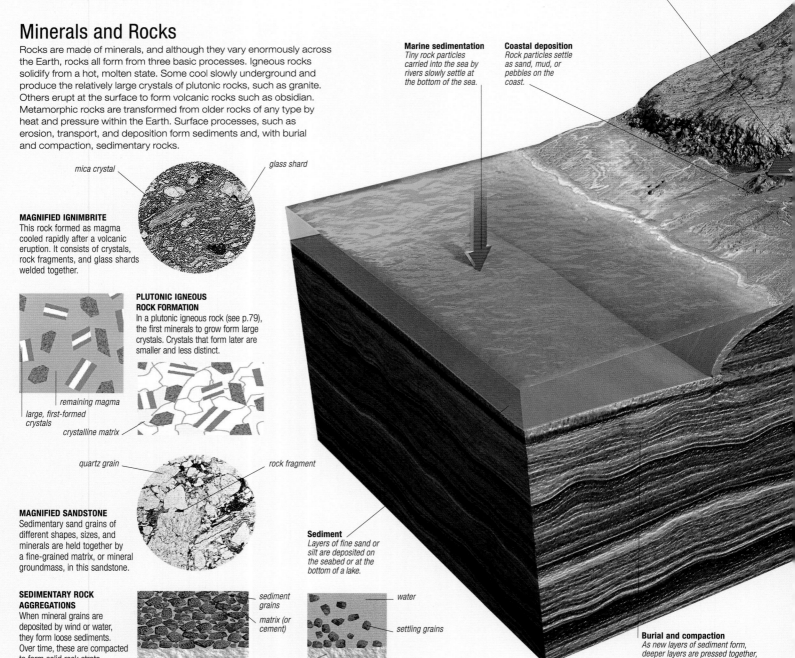

River erosion
Rivers transport silt and rock fragments created by weathering.

Marine sedimentation
Tiny rock particles carried into the sea by rivers slowly settle at the bottom of the sea.

Coastal deposition
Rock particles settle as sand, mud, or pebbles on the coast.

Sediment
Layers of fine sand or silt are deposited on the seabed or at the bottom of a lake.

Burial and compaction
As new layers of sediment form, deeper layers are pressed together, forming solid sedimentary rock.

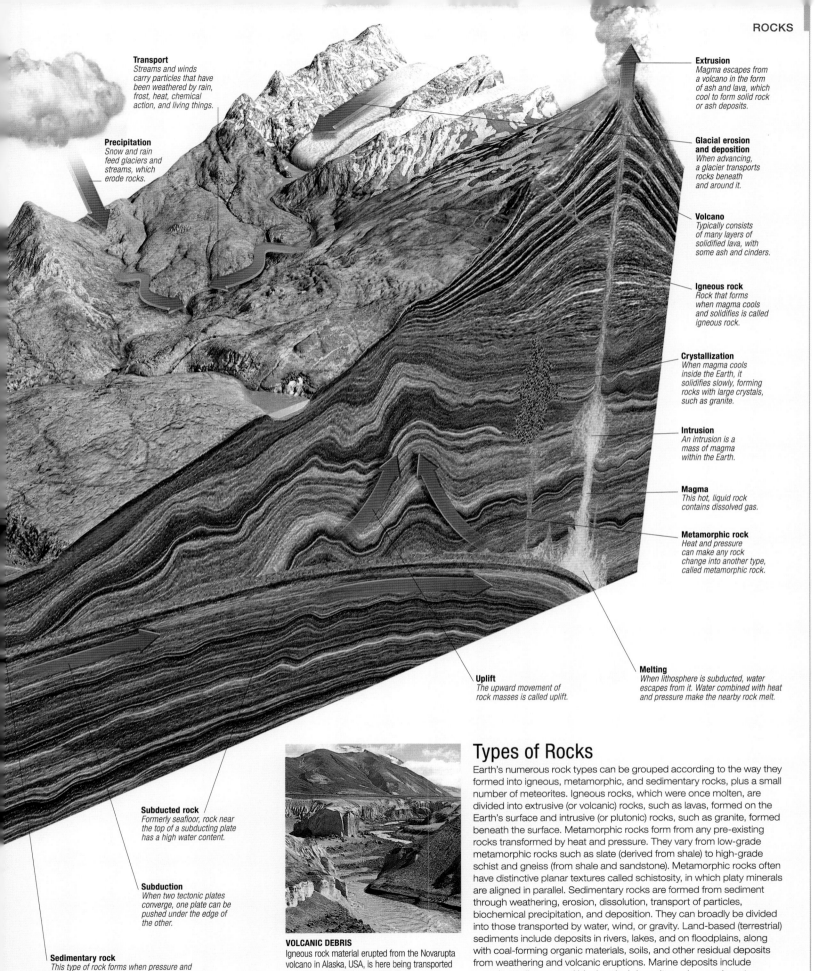

Transport
Streams and winds carry particles that have been weathered by rain, frost, heat, chemical action, and living things.

Precipitation
Snow and rain feed glaciers and streams, which erode rocks.

Extrusion
Magma escapes from a volcano in the form of ash and lava, which cool to form solid rock or ash deposits.

Glacial erosion and deposition
When advancing, a glacier transports rocks beneath and around it.

Volcano
Typically consists of many layers of solidified lava, with some ash and cinders.

Igneous rock
Rock that forms when magma cools and solidifies is called igneous rock.

Crystallization
When magma cools inside the Earth, it solidifies slowly, forming rocks with large crystals, such as granite.

Intrusion
An intrusion is a mass of magma within the Earth.

Magma
This hot, liquid rock contains dissolved gas.

Metamorphic rock
Heat and pressure can make any rock change into another type, called metamorphic rock.

Uplift
The upward movement of rock masses is called uplift.

Melting
When lithosphere is subducted, water escapes from it. Water combined with heat and pressure make the nearby rock melt.

Subducted rock
Formerly seafloor, rock near the top of a subducting plate has a high water content.

Subduction
When two tectonic plates converge, one plate can be pushed under the edge of the other.

Sedimentary rock
This type of rock forms when pressure and cementation bind particles of sediment together.

VOLCANIC DEBRIS
Igneous rock material erupted from the Novarupta volcano in Alaska, USA, is here being transported and redeposited to form river sediments.

Types of Rocks

Earth's numerous rock types can be grouped according to the way they formed into igneous, metamorphic, and sedimentary rocks, plus a small number of meteorites. Igneous rocks, which were once molten, are divided into extrusive (or volcanic) rocks, such as lavas, formed on the Earth's surface and intrusive (or plutonic) rocks, such as granite, formed beneath the surface. Metamorphic rocks form from any pre-existing rocks transformed by heat and pressure. They vary from low-grade metamorphic rocks such as slate (derived from shale) to high-grade schist and gneiss (from shale and sandstone). Metamorphic rocks often have distinctive planar textures called schistosity, in which platy minerals are aligned in parallel. Sedimentary rocks are formed from sediment through weathering, erosion, dissolution, transport of particles, biochemical precipitation, and deposition. They can broadly be divided into those transported by water, wind, or gravity. Land-based (terrestrial) sediments include deposits in rivers, lakes, and on floodplains, along with coal-forming organic materials, soils, and other residual deposits from weathering and volcanic eruptions. Marine deposits include granular sediments and biochemical deposits such as carbonates.

IGNEOUS INTRUSION
This vertical intrusive dolerite dyke in the Cuillin Hills, UK, acted as a feeder for the extrusive basaltic lavas that form the cliffs at top left.

Igneous Rocks

Rocks that solidify from a hot, molten state are known as igneous and are broadly divided into extrusive rocks, such as those produced by volcanoes, and intrusive bodies, such as plutons and dykes. They vary in composition from basalt to granite and in texture from rapidly cooled glasses and tiny crystals to slowly cooled coarse grains. Igneous rocks include the basalts (volcanic) and underlying gabbros (plutonic) that create all of the oceanic crust, plateau basalts (such as the Deccan Traps, see p.494), basaltic island volcanoes (such as those of Iceland), many hazardous subduction-zone volcanoes (such as Fuji, see p.174), and expansive plutons of granite (such as the Sierra Nevada, USA) and gabbro (such as the Skaergaard, see p.180). Plutonic rocks also form smaller bodies, which are classified (in decreasing size) as plutons, stocks, sills (roughly horizontal), and dykes (roughly vertical). The term pegmatite refers to silica-rich (granitic) plutonic bodies with giant crystals, which are sources of both gems and industrial minerals rich in rare elements.

IGNEOUS EXTRUSION
Large volcanic outpourings of basaltic magma shrink as they cool and develop roughly 5-sided columns that are perpendicular to the cooling surfaces, such as these near Aldeyarfoss, Iceland.

IGNEOUS ROCK FORMATION
Molten rock (magma) collects in underground chambers within the Earth's crust. It may solidify in place to form an intrusion, or it may erupt at the surface to form an extrusive igneous rock.

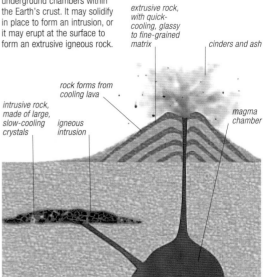

extrusive rock, with quick-cooling, glassy to fine-grained matrix

cinders and ash

magma chamber

rock forms from cooling lava

intrusive rock, made of large, slow-cooling crystals

igneous intrusion

Metamorphic Rocks

All rocks can be changed by heat and pressure. Such changes are called metamorphism. The most significant transformations occur at depth within the Earth, but local changes can be produced nearer the surface. Strong pressure creates dynamic metamorphism, typified by the compression of mud-rocks into slates. On a smaller scale, such pressure may be localized, for example along fault planes, where rocks are distorted by the Earth's movements. Intense heat produces thermal metamorphism, which ranges from localized heating of rocks around a small igneous body to more widespread heating that produces kilometre-wide zones of recrystallization, called thermal aureoles, around major, deep-seated plutonic intrusions. Together, heat and pressure on a large scale create regional metamorphism, which is graded from low to high, with increasing pressure and temperature triggering the formation of assemblages of new minerals, and so converting the original rock texture into a metamorphic texture, often dominated by planar foliation.

CHANGE OF ROCK TYPE
This tilted and frost-shattered layer of sandstone in Shropshire, England, was originally quartz sand on a sea bed, but it has since been cemented and recrystallized, and ultimately metamorphosed into quartzite.

THERMAL METAMORPHISM
Where large bodies of igneous rock such as granite are formed, they radiate sufficient heat into the surrounding rocks to alter their mineralogy, a process known as thermal metamorphism.

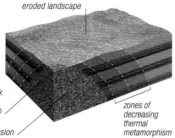

eroded landscape

existing rock changed by hot intrusion

zones of decreasing thermal metamorphism

granite intrusion

DYNAMIC METAMORPHISM
The Earth's movements can apply pressure to sedimentary rocks and transform them through dynamic metamorphism by orientating mica minerals to form slaty cleavage.

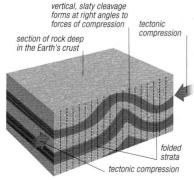

vertical, slaty cleavage forms at right angles to forces of compression

tectonic compression

section of rock deep in the Earth's crust

folded strata

tectonic compression

SPECTACULAR FOLDING
Intense folding of sedimentary rocks, seen here in the Ugab Valley, Namibia, also initiates dynamic metamorphism through compression and recrystallization.

CREATION OF SLATE
Clay minerals of sea-bed muds at Ingleton quarry, England, have been changed by dynamic metamorphism into a slaty cleavage.

Sedimentary Rocks

For at least 3,800 million years of the Earth's history, sediments have accumulated on the surface of the planet. Laid down layer upon layer and separated by bedding planes, they accumulate in sequences with the oldest layer at the bottom. With deep burial, they are compacted (lithified) into sedimentary rock, occasionally disturbed by folding or faulting, and returned to the surface again. These processes were first demonstrated by Nicolaus Steno (see panel, below). The Earth's depositional environments vary greatly, producing many kinds of sediments and sedimentary structures, such as ripples. The sands and muds from ancient rivers, deserts, and seas, along with the deposits associated with glaciers and volcanoes, are all preserved in the Earth's stratigraphic record. The original sediment is mainly mineral and rock debris from weathering, erosion, and transport by water (for example, sandstone), wind (loess), glacial and mass movement (tillite), biochemical activity (limestone), or evaporation (salt deposits). Once laid down, sediments may be altered by a variety of processes. Water is driven off, mainly due to compaction, and further changes over time transform the soft sediment into hard and brittle, layered rocks such as sandstone, limestone, and shale.

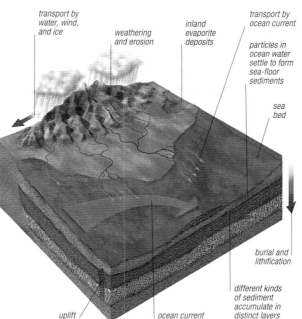

transport by water, wind, and ice

weathering and erosion

inland evaporite deposits

transport by ocean current

particles in ocean water settle to form sea-floor sediments

sea bed

burial and lithification

different kinds of sediment accumulate in distinct layers

uplift

ocean current

SEDIMENTARY ROCK FORMATION

The creation of sedimentary rocks occurs in overlapping stages. Rocks are weathered and eroded on land to create sediments that ultimately are transported to the sea. There, further transport may occur before deposition, burial, and lithification, the process that turns loose sediment into hard sedimentary rock.

NICOLAUS STENO

Although Danish, Nicolaus Steno (1638–86) studied in Italy and became physician to the Grand Duke of Tuscany. While there, he demonstrated the principle of superposition and showed that sedimentary rock strata were originally laid horizontally on top of one another by the deposition of particles from a fluid. Consequently, in any series of strata younger layers lie on older ones. He also ascertained that tilted and deformed strata result from displacement by the Earth's movements after deposition, and proved that fossils are the remains of once-living organisms.

QUARTZ SANDSTONE
At a magnification of x 25, the individual quartz grains of this sandstone can be seen to be held together by tiny but well-formed crystals of a quartz cement – the small, angular, surface protrusions.

WATER-LAID SEDIMENTARY ROCKS
The alternation of limestone and shale deposits at Lyme Regis, in southern England, show that conditions of sedimentation kept changing from carbonate to that of fine-grained mud deposition.

Meteorites

Natural objects from space made of iron and stone crash through the Earth's atmosphere all the time. They range from dust-sized particles to massive lumps weighing many tonnes. Such meteorites have fallen throughout the history of the Earth, and they are important as they record the birth of the Solar System 4,560 million years ago and its subsequent history. Records of meteorite falls first appear in cuneiform writings, in about 1900 BCE, while fossilized meteorites recently found in Sweden fell to the Earth more than 450 million years ago. Meteorites are made of the same elements as the Earth and Sun. These elements combine to form more than 300 different minerals, including many not found on the Earth. The abundance and types of minerals found in the meteorites are used to classify them into stones (both chondrites and achondrites), stony-irons (such as pallasites), and irons.

METEORITE SECTION
Cut in half, this spindle-shaped meteorite (left) with its black surface shows a grey metallic interior made entirely of intergrown crystals of nickel-iron alloy typical of iron meteorites.

alloy of nickel and iron metal

ANTARCTIC FIND
In recent decades, a large number of extremely well-preserved meteorites (left) have been collected from the surface of the Antarctic ice-sheet, where they have accumulated over hundreds of years.

ANCIENT FRESHWATER DEPOSITS
Many old sedimentary rocks (such as these ones in Bryce Canyon, Utah, USA) are derived from sediments originally deposited in rivers and lakes.

ROCK PROFILES

The pages that follow contain profiles of a selection of rock types. Each profile begins with the following summary information:

ORIGIN (igneous) Intrusive or extrusive; (metamorphic) dynamic, thermal, or regional; (sedimentary) marine, freshwater, wind-blown, glacial, organic, terrestrial, residual, or volcanic; (meteorites) extra-terrestrial

COLOUR A brief description, including common variations

GRAIN SIZE Fine grains are less than 0.1mm (0.004in) in diameter; medium 0.1–2mm (0.004–0.08in); coarse more than 2mm (0.08in)

Igneous and Metamorphic Rocks

Apart from the core, the vast bulk of the Earth is made up of igneous rocks and metamorphic rocks formed by the cooling and crystallization of molten silicate rocks and their transformation in the solid state to new textures and assemblages of minerals. Igneous rocks can be classified according to their composition (especially the relative proportions of silica and iron-magnesium minerals). They can also be grouped according to grain size (from fine to coarse, reflecting how quickly they cooled) and colour (silica-rich rocks, dominated by quartz and feldspar, tend to be pale coloured, while silica-poor rocks tend to be coloured darker from iron-magnesium minerals, such as olivine, pyroxenes, and amphiboles). Also invaluable is texture (how the crystal grains relate to one another). Other characteristics, such as the mode of formation and presence of certain minerals or assemblages of minerals (for example, chlorite, kyanite, and garnet), are particularly useful in categorizing metamorphic rocks, as they reflect the temperature and pressure conditions under which they were formed.

BUILDING WITH GRANITE

The mineral composition and structure of granite give it good load-bearing properties so that it can be used for a variety of building purposes, such as road building (shown here, in Egypt). Granite can be a massive rock without any planes of weakness, but cracks, called joints, may form as the rock cools and pressure is released. These joints can be exploited in quarrying; otherwise, granite has to be drilled and blasted to extract blocks. The surface of rough-hewn granite can be susceptible to chemical weathering, but polished granite is resistant to such weathering. The relative coarseness of the grain means that it is not suitable for fine carving.

IGNEOUS

Granite

ORIGIN Intrusive	**GRAIN SIZE** Mostly coarse
COLOUR White–red, pale green–blue, grey–black	

Typically, granite is a tough rock that forms pronounced topographical features and is resistant to erosion. It forms by the slow cooling of magma deep in the Earth's crust, and is composed of silica-rich minerals such as quartz, feldspar, and smaller amounts of mica and amphibole (generally hornblende).

MAGNIFIED SECTION
Brown biotite crystals set in a matrix of quartz and feldspar (both clear coloured) are typical of plutonic igneous rock granite.

GRANITE PORPHYRY
A large feldspar crystal set in smaller crystals gives this granite from south-west England a texture known as porphyritic.

Granite's colour varies from white to red, pale green or blue, even black, depending on the colours of the minerals and the texture. Most crystals are a few millimetres in size and interlock one another, but some minerals such as feldspar grow bigger – up to several centimetres. Granite is composed of about 70 per cent silica (SiO_2) and more than 10 per cent quartz, and forms large crystals. Overall, granite does not vary greatly in its mineral composition, except near the margins of granite intrusions, where blocks

TEXTURES AND COLOURS
Different mineral compositions cause the pink and red colouring in the two coarse-grained granites shown here, while a medium-grained granophyre has more amphiboles and dark micas than usual.

ferro-magnesian minerals create dark colour

biotite mica

GRANOPHYRE

PINK GRANITE

called xenoliths, which are derived from the surrounding rocks, can be found in various stages of assimilation. Granite forms large intrusive masses called batholiths, with volumes of tens to thousands of cubic kilometres, yet it can also occur as a lens, vein, or dyke on a metre-scale. Because granite forms between one and several kilometres beneath the Earth's surface, it is exposed only after long periods of erosion or tectonic movements of the Earth's crust. Feldspars in granite break down to form kaolin-rich clays on the Earth's surface.

orthoclase feldspar crystals

RED GRANITE

CHANGE OF SHAPE
Prolonged exposure can change the shape of jointed granite blocks to rounded boulders in a process known as spheroidal weathering, as here in the Alabama Hills, California, USA.

IGNEOUS
Pegmatite

ORIGIN Intrusive **GRAIN SIZE** Coarse

COLOUR White, pale pink, and pale grey

After most of a granitic intrusion has cooled and crystallized, the residual liquid part of the magma solidifies into extremely coarse-grained rocks called pegmatites. Pegmatites typically occur as sheet- or lens-shaped bodies at the margins of granitic plutons. Like granites, most pegmatites are composed of quartz, feldspars, and micas. Pegmatites are an important source of industrial minerals and rare elements such as boron, beryllium, and lithium. They also form gem minerals such as tourmaline and beryl.

pink feldspar

tourmaline

IGNEOUS
Granodiorite

ORIGIN Intrusive **GRAIN SIZE** Coarse

COLOUR White, grey, and black

dark amphibole

Granodiorite is the most voluminous of all intrusive igneous rocks in the continental crust, and its composition is thought to approximate the average composition of all rocks in the continental crust. With the addition of iron and magnesium, granite grades into granodiorite and then diorite. Like granite, granodiorite forms large intrusions within the continental crust, but it is slightly darker in colour as a result of its composition: it has more plagioclase (over 65 per cent) than potassic feldspar and slightly more (up to 40 per cent) dark minerals such as biotite mica. A tough rock, granodiorite is widely quarried, cut, and crushed for use as facings for buildings, curbstones, and road surfaces.

IGNEOUS
Dolerite

ORIGIN Intrusive **GRAIN SIZE** Medium

COLOUR Dark green, grey, and black

Broadly similar in composition to gabbro, dolerite typically comprises pyroxene, plagioclase feldspar (generally labradorite), and iron-titanium oxides such as ilmenite. Varieties of dolerite range from those with some quartz to those with large amounts of olivine or hornblende. Dolerite usually occurs as sheet-like or circular, plug-shaped intrusions related to a volcano. In some cases, the rocks surrounding such intrusions have been thermally metamorphosed. This suggests that many dolerite intrusions have acted as pathways for magma feeding extrusive lavas or further intrusive and extensive dykes and sills. Dolerite intrusions occasionally include exotic blocks (called xenoliths) of coarse-grained igneous or metamorphic rocks that have been raised from

abundant amphibole

MINERAL CONTENT
The characteristic medium-grained texture and dark colour of dolerite rocks are derived from their abundant amphibole and mica minerals.

DOLERITE DYKES
This road-cutting in Jordan reveals a network of branching and rejoining dark dolerite dykes in a surrounding pink host rock.

great depths. These are of particular interest because they provide samples of rocks from the base of the continental crust and possibly the upper mantle, which are not normally available. The mineral grains of dolerite interlock, to produce a very compact rock with high crushing strength. It is therefore often used as aggregate in load-bearing layers of road beds.

IGNEOUS
Gabbro

ORIGIN Intrusive **GRAIN SIZE** Coarse

COLOUR Grey–black

Named after a town in Tuscany, Italy, by Christian Leopold von Buch (see panel, right), gabbro is a coarse-grained, plutonic rock with a high percentage (20–65 per cent) of dark minerals, especially pyroxene and olivine. As these dark minerals increase at the expense of feldspar, gabbro grades into peridotite. The light-coloured minerals are predominantly plagioclase feldspar (35–80 per cent). Gabbro is found in continental and oceanic crusts, but it is a more important constituent of the

GABBRO IN GREENLAND
Layered gabbro intrusions in the Skaergaard intrusion are here are cut through by dykes, which are compositionally similar but finer grained.

CHRISTIAN LEOPOLD VON BUCH

One of the most eminent geologists of all time, Christian Leopold von Buch (1774–1852) studied mineralogy with Abraham Gottlob Werner (see p.45) at Freiberg, where one of his fellow students was the famous geologist and geographer Alexander von Humboldt (see p.141). Von Buch travelled widely and published on many geologically related topics. His early studies focused on volcanism, but his later interests were in fossils and stratigraphy. He effectively defined the Jurassic system of strata in his 1839 book *Über den Jur in Deutschland*.

latter, where it forms as part of what is known as the ophiolite sequence. Below ocean-floor sediments, there are pillow lavas, sheeted dykes, and then gabbro. This has developed from the slow cooling of basaltic magma at depth. Gabbro can form plutons, or sheet-like bodies. Gravity segregation of the crystallizing minerals has produced some layered gabbros such as the

Bushveld Complex of South Africa, which contains important economic minerals such as chromite. Gabbro is a tough, compact rock and is used as an ornamental facing stone, especially when it contains an iridescent variety of labradorite feldspar.

light plagioclase feldspar

dark pyroxene

TRANSFORMATION OF MINERALS
White plagioclase feldspar and dark minerals such as pyroxene (shown here) typically crystallize to form gabbro from slow-cooling, silica-poor magmas.

IGNEOUS
Diorite

ORIGIN Intrusive **GRAIN SIZE** Coarse

COLOUR Black and white

light plagioclase feldspar

scattered dark minerals

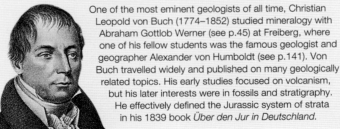

Less silica-rich than granodiorite, diorite is the intrusive equivalent of andesitic lava and has little or no quartz (less than 20 per cent). It is composed essentially of plagioclase feldspar and has more dark (mafic) minerals, such as amphibole and pyroxene (up to 50 per cent), than granodiorite, of which it is a compositional variant. As a result, diorite is darker in colour and can be difficult to distinguish from gabbro, which it also grades into. Diorite often contains fragments (xenoliths) of other rocks, especially gabbro. This tough rock is used for ornamental purposes and as road metal.

Lamprophyre

ORIGIN Intrusive and extrusive

GRAIN SIZE Fine–coarse   **COLOUR** Dark grey

The name lamprophyre is derived from the Greek and means "glistening porphyry", referring to its texture. The rapidly cooled matrix is typically very fine grained, but it is studded with larger phenocrysts of mica or amphibole (or both). These two water-bearing minerals with strong cleavages give lamprophyre its glistening porphyritic texture. The restriction of feldspars to the matrix and the presence of phenocrystic mica and amphibole attest to the exceptionally high water content of lamprophyric magmas. Lamprophyre typically forms late-stage dykes and sills cutting across granitic plutons. Volcanic types are also known. (These include the scoria shown below – this is a type of volcanic glass from a late Pleistocene cinder cone in Mexico.)

fine-grained matrix

phlogopite phenocryst

Lamproite

ORIGIN Intrusive and extrusive   **GRAIN SIZE** Coarse

COLOUR Dark grey

Lamproites are volcanic igneous rocks that contain diamonds. They are rich in potassium but silica-poor, containing 45–55 per cent by weight of silicon dioxide. Lamproite generally overlaps in composition with kimberlite and lamprophyre, and may be subdivided into olivine and leucite lamproite. Major minerals include olivine, leucite, and phlogopite mica, as well as accessory minerals such as calcite, apatite, serpentine, spinel, and diamond – but only olivine lamproite contains diamonds. Lamproite typically occurs as intrusive dykes or extrusive flows, but may also form in diatremes or pipes on the margins of ancient shields or cratons. A lamproite pipe in the Kimberley region of Australia produced 40 per cent by volume of the world's diamonds at its peak.

Kimberlite

ORIGIN Intrusive   **GRAIN SIZE** Medium–coarse

COLOUR Bluish grey to greenish grey and black

The world's primary source of diamonds is kimberlite pipes. These steep-sided, deep volcanic conduits often occur in clusters, which merge at depth. Kimberlite is a dark-coloured, originally coarse-grained, ultramafic rock with variable composition. It is often fragmented during the process of intrusion. The predominant minerals are olivine, mica, pyrope garnet, orthopyroxene, and a variety of accessory minerals, including diamond. The matrix is generally altered to serpentine, chlorite, and carbonate minerals. Kimberlite pipes also often include exotic rocks, known as xenoliths, which are derived from the Earth's crust and upper mantle.

dark matrix

xenolith

MINING KIMBERLITE

Diamonds are rare but exceedingly valuable minerals within kimberlite, so it can be worth processing many tonnes of rock to extract a few diamonds. Because of the near-vertical walls of kimberlite pipes, the process of mining them for their diamonds, as here in Kimberley, South Africa, has resulted in some of the biggest and deepest artificial holes in the world.

A SOURCE OF DIAMONDS
Diamond-bearing kimberlite forms volcanic pipes less than a kilometre wide, which extend to great depth.

Peridotite

ORIGIN Intrusive

GRAIN SIZE Coarse   **COLOUR** Dark green to black

Peridotite is the main rock of the Earth's upper mantle. Most peridotite is a dark-coloured, dense rock that slowly cooled at depth, and it is often found layered with gabbro. It is mainly composed of olivine and pyroxene, along with garnet or spinel, and more rarely amphibole and phlogopite. Peridotite occurs as sheet-like rocks associated with other mafic to ultramafic rocks typical in oceanic crust sequences, or in large massifs within mountain belts. It is also found as xenoliths brought up by alkali-rich basaltic magmas.

CANADIAN COASTLINE
These layered ultramafic rocks at Lobster Cove, Canada, include peridotite.

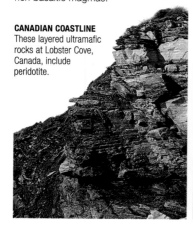

Basalt

ORIGIN Intrusive and extrusive   **GRAIN SIZE** Fine

COLOUR Greyish black to black when fresh

The most common igneous rock on the Earth's surface, basalt forms the rock floor of most of the oceans. It also occurs in continental settings as extensive plateau basalts, such as in the Columbia River region of the northwest USA (see p.158). In both settings, basalt is found as intrusive and extrusive bodies of rock. This fine-grained, dark-coloured equivalent of gabbro is dominated by pyroxene, olivine, plagioclase, and accessory iron-titanium oxide minerals such as magnetite and ilmenite. Basalt may include some large crystals, typically olivine or plagioclase. It also commonly contains former gas bubbles, known as vesicles, which may become filled with zeolite, carbonate, or

fine-grained matrix

small crystals

FINGAL'S CAVE
The dramatic scenery surrounding Fingal's Cave in Scotland inspired the classical composer Felix Mendelssohn to write his now-famous Overture.

FINE-GRAINED TEXTURE
The crystalline texture of basalt is so fine grained that individual crystals are not visible to the unaided eye, but they give basalt a tough, compact structure.

agate by a secondary process. Basalt weathers to pale green, brown, or grey or, where oxidized, to red. It occurs mostly as extrusive lavas, but also forms intrusions such as dykes and sills. On cooling, it may form distinctive, jointed columns. The occurrence of intrusions at sites such as the basaltic lavas of the Giant's Causeway in Northern Ireland and the nearby Fingal's Cave on the Scottish island of Staffa (see left) has fascinated scientists and artists alike for hundreds of years. When fresh, basalt makes a high-grade aggregate, and it has been used as an ornamental stone since ancient Egyptian times.

fine-grained matrix

mineral-filled vesicles

AMYGDALOIDAL BASALT
Amygdaloidal means almond-shaped and refers to the gas bubbles commonly found in basaltic lavas, which are often infilled with secondary minerals.

IGNEOUS

Dacite

ORIGIN Mostly extrusive

GRAIN SIZE Fine–coarse

COLOUR Pale to medium shades of grey and pink, through to brown

abundant plagioclase

Named after a Roman province in Romania, dacite is a light- to medium-coloured volcanic rock with about 65 per cent by weight of silica. It occurs as extrusive lavas, small intrusions such as domes and plugs, or pumice and ash, having been formed by the rapid cooling of viscous magma extruded at 800–900°C (1,500– 1,700°F). In 1980, Mount St. Helens, USA, erupted dacite pumice and ash, as did Mount Pinatubo, in the Philippines, in 1991. Typically porphyritic with phenocrysts of plagioclase, quartz, hornblende, or biotite mica, some coloured varieties have been used for centuries as ornamental stones, and currently as aggregates.

IGNEOUS

Obsidian

ORIGIN Mostly extrusive   **GRAIN SIZE** Fine

COLOUR Greenish black to black when fresh

Rapid cooling of highly viscous, hot, rhyolitic lava before individual minerals have had time to form crystals gives obsidian its overall glassy texture. Typically, this dense, opaque to transparent rock is dark coloured, due to the presence of very fine-grained iron-titanium oxide minerals such as magnetite. Small crystals of feldspar or quartz may form, and the rock sometimes shows flow banding with alternating glassy and devitrified layers. Obsidian occurs in a number of different volcanic settings, ranging from glassy surfaces to voluminous

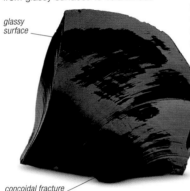

glassy surface

concoidal fracture

OBSIDIAN TOOLS

For prehistoric peoples, obsidian was ideal for making very sharp blades and points, because of its glassy texture, conchoidal fracture, and considerable hardness. Broken shards of obsidian were extremely sharp – although fragile – but if damaged they could be reworked to give new, equally sharp edges. Obsidian, for example, was often used in the technically sophisticated, Clovis-point weapons adopted by hunters in the Americas from the tenth millennium BCE.

lavas (such as the 1,300-year-old Big Obsidian Flow at Newberry, Oregon, USA), small intrusions, and the outer layers of lava domes. Obsidian has been used in jewellery since ancient times: for example, the black eyes of the young King Tutankhamen's gold funerary mask were made from polished obsidian.

REFLECTIVE IMAGE
The smooth, rounded, cracked surface of obsidian is typically dark coloured and slightly translucent, with a glassy texture similar to that of manufactured glass.

IGNEOUS

Andesite

ORIGIN Mostly extrusive

GRAIN SIZE Fine–coarse

COLOUR Variable, including brown–purple to grey

Named after the Andes Mountains of South America, andesite is a fine-grained to porphyritic volcanic rock common in many of the world's subduction-related volcanic arcs. Andesite typically has about 60 per cent silica by weight and erupts at temperatures of 950–1,000°C (1,750–1,800°F). Plagioclase feldspar commonly forms prominent phenocrysts accompanied by pyroxenes, iron-titanium oxides, and, in some andesites, the hydrous minerals amphibole or biotite (or both). Andesitic magma can erupt explosively, but typically the gas propellant bleeds away from the magma prior to eruption as sluggish and viscous block-lava flows.

white plagioclase

IGNEOUS

Rhyolite

ORIGIN Mostly extrusive   **GRAIN SIZE** Fine

COLOUR White to pale grey, green, pink, and brown

This light-coloured, often banded rock is rich in sodium, potassium, and silica (70–78 per cent by weight). It is primarily found in continental settings. Rhyolite contains a fine-grained matrix, which may be glassy and sometimes includes visible phenocrysts (larger crystals) of feldspar, quartz, or mica. Colour banding and parallel phenocrysts within the rhyolite indicate the direction of the magma flow in the conduit through which the lava was released. Silica-rich rhyolite magma is relatively cool, emerging at temperatures as low as 700–800°C (1,300–1,500°F). Due to its high silica content, rhyolitic magma

is extremely viscous, which causes it to flow slowly and to retain its gases, because gas bubbles cannot effectively rise through the viscous liquid. If rhyolite has a low gas content as it nears the surface, it erupts to form thick, sluggish lava flows that tend to pile up around the vent, forming dome-like onstructions. More commonly, rhyolite is rich in gas as it nears the surface and erupts explosively. The once-dissolved

volatiles, mainly water and carbon dioxide, expand rapidly into the atmosphere. Expansion froths the magma and rips it apart to form pumice and ash, and propels the fragmented magma out of the vent at high speed. The hot eruption column entrains and heats surrounding air, which can cause the plume to rise buoyantly into the stratosphere. In other cases, the margins of the plume collapse downwards to form destructive

pyroclastic flows. Prehistoric rhyolitic eruptions, such as those that have occurred at Yellowstone, USA, in the past 2.1 million years, have ejected up to several thousand cubic kilometers of magma.

THICK LAVA FLOWS
Silica-rich rhyolite lavas are viscous and so do not easily flow any great distance after expulsion. Here, in the Rhyolite Hills, Iceland, they have therefore formed very thick flows.

fine, pale matrix

large, dark mica

TWO CRYSTAL SIZES
Although characteristically fine grained and pale coloured, rhyolite may contain a variety of slightly larger crystals within its matrix.

IGNEOUS

Volcanic deposits

ORIGIN Extrusive   **GRAIN SIZE** Fine–coarse

COLOUR Very variable, including white, grey to black

The material erupted by volcanoes varies according to the magma from which it was derived. As recognized by James Hutton (see panel, right), these deposits range from effusive lavas to a variety of other forms. These other forms are collectively known as volcanic deposits, and they include scoria-, pumice-, ash-fall, pyroclastic-flow, and debris-flow deposits. The nature of each volcanic deposit depends on the chemical composition, gas content, and temperature of its silica-poor magma during eruption. Silica-poor magmas such as basalt have low viscosities and high temperatures. Consequently, basaltic magmas are very fluid and, on eruption, develop a wide range of fluidal shapes. Clots of basalt thrown through the air, called bombs, can develop aerodynamic

gas-bubble vesicle

PUMICE
This gas-frothed laval rock is filled with numerous cavities, or vesicles, and is so light that it can float.

VOLCANIC DEBRIS
Fine ash through to pebble-sized lapilli are among debris erupted from volcanoes that falls back to Earth.

LAPILLI (PUMICE)

COARSE ASH   **FINE ASH**

shapes (see below). Fast-moving basaltic lava flows can form blocks with jagged, spiny surfaces known by the Hawaiian name a'a, whereas slower-moving basalts have smooth and ropey textures known by the Hawaiian name pahoehoe. When basalt erupts below water, it forms stacked, bulbous masses called pillow basalts. Magmas richer in silica (andesite, dacite, and rhyolite) are more viscous, which inhibits their ability to degas. Nonetheless, many silica-rich magmas do lose their gases during ascent, and erupt to form sluggish and

PYROCLASTIC LAYERS
These layers of ash and pebble-sized rock and pumice pieces have piled up around the volcano on the Italian island of Lipari.

PYROCLASTIC FLOWS

Our knowledge of volcanic phenomena such as pyroclastic flows has generally come from observations made during or following important recent eruptions. These dramatic and deadly phenomena were only dimly perceived prior to the eruptions of the Caribbean volcanoes Mount Pelée and Soufrière St. Vincent in 1902, when pyroclastic flows (also known by the French term *nuées ardentes*, or "glowing clouds") were observed and photographed as they descended canyons on the flanks of the volcanoes. Volcanologists came to realize that pyroclastic flows are hot mixtures of gas, pumice, ash, and old rock fragments that move downhill as turbulent, gravity-driven flows. Insight from these eruptions led to the realization that many ancient deposits of pumice and ash (such as those from the eruption of Vesuvius in 79 CE, see p.170), which were formerly thought to have fallen through the air, were actually pyroclastic-flow deposits.

PYROCLASTIC-FLOW DEPOSIT
Layers of ash with different-sized particles, such as these from around Herculaneum, Italy, are now interpreted as deposits of pyroclastic flows.

thick block-lava flows. Other silica-rich magmas, however, reach the surface with high contents of water, carbon dioxide, and other gases. These explosively fragment the magma into pumice and ash, sending eruptive clouds high into the atmosphere, which are blown downwind and begin to fall back to Earth. The largest, densest particles fall closest to the vent, and finer and less dense particles drop at progressively greater distances, forming thinner deposits. The margins of the eruptive column can collapse to form pyroclastic flows (see panel, above). Such flows can travel rapidly downhill, coming to rest as accumulations of pumice, ash, and rock fragments called pyroclastic-flow deposits. When stacked, the pressure in the hot interior bonds the deposit into welded tuff.

VOLCANIC BOMB
Molten, low-viscosity basaltic magma blasted through the air can assume aerodynamic forms, such as shown by this streamlined volcanic bomb.

spindle shaped

piled droplets

LAVA STALAGMITE
When fluid, molten basaltic lava drips into a lava tube, it can cool drop by drop. As the drops land on the ground, they accumulate to form a stalagmite-like structure.

JAMES HUTTON

One of the most important Earth scientists of the 18th century, James Hutton (1726–79) was born in Scotland and educated in chemistry and medicine in Edinburgh, Paris, and Leiden. He studied volcanic rocks in particular and showed that both intrusive and extrusive rocks are of igneous origin. His *Theory of the Earth*, published in 1785–88, was particularly influential.

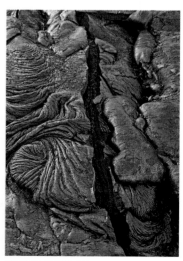

PAHOEHOE LAVA FLOW
Hot, fluid basalt forms a plastic skin, which can stretch and fold as the lava moves within it. The skin gradually hardens to a ropey-textured, glassy surface, as seen here at Kilauea volcano, Hawaii.

METAMORPHIC

Skarn

ORIGIN Thermal   **GRAIN SIZE** Fine–coarse

COLOUR Most commonly green and red, but also grey, brown, black, and white

Formed from high-temperature metamorphism and alteration of carbonate rock, skarn contains a variety of minerals rich in calcium, magnesium, and iron (such as garnet and pyroxene), as well as sulphide minerals of copper, lead, iron, zinc, and tungsten in commercially valuable quantities. Skarn deposits are found adjacent to porphyry deposits within the thermal contact zone of large igneous intrusions (such as granites) that have invaded carbonate rocks. Mineral-enriched solutions react with the surrounding rocks to form new silicate minerals and ores.

banded structure

METAMORPHIC

Serpentinite

ORIGIN Dynamic   **GRAIN SIZE** Fine–coarse

COLOUR Mainly green with black streaks

This generally green rock is made of the mineral serpentine and is commonly banded, blotched, or streaked with colours ranging from black to bright green and occasionally red. Serpentinite is also typically transected by veins of chrysotile asbestos. It is mainly composed of a silica-poor, hydrated magnesium-silicate, which is in turn the alteration product of ferro-magnesian minerals, especially olivine – serpentinite being derived from the igneous rock peridotite, which is rich in olivine. Other minerals present include pyroxene and iron oxides. Serpentinite is a relatively dense but soft metamorphic rock, and it is often fibrous in texture and readily deformed. The presence of serpentinite within mountain belts provides evidence of major tectonic events in the

patchy colour

ORNAMENTAL STONE
The attractive coloration of serpentinite rock such as this, with its greens and black, has led to its widespread use as an ornamental stone.

geological past, which brought slices of ocean-floor rocks or even the mantle itself into contact with the continental crust. Most serpentinite is found in convergence zones between tectonic plates, such as within the Alpine–Himalayan mountain belt. These are often the only remaining traces of the ocean basin that once existed between two continental masses that are now firmly

DEEP-SEA CHANGES
These serpentinite rocks at the Lizard Peninsula, in the far southwestern part of the UK, were metamorphosed 400 million years ago.

joined together (or sutured). Substantial exposures of serpentinite include those found at the Bay of Islands, in Newfoundland, Canada, and those in Oman, on the Arabian Peninsula. For example, the parent rocks of the Oman serpentinites were originally formed on the Tethys Ocean floor in mid-Cretaceous times. The closure of the Tethys Ocean resulted in part of the sea-floor rocks being thrust by faults onto the Arabian continent.

METAMORPHIC

Marble

ORIGIN Thermal   **GRAIN SIZE** Fine–coarse

COLOUR Mainly white, pink, green, brown, and black

Over millennia, marble has been valued for its smooth texture, colour, and ease of working for both sculpture and building purposes. It is the product of thermal or regional metamorphism of limestone. Heat and pressure cause the carbonate minerals to recrystallize into compact and often homogenous marble of various colours. Superficially, it may be confused with quartzite (another metamorphic rock), but marble can be scratched by a steel point, and it also reacts to weak hydrochloric or acetic acid. Recrystallization tends

to remove any cracks or cleavage, but traces of folds – often highly contorted – may be preserved, indicating that the rock flowed during metamorphism. Also, marble that is the product of low-grade metamorphism may preserve traces of fossils and sedimentary rocks. The close intergrowth of the carbonate minerals gives some marble a sugar-like texture that is suitable for carving (see panel, right). It is composed mainly of calcite but may contain other minerals rich in magnesium, calcium, and iron, such as serpentine, dolomite, phlogopite mica, and the amphibole tremolite,

which can produce distinctive colours and textures. Decorative marble is often particularly valued for specific colour and textural variations as a result of such mineral "impurities" in the matrix. As proportions of these accessory minerals increase, marble grades into other calcsilicate metamorphic rocks such as skarn. Marble is widespread through regional metamorphic terrains wherever there were original limestones. It is therefore often interlayered or found close to phyllite, quartzite, and schist. In addition, it may be occur in the thermal metamorphic zones around large igneous intrusions.

MARBLE TEXTURE
The crystalline calcium carbonate in marble gives it a compact texture that is suitable for intricate carving.

sugar-like texture

MARBLE SCULPTURE

Because marble can readily be carved, it has been used since preclassical times for sculpture and in buildings such as the Parthenon, in Greece, and the Taj Mahal, in India. The pure white marbles of Greece and Italy have been reshaped by the greatest sculptors from Praxiteles to Michelangelo, to produce world-renowned statues. Unfortunately, marble is prone to attack by acid rain, which, since the Industrial Revolution, has seriously damaged marble stonework on many buildings worldwide.

RENOWNED
The Taj Mahal was built of white marble and semi-precious stones in 1632–54.

FLAWLESS ROCK
Some of the best-quality marble such as that found in Carrara, Italy, has a pure white colour and can be cut into almost any form.

Hornfels

ORIGIN Thermal   **GRAIN SIZE** Fine–medium

COLOUR Variable, including white and shades of grey, green, blue, and black

The heat given off by large igneous intrusions can metamorphose surrounding sedimentary rocks into a new rock type called hornfels. This hard, compact stone has a fine matrix of interlocking mineral grains. Some varieties, such as pyroxene hornfels, however, are medium grained and completely recrystallized. The colour of hornfels depends on the composition of the original rock, but it generally looks spotted, because of the random growth of the metamorphic minerals. The new

red garnet crystals

GARNET HORNFELS
Rounded garnet crystals here have grown larger than the other minerals of the matrix and so have given this hornfels a distinctive spotted appearance.

mineral growths (the spots) on, for example garnet hornfels, are technically known as porphyroblasts, and the texture they produce is termed porphyroblastic. Their size and shape depend on the extent of metamorphism and the minerals growing within them. Generally, the rocks are too fine grained to allow individual minerals to be distinguished in a hand specimen. Some porphyroblasts have diagnostic shapes: for example, chiastolite grows in long prisms

chiastolite crystals

even-sized grains

CHIASTOLITE HORNFELS
Long, square-sectioned needles of chiastolite here grow randomly through a fine-grained matrix.

with a distinct cross shape of inclusions within each grain. These inclusions are seen in the direction perpendicular to the prism's long axis. Heating from an intrusion to temperatures within the range 300–1,000°C (600–1,800°F) causes solid-state changes to occur, with the progressive development of metamorphic minerals depending on the temperature. A succession of hornfels is therefore found, from low-temperature, spotted slate with porphyroblasts, which cannot be identified by eye in a hand specimen, through cordierite hornfels, which may be in direct contact with the intrusion or pass into a higher-grade pyroxene hornfels, which is in direct contact. The thermal contact zone around even very large granite intrusions may be surprisingly narrow (1m/3ft or so thick), showing that the heat from the intrusion dissipated rapidly away from the contact. The name hornfels is an old German miners' term meaning "horn rock".

dark pyroxenes

PYROXENE HORNFELS
The pyroxene crystals in this hornfels specimen can be distinguished only with the aid of a microscope.

Migmatite

ORIGIN Regional   **GRAIN SIZE** Medium–coarse

COLOUR Dark grey, white, and pink

banded structure

mixture of quartz and feldspar

Migmatite is formed at great depth in the continental crust and consists of light-coloured (leucocratic) and dark-coloured layers. Unlike gneiss, which it superficially resembles, migmatite shows evidence of partial melting either by the *in situ* production of granite from a pre-existing rock or by the fine-scale injection of granite veins. The latter appears as small clusters of crystals, bigger lenses, or veins, which may be layered or highly folded. Folded structures in migmatites indicate that deformation is common in partially molten bodies of rock. In 1907, the Finnish geologist Jakob Johannes Sederholm gave migmatite its name, which is derived from the Greek *migma* ("mixture").

Slate

ORIGIN Regional   **GRAIN SIZE** Fine

COLOUR Generally grey, also tinged green or purple

This compact rock is characterized by the way in which it may be split into thin sheets along parallel planes of cleavage. This property – along with slate's waterproof quality – have led to its widespread use as a building material, especially as roofing slate (see panel, below). The formation of slates occurs during low-grade metamorphism of mud-rocks (fine-grained, clay-rich sediments such as mudstones, shales, siltstones, and fine-grained volcanic deposits such as layered tuffs). When mud-rock

poorly developed cleavage

BLACK SLATE
Two successive deformations have produced this slate – the second, weaker one having realigned some of the minerals diagonally.

is subjected to compression, it is reduced in volume as water is driven out and the minute clay mineral grains are all reorientated at right angles to the pressure direction. This can happen irrespective of any original bedding planes within the sediment or any shape they are folded into. Cleavage formation can destroy any original sedimentary structures or fossils entombed within the rock. When the cleavage is closely parallel to the original sedimentary bedding, however, fossils may be preserved even if considerably distorted. Since the cleavage is generally a product of the same forces that produce folding, there is an important geometrical relationship between the cleavage and fold geometry. Like minerals of the mica group, the clay minerals have a tabular (or platy) shape and a well-developed plane of cleavage. When they are orientated parallel to one another, the whole rock acquires the same parallel slaty cleavage, as it is known. Mica can be a constituent of slate; both primary and synmetamorphic mica grains define some slaty cleavage. Slate may vary enormously in colour but typically is grey, purplish, or greenish grey. Quarrymen have used the word slate in a much wider sense

COLOUR BANDING
Cleavage surfaces of slates are often coated with mineral deposits such as the metals iron and copper, as here in Pembrokeshire, UK.

to describe any fissile rock that can be used as roofing material: for example, Stonesfield slate from Oxfordshire, England, is an unmetamorphosed platy Jurassic limestone.

brachiopod shell

FOSSIL-IMPRINTED SLATE
Occasionally, fossils like this Paleozoic shell are preserved within slates. They can be difficult to identify if they are highly flattened and distorted.

SLATE ROOFS
When highly compacted and cleaved into slates, mud-rocks make ideal waterproof roofing materials. They can be split into thin sheets, even as fine as 5mm (³⁄₁₆in), which are not too heavy and can be trimmed to shape. Good-quality slates can last for hundreds of years. Impurities such as pyrite (iron sulphide), which are common and can be hard to detect in the slate, gradually erode and can make the slate porous. The craft of hand splitting and trimming slates is arduous, labour intensive, and unhealthy (because the dust is harmful), and is no longer practised to the extent it once was. Most commercial "slates" are now therefore preformed and made of reconstituted mineral material.

WEATHER RESISTANCE
Well-fitting slates protect this mountain house in Piedmont, Italy, against rain and snow.

HARD-WEARING
Quartzite is a hard, tough, and compact rock because it is almost entirely composed of quartz grains that are interlocked.

Phyllite

ORIGIN Dynamic, low-grade regional

GRAIN SIZE Fine–medium

COLOUR Greenish grey to plain grey

With progressive metamorphism, slate grades into cleaved phyllite with a slight increase in grain size, and then into schist. Phyllite's cleavage planes have a characteristic sheen resulting from the reflection of light from the individual grains, which are barely detectable by the unaided eye. A single cleavage is generally dominant but is sometimes corrugated or folded by subsequent deformation. Phyllite therefore does not split as easily or uniformly as slate and cannot be used in the same way as a building material. As in slate, the cleavage is produced by the reorientation of clay mineral grains in response to confining pressure. Phyllite represents a greater amount of recrystallization of original clay minerals than slate to form minerals such as chlorite and mica (especially muscovite). The name phyllite is derived from the Greek for "leaf", referring to the parallel cleavage.

Quartzite

ORIGIN Thermal and dynamic

GRAIN SIZE Fine–medium

COLOUR Mostly white to grey

Superficially, quartzite may be confused with marble, although it is considerably harder than both the latter and most other metamorphic rocks. Consequently, quartzite often forms resistant features such as ridges in a landscape. Produced by the metamorphism of sandstone, it is generally pale coloured or white but may be in various shades of brown or grey, depending on its minor mineral constituents, especially magnetite and pyrite. Quartzite is predominantly made of quartz grains that have recrystallized. With high-grade metamorphism, however, individual grains within sandstones may be lengthened by the combination of pressure, heat, and stress, which may flatten or shear the rock. This is best seen in conglomerates,

SHATTERED QUARTZITE
These blocks of quartzite in Israel formed when sea sand partially melted during a volcanic eruption, before cooling to form quartzite prisms.

where the grains are very large. When sandstone grains recrystallize, they overgrow their original grain boundaries and closely intermesh to form compact and brittle quartzite. Original sedimentary structures and fossils may be destroyed in the process, although, with low levels of metamorphism, traces of some sedimentary structures may persist, as well as outlines of some fossils. The main variation in quartzite is due to original compositional variations such as the presence of feldspar, mica, or pyrite. Because it is brittle, quartzite is often

fractured by a series of joints that may act as conduits for mineral-rich fluids. Due to their relative hardness and chemical stability, eroded quartzite fragments (clasts) persist longer in sediments than most other rocks.

Schist

ORIGIN Regional   **GRAIN SIZE** Medium–coarse

COLOUR Variable, including white and shades of grey, green, blue, brown, and black

The essential feature of schist is its parallel planes of similarly orientated minerals within the rock. This originates in the same way as cleavage, with minerals growing with a preferred orientation in response to a powerful

HIGHLAND SCHIST
Alternating layers of light and dark minerals characterize this Scottish metamorphic rock in Glen Nevis as a schist.

FOLDED SCHIST
The parallel planes in this specimen have been compressed into corrugations by later movements of the Earth.

shiny surface

crumpled cleavage

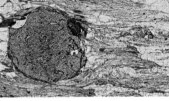

pressure and deforming stress on the rock. Minerals that are tabular (such as mica or chlorite) or blade- or needle-shaped (such as kyanite or sillimanite) can all contribute to the formation of schistosity. Clay minerals are absent, having been recrystallized into mica and chlorite. In general, schists are all produced by relatively high pressures and temperatures typical of large-scale

regional metamorphism. The result is that schist typically has medium- to coarse-grained texture, but may frequently contain larger grains of particular minerals called porphyroblasts. Some minerals tend to grow within specific pressure and temperature ranges, and consequently can be used as indicators of the conditions under which their host schist originally developed. The schist

MICROSCOPIC VIEW
As they grow, rounded garnet crystals form islands around which mica grains wrap, as seen in this highly magnified thin section.

rock name is then modified by the indicator mineral – for example, biotite or garnet schist (see below). The dominant minerals for most common schists include quartz and mica plus various other metamorphic minerals such as garnet, kyanite, and staurolite. The name schist is an old German miners' term and is derived from the Greek *skhizein* (" split" or "divide"). In some mountainous regions where slates are not available, schist is used as roofing material, as it can be finely split.

GARNET SCHIST
Large, rounded garnets indicate that this schist developed at relatively high temperatures and pressures deep within the continental crust.

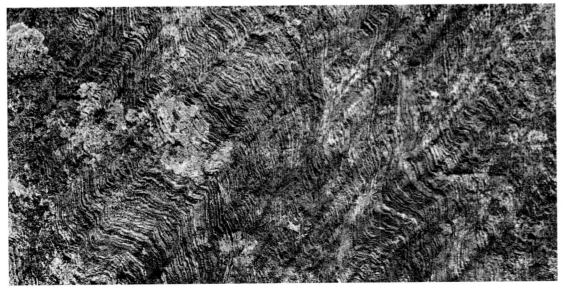

METAMORPHIC

Gneiss

ORIGIN Regional   **GRAIN SIZE** Medium–coarse

COLOUR Pink-grey

With its characteristic alternation of centimetre-scale light and dark layers, this silica-rich metamorphic rock has a feldspar content of more than 20 per cent. It is the product of high grade metamorphism at temperatures approaching the melting-points of the

large feldspar crystals

tight folds

BIOTITE GNEISS

MULTI-LAYERED
Gneiss rocks may appear different depending on the rocks they originated from, yet they share characteristic, alternating light and dark layers.

PRECAMBRIAN GNEISS

constituent minerals, thus allowing some of them to recrystallize and produce the porphyroblastic textures seen in gneiss. Although gneiss is typically an overall pink or grey, its colouring may vary within the rock, depending on its composition. It is composed of layers, often in different colours and sometimes with a wavy, planar pattern of orientated biotite grains. Pale layers are dominated by white quartz and plagioclase feldspar, although pink feldspar may also be present. Darker bands include biotite and hornblende and tend to have parallel orientations. Some minerals, such as garnet or clustered associations of feldspar and quartz, form eye-shaped lumps as gneiss is strongly deformed (see panel, right). Gneiss may be derived from sedimentary or igneous rocks.

The more compact varieties of gneiss are often used as building stone, especially those that display spectacular folds and colour bands and other metamorphic textures when cut and polished.

Because of their high crushing strength, these compact varieties of gneiss are also used as aggregates. The name gneiss is very ancient and probably derives from a slavonic word meaning "nest"; it was certainly in use in medieval times by miners in Central Europe. The word was adopted – along with schist – by the German mineralogist and teacher Abraham Gottlob Werner (see p.45) and his followers in the 18th century. Thus it became standard geological terminology.

FOLDED GNEISS
Alternating, pale- and dark-coloured layers are intensely metamorphosed and folded in this ancient gneiss from Namibia, in southwest Africa.

GARNET AUGEN GNEISS

Garnet is a common mineral in rocks, such as schist and gneiss, found in areas of regional metamorphism. Generally, the garnet is the iron-rich, pink variety almandine. It tends to grow as distinctive porphyroblastic "augen" (eye-shaped grains). Sometimes the

garnet crystals rotate during growth to produce what are known as snowball garnets, which have internal spiral trails of small inclusions of other minerals that are visible with a microscope. Garnet minerals are useful to geologists because their occurrence within regionally metamorphosed rocks indicates that the rocks formed within certain pressure and temperature ranges. At pressures and temperatures equivalent to depths of 100km (60 miles) beneath the Earth's surface, redistribution of aluminium among the minerals of upper-mantle peridotites leads to a progressive conversion of spinel and pyroxene into garnet.

METAMORPHIC

Eclogite

ORIGIN Thermal and dynamic

GRAIN SIZE Medium–coarse

COLOUR A mixture of green and red

One of the densest silicate rocks known, eclogite is typically massive and formed by very high pressures and temperatures in excess of 500°C (900°F) associated with regional metamorphism at considerable depth in the Earth's crust. Although rare at the surface, it occurs as exotic blocks within serpentinite or kimberlite, as well as in mountain belts that preserve metamorphic rocks that formed

in ancient subduction zones. Eclogite is formed when slabs of dense ocean-floor rocks such as basalt subduct beneath the edge of a continent or another oceanic slab. On reaching depths of 80–90km (50–56 miles), the pressure and temperature conditions cause the silica-poor, dark silicates such as amphibole to lose their water and so transform into the denser minerals of eclogite. Eclogite provides important information about the process of subduction. This rock can be striking when cut and polished, showing bright green and red from its main constituent minerals – green pyroxene (omphacite) and red garnet, and other high-pressure minerals such as kyanite.

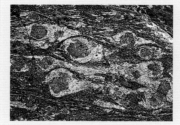

high-pressure minerals

DEEP SAMPLE
Dark red garnet crystals have grown as large porphyroblasts within the green pyroxene matrix of this eclogite hand specimen.

red garnet

green pyroxene

TYPICAL COLORATION
The characteristic and attractive eclogite mixture of red garnet and green omphacite pyroxene is clearly visible in this hand specimen.

BROUGHT TO THE SURFACE
Eclogite forms the dark bands in this outcrop of regionally metamorphosed gneiss in western Norway.

METAMORPHIC

Granulite

ORIGIN Regional   **GRAIN SIZE** Medium–coarse

COLOUR Brownish grey to dark grey

This tough, massive rock is formed at the base of the continental crust under high pressure and temperature. Typically, granulite may be banded and display what is known as a granoblastic texture, in which most of the minerals form a mosaic of roughly rounded, interlocking crystals. They have variable compositions including feldspar, quartz, and anhydrous minerals (meaning those without water) rich in iron and magnesium, such as pyroxene and garnet. Granulite forms at temperatures above 700°C (1,300°F) and over a wide range of pressures equivalent to crustal depths of 7–45km (4–28 miles) below sea-level. It is thought that much of the lower continental crust is made of granulite, which surfaces only in mountain belts and xenoliths in volcanic eruptions.

lava-coated granulite xenolith

STAR-SHAPED AMPHIBOLITE
Deep erosion of regionally metamorphosed rocks in Switzerland has exposed eclogites, which form at high pressures deep within the continental crust.

METAMORPHIC

Amphibolite

ORIGIN Regional   **GRAIN SIZE** Coarse

COLOUR Green-black

Amphibolite is one of the few metamorphic rock types that are named after a single mineral (amphibole). This mineral is extremely important in metamorphic rocks that are derived from extrusive rocks (lavas), such as basalt or andesite, and intrusive rocks, such as gabbro and diorite. This is because the mineral structure of amphibole can incorporate a large variety of elements – indeed, some geologists have described amphibole as a "waste bin" or "sponge" because it may contain so many elements. Despite changes in its composition, amphibole remains a stable mineral during the metamorphism of mafic to intermediate igneous rocks. This property also explains why amphibolite is found in the higher-temperature parts of both thermal and regional metamorphic terrains worldwide, provided the original rocks were of basaltic to andesitic composition. Amphibolite is formed at moderate temperatures (500–700°C/ 900–1,300°F) and across a wide range of pressure conditions, which typically occur 10–30km (6–19 miles) below sea-level. The major mineral components of amphibolite are 50 or more per cent hornblende amphibole (and sometimes other amphiboles, such as actinolite or tremolite) and plagioclase feldspar, as well as a variety of other minerals such as biotite and garnet. Some of these, especially garnet, may grow as porphyroblasts – relatively large crystals surrounded by smaller ones. The presence of large numbers of amphibole crystals aligned in parallel planes may give the rock a distinctive, cleavage-like pattern known as foliation. Sometimes amphibolite is divided into alternating layers of dark amphibole and light feldspar. Amphibolite's strength and compactness mean that it can also be used as high-grade aggregates with good load-bearing properties for road construction.

red garnet   matrix of amphibole and plagioclase

MINERAL COMPOSITION
Garnet crystals set in a matrix of amphibole and plagioclase crystals help identification of this metamorphic rock as an amphibolite.

METAMORPHIC

Mylonite

ORIGIN Dynamic   **GRAIN SIZE** Fine

COLOUR Variable, including grey, brown, and black

fractured rock

Mylonite is a banded or laminated metamorphic rock in which the original texture or fabric of the parent rock is destroyed by intense shearing deep within a fault zone. The parent rock can be of almost any kind, and its fabric is replaced by a kind of roughly developed foliation – a deformed streaky pattern known as cataclastic texture. The grains are reduced in size from those of the parent rock, and may recrystallize into a hard, compact, and often splintery rock with orientated crystal growth and occasionally eye-shaped augen porphyroblasts.

METAMORPHIC

Jadeitite

ORIGIN Dynamic   **GRAIN SIZE** Fine

COLOUR Shades of green, grey, white, pink, and mauve

Jadeitite is a very hard and tough metamorphic rock formed largely from the pyroxene-group mineral jadeite, which is a sodium alumino-silicate mineral. It is usually massive and coloured in various shades of green from white to almost black, and including a rare mauve-coloured variety, as well as the very rare emerald-green Imperial, Kingfisher, or quetzal jade. Impurity-free jadeitite is white. The finest-quality jadeitite is translucent. Jadeitite's toughness is

compact structure

COLOUR VARIATIONS
Although commonly green, the colour of jadeitite is extremely variable. The lavender colouring is here caused by manganese and iron impurities.

JADE ARTEFACTS

Because of its great toughness and beauty, the rock jadeitite, which is made of the mineral jadeite, has been valued for thousands of years, especially in China and Mexico. Wood found with jadeite artefacts in Mexico, for example, has been carbon dated to 1500 BCE. Initially, jadeitite was probably used for utilitarian purposes because it was made into high-quality axes. Its rarity, combined with its durability, attractive colours, and ability to be carved into intricate, three-dimensional forms without breaking, have since resulted in jadeitite being adopted more extensively, for ornamental and ceremonial purposes. Jadeitite is mined commercially in Myanmar, Russia, and Guatemala. The name jadeite is a derivation of l'jade, which is a mistranslation of the Spanish *pietra de ijada* ("stone of the loin or kidney").

ENHANCING THE COLOURS
Expert carvers of jade combine natural colour variations in their designs, such as the pale green and lavender seen here.

due to a strong network of closely interlocking prismatic crystals that formed at high pressure in crustal depths of 15–30km (9–18 miles), that is, in the lower part of the continental crust – but at relatively low temperatures (about 400°C/750°F). These conditions, which can be attained only in subduction zones, transform basalt into a blue-coloured schist that contains the blue-coloured amphibole glaucophane, plus lawsonite or epidote. Jadeite was first distinguished as a mineral in 1863, when the French chemist Augustine Damour analyzed an object from the New World, which he thought was the same mineral as Chinese jade. In fact, it transpired that Chinese jade was not the sodium alumino-silicate mineral jadeite. Instead, it is a massive form of the calcium-magnesium-iron silicate mineral amphibole, sometimes referred to as nephrite actinolite.

TRACE ELEMENTS
The green colour here is due to the presence of iron and chromium.

Sedimentary Rocks

These rocks are formed from deposits that accumulate under the influence of gravity, layer by layer, on the Earth's surface, and are laid down by water, wind, ice, and mass movement. Those deposited in seas and oceans can be grouped as marine deposits, from the shoreline to abyssal depths. Those formed on land are terrestrial, including freshwater lakes and rivers, upland areas, deserts, and glacial terrains. A broad division is made into: clastic (or fragmental), biogenic, and chemogenic rocks. Clastic rocks are subdivided by grain size from coarsest (conglomerates) to finest (claystone, mudstone, and shale), with further identification by composition. Sandstones, for example, are divided into quartz-rich, lithic-rich and feldspar-rich sandstones (arkoses). Biogenic sedimentary rocks are categorized into limestones (calcium carbonate), cherts (silica), and peat and coal (organic carbon). Limestones can be formed from many different organisms – shells, corals, marine plankton – or from chemical precipitation, for example travertine and tufa. Dolomites are mostly formed from limestones that have been chemically altered after deposition. Chemogenic rocks include evaporites (such as halite and gypsum), ironstones, and phosphorites.

SEDIMENTARY
Breccia

ORIGIN Marine, terrestrial, and glacial

COLOUR Variable, depending on mineralogy

GRAIN SIZE Coarse

Fragments of angular rock distinguish breccia from round-grained conglomerate. In breccia, fossils are rare, and layering is not generally evident. Although this rock has a variable composition, it typically comprises immature sediment that has not been transported far. Consequently, it tends to form from sediments derived from terrestrial erosion or mass movement such as high-angle scree slopes or cliff bases (known as talus). Some frost-shattered or glacial moraine debris is made up of breccia, but the former is not normally consolidated as a rock and the

latter tends to show signs – such as ice scratches – of its glacial origin. The term breccia is also used to describe rock fragments produced by fault movement, and these are often cemented with minerals derived from hydrothermal fluids and may be of some economic value.

fine matrix

angular fragments

SEDIMENTARY
Conglomerate

ORIGIN Marine, terrestrial, and glacial

COLOUR Variable, depending on mineralogy

GRAIN SIZE Coarse

rounded pebbles

Typically containing pebble-sized grains but also including boulders more than 1m (3ft) across, conglomerates are the coarsest-grained sedimentary rocks. They are formed through high-energy processes of sedimentation such as glaciation, gravitational movement, and rapid water flow in environments such as mountain streams and wave-battered beaches. The grains are rounded by friction and generally set in a finer-grained matrix. Composition varies enormously and depends on the parent rocks and the amount of weathering and erosion – immature conglomerates can include some soft rocks, such as mudstones or limestone. Fossils are rarely present.

SEDIMENTARY
Arkose

ORIGIN Marine, terrestrial, and glacial

COLOUR Variable, including white, yellow, and brown

GRAIN SIZE Medium–coarse

This type of sandstone contains a mixture of quartz, lithic fragments, and feldspar (the latter comprising over 25 per cent of the rock), sometimes with a clay matrix or carbonate cement. Arkose generally forms where feldspar grains are particularly abundant – in deposits derived from the weathering and erosion of metamorphic and igneous rocks. Compared with quartz, the survival of feldspar grains in sedimentary environments is relatively limited, since the feldspar is chemically weathered to form clay minerals, especially in hot and humid climates. Arkose therefore forms around high-latitude mountain belts, where uplift and high rates of erosion expose feldspar-rich rocks such as granite. It is found in rivers or alluvial plains at the base of mountains, too.

visible grains

SEDIMENTARY
Lithic Sandstone

ORIGIN Marine, freshwater, and terrestrial

COLOUR Mostly grey

GRAIN SIZE Fine–coarse

Lithic sandstones are characterized by a relatively high proportion (over 25 per cent) of rock fragments – also known as lithic grains. The sandstones may also contain quartz, feldspar, and other minerals, as well as a clay matrix that is derived from the alteration of chemically unstable minerals and/or rock fragments. The lithic material has mostly been worn down to sand-sized

grains, which are more or less angular and irregular in shape. Several different lithic types occur depending on the type of rocks eroded from the source area – for example volcanic, metamorphic, limestone, or shale. They are commonly found quite close to the source area, deposited by rivers or on alluvial fans, but can also be transported long distances into the deep sea by turbidity currents. Some lithic sandstone turbidites (the

deposits of these currents) with a high proportion of clay matrix were formerly called greywacke sandstones, but this term is no longer in common use.

PEMBROKESHIRE COASTLINE
These Paleozoic lithic sandstones in Abereiddy Bay, Wales, were originally laid down on the sea bed some 440 million years ago.

MIXED COMPOSITION
A typical lithic sandstone, such as this one, consists of mineral grains and rock fragments with a matrix of clay particles.

fine clay matrix

SEDIMENTARY
Siltstone

ORIGIN Marine, freshwater, and glacial

COLOUR Variable, including grey, brown, reddish brown

GRAIN SIZE Fine–medium

This clastic sedimentary rock typically has fine internal layering and sometimes small ripples. Its particles are smaller than fine sand but larger than clay grains. Most silt is quartz, but clay is often present. As a sediment, siltstone tends to behave more as sand than clay, and is produced by prolonged weathering, erosion, and transport of sediment in both marine and terrestrial environments, including glacial ones. Being relatively fine grained, siltstone often contains fossils.

silt grains

SEDIMENTARY
Sandstone

ORIGIN Marine, terrestrial, and glacial

COLOUR Variable, including white, yellow, brown to red-black

GRAIN SIZE Medium

Accounting for 10–15 per cent of all sedimentary rock in the Earth's crust, sandstone typically occurs as stratified layers of sand-sized particles, held together by various mineral cements. Typically porous, sandstone forms economically important reservoirs for water and hydrocarbons. Sandy sediment accumulates particle by particle, through the actions of wind and water, on land and at sea. It is deposited in different environments, from high mountain rivers and lakes to the bottom of the deep ocean. The environment in which a sandstone deposit formed is reflected in many of its characteristics, from its large-scale geometry to smaller structures and its composition. For example, river-lain sandstones are tens to hundreds of kilometres long, up to several kilometres wide, and up to hundreds of metres thick. Locally, they appear as meandering channels, with curved lower boundaries and flat upper surfaces. Sequences of rippled layers with well-rounded, similar-sized grains reflect the original river's current strength, varying from coarse, high-energy basal layers passing up into finer, low-energy tops and then silts and muds. Technically, the term sandstone refers to the grain size of the particles, which may be composed of any rock fragment or mineral combination. In practice, most sandstones are rich in quartz, because quartz is the most common residual mineral in sedimentary environments. Sandstone can be classified according to its grain size (fine, medium, and coarse) and composition: for example, arkose (or arkosic sandstone) is feldspar rich. The size and roundness of the grains reflect how far they have travelled, and their composition may indicate the geology of their source rocks.

DINOSAUR FOOTPRINTS
The fossil record of many organisms ranging from worms to dinosaurs is preserved by their trace fossils, such as burrows or footprints, left in soft, sandy or muddy sediment, here in Namibia. Where the traces are quickly infilled by younger sediment, they can be protected from destruction, and on burial and lithification may be preserved. Subsequent uplift, erosion, and weathering of ancient sedimentary strata may then reveal the fossil traces once again. Analysis of these can tell scientists a great deal about the behaviour of the long-vanished animals.

plant fossil

sand grains

FOSSILIFEROUS
An ancient, sand-laden river carrying plant debris, which has been buried, will later turn into sandstone of the kind shown here.

CLIFF-TOP RETREAT
These Greek monasteries at Metéora perch on top of a cliff made of sandstone, which has resisted erosion over several millennia.

SEDIMENTARY
Claystone

ORIGIN Marine, terrestrial, and glacial

COLOUR White through to pale shades of grey, yellow, brown, and red

GRAIN SIZE Fine

This common sedimentary rock is typically composed of particles less than 0.004mm (0.0002in) across – mostly derived from the weathering of feldspar and other minerals. The particles in claystone are predominantly silicate clay minerals, such as kaolinite, organic material, and minute particles of other types, such as carbonate, pyrite, and gypsum – hence the name of the rock. Claystone is similar in grain size and composition to shale but lacks its layered structure and ability to be split. Fine-grained clay particles can be transported great distances by wind and water.

clay matrix

SEDIMENTARY
Mudstone and shale

ORIGIN Marine, terrestrial, and glacial

COLOUR White, also pale shades of grey, yellow, brown, and red

GRAIN SIZE Fine

mud matrix

curved surface

The most abundant of all sedimentary rocks, mudstone and shale are variable in composition but commonly include silt and clay minerals, organic material (of economic value in oil shale), iron oxides, and minute crystals of minerals such as pyrite and gypsum. When the grains occur randomly within the sediment, the resulting rock is a blocky mudstone with little or no layering. However, where the grains have been orientated by the steady flow of a current or by compression to produce layering along which the rock splits, the resulting rock is called a shale. Mudstone and shale generally contain well-preserved fossils.

SEDIMENTARY
Loess

ORIGIN Wind-blown and terrestrial

COLOUR Pale grey to brown

GRAIN SIZE Fine

Lifted by the wind from dry land surfaces, loess is a fine-grained silt and clay-sized material that is carried as airborne dust, frequently over very great distances, before being redeposited and eventually compacted to form sedimentary rock. Dry, windy landscapes lacking in vegetation, such as arid and semi-arid continental interiors, are typical source regions for loess, which is finer grained than sand. There are huge loess deposits in eastern Europe and much of Asia, all of which are derived from areas that were on the edges of ice-sheets during the last ice age. Fluctuations in loess sediment can help provide a detailed record of climate change.

fine-grained texture

SEDIMENTARY
Dolomite

ORIGIN Marine and terrestrial

COLOUR Variable, but mostly white or pale shades of yellow, grey, or brown

GRAIN SIZE Fine–coarse

compact carbonate rock

To distinguish it from the mineral of the same name, dolomite is also known as dolostone. This carbonate rock is mostly formed by post-depositional chemical replacement of original limestone. Calcium ions are exchanged for magnesium ions by percolating magnesium-enriched water. The original carbonate mineral is converted by this process of dolomitization into the magnesium carbonate mineral dolomite. Typically, dolomite forms massive rocks in which the original bedding and other sedimentary structures are destroyed by the process of dolomitization.

SEDIMENTARY

Limestone

ORIGIN Marine and freshwater

COLOUR Variable, but mostly white or pale shades of yellow, grey, or brown

GRAIN SIZE Fine—coarse

After mudstone and sandstone, limestone is the next most abundant sedimentary rock. It forms extensive, thick, multiple layers within the continental crust, and contains carbonate minerals, especially calcite (50 per cent or more), which recrystallize during lithification into a hard and brittle rock. Carbonate sedimentation often involves animals or plants, and especially micro-organisms. These change the local chemical environment, transforming tiny carbonate crystals (generally the mineral

oolitic grains

OOLITIC LIMESTONE
Rounded grains of calcium carbonate rolled by sea-bed currents (oolites) have here been cemented by carbonate mud to form rock.

FOSSILIFEROUS LIMESTONE
A deposit of shells that has accumulated in sufficient quantity may turn into limestone rock. This one has a matrix of carbonate minerals.

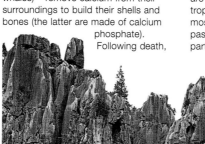

fossil shell

aragonite) into carbonate muds on shallow sea beds. A wide range of organisms – from the microscopic, single-celled foraminifera to reef builders (such as corals) and gigantic vertebrates (for example, dinosaurs and whales) – remove calcium from their surroundings to build their shells and bones (the latter are made of calcium phosphate).
Following death, carbonate remains may accumulate in sufficient quantities to form sediments such as ocean-bed muds and reef limestones. In shallow tropical seas with high evaporation rates, carbonate may also be deposited directly on the sea bed. There, the currents roll the carbonate-accreting grains around, like snowballs (up to several millimetres in diameter), forming carbonate sands called oolites, which are later lithified to form rocks. Shallow tropical sea carbonates build up the most important limestone strata. In the past, such seas have flooded extensive parts of the Earth's continents, building huge reefs and associated limestone sediments: for example, during Lower Carboniferous (Mississippian) times, they covered much of the ancient continent of Laurentia (now roughly North America) as well as Europe. Limestone rocks can dissolve to form karst and cave features and then, on evaporation, be redeposited to form travertine deposits around hot springs and geysers (see p.185), and cave carbonate deposits such as flowstones and stalagmites (see p.263).

PAGODA PILLARS
This limestone in Lunan, southern China, which dates from the Permian Period, has been largely dissolved away by chemical weathering.

SEDIMENTARY

Evaporites

ORIGIN Marine and terrestrial

COLOUR Generally white or pale shades of yellow and grey, through to red

GRAIN SIZE Fine—coarse

Minerals such as halite (rock salt) and gypsum form extensive sedimentary deposits beside shallow seas or, in arid areas, within saltwater lakes and seas. Evaporation triggers the growth of salt crystals from brines derived from the weathering of local rocks. Thus, ancient evaporite deposits can be good indicators of past climates, geography, and seawater composition. Seawater evaporation produces a sequence of

DEVIL'S GOLF COURSE
In Death Valley, California, salt evaporites have formed pressure-rimmed pans as growing crystals push against one another while they develop.

crystalline texture

ROCK SALT
Although crystalline halite formed by evaporation of seawater can look like marble or quartzite, it differs in being soluble and much softer.

evaporite minerals regulated by their solubility, from calcite through gypsum to halite and potassium and magnesium salts such as sylvite (bitter salt) and carnallite. Some evaporite deposits are economically important for the production of salt, gypsum, and potash, which is an important constituent of fertilizer. Because of their solubility, evaporite deposits may be dissolved and redeposited in other layers.

SEDIMENTARY

Iron formations

ORIGIN Marine, freshwater, and terrestrial

COLOUR Reddish brown to red   **GRAIN SIZE** Fine

Sedimentary iron formations are of great economic importance, with their iron minerals deposited in a variety of environments (see panel, right). The commonest sediment-related iron minerals include oxides (such as hematite, magnetite, and goethite), carbonates (such as siderite), and sulphides (such as pyrite and marcasite). Iron-rich strata may show sedimentary features such as rounded grains and current ripples. Fossil remains, especially plants, are common in Phanerozoic iron deposits. Most iron deposits are chemical in origin and most important are the banded iron formations of the Precambrian Eon, which form the bulk of the world's iron reserves, with iron contents in excess of 50 per cent by weight. Typically hundreds of metres thick, they first occur around 3,800 million years ago, and, apart from those associated with the late Precambrian ice ages, are rare in the last 1,800 million years of the geological record. The most extensive have thinly bedded layers of chert alternating with layers of magnetite and hematite, and are products of chemical sedimentation processes. Others are associated with volcanism.

specular hematite

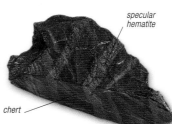

chert

BANDED IRONSTONE
Such alternating layers of red chert and black hematite, which is rich in iron, are typical of marine deposits dating from the Precambrian Eon.

ORE DEPOSITS

Iron is the most important metal in the modern world, and is mostly sourced from sedimentary ores. Precambrian, banded-iron deposits in the Lake Superior region have produced most of the iron ore mined in North America over the last 120 years. Iron's availability was an important factor in the rapid industrialization of that continent at the end of the 19th century. Australia is now a major producer of iron ore, from similar deposits, as here in the Hamersley Range.

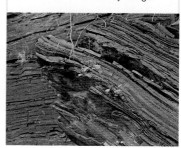

SEDIMENTARY
Organic deposits

ORIGIN Marine, freshwater, and terrestrial

COLOUR Generally dark brown to black

GRAIN SIZE Fine

This group of sedimentary deposits includes the economically important hydrocarbons (such as peat, coal, oil, and asphalt). In addition, it includes other related organic products that are preserved in rocks, such as plant-derived amber resin. Essentially, organic deposits are made of carbon-related materials derived from the post-mortem breakdown of organic tissues – both plant and animal. All life consists of complex organic compounds, mainly long chains of carbon atoms (aliphatic molecules) such as carbohydrates, proteins, and lipids. In the process of photosynthesis, plants on land and algae in the sea convert carbon dioxide

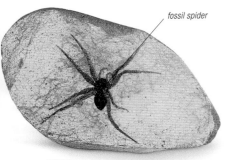

fossil spider

AMBER
Resin from some kinds of ancient pines trapped invertebrates and, on hardening, became part of the sedimentary rock record known as amber.

COASTAL COALS
In Portugal, erosion and open-cast mining have exposed these thin dark coal seams with paler soil layers and other mixed sediments.

TAR PITS
Natural seepages of oil occasionally appear at the surface, where their volatile compounds are lost, leaving heavy tars, as here in Trinidad.

into organic tissue using light energy from the Sun. In recent decades, however, an independent process of chemosynthesis, provided by hydrothermal fluids, has been found to be widespread within the deep ocean. Following the death of an organism, its tissues decay and are broken down, often forming rings of carbon atoms (aromatic molecules). The carbon molecules may be dissolved, forming humic or fulvic acids, or be stored in sedimevnts. The latter may develop into petroleum and natural gas (methane) in marine sediments, depending on a succession of geological events. On land, organic matter is stored in soils and aquatic deposits such as peat, coal, and gas (methane), depending on geological constraints of burial pressure and temperature, as realized by William Smith (see panel, above). Having low densities, petroleum and gas often migrate from their source rocks and may return to the surface and be lost, or they may be stored underground in porous

reservoir rocks such as sandstone. Peat and coal, however, are made of accumulated plant materials in layers up to several metres thick, interbedded with fossil soil layers (called seat-earths) and other rocks such as sandstone, shale, and limestone, depending on the original depositional site. Since the Carboniferous Period, tropical forests and swamps have been extensive

PEAT FUEL
Historically, plant deposits of bogs and swamps, or peat, were cut and dried, as here in County Kerry, Ireland, for use as domestic fuel, garden compost, and in power stations. It is now being substituted with biomass, a cleaner source of energy.

WILLIAM SMITH

The son of a rural Oxfordshire blacksmith, William Smith (1769–1839) earned his living as a civil engineer and land surveyor, building canals and sinking shafts for coal. From this practical experience of rock sequences, Strata Smith, as he was nicknamed, independently developed techniques for mapping and correlating strata. Single-handedly, he produced the first detailed geological map of England, Wales, and parts of Scotland in 1815, and provided cross-sections of the correct vertical succession of strata, which he identified from the colour, texture, and composition of the rocks. In doing so, Smith laid down the principles of geological mapping.

enough to form coal deposits, the grade of which depends on the burial depth of the plant-bearing layer. With increasing pressure and temperature, peat-type deposits are progressively transformed into lignite, bituminous coal, and anthracite, which have progressively higher carbon contents and calorific values and lower moisture contents and volatility.

vitreous surface

compression fractures

ANTHRACITE
Plant remains that have accumulated in brown, earthy peats are metamorphosed into anthracite by deep burial and elevated temperature.

PLANET EARTH

FLOODED FIELD
Drenched by crude oil, two men struggle to cap a
Kuwaiti oil well that was set on fire during the Gulf
War of 1991. The Middle East contains over half
the world's oil reserves.

FOSSIL FUELS

Despite the growing use of renewable resources, fossil fuels still supply more than four-fifths of the world's energy needs. Oil and gas often occur together, and are extracted using sophisticated drilling technology that maximizes shrinking reserves. The third fossil fuel – coal – is much more abundant than oil and gas, although it is harder to extract, and also more polluting when used. The after-effects of coal mining dominate the landscape in some parts of the world.

Oil and Gas Extraction

When the world's first oil well opened, in Pennsylvania, USA, in 1859, it struck oil at a depth of just 21m (69ft). A century and a half later, such easily accessible deposits are extremely rare. In the relentless quest for more oil, today's oil wells can be more than 5,000m (16,500ft) deep, and, instead of drilling directly downwards, drill bits can be steered underground at almost any angle. Called directional drilling, this enables a single oil rig to collect oil from an area of more than 100 square km (40 square miles).

Compared to coal mining, oil extraction creates very little surface spoil. One of oil's few waste products is natural gas, which has now become an important fuel in its own right. But with great distances often separating production and consumption, the overall impact of oil and gas extraction is high. Oil installations can pollute groundwater on land, and numerous accidents by oil-carrying tankers underline the pollution hazard at sea. Oil and gas extraction can also intrude into environmentally sensitive areas.

Coal Mining

Coal was the fuel that powered the Industrial Age. In energy terms, its place has since been taken by oil, yet it remains an essential resource, particularly in the developing world. Of all the fossil fuels, coal is the simplest to use, but, because it is a sedimentary rock, it is also the hardest and most dangerous to extract.

The earliest coal mines were shallow pits, excavated where coal seams were near the surface. When these became depleted, miners were forced to dig downwards to find seams, which they could follow underground. By the early 1600s, some mines in Europe were already more than 1km (⅗ mile) long, and they generated spoil heaps that have become a hallmark of mining country worldwide. With underground mining, subsidence and flooding are constant risks, and the older a mine is, the greater these problems become.

Much of today's coal is produced by an alternative method: surface or open-cast mining. In this, the excavation is typically carried out by power shovels or self-propelled augers, which remove the coal by digging or boring through it. Compared to underground mines, surface mines are cheaper and safer, but they alter the landscape even more. In many mines, coal makes up less than 10 per cent of the extracted rock. The rest is waste, which is

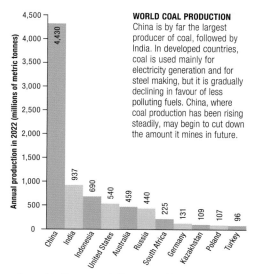

4,500 — 4,000 — 3,500 — 3,000 — 2,500 — 2,000 — 1,500 — 1,000 — 500 — 0

Annual production in 2022 (millions of metric tonnes)

China 4,430 | India 937 | Indonesia 690 | United States 540 | Australia 459 | Russia 440 | South Africa 225 | Germany 131 | Kazakhstan 109 | Poland 107 | Turkey 96

WORLD COAL PRODUCTION
China is by far the largest producer of coal, followed by India. In developed countries, coal is used mainly for electricity generation and for steel making, but it is gradually declining in favour of less polluting fuels. China, where coal production has been rising steadily, may begin to cut down the amount it mines in future.

returned to the ground. Because of its lack of soil and high mineral content, such waste is a difficult habitat for plants. Fortunately, in recent years, researchers have bred plant strains that can tolerate this man-made environment. With these, and careful landscaping, spoil heaps can be stabilized and their visual impact much reduced.

Future Supplies

The Earth's accessible coal reserves amount to at least 1,000 billion tonnes, which is enough to last several centuries if consumption rates remain unchanged. However, coal is a highly polluting fuel, so environmental considerations are likely to limit its use. There is enough natural gas to last at least another century, but for oil the picture is less clear. Oil exploration constantly finds new reserves, while new technology increases the efficiency of extraction. This century, the world's estimated oil reserves have been increased greatly by the inclusion of new "unconventional" sources, such as tar sands and oil shale. Oil and natural gas supplies have also increased through the use of fracking – a process to extract trapped oil and gas by cracking open rocks with water pressure. Including these resources, the USA has the world's biggest reserves.

SURFACE MINING
The Arkwright coal mine, near Derby, is one of England's largest open-cast pits. Coal from this region powered the world's earliest locomotives.

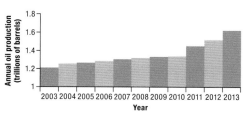

1.8 — 1.6 — 1.4 — 1.2 — 1

Annual oil production (trillions of barrels)

2003 2004 2005 2006 2007 2008 2009 2010 2011 2012 2013
Year

WORLD OIL PRODUCTION
Talk of "Peak Oil" – a time in the near future when oil production would reach a peak then decline as resources dwindled – has waned with the discovery of new reserves and extraction techniques.

SEDIMENTARY

Bauxite

ORIGIN Terrestrial

COLOUR Variable, including white, grey, yellow, and red

GRAIN SIZE Fine–medium

When exposed at the surface, many minerals are not stable to normal processes of weathering, especially in humid tropical regions. Decaying vegetation releases organic acids, which percolate down to the bedrock, causing chemical changes known as

VARYING LEVELS
Hydrated aluminium oxide minerals make up the large grains and the intervening cement in this bauxite ore hand specimen.

leaching. This removes soluble constituents of minerals, leaving a residue to form new minerals. The most immobile residual elements are silicon, aluminium, and iron, usually in the form of oxide or hydroxide minerals in the rock bauxite. Bauxite is an ore containing more than 50 per cent aluminium oxide and is found in tropical limestone areas, where all the limestone has been dissolved away. Small mineral

rounded fragments

residues from the original limestone are left behind, concentrated, and form the insoluble bauxite minerals by additional chemical leaching. Bauxite also occurs as a residual deposit through the leaching of alumina and silica-rich minerals from igneous, metamorphic, and sedimentary rocks. Bauxite is a rock composed primarily of aluminium hydroxide minerals (for example, gibbsite) along with clay minerals, silica, and iron hydroxide minerals (for example, limonite). Layers of bauxite may be tens of metres thick and take millions of years to develop. It is the principal ore of aluminium, with Australia producing nearly a third of the global ore production. Bauxite is a useful indicator of past climates, because it records periods of tropical climate in the geological past.

SURFACE EXTRACTION
Bauxite deposits in Jamaica are formed by the intense tropical weathering of surface rocks, and can therefore conveniently be quarried.

SEDIMENTARY

Phosphate deposits

ORIGIN Marine and freshwater

COLOUR Greenish grey to brownish grey

GRAIN SIZE Fine–medium

Animal bones and the faeces of carnivorous animals are the source of most phosphate deposits. Such deposits (known as phosphorites) are typically layered accumulations (with more than 20 per cent of phosphate minerals, especially apatite). Present-day phosphate deposits formed when animal remains accumulated relatively quickly in marine sediments and were then altered by chemical reactions. Phosphate deposits are commonly associated with limestones, and some are of economic value.

GUANO DEPOSITS
Guano deposited by fish-eating birds on islands, such as this one in the Galápagos, has in the past proved to be an economic source of phosphate.

SEDIMENTARY

Septarian concretion

ORIGIN Marine and terrestrial

COLOUR Greyish yellow to brown

GRAIN SIZE Fine

shrinkage cracks

calcite

Concretions are typically formed within sediments. They vary in their composition, shape (round or disc-shape to irregular), and size (metres to millimetres), but are essentially the same material as the host sediment, which has been cemented (concreted) by other minerals, commonly carbonate, silica, or iron oxide. A septarian concretion has internal cracks filled with crystals such as calcite. The infilling mineral also hardens or cements the concretion, often before compression and flattening, and it is therefore more resistant to weathering and erosion than the surrounding sediment.

SEDIMENTARY

Chert

ORIGIN Marine and terrestrial

COLOUR Variable, but mostly grey to black

GRAIN SIZE Extremely fine

Chert is an extremely fine-grained (cryptocrystalline) form of silica. Also known as flint, it develops in a variety of sedimentary environments both as isolated nodules and in more continuous layers and massive beds, especially in limestone, banded ironstone, and above ocean-floor basalts. Most chert is formed by the accumulation of silica in the sediment by the decay of silica-rich organisms (such as glass sponges and radiolarians), hydrothermal activity on the ocean floor, and – during the Precambrian Eon – due to changes in ocean chemistry, which created

curved fracture

compact silica

FLINT TOOLS

The glassy texture of chert, along with its hardness, toughness, and brittleness, have proved to be suitable qualities for stone tools such as axes, blades, and points, and so this rock was an important resource in prehistoric times. First crafted by early peoples around 2.6 million years ago, chert has continued to be used even in more recent times, as part of the firing mechanism of flintlock muskets.

sedimentary iron formations. In some cases, further chemical changes within the sediment form silica gels, some of which harden into chert, often around fossil remains and before compaction of the wet sediment. Due to its hardness, and resistance to erosion, chert may accumulate to form significant deposits such as beach conglomerates. It has also provided an invaluable source of weapons and tools for humans (see panel, above).

FLINT CORE
Flint nodules can form in sedimentary rocks, especially limestone, and may sometimes include organic fossils not normally preserved.

SEDIMENTARY

Nodules

ORIGIN Marine, freshwater, and terrestrial

COLOUR Variable, depending on mineralogy

GRAIN SIZE Fine–coarse

Minerals such as calcite, quartz, pyrite, and apatite may crystallize in spherical aggregates, to develop nodular growths in many sedimentary environments. Most nodules form within sediment, while others grow near or at the sediment–water interface. Abundant ferro-manganese nodules containing valuable metals such as titanium, chromium, cobalt, copper, and zinc form on the deep sea-floor, but so far are too expensive to recover. Other nodules may preserve rare fossils or soft tissues.

spherical nodule

radiating structure

Meteorites

Meteorites are rocks found on the Earth's surface that have an extra-terrestrial origin – they continuously fall to the Earth from space, with a small number derived from the Moon and Mars. They can be divided into three broad groups: stony meteorites, iron meteorites, and a mixed stony-iron group. By far the most common are stony meteorites (subdivided into chondrites and achondrites), which are made of pyroxene, olivine, and plagioclase silicate minerals with some nickel-iron metal. By contrast, the irons are mainly nickel-iron alloys. Stony irons such as pallasites are rare; they contain mixtures of silicates and nickel-iron alloy. Dating from the early stages of the formation of the Solar System, meteorites provide scientists with invaluable evidence about the minerals that were originally aggregated in the Earth.

Pallasites

ORIGIN Extra-terrestrial   **GRAIN SIZE** Coarse

COLOUR Yellow to green and metallic grey

When sliced open, pallasites reveal a mixture of rounded, yellowish green olivine crystals in a shiny grey

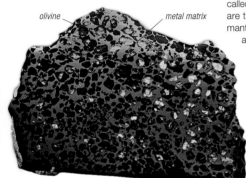
olivine   metal matrix

groundmass. This kind of stony-iron meteorite, which accounts for about 1 per cent of meteorite falls, was named after the German naturalist Peter Pallas who, in 1776, provided the first detailed description of a meteorite that he had found in Siberia. Pallasites typically comprise one part magnesium-rich olivine crystals to two parts metal. Where the metal is even more abundant, it shows a pattern called Widmanstätten structure. Pallasites are thought to originate at the core–mantle boundary of long-destroyed asteroids, where the thick, olivine-rich silicate mantle material mixes with the underlying metallic iron core. Similarities in metal composition (among elements such as gallium, germanium, and nickel) suggest there is a link between some pallasites and some iron meteorites. However, compositional differences between pallasites indicate that they are derived from more than one asteroid.

Chondrites

ORIGIN Extra-terrestrial

GRAIN SIZE Medium–coarse

COLOUR Brownish grey with black crust

First analysed by the French chemist Antoine Lavoisier in 1772, chondrites are by far the most common type of meteorite, forming about 86 per cent of meteorites that have fallen to the Earth. Whether this reflects their true abundance in space is less certain.

rounded silicate grains

CHANGES DURING TRANSIT
As chondrites move through the atmosphere, their exteriors melt to form black, glass-like crusts. The insides remain a mass of silicate mineral grains.

Chondrites have provided the oldest and most precise dates (of 4,560 million years) for rocks of the Solar System. As they contain the same ratio of elements as the Sun, it is likely that they formed within a few million years of the birth of the Solar System. Chondrite meteorites are essentially cosmic sediments. They take their name from their major constituent, chondrules. These are rounded spheres formed when molten silicates cooled rapidly in the early solar nebula. Chondrites comprise olivine, pyroxene, and plagioclase from chondrules, along with metal and sulphide. A long history is recorded in chondrite meteorites after their initial formation. This history includes more than 60 million years of heating within asteroids, subsequent impacts that heated parts of asteroids, fragmentation that sent chondrites on their journey towards the Earth, and their eventual arrival on our planet.

Achondrites

ORIGIN Extra-terrestrial   **GRAIN SIZE** Coarse

COLOUR Grey

Achondrites are rare, constituting only about 3 per cent of all known meteorites. Despite their scarcity, they are made of the types of igneous rocks most familiar to humans. Achondrites formed in the outer layer (or crust) of an asteroid, which is similar to the crust of the Earth. Like the Earth, many asteroids melted in the early part of the history of the Solar System. During this melting and differentitation, dense metal sank to the centre to form a core, and lighter silicate melts rose to the surface,

light feldspar

fusion crust

ROUGH SURFACE
The apparently chaotic mixture of medium and coarse grains on the surface of this achondrite in fact conceals its crystalline igneous texture.

erupted in lava flows and fountaining volcanoes, and eventually solidified and cooled to form the achondrites. The igneous origin of achondrites was first recognized in 1825 by the German Gustav Rose, who was a professor of mineralogy at Berlin University. With Alexander von Humboldt (see p.141), Rose travelled widely in Asiatic Russia and Siberia, collecting minerals and rocks including meteorites. Composed of a variety of silicate minerals, including olivine, pyroxene, and plagioclase, achondrites typically contain far less metal than the more common chondrites (see below). Achondrites are divided into several groups. The eucrites – the most prevalent group – are basalts. Composed essentially of pyroxene and plagioclase, they form first when chondrites melt. They resemble basalts found in abundance on the Earth, particularly on the floors of the oceans. The related diogenites formed when the basaltic melt stagnated in large magma chambers and tiny pyroxene crystals grew and sank through to melt to form a layer of pyroxene. Howardites – rocks composed of eucrites and diogenites mixed together by impact – testify to a long history of bombardment in space. A variety of other meteorites are also classified as achondrites, including angrites, aubrites, ureilites, and acapulcoites. While these groups are considerably smaller, they suggest that melting was a common process that modified many asteroids during the early history of the Solar System.

Irons

ORIGIN Extra-terrestrial   **Grain Size** Coarse

COLOUR Metallic grey with brown-black crust

About 5 per cent of meteorites are irons, which are composed of iron-nickel metal alloy (which is 5–20 per cent by weight nickel). When the cores of ancient asteroids cooled, the metal separated into two minerals – kamacite and taenite – whose intergrowth forms this meteorite's Widmanstätten structure (see Pallasite, above). Because Earth's metallic core cannot be reached, iron meteorites provide the only material evidence for its probable composition.

MARTIAN METEORITES

The possibility of life on Mars cannot be discounted even though recent claims of microscopic organisms and hydrocarbon residues being preserved in a Martian meteorite known as ALH 84001 (shown here) have now been questioned. There is no doubt, however, that extra-terrestrial hydrocarbons are preserved in some chondrites. Magnetite grains and a microbe-like tubule were found within carbonate minerals from ALH 84001, and the tubule was thought to be a minute, rod-shaped, bacterial fossil. Nevertheless, some critics believe that its structure resulted from mineral formation at high temperature and was therefore unlikely to have occurred within bacterial cells. Also, the complex structured hydrocarbons had Earth-derived, radiocarbon isotopes and so may be contaminants rather than from the planet Mars.

RIVER FLOW
Water is a powerful force of change.
Rivers such as the Thjorsa in Iceland
(shown here) move rock fragments and
organic material from land to the sea,
where it may eventually form new rock.

THE CHANGING EARTH

The surface of the Earth is constantly changing under the influence of forces within the planet, and surface processes, and even the impact of objects from outer space. The effects of such change are often discernible only after many thousands of years, although some changes such as earthquakes and volcanic eruptions have an instant effect. The relentless movement of the tectonic plates, which form the Earth's surface, can open new oceans, create mountains, and change continents. Wherever the Earth's materials are exposed at the surface, they are subjected to the processes of weathering, erosion, and transport by wind, water, and ice. These processes can sculpt rock and create sediment, which is deposited on land and beneath the sea. Water covers most of the Earth's surface and is a significant agent of change. The Earth is also the only planet known to support life. Living things both adapt to the conditions on the surface and shape the environment around them.

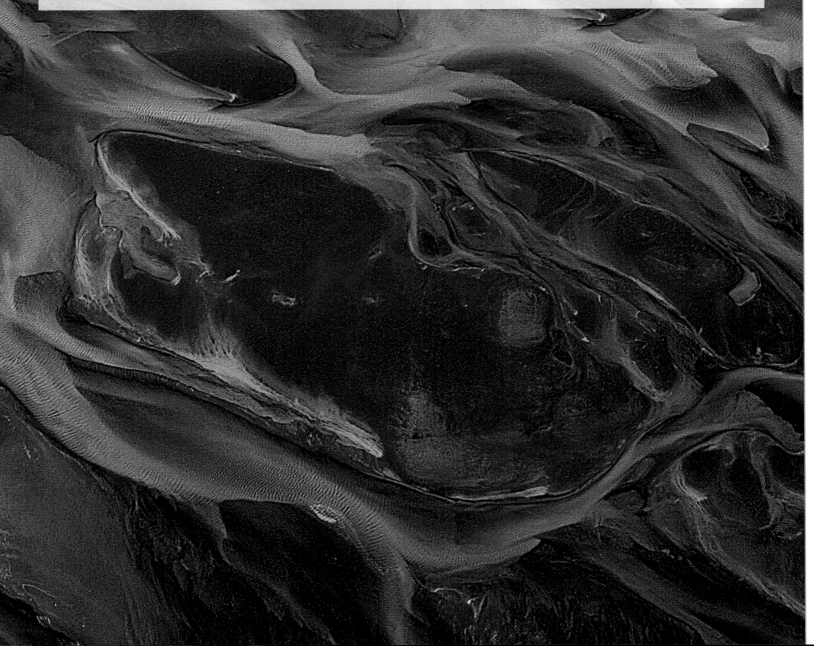

Tectonic Plates

12–13 The Earth's past

40 The crust

Plate boundaries 88–89

Fault systems 122–33

Mountains 136–53

Volcanoes 154–55

Ocean tectonics 386–87

Tectonic Earth 486–503

The cool, brittle outer rock layer of the Earth is called the lithosphere, and it is broken into seven large, continent-sized tectonic plates and about a dozen much smaller plates. These irregularly shaped segments fit together like a jigsaw puzzle and cover the entire surface of the Earth. Tectonic plates have an outer layer of crustal rocks that is inseparable from an underlying layer of upper-mantle rocks. Throughout geological time, the growth and reduction of plate size has also moved them across the Earth's surface. Oceans have opened and closed, forming mountain ranges and volcanoes, and shuffled the positions and orientations of the continents.

ALFRED WEGENER

The German meteorologist and geophysicist Alfred Wegener (1880–1930) was responsible for the theory of continental drift, which was the forerunner of current ideas about plate tectonics. Impressed by the complementary shapes of the coastlines of Africa and South America, he believed that all the continents once joined together in a supercontinent called Pangea, which split in the Mesozoic Era.

The Earth's Plates

Seven major plates, which are up to 100km (60 miles) thick, cover 94 per cent of the Earth. They carry the major continents, most of which were once part of the ancient landmass of Pangea (see p.20). The Pacific Plate is by far the largest in area at 108 million square kilometres (42 million square miles). It is followed in size by the African Plate; the Eurasian Plate, which includes Europe and Asia; the Australian Plate; the North American Plate; Antarctica; and the South American Plate. The remaining surface area is accounted for by about a dozen much smaller plates, such as the Juan de Fuca Plate, which is located off the coast of the northwest USA. The Pacific Plate is constructed entirely of oceanic crust and mantle, while, at present, the rest of the major plates contain both continental and oceanic material. Plate size is constantly changing as some expand and others get smaller. These processes occur at plate boundaries (see pp.88–89). A balance between the two processes is maintained as the size of the Earth is constant. Most plate boundaries occur within the oceans or adjacent to continents, but some are marked on land by mountain belts, such as the Himalayas, which is where the Eurasian and Indian plates meet, or volcanic island arcs, such as the Aleutians in Alaska, which marks the subduction of the Pacific Plate beneath the North American Plate.

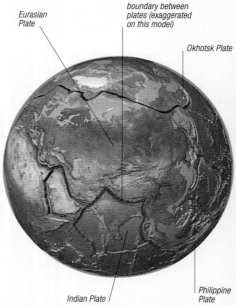

Eurasian Plate

boundary between plates (exaggerated on this model)

Okhotsk Plate

Indian Plate

Philippine Plate

TECTONIC PLATES TODAY
At present the Earth's surface is broken into seven major tectonic plates and about a dozen minor ones. The boundaries of the plates are marked by features such as mountains, ocean ridges, and deep-sea trenches.

THE EUROPEAN ALPS
The division between the African and Eurasian plates is marked by the high peaks of the European Alps.

CASCADE RANGE, USA
When two plates converge, mountains rise. Subduction of the Juan de Fuca Plate beneath the North American Plate has generated the Cascade Range in northwestern USA, including Mount Adams on the horizon.

WESTERN AUSTRALIA
The Precambrian rocks of Western Australia (above) are more than 3,400 million years old. They form a stable continental mass that is part of the Australian Plate.

How Plates Move

The Earth's tectonic plates are known to be in constant motion, shifting by up to 15cm (6in) each year. However, the mechanism that causes them to move has been hotly debated. Some scientists have argued that organized convection cells in the mantle move the tectonic plates but, new imaging techniques have failed to find evidence of such convection cells. It is now thought that it is a process known as slab pull helps to drive motion in the mantle rather than the other way around. New oceanic lithosphere formed at mid-ocean ridges is less dense than the underlying asthenosphere but becomes denser as it cools and thickens with age. This increasing density causes it to sink into the mantle at subduction zones and pull the lithosphere apart at divergent boundaries resulting in sea floor spreading or rifting. However, there is another possibility called ridge push in which newly formed, "hot", lithospheric plates slide sideways from spreading ridges and push the plate in front of them.

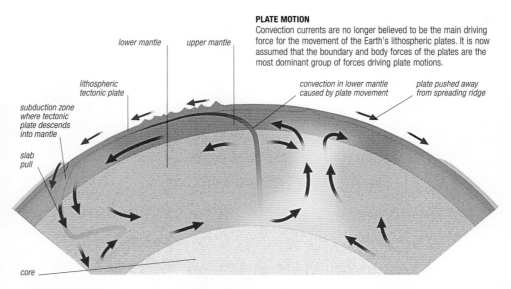

lower mantle
upper mantle

lithospheric
tectonic plate

subduction zone
where tectonic
plate descends
into mantle

slab
pull

core

PLATE MOTION
Convection currents are no longer believed to be the main driving force for the movement of the Earth's lithospheric plates. It is now assumed that the boundary and body forces of the plates are the most dominant group of forces driving plate motions.

convection in lower mantle
caused by plate movement

plate pushed away
from spreading ridge

Past Plate Movement

Several scientific breakthroughs have led to the reconstruction of past movements of the Earth's tectonic plates. Mapping of plate boundaries, the discovery of sea-floor spreading (see p.88), and measurement of the direction and rate of movement of the plates has enabled scientists to recreate their history. There is also much geological evidence to give clues to the positions of the plates in the past. For instance, the distribution of ancient mountains, such as the Caledonian belt – which today has remnants on either side of the Atlantic (see pp.140 and 142), and unusual deposits and fossils, have helped to pin down past plate positions. Fossils of the marine reptile *Mesosaurus* have been found in both Africa and South America, suggesting that the two continents were once joined. Mapping and drilling of the ocean floor beginning in the 1950s and 1960s provided further evidence of plate movement, as it was discovered that the oldest-known sea-floor rocks were just 180 million years old. Thus the ocean floor is relatively young, and any older rocks have disappeared, swallowed back within the Earth. Plate movements from the present back to the Permian Period show that over the course of about 200 million years, today's continents evolved from a continental cluster called Pangea, which stretched from pole to pole (see p.20). Late Precambrian models have also been reconstructed, and include a group of continents called Rodinia. Little information exists for earlier tectonic distributions.

FOSSIL FOUNDATIONS
Fossils are a key to past plate reconstructions. Identical fossils have been found on either side of the Atlantic. Other species, such as this *Gasosaurus*, are isolated within stable continents.

MADAGASCAN PRIMATE
The lemurs of Madagascar, including this Ring-Tailed Lemur, were isolated as a species when the island split from Africa in the late Cretaceous Period.

MEASURING PLATE MOVEMENT

Data about current plate movement is provided by both ground-based or space-based geodetic measurements, which constantly monitor Earth's size and shape. Plate motion is best measured by satellite-based space geodesy with repeated measures of points located on the different tectonic plates. Currently some 21 satellites in space orbits some 20,000 km (12,400 miles) above Earth form the NavStar system of the U.S. Department of Defense. The image here is an artist's concept of a NavStar GPS satellite.

PLANET EARTH

Plate Boundaries

17	The first continents
34–35	The Earth's Structure
40–41	The crust
86–87	Tectonic plates
Fault systems	122–33
Mountains	136–53
Volcanoes	154–75
Ocean tectonics	386–87
Tectonic Earth	486–503

There are three main types of boundaries between the Earth's tectonic plates. At two of these types of boundaries, plates change in size. Plate growth occurs at divergent boundaries along sea-floor spreading ridges, while plate consumption occurs at convergent boundaries, which are marked by deep-ocean trenches and subduction zones. Divergent and convergent zones form 80 percent of plate boundaries; the remaining 20 percent are transform boundaries, where two plates slide past one another with no significant change in the size of either plate.

(see p.138)

Divergent Boundaries

Where plates move apart, they form divergent (or constructive) boundaries. An example of a divergent boundary is the Mid-Atlantic Ridge, which stretches beneath the ocean between the North and South Poles, although it rises above sea-level in Iceland. Divergent boundaries are marked by slowly spreading rift valleys (see p.138). As two plates move away from each other, the mantle rises and partially melts to form basaltic magma. When the molten basalt erupts on the sea floor, it solidifies as it comes into contact with ocean water, forming oceanic crust, a process that is repeated as the plates continue to diverge. New oceanic crust is created at divergent boundaries at a rate of a few centimetres a year, a process called sea-floor spreading. Over many millions of years, this widening process can lead to the opening of new oceans. Because the Earth remains the same size, lithosphere is consumed by subduction (see opposite page) at the same rate that it is created. The straight lines of divergent boundaries are disrupted by the Earth's curvature, and transform boundaries (where two plates slide past each other) form. These are neither constructive nor destructive boundaries.

VOLCANIC ACTIVITY
At Geysir in Iceland (left), groundwater, superheated by hot rocks below the Mid-Atlantic Ridge, regularly blasts out at the surface in the form of a steam geyser.

VOLCANIC CHAIN
Iceland is split by the divergent plate boundary between the North American and Eurasian plates. Volcanoes have formed parallel to the boundary.

DIVERGENT AND TRANSFORM BOUNDARIES
Divergent plate boundaries are offset by transform boundaries. Plates slide past one another along transform boundaries, and, since the lithosphere is neither destroyed nor created, they are also known as conservative plate boundaries.

TRANSFORM FAULT
Horizontal sliding movement along the boundary between the North American and Pacific plates occurs either side of the San Andreas fault.

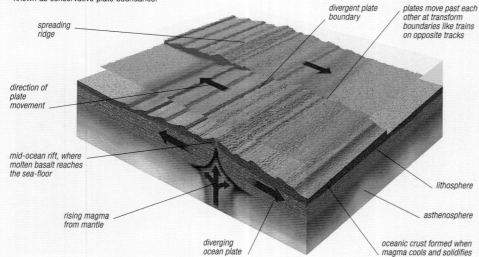

- spreading ridge
- direction of plate movement
- mid-ocean rift, where molten basalt reaches the sea-floor
- rising magma from mantle
- diverging ocean plate
- divergent plate boundary
- plates move past each other at transform boundaries like trains on opposite tracks
- lithosphere
- asthenosphere
- oceanic crust formed when magma cools and solidifies

MAGNETIC REVERSALS

The rate at which new crust is created along a spreading ridge can be measured by studying the magnetic properties of the rocks on either side. As upwelling magma cools, iron-rich minerals within it, orientated by the Earth's magnetic field, are set in position. These reflect the direction and inclination of the magnetic poles at the time the rocks formed. Their magnetism can be mapped by magnetometers, and these reveal symmetrical patterns on either side of a ridge, which show the periodic reversals of the Earth's magnetic field. In this image, bands of colour identify rocks of normal and reversed magnetization.

Convergent Boundaries

Convergent boundaries (also known as destructive boundaries) occur where plates move towards one another. Continental crust is thicker and less dense than oceanic crust. If continental and oceanic plate boundaries converge, for example, as is happening off the coast of South America, a subduction zone forms, at which the denser oceanic crust descends beneath the continental margin. This leads to uplift of the continental plate, and the formation of volcanic mountains, such as the Andes. Where two oceanic plates converge (see p.137), the older of the two will be subducted beneath the younger, as it is cooler and denser. At these boundaries, volcanic island arcs form (see p.387). Convergent boundaries involving oceanic plates are marked by deep-sea trenches, where the ocean floor reaches depths of up to 11km (7 miles). At subduction zones, earthquakes occur deep within the Earth as the tectonic slab descends into the mantle. The plate also begins to bleed fluids, which trigger partial melting in the mantle. The resulting magma can ascend to the surface in a volcanic eruption. When two continental plates converge, no subduction occurs as the plates are of a similar density. Instead the pressure leads to the formation of mountain ranges such as the Himalayas (see pp.136–37).

VALLE DE LA LUNA, ANDES
The Andes have been formed by crustal compression as a result of the convergence of the Nazca and South American plates.

SUBDUCTION ZONE
When an oceanic plate collides with a continental plate, it sinks into the mantle and eventually melts. The collision leads to the formation of an ocean trench and a chain of volcanic mountains on land.

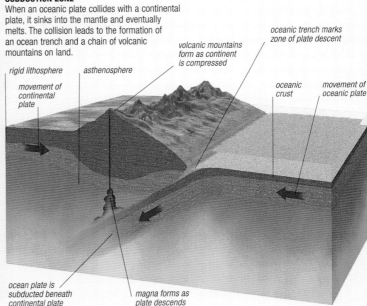

rigid lithosphere

asthenosphere

volcanic mountains form as continent is compressed

oceanic trench marks zone of plate descent

movement of continental plate

oceanic crust

movement of oceanic plate

ocean plate is subducted beneath continental plate

magna forms as plate descends

World Boundaries

Six of the seven largest tectonic plates are predominantly continental, but the largest of all, the Pacific Plate, is oceanic. Asia and North America, in particular, have grown by the addition of numerous smaller plates, and are known as accretionary terranes. These are geologically welded to the main plates. Well-defined zones of seismic and volcanic activity are distributed along the current plate boundaries, as shown here.

— Divergent or transform plate boundary

— Convergent plate boundary

— Deep-sea trenches

JOHN TUZO WILSON

In 1965, the Canadian geophysicist John Tuzo Wilson (1908–93) demonstrated that sea-floor spreading ridges are segmented and offset by what he called transform-fault boundaries. Wilson also put forward the theory of hotspots (see p.387) and their part in the formation of volcanic island chains such as Hawaii. His discoveries were a key contribution to the acceptance of the theory of plate tectonics.

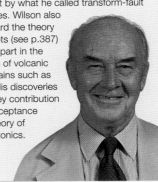

Weathering

62–83	Rocks
Erosion	92–93
Mass movement	96–97
Karst landscapes	241
Deserts	278–79
Precipitation and clouds	472–73

When rocks are exposed at the Earth's surface, they react to a group of physical, chemical, and biological processes collectively known as weathering. Over long periods – sometimes millions of years – weathering can lead to the disintegration of rocks, making them susceptible to later transport and erosion. Many rocks are formed in conditions that are very different to those on the Earth's surface. When rocks are exposed, they react in various ways with the atmosphere, water, and living organisms, depending on their structure and composition. The speed at which weathering occurs also depends on climatic conditions – rocks in warm, humid environments tend to weather more quickly than those in arid or cold conditions.

Physical Weathering

The process whereby rocks break apart without substantial change to their chemical structure is known as physical or mechanical weathering, and is largely due to changes in pressure or temperature. All rocks have natural areas of weakness, along which cracks can appear. During the processes of uplift (see p.63) and erosion (see pp.92–93), which expose rocks at the surface, much of the heat and pressure that surrounded them is removed, so they cool and shrink, causing them to fracture. Water invades the resulting cracks and when it freezes, it expands by almost one-tenth of its volume. This exerts immense pressure on the surrounding rock and enlarges the fracture. If a cycle of freezing and thawing is established, the rock eventually breaks up. Called frost shattering or heaving, such freeze–thaw processes are particularly visible on exposed mountainsides. Similar processes of expansion and contraction may also occur in deserts, where the fluctuation between daytime rock surface temperatures, which can reach up to 60°C (140°F), and night-time temperatures, which can be freezing, may result in rock fracture.

SHATTERED ROCK
Exactly how desert boulders, such as this one in Arizona, shatter is debated. Cycles of heat expansion and contraction and frost shattering may play a role in this type of weathering.

FROST-SHATTERED ROCK
The fracture of rocks by freeze–thaw action in exposed mountain areas, such as this one in Wales, UK, is a physical weathering process.

Chemical Weathering

Physical weathering exposes rock surfaces to other processes, including chemical weathering. This type of weathering can lead to the breakdown of some minerals within the rock, and the formation of new ones. In warm, wet climates, chemical weathering occurs far more rapidly than in arid conditions. The process is enhanced by rainfall, which combines with carbon dioxide in the atmosphere to form an acidic solution. As this oozes through soils it takes some minerals, particularly calcium carbonate, into solution in a chemical reaction that is reversible. For example, limestone readily dissolves in acidic water, which can hollow out underground passages in the rock. Inside these passages, the limestone may be redeposited from solution to create rock formations such as stalactites or flowstone (see pp.240–41). The chemical weathering of limestone produces a distinctive type of landscape called karst. Groundwater and seawater can also lead to reactions with other minerals. For instance, many rock-forming silicate minerals, such as olivine, combine with water. At or just below the Earth's surface, micas, feldspars, and some amphiboles may also combine with water to form clay minerals, such as kaolinite (china clay). These are a principal component of the Earth's soil and sediment, so chemical weathering is a hugely important geological process. Many rocks also contain iron-rich minerals, which react in the atmosphere to form iron oxide residues that are characterized by their rust colour.

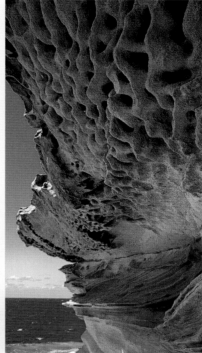

SALT WEATHERING
Seawater has lashed against this rock, penetrated it, and then evaporated to leave behind salt crystals, which have expanded to produce holes in the surface.

KARST LANDSCAPE
Extensive weathering of limestone has led to the creation of spectacular scenery in Guilin, China. The hills and towers are all that remain of the original rock.

ACID RAIN

Chemical weathering by acid rain, which is so effective in forming cave systems, also attacks limestone buildings and sculptures (see below). Since the Industrial Revolution, the process has accelerated, particularly in areas of atmospheric pollution. Nitrogen oxides and sulphur dioxide are released as a result of burning fossil fuels and vehicle emissions, and compounds react in the atmosphere to form acid rain. Damage is also caused by acid-rain run-off into groundwater, rivers, and lakes.

Biological Weathering

Organisms play an important role in the weathering of rocks, reacting with them to cause biochemical decay. Lichens tolerate extreme environmental conditions, and were probably the first organisms to colonize the land. Their acids dissolved the minerals within rocks, breaking them down in nutrients and forming soil. This addition of organic material to inorganic rock debris was an important stage in the formation of materials that could support the growth of the first land plants. In recent years it has been discovered that microbial life exists far below the ground, living on the surface of mineral grains. This evidence suggests that biological weathering may play a far more important role in the alteration of rock-forming minerals than has been recognized previously. Macroscopic plants, such as trees, also play a significant role in the formation of soils as they release organic acids that attack rocks. They also contribute to physical weathering as their roots and seeds penetrate and grow within the cracks in rocks, forcing them to open further. It could also be argued that humans have been one of the most influential organisms to have brought about weathering of landscapes, by clearing land for agriculture and construction.

LICHEN
In the soil-free Antarctic landscape, primitive lichens obtain nutrients from rock surfaces. Acids within the lichens dissolve the outer parts of the rock.

PLANT COLONIZATION
Even recent lava flows, such as this example in Hawaii, can be colonized by plants. These succulents derive nutrition from their harsh environment.

TREE INVASION
Angkor Wat, the 12th-century capital of the Khmer kingdom (present-day Cambodia), was abandoned in 1443. Since then it has been overtaken by the growth of the encroaching jungle.

Erosion

◄ 62–83 Rocks

◄ 90–91 Weathering

Deposition 94–95 ►

Erosion and transport 199 ►

Underground erosion 240 ►

Glacial erosion 255 ►

Wind and dunes 279 ►

Erosional and depositional coastlines 436 ►

The processes by which soil and rock material are loosened, dissolved, and removed from any part of the Earth's surface are known collectively as erosion (from the latin *erodere*, which means "to gnaw"). They include corrasion, which is the mechanical erosion of rocks by materials carried within water, wind, and ice, and the movement of rock material as sediment or in solution. Erosion follows vertical uplift of rocks. Together with gravity, the processes of erosion combine to lower and flatten the Earth's surface. Without erosion and sediment transport, the rock debris that results from weathering (see pp.90–91) would accumulate where it formed. There are two primary ways in which material is transported: by the movement of air, water, or ice; and by the forces of gravity.

W. M. DAVIS

The American physical geographer William Morris Davis (1850–1934) was a pioneering student of landscape erosion and erosive cycles. He envisioned a cycle in which juvenile landscapes evolved to maturity and old age, with characteristic landforms for each stage. Davis's model saw rivers cutting steep V-shaped valleys, which matured into broader valleys with meandering rivers and floodplains, which then became more prolonged "old-age" lowland landscapes close to sea-level. Over the last few decades, the reality has been shown to be more complex.

Water Erosion

Flowing water is continually reshaping the landscape, and the effects of its erosive power are evident all around us. Water is capable of lifting and transporting loose rock fragments (sediment), which then act abrasively on rock, wearing it away. It can also wash away soil and dissolve minerals, carrying them downstream before depositing them elsewhere (see pp.94–95). When rainwater falls on slopes, it runs downhill, cutting small channels called rills, which can deepen with further rainfall to form gullies. These channels can eventually join up with streams and rivers. Rivers are important erosive agents, carving their way through the Earth's surface to create valleys and canyons, and carrying sediment towards the sea. The volume of sediment, and the distance that it is transported, depends upon the size of the particles a river carries and the energy of the flow. A major flood has huge erosive potential. Seawater is also a powerful erosive force, as waves constantly bombard coastlines, re-shaping them and creating dramatic rock formations, such as caves and stacks (see p.436).

DELICATE ARCH
The flow of seawater millions of years ago, ice wedging, and collapse along joints, contributed to the formation of this remarkable stone arch in Utah, USA.

AUSABLE RIVER
A powerful, fast-flowing river, such as this one in New York State, USA, transports abrasive particles, and can quickly cut a steep gorge through rock. In other settings, rivers may erode the rock more slowly.

CLIFF EROSION
Exceptionally high tidal ranges in the Bay of Fundy, Canada, and sediment-laden attack by waves, have cut through the cliff face to produce these free-standing rocks.

ACCELERATED EROSION

High winds, the shifting course of the Yellow River (see p.221), and overcultivation of land in northwest China's arid interior have led to the erosion of the vast layer of fertile loess (silt) that blankets the landscape. The loess was deposited by winds from the arid Arctic regions of northern Asia during the last ice age. Today, lack of vegetation to bind the surface allows strong winds to lift these light particles into the air, where they are carried in suspension as dust storms over large distances. Millions of tonnes of China's loess are stripped away by wind and water erosion every year.

Wind Erosion

Moving air has the power to carry huge volumes of sediment, especially under arid conditions. When there is a lack of vegetation or moisture to hold surface deposits together, the wind picks up loose grains and transports them. Wind carried material is itself an effective agent of erosion, sculpting rocks into a variety of shapes. These include ventifacts, which are rocks that have been blasted by particles carried within the air flow, and yardangs, which are wind-shaped desert ridges (see p.278). By removing loose material, the wind exposes rock to further weathering and erosion, with the result that land surfaces are actively transformed by these two processes. Until the Earth's landscapes were covered by vegetation in the Permian Period, wind was probably as significant an agent of erosion as water. Today, wind erosion occurs primarily in arid or desert regions. Only the tiniest grains, about the size of a particle of silt or clay, are lifted high into the wind flow. The transport of larger grains takes place in a process called saltation, in which grains are lifted up to 1m (3ft) into the air, and carried along briefly, before landing back on the ground. As each particle bounces back to the surface, it collides with other fragments that are too large to be lifted, and shunts them along in a creeping motion. This results in the build up of ripples and dunes.

ERODED ROCK
Wind erosion has stripped away the softer strata of this sedimentary rock in the Sahara Desert, emphasizing its constituent layers.

VENTIFACT
Grains carried by wind have sand-blasted this erratic Antarctic boulder. Known as ventifacts, such structures are normally associated with hot deserts.

Glacial Erosion

Glaciers form in areas where ice and snow remain all year round and where, over the course of a year, more accumulates in winter than melts in summer. At present, 10 per cent of the Earth's land surface, about 14.9 million square km (5.75 million square miles), is covered by diminishing amounts of glacial ice as global climates continue to warm. However, 18,000 years ago, during the last ice age, this proportion was closer to 30 per cent. Rock-laden ice is one of the most effective abrasive forces in the natural world. The ice removes loose rock debris and incorporates it into its flow, which then carves, scours, and scrapes the surfaces it creeps over, dramatically modifying landscapes. These effects become visible only when the ice melts. The classic land-forms of glacial erosion include pyramid-shaped mountain peaks with steep rock walls called glacial horns – one of the best examples of which is Switzerland's Matterhorn (see p.145); scooped-out cirques separated by sharp rock ridges called arêtes (see p.255); and broad U-shaped valleys (see p.255). Many rock surfaces are covered with tell-tale traces of glacial erosion in the form of scratches and grooves.

GLACIAL VALLEY
This broad valley in northern England's Lake District, surrounded by steep, bare rock walls that contain a flat valley floor without a major river, is a typical result of glacial erosion.

GLACIAL GROOVE
Smooth, rounded rock outcrops covered in parallel scratches and grooves, such as these in northern Norway, are typical of erosion by rocks embedded in glacial ice.

PLANET EARTH

PLANET EARTH

Deposition

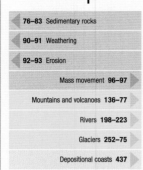

76–83 Sedimentary rocks

90–91 Weathering

92–93 Erosion

Mass movement 96–97

Mountains and volcanoes 136–77

Rivers 198–223

Glaciers 252–75

Depositional coasts 437

The solid product of the processes of weathering and erosion is sediment, which is moved across the Earth's surface by the fluid motion of water, wind, ice, or mass movement (see pp.96–97). As this movement slows down – for example, if a river reaches flatter ground – particles carried within the flow are deposited, with denser fragments falling out first. Sediment is deposited in layers in natural "traps", such as river deltas, or basins on land or in the sea. This accumulation of layers of sediment represents the initial stage of sedimentary rock formation (see p.65). Deposition is also an important process in the formation of fertile land for cultivation.

Water-Borne Deposition

Water plays a major role in the transportation and deposition of sediment. The particles carried by water can range from tiny grains of clay to boulders, depending on the speed and depth of the flow. Water is also capable of dissolving certain minerals and rocks, especially limestones, and transporting the mineral constituents in solution. Some minerals may subsequently come out of solution to form new minerals. Rivers carry sediment from the steeper parts of their courses, depositing it within lakes and on floodplains or in deltas. Floodplains are important sites of deposition: for instance, the Mississippi floodplain in the USA (see p.205) is 15km (9 miles) wide, and is characterized by fertile deposits of sand, silt, and mud. Much river-borne material is also carried towards the sea and deposited in deltas (see p.201). The deltas of large rivers, such as the Nile (see p.217) and the Ganges (see p.222), are sites of significant deposition. As it enters the sea, river-borne sediment is redistributed, contributing to the formation of continental shelves. Sediment can also be transported into the ocean by a high-energy flow called a turbidity current (see p.392).

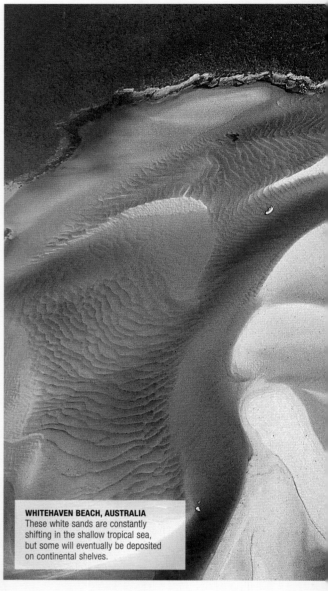

WHITEHAVEN BEACH, AUSTRALIA
These white sands are constantly shifting in the shallow tropical sea, but some will eventually be deposited on continental shelves.

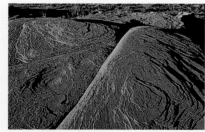

ANCIENT SANDFLATS
Washed by waves and currents, layers of sand were deposited on an ancient beach. The layers were buried and compacted to form this sandstone in Utah, USA.

RIVER TRANSPORT
Pebbles and boulders of various sizes are transported downstream by high-energy waters that also smoothe their shape.

SEDIMENT SAMPLING

By taking samples of the layers of sediment at a deposition site, scientists can reconstruct the conditions that prevailed in the past. From these samples, it is possible to determine the structure of the deposits. This is done by measuring the thickness of each layer of sediment, identifying fossils, measuring the size and shape of grains, and analysing mineral composition, all of which can be used to track the site's geological history.

LAND RECLAMATION

Where land is at a premium for human habitation, such as around Hong Kong Harbour, materials are removed from other locations and deposited to form new land. In Hong Kong, material has been removed from mountain tops and deposited in the sea, supplemented by rock that has been transported from elsewhere in the world. Land reclamation has increased the space for building in Hong Kong, but has also narrowed one of the world's finest natural harbours.

Wind-Borne Deposition

Air cannot transport large particles of sediment. However, atmospheric wind and storm systems can be so energetic that they carry enormous volumes of dust, silt, and fine sand. As wind subsides, it deposits the material it is carrying. A strong wind can carry sediment over vast distances – particles from the Sahara Desert in Africa have been found deposited on the opposite side of the Atlantic, in Barbados.

Most wind-formed landscapes are situated within the world's hot, arid equatorial deserts, although there are significant mid-latitude deserts, such as the Gobi in Asia (see p.294). The most striking desert landforms caused by wind-borne deposition are dunes, which are usually between 5 and 30m (15 and 100ft) high. Some dunes are concentrated in huge sand seas called ergs, which have an area of more than 32,000 square km (12,000 square miles). Dunes vary in shape, and include ripples and ridges, which are transverse and parallel to wind direction, crescent-shaped barchans, domes, and star-shaped formations (see p.279). Narrow zones of sand dunes also form along coastlines, as beach sands are blown inland and deposited. One of the Earth's most significant wind deposits is a thick layer of fine, fertile silt called loess. It covers large areas of the Earth's surface, particularly in China, and was deposited during the last ice age (see p.92).

AFRICAN SANDSTORM
Wind-driven sandstorms, such as this massive system over Chad in central Africa, can transport and deposit desert particles over many thousands of kilometres.

LAYERED SANDSTONE
Distinctive layers of large-scale dune deposits, dating back to the Jurassic Period, are evident in these sandstones in Utah, USA.

Glacial Deposition

Rock material eroded by mountain glaciers and ice sheets is called till, and consists of particles that range from clay and silt to boulders. This detritus is carried downhill by the glacier, and then deposited over surrounding landscapes. The features produced by till deposits are called moraines, which range from linear trains of debris to wide sheets of rock that can be moulded into many different shapes (see p.255). Till that is deposited at the sides of glaciers is known as a lateral moraine, and creates ridges that are parallel to ice flow. If two glaciers merge, a medial moraine is formed from the till, while ridges formed at right angles to the front edge of a glacier are called terminal moraines. These deposits remain as distinctive landforms when glaciers recede. Melt-waters re-shape till and generate sediments called fluvioglacial deposits that consist primarily of sand and gravel. Rivers that flow beneath glaciers channel these deposits into sinuous ridges called eskers.

MEDIAL MORAINE
A medial moraine, such as this one in Switzerland, is formed when two glaciers merge. Their lateral moraines combine to form a central band of till deposits.

Volcanic Deposition

Material ejected by volcanoes into the atmosphere includes ash and aerosol particles, which are deposited in layers over surrounding landscapes. The finest material can be spewed high into the atmosphere and distributed over great distances before it is finally deposited. The longest-lasting and most widely dispersed volcanic particles are droplets of sulphuric acid. Clouds of these droplets absorb incoming solar radiation, preventing it from reaching the Earth's surface, which then affects climate (see p.455). Ultimately, these droplets fall out of the atmosphere at the poles, where they are buried by snow. These deposits have been detected within cores drilled from polar ice sheets.

 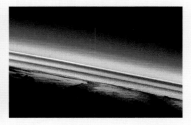

DEPOSITS IN THE ATMOSPHERE
The left-hand image, taken from the Space Shuttle in 1984, shows a clear atmosphere. The right-hand image was taken less than two months after the eruption of Mount Pinatubo in 1991, which released nearly 20 million tonnes of aerosol particles into the atmosphere. Two distinct dark layers of aerosol particles are visible, some of which eventually fell to Earth.

Mass Movement

92–93 Erosion

Landslides 98–99

Fault systems 122–23

Earthquakes 130–31

Volcanoes 154–77

Glaciers 250–75

Tsunamis 388–89

Rock materials may appear to be strong and capable of maintaining mountains and cliffs high above their surroundings. However, all rocks have finite strengths, and the pull of gravity, combined with the processes of weathering (see pp.90–91) and erosion (see pp.92–93), alters the shape of the Earth's surface. A steep cliff may survive for millennia, but can quickly collapse if it is undercut by a river. A hillside may seem to have stable, shallow slopes, but will fail suddenly and collapse, sometimes with disastrous results, if the forces that bind it are altered by rain, vegetation clearance, or a geological event such as an earthquake (see pp.98–99). Known as mass movement, the speed of the motion depends on the angle of the slope and the material involved.

Slope Stability

The slope of a pile of material has a natural angle of repose that is maintained by friction between the individual particles. This is true both of slopes that are consolidated (in which the particles are joined) and those that are unconsolidated (in which they are loose). If the angle is steepened or disturbed, the pile will collapse until it reaches a more stable position. Masses of material will also start to move, under the influence of gravity, if the cohesive strength is altered by the presence of water, additional weight, or sudden movement such as an earthquake. Any of these factors can drastically alter the equilibrium of the mass, and cause it to collapse. Vast amounts of unstable unconsolidated and consolidated material is found all over the Earth's surface both on land and beneath the sea. Much of this is in slopes that are close to their maximum angle of repose and, if triggered, will fail. The inhabitants of high mountains and coasts with steep cliffs are only too familiar with the processes of mass movement and the potential instability of slopes.

SAND DUNES
If disturbed, sand grains will roll down the slope of a sand dune until they reach their natural angle of repose.

slope is stable at 35°

FINE SAND

slope is stable at 40°

COARSE SAND

slope is stable at 45°

ANGULAR PEBBLES

ANGLES OF STABILITY
When loose materials are poured, they form natural piles. Large particles form slopes with higher angles of repose than those made of fine particles.

Rock Mass Movement

Several factors and processes contribute to the downward movement of rock material under the force of gravity. Rock that is weakened by weathering, particularly the process of frost shattering (see p.90), can break away from a slope and tumble down it. Weathering can decrease the strength of a rock mass, until it, too, becomes unstable and moves down a slope. A rock slide occurs when a block breaks off and moves along a bedding plane or a joint. The most devastating rock mass movement is a rock avalanche, which can be triggered by an earthquake or torrential rain. This type of mass movement is violent, sudden, and rapid, moving at speeds of up to 320kph (200mph), and contains many pieces of rock. The largest rock avalanches are the result of the collapse of volcanoes, such as the one that occurred during the eruption of Mount St. Helens in 1980 (see p.159). Evidence of rock mass movement can be seen in talus or scree slopes, which accumulate at the bottom of cliffs or mountains. Rocks can also form part of unconsolidated mass movement (see opposite page).

ROCK FALL
Avalanches of rock debris, such as this one above the Hood River in Oregon, USA, are common in steep, hilly areas prone to heavy rainfall.

CLIFF SLIDE
Cliffs that are weakened by the bombardment of waves are prone to rock falls, especially in areas where movement along active faults can trigger them, such as the coast of California, USA.

PLANET EARTH

AVALANCHES
An avalanche is a downwards movement of snow, ice, or debris in a mountainous area, such as Alaska (seen here). It can be triggered by sudden noises or thawing.

Unconsolidated Mass Movement

Soil and sediment consisting of uncemented grains can move downhill in various ways. The addition of water further affects this movement. Unconsolidated mass movements include creep, earth-flows, mudflows, and debris avalanches. Soil creep and solifluction (see below) are slow movements (a few millimetres a year) that usually occur over gentle slopes. These processes are common in permanently frozen ground (permafrost), particularly during the seasonal thaws, when saturated debris moves downhill over surfaces lubricated by water, fine sediment, or a combination of the two. Debris flows and mudflows containing soil, rocks, and water, can occur suddenly and move very rapidly. Lahars occur on the sides of volcanoes, and can be triggered by an eruption or rainfall that remobilizes ash and volcanic debris to create a fast-moving and often catastrophic flow. Such rapid movements can pose substantial threat to life, as they have the potential to envelop everything in their path. Mass movements of unconsolidated material can include simple, unconfined hillside flows, those that are confined by valleys and eventually spread out on to open ground, and highly destructive large-scale flows with enough energy to over-ride topography.

MUDFLOW
Heavy rainfall on this cliff on the Isle of Wight, UK, has saturated and liquefied the clay within it, causing it to collapse and resulting in a mudflow that has spread across this beach.

SOLIFLUCTION
The slow flow of water-saturated soil from higher to lower ground is known as solifluction. The tilt of these old building piles in Svalbard, in the Arctic, is evidence that solifluction has occurred.

SUBMARINE MASS MOVEMENT

The large blocks of rock detected on the submarine slopes of oceanic volcanoes (such as those of the Hawaiian Islands in the Pacific and the Canary Islands in the Atlantic) suggest that these unstable edifices have repeatedly collapsed and swept down their submarine slopes. Imagery of the ocean floor surrounding the Canary Islands has revealed blocks of rock 1km (⅔ mile) wide on the underwater slopes, extending 80km (50 miles) out from the islands. Geologists have suggested that the Cumbre Vieja volcano on La Palma in the Canary Islands (see right), which is the most active volcano in the group, could collapse and cause a huge submarine mass movement, triggering a tsunami with the potential to devastate the eastern seaboard of the USA.

Types of Slope

The development, form, and decline of slopes is a fundamental feature of landscape evolution and depends on a number of elements. These include the relative hardness of the materials forming the slope, and the elements that impact upon it such as wind and rain. For example, resistant rocks such as granite can form perpendicular cliffs, while softer rock such as shale forms slopes that are more gradual. Free-standing cliffs may form as a result of undercutting by waves or stream erosion, glacial excavation, differential weathering, or collapse along vertical joints. Then, sloping pediments of debris build up and protect the lower parts of the cliffs from further weathering or erosion. In arid and semi-arid environments, prominent rock cliffs such as mesas (see p.279) are common, but in areas where the climate is more varied and there is more rainfall, fragmented debris may dominate the slope – unless another geological process, such as glaciation, removes it.

BUTTES
Vertical-faced mesas or smaller buttes, such as these at Monument Valley, USA, rising above gently sloping pediments, are typical of slope development under arid conditions.

SCREE
Rock debris that has been loosened by freeze–thaw action (see p.90) tumbles down a hillside, piling up to form scree slopes on the sides of valleys, such as this one in England.

LANDSLIDES

On 8 August 2010, nearly two million cubic metres (70 million cubic ft) of mud and rocks tore through the city of Zhugqu in Gansu province, China, in a surge up to five storeys high. The landslide began when heavy rain piled up water behind a dam of debris that had built up in a small river – then suddenly swept it away. Many landslides have entirely natural causes, but excessive tree-felling and dam construction may well create the instability that triggers landslides too.

Slopes That Fail

Landslides are mass movements (see p.96) that occur when gravity overcomes friction and pulls soil or rock downhill. During a landslide, part of a slope shears away along a basal failure plane, and initially moves as a coherent block. In some landslides, the moving material breaks up, but in others it stays more or less intact as it shifts. An example of this occurred on the coast of Dorset, England, in 1839. Coastal cliffs gave way, carrying clifftop farmland with them, but the fields remained usable, and crops were grown on them for several years.

One of the most common causes of landslides is heavy rainfall, which saturates the ground. Earthquakes and volcanic eruptions are much more violent triggers. Human activities can increase both the risk of landslides and the danger that they present.

In mountainous regions, forest clearance is a frequent cause of landslides, because it increases water run-off and accelerates erosion. Road construction is another hazard, because it cuts into hillsides, destabilizing rock. Urbanization also causes problems. In Malaysia, a landslide demolished a 12-storey apartment block in 1994, while in Hong Kong, landslides have killed nearly 500 people in the last 50 years. Most of these fatalities have been due to poor construction techniques on steep slopes, aggravated by the region's heavy summer rainfall.

Although some landslides strike without any warning, in many cases ground movements show that an area is at risk. During the early 1960s, several small landslips occurred during the construction of the Vaiont Dam, in the Italian Alps. During the autumn of 1963, rising water levels saturated a clay layer, triggering a massive landslide that plunged into the reservoir below. The dam held, but displaced water surged over its crest, devastating a town in its path. Altogether, more than 2,000 people lost their lives.

The most catastrophic landslides in recorded history took place during a major earthquake in the Gansu region of China in 1920, which triggered collapses of steep cliffs of fine-grained loess soils. Ten large cities and many villages were buried, and an estimated 200,000 people lost their lives during a single day.

Mud on the Move

Landslides move at speeds ranging from a slow, barely perceptible creep to more than 100kph (60mph) on steep slopes. Both debris avalanches and water-saturated mass movements (mudflows) can travel many tens of kilometres, causing widespread devastation and smothering everything in their paths. Mudflows are common in the western USA during El Niño years (see p.453), and they were responsible for many of the fatalities caused by Hurricane Mitch, when it stalled over Central America in 1998, shedding considerable quantities of rain.

Mudflows in volcanic terrain, called lahars, can occur in association with eruptions, or later, as rains remobilize loose ash deposits. Fine ash is particularly dangerous, because it easily forms a fluid mass. When a lahar comes to a halt, the transported mud or ash sets like cement, making escape almost impossible. In 1985, a lahar poured down the flanks of Nevado del Ruiz, a volcano in Colombia (see p.162), and struck the town of Armero, 50km (30 miles) to the east. Most of its 29,000 inhabitants were asleep, and fewer than a third survived.

Mudflows on this scale are impossible to prevent, but, in the future, improved communications may reduce the cost in human lives. A satellite-based system, now under trial in the Caribbean, will gather rainfall data and flash warnings to home computers, or even mobile phones.

VAIONT DAM DISASTER
Following a landslide, water from the Vaiont reservoir destroyed the town of Longarone, in 1963. A second overspill – also caused by a landslide – occurred in 1966.

LARGEST LANDSLIDES

Most of the largest landslides and debris flows in the last 120 years (listed below) have occurred in tectonically active zones.

Location	Date	Cause	Fatalities
Gansu, China	1920	Earthquake	200,000+
Vaiont, Italy	1963	Dam construction	2,600
Mount Huascarán, Peru	1970	Earthquake	18,000
Mount St Helens, USA	1980	Volcanic eruption	60
Nevado del Ruiz, Colombia	1985	Volcanic eruption	23,000
Huaraz, Peru	1991	Lake drainage	5,000
Honduras and Nicaragua	1998	Hurricane	2,000+
Gansu, China	2010	Rainfall	1,471
Glacier Bay, USA	2012	Snowmelt	0

EXPOSED TO THE ELEMENTS
Completely clearing a forested area aggravates natural erosion, leading to frequent landslides, as has happened here in Cabin Creek, in the Cascade Range, USA.

TRAIL OF DESTRUCTION
The mudslides that swept through Zhugqu in China covered 300 homes, including an entire village, and claimed over 1,400 lives.

PLANET EARTH

Meteorite Impacts

10–31 The history of the Earth

34–35 The Earth's structure

83 Meteorites

Extinctions 113

The Earth is regularly struck by debris from space. In the last billion years, it has been hit by about 130,000 meteorites large enough to produce a crater at least 1km (⅔ mile) wide. Although this kind of impact site is visible on the Moon, the surface of the Earth is so geologically active that evidence of many impacts has disappeared. Until recently, it was believed that the Earth was hit by large meteorites only very early in its history. However, scientists now realize that the Earth is being struck continually, and that a number of giant craters have been preserved. So far, more than 190 impact sites have been confirmed worldwide, with new craters discovered every year.

NAMIBIAN METEORITE
This 60-tonne rock discovered at Hoba West, Namibia, is the largest meteorite to have been found on Earth.

Causes of Impacts

Most meteorites are fragments of asteroids, although some were produced when asteroids hit the surfaces of the Moon or Mars, flinging debris into space. While they are outside the Earth's atmosphere, these rocks are called meteoroids; those that enter the atmosphere are known as meteors. These are usually so small that they burn up as they plunge towards the Earth, leaving a bright trail in the night sky. Meteor showers are caused when the Earth passes through the tail of a comet. Meteorites are objects large enough to travel through the atmosphere and hit the ground. Impacts may also be caused by debris from comets. If a meteorite is the size of a house, or even larger, the force of the impact blasts out a crater.

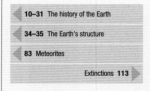

METEOR TRAIL
Meteors, which are also called shooting stars, are visible in the night sky when their trails burn up as they travel through the Earth's atmosphere.

EUGENE SHOEMAKER

American scientist Eugene Shoemaker (1928–97) was an expert on meteorites. His work confirmed the first identification of an impact crater on the Earth: Meteor Crater in Arizona, USA (see p.103). In the process, he and Edward Chao discovered coesite, a silica mineral produced under the extreme pressure and temperature of an impact. The presence of coesite is now taken as a key indicator of an impact. With his wife Carolyn and friend David Levy, he discovered the Shoemaker-Levy 9 comet, which hit Jupiter in 1994.

Identifying Impact Sites

The most obvious indication of a meteorite impact is a large, circular crater, and geologists may begin their search for new sites by scouring satellite images of the Earth's surface for such features. Although craters erode over time, scars called astroblemes ("star wounds") remain. Once spotted, each site is investigated for evidence to confirm impact. Usually a meteorite vaporizes on impact, but shattered or melted remnants may still exist. These fragments can resemble ordinary rock, but if rich in iridium, osmium, or platinum they probably originate from a meteorite. Geologists also look for damage to rocks around the impact site. These effects are called shock metamorphism, and include shatter cones, tektites, diaplectic glass (a natural glass formed during a meteorite impact), and high-pressure silica minerals such as stishovite and coesite.

TEKTITE
Tektites are beads of glass that are formed when droplets of silica-rich molten material are expelled from a crater during a meteorite impact.

SHOCKED QUARTZ
This image of a quartz grain reveals a key sign of a meteorite impact – shock lamellae. These are deformed layers within the quartz, indicated by bright colours.

METEOR CRATER, ARIZONA
With its distinctive simple bowl shape, Meteor Crater is the result of the most recent significant meteorite impact on Earth, about 50,000 years ago. It was the first major impact structure to be identified (see p.103).

WOLFE CREEK CRATER, AUSTRALIA
This Australian impact structure shows features typical of meteorite craters: the near-circular depression is surrounded by a well-preserved blanket of debris.

Impact Craters

When a meteorite strikes the Earth, the impact sends shock waves through the ground, squeezing the surrounding rock to two or three times its usual density. The compressed rock then springs back and shatters into fragments, hurling chunks upwards and outwards, along with pieces of the meteorite that have not vapourized. The result is a bowl-shaped crater that is much larger than the meteorite. Rock fragments called ejecta are blasted far beyond the crater. Other fragments fall back into the hollow to form rocks called breccia. Small craters form in just a few seconds, and even giant ones form in a few minutes. Meteorite craters vary widely in appearance from small, cup-shaped bowls to giant depressions filled with multiple ridges and mounds. They are divided into two groups: simple and complex (see below).

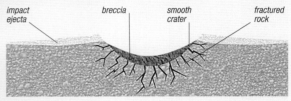

impact ejecta   breccia   smooth crater   fractured rock

SIMPLE CRATER
Typically no more than 4km (2½ miles) in diameter, simple craters are smooth bowls that are wider than they are deep. They often have a steep, well-defined rim, and there may be a layer of breccia at the bottom of the crater that is thicker at the centre.

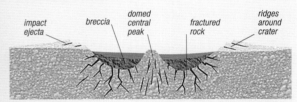

impact ejecta   breccia   domed central peak   fractured rock   ridges around crater

COMPLEX CRATER
Complex craters are larger than simple craters and they are usually shallower. This type of crater often has rings of ridges. Many also have a domed central peak, which was raised as the compressed rock sprang back after impact.

After Impact

The potential energy of a significant meteorite impact on the Earth is over 100 million megatons – more than the world's entire nuclear arsenal. Studying nuclear explosions has given scientists some idea of what might happen after the impact of a large meteorite about 10km (6 miles) in diameter. At the moment of collision, an immense wave of heat and pressure roars outwards, flattening and incinerating everything over a vast area. Massive amounts of debris are blasted high into the atmosphere and blown around the world by strong winds. Hot ash rains down, starting forest fires, while dust clouds linger in the atmosphere for months, blocking out the Sun and turning the Earth into a dark and frozen planet. When the sky finally clears, the carbon dioxide that has flooded the atmosphere creates a greenhouse effect, warming the global climate by an average of 15°C (27°F).

SHATTER CONES

Among the most persuasive signs of a meteorite impact are shatter cones. These are rock structures that have distinctive fractures resembling horsetails. The fractures converge in a cone shape, which can range in size from 2.5cm (1in) to over 2m (6ft) in length. Fractures such as these are only caused by sudden, intense pressure on existing rock. In impact craters, most shatter cones point upwards, which indicates that the impact on the rock came from above.

DINOSAUR EXTINCTION
The global effects of a large meteorite impact may be devastating and scientists have evidence that they have been the cause of some of the mass extinctions that have occurred, in particular the death of the non-avian dinosaurs 66 million years ago.

METEORITE IMPACT PROFILES ▶

The pages that follow contain profiles of the most significant impact craters in the world. Each profile begins with the following summary information:

DATE OF IMPACT Approximate date at which the meteorite hit the Earth

TYPE Simple or complex

DIAMETER Width across the crater

PLANET EARTH

PLANET EARTH

Haughton

LOCATION On Devon Island, Nunavut Territory, in the Arctic region of northwest Canada

DATE OF IMPACT 23 million years ago

TYPE Complex

DIAMETER 24km (15 miles)

Most craters that are as old as the Haughton impact site have undergone significant erosion. However, because it lies within the Arctic Circle where much of the water is frozen solid, Haughton has escaped the worst effects of weathering. The lack of vegetation covering the site makes it easy for geologists to study the crater both on the ground and from space using satellites. Studies have shown that the giant meteorite that

MARS TEST

Rocky, arid, and icy cold, Haughton Crater resembles conditions found on the surface of Mars. Scientists have been studying the crater to see what it can tell them about the geology of the Red Planet and also how astronauts could survive on Mars.

struck here penetrated 1.7km (1 mile) into the ground – as far as the buried ancient rocks of the continental crust. As it crashed into the Earth, it threw a huge shower of shattered fragments of this ancient crust into the air, which can now be found on the surface.

ARCTIC IMPACT
The Haughton meteorite struck within the hilly landscape of Devon Island in Arctic Canada.

NORTH AMERICA *northeast*

Manicouagan

LOCATION Within Réservoir Manicouagan, near the Laurentian Mountains, Quebec, Canada

DATE OF IMPACT 215.5 million years ago

TYPE Complex

DIAMETER 100km (60 miles)

The Manicouagan Crater is one of the largest and most intact impact structures on the Earth's surface. Its width is exceeded by only four other craters in the world and matches that of the Popigai Crater in Russia (see p.105). It also one of the Earth's oldest craters, caused by the impact of a huge meteorite in the Triassic Period. The site is notable for its circular central plateau, which is surrounded by the ring-shaped Lake Manicouagan. The crater has been the subject of a great deal of speculation about whether the impact caused a wave of global extinction. Paleontologists identify the boundary of the Triassic and Jurassic periods as one of the times in the Earth's history when vast numbers of species disappeared in

RIVERS AND LAKE
Manicouagan Crater has become a focus for numerous surrounding rivers and streams, which flow into its lake.

a short span of time, an incident known as the TJ event. One theory to explain this is that the Manicouagan meteorite impact threw so much rock and dust into the atmosphere that its effect on the planet's climate destroyed many living organisms. However, new dates for the impact predate the end Triassic by 14 million years, making such a causal connection unlikely.

NORTH AMERICA *northeast*

Sudbury

LOCATION North of Lake Huron and the town of Sudbury, Ontario, eastern Canada

DATE OF IMPACT 1,840 million years ago

TYPE Complex

DIAMETER 200km (125 miles)

The Sudbury Crater is such an unusual shape that it took geologists a long time to identify it as a meteorite crater. The area is one of the Earth's richest sources of nickel-copper sulphide minerals, and

was initially thought to be of volcanic origin. Now, however, scientists are certain that it is an impact site. The object that smashed into the Earth here produced a crater about 20km (12 miles) deep. What makes the crater so unusual is the fact that it is elliptical and is nearly twice as long as it is wide. The crater is now believed originally to have been circular – its elliptical shape was produced by post-impact deformation of the Canadian Shield, the stable continental rocks in which the impact took place.

NICKEL MINING
The highly productive mines that lie within the Sudbury Crater yield over 1,000 tons of nickel and copper ores every day.

NORTH AMERICA *central*

Manson

LOCATION Around the town of Manson, northwest of Des Moines, Iowa, midwestern USA

DATE OF IMPACT 74 million years ago

TYPE Complex

DIAMETER 37km (23 miles)

Completely buried up to 90m (295ft) beneath glacial deposits, the Manson impact site was thought to be a volcanic structure, but studies in the 1960s revealed its true origin. Initially, the crater was dated at about 66 million years old, so scientists began to wonder if it might be linked to the extinction of the dinosaurs, which occurred at a similar time. In the 1990s, scientists drilled about a dozen research cores to find out more about Manson. They discovered the structure consists of a ring-shaped moat around a central peak. They also revised the date of impact to 74 million years ago, well before the demise of the dinosaurs. The effects of the impact must been enormously destructive, killing most wildlife within 1,000km (620 miles).

NORTH AMERICA *east*

Chesapeake Bay

LOCATION On the Atlantic coast of the USA, beneath Chesapeake Bay, Maryland, and Virginia

DATE OF IMPACT 35 million years ago

TYPE Complex

DIAMETER 85km (53 miles)

One of the largest impact craters on Earth, Chesapeake Bay was identified by chance in the 1980s when teams drilling the sea bed for water sources found samples of jumbled rocks, and an ocean survey ship found shattered quartz and microtektites. Subsequent seismic and gravity surveys then revealed the huge extent of the crater.

SATELLITE IMAGE OF CHESAPEAKE BAY

Meteor Crater

LOCATION East of Flagstaff and west of Winslow, in the Painted Desert, Arizona, USA

DATE OF IMPACT 50,000 years ago

TYPE Complex

DIAMETER 1.2km (³/₄ mile)

A vast dish carved into the Arizona desert, Meteor Crater is perhaps the Earth's most famous meteorite crater. Its shape has been remarkably well preserved, and it is comparatively young for an impact site. Scientists estimate that the meteorite that crashed here was about 50m (150ft) across and weighed about 300,000 tonnes. It probably hurtled into the ground at a speed of 65,000kph (40,000mph), creating an immense force as it hit the surface – the equivalent of 20 million tonnes of TNT exploding. Fragments of the nickel-iron meteorite that impacted are scattered all around the crater, as are tiny metallic balls, which were created as the meteorite vaporized on impact. It was the presence of nickel-iron that led Daniel Moreau Barringer (see panel, right) to conclude that the crater was the result of a meteorite impact. Further crucial evidence was discovered in 1960 by Eugene Shoemaker, Edward Chao, and Daniel

CRATER AND CANYON
The Colorado Plateau is scarred by two dramatic geological structures: Meteor Crater (left) and the Grand Canyon (right).

Milton, who found coesite and stishovite at the site. These minerals form only under enormous pressure and at high temperatures, and are key signs of an extra-terrestrial impact. Shoemaker's demonstration three years later of the similarities between Meteor Crater and craters made by nuclear tests in Nevada clinched the argument.

DANIEL MOREAU BARRINGER

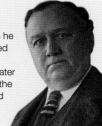

Meteor Crater is also called Barringer Crater after the mining engineer who was determined to prove its impact origin, Daniel Moreau Barringer (1860–1929). He found it hard to convince geologists of his theory, which he put forward in 1902, and lost a fortune searching for the meteorite he was convinced was buried there. It was later realized that the meteorite had vaporized on impact.

ARIZONA'S GIANT IMPACT SITE
Meteor Crater is a dramatic, cup-shaped bowl in the Arizona desert. It was the first meteorite impact site on the Earth's surface to be positively identified.

Chicxulub

LOCATION Centred on the town of Chicxulub, beneath the Yucatan Peninsula, Mexico

DATE OF IMPACT 65 million years ago

TYPE Complex

DIAMETER 160–240km (100–150 miles)

Buried deep beneath the limestone of Mexico's Yucatan Peninsula, Chicxulub is one of the largest meteorite craters yet found on the Earth. Some estimates put its width at over 300km (185 miles). At the end of the Cretaceous Period, 65 million years ago, a comet or an asteroid that may have been more than 10km (6 miles) wide crashed into the Earth at this site. The impact was cataclysmic: fires raged over the surface, giant tsunamis radiated across the oceans, and the planet was rocked by massive earthquakes. Many scientists believe that the devastating global effects of the Chicxulub impact caused the extinction of the dinosaurs (see p.24) and two-thirds of other animal species. Both the size of the catastrophe and its timing are right, and the evidence is very convincing. The Chicxulub Crater is a complex crater with several ring-shaped

CATASTROPHIC CHICXULUB
Magnetic and gravity surveys were used to create this vertically exaggerated computer image of the crater hidden about 1km (⅔ mile) beneath the Yucatan Peninsula. The top of the image represents north.

ridges and a central mound like that of the Manicouagan Crater (see opposite page). To explain its shape, scientists have suggested that the impact deformed the solid rock of the Yucatan Peninsula rapidly, almost like a fluid. Rings rippled out from the point of impact, and at the centre a towering mound, at least twice the height of Mount Everest, piled up and then subsided gradually. Because it is buried so deep beneath the surface, Chicxulub was discovered almost by accident when an oil company was prospecting in the area in the 1980s.

THE K-PG BOUNDARY

About 66 million years ago, two-thirds of the Earth's animal species died out suddenly. Scientists link this event to a thin layer of clay that dates from the same time, which is sandwiched between Cretaceous and Tertiary rocks. Called the K-PG boundary, this clay contains high concentrations of iridium, an element rare on the Earth but abundant in meteorites. It is likely that this layer represents debris from the Chicxulub impact that was suspended in the atmosphere. This would have blocked out the Sun's rays, turning the world ice cold and spelling doom for the non-avian dinosaurs.

PLANET EARTH

Bosumtwi

LOCATION	Beneath Lake Bosumtwi, northwest of Accra and south of Kumasi, Ghana, west Africa
DATE OF IMPACT	One million years ago
TYPE	Complex
DIAMETER	10km (6 miles)

Hidden deep within rainforest, and drowned beneath the waters of the lake that fills its crater, Bosumtwi has been more awkward to study than other impact sites. Moreover, the heat and humidity of the region have resulted in rapid erosion of many of its features. However, scientists have managed to confirm that the crater is complex, with a raised central dome buried beneath the lake floor. Tektites of a similar age found along the Ivory Coast are now thought to have been flung out by the Bosumtwi impact.

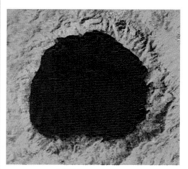

LAKE BOSUMTWI

Ries

LOCATION	In southern Bavaria, northwest of Munich, western Germany
DATE OF IMPACT	15 million years ago
TYPE	Complex
DIAMETER	24km (15 miles)

The Ries Crater is perhaps the most studied impact structure in Europe. It is a flat, circular basin with a narrow rim around the edge. The Ries meteorite is thought to have been about 1km (⅔ mile) across. Like some other impact sites, the Ries Crater is not alone in the landscape. Just 36km (22 miles) to the southwest is another, less perfectly shaped basin called the Steinheim Crater, which measures only 3.8km (2.4 miles) across. This was created by a much smaller meteorite that was barely 100m (300ft) long. Radioactive dating suggests that both these meteorites struck at the same time. It is possible that they were the result of a single meteorite that disintegrated as it came through the atmosphere, but the craters are so far apart that this seems unlikely. Some scientists think it more probable that two meteorites swept into the atmosphere together. When the larger Ries meteorite struck, it penetrated the limestone and marl sediments at the surface and plunged right through to the crystalline crustal basement over 600m (2,000ft) below

TOWN BUILT WITHIN A CRATER
Within the Ries Crater lies the medieval walled town of Nördlingen, which was built on the dried sediments of the lake that once filled the impact site.

the surface of the Earth. The shock melted surface rocks, which were ejected from the site, cooling into spheres of transparent green glass (tektites) that were scattered across the landscape of Bohemia and Moravia in the Czech Republic. Clear examples of this green glass, called moldavite, are today cut to make semi-precious stones. Soon after the impact, both the Ries and the Steinheim craters filled with water to become lakes, but over the next 2 million years they became choked with sediment and eventually formed dry land.

Vredefort

LOCATION	Southwest of Johannesburg, east of the Vaal River, Witwatersrand Basin, South Africa
DATE OF IMPACT	2,020 million years ago
TYPE	Complex
DIAMETER	300km (180 miles)

The Vredefort impact site is marked by the oldest and largest known crater on the Earth's surface. It is also one of the few multi-ringed craters in the world. Such structures are common on the Moon but are rare on the Earth, as most would have formed in the early days of the planet's history and have been destroyed by geological processes. It was initially thought that the dome in the centre of the Vredefort Crater was the result of a volcanic explosion. However, since the mid-1990s, mounting mineralogical evidence, accompanied by the discovery of dramatic distortion of rocks within the dome, has convinced scientists that it is the site of a meteorite impact.

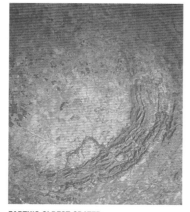

EARTH'S OLDEST CRATER
This image taken by a satellite over southwestern South Africa clearly reveals the circular wrinkles at the edge of the giant Vredefort structure.

Shatter cones are found frequently in the bed of the nearby Vaal River, while rocks that lie flat outside the crater have been forced into almost vertical positions within the basin. The extra-terrestrial body that hit Vredefort was one of the biggest meteorites ever to have hit the Earth, measuring about 10km (6 miles) across.

MINING CRATERS
Many impact sites are also the locations of valuable mineral ores, which have been subsequently exploited by mining companies. At Sudbury Crater in Canada (see p.102), the rich nickel and copper ores were formed by the heat of the meteorite's impact. The older sediments beneath Vredefort are rich in gold, which is preserved beneath the crater. Meteorite craters also make very good traps for oil, as it works its way towards the surface through cracks formed by an impact.

GOLD MINING IN VREDEFORT
These mines yield over 7 billion dollars' worth of gold annually, and contain the largest concentration of the precious metal in the world.

Kara and Ust Kara

LOCATION	Close to the mouth of the Kara River, near the coast of the Kara Sea, northwest Siberia, Russia
DATE OF IMPACT	70 million years ago
TYPE	Complex
DIAMETER	65–120km (40–75 miles)

Discovered in the 1970s, the Kara Crater is situated in an area of bleak, barely inhabited tundra. It was initially believed that there was just one crater here, about 65km (40 miles) wide. Then more impact outcrops were spotted to the west by the coast of the Kara Sea (see p.404). Some geologists suggested that this was another crater, named Ust Kara, which formed at the same time. However, the results of scientific expeditions in 2001 concluded that the two sites probably form one vast crater. Its age once led scientists to suggest the crater as an alternative candidate to Chicxulub (see p.103) for the K-PG event that might have wiped out the dinosaurs, but it is too old.

ASIA *west*

Tunguska

LOCATION Near the Tunguska River in the forests of eastern Siberia, north of Lake Baikal, Russia

DATE OF IMPACT 1908

TYPE No crater has been discovered

DIAMETER Unknown

Tunguska is one of the Earth's great mysteries. On 30 June 1908, local people witnessed a fireball streaking through the sky before vanishing below the horizon. A huge explosion followed, felling trees over an area of 2,200 square km (850 square miles). There have been several expeditions to the Siberian site to try and discover the cause of the explosion; one theory is that it was due to a comet vaporizing before it hit the surface.

DEVASTATION AT TUNGUSKA

ASIA *northeast*

Popigai

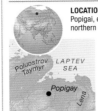

LOCATION Near the town of Popigai, east of the Popigai River, northern Siberia, Russia

DATE OF IMPACT 35 million years ago

TYPE Complex

DIAMETER 100km (62 miles)

ASIA *southwest*

Lonar

LOCATION Northeast of Mumbai, in the Buldana district of Maharashtra, western India

DATE OF IMPACT 50,000 years ago

TYPE Simple

DIAMETER 1.8km (1¹/₁₀ mile)

Filled with a lake and surrounded by temples and dense woodland, Lonar is undoubtedly one of the world's most spectacular meteorite impact sites. According to Hindu legend, the crater was created by Lord Vishnu when he destroyed the demon Lavanasur. The site was first recognized as a meteorite impact crater by the American geologist

The vast Popigai Crater is one of the Earth's largest impact sites, and was probably caused by a 5-km- (3-mile-) wide meteorite. The force of the impact was sufficient to turn graphite into microscopic diamonds, and ejecta discovered as far away as Massignano, Italy, may have originated there. The timing of the Popigai collision coincides closely with the impact at Chesapeake Bay (see p.122). This has led to the conjecture that at the time of the impact the Earth may have been subjected to a comet shower.

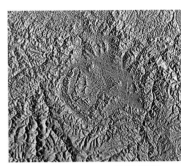

POPIGAI IMPACT SITE SEEN FROM SPACE

G.K. Gilbert, in 1896. The crater is about 170m (560ft) deep, with a rim that is 20m (65ft) high. Estimates suggest the meteorite that struck here was about 60m (200ft) long. The Lonar Crater is remarkable in that the impact occurred in hard basalt, therefore the shape of the crater is exceptionally well preserved. The salty, alkaline

SALT LAKE
Surrounded by a rim produced by the meteorite's impact, Lonar's lake is one of the most saline on Earth, and is an almost perfect circle.

chemistry of the lake means that it is inhabited by unique species of flora and fauna.

AUSTRALASIA *Australia*

Woodleigh

LOCATION South of Shark Bay and the town of Denham, Western Australia

DATE OF IMPACT 250–364 million years ago

TYPE Complex

DIAMETER 60km (37 miles)

Woodleigh was discovered by mining prospectors in the 1970s, but it wasn't until a 1997 survey revealed a tell-tale dome that a meteorite impact was suspected. Shocked quartz, which was drilled from the dome in 1999, confirmed the site's origins. An impact this large, from a meteorite about 3km (2 miles) wide, probably caused major devastation to the planet's living organisms.

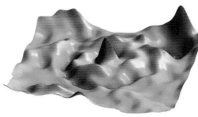

COMPUTER IMAGE OF WOODLEIGH CRATER

AUSTRALASIA *Australia*

Gosses Bluff

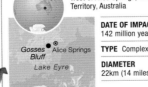

LOCATION Near Hermannsburg, west of Alice Springs, Northern Territory, Australia

DATE OF IMPACT 142 million years ago

TYPE Complex

DIAMETER 22km (14 miles)

One of the best-preserved meteorite craters in the world, Gosses Bluff is an imposing sandstone ring in the northern Australian desert. The site is very important to the Western

Arrernte Aboriginal people, featuring in some of their Dreamtime stories. The crater formed when an asteroid 1km (⅔ mile) wide smashed into the surface of the Earth. The meteorite is thought to have penetrated 5km (3 miles) into the ground and vaporized. The ground then recoiled to produce the crater's central dome. Over millions of years, some of the crater's original features have eroded and weathered, and the entire land surface of the region is about 2km (1¼ miles) lower than it was at the time of impact. However, in addition to the striking outer ring, there remains a 4.5km- (2¾-mile-) wide inner ring, which is overlain with breccias and the remnants of the central dome.

SANDSTONE CIRCLE
Rearing out of the desert like a castellated wall, the outer ring of the Gosses Bluff Crater is a striking reminder of the massive asteroid impact millions of years ago.

CENTRAL UPLIFT
The true shape of the Gosses Bluff Crater, with its ring of sandstone hills and its slightly raised central area, becomes clearly visible in this photograph from space.

Water

Water resources 110–111
Rivers 198–99
Lakes 224
Freshwater quality 230–31
Glaciers 253–75
Ocean water 384–85
Polar oceans 398–99
Precipitation and clouds 472–81

The Earth's outer layers are dominated by water. More than two-thirds of its surface is covered with liquid water; if frozen water, in the form of ice, is also included, this proportion rises to more than four-fifths. Water is essential to life because it is an excellent solvent and it can move or flow easily. Living organisms require not merely the presence of water, but a continuing supply for the maintenance of life. Humans are typical of most animals, being 62 per cent water; soft-bodied aquatic creatures such as jellyfish are more than 98 per cent water.

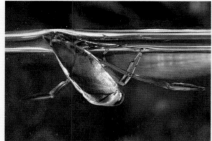

SURFACE TENSION
The film that forms on water is strong enough to support the weight of insects such as the Common Back-swimmer. This predatory bug can hang down from the surface of ponds.

Properties of Water

Pure water has no colour, taste, or smell. Its freezing point, at which it turns into solid ice, is 0°C (32°F), and its boiling point, when it changes into water vapour, is 100°C (212°F). Water's density, or mass per volume, is 1kg per litre (1¼lb per pint). This is high for a liquid – that is, water is relatively heavy – and the figure is used as a standard for comparing densities of other substances. Each molecule of water is made of two hydrogen atoms and one oxygen atom (written H_2O). A pinhead-sized drop of water contains about one billion billion molecules. These attract each other powerfully, especially at the surface of the water, where their mutual attraction forms a strong "skin", known as surface tension. As water is difficult to break by tension, it can "creep" into small holes and along narrow cracks. Its ability to change state easily, and to spread or flow even through rock, means that water is always on the move through global and local cycles (see p.109).

Salt Water

Salt water is found in seas and oceans, coastal lagoons, and inland lakes with little or no outlet, such as the Dead Sea (see p.239). In the latter, minerals are washed into drainage basins, where they accumulate at high concentrations because pure water evaporates relatively quickly. The main dissolved minerals in seawater are sodium (10.8 parts per thousand) and chloride (19.35ppt), with smaller amounts of other salts (see p.384). The freezing point of salt water is -1.9°C (28.6°F) at 35ppt. Overall salinity (or salt concentration) is highest in the tropics (above 35ppt), where evaporation occurs fastest, and lowest near the poles (below 30ppt), where there is slower evaporation and melting ice. Seawater is slightly less saline than the body fluids of most marine invertebrates, so water tends to enter their bodies by osmosis; to counteract the inflow, they have body structures such as gills and nephridia (tiny tubes) that are able to pump water out. Marine vertebrates, including most fish, face the opposite problem. Seawater is two to three times more concentrated than their body fluids, so they lose water through their gills and other surfaces. To counteract this, they drink seawater and rapidly filter out the salts, mainly through the kidneys and other specialized organs, to produce small volumes of highly concentrated urine.

SOCKEYE SALMON
Salmon hatch in fresh water, migrate to the sea to mature, then return to their home river to breed. Their physiology changes to enable them to cope with varying salinity.

HIGH-SALINITY SEA
In the landlocked Dead Sea (above) between Israel and Jordan, fluctuating water levels and rapid evaporation lead to the deposition of minerals in pillar-like shapes.

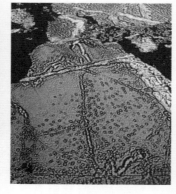

LOW-SALINITY PACK ICE
Pack ice is a typical feature of low-salinity seas, such as the Beaufort Sea off northern Canada (left). In winter, the seawater freezes easily to form a solid sheet, which breaks up in summer.

RJUKANDI WATERFALL
A mass of water flows over this Icelandic waterfall. As water moves in this way, it releases a large amount of erosive energy.

Fresh Water

Of all the water on Earth, about 3 per cent is fresh – that is, it contains far lower concentrations of dissolved minerals than salt water. Almost four-fifths of this fresh water is locked up in polar regions as glacier ice; the other fifth lies under the surface, as groundwater in rocks. Only 0.3 per cent of fresh water is in liquid form at the surface, in rivers, lakes, and wetlands, yet this is the form most useful for land animals and many aquatic creatures and plants. Although it may look pure, natural fresh water usually holds measurable amounts of at least 25 dissolved minerals, such as silica, calcium, potassium, magnesium, and iron. The relative terms "hard" and "soft" are used to describe water with higher or lower levels of these minerals, respectively. Rainwater is naturally slightly acidic. It absorbs carbon dioxide gas from the air and from the soil, and becomes a weak form of carbonic acid. This is almost insignificant compared to acid rain formed by atmospheric pollution (see p.91), but it is enough to dissolve chemically rocks such as limestones, where cracks gradually widen to form potholes and caverns (see p.240). The temperature of fresh water varies hugely. Far underground, it may be warmed by hot rocks until it is superheated. It may then burst forth as a geyser.

ACIDIC WATER
Bogs are isolated from alkaline groundwater and depend entirely on rainwater, which is relatively mineral poor. As a result, they are naturally acidic, a condition that favours the growth of sphagnum moss, but few other plant species.

MINERAL TERRACE
At Pamukkale in southwest Turkey, hard water reaches the surface as hot springs. Minerals in the water precipitate out, or become solid, forming terraces (below).

HOT WATER
The temperature of hot springs (above) may exceed boiling point. Yet life survives here, especially in the form of thermophilic, or heat-loving, micro-organisms such as blue-green algae.

Ice

At any one time, almost four-fifths of all fresh water is frozen solid as ice. Ice exists in numerous forms, including mountain glaciers, polar ice-sheets, icebergs, periglacial landforms (see p.338), and mountain-top coverings. The mass of floating ice in the Arctic Ocean is, on average, 5–7m (16–23ft) thick. The ice covering the landmass of Antarctica has accumulated over thousands of years to become more than 4.5km (2.8 miles) thick in places. Ice is slightly less dense than liquid water, so that icebergs and ice cubes float (although only one-eighth to one-tenth of their bulk is exposed above the surface). As water cools towards its freezing point, it becomes most dense at 4°C (39°F). Below this temperature, it starts to expand again. This property is helpful to aquatic life. In a pond or lake, the densest water at 4°C (39°F) sinks to the bottom, which enables animals to survive beneath the ice that forms at the colder surface.

ICEBERG
As an iceberg warms, chunks break off and it floats at different levels in the water, leaving a vertical series of grooves eroded by the waves.

MELTWATER
Rising temperatures in spring melt ice and snow on higher ground, producing a surge of water along rivers. The tips of glaciers also release meltwater.

Water in the Atmosphere

The proportion of water vapour in the atmosphere is known as absolute humidity. It varies with air temperature and pressure, from almost zero to four per cent by volume. Often a more useful measure is relative humidity. This compares the actual amount of water vapour in a given volume of air to the maximum amount which that air could hold. When relative humidity is 100 per cent, the air is completely saturated and cannot hold any more moisture. At this point, no further evaporation can take place. Water also exists in the atmosphere in both liquid and solid states. When water vapour condenses around particles of dust, it forms either tiny droplets or ice crystals, depending on the air temperature. Large masses of droplets or crystals become visible in the form of clouds (see pp.472–81). Another feature of water droplets and ice crystals is that they float easily in even the weakest air currents, and so are continually colliding and merging. If they grow sufficiently large, they fall as either rain or snow. Sometimes, powerful winds create updraughts that blow raindrops upwards into supercooled clouds, where they rapidly freeze, only to fall again as pellets of frozen rain (or hailstones).

SNOWSTORM
Windblown snow collects on the sides of the Grand Canyon, Arizona, USA, during a storm. In low atmospheric temperatures, for example, in winter in temperate regions or at high altitudes, clouds may be made of ice crystals and precipitation takes the form of snow.

ALTOCUMULUS
These mid-altitude clouds occur when a large air mass is forced to rise, leading to cooling and condensation over a wide area. They can be grey or white or both.

CIRRUS
The wispy strands of cirrus clouds are made of ice crystals blown into shape by the wind. They form when an air mass cools at high altitude and becomes saturated, forming ice instead of water droplets.

RAINSTORM OVER BOHOL SEAD
Banks of moisture-laden clouds cause a squall off the southern Philippines. Clouds contain the same mixture of water vapour and droplets as steam.

The Global Water Cycle

Water represents only 0.2 per cent of the Earth's weight, but the global water cycle, or hydrological cycle, is one of the world's most important systems. It is responsible for the continuous circulation of water around the planet. The Sun provides a massive input of energy to the Earth in the form of solar radiation, especially infrared radiation or heat (see p.446). This warms liquid water, which evaporates into invisible water vapour and then dissipates into the atmosphere. About 22 per cent of the solar radiation that reaches the Earth heats oceans, seas, lakes, and rivers, changing liquid water into water vapour. Only pure water evaporates; any dissolved minerals and other substances in it are left behind, mainly as salts in the sea. This process is, in effect, solar-powered distillation (purification) of water. Water vapour is also given off by plants during transpiration, and by animals in exhaled air and perspiration. As water vapour rises into higher, cooler regions of the atmosphere, it condenses to form clouds. Eventually, this returns to the land as precipitation. Here, it is warmed by the Sun – and so the global water cycle continues.

PRECIPITATION
Rain, hail, sleet, snow, frost, and dew form a vital part of the global water cycle. They return water in the atmosphere to the Earth's surface, and in the process release the latent energy stored in water vapour.

BLUE-AND-WHITE PLANET
From space, the Earth appears mainly blue and white. These areas are the visible forms of water, in seas and oceans, and in clouds. Water is the chief weather-determining component of the atmosphere.

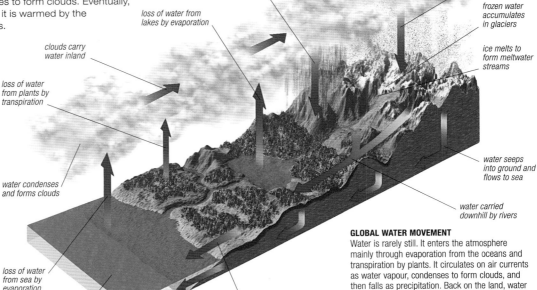

water returns to land as rain

water returns to land as snow

loss of water from lakes by evaporation

frozen water accumulates in glaciers

clouds carry water inland

ice melts to form meltwater streams

loss of water from plants by transpiration

water condenses and forms clouds

water seeps into ground and flows to sea

water carried downhill by rivers

loss of water from sea by evaporation

water returns to sea via rivers and streams

water stored in seas and oceans

GLOBAL WATER MOVEMENT
Water is rarely still. It enters the atmosphere mainly through evaporation from the oceans and transpiration by plants. It circulates on air currents as water vapour, condenses to form clouds, and then falls as precipitation. Back on the land, water moves as rivers and glaciers, and it soaks into the soil and rocks; it is also stored in lakes, oceans, and the tissues of living things.

FRESHWATER SPRING
The Ozark Plateau, USA, has many springs, including Greer Springs, Missouri (above), which releases an average flow of a billion litres (222 million gallons) of water each day.

The Local Water Cycle

Many interconnected local water cycles together make up the global water cycle. On a local scale, water movement is determined by factors such as the amount and type of precipitation, the contours of the landscape, and the geology (the composition and layering of soil and rock). Water can percolate through porous or permeable rocks such as chalks, but not through impermeable types such as clay. So regions with porous soils and rocks tend to have less surface drainage, in the form of streams and rivers, because rain and other precipitation soaks straight down. Percolating water accumulates above poorly permeable layers, saturating rocks and creating groundwater deposits. A rock formation that conducts groundwater is known as an aquifer, which people can draw upon for water wells. The upper level of groundwater saturation is called the water table, and usually rises and falls with varying precipitation through the seasons. Where the surface of the land intersects an aquifer, water may escape from it as a spring. Alternatively, the water can percolate unseen below the surface, through to rivers, lakes, and wetlands.

PAVED LANDSCAPE
Today, vast areas of land are covered with waterproof layers, such as tarmac, paving, and concrete, which greatly alter local water cycles. Water can no longer easily percolate into porous soils and rocks to replenish groundwater supplies. Instead, it flows into drains and channels and rushes away into larger watercourses. This can lower the water table, cause springs to dry up, and deplete local groundwater supplies needed for irrigation and human use. The sudden rush of surface water after a storm may also cause local flooding.

LOCAL WATER MOVEMENT
In a given area, water may exist at the surface, for example, in a stream, marsh, or lake. Other water is underground. In the zone of aeration, air fills the spaces between particles of soil and rock. Beneath this lie zones of temporary and permanent water saturation, which vary according to the rainfall. The water table is the level below which the ground is saturated.

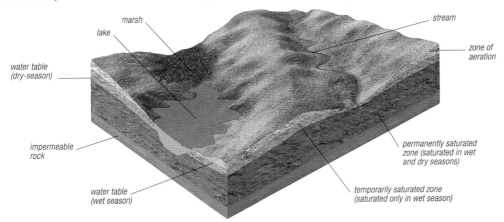

marsh

lake

stream

water table (dry-season)

zone of aeration

impermeable rock

permanently saturated zone (saturated in wet and dry seasons)

water table (wet season)

temporarily saturated zone (saturated only in wet season)

WATER RESOURCES

Every year, about 40,000 cubic km (9,500 cubic miles) of water evaporates from the oceans and falls on land. This yearly flow is about ten times that of the River Amazon, and – in theory – it is enough to support at least five times the current number of people on Earth. Most of the world's fresh water, however, is difficult to reach, and the remainder is unevenly spread. As demand for water grows, the world's resources are coming under increasing strain.

Surface Water

Water supplies come from two major sources. Some is removed from rivers and lakes (surface water), and some from aquifers – natural reservoirs of water in porous rock (groundwater). A small amount is also produced by desalinization, but this is an expensive procedure that relatively few countries can afford. Surface water has the advantage of being accessible, and, because it is quickly replenished, it can be treated as a renewable resource. Unfortunately, it is also easily polluted – a growing problem in regions with high populations and poor hygiene.

Globally, domestic use accounts for less than a tenth of total water consumption, and industry uses about 25 per cent. The biggest consumer by far is agriculture, which takes about 60 per cent, much of it from rivers. Most of this water is for irrigation, and once it is has been used it often evaporates instead of draining back into natural waterways. In arid parts of the world, from the American West to China, irrigation has – at times – brought major rivers to a halt.

Cutting household consumption helps to conserve water, but cutting agricultural consumption can have an even greater impact. By drip-feeding plants with moisture, micro-irrigation can reduce water consumption by more than 250 per cent. However, this is financially feasible only for high-value crops. For large-scale cultivation – such as cereal farming – growing drought-resistant varieties can yield much greater savings.

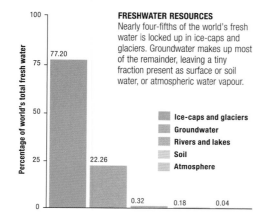

HUMAN-MADE OASIS
A golf course contrasts with desert scenery near Las Vegas, Nevada, USA. The fairways are irrigated from the Colorado River – an example of recreational water use.

FRESHWATER RESOURCES
Nearly four-fifths of the world's fresh water is locked up in ice-caps and glaciers. Groundwater makes up most of the remainder, leaving a tiny fraction present as surface or soil water, or atmospheric water vapour.

Chart: Percentage of world's total fresh water
- 77.20 — Ice-caps and glaciers
- 22.26 — Groundwater
- 0.32 — Rivers and lakes
- 0.18 — Soil
- 0.04 — Atmosphere

Groundwater

Some of the world's earliest civilizations depended on groundwater for their survival, and aquifers now underpin life in many dry parts of the world. Groundwater is harder to reach than surface water but is less prone to pollution, and it is often present in places where surface water is scarce. The drawback is that it replenishes itself very slowly. If too much water is used, the water-table sinks and wells run dry. The drop in the water-table can be precipitous: in southern India, falls of more than 25m (82ft) have been recorded in a single year.

In the American Great Plains, declining water-tables threaten some of the most productive farmland in the world. Here, farming relies on the Ogallala Aquifer, an immense reservoir of fossil melt water dating back to the last ice age. Large-scale use of this water began in the 1970s, but with current usage it should be considered as a nonrenewable resource. Australian farmers face similar problems in the Great Artesian Basin – a huge aquifer where boreholes currently tap about 1.3 billion litres (290 million gallons) a day.

TESTING TIME
In Pakistan, a technician and a farmer check water from a newly sunk artesian well. In such wells, pressurized groundwater flows up to the surface.

Water Wars

When water is in short supply, the potential for conflict grows. Within national borders, this can lead to protracted legal disputes, but when it develops between countries it may even cause wars. More than 250 of the world's largest river basins straddle more than one country, and many experts believe that the threat of "water wars" is likely to increase as the human population grows. According to international law, water cannot be owned outright, but at present there is little to prevent upstream nations from exploiting supplies in ways that harm their neighbours. Likely flash-points for water conflict include the Middle East, southern Africa, and Central Asia – all regions where low rainfall brings about water stress.

In 2003 – the International Year of Fresh Water – a UN report predicted that the average supply of water per person would fall by over a third in 20 years, and that 7 billion people could face a shortage of water by 2050. In 2023, according to the UN, an estimated 2.2 billion people already lacked access to safely managed drinking water. With diminishing water resources, an even bigger challenge will be to ensure that existing supplies are shared fairly.

NATURAL CYCLE
In many parts of India, the monsoon climate means that most of the year's rain falls in a four-month period, after which reservoirs dry out all too quickly.

PLANET EARTH

Life

10–31	The history of the Earth
Threats to biodiversity	116–17
Deserts	276–97
Forests	298–317
Wetlands	318–27
Grasslands and Tundra	328–39
Reefs	396–97

Living organisms have been present on Earth for about 3.8 billion years – or some two-thirds of the Earth's history. During the first 2 billion years, aquatic micro-organisms dominated the planet. As they released oxygen, they changed the composition of ocean waters and the atmosphere, and made them more habitable. Evolving organisms adapted to the changing conditions and spread from seas to land and the air. Their growing presence affected the very nature of the planet, so that the Earth today is the result of myriad intimate interactions between the non-living and the living.

What is Life?

Although some organisms are easy to identify as living, the boundary between the living and the non-living is hard to discern given that many micro-organisms and organic structures, such as viruses or seeds, appear inanimate. However, some key features can help define a living organism, such as its ability to function, grow, change, repair, and reproduce. Energy and raw materials from the environment are required to construct self-sustaining life forms, maintain their function, and carry out essential processes. The one element that can perform all these tasks is carbon. It can combine with other elements, such as oxygen and nitrogen, to build diverse and complex molecular structures, such as the DNA (deoxyribonucleic acid) molecule. DNA encodes the genetic instructions for the development, function, and reproduction of all known living organisms. Even the most primitive and simple life forms – single-celled prokaryotes – have organized structures that can transform energy to regulate their function, build body components, respond to the environment, and reproduce. Many prokaryotes can survive extreme conditions and are thought to have been the earliest life forms to evolve over 3,500 million years ago.

CO-EXISTING SPECIES
All life relies on water. Here, Ostrich, Springbok, Giraffe, and Burchell's Zebra gather at a drying waterhole at Etosha National Park, southwest Africa.

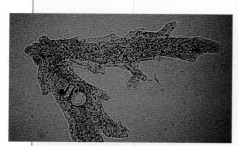

SIMPLE REPRODUCTION
Simple organisms and some plants can produce offspring identical to themselves. The single-celled amoeba shown here can divide into two cells, both of which can feed and grow back to full size, and split again, all in a few hours.

COMPLEX REPRODUCTION
Complex organisms require another member of the same species to reproduce. Many animals care for their offspring.

DOMAINS OF LIFE

The classification of living things has changed considerably in recent decades. Major groups of familiar organisms are now assigned to larger groups called domains. The most primitive, but still diverse organisms, the prokaryotes, form two domains called archaea and bacteria. A third domain, the eukaryotes, comprises the more complex life forms – such as plants, animals, and fungi – as well as numerous less familiar single-celled protists.

Animals
Ranging from ants to whales, animals may seem a distinctive group, but most fundamental animal characteristics are also present in other groups. The most distinctive features are found at the molecular level of various proteins, such as collagen, and in the molecules that bind animal cells together in distinctive tissues, such as epithelia, and help in the transformation of one cell type to another.

INVERTEBRATE / VERTEBRATE

Plants
One of the most diverse groups of organisms known, with over 290,000 species, plants are essential for life on land and some watery environments. Through a process called photosynthesis, they capture light energy from the Sun to build sugars, starches, and carbohydrates, and so produce the basic food substances for life on land.

FLOWERING PLANT

NON-FLOWERING PLANT

Fungi
This domain includes some of the most important organisms for life on land. They break down organic material and make nutrients available for other organisms. They often enter into close symbiotic relationships without which many plants could not flourish.

MUSHROOM

Protists
This vast, diverse group of eukaryotes includes many organisms that are not closely related, making this a grouping born more out of convenience than natural order. It includes flagellates, amoebae, algae, and others. While most are single-celled micro-organisms, some, such as algae, include multicellular forms of considerable size.

GIANT AMOEBA

Bacteria
Single-celled bacteria are often associated with disease, but they occupy an extremely important position in the chain of life. They grow and reproduce rapidly and can live in many different environments, from soil to the guts of animals. They perform beneficial tasks, such as breaking down organic matter, and often form the base of food chains.

BACTERIUM

Archaea
Superficially bacteria-like, archaea can be distinguished by their genetic and chemical makeup, which allows many of them to survive in extreme conditions of temperature, acidity, and salinity. Often referred to as "extremophiles", they may have been some of the first life forms to evolve on Earth.

NATRONOCOCCUS

Diversity

There are at least 9 million species of animal alive today, most of which are insects. The total number of eukaryotes is close to nine million, and then there are unknown numbers of prokaryote species. When fossil records are included, the total number of species rises to hundreds of millions; however, most of these are extinct. This immense diversity is defined using the concept of biological species, according to which, a species is a member of a population that can produce viable young that can further reproduce. Traditionally, species were grouped in a hierarchical system of family, order, class, phylum, and kingdom. However, this has been replaced with a branching (cladistic) system of groups, in which members share unique characteristics resulting from descent through evolution.

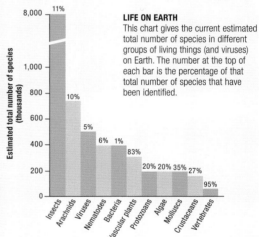

BROWN KIWI
The number of species fluctuates with scientific knowledge. Kiwis were once regarded as two species, brown and spotted, but recent studies suggest that there are actually four.

LIFE ON EARTH
This chart gives the current estimated total number of species in different groups of living things (and viruses) on Earth. The number at the top of each bar is the percentage of that total number of species that have been identified.

Estimated total number of species (thousands)

Group	Value	% identified
Insects	8,000	11%
Arachnids		10%
Viruses		5%
Nematodes		6%
Bacteria		1%
Vascular plants		83%
Protozoans		20%
Algae		20%
Molluscs		35%
Crustaceans		27%
Vertebrates		95%

Evolution

Over time, organisms change through the process of natural selection as a result of reproduction and genetic modifications inherited by their offspring. These modifications are then subjected to environmental influences, such as climate, soils, and biological factors such as predation. Charles Darwin referred to the whole process as "descent with modification". Consequently, those offspring with inheritable modifications that make them better adapted to the environment have an increased chance of survival and passing on their genes. Over generations, the useful characteristics spread and the species itself changes, or evolves. If a species is found in two different habitats, each subgroup becomes adapted to its own habitat, and one species eventually becomes two, in an evolutionary process known as speciation.

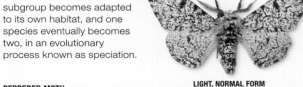

LIGHT, NORMAL FORM

PEPPERED MOTH
The Peppered Moth has pale, speckled wings for camouflage on tree trunks. As industry spread through England in the 1800s, sooty city air darkened tree bark. The few dark moths produced by natural variation were better camouflaged and survived to leave dark offspring, which came to dominate.

DARK, OR CITY, FORM

Extinctions

The vast majority of species that ever lived on Earth are now extinct – they have died out completely. The fossil record shows that creatures such as ammonites, trilobites, and non-avian dinosaurs once thrived in huge numbers but no longer exist. However, background (ongoing) extinction is also a natural part of evolution, as is the turnover of species, which have an average duration of about 5 million years. As environments change, some species adapt and thrive, while others perish. Life on Earth has been disrupted by five major environmental changes. The most famous extinction occurred 66 million years ago at the end of Cretaceous times when non-avian dinosaurs and over 40 per cent of marine invertebrates died out. More drastic was the Permian extinction, 252 million years ago, when 96 per cent of marine species, 70 per cent of terrestrial vertebrate species, and 83 per cent of all insect species died out. Since humans have inhabited the Earth and altered it for their own needs, the rate of extinction has accelerated to between 100 and 1,000 times the background rate, with between one species a day and one every 20 minutes being lost.

EXTINCT IN THE WILD
The Hawaiian Koki'o, a species of tree unique to the island of Molokai, has been extinct in the wild since 1918. Ten grafted plants remain in cultivation, but the species of Hawaiian honeycreeper that fed on the plant's nectar is now extinct.

Biomes and Ecosystems

Ecologists recognize several levels of organization among living things. A biome is a broad category, taking in similar assemblages of plants and animals, across whole regions and continents. For example, the coniferous forest biome, in which the trees have needle-like leaves that can withstand long, cold winters, stretches over vast areas of the northern continents. A biome can be divided into several smaller-scale, more specific communities of plants and animals living in a certain area (known as a habitat). A community could be as large as a huge lake, or as small as a rotting tree stump (a microhabitat). So the temperate forest biome includes communities such as woods of oak, beech, birch, maple, and many other deciduous trees. Biomes are typically named after their dominant plants – for example, the grassland biomes include communities based on Elephant Grass, pampas grasses, cotton-grasses, and others. Biomes are determined mainly by environmental factors such as temperature, rainfall, and wind or water currents. Also important are the topography and soil type. Each biome has not only its characteristic plants, but also typical species of animals. An ecosystem is a functional rather than descriptive term. It refers to the ways living organisms interact with each other and with the physical and chemical factors of their surroundings. An ecosystem is never totally self-contained – a foraging animal can move from a forest to nearby grassland, for example.

SEA OTTER COMMUNITY
Sea Otters live in the forests of kelp that lie off the Pacific coast of Alaska and Canada south to California. The Sea Otter's diet includes red and purple sea urchins, whose main food is the kelp. If the Sea Otter population declines, then the sea urchin population grows unchecked and this prevents the kelp fronds from developing.

GROUND FINCH

CACTUS FINCH

DARWIN'S FINCHES
The Galápagos Islands were colonized a few million years ago by finches, blown from South America. They evolved into several species that exploit different food sources.

Biogeography

The study of how living things are distributed around the world, why they live there, and especially how they got there, is known as biogeography. Each species is adapted to its own habitat, which may be continuous, like a vast tract of grassland, or discontinuous, like a group of islands in the ocean. Species can spread through a continuous habitat relatively easily. They have more difficulty overcoming geographic or habitat barriers – as when a lowland species tries to cross a mountain range, although this may not be a problem to animals that can fly. The geographical history of the Earth, with its changing pattern of bridges and barriers caused by plate movements and volcanic activity, has greatly affected present-day species distribution. In some places, what is now the ocean bed was exposed long ago by lower sea-levels. It was colonized by terrestrial plants and animals, and used as a land bridge for species to spread from their original area to another region with a similar habitat. Other types of geographical barriers encourage new species to evolve. This occurs, for example, when members of a terrestrial species arrive at an oceanic island on a floating mass of matted vegetation. The island has its own specialized habitat, to which some members of the species adapt. They gradually evolve into a new species, distinct from the founder species. Continued island-hopping may lead to a group of related but distinct species that are adapted to different habitats.

WOOLLY MONKEY   **GUINEA BABOON**

OLD WORLD VERSUS NEW WORLD
Evolution of separate populations into different species occurs on all geographic scales. It has produced distinctive sets of monkey species in the New and Old Worlds, such as the Woolly Monkey in South America and the baboons of Africa.

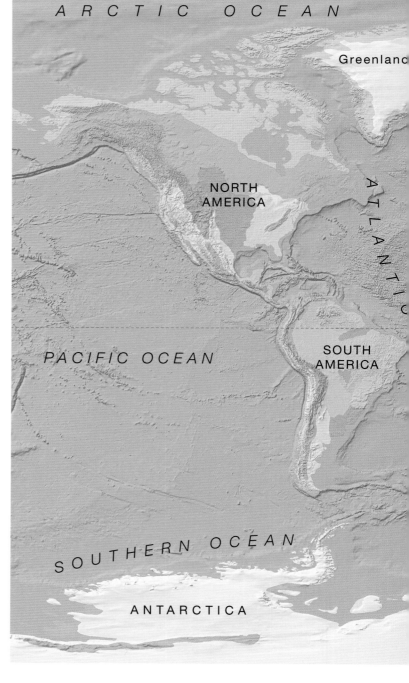

A R C T I C   O C E A N

Greenland

NORTH AMERICA

ATLANTIC

PACIFIC OCEAN

SOUTH AMERICA

S O U T H E R N   O C E A N

ANTARCTICA

ACORN BANKSIA   **COMMON WOMBAT**

ENDEMIC SPECIES
Australia has been an island for over 40 million years. Its wildlife has evolved in unique directions with thousands of endemic species (those that occur naturally nowhere else). These include more than 170 species of marsupials, such as the Common Wombat, and 75 species of the plant genus *Banksia*.

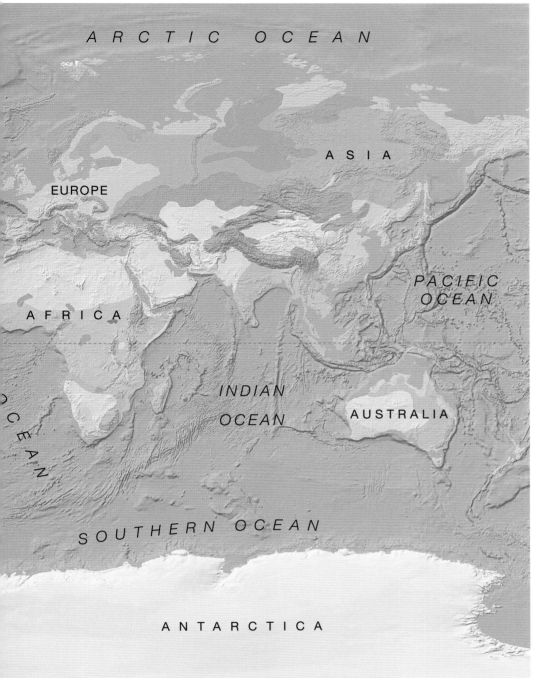

ARCTIC OCEAN

ASIA

EUROPE

PACIFIC OCEAN

AFRICA

INDIAN OCEAN

AUSTRALIA

SOUTHERN OCEAN

ANTARCTICA

BIOMES OF THE WORLD

This map shows the distribution of the Earth's major biomes. The biome distribution shown here is the pattern that would exist if man-made changes to the planet, resulting from urbanization, industrialization, the spread of agriculture, and logging, had not occurred. Only the major wetland areas are shown here, as the rest are too small to show up on a map of this scale, and the size of many of them varies with the season. Similarly, only the largest lakes and inland seas have been shown, and no rivers have been mapped.

TROPICAL FOREST	TUNDRA
TEMPERATE FOREST	WETLAND
CONIFEROUS FOREST	DESERT
TROPICAL GRASSLAND	MOUNTAIN
TEMPERATE GRASSLAND	POLAR

Nutrient Cycles

Certain chemicals are vital to living things, being common constituents of the complex organic molecules found in living cells. One of the most important is the element carbon, on which all life is based. Other important chemicals include nitrogen and phosphorus for plant growth, and the metals magnesium (which is part of the pigment chlorophyll that captures sunlight energy in plants) and iron (found in the blood of many animals). But these chemicals are finite and must be reused. So the Earth can be regarded as a giant self-contained ecosystem where the same nutrients are recycled, going round on an endless variety of pathways through countless organisms. In the carbon cycle, carbon dioxide from the air is incorporated into carbon compounds in the living tissues of plants and phytoplankton by photosynthesis. The carbon is subsequently returned to the atmosphere as a by-product of respiration by the plants and phytoplankton, and the animals that ate them. The carbon cycle also includes important inorganic components and processes (see below).

NITROGEN FIXATION

Some species of plant, such as the acacias shown here, have nitrogen-fixing bacteria living in their roots. The bacteria take nitrogen gas from the air and reduce it to nitrate, a nutrient for the plant.

THE CARBON CYCLE

The pathways by which carbon is cycled through ecosystems are complex. As organisms die and their cells rot, some of the carbon in them enters the soil on land or the sediments on the sea floor. Here it can be used by micro-organisms, or taken in as mineral substances such as carbonates through roots, for new plant growth. In some cases, carbon is transformed into fossil fuels such as coal or oil. In addition to the natural routes, which include volcanic eruptions, humans add carbon dioxide (CO_2) to the atmosphere, mainly through the burning of fossil fuels.

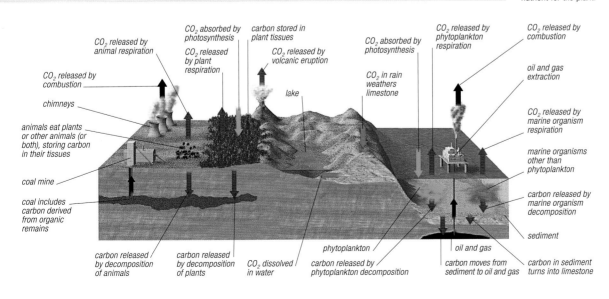

CO_2 released by animal respiration

CO_2 absorbed by photosynthesis

CO_2 released by plant respiration

carbon stored in plant tissues

CO_2 released by volcanic eruption

CO_2 absorbed by photosynthesis

CO_2 released by phytoplankton respiration

CO_2 released by combustion

CO_2 released by combustion

chimneys

CO_2 in rain weathers limestone

lake

oil and gas extraction

animals eat plants or other animals (or both), storing carbon in their tissues

CO_2 released by marine organism respiration

coal mine

marine organisms other than phytoplankton

coal includes carbon derived from organic remains

carbon released by marine organism decomposition

sediment

carbon released by decomposition of animals

carbon released by decomposition of plants

CO_2 dissolved in water

phytoplankton

carbon released by phytoplankton decomposition

oil and gas

carbon moves from sediment to oil and gas

carbon in sediment turns into limestone

CARBON MOVEMENT

PHOTOSYNTHESIS

WEATHERING AND EROSION

HUMAN CARBON TRANSFORMATION

THREATS TO BIODIVERSITY

Although no one knows how many species of living things exist on the Earth, one fact is certain: the Earth's biodiversity – or biological richness – is currently undergoing a steep decline. According to some estimates, more than 5 per cent of the planet's species are disappearing each decade, which is the highest rate for 66 million years. Mass extinctions happened long ago on several occasions, triggered by natural events. In today's case, humans are largely to blame.

A Downward Trend

The current decline in biodiversity affects life on all fronts. According to the International Union for the Conservation of Nature and Natural Resources (IUCN), nearly one in four species of mammal is under threat, while the figure is one in eight for birds. Plants fare less badly, but even so more than 7,000 species are listed as endangered. However, these figures represent only the tip of the iceberg, because data are far from complete. Every one of the world's 6,495 known mammal species has been assessed, for example, yet less than 0.1 per cent of invertebrate animals, which number more than 1 million species, has been evaluated in the same detail.

The forces threatening biodiversity are much easier to establish. Habitat destruction is by far the most important cause, because it sweeps away not only individual species but also the ecosystems on which they depend. Habitats can also be disrupted by introduced species, by pollution, and by the direct exploitation of living things – particularly through hunting and collecting. Pollution can have localized effects, but it can also cause habitat disruption on a worldwide scale.

LIFE IN ISOLATION
Giant Tortoises evolved on remote islands, such as the Galápagos and Seychelles. Hunting, and competition from introduced animals, have made many island races extinct.

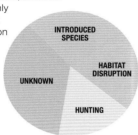

CAUSES OF RECENT EXTINCTION
Three main factors are behind the more than 400 cases of animal extinction, dating back to the year 1600. In the last 50 years, habitat destruction has become the most important cause of biodiversity decline.

(pie chart labels: INTRODUCED SPECIES / HABITAT DISRUPTION / UNKNOWN / HUNTING)

Biodiversity Hotspots

For climatic and geographic reasons, biodiversity is unevenly spread. Some large regions, such as the Arctic, contain relatively few species, while some much smaller regions boast considerable biological riches. These areas, known as biodiversity hotspots, include ecologically isolated parts of continents, as well as islands such as the Galápagos. Together, two dozen hotspots – making up less than a fiftieth of the Earth's land surface – are home to over a third of the world's vertebrate animals and flowering plants.

Hotspots are concentrated in the tropics, where habitat destruction can do disproportionate damage. Tropical rainforests and coral reefs are

particularly vulnerable, and are home to many declining species. But the fall in biodiversity is not just a tropical problem: after two centuries of economic growth, many of the world's industrialized countries have pushed their natural inhabitants to the very edge of survival.

Preserving Biodiversity

Biodiversity matters because it is a reservoir of resources on which we all depend. Today's crop plants and farm animals originally came from the wild, and their wild relatives still contain a huge reserve of potentially useful genes. In the past, plants have supplied us with hundreds of useful drugs, and it is likely that they will offer many more in the future. A high level of biodiversity is generally believed to stabilize ecosystems and to provide vital environmental services, such as water purification and soil formation. Without these, human survival may be put at risk.

Where species are facing the immediate risk of extinction, emergency methods can be used to help them survive. These include breeding species in captivity, so they can be released at a later date. However, this form of protection is extremely expensive

WILDLIFE CRISIS
In the last 30 years, the number of Black Rhinoceros has fallen by 96 per cent – a decline that is entirely due to unauthorized hunting.

and raises difficult questions about which species should be selected. It is easier, for example, to raise funds for saving tigers than vultures, even though vultures play an equally important part in natural food chains. Most ecologists believe that the key to stabilizing biodiversity lies in preserving fully functioning habitats. Many also advocate that hotspots should have priority when resources are limited.

ENDEMIC SPECIES
The Cirio or Boojum Tree, from Baja California, Mexico, is a typical endemic species – one that is found in a small geographical area and nowhere else.

MANY VARIETIES
This collection of crickets and katydids is from one of the world's richest habitats – the tropical forests of Costa Rica. Despite decades of study, new species are discovered every year.

LAND

LAND

MOUNT ST. HELENS, CASCADE RANGE
In May 1980, a violent eruption tore apart the
snow-capped peak of Mount St. Helens in
Washington State. Such events are a reminder
of the hugely powerful, and often devastating,
internal workings of the Earth.

MOUNTAINS AND VOLCANOES

The Earth's rocky elevations, mountains and volcanoes, have an enduring fascination. Mountain belts form imposing barriers and are liable to violent extremes of weather. They are also associated with catastrophic events such as landslides and avalanches. Until recently, some people saw mountains and volcanoes as the homes of wrathful gods, who vent their anger without warning, shaking the ground, and spewing fire, rocks, and ash into the air. But mountains and volcanoes are just the most visible sign of the tectonic forces shaping the Earth – elsewhere fault systems and earthquake zones follow the outlines of the tectonic plates themselves, and igneous intrusions mark areas where molten material has penetrated existing rock. Since scientific investigation of these phenomena began in the 18th century, they have lost many of their superstitious associations, but our fascination with them continues, and they remain impressive reminders of the often spectacular power of Earth's continuing evolution.

Fault Systems

36–37 The core

86–87 Tectonic plates

88–89 Plate boundaries

Living with earthquakes 130–31

Mountains 136–53

Ocean tectonics 386–89

Tsunamis 433

Over geological time, the relentless movement of the Earth's tectonic plates puts the brittle rock of the crust under so much strain that it buckles, fractures, and is pushed deep down into the mantle. When rock fractures, huge blocks slip past one another. These cracks in the Earth's surface are called faults, and often extend through the crust for many kilometres. Many faults are the result of past movement and are currently inactive. However, many fault systems are still active, and massive amounts of energy can be released in a few seconds as rocks move past each other in an earthquake.

Faults and Joints

Faults are cracks in rocks across which there has been displacement. They range from tiny fissures to sets of interconnecting faults that are many kilometres long. Some active faults move in a slow, continuous manner called creep. More often, they judder in a series of spasmodic jumps, which can release energy in the form of an earthquake (see opposite page). Over millions of years, the cumulative effect of countless small movements along the fault can displace rocks across a fault a substantial distance up, down, or horizontally. Not all rock fractures are faults: joints are cracks along which there is no evidence of movement. These are usually caused by tectonic activity, but may also be the result of erosion (see pp.92–93), which strips away the overlying layer of rock, releasing pressure and allowing the surface of rocks to expand and crack. They can also be formed as igneous rock cools. Joints are exploited by the processes of weathering (see pp.90–91), which may exaggerate the shape of the original fracture. They usually occur close to the surface.

LIMESTONE JOINTS
Sedimentary rocks, such as this limestone in England, have particularly marked joints. When the limestone is exposed at the surface, its strong pattern of criss-cross joints is further etched by acidic rainfall. Such striking formations are called limestone pavements.

MEASURING EARTHQUAKES

Earthquakes are recorded on a seismograph, which charts the magnitude of each vibration. They were traditionally measured on the Richter Scale. Each step in the scale, which begins at 0 and has no upper limit, represents a 30-fold increase in the energy released by an earthquake. The greatest magnitude recorded, 9.5, was from the 1960 Chilean earthquake. Many scientists now use the moment-magnitude scale, which combines seismograph readings with measurements of rock movement.

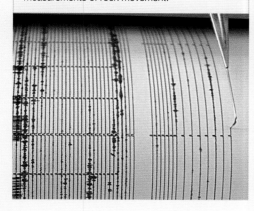

FAULT PLANE
The huge uplift of rock on one side of a fault over a period of thousands of years has created this dramatic precipice in Greece. Scratch marks called slickensides can be seen on the face of the cliff.

THINGVELLIR RIFT, ICELAND
This exposure of fault lines parallel to the Thingvellir Rift (see right) provides dramatic evidence of tectonic activity. The rift lies on the boundary separating the American and Eurasian plates.

CHARLES RICHTER

In 1935, the American seismologist Charles Richter (1900–85) devised the scale for measuring earthquakes that bears his name. Before this, seismologists had been able to compare earthquakes only by using the Mercalli scale, a rule-of-thumb guide to the severity of their effects. The Richter Scale, devised with Beno Gutenberg (1880–1960), measured the magnitude of each earthquake precisely, from the variations in movement it produced on a seismograph (see below). Richter also mapped out the zones in North America that were vulnerable to earthquakes, but he was scathing about the idea of predicting them.

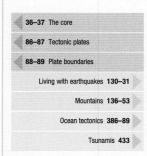

STRIKE-SLIP FAULT, NEVADA, USA
The displacement of the two sides of this right-lateral strike-slip fault can be seen in the white areas on either side of the crack, which were once continuous vertical layers.

FAULT MOVEMENT
Faults are classified by their direction of movement, which reflect surrounding tectonic stresses and whether they shorten or lengthen the original surfaces.

Types of Fault

Faults are created by the compression or extension of the Earth's crust as its tectonic plates move (see pp.88–89). The slope or angle of a fault to the horizontal is called the dip. Faults that show vertical movement are called dip-slip faults, and may be either normal or reverse. Normal faults occur where tension in the Earth's crust pulls the rock apart, and allows a block to slip down the fault plane, which is the surface along which the crack occurs. Parallel sets of normal faults form rift valleys (see p.138), and are associated with divergent plate boundaries. Reverse faults, known as thrust faults if they are at a shallow angle, occur where compression in the crust pushes one block of rock up over another. If two blocks slide past each other horizontally, the fracture is classified as a strike-slip fault, an example of which is the San Andreas Fault (see pp.124–25). If the opposite block in a strike-slip fault moves left, it is left-lateral; if it moves right, it is right-lateral. If a strike-slip fault is combined with compression or tension, the blocks can slide diagonally, creating an oblique-slip fault.

FAULT LINE, UZBEKISTAN
The dramatic effects of fault displacement on the landscape are visible in this escarpment on the outskirts of Tashkent in Uzbekistan, which is the result of ancient earthquakes.

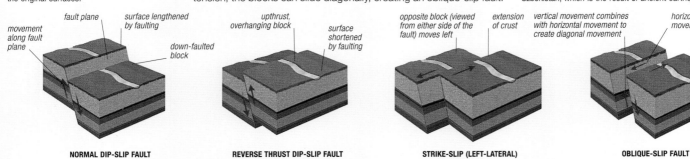

movement along fault plane — fault plane — surface lengthened by faulting — down-faulted block

NORMAL DIP-SLIP FAULT

upthrust, overhanging block — surface shortened by faulting

REVERSE THRUST DIP-SLIP FAULT

opposite block (viewed from either side of the fault) moves left — extension of crust

STRIKE-SLIP (LEFT-LATERAL)

vertical movement combines with horizontal movement to create diagonal movement — horizontal movement

OBLIQUE-SLIP FAULT

Earthquakes

A sudden slip on a fault releases energy in the form of an earthquake, which radiates outwards as seismic waves (see p.36). The Earth's major earthquake zones coincide with faults between tectonic plates. The immense forces generated as these plates grind together trigger most large earthquakes. This happens as the rocks on either side of a fault lock, and stress builds up. The fault eventually ruptures, sending shockwaves shuddering through the planet. The Pacific Ocean's subduction zones are the source of 80 per cent of the world's earthquakes. Almost every day, several hundred minor earthquakes occur around the world. Massive disturbances, such as the 9.1 magnitude Indonesian earthquake of 2004, which triggered tsunamis across the Indian Ocean and killed over 230,000 people, are rare. However, there have been 10 earthquakes of 8.6 and higher magnitude since 1950.

EARTH MOVEMENT
This radar image shows the ground displacement from an earthquake. The closely packed contours near the black line of the fault indicate the greatest movement.

FAULT SYSTEM PROFILES

The pages that follow contain profiles of a selection of the world's most significant fault systems. Each profile begins with the following summary information:

TYPE Normal dip-slip, reverse dip-slip, reverse thrust dip-slip, strike-slip (left-lateral), strike-slip (right-lateral), or oblique slip

LENGTH/AREA Extent of fault line or complex

ACTIVITY Active or inactive

LAND

NORTH AMERICA *west*

San Andreas Fault

LOCATION Extending from Cape Mendocino, northern California, to the Gulf of California, western USA

TYPE Strike-slip (right-lateral)

LENGTH 1,200km (800 miles)

ACTIVITY Active

Slicing across California's coastal region, the San Andreas Fault is one of the Earth's most famous faults. It is a strike-slip fault in which the rocks move sideways in opposite directions, although, because it forms a plate boundary, the San Andreas is also a transform fault (see p.88). To the west of the fault is the Pacific Plate, which extends from the edge of California almost as far as the eastern edge of Asia. To the east is the North American Plate, which forms the bulk of the continent. As the Pacific Plate rotates, coastal California is sliding slowly northwest past the rest of North America. Over the past 20 million years the Pacific Plate has moved about 560km (350 miles), relative to North America, averaging about 1cm (³/₈in) a year. Plate movement seems to be accelerating, and over the past century the fault has shifted almost 5cm (2in) each year. The San Andreas Fault is not one long crack in the Earth's crust, but consists of four major segments and several minor sections. Along some segments of the fault, the strain of the plate movement is released as minor tremors. In major sections, the

FLOODED FAULT
Water exploits the cracks in the landscape created by faults. Rivers flow along them and lakes form in them. In the case of Tomales Bay, the sea has flooded an inlet created by the San Andreas Fault.

plates lock together for many years, allowing stresses to build up that are eventually unleashed as massive earthquakes. These disturbances typically occur in one section of the fault at a time. In 1857, a sudden

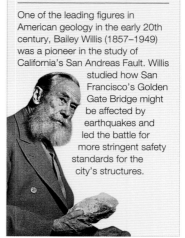

BAILEY WILLIS

One of the leading figures in American geology in the early 20th century, Bailey Willis (1857–1949) was a pioneer in the study of California's San Andreas Fault. Willis studied how San Francisco's Golden Gate Bridge might be affected by earthquakes and led the battle for more stringent safety standards for the city's structures.

SLIPPING SIDEWAYS
Images from space reveal the gash across the land created by the San Andreas Fault as the Pacific and North American plates move in opposite directions.

THE LAST BIG ONE

At 5.12 a.m. on 18 April 1906, the Pacific Plate lurched approximately 6m (20ft) northward along a 430-km- (267-mile)- long stretch of the northern San Andreas Fault. In just a few seconds, centuries of pent-up energy were released as a massive earthquake that devastated San Francisco. Although the earthquake destroyed many buildings, the worst damage resulted from the fires that raged through the city in the aftermath.

movement along the segment of the fault in the Transverse Ranges, which separate Central and Southern California, resulted in a severe earthquake that opened a crack 350km (218 miles) long. In 1906, movement in the northern section of the fault caused an earthquake that is estimated to have measured 8.3 on the Richter Scale, and which devastated San Francisco (see panel, left). An earthquake, with its epicentre at Loma Prieta, rocked San Francisco again in 1989. It is feared that pressure is building up in the southern section of the fault, and when it finally unlocks it will unleash an earthquake known as the Big One. This earthquake is now expected to happen before the year 2032. Most of California's population lives close to the San Andreas Fault, and some communities are built right over it.

EARTHQUAKE WATCH

In 1985, the US Geological Survey predicted that an earthquake would strike the small Californian community of Parkfield in 1993. The village was fitted with equipment to monitor the potential disturbance, although it did not happen. However, Parkfield remains the most closely observed earthquake zone in the world, as scientists probe the ground to measure factors such as strain in the rocks, heat flow, and geomagnetism.

VISIBLE FAULT LINE
The San Andreas Fault scars the Californian landscape. In addition to the very obvious fault line itself, it is often possible to see hills, valleys, and streams truncated and offset on opposite sides of the fault.

NORTH AMERICA *west*

Basin and Range

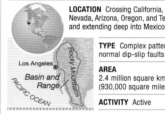

LOCATION Crossing California, Utah, Nevada, Arizona, Oregon, and Texas, and extending deep into Mexico

TYPE Complex pattern of normal dip-slip faults

AREA 2.4 million square km (930,000 square miles)

ACTIVITY Active

The Basin and Range country of the southwest USA and northern Mexico has one of the most distinctive landscapes in the world. More than 100 long, roughly parallel mountain ranges run from north to south through the region, each separated from the next by a wide, flat, dry desert basin. On each side, the mountain ranges are bounded by faults that slope into the valleys. This extraordinary landscape is the result of extensive stretching of the North American Plate, which began about 20 million years ago. The crust was pulled, cracked, and thinned, and it is now almost twice its original width. As this happened the crust fractured, and fault

DEATH VALLEY, CALIFORNIA
The lowest point in the Basin and Range – and the US – is Death Valley, at 86m (282ft) below sea level. Surrounded by sharp mountain peaks, the basin contains several extremely active faults.

after fault opened up. Basins were created as the land dropped down between parallel faults. As the basin blocks slid down, the mountains were pushed up, reaching heights of 3,500m (11,500ft) in places. Although there are a variety of fault types in the region, the majority of the faults are normal dip-slip faults, created as the Earth's crust was wrenched apart. The faults are all steep, typically descending into the crust at an angle of 60° but flattening at depth. As the mountain ranges rose, they were immediately affected by weathering and erosion (see pp.90–93). Water, wind, and ice have worn down the peaks and deposited debris in the intervening valleys. Sometimes, so much debris has collected in the basins that the solid bedrock is buried thousands of metres below the surface.

LINEAR VALLEYS
In the Great Basin of southern Nevada, the pattern of alternating parallel mountain ridges and deep basins is clearly visible.

SNOW-CAPPED SIERRA NEVADA
The uplifted blocks forming California's Sierra Nevada tower up to 3,300m (11,000ft) above Owens Valley to the east.

LAND

Midland Valley

LOCATION Running from east to west between the Southern Uplands and Scottish Highlands, Scotland

TYPE Strike-slip (right-lateral)

LENGTH 90km (56 miles)

ACTIVITY Inactive

The Midland Valley is bounded on either side by two very ancient strike-slip faults that divide Scotland in two. On the southern side of the valley lies the Southern Upland Fault, while to the north is the Highland Boundary Fault. The origin of these faults dates back more than 540 million years to a time when northern Scotland was part of the ancient continent of Laurentia and was separated from the rest of the British Isles, which were

situated on the ancient continent of Avalonia. During the course of the next 200 million years, Laurentia moved gradually towards Avalonia, carrying northern Scotland towards the rest of Britain. As the continents converged, the enormous pressure formed the Scottish mountains in an event called the Caledonian Orogeny (see p.138). About 400 million years ago, Laurentia began to swing away to the southeast, ripping Scotland apart again, and creating not only the two great faults on either side of Midland Valley but also another

LOCH LOMOND
Seen from Duncryne Hill on its southern shore, Loch Lomond is the largest of Scotland's lochs, and is crossed by the Highland Boundary Fault.

fault north of the Highlands, Glen Mor (also known as Great Glen), which forms the long basin filled by Loch Ness.

COAL MINING

About 320 million years ago, Scotland sat on the Equator, and had a tropical climate. Swampy forests flourished on the ancient land where the Midland Valley now lies. Over hundreds of millions of years, the remains of the plants from these forests compressed to form Scotland's richest coal deposits, which were exploited for years. Coal mining in Scotland is now over and the last mine shut down in 2022.

HIGHLAND BOUNDARY
The Scottish Highlands rise along the northwestern margin of the Midland Valley, where the Highland Boundary Fault creates a dramatic shift in the region's geology.

Moine Thrust

LOCATION Extending northwest from the south of the Isle of Skye to the Moine Peninsula, Scotland

TYPE Reverse thrust dip-slip

LENGTH 180km (112 miles)

ACTIVITY Inactive

NORTHWEST HIGHLAND MOUNTAIN
At 1,062m (3,484ft) high, An Teallach is one of Scotland's many spectacular mountains. It is the site of intense geological study, because layers of the Moine Thrust belt are well-exposed here.

Thrust faults are shallow reverse faults that are created as Earth's upper crust is squeezed, forcing one block up to overhang another. The Moine Thrust is not a single thrust fault, but a formation called a thrust belt. These are created when plate movement forces layers of crust up and over each other, and then pulls them back, a motion that happens again and again. The discovery of the Moine Thrust in 1907 was a milestone in the history of geology, as it was the first thrust belt to be identified. It formed between 410 and 430 million years ago, as Scotland was compressed by tectonic movement. Thrust belts are now recognized as features at the edges of mountain ranges all around the world.

North Sea Basin

LOCATION Lying beneath the North Sea between the British Isles and Scandinavia

TYPE Normal dip-slip

LENGTH 250km (155 miles)

ACTIVITY Inactive

As geologists prospected the rocks beneath the North Sea for oil in the 1960s, they discovered not only a rift valley but a whole series of faults. Rift valleys form when two normal faults pull in opposite directions, so that as the Earth's crust stretches apart, a block drops down between them (see p.138). The rift valley under the North Sea dates from the Jurassic Period, about 200 million years ago, when the supercontinent of Pangea began to break up, opening up the Atlantic Ocean and the North Sea. It overlies a complex series of much older faults

dating from about 50 million years earlier. By studying these structures using seismic surveys that probe into the solid bedrock, geologists are discovering more about how continents diverge, and also the way that rich oil deposits accumulate in down-thrown fault features such as those in the North Sea Basin.

DRILLING FOR OIL
The rich oil fields beneath the North Sea are the largest natural oil and gas deposits in Europe.

EUROPE *eastern*

North Anatolian Fault

LOCATION Extending across northern Turkey and beneath the Sea of Marmara

TYPE Strike-slip (right-lateral)

LENGTH 1,000km (600 miles)

ACTIVITY Active

The North Anatolian Fault is one of the most energetic earthquake zones in the world. Turkey is set on a minor tectonic plate that is being squeezed westwards as the Arabian and Eurasian plates move together. The geological setting is similar to America's San Andreas Fault, with the Anatolian Plate grinding past the Eurasian Plate at 1–20 cm (½–8 in) each year and triggering earthquakes as it moves. Since the 1939 Erzincan earthquake, which killed 33,000 people, there have been nine earthquakes measuring over 7.0 on the Richter Scale and located progressively westwards. Seismologists studying this pattern believe that earthquakes occur in "storms" over a number of decades, and that one earthquake triggers the next. By analysing the stresses caused along the fault by each earthquake, they were able to forecast a disturbance that struck the town of Izmit with such devastating effect in August 1999. More recently, on 6 February 2023, an earthquake measuring over 7.8 on the Richter Scale struck southern and central Turkey and northern and western Syria.

SEEN FROM SPACE
The great scar across the Turkish landscape made by the North Anatolian Fault is clearly visible when viewed from space.

IZMIT EARTHQUAKE DEVASTATION
In 1999, an earthquake measuring 7.0 on the Richter Scale struck on the North Anatolian Fault. It killed 17,000 people, injured 50,000, and left 500,000 homeless.

EUROPE *central*

Rhine Rift

LOCATION Extending through the middle of northwest Europe, from the Swiss Alps to the North Sea

TYPE Reverse thrust dip-slip

LENGTH 1,320km (820 miles)

ACTIVITY Active

The Rhine Rift is a striking reminder of the tectonic forces that formed the European continent. Rift valleys are usually created when continents diverge, causing land to drop down within the gap as the crust is pulled apart. However, the Rhine Rift is thought to have formed as the two halves of what is now northwestern Europe knitted together between 40 and 50 million years ago. As this happened, a long strip of rock called a graben (from the German word for "ditch") dropped between the two continental massifs: the Vosges in France and the German Black Forest. It was initially assumed that the Rhine Rift was inactive. However, scientists have recently detected signs of movement beneath the surface. Faults within stable continental interiors rarely rupture, but if they do, the subsequent earthquake can be sudden and extremely violent. Northwest Europe's worst earthquake occurred in 1356 in the Upper Rhine Valley, destroying the Swiss city of Basle, and knocking down buildings as far as 200km (125 miles) away. Some geologists believe the cause of the earthquake was not the Rhine Rift, but a fault in the Alps. However, Basle may yet fall victim to an earthquake unleashed by the Rhine Rift, as an active fault has been found southwest of the city. In 1992, an earthquake along the Peel Fault, which is further down the Rhine, rocked the town of Roermond in the Netherlands.

FAULTED FOREST
Germany's Black Forest lies on a horst block, which is a section of rock that has been left upstanding after the land on either side of it faulted and dropped downwards.

KAISERSTUHL, GERMANY
To the north of Basle is Kaiserstuhl, which is an ancient complex of volcanoes. These were created along the plate margin that also caused the Rhine Rift to form.

AFRICA *east*

Great Rift Valley

LOCATION Extending from the southern Red Sea through East Africa to Beira in Mozambique

TYPE Normal dip-slip

LENGTH 6,400km (4,000 miles)

ACTIVITY Active

Great Rift Valley · Lake Victoria · ATLANTIC OCEAN

One of the longest fault systems in the world, the Great Rift Valley is part of a huge set of fissures in the Earth's crust called the East African Rift System, which threatens to split Africa in two. The Great Rift Valley is the primary

STEEP VALLEY CLIFFS
At Losiolo Maralal, Kenya, the valley floor has dropped away to leave sheer escarpments above the Suguta Valley, one of the hottest places on the Earth.

branch of the system, which extends from Jordan in the north, through the Dead Sea and the Red Sea (see p.418), and along the length of East Africa to the mouth of the Zambezi River (see p.219). The average width of the valley is 50km (30 miles), although at its widest point in the Danakil Desert it is nearly 480km (300 miles) across. Its steep walls rise 900m (2,955ft) above the valley floor, but in some places, such as Mau Escarpment, Kenya, the cliffs soar to 2,700m (8,860ft). The valley splits into two branches in Kenya and Tanzania. The western arm extends from the north shore of Lake Nyasa (see p.235) along the western borders of Tanzania, forming an arc that is dominated by Lake Tanganyika (see p.236). The eastern branch includes lakes Manyara (see p.235) and Natron. Most of the lakes in the Rift Valley are extremely deep, with Tanganyika descending to 1,471m (4,800ft). The East African Rift System lies along the boundaries of three

EARLY HUMANS

In 1974, American paleontologist Don Johanson found some of the most important hominid remains ever discovered, an almost complete skeleton of a 3-million-year-old *Australopithecus afarensis*. The skeleton, nicknamed Lucy, is that of a woman drowned. It was preserved in volcanic mud, where minerals from the surrounding liquid slowly replaced the calcium in the bones. The land turned to desert, and the ground was then washed away by rain to reveal the skeleton millions of years later.

RIFT VALLEY WILDLIFE
The Great Rift Valley is home to a huge array of endemic species of African wildlife, among which are leopards.

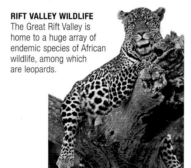

RED SEA CORAL REEFS
The Red Sea was formed by the divergence of the African and Arabian plates. Stunning coral reefs have formed in the sea's warm shallows.

tectonic plates, the Arabian, the African–Nubian, and the African–Somalian. It started to form when the plates began to diverge about 100 million years ago. As the Earth's crust stretched, volcanoes erupted; they dot the length of the rift today. About 20 million years ago a series of huge faults opened, and the land dropped between the three plates to form the Great Rift Valley. The plates meet at a triple junction under the Afar Triangle in Ethiopia, where the Red Sea merges with the Gulf of Aden. This may be the beginning of the next major ocean, as hot mantle material doming under the junction splits the plates apart. It is thought that the plates will eventually separate, and that the Horn of Africa will break off from the rest of the continent to form an island. Recent activity includes a magnitude 6.8 earthquake on 5 December 2005, in Lake Tanganyika, the oldest and deepest of the Rift Valley lakes.

THE LEAKEY FAMILY

No name is more closely linked to the wealth of early hominid discoveries in the Great Rift Valley than that of Leakey. Louis Leakey (1903–72) and his wife Mary (1913–96) discovered the first skull of an *Australopithecus* in Olduvai Gorge, Tanzania, in 1959. They also discovered that the remains of the earliest humans *Homo habilis* and *Homo erectus*. Later, Mary Leakey found 3.5-million-year-old footprints in Tanzania proving that these human ancestors walked upright. The couple's son, Richard, has also made significant discoveries in the same field.

VOLCANIC LANDSCAPE
Where the Great Rift Valley meets the Red Sea, at the Ardoukoba Volcanic Field in Djibouti, the terrain is dominated by volcanoes. A fault-bounded rift extends seaward in the centre of the picture.

GREAT RIFT VALLEY AT THE RED SEA
This image shows the northern edge of the
Great Rift Valley and its extension between
the African and Arabian plates. The resulting
rifts have been filled by the Gulf of Suez (left)
and the Gulf of Aqaba (right).

FALLING DOWN
New Zealand's 2011 Christchurch earthquake had a magnitude of 6.3. It caused extensive damage to the city's centre and toppled two bell towers on the Christchurch Catholic Cathedral.

LIVING WITH EARTHQUAKES

Earthquakes are the most dangerous of all natural hazards. Unlike storms and volcanic eruptions, they strike within seconds, giving no opportunity for escape. Quite apart from the toll in lives, they cause damage that can take years to repair. Earthquakes cannot be prevented, but current research aims to make life safer in earthquake zones. The chief goal – which still remains elusive – is a reliable way of forecasting exactly when and where future earthquakes will strike.

Predicting Earthquakes

Forecasting earthquakes is notoriously difficult, because they are so complex. Hundreds of variables are involved, and no two fault systems are exactly the same. However, two kinds of data are particularly revealing: the past record of earthquakes in a given area, and the present build-up of stress within a fault system's rocks.

The past record gives average frequencies for earthquakes, which can be used as a guide to future earthquakes. For example, if there has previously been a major earthquake every 100 years, there is a 10 per cent probability of one in any given decade, if earthquakes are randomly spaced. But research into fault-system stress suggests that earthquakes are not random. Far from relieving stress, one earthquake can actually increase the likelihood of another one elsewhere. In 1999, this discovery was successfully used to predict a major aftershock in Turkey, three months after the devastating earthquake at Izmit, which claimed 25,000 lives (see p.127).

LARGEST EARTHQUAKES

The ten most powerful earthquakes of the last 25 years are here ranked by moment magnitude – a measurement that reflects seismic energy more accurately than the Richter scale.

Location	Date	Magnitude
Off coast of North Sumatra	26 December 2004	9.1
Off east coast of Honshu, Japan	11 March 2011	9.0
Offshore Maule, Chile	27 February 2010	8.8
North Sumatra, Indonesia	28 March 2005	8.6
Southwest Sumatra, Indonesia	11 April 2012	8.6
Southwest Sumatra, Indonesia	12 September 2007	8.5
Southern Peru	23 June 2001	8.4
Illapel, Chile	16 September 2015	8.3
Okhotsk Sea, Russia	24 May 2013	8.3
Tokachi, Japan	25 September 2003	8.3

Earthquake Proofing

The Izmit earthquake vividly illustrated a well-established fact: it is not earthquakes that kill, but buildings. Earthquakes have a relatively minor impact in the open – many people even fail to notice them – but when they shake rigid structures, the results can be catastrophic. In Izmit (see below), where the initial shake had a magnitude of 7.4, many new apartment blocks collapsed completely,

SELECTIVE DESTRUCTION
The Izmit 1999 earthquake in Turkey destroyed many new apartments, and reinforced the need for better-designed buildings in earthquake-prone areas.

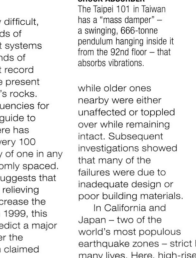

SHOCK ABSORBER
The Taipei 101 in Taiwan has a "mass damper" – a swinging, 666-tonne pendulum hanging inside it from the 92nd floor – that absorbs vibrations.

while older ones nearby were either unaffected or toppled over while remaining intact. Subsequent investigations showed that many of the failures were due to inadequate design or poor building materials.

In California and Japan – two of the world's most populous earthquake zones – strict building codes have saved many lives. Here, high-rise buildings are designed to be slightly elastic, so that the energy from a shock is dissipated harmlessly, instead of fracturing supporting columns and floors. Foundations are fitted with mechanical dampers – structures that function in much the same way as shock-absorbers in cars. Using this kind of technology, skyscrapers such as San Francisco's Transamerica Building have emerged from major shocks unscathed.

Almost as important is the prevention of fire. After San Francisco's 1906 earthquake, fire swept through its central business district (see p.124); similarly, in 1923, thousands of houses were burned in Tokyo, causing almost as much loss of life as the earthquake itself. Although the use of non-flammable building materials greatly reduces this danger, risks still remain. In 1995, more than 5,000 people died when an earthquake hit Kobe (see p.132) – a city widely believed to be one of the most earthquake-proof in the world.

Tsunamis

When earthquakes occur beneath the sea, the sudden slippage of the sea bed can produce tsunamis, or tidal waves. These radiate outwards from the earthquake site, travelling across oceans at speeds of up to 800kph (500mph). In the open ocean, tsunamis are rarely more than 1m (3ft) high, but as they approach coasts their amplitude increases, creating a wall of water that can tower 20m (65ft) above the normal high-tide level.

Like earthquakes, tsunamis cannot be prevented, but monitoring systems can warn when a tsunami is on its way. In the Pacific Basin – by far the world's most tsunami-prone region – an ocean-wide warning system was established in 1948. Based in Hawaii, the system receives information about earthquakes around the Pacific Rim and issues alerts when tsunamis are detected. After the disastrous tsunami on 26 December 2004, a similar system was set up in the Indian Ocean.

LAND

Nojima Fault

LOCATION Extending the length of Awaji-shima island, Osaka Bay, ending beneath Kobe, Japan

TYPE Strike-slip (right-lateral).

LENGTH 60km (37 miles)

ACTIVITY Active

Early on the morning of 17 January 1995, Kobe, Japan's sixth-largest city, was hit by one of the worst earthquakes to strike the country since the 1923 Tokyo disaster. The earthquake lasted for barely 20 seconds, but it was so powerful, reaching 6.9 on the Richter Scale, and so close to the city, that its effects were devastating. Some 5,500 people were killed, 37,000 were injured, and 300,000 people were left homeless. Scientists began an urgent search for the origins of the earthquake, as the disaster had occurred in a region of the country that was not prone to serious seismic disturbances. The major fault in the region is the Median Tectonic Line (MTL), which extends beneath the sea off the eastern coast of Japan. The Nojima Fault, which is part of the Arima Takatsuki Line, which

KOBE EARTHQUAKE

The Kobe earthquake was traced to the Nojima Fault on Awaji-shima island, which is one of a series of faults. Southern Japan sits on the eastern edge of the Eurasian Plate, where the Philippine Plate is being subducted along the Nankai Trough. However, the Eurasian Plate is moving in a different direction. The plate boundary is marked by strike-slip faults that trigger seismic disturbances.

in turn is a branch of the MTL, was found to be the cause of the Kobe disturbance. The earthquake began 22km (14 miles) beneath the surface of the Earth, and within 20km (12 miles) of the city itself, which is why its effects were so catastrophic. During the earthquake, the ground on the south side of the Nojima Fault moved 1.5m (5ft) to the right, and slipped 1.2m (4ft) downwards.

CRACKED EARTH
The movement along the Nojima Fault during the Kobe earthquake was marked at the surface by a visible rupture and cracks that scarred the surrounding landscape.

REBUILDING

The Kobe earthquake destroyed 100,000 buildings and damaged an equal number. The repair cost for the infrastructure alone was estimated at US$150 billion. Yet rebuilding work began immediately, and within a year all railways and almost all roads had opened again. The people who suffered the most were the poor, who could not afford to rebuild their wooden houses, which were destroyed by the earthquake.

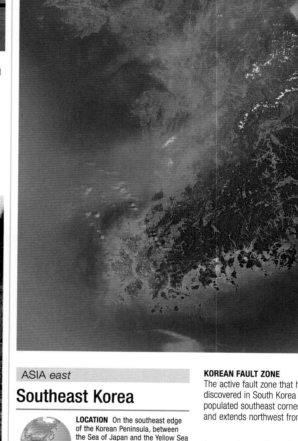

Southeast Korea

LOCATION On the southeast edge of the Korean Peninsula, between the Sea of Japan and the Yellow Sea

TYPE Strike-slip (left-lateral).

LENGTH 200km (124 miles)

ACTIVITY Active

Until recently, Korea was not believed to be a major earthquake zone. However, over 30 active faults have now been mapped in the southeast of the peninsula and beneath the surface of the surrounding ocean. Unlike neighbouring Japan, Korea is not located next to any plate boundaries, situated instead within the relatively stable interior of the Eurasian Plate. Nevertheless, there has been at least one minor earthquake here every year for the past 2,000 years. The possibility that Korea may unexpectedly be hit

KOREAN FAULT ZONE

The active fault zone that has recently been discovered in South Korea begins in the heavily populated southeast corner of the peninsula and extends northwest from its edge.

by a major earthquake like the one that devastated Kobe (see Nojima Fault, left) has focused attention on the peninsula's fault network – particularly as many nuclear and industrial facilities are situated in the region. The major fault is the Yangsan, which runs 200km (125 miles) inland in a northeasterly direction from the peninsula's tip. The Yangsan Fault is a strike-slip fault that is moving by up to 1mm (¹⁄₅₀in) every year. Most of the Korean faults were originally opened up by the way the Pacific Plate interacts with the Eurasian Plate and the Okhotsk Plate. The resulting Korean faults were all originally right-lateral strike-slip faults. Yet the way the Pacific Plate is now being subducted under the Okhotsk and Eurasian plates has reversed the movement of the faults, and they are now left-lateral faults.

Great Alpine Fault

LOCATION Extending along the west of South Island from Fiordland to Blenheim, New Zealand

North Island
Wellington
South Island
Great Alpine Fault

TYPE Strike-slip (right-lateral)

LENGTH 500km (310 miles

ACTIVITY Active

The Great Alpine Fault of New Zealand's South Island is one of the Earth's most remarkable faults; its curving trace is easily visible from space and readily visible on the ground. Over the last 12 million years, the mountains of South Island's Southern Alps have been uplifted about 20km (12.4 miles) along the eastern side of the fault. However, rapid erosion of the mountains, especially by glaciers, has kept their highest point below 4km (2.4 miles). New Zealand is one of the most seismically active regions of the world with over 15,000 earthquakes a year, but only a 100 or so are large enough

to be felt. However, there have been four major earthquakes of magnitude 8 or more and 15 of magnitude 7.5 or more since 1900. New Zealand straddles the Australian and Pacific plates. At the southern end of South Island, the Australian Plate is subducted (pushed down) below the Pacific Plate.

FRACTURED EARTH
The activity of the Great Alpine Fault is clearly revealed in this field on South Island, where a large crack has opened across the Earth's surface.

At the other end of the fault system, in North Island, the opposite occurs – the Pacific Plate is subducted over 300km (186.4 miles) below the Australian Plate. This generates deep earthquakes, and the magma that surfaces forms North Island's active volcanic zone. In the middle of the fault system, the two plates shake past one another, with the Pacific Plate rotating anticlockwise and moving south at a rate of 35–45mm (1¼–1¾ in) a year relative to the Australian Plate. This forms South Island's Alpine fault, which is consequently known as a right-lateral fault. The origins of the fault, and so too those of the Southern Alps, date back about 26 million years to a time geologists call the Kaikoura Orogeny. At this time, New Zealand was barely above sea-level and there were no mountains, but as the Pacific and Australian plates twisted and thrust together, the Southern Alps were forced up and the Great Alpine Fault opened.

HAROLD WELLMAN

Harold Wellman (1908–99) was an English geologist who discovered the Great Alpine Fault. During World War II, the New Zealand Geological Survey sent him to South Island to look for the mineral mica. While there, he discovered a line across the ground where rocks changed from granite to schist. Wellman realized the rocks had been displaced over 500km (310 miles) sideways along this line. He had found one of the Earth's major faults.

MOUNTAIN FAULT
The straight line of the Great Alpine Fault strikes across the landscape of New Zealand's South Island. The line of the fault becomes visible as it emerges in the Southern Alps.

Mountains

86–89 Tectonic plates

122–33 Fault systems

Volcanoes 154–77

Glaciers 252–75

Forests 298–317

Ocean tectonics 386–89

Mountain climates 463

Precipitation and clouds 472–81

Mountains are rock masses that have been elevated high above their surroundings, mostly by the Earth's plate-tectonic processes. The origin of a mountain belt can generally be traced to one of the Earth's plate boundaries, past or present. Most mountains can be classified by the type of plate boundary involved in their formation (see pp.88–89). A broad classification can be based on whether the plate boundary is divergent or convergent. The plates involved may be both continental, both oceanic, or of differing types. Once they have been uplifted by tectonic forces, mountains are sculpted into peaks, valleys, and other features by the processes of weathering and erosion.

ANTARCTIC FOLD
Intense folding and metamorphism have transformed the white layers seen in this picture of Antarctica's Royal Society Range from limestone into marble, and deformed them into an overturned anticline.

Folding

Rocks near the Earth's surface are usually hard and brittle, and when subject to pressure they may break, as happens during an earthquake. However, when rocks are deeply buried, heated, and compressed over a long period, they become ductile and can deform, leading to folding, and making them susceptible to the geological processes associated with mountain building. Horizontal sedimentary rock strata buckle and dislocate into a variety of curved shapes. Upfolds, known as anticlines, are separated by downfolds known as synclines. These processes can produce folds from a few millimetres to many kilometres long. The hills and valleys flanking mountain belts are typically symmetrical folds with vertical planes of symmetry. By contrast, deep within mountain belts, where temperatures are greater, folds may become overturned to such an extent that the fold becomes almost flat (recumbent), and the tip may become detached and slide or be forced further along a fault plane.

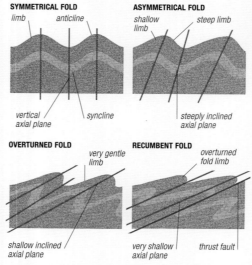

SYMMETRICAL FOLD
limb   anticline
vertical axial plane   syncline

ASYMMETRICAL FOLD
shallow limb   steep limb
steeply inclined axial plane

OVERTURNED FOLD
very gentle limb
shallow inclined axial plane

RECUMBENT FOLD
overturned fold limb
very shallow axial plane   thrust fault

VERTICAL STRATA, NAMIBIA
The effect of differential stress (unequal pressure in different directions) on rock strata can be powerful enough to rotate them from a horizontal to a vertical position.

FOLD SHAPES
Folds with limbs (sides) whose axial planes are inclined at the same angle are called symmetrical folds. If the axis of a fold is tilted to one side, and one limb has deformed so that it is steeper than the other, the fold is asymmetrical. When the fold axis is inclined at 45° or less to the horizontal, the fold is overturned. If the fold axis lies at a shallow angle, the fold is known as recumbent and is often accompanied by low-angle thrust faulting.

ANDEAN PEAKS
Although they began forming millions of years ago, the Andes, seen here in Patagonia, are among the Earth's youngest mountains.

Continents in Collision

Some of the Earth's most impressive mountain ranges, such as the Himalayas (see pp.150–51), have been formed by continents colliding directly with each other. Most tectonic plates are a combination of upper mantle rocks with dense oceanic crust and lighter but thicker continental crust. When an ocean basin is consumed by subduction, two continental masses move towards each other and meet, forming a convergence zone. Their similar density means neither will readily subduct into the mantle. Subduction cannot continue, even though the plates carrying the continents continue to converge. Instead the continental margins, with their thick layers of sediments washed off the continents, and then the continents themselves collide to form vast folded and faulted mountain belts in which crustal rocks are thickened, deformed, and metamorphosed. Within such a belt, slivers of oceanic crust and even upper mantle rocks may be preserved, testifying to the earlier subduction.

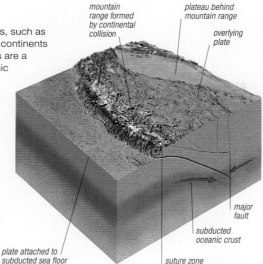

mountain range formed by continental collision
plateau behind mountain range
overlying plate
major fault
subducted oceanic crust
plate attached to subducted sea floor
suture zone

HIMALAYAN PEAKS
The mass of rock is supported by a root of thickened continental crust below it that floats on the mantle. This balancing act is called isostasy.

PLATE MOVEMENT
As continents come together, sea floor from one plate is subducted beneath the other, creating volcanic mountains. As the continental crust blocks finally collide, a second range is created by compression and uplift, with the previous mountains forming a plateau.

Ocean–Continent Collisions

Where the oceanic part of a plate converges with the edge of a continent, a subduction zone is formed, as the denser oceanic slab of crust and mantle is driven beneath the less dense continental crust. This type of subduction forms a continental volcanic arc, such as the Cascade Range of the western United States. Sediments scraped from the ocean floor by the subduction process are faulted onto the continental margin, which is further compressed and thickened by folding. Such subduction is also responsible for the Ring of Fire around the Pacific Ocean, and contributes to the building of the young mountain belt extending from the South American Andes (see pp.140–41) through the Western Cordillera of North America and the mountains of the Aleutian Islands to Japan and beyond.

ANDES AND PACIFIC OCEAN
This image, taken from space, shows the Chilean Andes, with the Pacific Ocean to the right of the range.

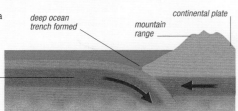

deep ocean trench formed
continental plate
mountain range
subducting oceanic plate

ANDEAN-STYLE MOUNTAINS
An oceanic plate subducts below the overriding continental plate. The continent's edge thickens, buckles, and is built up further by volcanism.

Ocean–Ocean Collisions

When converging plates are both capped by oceanic crust of similar density, a volcanic island arc (see p.388) can be produced. Typically, this consists of mountainous islands with a parallel trench running offshore along one side of the chain. The trench marks where the plate that is older, cooler, and therefore slightly denser is subducting beneath the other one. Hot, molten rock (magma) rises through the overriding plate to erupt at the surface, building up gradually into a chain or arc of mountainous volcanic islands such as the Marianas, Vanuatu, or Solomon Islands in the Pacific and the Antilles and Scotia arc in the Atlantic. Island arcs and their associated sediments can sometimes become joined or accreted to continental margins, forming so-called exotic terranes. Such terranes are thought to have contributed to new mountains on the Pacific margin of North America's Western Cordillera.

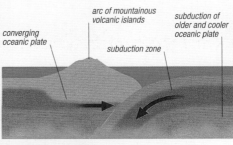

arc of mountainous volcanic islands
subduction of older and cooler oceanic plate
converging oceanic plate
subduction zone

VOLCANIC ISLAND ARC
Subduction of the older plate triggers volcanic activity, creating an island arc and forming new continental crust.

JAPANESE ISLANDS
Plate convergence below Japan generates frequent and often damaging earthquakes, fault movements, and occasional volcanic eruptions.

Rift Mountains

Where continental crust is pulled apart at divergent boundaries, the brittle surface rocks do not stretch very far before fracturing along parallel fault lines. Between the faults, the central block subsides, forming a rift valley flanked by cliff-like escarpments that can reach mountainous dimensions, as in parts of Africa's Great Rift Valley (see pp.128–29). If rifting continues, plates may split in two, with a new ocean between them. Where two oceanic plates diverge, they generate sea-floor spreading (see p.386), leading to the formation of submarine mountain ranges quite different in structure to those on land. As the plates are pulled apart, molten rock rises from the upper mantle to erupt along the axis of the sea-floor spreading ridge. These ridges form the Earth's longest mountain chain.

RIFT VALLEY
The Great Rift Valley is flanked by mountains, including volcanoes, that rise over 2,000m (6,600ft) above the valley floor.

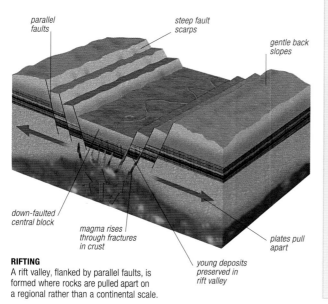

RIFTING
A rift valley, flanked by parallel faults, is formed where rocks are pulled apart on a regional rather than a continental scale.

parallel faults

steep fault scarps

gentle back slopes

down-faulted central block

magma rises through fractures in crust

plates pull apart

young deposits preserved in rift valley

EDUARD SUESS

Austrian geologist Eduard Suess (1831–1914) pioneered studies of the folds, faults, and geological structure of the European Alps. Suess's acclaimed four-volume book *The Face of the Earth (Das Antlitz der Erde,* 1885–1909) was an encyclopedic overview of the Earth's mountains, their structure and formation, and a history of the oceans. He was also the geologist who first put forward the idea – later disproved – that there could have been land bridges that joined the continents of South America, Africa, India, and Australia. Suess named this once-vast continent Gondwanaland – the name is still used for an ancient supercontinent.

Epeirogenic Movements

The Earth's crust can be subjected to large-scale movement without extensive deformation and folding. This movement can be either downwards, which results in the creation of large sedimentary basins, or upwards to form elevated domes or plateaus. Both trends are referred to as epeirogenic movements. These events tend to be gradual, and usually affect large regions, such as the Colorado Plateau in the USA. An area of about 40,000 square km (15,500 square miles), the plateau is cut off from the Great Plains by the Rocky Mountains, and has been stable for at least 500 million years. Epeirogenic uplift in the last 65 million years has led to increased erosion by the Colorado River, excavating deep canyons in the plateau.

UTAH GOOSENECKS
At Goosenecks State Park, Utah, epeirogenic uplift has led to the formation of a deep canyon by the San Juan River.

Orogenic Belts

Over geological time, plate movements have produced innumerable episodes of mountain building around the world, a process known as orogenesis. The resulting linear arrays of mountains, shown below, are called orogenic belts, with the events that form them known as orogenies. They are usually identified by a geographical name, such as the Caledonian orogeny, which was first identified in Scotland (known as Caledonia to the Romans). Due to later plate movements, most ancient orogenic belts are no longer located where they formed; for instance, the Caledonian belt was originally an extension of the North American Appalachians.

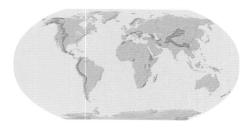

ANCIENT MOUNTAIN ROOTS
The Scottish hills now barely qualify as mountains, but they are the eroded remnants of once higher mountains.

MOUNTAIN PROFILES

The pages that follow contain profiles of some of the world's mountain belts. Each profile includes the following summary information.

HIGHEST PEAK Name and height above sea-level

LENGTH Distance along the belt's longest axis

AGE Geological era within which the mountain belt formed

FORMATION TYPE Continent–continent, ocean–continent, ocean–ocean, or rift

Rocky Mountains

LOCATION Stretching along the western side of North America from Alaska to New Mexico

HIGHEST PEAK Mount Elbert, USA 4,400m (14,400ft)

LENGTH 4,800km (3,000 miles)

AGE Mesozoic– Cenozoic

FORMATION TYPE Ocean–continent

The Rocky Mountains are part of one of the largest mountain belts on Earth, the great Western Cordillera of North America, which extends from above the Arctic Circle to the Sierra Madre Occidental in Mexico. Strictly speaking, the Rocky Mountains are the high peaks of the cordillera's eastern ranges, which also include the Canadian Rockies, and the Northern, Central, and Southern Rockies in the USA. Most of the mountains peak above 1,000m (3,000ft), with more than 100 summits above 3,000m (10,000ft). Within the belt is North America's continental divide, to the west of which rivers flow into the Pacific, while in the east they flow into the Arctic and Atlantic oceans. Geologically, the history of the cordillera is extremely complex, with large amounts of uplift taking place in the last 15 million years. The range

includes intensely deformed zones with extensive folds and faults, such as the Canadian Rockies; belts of volcanic activity and lava plateaus, which are evident in Yellowstone, Wyoming; and highly metamorphosed areas with granite intrusions, including Grand Teton. Features such as Wyoming's Great Divide Basin are also part of the chain. the Alaskan ranges are snow-covered and

heavily glaciated. An abundance of wildlife inhabits the cordillera, including mountain goats and sheep, bears, wolves, cougars, and moose.

FEATURES OF THE WESTERN CORDILLERA

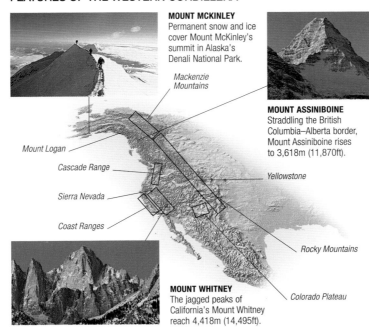

MOUNT MCKINLEY
Permanent snow and ice cover Mount McKinley's summit in Alaska's Denali National Park.

Mackenzie Mountains

Mount Logan

Cascade Range

Sierra Nevada

Coast Ranges

MOUNT ASSINIBOINE
Straddling the British Columbia–Alberta border, Mount Assiniboine rises to 3,618m (11,870ft).

Yellowstone

Rocky Mountains

Colorado Plateau

MOUNT WHITNEY
The jagged peaks of California's Mount Whitney reach 4,418m (14,495ft).

AMERICA'S ICON
The bald eagle is the only species of eagle endemic to North America.

ALASKAN OIL PIPELINES

The discovery of oil and gas in Alaska has led to the construction of huge pipelines that overcome the barrier presented by the high mountains of Alaska. However, the pipelines cross environmentally sensitive wilderness areas, creating a physical obstacle for migrating animals such as Caribou. Although burying pipelines might appear to solve such difficulties, it can actually cause further problems, as they may fracture. This is also a region of permafrost, which is unstable. In places, pipelines are elevated to allow wildlife to pass safely beneath.

ROCKY MOUNTAIN LANDSCAPE
The head waters of the Snake River, with its flat alluvial plain and terraces, lie in the Teton Range, northwest Wyoming. They are overlooked by Grand Teton peak, which reaches a height of 4,197m (13,770ft).

LAND

LAND

NORTH AMERICA *east*

Appalachians

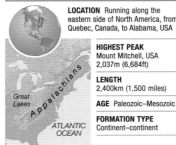

LOCATION Running along the eastern side of North America, from Quebec, Canada, to Alabama, USA

HIGHEST PEAK
Mount Mitchell, USA
2,037m (6,684ft)

LENGTH
2,400km (1,500 miles)

AGE Paleozoic–Mesozoic

FORMATION TYPE
Continent–continent

A much-eroded, ancient, and complex mountain belt, the Appalachians were formed during a series of plate separations and collisions. Today, the chain consists of a series of ranges that increase in geological complexity from

APPALACHIAN AMPHIBIAN
There are 35 native varieties of salamander resident in the Appalachian forests, including this Marbled Salamander.

west to east. The slightly folded and uplifted plateaus in the west give way to parallel ridges and valleys, followed by the highly deformed Blue Ridge. The belt becomes less elevated closer to the Atlantic coast, where metamorphosed sediments and volcanic rocks are overlain by the younger accumulations of the coastal plain. The Appalachians are the North American continuation of Europe's Caledonian range (see p.142).

TENNESSEE MOUNTAINS
The forest-clad slopes of much of the Appalachian range disguise a geologically complex area of ancient metamorphosed rocks.

NORTH AMERICA *south*

Sierra Madre Oriental

LOCATION In eastern Mexico – extending from north to south, and bending west at the northern end

HIGHEST PEAK
Cerro Peña Nevada,
Mexico 3,644m (11,955ft)

LENGTH
1,100km (700 miles)

AGE Mesozoic–Cenozoic

FORMATION TYPE
Ocean–continent

Much of Mexico is elevated above 1,000m (3,000ft), with the three Sierra Madre ranges and the Mexican Volcanic Belt forming the country's mountain backbone. A continuation of the Western Cordillera, the Sierra Madre Oriental was formed by compressional tectonic forces and folding about 50–70 million years ago, which produced a long, dramatic series of alternating anticlines and synclines in limestone.

LIMESTONE CRESTS
The sharp peaks of the Sierra Madre Oriental are formed of limestone, which was elevated by tectonic forces.

VIEW FROM ORBIT
This view from the space shuttle shows the Sierra Madre Oriental, close to the city of Monterrey.

SOUTH AMERICA *west*

Andes

LOCATION Running down the western side of South America from the Caribbean Sea to Cape Horn

HIGHEST PEAK
Aconcagua, Argentina
6,959m (22,834ft)

LENGTH
7,200km (4,500 miles)

AGE
Mesozoic–Cenozoic

FORMATION TYPE
Ocean-continent

The longest mountain chain in the world (apart from the Mid-Atlantic Ridge beneath the Atlantic Ocean), the Andes is also one of the Earth's most spectacular and active mountain belts. Within it lies one of the most arid places on Earth, Chile's Atacama Desert (see p.284); the world's highest navigable body of water, Lake Titicaca (see p.229); the source of the world's greatest river, the Amazon see pp.210–11); and the highest city in the world, La Paz, in Bolivia, which is situated at an altitude of 3,630m (11,910ft). The whole range is frequently affected by earthquakes and eruptions from 183 active volcanoes, including the highest on Earth. The mountains ascend

FEATURES OF THE ANDES

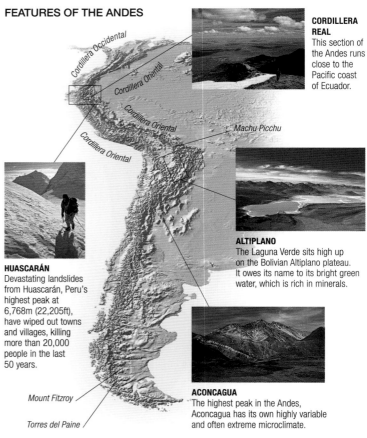

Cordillera Occidental

Cordillera Oriental

Cordillera Oriental

Cordillera Oriental

Cordillera Oriental

Machu Picchu

HUASCARÁN
Devastating landslides from Huascarán, Peru's highest peak at 6,768m (22,205ft), have wiped out towns and villages, killing more than 20,000 people in the last 50 years.

Mount Fitzroy

Torres del Paine

CORDILLERA REAL
This section of the Andes runs close to the Pacific coast of Ecuador.

ALTIPLANO
The Laguna Verde sits high up on the Bolivian Altiplano plateau. It owes its name to its bright green water, which is rich in minerals.

ACONCAGUA
The highest peak in the Andes, Aconcagua has its own highly variable and often extreme microclimate.

CITY OF INCAS
Situated high in the Peruvian Andes, Machu Picchu is a remarkably preserved Inca city. The exact function of the settlement is unknown, but it may have been a royal estate built by the Inca emperor Pachacuti in the 15th century. Located on a prominent ridge, partly surrounded by steep cliffs, the site includes houses, religious architecture, tombs, and agricultural terraces. When the Spanish conquistadors arrived in South America in the 16th century, Machu Picchu was abandoned. The city remained hidden for centuries until 1911, when it was found by American archaeologist Hiram Bingham.

HIGH FLIER

With the largest wing area of any bird, the Andean Condor can fly at altitudes of 5,500m (18,000ft).

abruptly from sea-level on the western, Pacific coast of South America to altitudes of over 6,500m (21,300ft). The belt forms a formidable physical, meteorological, and biological barrier, ranging from 100km (60 miles) wide at the southern tip of the continent, to 700km (435 miles) wide in the central region. In Peru, Bolivia, and Ecuador, between the Cordillera Occidental range in the west and the Cordillera Oriental in the east, lies the Altiplano (also known as the Puña), which is the world's most densely populated high plateau. The farmers who inhabit the Altiplano exist in one of the most hostile environments for human life on Earth: the high altitude means that the air is thin, and the weather is extreme. The southern sector of the range, which runs through Patagonia and Tierra del Fuego, has been glaciated as far as sea-level and is characterized by numerous fiords. Geologically, the mountains are the result of the continuing eastward subduction of the Nazca Plate beneath the South American Plate. The subduction zone is marked by the Peru–Chile Trench off South America's Pacific coast, which plunges to over 8,000m (26,250ft) below sea-level. The present topography of the Andes is largely the result of deformation that has taken place during the past 50 million years, with significant elevation of the mountains occurring over the past 10 million years. As such, the belt is young in geological terms, and

GRANITE HORNS
The granite peaks of the Torres del Paine soar to 3,050m (10,006ft) above the glacial lakes of Patagonia in southern Chile.

mountain building continues in the present. Minerals such as gold, silver, platinum, and copper are all mined in the Andes.

JUTTING PEAKS
The tooth-shaped peaks of the Fitzroy Range in Patagonia are the stumps of much higher mountains that have been steadily eroded by glaciers.

LAND

Urals

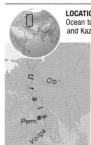

LOCATION Running from the Arctic Ocean to the border between Russia and Kazakhstan

HIGHEST PEAK
Narodnaya, Russia
1,894m (6,214ft)

LENGTH
2,400km (1,500 miles)

AGE Paleozoic

FORMATION TYPE
Continent–continent

The ancient, extensively eroded Urals form a wall-like, linear mountain chain traditionally regarded as the boundary between Europe and Asia. Only 200km (125 miles) wide at its broadest point, it is a narrow range. The Urals represent the seam along which the Baltic and Russian continents collided about 300 million years ago, after the vast Uralian Ocean had been completely subducted under the Russian continent. The movement that formed the mountains was part of the assembly of the supercontinent Pangea, which once covered two-fifths of the world's surface. Following subduction of the ocean crust, the compressive forces of convergence crumpled sediments on the leading edges of the two colliding continents, thickening the crust and elevating the mountains. In 1841, British geologist Roderick Murchison mapped the stratification of deposits in the western Urals. In doing so he defined the Permian Period (named after the Russian region of Perm), which extended from approximately 290 million to 252 million years ago and saw the extinction of much of the marine life of the period. Sulphide mineral deposits, including copper, iron, and zinc, were formed by hydrothermal vents on the ancient ocean floor. These are now preserved in the southern Urals and form an ore belt that is 2,000km (1,200 miles) long. The area has been one of the most important industrial regions of modern times. The central and southern Urals are heavily forested, while further north the mountains feature alpine meadows and tundra.

PLATINUM MINING

Pure platinum is an extremely rare and valuable natural mineral. It was once confused with silver, but it actually has a much higher melting point of 1,772°C (3,222°F). With a specific gravity of 21.5, platinum is the densest naturally occurring mineral. Nuggets of platinum have been found in placer deposits (see p.45), especially at Nizhniy Tagil in the Urals, where one of the largest accumulations of the mineral in the world is found.

THICKLY FORESTED MOUNTAINS
Much of the Urals are covered in taiga, the Russian name for the coniferous forest that is indigenous to Eurasia, and stretches from the Baltic Sea to the Pacific Ocean (see p.312).

Caledonian Mountains

LOCATION Running from the Arctic Circle through Scandinavia and Scotland to northwest Ireland

HIGHEST PEAK
Galdhøpiggen, Norway
2,469m (8,100ft)

LENGTH
2,600km (1,625 miles)

AGE Paleozoic

FORMATION TYPE
Continent–continent

The Caledonian Mountains are the European section of a much longer mountain chain that continues as the Appalachians (see p.140) on the opposite side of the Atlantic Ocean. This ancient mountain belt is one of the most intensely studied in the world. This is especially true of its British ranges, studies of which generated many formative ideas about mountain building in the late 19th century. However, none of the pioneers of geology could have guessed at the extraordinary history of the Caledonian Mountains. Five hundred million years ago, North America, Scotland, and northwest Ireland were part of the ancient continent of Laurentia, and were separated from Avalonia (England, Wales, and southwest Ireland) by the Iapetus Ocean, all of which were located in the southern hemisphere. About 490 million years ago, Avalonia began to move north, subducting the Iapetus Ocean in the process. Eventually, Avalonia impacted upon Laurentia, a convergence that resulted in a period of mountain building. Much later, as the North Atlantic Ocean began to open 80 million years ago, North America split from northwest Ireland and Scotland, which converged with the rest of what is now the British Isles. The Caledonian Mountains are continuing to rise by a

BUACHAILLE ETIVE MOR
This striking triangular peak descends to Glen Coe, in the Scottish Grampian Mountains.

tiny amount each year. The range includes the North West Highlands and Grampian mountains of Scotland (which are the most elevated peaks in the British Isles) and the Jotunheimen of Norway. The rugged landscape of the range today features craggy, snow-capped peaks, glens, lochs, and fiords.

NORWEGIAN FIORDLAND
With their precipitous mountainsides, the deep fiords of Norway's Atlantic coast are evidence of the incredible erosive power of glaciers.

Pyrenees

LOCATION Between France and Spain, extending from the Atlantic Ocean to the Mediterranean Sea

HIGHEST PEAK
Aneto, Spain
3,404m (11,168ft)

LENGTH
435km (270 miles)

AGE Late Paleozoic–Cenozoic

FORMATION TYPE
Continent–continent

WATCHING SPACE

The Earth's increasingly polluted atmosphere mars our view of the Universe. Building astronomical observatories as high as possible, and away from heavily populated areas, minimizes the effects of atmospheric pollution and light interference. The Pic du Midi Observatory is situated above the clouds in the Pyrenees.

FRINGED PINKS
Thousands of varieties of wild flower grow in the Pyrenees, including these delicate Fringed Pinks, which are found in the Spanish mountains.

The Pyrenees form a natural barrier between France and the Iberian Peninsula (Spain and Portugal). Two-thirds of the belt is situated in Spain, and there are relatively few natural passes between the mountains. Nevertheless, human predecessors (*Homo antecessor*) managed to cross the mountains into northern Spain approximately 800,000 years ago.

The limestone caves of the Pyrenees are particularly famous for preserving some of the most recent Neanderthal remains in Europe, which are approximately 29,000 years old. The walls of the caves also contain paintings by the earliest modern humans, dating back about 15,000 years. The geological history of the Pyrenees is long, complex, and intimately associated with that of the Iberian Peninsula. During the Cretaceous Period, approximately 142 million years ago, Iberia became a micro-continent, rupturing from the major landmasses during the break-up of the supercontinent of Pangea. As the Atlantic Ocean opened, the Iberian plate was wedged between Europe and North Africa, leading to the initial development of the Pyrenees. Then, about 60 million years ago, rotation of the African and Eurasian plates meant that the Iberian Plate again converged

PYRENEAN IBEX
Once common throughout Europe, wild goats were depicted by early humans. The Pyrenean ibex is a surviving species though it is declining dramatically.

with southern France. The resulting thickening and folding of the crust further uplifted the Pyrenees. The peninsula's motion ended about 10 million years ago. During the last ice age, the Pyrenees experienced extensive glaciation. The mountains are characterized by jagged peaks that are steepest on the French side of the range. In the west, the slopes are heavily wooded, but the granite peaks of the east are much starker with less vegetation. Dramatic, semicircular glaciated hollows known as cirques (see p.252) are a prominent feature of the Pyrenees, and the mountains also contain some of Europe's most spectacular waterfalls, the highest of which, at the Cirque de Gavarnie, plunges 422m (1,385ft). The wildlife of the Pyrenees includes abundant bird species, among them vultures (including the rare Lammergeier), eagles, capercaillie, and Ptarmigan. Mammals include Chamois, wild boar, lynx, and a tiny population of brown bears.

FEATURES OF THE PYRENEES

Pyrénées-Atlantiques

CIRQUE DE GAVARNIE
This sheer rock wall is the site of the highest waterfall in Europe.

Mount Aneto

HAUTES-PYRÉNÉES
Straddling the French–Spanish border, the Hautes-Pyrénées include the highest peaks in the belt.

Pyrénées-Orientales

ORDESA NATIONAL PARK, SPAIN
Rising high above the tree-line, horizontal rock strata form precipitous cliffs above valleys that were once filled with glacial ice. Only the high peaks stood out above the glaciers at the height of the last ice age.

LAND

EUROPE *south*

Alps

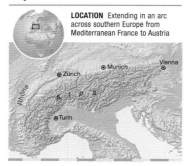

LOCATION Extending in an arc across southern Europe from Mediterranean France to Austria

HIGHEST PEAK Mont Blanc, France 4,807m (15,771ft)	
LENGTH 1,050km (650 miles)	
AGE Mesozoic–Cenozoic	
FORMATION TYPE Continent–continent	

The largest mountain chain in Europe, the Alps contains some of the most intensely studied mountains in the world. Studies of the Alps in the 18th and 19th centuries led to early theories about mountain formation and the nature of large-scale folding and thrusting. Between 100 and 200km (60 and 120 miles) wide, the Alps form a curved fold belt, with several peaks rising to over 4,000m (13,000ft). The west of the chain includes the Maritime and Pennine Alps; the central ranges include the Bernese (or Central) Alps; and the eastern ranges include the Dolomites and the Austrian Alps. In the past, the mountains have formed a physical obstacle to human migrations, although roads were

first built through the belt's natural passes in Roman times. The Alps were generated by the convergence of the African and Eurasian plates. This process began approximately 90 million years ago, when the African plate began to move north, subducting the ancient Tethys Ocean and uplifting the sediment on the sea floor to form faulted recumbent fold structures called nappes. Folding and thrusting continued, with a major period of mountain building taking place 33–15 million years ago. The current landscape of the Alps is the result of glaciation during the last 2 million years. Primary features include U-shaped and hanging valleys (see p.255), and lakes such as Como (see p.233) and Garda, which have been naturally dammed by glacial moraines. The longest remaining glacier in the alps is the Aletsch (see p.267), which extends for 25km (16 miles).

ALBRECHT PENCK

German geomorphologist Albrecht Penck (1859–1945) was director of the Berlin Institute of Oceanography and Geography. Penck studied the development of landforms and surface deposits in the Alps and the boulder clay of the North German Plain to determine the extent to which glaciation had affected them. This led him to propose that four main episodes of glaciation had taken place in the area. He also argued that the elevation of mountains is limited by the balance between uplift and erosion, a theory that is well illustrated by the Alps. The mountains continue to be uplifted by approximately 0.5mm (1/50in) every year, a rise that is compensated for by similar levels of erosion. His most famous work is *The Morphology of the Earth's Surface* (1894).

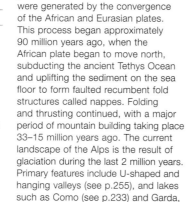

EDELWEISS
This perennial alpine flower has tufted leaves and flowers protected by thick, woolly bracts. These protect the Edelweiss from the freezing temperatures that are common close to the snow-line where it grows.

ALPINE MEADOW
Retreating glaciers allow specialized alpine plants, especially grasses and saxifrages, to colonize newly exposed rocky soils within a few years.

ICE MUMMY

In 1991, on the Alpine border between Italy and Austria, the body of Otzi, as he was later named, was discovered. He was so well preserved in ice that he was thought to have died recently. However, the discovery of his bow, arrows, and copper axe suggested that he was a Neolithic hunter. A radiocarbon age of 5,200 years confirmed this. An arrow embedded in Otzi's back suggests that he was attacked while climbing over the mountains.

FEATURES OF THE ALPS

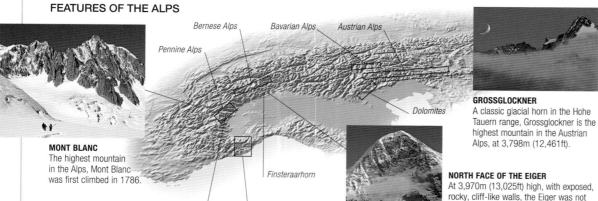

Bernese Alps　*Bavarian Alps*　*Austrian Alps*

Pennine Alps

Dolomites

MONT BLANC
The highest mountain in the Alps, Mont Blanc was first climbed in 1786.

Finsteraarhorn

Matterhorn　*Maritime Alps*

GROSSGLOCKNER
A classic glacial horn in the Hohe Tauern range, Grossglockner is the highest mountain in the Austrian Alps, at 3,798m (12,461ft).

NORTH FACE OF THE EIGER
At 3,970m (13,025ft) high, with exposed, rocky, cliff-like walls, the Eiger was not climbed until 1858, and its north face defied all attempts until 1938.

FORMIDABLE MATTERHORN
With its pyramid-shaped peak and precipitous rock walls sculpted by glacial ice, the Matterhorn, 4,478m (14,692ft) high, is seen as the epitome of the alpine mountain form.

LAND

EUROPE *southwest*

Sierra Nevada

LOCATION In southern Spain, running west–east between Granada and the Mediterranean Costa del Sol

HIGHEST PEAK
Mulhacén, Spain
3,481m (11,421ft)

LENGTH 100km
(63 miles)

AGE Mesozoic–Cenozoic

FORMATION TYPE
Continent–continent

Southern Spain's snow-capped Sierra Nevada mountain range forms a dramatic backdrop to the Mediterranean holiday beaches of the Costa del Sol. The range is the most southerly part of the more extensive Betic Cordillera, which run southwest to northeast from Cadíz to Alicante, and continue offshore to the Balearic Islands of Majorca, Minorca, and Ibiza. The formation of the Sierra Nevada was contemporary with that of the Atlas Mountains (see opposite) and the European Alps (see pp.144–45). It resulted from the northwards movement of the African Plate during the Cenozoic Era. The Sierra Nevada continues to rise by about 3mm (⅛in) every 100 years. The southern folds of the range, known as Las Alpujarras, are deep valleys that provide fertile farming land.

ARID BADLANDS
Deforestation and overgrazing of fragile, semi-arid lands have resulted in erosion and the formation of semi-desert badlands in the Sierra Nevada.

EUROPE *central*

Carpathians

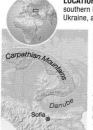

LOCATION Forming an arc from southern Poland through Slovakia, Ukraine, and into central Romania

HIGHEST PEAK
Gerlachovka, Slovakia
2,655m (8,711ft)

LENGTH
1,450km (900 miles)

AGE Mesozoic–Cenozoic

FORMATION TYPE
Continent–continent

The Carpathians are an extension of the European Alps, and they contain a complex mixture of ancient rocks covered by younger sediments that are often highly folded and faulted. Many of the rocks that form the range were once deposits in the Mesozoic Tethys Ocean.

APOLLO BUTTERFLY
A huge conservation programme was undertaken to preserve this species from extinction in the Pieniny range of the Polish Carpathians.

The highest Carpathian peaks are the granitic Tatra Mountains in Slovakia, where steep, rocky walls rise above U-shaped valleys. In Romania, where more than half of the mountains are situated, the chain breaks up into several ranges, the loftiest being the Transylvanian Alps. Broad, fertile valleys, lakes, and limestone plateaus are all features of the range.

STEEP-SIDED VALLEY
Pasture land is grazed by sheep at the base of sharp, towering Carpathian peaks in Slovakia.

EUROPE *south*

Dinaric Alps

LOCATION Running along the coast of the Adriatic Sea from Slovenia into northern Albania

HIGHEST PEAK
Durmitor, Yugoslavia
2,522m (8,274ft)

LENGTH
750km (470 miles)

AGE Mesozoic–Cenozoic

FORMATION TYPE
Continent–continent

The northwest–southeast trend of the Dinaric Alps parallels Italy's Apennines, from which they are

DURMITOR, YUGOSLAVIA

separated by the Adriatic Sea. Deep gorges have been cut into the mountains by rivers such as the Neretva. The soluble nature of the limestone that forms the mountains has led to the development of the region's famous karst landscape (see p.241). Below ground, extensive cave systems have developed, and in places cavern collapse has formed large hollows called poljes.

EUROPE *east*

Caucasus Mountains

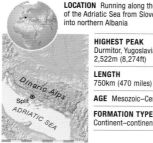

LOCATION Running between the Black Sea and Caspian Sea along the border of Russia and Georgia

HIGHEST PEAK
El'brus, Russia
5,642m (18,510ft)

LENGTH
1,200km (750 miles)

AGE Late
Paleozoic–Cenozoic

FORMATION TYPE
Continent–continent

One of Europe's natural boundaries in the east, the Caucasus Mountains consist of two parallel ranges separated by a central trough. The southeastern peaks are mainly limestone, which has formed karst landscapes. A higher range, formed of igneous rocks, dominates in the northwest. The mountains contain one of the world's deepest caves, Snezhnaya, Georgia, which is 1,470m (4,823 ft) deep.

PONTIC MOUNTAINS, TURKEY
On the southern shore of the Black Sea, the Lesser Caucasus fold structure continues westwards as the Pontic Mountains.

BELOLAKAYA
This peak attracts alpinists, trekkers, and tourists. It is one of several high mountains in the western Caucasus, the tallest of which is Dombay-Ul'gen, at 4,042m (13,262ft).

Atlas Mountains

LOCATION Extending from the Atlantic coast of Morocco to the Mediterranean east coast of Tunisia

HIGHEST PEAK
Toubkal, Morocco
4,165m (13,665ft)

LENGTH
2,400km (1,500 miles)

AGE Mesozoic–Cenozoic

FORMATION TYPE
Continent–continent

The Atlas Mountains form a barrier between the Mediterranean Sea and the Sahara. Many of the tallest peaks preserve evidence of glaciation during the last ice age, with cirques (see p.252) and periglacial features (see p.338) evident. The mountains do not form a continuous chain, but are a series of ranges that are interspersed with plateaus, basins, and gorges. The most prominent range is the High Atlas, which runs inland from Agadir on Morocco's west coast for 650km (400 miles) and contains several snow-capped peaks, including the highest, Toubkal. The eastern ranges of the Atlas form less elevated plateaus. In the north, the

ATLAS MONKEY
The Barbary Ape is actually a monkey that has become adapted to living in the rocky cliffs and high-altitude oak and cedar forests of the Atlas range.

WATER SOURCES

In the arid regions of the Atlas Mountains, water is a scarce commodity, and many communities depend upon meltwater streams formed by the winter snows. In places, the water is collected in deep cisterns cut into the rock centuries ago. With the advent of global warming there is a risk that diminishing snowfalls will result in water shortages for the inhabitants.

FEATURES OF THE ATLAS MOUNTAINS

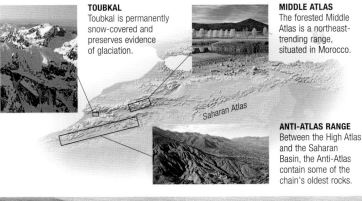

TOUBKAL
Toubkal is permanently snow-covered and preserves evidence of glaciation.

MIDDLE ATLAS
The forested Middle Atlas is a northeast-trending range, situated in Morocco.

ANTI-ATLAS RANGE
Between the High Atlas and the Saharan Basin, the Anti-Atlas contain some of the chain's oldest rocks.

mountain slopes receive ample rainfall and are forested with cedar, pine, and oak. The more southerly ranges are more arid, with salt flats occurring in those areas of the Atlas close to the Sahara. An earthquake in 1960 destroyed Agadir and confirmed the presence of major active faults separating the Atlas from the North Saharan Basin. Like the European Alps (see pp.144–45) and Spain's Sierra Nevada (see opposite), the Atlas Mountains were built by plate convergence between Africa and Europe, beginning about 60 million years ago. However, as is also true for Central Europe, there are folds and faults in the Moroccan ranges that were created during an era known as the Variscan orogeny, a period of mountain building that occurred about 300 million years ago. Younger sedimentary rocks, including many limestones that were originally deposited on the floor of the Tethys Ocean, eventually covered these folds. The northernmost ranges were formed by folding that began 60 million

years ago. Further movement of the African Plate, approximately 20 million years ago, deformed rocks in central Tunisia and Algeria, and elevated the central plateaus. The mountains are rich in minerals, including coal, iron, gold, silver, and phosphate minerals.

BARREN SLOPES
The inhabitants of ancient settlements continue a precarious existence in the arid parts of the Atlas. Survival depends on local microclimates and the availability of water for plant growth and domestic uses.

LAND

AFRICA *south*

Drakensberg Plateau

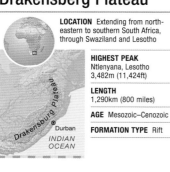

LOCATION Extending from north-eastern to southern South Africa, through Swaziland and Lesotho

HIGHEST PEAK
Ntlenyana, Lesotho
3,482m (11,424ft)

LENGTH
1,290km (800 miles)

AGE Mesozoic–Cenozoic

FORMATION TYPE Rift

A vast, westward-dipping escarpment, the Drakensberg range is one of the world's largest elevated plateaus. It separates a narrow coastal plain from southern Africa's vast interior of savanna and desert. The east of the plateau is dominated by the Great Escarpment, which soars to over 3,300m (10,000ft), high enough for frost action to fracture the rocks. In places, the sheer scarp face rises for over 2,100m (7,000ft) from its base to the crest. The Drakensberg Plateau consists of sedimentary rocks capped with volcanic deposits. The base of the plateau was built up by the non-marine Karoo sediments that were laid down about 250–300 million years ago. They begin with glacial deposits (the Dwyka tillite) and pass up into terrestrial coals. The summit of the plateau contains remnants of the ancient continent called Gondwanaland, which were originally close to sea-level before being elevated prior to the break-up of the landmass 200 million years ago. A rising hotspot domed the region, stretching and rifting the crust as Africa and South America separated. Subsequently, basalts erupted through volcanic fissures in the underlying sediment, a process that continued for about 50 million years. These layers of basalt originally covered an area of approximately 2 million square km (800,000 square miles), a region that includes much of southern Africa and parts of what is now Antarctica. The basaltic deposits were originally 1,500m (4,900ft) thick, but erosion has heavily dissected them. The resulting landscape of the Drakensberg Plateau is incredibly varied, including steep sandstone cliffs, free-standing pinnacles, torrential waterfalls, and extensive caves. Major peaks include Mont-aux-Sources (which is the origin of three rivers, including the Tugela) and Giant's Castle, so named because its steep, rocky cliffs resemble battlements. Rivers rising in the mountains, such as the Tugela and Orange, have carved impressive gorges through the rock. The diverse and inaccessible landscape, which also includes extensive high grasslands, provides habitats for an astonishing array of flora and fauna, with about 300 species of indigenous plant life. Drakensberg's name is taken from the Afrikaans for "dragon's mountain", but it is also known by its Zulu name of Ukhahlamba.

FEATURES OF THE DRAKENSBERG PLATEAU

Mont-aux-Sources

Giant's Castle

NTLENYANA
The highest point of the plateau is Mount Ntlenyana.

Great Karoo

GRASS PEAKS
Lesotho is located in the highest part of the Drakensberg Plateau.

ABOVE THE CLOUDS
The Great Escarpment of the Drakensberg rises high above the low clouds of the coastal plains. The clouds bring moisture to the lower slopes, while the top of the plateau is much drier.

ROCKY MAMMAL
Related to early hoofed mammals, the Rock Hyrax inhabits rocky slopes in colonies of up to 40 for protection against predators.

ASIA *west*

Zagros Mountains

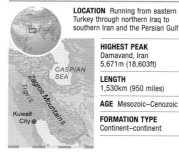

LOCATION Running from eastern Turkey through northern Iraq to southern Iran and the Persian Gulf.

HIGHEST PEAK
Damavand, Iran
5,671m (18,603ft)

LENGTH
1,530km (950 miles)

AGE Mesozoic–Cenozoic

FORMATION TYPE
Continent–continent

The Zagros Mountains run parallel to the valley of the Euphrates River and the Persian Gulf. Part of an unstable fold belt, the range is vulnerable to earthquakes, as it includes a complex zone of thrust faulting and volcanism. In the centre of the region lies an area of salt domes up to 1,500m (5,000ft) high. Many of the Zagros Mountains are perennially snow-capped.

SNOW FIELDS
Although on the same latitude as southern Turkey in the Mediterranean, some regions of the Zagros Mountains rise to over 5,000m (16,400ft), and are covered with snow in winter.

OIL-RICH HILLS

The sediments of the western foothills of the Zagros Mountains are a substantial source of hydrocarbons. Folding has produced anticlines, which trap petroleum in reservoirs beneath impermeable rock. However, some oil seeps to the surface where volatile gases burn off.

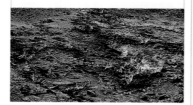

ASIA *central*

Altai Mountains

LOCATION Running from eastern Kazakhstan through Russia and Mongolia to western China

HIGHEST PEAK
Belukha, Russia
4,506m (14,783ft)

LENGTH
2,000km (1,250 miles)

AGE Late Paleozoic

FORMATION TYPE
Continent–continent

The Altai mountain system forms the northern rim of the Dzungarian Basin depression in western China. Much of the range exceeds 3,000m (10,000ft), and there are numerous glaciated peaks over 4,000m (13,000ft) in the Mongolian sections. The highest peak, Mount Belukha, lies in the north at the headwaters of the rivers Ob' (see p.221) and Irtysh in Russian Siberia

SNOW LEOPARD
This endangered species of big cat is ideally suited to the cold mountains of the Altai, where it hunts small mammals and birds.

MONGOLIAN HIGHLANDS
The Altai Mountains rise above the cold steppe grasslands of Mongolia, home to nomads and their herds of horses, camels, goats, and sheep.

ANCIENT INHABITANTS

Southern Africa's rock art was first described in the mid-18th century by European travellers. Engravings and paintings of hunting scenes, containing beautiful representations of animals and human hunters, are widespread throughout the continent. It is difficult to date the rock art, but it was probably produced less than 40,000 years ago by modern humans who first evolved in Africa around 200,000 years ago.

BLYDE RIVER CANYON
The Three Rondavels (also known as the Three Sisters) form part of the Blyde River Canyon in the Transvaal, South Africa. The cliffs of the canyon rise up to 800m (2,600ft).

Tien Shan Mountains

LOCATION To the north of the Tarim Basin between western China, Kyrgyzstan, and Kazakhstan

HIGHEST PEAK Pik Pobedy, Kazakhstan 7,439m (24,407ft)

LENGTH 2,250km (1,400 miles)

AGE Late Paleozoic–Cenozoic

FORMATION TYPE Continent–continent

With more than 30 peaks of 6,000m (20,000ft) or more in height, the Tien Shan mountain belt is an imposing physical barrier. In 1856, Russian explorer Peter Semonyov became the first

KHAN TENGRI
Engilchek, the largest glacier in the Tien Shan, flows for 62km (39 miles) down the western slopes of the Khan Tengri massif.

European to survey the Tien Shan, whose Chinese name translates as "celestial mountains". The belt consists of a series of parallel ranges that extend for about 2,800km (1,740 miles), with Ozero Issyk-Kul' – one of the largest mountain lakes in the world – located in the northeastern end of the range. The western Alai range includes many lakes, and the surrounding

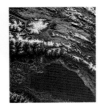

GLACIER COVER
Peaks that rise above 4,000m (13,000ft) in the Tien Shan are capped by glaciers, covering an area of about 9,500 square km (3,670 square miles).

peaks average about 4,600m (15,000ft) in height. In the east, the mountains of the Tien Shan become lower and narrower, but then rise again to over 5,000m (16,400ft) in the Bogda Shan range, with Bogda Feng in China ascending to 5,445m (17,864ft). The east–west trending folds of the range expose lower Paleozoic rocks that have been folded, faulted, and uplifted in a series of tectonic events that included early subduction and subsequent continent–continent collision. These disturbances continue in the present, with massive earthquakes associated with major faults in the region. The Tien Shan range separates two major topographical depressions, the southerly Turpan Depression and the northern Dzungarian

MOUNTAIN FARMS
Fertile pastures flank the Tien Shan Mountains of Kyrgyzstan and Kazakhstan. Herds of goats, sheep, and horses are farmed in the region.

Basin. The floor of Turpan Pendi is 154m (505ft) below sea-level. When the English explorer Charles Howard-Bury visited the area in June 1913, he marvelled at the diversity and profusion of wild flowers, commenting that he had never "seen such a luxuriant flora" as he discovered in the alpine meadows of the Tien Shan. The mountains are the natural habitat of many domesticated ornamental and edible plants, including fruit trees, roses, tulips, and onions. The Tien Shan's animal life includes the endangered Snow Leopard.

ASIA *central*

Himalayas

LOCATION Running southeast from northern Pakistan and India across Nepal to Bhutan

HIGHEST PEAK Mount Everest, Nepal 8,850m (29,035ft)

LENGTH 2,300km (1,400 miles)

AGE Cenozoic

FORMATION TYPE Continent–continent

In addition to being the highest mountains on Earth, the Himalayas are also one of the youngest mountain belts. Formed within the last 50 million years, the belt is largely composed of deformed and metamorphosed rocks from the Indian crustal plate. Mountain building began with the collision of the Indian continental plate and Southeast Asia. This convergence caused India's northern edge to shorten by about 2,000km (1,200 miles), thickening the Indian crust to 55km (34 miles) beneath the Himalayas and 70km (43 miles) below Tibet. The elevation of the Himalayas transformed the region's climate, establishing the annual Southeast Asian monsoon, with its regular torrential rains. Between 250 and 350km (155 and 218 miles) wide, the mountains form a virtually impassable barrier, although one species, the Bar-headed Goose, regularly flies across the peaks from Central Asia to the Indian subcontinent. A series of ranges run parallel along the length of the Himalayan chain. In the south, the mountains rise from the flat, fertile Ganges Plain to the low, gently folded Siwalik Range. This range contains numerous fossils of early land-living mammals, including the first apes, which have been found to be 5–25 million years old. A major thrust fault separates the Siwaliks from the central Himalayas, which rise up to 5,000m (16,400ft) high. Geologically complex, this range includes slices of Precambrian and Paleozoic rocks covered with younger sediments. Brought to the surface by great thrust faults, these rocks have been deeply eroded, and frequent landslides still occur in the region. To the north are the Great Himalayas, which soar to more than 7,000m (23,000ft) above sea-level. These mountains are formed of Upper Paleozoic rocks, slices of the Tethys Ocean floor,

FEATURES OF THE HIMALAYAS

Nanga Parbat · Karakoram Range · Plateau of Tibet · Lhotse · Makalu · Kanchenjunga

K2
At 8,611m (28,253ft) above sea-level, K2, the second highest mountain in the world, was first conquered in 1954 by a team of Italian climbers.

ANNAPURNA
At 8,091m (26,545ft) high, the north face of Annapurna was first climbed by a French team led by Maurice Herzog in 1950.

EVEREST'S PEAK
Only when first surveyed in 1852 was Everest confirmed to be the world's highest peak.

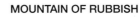

NORGAY AND HILLARY

New Zealander Edmund Hillary (1919–2008) and Nepali Tenzing Norgay (1914–86) were the first people to reach the summit of Everest, on 29 May 1953. Hillary went on to lead the first overland crossing of Antarctica, and Tenzing to train guides at the Mountaineering Institute in Darjeeling, India. The pair carried oxygen tanks to counteract the thin air at high altitude.

PLATEAU OF TIBET
With an average altitude of 5,000m (16,400ft), the Tibetan Plateau, seen in this picture taken from the Space Shuttle, is the highest and largest plateau in the world.

MOUNTAIN OF RUBBISH

Since the summit of Mount Everest was first reached in 1953, more than 6,300 people have ascended the mountain, and over 320 have perished in the attempt. For many, climbing Everest is now merely another extreme sport, but the business provides a valuable income for the local population. However, the disposal of rubbish from climbing parties is now a major concern. Debris, including empty gas bottles, does not rot away at altitude and is difficult to bury.

and granite intrusions that have been thrust over early Cenozoic sediments. Many peaks in this range reach over 8,000m (26,000ft) and are covered with permanent snow and glaciers. Two major rivers, the Indus (see p.223) and the Brahmaputra, have their sources in these mountains. Beyond and to the north of the Great Himalayas lies the vast Plateau of Tibet. The Himalayas are still rising, at rates of up to 4mm (1/16in) a year, growth that is offset by the processes of weathering and erosion. Because the belt is so vast, plant species are varied, with tea cultivated on the lower slopes and oak,

HIMALAYAN POPPIES
Many varieties of poppy collected in the Himalayas in the 19th century are today familiar garden plants all over the world.

rhododendrons, and pine forests growing on the higher slopes. The highest peaks are inhospitable, with little or no vegetation.

FROM OCEAN TO SKY
Many Himalayan rocks were originally sediments on the Tethys Ocean floor. Now they lie within the highest mountains in the world, many of which rise over 8,000m (26,000ft) above sea-level.

LAND

LAND

AUSTRALASIA *Australia*

Great Dividing Range

LOCATION Running from the Cape York Peninsula, Queensland, along Australia's eastern coast to Tasmania

HIGHEST PEAK Mount Kosciuszko, Australia 2,228m (7,310ft)

LENGTH 3,600km (2,250 miles)

AGE Paleozoic

FORMATION TYPE Ocean–continent

To the first Europeans who arrived there, the elevated eastern margin of the Australian continent, now known as the Great Dividing Range, seemed to be a single mountain chain rising abruptly from the narrow coastal plain. However, approached from Cairns in the north, the chain forms an uplifted plateau with a great scarp face set back inland. The Great Dividing Range actually consists of a complex of tablelands and lowlands, rising to higher, alpine-like peaks in the south. The landscape ranges from green hills covered with lush tropical rainforest in Queensland, to snow-covered mountains in New South Wales. The mountains are the source of several major rivers, including the Murray (see p.223) and the Darling. The wildlife inhabiting the range is as diverse as the landscape, and includes kangaroos and platypuses. The highest peak, Mount Kosciuszko, was first noted by Polish-born explorer Paul Strzelecki in 1829, and is named after a Polish national hero.

BLUE MOUNTAINS
Trees cling to the rugged slopes of the Three Sisters, an outcrop in the Blue Mountains, New South Wales.

BREAD KNIFE, WARRUMBUNGLE
The peaks of the Warrumbungle Range, including the Bread Knife (shown here), are the remnants of ancient volcanoes.

AUSTRALASIA *New Zealand*

Southern Alps

LOCATION Running the length of New Zealand's South Island, from northeast to southwest

HIGHEST PEAK Mount Cook, New Zealand 3,744m (12,284ft)

LENGTH 500km (310 miles)

AGE Cenozoic

FORMATION TYPE Ocean–continent

A spectacular range of glaciated mountains, valleys, and fiords, the Southern Alps dominate New Zealand's South Island. The range is at its highest near the centre, where the snow fields around Mount Cook feed glaciers such as the Tasman Glacier (see p.274), which flows for 23km (14 miles). The Southern

ALPINE PARROT
The Kea, a large, opportunistic parrot, has prospered better than its relative, the endangered and flightless Kakapo.

Alps were formed by the convergence of the Australian and Pacific plates. The relatively young range – it was formed in the last 60 million years – is still tectonically active and continues to experience uplift. Year-round rain from westerly winds is among the heaviest in the world, with up to 8m (26ft) falling in a year. Thus the western slopes of the range are clad in forest, while in the more arid rain-shadow to the east, tussocky grasses grow, providing pasture for summer grazing by highland sheep.

SOUTHERN PEAK
Mount Cook is just one of many peaks above 3,000m (10,000ft) in the Southern Alps.

MITRE PEAK, FIORDLAND
South Island's southwest coast is deeply indented with glacial fiords and valleys flanked by steep, triangular peaks.

Transantarctic Mountains

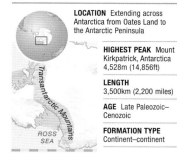

LOCATION Extending across Antarctica from Oates Land to the Antarctic Peninsula

HIGHEST PEAK Mount Kirkpatrick, Antarctica 4,528m (14,856ft)

LENGTH 3,500km (2,200 miles)

AGE Late Paleozoic–Cenozoic

FORMATION TYPE Continent–continent

FEATURES OF THE TRANSANTARCTIC MOUNTAINS

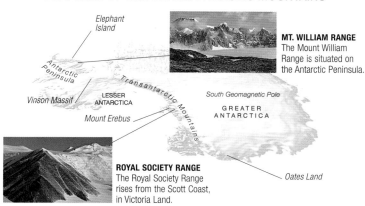

MT. WILLIAM RANGE
The Mount William Range is situated on the Antarctic Peninsula.

ROYAL SOCIETY RANGE
The Royal Society Range rises from the Scott Coast, in Victoria Land.

PAST CLIMATES

On Antarctica's Alexander Island, fossilized conifer plants (see image, below) have been preserved that were alive in the mid-Cretaceous Period, approximately 100 million years ago. One tree is surrounded by fossils of its foliage and those of liverworts and ferns. It was entombed as it grew by sediments from floods. At the time, Antarctica was in a polar position, and the presence of such coniferous woodlands is convincing evidence of a much warmer climate during the period.

The curved belt of the Transantarctic Mountains separates the highly elevated, ancient region of Greater Antarctica in the east from the much younger and lower Lesser Antarctica in the west. More than three-quarters of the continent is covered with a deep ice-sheet (see pp.272–73) but several mountain peaks over 4,000m (13,000ft) high emerge from it. The belt extends northwest through the continent and

DEEP IN SNOW
Forming the northeast coast of the Ross Ice Shelf, the mountains of Queen Maud Land hold back the East Antarctic ice-sheet except where glaciers spill through.

DRY VALLEY
With temperatures permanently below freezing and lying in a rain-shadow, some parts of Antarctica's Dry Valley have seen no rainfall for more than 2 million years (see right).

includes many volcanoes, some still active, including Deception Island. It is thought to mark a region along the western edge of Greater Antarctica that has been subjected to repeated compression and extension events over very long periods. The mountains continue offshore to form the South Sandwich volcanic island arc and the southwestern end of the Pacific Ring of Fire.

LAND

Volcanoes

12–31 The history of the Earth

34–35 The Earth's structure

64 Igneous rocks

88–89 Plate boundaries

Living with volcanoes 168–69

Igneous intrusions 178–84

Geothermal energy 194–95

Ocean tectonics 386–89

Climate change 454–57

Volcanic eruptions of molten rock from within the Earth are awesome demonstrations of the pent-up heat energy stored within our planet. The world is dotted with volcanoes and their rock products, both ancient and modern – reminders of the Earth's long history of volcanism. The variety of materials volcanoes produce, their styles of eruption, and the edifices they create all reflect the geological environments within which they develop. Active volcanoes are found on land and, above all, in the deep sea, and their distribution – which is far from random – reflects the continuing dynamism of the planet.

Erupting Magma

Volcanism is the process by which magma – molten rock from inside the Earth – rises through the Earth's crust onto the surface. Magma chambers within the crust are created by localized melting and upward migration of partially molten crust and mantle rock. This melting requires special conditions, such as the ascent through convection of solid mantle rock beneath oceanic spreading ridges, the addition of seawater at subduction zones, or the presence of rising mantle plumes beneath hotspot volcanoes. Magma travels from the chamber up to the Earth's surface either through fractures in the crust or by literally melting a path through the surrounding rock. A volcano is both the vent at the Earth's surface through which volcanic materials erupt and the edifice created by their accumulation around the vent. As well as magma that erupts as lava, volcanic products include gas, ash, and fragmented rocks (see p.157). These can pour gently out onto the Earth's surface or be blasted through the atmosphere, to be spread globally by high-level winds (see p.156).

VISIBLE FROM SPACE
The scale and power of volcanic eruptions, such as this one at Rabaul, blast gas and debris high into the stratosphere. They can be seen even from space.

VOLCANISM
A typical volcanic system is based around a central "plumbing" network that transfers magma from its crustal chamber to the surface.

spreading eruptive cloud

erupting volcanic vent

volcanic bombs

layers of lava and pyroclastic materials

pyroclastic flow

magma-filled cracks and fissures

surrounding country rock

magma chamber

LAND

Lavas and Volcanoes

Volcanic material erupted at the Earth's surface varies according to the properties of the magma from which it is derived. Differences in chemical composition, gas content, and temperature all affect the way in which magmas erupt, and therefore the type of landforms and volcanoes they produce. Magmas that have high silica content and low temperature are very viscous and slow-moving. If they reach the surface with high gas content, their eruptions are highly explosive. If they lose their gas during ascent, they instead erupt as lava. Silica-rich lavas are thick, slow-moving block lavas that do not travel far from their source. They are associated with subduction at destructive plate margins and the assimilation of rock materials from continental crust. Basaltic lavas, including pahoehoe and 'a'a, have low silica content and high temperatures. They are extremely fluid and can travel at high speeds – up to 100kph (60mph) – over great distances. They are associated with hotspot activity, such as that under Hawaii, as well as ocean-floor spreading, where they produce pillow lavas (see pp.386–87).

PAHOEHOE (KILAUEA)
Known by its Hawaiian name, this fluid basaltic lava cools to form a glassy skin. It can have a flat, hummocky, or ropy surface produced by further flow after the skin has formed.

BLOCK LAVA (VOLCAN DE COLIMA)
Silica-rich, viscous lavas, such as andesites and rhyolites, typically flow as piles of large, fractured blocks.

A'A (KILAUEA)
Pronounced "ah-ah", this fast-flowing basaltic lava has a rough texture, formed as the surface cools and fractures.

SHIELD VOLCANO (MAUNA KEA)
Fluid basaltic lavas that can flow easily across the ground tend to build massive volcanoes with low, angled slopes and a shield-shaped profile.

STRATOVOLCANO (MOUNT FUJI)
The flanks of these large, often steep-sloped volcanoes are built up of alternating layers of more viscous lava and pyroclastic deposits (see p.157).

DOME VOLCANO (MOUNT ST. HELENS)
Lava domes are rounded, steep-sided mounds of silica-rich lava that is too viscous to flow very far from the vent before it cools and crystallizes.

CINDER CONE (CERRO NEGRO)
Steep, straight-sided cinder cones are made of loose, angular, volcanic rock fragments that fall from the eruption cloud and accumulate above a vent.

CALDERA FORMATION
A caldera is a broadly circular depression formed when a volcano's magma chamber is emptied. As the chamber becomes depleted, and support of the overlying rock is removed, the volcano's summit collapses, forming a caldera.

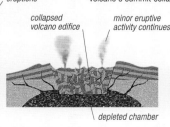
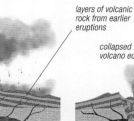
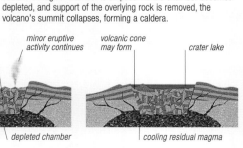

layers of volcanic rock from earlier eruptions

fissures in rock

collapsed volcano edifice

minor eruptive activity continues

volcanic cone may form

crater lake

continuing eruption

emptying magma chamber

depleted chamber

cooling residual magma

Eruption Styles

Volcanologists traditionally have recognized different styles of volcanic eruption, although these styles can often mark different phases in a single event. Strombolian eruptions are spasmodic and discrete explosive bursts, a few seconds long, which eject pyroclastic materials (see opposite page) a relatively short distance into the air. High gas pressures fragment magma within the vent, but with frequent release no sustained eruption cloud develops. Vulcanian eruptions are more violent, producing highly fragmented ash and rock. Eruptive columns can rise 10–20km (6–12 miles), with ash spread over huge areas. Plinian eruptions expel large columns of pyroclasts and gas at high velocities (hundreds of metres per second) up to 45km (30 miles), well into the stratosphere. Collapse of such columns can produce deadly pyroclastic flows, containing gas and pyroclasts of all sizes, travelling at high speeds and covering large distances. In contrast, Hawaiian eruptions produce copious low-viscosity lavas, associated with shield volcanoes. Where similar eruptions occur in shallow water, they are known as Surtseyan. These eruptions can be highly explosive as a result of water invading the vent, expanding as it turns to steam, and fragmenting the magma to blast out glassy fragments.

PLINIAN ERUPTION (MAYON)
These eruptions produce massive clouds of expanding gas that carry volcanic materials high into the atmosphere. They are named after Pliny the Younger, who described the eruption of Vesuviusin 79 CE.

PHREATIC ERUPTION (SURTSEY)
Phreatic eruptions are steam-driven explosions produced when water, either underground or on the surface, is heated by magma or hot rocks, and expands rapidly.

LAHAR (MT. PINATUBO)
The saturation of volcanic ash by water, from heavy rain, glacial meltwater, or drainage of crater lakes, can generate catastrophic mud or debris flows, called lahars, that devastate surrounding landscapes.

FISSURE ERUPTION (KILAUEA)
In 1971, lava erupted from this fissure, which extends southwest from Halemaumau crater at Kilauea volcano, Hawaii, and covered the surrounding landscape in basalt.

STROMBOLIAN ERUPTIONS
Intermittent, small-scale explosive eruptions like this one in Hawaii are produced as gas bubbles through a magma column in a vent.

Volcanic Deposits

Many different types of volcanic material can be preserved in the Earth's rock record, and their analysis reveals a great deal about the nature of past volcanism. Although lavas are the best known volcanic products and can vary considerably in composition, texture, and volume (see p.155), fragmented volcanic rock materials, or pyroclasts, are in many ways the most interesting and have the greatest impact on life. They are produced by explosive eruptions of parent magmas that evolve into gaseous columns containing fragmented rock, minerals, and glassy shards. Deposits are classified by particle size. The smallest, usually called ash, is fine enough to stay aloft for days to weeks and can be distributed

ASH BURIAL
Volcanic deposits of all sizes can smother and bury landscapes and buildings, as here on Montserrat, and totally extinguish life.

LAYERED DEPOSITS
Pyroclastic eruptive material is deposited in a similar way to sediments and accumulates as stratified layers.

globally by high-altitude winds. Fragments can also combine with hot gases to produce glowing pyroclastic flows rolling across landscapes at up to 200kph (125mph). At the other end of the scale, some volcanoes have been known to

LAVA BOMBS
Large fragments of ejected molten rock take on aerodynamic shapes as they travel through the air.

eject rock fragments bigger than a house up to several kilometres. Tephra is a collective term for fragments of volcanic rock and lava of all sizes that are blasted into the air.

MONITORING VOLCANOES

Many millions of people live near dangerously active volcanoes, often because their agriculture relies on rich volcanic soils (see pp.168–69). Advance notice of when eruptions might occur is highly desirable but it also has to be consistently accurate for local populations to heed the warnings. Unlike major earthquakes, which can happen with little or no warning, large-scale volcanic eruptions are often preceded by warning signs. Volcanologists have developed a number of techniques for monitoring the evolution of magmas beneath volcanoes and anticipating when volcanic activity is likely to develop into a potentially dangerous phase. These include measuring and interpreting the seismic activity that accompanies eruptions, analysis of changes in gases expelled from volcanic vents, and remote sensing by satellite to measure any changes in the elevation or shape of the volcanic edifice.

SEISMIC READINGS
Seismometers are used to detect vibrations caused by fracturing rock, magma movement, or bursting gas bubbles prior to eruptions.

Volcano Distribution

The global distribution of volcanoes clearly relates to the Earth's tectonic plates. About 80 per cent of volcanoes above sea-level occur at convergent plate boundaries, 5 per cent occur at divergent boundaries, and mantle hotspots account for the rest, found away from plate boundaries. The volcanic chains around the Pacific (called the Ring of Fire) include volcanic island arcs, formed by ocean–ocean plate convergence, and Andean-style continental arcs, produced by ocean–continent convergence. Continental plate divergence generates rift volcanoes, such as those of eastern Africa, while oceanic plate divergence forms ridges, such as that of the mid-Atlantic, and isolated volcanic islands.

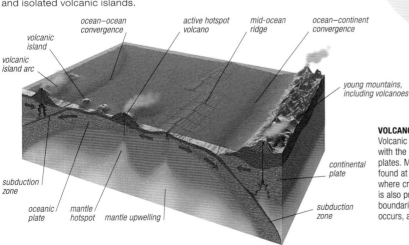

ocean–ocean convergence · active hotspot volcano · mid-ocean ridge · ocean–continent convergence · volcanic island · volcanic island arc · young mountains, including volcanoes · continental plate · subduction zone · oceanic plate · mantle hotspot · mantle upwelling · subduction zone

VOLCANOES AND PLATES
Volcanic activity is closely associated with the movement of tectonic plates. Most active volcanism is found at undersea mid-ocean ridges where crust is being pulled apart. It is also produced by convergent plate boundaries, where subduction occurs, and rising mantle hotspots.

VOLCANO PROFILES

The pages that follow contain profiles of many of the world's most significant and best-known volcanoes. Each profile begins with the following summary information:

HEIGHT Elevation above sea-level of each volcano's highest point. Due to volcanic activity this figure is likely to change over time

TYPE Stratovolcano, shield volcano, cinder cone, caldera, fissure vent, maar (vent formed by steam explosions), tuff cone, lava dome, flood basalt, complex

LAST ERUPTION Date of the last confirmed eruption, or, in the case of ancient volcanoes, an estimate based on rock records

LAND

NORTH AMERICA *northwest*

Novarupta

LOCATION In the Katmai region of the Alaskan Peninsula, in the northwest USA

HEIGHT 841m (2,760ft)

TYPE Caldera, lava dome

LAST ERUPTION 1912

Although Novarupta is the lowest volcano in the Katmai area of Alaska, it was formed, in 1912, by the biggest volcanic eruption of the 20th century. Within 60 hours, some 30 cubic km (7 cubic miles) of silica-rich rhyolitic magma blasted out of Novarupta. Most of it fell back to Earth as ash, while the remainder formed a pyroclastic flow – the renowned Valley of Ten Thousand Smokes (VTTS) ash flow (see p.188). At the end of the eruption, two calderas formed. An indistinct shallow depression about 2km (1.2 miles) wide formed at Novarupta and was followed by the extrusion of a lava dome, 65m (215ft) high and 380m (1250ft) wide, at the vent of the VTTS ash flow. A much larger collapse caldera, about 5km (3 miles) wide, was formed at Katmai volcano 16km (10 miles) to the east, as its underground magma reservoir drained towards Novarupta. Traces of ash that settled after the 1912 eruption have been found in Greenland ice cores.

PLUG DOME
Novarupta's lava dome was formed as masses of viscous lava piled over and around the vent. As it expanded from within, its surface hardened and cracked.

NORTH AMERICA *west*

Mount Rainier

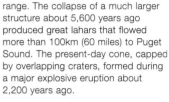

LOCATION In the northern Cascade Range, Washington State, west USA

HEIGHT 4,392m (14,410ft)

TYPE Stratovolcano

LAST ERUPTION 1894

Heavily glaciated, with an ice-filled summit crater complex, Mount Rainier is the highest peak in the Cascade range. The collapse of a much larger structure about 5,600 years ago produced great lahars that flowed more than 100km (60 miles) to Puget Sound. The present-day cone, capped by overlapping craters, formed during a major explosive eruption about 2,200 years ago.

MOUNT RAINIER CAPPED WITH SNOW

NORTH AMERICA *west*

Columbia River Plateau

LOCATION Ranging across southern Washington, northern Oregon, and Idaho

HEIGHT Up to 2,000m (6,600ft)

TYPE Flood basalt

LAST ERUPTION About 15 million years ago

Between 17 and 15 million years ago, about 170,000 cubic km (40,000 cubic miles) of basaltic lava poured out of deep crustal fissures to create the Columbia River Plateau. Many of the flows are over 100km (160 miles) long and up to 30m (100ft) thick, and some cooled to form distinctive shrinkage columns. More than 95 per cent of the basalts may have been erupted in less than 2 million years. The origin of the eruption is thought to have been a rising mantle plume or hotspot beneath the continental crust, which produced partial melting in the upper mantle. The increased heat flow caused the crustal rocks above to dome and stretch, creating deep cracks that tapped the magma below and allowed it to flow out at the surface.

TWIN SISTERS
These basalt pillars were isolated about 15,000 years ago when catastrophic floods carved great channels in the Plateau.

NORTH AMERICA *west*

Crater Lake

LOCATION In the southern part of the Cascade Range, Oregon, west USA

HEIGHT 2,487m (8,159ft)

TYPE Caldera

LAST ERUPTION About 2290 BCE

Crater Lake, the second deepest lake in North America at 589m (1,934ft) deep, fills a caldera 14km (9 miles) wide and is surrounded by a spectacular rock wall. The caldera marks the site of a complex of overlapping volcanoes, known as Mount Mazama. This complex was built up 420,000–40,000 years ago and erupted 6,850 years ago in one of the biggest eruptions of the Holocene (the last 10,000 years). A huge Plinian blast showered ash as far away as Alberta, Canada, and pyroclastic flows travelled up to 40km (25 miles) from the volcano. A series of now-lake-covered lava domes formed on the caldera floor a few hundred years after its formation.

VOLCANIC DEPOSITS
Crater Lake's "Pinnacles" are pyroclastic flow deposits that have eroded into dramatic spires.

FLOODED CALDERA
Crater Lake fills an almost perfectly circular caldera. The lake's surface is broken by Wizard Island, a cinder cone that erupted out of the caldera floor.

Mount St. Helens

LOCATION In the northern region of the Cascade Range, Washington State, west USA

HEIGHT 2,549m (8,363ft)

TYPE Stratovolcano

LAST ERUPTION 2008

The strikingly symmetrical cone of Mount St. Helens had long been admired as America's Mount Fuji. However, on 18 May 1980, in one of the world's best-documented eruptions, the top 1,400m (4,600ft) of the volcano was destroyed, leaving a horseshoe-shaped caldera. Magma intrusion caused the north flank to bulge outward and then collapse in a debris

PLINIAN ERUPTION

In 1980, Mount St. Helens produced a Plinian-style eruption, blasting an eruptive column of gas, ash, and pumice fragments 24km (15 miles) into the atmosphere.

BEFORE ERUPTION
The volcano's conical peak was flanked by evergreen forests and grassy meadows.

avalanche, uncorking an explosive eruption. The avalanche travelled about 25km (16 miles) at speeds of up to 75m (250ft) per second, filling a valley with debris up to 195m (640ft) deep. An explosive blast of hot gases and magma travelled ahead of the debris avalanche, flattening more than 10 million trees over about 600 square kilometres (230 square miles). Later, collapse at the base of the eruptive column produced numerous pyroclastic flows, at temperatures of about 700°C

AFTER ERUPTION
Seared by a lethal combination of red-hot gases, rocks, and ash, the landscape to the north of the volcano resembled a desolate moonscape.

(1,300°F). Formed in nine eruptive phases beginning 50,000–40,000 years ago, Mount St. Helens has been the most active volcano in the Cascade Range during the Holocene Epoch (the last 10,000 years). The modern volcano has been constructed within the last 2,500 years as basaltic, andesitic, and dacitic magma erupted from both summit and flank vents.

WILDLIFE RETURNS
Although about 2,000 elk died in the eruption, large numbers have since returned to the area to graze.

RATES OF RENEWAL

Catastrophic volcanic eruptions can destroy the plant and animal life of a region. Mount St. Helens provided an invaluable opportunity to monitor the regeneration of wildlife after an eruption. Studies revealed that recolonization occurs at a much faster rate than scientists had estimated. Mats of river algae (see below) and sprouting shrubs were harbingers of regrowth, followed by birds, small mammals, and, eventually, elk.

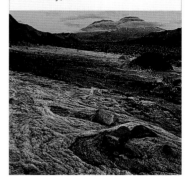

08:32:47 08:32:53 08:33:00 08:33:19

1 2 3 4

THE MOUNTAIN EXPLODES
As the north side of the volcano collapsed, a rock avalanche was set in motion (1). Pressure inside the volcano was released, and clouds of ash blasted out (2–3). These eventually caught up with and overtook the avalanche (4).

PACIFIC OCEAN *central*

Kilauea

LOCATION On the southeast coast of Hawaii, the most southeasterly of the Hawaiian Islands

HEIGHT 1,222m (4,009ft)

TYPE Shield volcano

LAST ERUPTION 2023

Kilauea is the most active of the overlapping volcanoes that have built up the island of Hawaii over the past million years. With its massive low-angle slope profile and wide caldera, it is a typical shield volcano built up by frequent summit and flank basaltic lava flows interspersed with lava-lake

LAVA FIELD
Low-viscocity, basaltic lava streams down the flanks of Kilauea. Lava flows can eventually reach the coast, where they cascade into the sea.

activity within the crater. About 90 per cent of Kilauea's surface is formed by lava that is less than 1,100 years old. Since 1983, during a period of constant eruption, flows have covered over 100 square kilometres (40 square miles) and destroyed up to 200 houses. Most of Kilauea is hidden beneath the sea. Hawaii is part of a chain of volcanic islands built up over the past 75 million years as the Pacific Plate passed over a mantle hotspot at a rate of about 9cm (3½in) per year.

SAMPLING LAVA

Sampling a volcano's lava is one way to monitor changes in its behaviour. Volcanologists studying Kilauea throw hammer-headed cables into lava tubes or, as shown here, take hand samples directly from lava flows or vent spatter. It is important to take hot lava samples, to obtain rapidly quenched glass for analysis.

PACIFIC OCEAN *central*

Mauna Loa

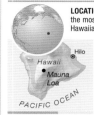

LOCATION In the centre of Hawaii, the most southeasterly of the Hawaiian Islands

HEIGHT 4,170m (13,681ft)

TYPE Shield volcano

LAST ERUPTION 2022

SUMMIT CRATERS

Rising nearly 9km (5½ miles) from the ocean floor, and covering about half the island of Hawaii, Mauna Loa is the largest active volcano in the world. Almost 90 per cent of its surface is basaltic lava less than 4,000 years old, erupted from flank fissures and overflowed from a summit lava lake. There have also been large-scale submarine avalanches, one of which travelled nearly 100km (60 miles).

NORTH AMERICA *south*

Parícutin

LOCATION In the Mexican Volcanic Belt, of southwestern Mexico, inland from the Pacific coast

HEIGHT 2,809m (9,216ft)

TYPE Cinder cone

LAST ERUPTION 1952

Parícutin is the best-known volcano of about 1,000 volcanic centres in the Michoacán–Guanajuato volcanic field. It is a cinder cone that started to erupt from a cornfield in 1943 and grew to 336m (1,102ft) within a year. The eruption continued until 1952, adding an extra 88m (288ft) to its height. Parícutin provided volcanologists with the rare chance to witness and document the birth, growth, and death of a volcano.

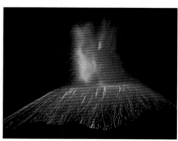

VOLCANO ABLAZE

NORTH AMERICA *south*

Colima

LOCATION In the Mexican Volcanic Belt, of southwestern Mexico inland from the Pacific coast

HEIGHT 3,860m (12,664ft)

TYPE Stratovolcano

LAST ERUPTION 2019

SHEER FLANKS
The steep profile of the Volcán de Colima has been shaped by viscous block-lava flows.

Colima is the most prominent complex in the western Mexican Volcanic Belt. It is made up of the 4,320m (14,173ft) Nevado de Colima and the younger 3,860m (12,664ft) Volcán de Colima, an active stratovolcano that grew on the south flank of Nevado in a southward-moving pattern typical of the belt. Volcán de Colima is Mexico's most active volcano – its active cone lies within a 5-km- (3-mile-) wide caldera formed by the collapse of an ancestral cone. Major slope failures at both Nevado and Volcán de Colima have formed a thick apron of debris-avalanche deposits. One of the biggest known occurred about 18,000 years ago, travelling for 120km (70 miles)

to the Pacific coast. The volcanic complex began to grow in late Pleistocene times, and recorded eruptions date back to the 16th century with occasional major explosive eruptions. The most recent, in 1913, destroyed the summit and left a steep-walled crater that was refilled and then overtopped by the growth of a lava dome that has fed five block lava flows since 1961.

GAS SAMPLING
Volcanologists collect samples from a fumarole field at the rim of Colima's crater. Monitoring composition and temperature of the gases can help scientists understand and predict the volcano's behaviour.

NORTH AMERICA *south*

Popocatépetl

LOCATION In the Mexican Volcanic Belt, within the Puebla region of south-central Mexico

HEIGHT	5,452m (17,887ft)
TYPE	Stratovolcano
LAST ERUPTION	2023

FERTILE SOILS

Volcanic products are rich in minerals and glass that can produce very fertile and well-drained soils. In highly populated areas, such as here in the shadow of Popocatépetl, the fertile soil is of great benefit to local farmers, despite the potential risks.

Popocatépetl is North America's second highest volcano, towering over Mexico city to its northwest. With its symmetrical shape, summit glaciers, and frequent eruptions during recorded history, it has an imposing presence. From the Pleistocene Epoch onwards, the volcano has grown to 28km (17½ miles) across. The summit crater is 670m (2,200ft) in diameter and up to 450m (1,480ft) deep. Since Popocatépetl's formation, at least three previous cones have been destroyed by gravitational collapse. These have produced massive debris-avalanche deposits to the south. Three major Plinian eruptions have occurred over the last few thousand years, most recently in 800 CE, producing pyroclastic flows and large-volume lahars. The Aztecs recorded frequent eruptions before the arrival of Western explorers in the 15th century, and gave the volcano its name, meaning "smoking mountain".

STEEP-WALLED CRATER
Popocatépetl's deep, oval-shaped crater has near-vertical walls. On the crater floor, a succession of sporadically erupting lava domes has evolved.

NORTH AMERICA *south*

El Chichón

LOCATION In an isolated part of the Chiapas region of southeastern Mexico

HEIGHT	1,060m (3,478ft)
TYPE	Lava dome
LAST ERUPTION	1982

This lava-dome and tuff-cone complex was relatively unknown until 1982, when it produced a series of highly explosive eruptions. Pyroclastic flows swept outward for more than 8km (5 miles), destroying 9 villages and killing more than 2,000 people. The magma was extremely sulphur-rich, and sulphuric acid droplets forming in the stratosphere produced brilliant sunsets around the globe.

ACIDIC POST-ERUPTION CRATER LAKE

CENTRAL AMERICA

Masaya

LOCATION Between the west coast of Lake Nicaragua and the Pacific coast, southwestern Nicaragua

HEIGHT	635m (2,083ft)
TYPE	Caldera
LAST ERUPTION	2023

ESCAPING GASES

One of Nicaragua's most active volcanoes, Masaya is an 11-km- (7-mile-) wide caldera, with more than a dozen vents, surrounded by walls 300m (985ft) high. The twin cones of Masaya and Nindiri have been the source of frequent recorded eruptions. A major Plinian eruption of basaltic tephra occured about 6,500 years ago. Since then, frequent lava flows have covered much of the caldera floor, and in 1670 CE one overtopped the northern rim.

CENTRAL AMERICA

Arenal

LOCATION On the southeastern shore of Lake Arenal in central Costa Rica

HEIGHT	1,657m (5,436ft)
TYPE	Stratovolcano
LAST ERUPTION	2010

Arenal is the youngest and one of the most active volcanoes in Costa Rica. The 1968 Vulcanian eruption signalled the start of the current long-term eruptive period. It threw large bombs up to 5km (3 miles) from the crater and generated pyroclastic flows (one of which killed 70 people) before releasing viscous lava flows from vents at the summit and the upper part of its western slope.

INCANDESCENT ROCK AVALANCHES

CENTRAL AMERICA

Soufrière Hills

LOCATION Occupies the southern half of the island of Montserrat in the eastern Caribbean Sea

HEIGHT	915m (3,002ft)
TYPE	Stratovolcano
LAST ERUPTION	2013

Soufrière Hills is a complex volcano, with a series of lava domes forming its summit area. Although swarms of seismic shocks had been felt at 30-year intervals throughout the

NEW DELTA
Some pyroclastic flows have travelled all the way to the coast and discharged over and into the sea, creating new deltas.

ERUPTION CLOUD
Soufrière Hills produces Plinian eruptions, sending huge clouds of ash high into the air and smothering the surrounding area.

20th century, the volcano was thought to be inactive. Then, in an eruption beginning in 1995, pyroclastic flows and mudflows overran and destroyed Plymouth, Montserrat's capital, burying it under several metres of pyroclastic deposits. The southern end of the island was evacuated and, inevitably, the population suffered major disruption. As the activity has continued, Soufrière Hills has become one of the world's most closely monitored volcanoes.

LAND

SOUTH AMERICA *northwest*
Nevado del Ruiz

LOCATION In the northern region of the Cordillera Central range, central Colombia

HEIGHT 5,321m (17,457ft)

TYPE Stratovolcano

LAST ERUPTION 2023

The modern volcano of Nevado del Ruiz sits within the caldera of the older Ruiz volcano. Its cone consists of a series of lava domes, with the summit occupied by the Arenas crater, 1km (3,300ft) wide and 240m (790ft) deep. The volcano's flanks are shaped by large landslides.

Although Nevado del Ruiz is only about 500km (300 miles) north of the equator, its position high in the Andes means that it is capped throughout the year by large volumes of snow and ice. Because of this, heat released by eruptions has, during recorded history, caused summit glaciers to melt, creating devastating lahars. In 1985, Nevado del Ruiz was the scene of South America's deadliest volcanic eruption and one of the worst volcanic disasters of modern times (see panel, right). A relatively small volume of hot ejecta, spewed across the snow- and ice-covered summit, proved to be a lethal combination.

BROAD MASS
Nevado del Ruiz is a wide, sprawling volcano that covers over 200 square km (77 square miles).

LAHAR DISASTER
On 13 November 1985, pyroclastic flows, released by a small eruption, melted the summit ice-cap, triggering a series of devastating lahars. Channelled down narrow valleys for up to 100km (60 miles), they wiped out the town of Armero and killed more than 23,500 people. When the flows stopped, they set like concrete, offering victims little chance of escape.

SOUTH AMERICA *northwest*
Galeras

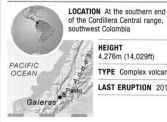

LOCATION At the southern end of the Cordillera Central range, southwest Colombia

HEIGHT 4,276m (14,029ft)

TYPE Complex volcano

LAST ERUPTION 2014

Now a modern cone positioned within an older caldera, Galeras has been active for more than a million years. Major eruptions and weakening of the edifice on many occasions have led to debris avalanches, pyroclastic flows, and widespread air-fall deposits. Situated next to the city of Pasto, the volcano is closely monitored.

PASTO'S VOLCANIC BACKDROP

SOUTH AMERICA *northwest*
Cotopaxi

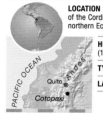

LOCATION At the southern end of the Cordillera Central range, northern Ecuador

HEIGHT 5,911m (19,393ft)

TYPE Stratovolcano

LAST ERUPTION 2023

The glacier-clad Cotopaxi is the highest active volcano on Earth. One of Ecuador's most active volcanoes, it has a steep-sided, almost perfectly symmetrical cone capped with deep craters nested within the summit.

ICY CRATER
Situated in the northern Andes, Cotopaxi is the highest active volcano in the world. Its peak is clad with glaciers and a constant snow covering.

VOLCANIC DEBRIS
A huge block of volcanic rock lies on a flank of Cotopaxi where it was probably deposited by a large debris avalanche or powerful lahar.

The present cone, scarred by lava flows and lahars, has been constructed over the last 5,000 years. Cotopaxi has a long history of explosive eruptions, including that of 1534, which put an end to a battle between the Incas and Spaniards. In one of its most violent recorded eruptions, in 1877, an eruptive column collapsed, generating pyroclastic flows and lahars that devastated nearby valleys and flowed over 100km (60 miles) into the Pacific Ocean.

SOUTH AMERICA *southwest*
Cerro Azul–Quizapu

LOCATION In the southern Andes, central Chile, close to the border with Argentina

HEIGHT 3,788m (12,428ft)

TYPE Stratovolcano

LAST ERUPTION 1967

The steep-sided Cerro Azul volcano is located at the southern end of a volcano chain called the Descabezado Grande–Cerro Azul eruptive system. It has a 500-m- (1,640-ft-) wide summit crater and several flank cinder cones, including the three La Resoloma craters to the west and Los Hornitos to the southwest. Quizapu is a vent on the northern flank that formed in 1846 with the emission of a large volume of lava. In 1932, Quizapu was the site of one of the largest explosive eruptions of the 20th century, ejecting a large volume of tephra and creating a crater up to 700m (2,300ft) wide and 150m (490ft) deep.

QUIZAPU CRATER

SOUTH AMERICA *east*
Paraná Plateau

LOCATION Across southern Brazil, eastern Paraguay, northern Argentina, and northwest Uruguay

HEIGHT Up to 1,000m (3,300ft)

TYPE Fissure vent

LAST ERUPTION About 120 million years ago

The eruption about 120 million years ago of more than 750,000 square km (290,000 square miles) of basaltic lava in the Paraná region of eastern South America coincided with the splitting of that continent from Africa and the opening of the south Atlantic Ocean. Heat from within the mantle formed the Walvis hotspot under the Paraná region of eastern South America and Namibia in southwest Africa, which at the time were joined together. Doming of the crust over a 1,000-km- (625-mile-) wide zone developed deep and extensive fractures, through which lava poured onto the Earth's surface. Continuing volcanic activity led to further rifting between the continents and the formation of a new ocean basin. The hotspot is today marked by the volcanic island of Tristan da Cunha, which last erupted in 1961.

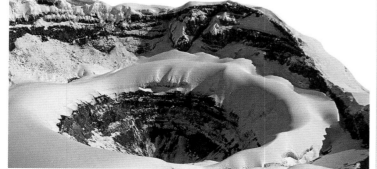

ATLANTIC OCEAN *north*

Grímsvötn

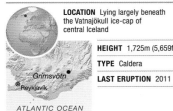

LOCATION Lying largely beneath the Vatnajökull ice-cap of central Iceland

HEIGHT 1,725m (5,659ft)

TYPE Caldera

LAST ERUPTION 2011

Iceland's most active volcano in recorded history, Grímsvötn is largely covered by the Vatnajökull ice-cap. Only the southern rim of its 8-km- (5-mile-) wide caldera is exposed above the ice, and heat from the volcano frequently causes floods of melted glacier ice (jökulhlaups). Long fissures extending from the volcano

LAKI FISSURE
The Laki fissure system is 27km (17 miles) long and contains over 140 craters.

include the well-known Laki fissure, which produced the largest lava flow in recorded history. In 1783 about 15 cubic km (3½ cubic miles) of lava were erupted over a period of seven months. More than 9,000 people died from starvation and disease in what became Iceland's worst natural disaster.

ERUPTING DOME
During an eruption in 1998, a plume 10km (6 miles) high erupted from the caldera.

EUROPE *west*

Antrim Plateau

LOCATION On the coast of County Antrim in northeastern Northern Ireland

HEIGHT Up to 550m (1,800ft)

TYPE Fissure vent

LAST ERUPTION About 55 million years ago

About 55 million years ago, a rising mantle plume caused the northwest of the British Isles and southeast Greenland to dome upwards. Rifting and sea-floor spreading erupted a pile of basaltic lavas up to 2km (1¼ miles)

thick, with flows up to 8m (26ft) thick. The lava flows are interbedded with pyroclastic material, and plant fossils in the beds show a humid temperate climate. The Giant's Causeway, a group of thousands of mostly hexagonal basalt columns, is world-famous for its spectacular appearance.

THE GIANT'S CAUSEWAY

EUROPE *central*

West Eifel Field

LOCATION In the Rhine Valley, south of the city of Cologne, western Germany

HEIGHT 600m (1,968ft)

TYPE Maars, cinder cones

LAST ERUPTION 7,000–11,000 years ago

The West Eifel Field is associated with the Rhine Rift (see p.127). Over the last 730,000 years, it has seen alternating episodes of Strombolian and phreatic activity, producing some 240 cinder cones, maars, and small stratovolcanoes. They

MAAR MINING

cover an area of about 600 square km (230 square miles), with the last eruption occurring 7,000–11,000 years ago. More than 70 maars, up to 1km (3,300ft) in diameter and up to 200m (650ft) deep, were formed between 12,500 and 10,000 years ago, when the end of the last ice age produced abundant meltwater.

EUROPE *west*

Chaîne des Puys

LOCATION In the Auvergne region of central France, in the northern Massif Central

HEIGHT 1,464m (4,803ft)

TYPE Cinder cones, lava domes, maars

LAST ERUPTION About 4040 BCE

The Chaîne des Puys was keenly studied by 19th-century scientists and played an important role in the history of geology. It is a series of cinder

cones, lava domes, and maars aligned north to south. The maars are craters formed by explosive eruptions, rimmed by ejected debris, and often filled by lakes. Their formation began about 70,000 years ago, with pyroclastic flows and long lava flows, and finally ended about 6,000 years ago. The most recent volcanic activity occurred near Besse-en-Chandesse. It has been radiocarbon-dated at around 4040 BCE and included powerful explosions that created Lac Pavin Maar.

VOLCANO CHAIN
A series of about 90 cinder cones, maars, and lava domes extends across nearly 50km (35 miles) of France's northern Massif Central.

EUROPE *south*

Stromboli

LOCATION One of the volcanic Aeolian Islands off the north coast of Sicily in the Mediterranean

HEIGHT 926m (3,038ft)

TYPE Stratovolcano

LAST ERUPTION 2023

The small island of Stromboli is an active volcano rising about 3,000m (9,850ft) from the ocean floor. It is one of the few almost continuously erupting volcanoes on Earth, having been active since records began more than 2,500 years ago. Stromboli's active summit vents are located in the Sciara del Fuoco, a curved scarp that was formed by a slope failure about 5,000 years ago. The volcano gave its name to one of the main types of volcanic eruption (see p.156). Most of its activity is moderate in scale,

consisting of brief, but explosive, ejections of glowing lava fragments to heights seldom more than 150m (490ft) and, less frequently, lava flows. Periods of more intense activity can produce prolonged eruptions from fountains and ejection of large bombs.

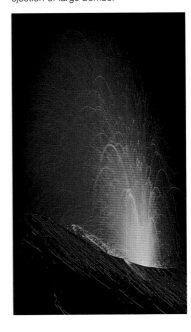

INCANDESCENT DISPLAY
This long-exposure photograph shows a typical small-scale explosive eruption ejecting glowing rock particles into the air.

UNPREDICTABLE ERUPTIONS
Predicting volcanic eruptions is difficult. Puyehue in Chile lay dormant for decades and then erupted in June 2011. Ash clouds from the eruption, seen here illuminated by lightning, darkened skies as far away as Argentina.

Mount Etna

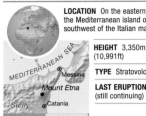

LOCATION On the eastern side of the Mediterranean island of Sicily, southwest of the Italian mainland

HEIGHT 3,350m (10,991ft)

TYPE Stratovolcano

LAST ERUPTION 2023 (still continuing)

Etna is Europe's highest volcano, and has been almost continuously active since observations were first recorded, about 2,500 years ago. About 2.5 million years old, Etna probably originates from a mantle hotspot. The volcano has a tall cone with a base circumference of about 150km (95 miles) and a surface area of about 1,600 square km (620 square miles). Etna's most prominent feature is the Valle del Bove, a 10km- (6-mile-) wide horseshoe-shaped caldera open to the southeast. The summit is a complex structure of old and new craters, and currently there are four active craters. The sides of the volcano are scored by deep fissures that radiate

SNOWY PEAK
Close to Etna's summit, the desolate wasteland of solid lava and ash is covered with snow for much of the year. Its fertile lower slopes are widely cultivated.

from the summit. Up to 200 small ash cones are present on its flanks, with three larger cones near the summit. Etna produces basaltic lava that has low viscosity, and lava flows extend to the foot of the volcano on all sides, reaching the sea over a broad area on the southeast flank. The summit craters are in a permanent state of low-level activity, with lava fountains erupting and phreatic explosions (see pp.156–57). Although Etna has not been considered an especially explosive volcano, at times it can be spectacular. Etna's largest and most famous eruption, in 1669, lasted 122 days. Earthquakes opened a 12-km- (8-mile-) long fissure and outpourings of basaltic lavas rapidly overwhelmed nearby villages (see panel, below). The 1986 summit eruption produced lava fountains that jetted up to 1.6km (1 mile) above the crater, and a column of ash and gas that rose 10km (6 miles) into the atmosphere. More recently, Etna erupted so much material over a six-month period that its height grew by 30 m (100 ft).

RIVER OF FIRE
Broad rivers of alkali basaltic lava flow down the lava field of the Valle del Bove on Etna's southeast flank in January 1992.

ERUPTIVE DISPLAY
On 28 July 2001, explosions from Etna's newly formed Lago cone hurled clouds of ash high into the air (1 and 2). Later, molten materials were discharged from two separate vents simultaneously (3 and 4).

DIVERTING LAVA FLOWS

Efforts to divert destructive lava flows on Etna date as far back as 1669, when lava overwhelmed a number of villages and threatened the town of Catania. A party was sent to try and breach a levee and allow the lava to spread in a different direction, away from the town. However, this action met with opposition from neighbouring villagers, and the lava flow was eventually left to enter the town. In recent years, methods have involved the use of industrial equipment to build earth dams and concrete barriers, and even the use of explosives. However, most lava-diversion attempts meet with only temporary success.

STROMBOLIAN ACTIVITY ON ETNA
Fountains of fluid lava and violent ejection of partially molten volcanic bombs, seen in this long-exposure photograph, are typical of Strombolian eruptions (see p.156).

OUT OF THE ASHES
Planted in pits designed to trap water, vines flourish in arid but fertile volcanic soil on Lanzarote, the easternmost of the Canary Islands.

LIVING WITH VOLCANOES

Volcanoes and the materials that they produce are essential for life on Earth. Without them, surface water would not have formed, and living things could not have evolved. Volcanoes also raise minerals and melts from deep in the Earth's crust, helping to create fertile soil. But this beneficial role comes at a price, because eruptions are among the most deadly hazards in the world. That danger cannot be removed, but with modern technology the risk it carries can be reduced.

Volcanoes as Neighbours

Although volcanoes have always been feared, people recognized their beneficial effects long before anything was known about the chemistry of soils. In Sicily, for example, the lower slopes of Mount Etna have been farmed for several thousand years, and were an important source of agricultural produce in classical times. On the Indonesian island of Java, weathered volcanic soil currently supports one of the densest populations on the Earth, with an average of nearly 1,100 people per square kilometre (2,850 per square mile). By way of contrast, the Amazon Basin, which is situated on almost exactly the same latitude as Java and enjoys a similar climate, is non-volcanic, and suffers from extremely poor soils, which are difficult to cultivate successfully.

Mount Etna has erupted more than 200 times in recorded history, but, despite this turbulent past, it has inflicted relatively little loss of human life. Etna's immense size means that most farms are distant from the summit, and the mountain's lava streams often follow predictable paths that people can avoid. Many volcanoes in populated regions, however, are prone to violently explosive eruptions (see p.156). Java's volcanoes fall into this category, as do those of Mount Pinatubo, in the Philippines. When the latter erupted in 1991, the death toll was high, and thousands of farms were buried by ash.

CLOSE ESCAPE
These houses in Shimabara, Japan, were damaged after Mount Unzen erupted in 1993 – about 200 years after the last devastation.

STANDING SENTINEL
A volcano rises high above rice-fields and buildings in Japan. More than 60 volcanoes are recorded as active in the Japanese archipelago.

Warning Signs

Modern volcanology began in the early 1900s, when an observatory was built on Kilauea – the world's largest active volcano – in Hawaii. Before this, people relied on a mixture of observation and superstition to judge when a volcano might erupt. Seismic tremors provided some clues, as did the sudden escape of gas, but just as much attention was paid to unusual behaviour in wild animals. Unfortunately, these portents often turned out to be unreliable – a fact vividly demonstrated in 79 CE, when Mount Vesuvius in Italy erupted, burying Pompeii and its inhabitants under several metres of ash (see p.170).

Volcanologists now gather several kinds of data to predict volcanic activity. One of the most useful is the physical deformation that occurs when magma rises up through a volcano's interior. Sudden expansion – such as that witnessed in Mount St. Helens, USA, in 1980 – is one of the surest signs that an eruption is imminent. Until recently, deformation measurements had to be taken manually, using instruments fixed to a crater's rim, but measurements can now be done remotely by satellite-based radar without any human presence at the crater.

Monitoring Volcanoes

Seismic activity is the most widely monitored parameter at active volcanoes – a warning sign is an increase or, just as ominously, a sudden stop in movement. Such information is gathered by automated seismic sensors and GPS, and is then transmitted to research stations for analysis. Seismic sensing provided advance warning of the Mount Pinatubo eruption, although, because of its immense scale, not everyone was able to escape.

Moving magma causes changes in a volcano's magnetic field, and vent-gas analysis provides a further line of evidence, because the concentration of gases changes as hot magma approaches the surface. Originally developed in Japan, vent-gas analysis can signal increased volcanic activity months before a volcano explodes, and, after an eruption, it also provides confirmation that volcanic activity is in decline. As yet, a fail-safe method of predicting eruptions does not exist, but with these monitoring techniques that goal may not be far off.

GAS ANALYSIS
A volcanologist collects gas samples on the island of Vulcano, off the north coast of Sicily. The last major eruption here was in 1890.

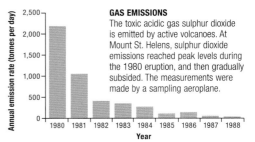

GAS EMISSIONS
The toxic acidic gas sulphur dioxide is emitted by active volcanoes. At Mount St. Helens, sulphur dioxide emissions reached peak levels during the 1980 eruption, and then gradually subsided. The measurements were made by a sampling aeroplane.

LAND

Vesuvius

LOCATION On the eastern coast of the Bay of Naples, 12km (7 miles) east of Naples in southwest Italy

HEIGHT	1,281m (4,203ft)
TYPE	Complex volcano
LAST ERUPTION	1944

Vesuvius is a frequently active volcano overlooking the Bay of Naples. Its cone, with a base circumference of about 70km (45 miles), sits within the caldera of an older volcano, Monte Somma, that formed about 17,000 years ago. Vesuvius is best known for its Plinian eruptions, such as the one in 79 CE that destroyed Pompeii (see panel, right) and Herculaneum. Since the modern cone formed, there have been eight major eruptions, accompanied by large pyroclastic flows and surges, and high eruptive columns capable of carrying pumice, ash, and bombs tens of kilometres into the air. In 1631, during the largest eruption since 79 CE, pyroclastic flows reached as far as the sea. More recently, Vesuvius erupted in 1906 and 1944, causing fatalities on both occasions. Today, about 3 million people live within the eruptive range of Vesuvius, and evacuation plans are in place.

LAST MAJOR ERUPTION
The 1944 eruption of Vesuvius claimed the lives of 27 people. As the average repose time between eruptions is 50 years, the next eruption is overdue.

GIANT PROFILE
The city of Naples sits in the shadow of Vesuvius, whose fertile slopes are dotted with villages and vineyards.

DESTRUCTION OF POMPEII

Most of the victims of the eruption that destroyed Pompeii in 79 CE died very suddenly, smothered by fast-moving, high-temperature pyroclastic surges. A town of about 20,000 people, Pompeii was buried under 3m (10ft) of materials that effectively "fossilized" the inhabitants and building remains.

Santorini

LOCATION One of the volcanic Greek Cyclades Islands in the eastern Mediterranean Sea

HEIGHT	367m (1,204ft)
TYPE	Caldera
LAST ERUPTION	1950

SANTORINI'S FLOODED CALDERA

Santorini is a complex of overlapping volcanoes and calderas forming a circular group of islands with steep, inward-facing walls. The most recent caldera-forming eruption occurred in about 1640 BCE, when about 58–68 cubic km (14–16 cubic miles) of material was ejected, and the current sea-filled caldera was formed. It is thought that this eruption might have caused the collapse of the Minoan civilization. New islands in the centre of the caldera were formed by a series of eruptions beginning in 197 BCE.

Vulcano

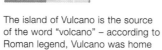

LOCATION One of the volcanic Aeolian Islands off the north coast of Sicily in the Mediterranean Sea

HEIGHT	500m (1,640ft)
TYPE	Stratovolcano
LAST ERUPTION	1890

The island of Vulcano is the source of the word "volcano" – according to Roman legend, Vulcano was home to Vulcan, the god of fire, and his forges. It also gives its name to the Vulcanian style of eruption, which is characterized by moderate-to-violent eruptive columns dominated by viscous and solid ejecta. Although it is only a small island (with an area of 22 square km/8.5 square miles), Vulcano is constructed of four volcanic complexes that have developed over the last 120,000 years. The Piano caldera, in the southern half of the island, is overlapped to the north by the Fossa cone, a modern volcano that last erupted in 1888–90. Vulcanello, on the northern tip of the island, is a partly submerged volcano with a lava platform, which now forms a low peninsula. The Vulcano complex also includes a combination of hot springs, mud pots, and active fumaroles.

NEARBY VILLAGE
A busy seaside town on the neighbouring island of Lipari sits in the shadow of the summit of the active Vulcano.

FUMAROLE GASES
Vulcano has become a popular site with volcanologists who monitor the gases emitted by the many steaming fumaroles on the volcano's flanks.

Pico de Teide

LOCATION On Tenerife, one of the Canary Islands in the Atlantic Ocean, off the northwest coast of Africa

HEIGHT
3,715m (12,188ft)

TYPE Stratovolcano

LAST ERUPTION 1909

The island of Tenerife was formed by a complex of overlapping volcanoes that remain active today. Pico de Teide lies within the 17-km- (10-mile-) wide Las Cañadas caldera. An eruption in 1492 was witnessed by Christopher Columbus on his way to the New World.

FLANK CONE

Erta Ale

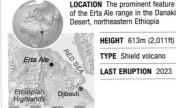

LOCATION The prominent feature of the Erta Ale range in the Danakil Desert, northeastern Ethiopia

HEIGHT 613m (2,011ft)

TYPE Shield volcano

LAST ERUPTION 2023

The remote Erta Ale is a shield volcano 50km (30 miles) wide that rises to over 600m (2,000ft) from below sea-level. Its 1.km-(1-mile-) wide elliptical summit crater contains circular, steep-sided pit craters. Another, larger, elongated depression parallel to the Erta Ale range is located to the southeast of the summit and is bounded by curvilinear fault ridges. Recent basaltic lava flows have erupted from these fault fissures into the caldera and have overflowed its rim. Erta Ale is Ethiopia's most active volcano, and its summit pit craters are renowned for their perpetually churning lava lakes. These have been active since at least 1967, but possibly since 1906, which would make it one of the longest known eruptions in recorded history. Recently, fissure eruptions have occurred on the volcano's northern flank.

PIT CRATER
The smaller of the two summit craters, Erta Ale's active south pit crater is about 150m (500ft) across its almost perfectly proportioned diameter.

VOLCANO CAULDRON
Within the summit crater, fountains of molten lava break through the churning lava lake's solidified, black surface crust.

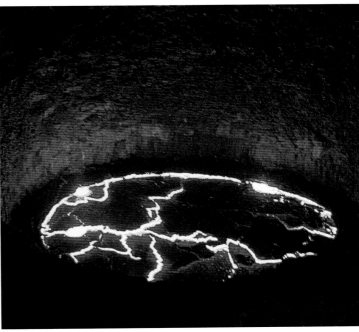

Mount Kilimanjaro

LOCATION At the southern end of the Great Rift Valley's eastern fork, in northeastern Tanzania

HEIGHT
5,895m (19,340ft)

TYPE Stratovolcano

LAST ERUPTION
Unknown

Africa's highest mountain, Kilimanjaro is one of several volcanoes whose formation is associated with the eastern part of the Great Rift Valley (see pp.128–29). An elongated complex cone, more than 50km (30 miles) long, it consists of three volcanoes. The mountain was mainly constructed during the Pleistocene Epoch (1.8 million–10,000 years ago) but includes a group of nested summit craters that are apparently younger. Kibo is the highest and central cone and erupted between the other two – Shira and Mawenzi. It is topped by a caldera, once covered by a glacier, with an inner crater called the Ash Pit. The 3.5-km- (2¼-mile-) wide caldera gives the summit its broad, elongated profile. Most of the 250 flanking lava and cinder cones are less than 100m (330ft) high. Recent volcanic activity has been mostly confined to fumaroles around Kibo's crater. Lying just 320km (200 miles) north of the equator and yet high enough to be snow-capped, Kilimanjaro supports a unique ecosystem adapted to a wide range in temperature.

MASSIVE MOUNTAIN
Rising from the savanna plains of eastern Africa, the massive Kilimanjaro is one of the most famous and most imposing volcanoes in the world.

UNIQUE VEGETATION
This water-holding cabbage in tussock grassland is typical of the specially adapted alpine vegetation.

GLOBAL WARMING

Global warming is transforming one of Africa's most iconic landmarks. Recent studies have suggested that since 1912 more than 80 per cent of the ice and snow at its summit has melted, and what remains is likely to vanish in the near future. As well as reshaping a dramatic skyline, the consequences for communities that rely on the meltwater could be huge.

LAND

Lake Nyos

LOCATION One of the craters on the Oku Volcanic Field in northwestern Cameroon

HEIGHT
3,011m (9,878ft)

TYPE Maar

LAST ERUPTION
Unknown

Lake Nyos, which lends its name to this volcano, lies within the crater of one of 29 maars in what is known as the Oku Volcanic Field. This, in turn, is part of a zone of crustal weakness, called the Cameroon Volcanic Line, which encompasses the Cameroon Mountain stratovolcano and extends for about

DEADLY GASES

In 1986, the release of a suffocating carbon-dioxide cloud from Lake Nyos killed around 1,800 people along with over 6,000 cattle. Up to 1 cubic km (¼ cubic mile) of the gas, travelling at nearly 50kph (30mph), was channelled down surrounding valleys for about 23km (14 miles). As it moved the heavy gas hugged the ground, displacing the air and asphyxiating all humans and animals in its path.

1,600km (1,000 miles), with half its length submerged in the Atlantic Ocean. The Lake Nyos crater might have been created by an explosive phreatic eruption about 500 years ago. Its rim, which is up to 1.8km (1 mile) wide, is made up

of fragments of basaltic ejecta containing large shattered blocks of granite. Lake Nyos was brought to the world's attention by a catastrophe in 1986 that killed about 1,800 people (see panel, above). This release of a great cloud of noxious gas was caused by the lake's water being saturated with carbon dioxide seeping from underground springs. It is thought that the gas could have been released from the water at the bottom of the lake by a landslide or earthquake. Since 1990, a team of French scientists has been working to degas the lake, and a series of pipes has been installed to try to prevent future build-ups of carbon dioxide.

CRATER LAKE
The crater lake is more than 200m (650ft) deep, and during the rainy season excess water often floods over the rim and down the nearby valleys.

Nyiragongo

LOCATION At the southern end of the Great Rift Valley's western fork, Democratic Republic of Congo

HEIGHT
3,470m (11,384ft)

TYPE Stratovolcano

LAST ERUPTION 2023

One of the most active volcanoes in the world, Nyiragongo is famous for the lava lake in its summit crater, first discovered by the German explorer Adolf von Gotzen, in 1894. On 10 January 1977, lava broke through the crater walls and travelled at up to 100kph (60mph) towards the city of Goma, killing 50 to 100 people. The lake started to build up again in 1982, and Goma was partially overrun by lava flows in January 2002.

CHURNING LAVA

Ol Doinyo Lengai

LOCATION At the southern end of the Great Rift Valley's eastern fork, northeastern Tanzania

HEIGHT 2,890m (9,482ft)

TYPE Stratovolcano

LAST ERUPTION 2023

The cone-building stage of this 370,000-year-old symmetrical volcano ended about 15,000 years ago. Its summit consists of two craters. The inactive older southern crater is covered

with vegetation and volcanic ash. Activity in the northern crater is centred around many hornitos (stacks pushed up from an underlying lava flow) and small cones. Ol Doinyo Lengai is most renowned for being the only volcano in recorded history known to have erupted compositionally unique carbonatite lavas

ACTIVE CRATER
The volcano's crater has gradually filled up with ash and rock. During a 1998 eruption, lava began to flow over the rim.

and tephra. With a viscosity close to that of water (due to low silica content), its lava is the most fluid in the world, and also the coolest, with temperatures up to only 590°C (1,100°F). This lava flows black during the day, glows a deep red at night, and when it comes in contact with water a chemical reaction turns it white. Long-term lava effusion has been punctuated with some ash eruptions, and strong explosive eruptions were recorded in 1917, 1940, 1960, and 1966.

MOUNTAIN OF GOD
Ol Doinyo Lengai is known to the Masai as the "Mountain of God". Local villages rely on tourists who visit the volcano.

STEEP PROFILE
The imposing, steep-sided form of Ol Doinyo Lengai rises from the flat, wide plain, south of Tanzania's Lake Natron.

AFRICA *east*

Piton de la Fournaise

LOCATION On the French island of Réunion in the Indian Ocean, off the east coast of Madagascar

HEIGHT 2,631m (8,632ft)

TYPE Shield volcano

LAST ERUPTION 2023

One of the biggest and most active volcanoes in the world, Piton de la Fournaise originated from a mantle hotspot under the Indian Ocean. Much of its 530,000-year eruption record has overlapped with that of the nearby Piton des Neiges shield volcano. Eastward slumping of the volcano produced three successive calderas, formed 250,000, 65,000, and less than 5,000 years ago. A 400-m- (1,300-ft-) high cone lies in the centre of the youngest of the three, Caldera de l'Enclos Fouque. The summit of the cone comprises two craters – Bory and the larger Dolomieu, from which most of the recent eruptions have originated. More than 150 eruptions have been recorded since the 17th century, producing fluid basaltic lava flows. The volcano's activity is closely monitored by an observatory on its slopes.

FURNACE PEAK
An active cone in the summit crater of the volcano belches fumes as molten lava wells up inside. In English, Piton de la Fournaise translates as "Furnace Peak".

VOLCANO TOURISM
Every year over 400,000 tourists are drawn to the isolated island of Réunion especially to visit Piton de la Fournaise. As one of the world's most active volcanoes, it offers the chance to witness at first hand the awe-inspiring scene of an eruption. Hiking paths wind through the dramatic landscape and lookout points offer safe viewing platforms.

ASIA *north*

Siberian Traps

LOCATION Centred on the town of Tura on the central Siberian Plateau, northern Russia

HEIGHT Up to 500m (1,650ft)

TYPE Fissure vent

LAST ERUPTION About 250 million years ago

One of the largest and most enigmatic of flood basalt effusions, the Siberian Traps cover more than 300,000 square km (115,000 square miles) of Arctic wilderness. About 1.5 million cubic km (360,000 cubic miles) of lava erupted through the Earth's crust within about a million years. This outpouring does not appear to be connected to rifting, although it may have been the result of a mantle plume. The timing of the formation of the Siberian Traps coincides with the Permo-Triassic extinction event – the largest extinction in the Earth's history (see pp.22–23). It has been argued that gases associated with such a massive eruption could have had enough effect on global climates to disrupt the food chain severely by damaging plant growth on a global scale.

ASIA *northeast*

Kliuchevskoi

LOCATION Near the eastern coast of the Kamchatka Peninsula, eastern Siberia, Russia

HEIGHT 4,835m (15,863ft)

TYPE Stratovolcano

LAST ERUPTION 2023

Symmetrical, snow-capped Kliuchevskoi is the highest and most active of a chain of volcanoes along the eastern side of the Kamchatka Peninsula. Since its formation about 6,000 years ago, it has frequently produced explosive and effusive eruptions without any extended periods of inactivity. Numerous cones and craters have been formed during the past 3,000 years by more than 100 flank eruptions. However, most recorded eruptions have originated from the 700-m- (2,300-ft-) wide summit crater, which is frequently reshaped. Many of Kliuchevskoi's eruptions have been viewed from space (see below).

SYMMETRICAL PEAK
Kliuchevskoi is a perfectly symmetrical stratovolcano formed by layers of pyroclastic materials and lava flows.

ERUPTION FROM SPACE
Photographed by astronauts on the space shuttle *Endeavour*, the eruption on 30 September 1994 sent a massive ash plume to a height of 18km (11 miles).

LAND

Mount Fuji

LOCATION About 90km (55 miles) southwest of Tokyo, on the Japanese island of Honshu

HEIGHT 3,776m (12,388ft)

TYPE Stratovolcano

LAST ERUPTION 1707

ANNUAL FESTIVAL

The Fujiyoshida Fire Festival, held at the end of August each year, is the highlight of Mount Fuji's tourist season. Shrines are dedicated to the goddess of Mount Fuji to give thanks for a safe climbing season. The mountain is the sacred epicentre of the country and climbing it is treated by many as a religious experience.

Japan's highest mountain, Mount Fuji is a symbol of its homeland around the world, and one of a chain of volcanoes along the western margin of the Pacific Ocean. It is a typical stratovolcano, with a steep-sided, symmetrical cone built up of layers of lava flows alternating with ash and other debris. Growth of the present volcano began 11,000 years ago on top of older volcanic remnants. Within 3,000 years, about 80 per cent of the current bulk of the volcano had built up in outpourings of basaltic lava. During the long eruptive history that followed, lava flows and violent eruptions were emitted from the 700-m- (2,300-ft-) wide summit crater and numerous flank vents, building up more than 100 cones. Some of these lava flows blocked river drainage to the north, forming the Fuji Five Lakes area, which has become a popular tourist area.

The last eruption, in 1707, was the largest in historical times. It formed a large new crater on the east flank and deposited ash on Tokyo, over 90km (55 miles) to the northeast.

ICONIC BEAUTY
The picturesque symmetrical cone of Mount Fuji has been celebrated by hundreds of artists. On a clear day, the mountain can be seen from over 150km (90 miles) away.

Baitoushan

LOCATION Straddling the mid-section of the border between China and North Korea

HEIGHT 2,744m (9,003ft)

TYPE Stratovolcano

LAST ERUPTION 1903

The Baitoushan (or Changbaishan) stratovolcano, 60km (38 miles) wide, was built up over an older shield volcano. In around 1000 CE, one of the largest eruptions in recorded history formed its summit caldera, 5km (3 miles) wide and 850m (2,790ft) deep. This is now filled by the scenic Lake Tianchi, or Sky Lake. Tephra were deposited more than 1,600km (1,000 miles) away in northern Japan. Only five eruptions have been recorded since the 15th century.

SKY LAKE IN THE BAITOUSHAN CALDERA

Mount Unzen

LOCATION On the Shimabara Peninsula, to the east of Nagasaki on the Japanese Island of Kyushu

HEIGHT 1,500m (4,921ft)

TYPE Complex volcano

LAST ERUPTION 1996

The huge Unzen volcanic complex includes three large stratovolcanoes within a 40-km- (25-mile-) long rift valley. The Mayu-yama lava dome complex formed about 4,000 years ago and was the source of a massive debris avalanche on 21 May 1792. The avalanche generated a tsunami up to 55m (180ft) high that devastated

SCENE OF DEVASTATION
In May 1993, a massive pyroclastic flow, travelling at great speed, destroyed everything in its path, including houses along the coast a few kilometres away.

the coastline and killed more than 14,500 people. Subsequent eruptive activity has been restricted to the summit and flanks of the Fugen-dake volcano. The most recent period of sustained volcanic activity, during 1990–95, was centred around the growth of a new summit lava dome. It has been estimated that during this time more than 10,000 pyroclastic flows were generated by the dome collapsing. Many of these flows swept up to 5km (3 miles) down the local river valleys, destroying hundreds of homes and causing many fatalities.

MAURICE AND KATIA KRAFFT

French volcanologists Maurice and Katia Krafft became famous for filming the hazardous nature of pyroclastic flows. Tragically, in 1991, while filming on Mount Unzen, they were killed by an incandescent ash flow that swept suddenly down the mountain.

ASIA *southeast*

Mount Pinatubo

LOCATION On the west coast of the island of Luzon in the Philippines

HEIGHT	1,486m (4,875ft)
TYPE	Stratovolcano
LAST ERUPTION	2021

In June 1991, Mount Pinatubo, a relatively unknown complex of lava domes, erupted in the second most violent volcanic explosion of the 20th century. A series of Plinian eruptions sent about 10 cubic km (2 cubic miles) of rock and ash 40km (25 miles) into the atmosphere. Huge pyroclastic flows incinerated land up to 17km (11 miles) from the volcano, and lahars destroyed villages up to 60km (38 miles) away. The top 250m (820ft) of the mountain collapsed, leaving a summit caldera. Although monitoring had given enough warning for a timely evacuation of the local population, 300 people were killed directly by the eruption, partly due to the simultaneous arrival of Typhoon Yuna.

CALDERA LAKE
The 2.5-km- (1½-mile-) wide summit caldera created by the 1991 eruption is partially filled by a lake.

As well as causing widespread disruption (see panel, right) the ash darkened the sky for a number of days. A month after the eruption, fine aerosol particles had circled the Earth, and had even reduced global temperatures. Lahars redistributing the deposits of the eruption continue to pose a threat to local populations.

ESCAPING THE ASH

Volcanic ash causes severe disruption both on the ground and in the air. At least 14 airliners flew into emissions from the 1991 eruption and several had to make emergency landings with badly damaged engines. Fine ash can also cause health problems for local populations. The eruption filled the air and covered the ground with huge amounts of ash. A major increase in respiratory illnesses such as bronchitis was reported at the time, and some believe that ash contamination could still be killing people today.

RACING THE FLOW
A lucky individual makes a dramatic escape from a pyroclastic flow, travelling at speeds of more than 80kph (50mph), released by the 1991 eruption.

LAND

ASIA *southeast*

Tambora

LOCATION On the island of Sumbawa, one of the Lesser Sunda Islands, Indonesia

HEIGHT	2,850m (9,350ft)
TYPE	Stratovolcano
LAST ERUPTION	2011

The 1815 eruption of Tambora was the largest in recorded history. It ejected more than 150 cubic km (30 cubic miles) of volcanic materials, producing a caldera 6km (4 miles) wide and 1,250m (4,100ft) deep. More than

MASSIVE CALDERA

10,000 people were killed by pyroclastic flows, and many thousands more were victims of subsequent famine and disease. Worldwide circulation of the dust and gas produced global climate change and a series of highly coloured sunsets famously painted by William Turner. The year 1816 was to be known as the "year without a summer".

ASIA *southeast*

Taal

LOCATION South of Pinatubo on the west coast of the island of Luzon in the Philippines

HEIGHT	400m (1,312ft)
TYPE	Caldera
LAST ERUPTION	2022

One of the most active and powerful volcanoes in the Philippines, Taal has erupted more than 30 times since records began in 1572. Its caldera, 20km (12½ miles) wide, is filled by Lake Taal. The 5-km- (3-mile-) wide Volcano Island in the north of the lake has been the focus of all of the volcano's recorded eruptions. This complex of stratovolcanoes, tuff rings, and cones has produced massive pyroclastic flows and surges. A 1911 eruption killed 1,300 people, and 200 died in 1965. Since 1991, Taal has showed signs of increased activity.

LAKE TAAL, THE FLOODED CALDERA

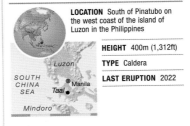

LAND

Merapi

LOCATION On the Indonesian Island of Java, just north of the city of Yogyakarta

HEIGHT 2,968m (9,737ft)

TYPE Stratovolcano

LAST ERUPTION 2023

Lying in one of the world's most densely populated areas, Merapi is one of Indonesia's most active volcanoes. The upper part of the volcano is unvegetated due to frequent eruptions of pyroclastic flows and lahars, which have accompanied the growth and collapse of the steep-sided summit lava dome. The dome occupies an unsupported position, at the western edge of the summit, making it prone to collapse, especially during periods of active dome growth. The collapses release pyroclastic flows that have devastated cultivated lands and caused many fatalities throughout recorded history. Eruptions of Merapi frequently produce hot, pyroclastic flows, formed by

gravitational collapse from lava flows and domes, and sometimes known as Merapi-type pyroclastic flows. Merapi was one of a dozen

"decade" volcanoes chosen in 1991 for scientific monitoring as part of the United Nations International Decade for Natural Disaster Reduction.

2010 ERUPTION
This satellite image shows one of a series of eruptions of Merapi in late 2010 that killed more than 250 people.

EVACUATION

Merapi's proximity to one of the world's most densely populated areas poses a constant threat. In such situations, advance warning of eruptions and effective crisis management are important. In 2018 and 2021, volcanologists monitoring the volcano issued a maximum alert, and thousands of villagers, already affected by ash falls, had time to flee to safety.

Krakatau

LOCATION A volcanic island in the Sunda Strait between Sumatra and Java, Indonesia

HEIGHT 813m (2,667ft)

TYPE Caldera

LAST ERUPTION 2022

An eruption in around 416 CE caused the ancestral Krakatau volcano to collapse, and formed a 7-km- (4-mile-) wide caldera. The regrown volcano was dramatically transformed in 1883 by one of the most notorious volcanic eruptions in recorded history. Earthquakes during the late 1870s preceded a series of eruptions, culminating in a massive explosion on 27 August 1883 that was

heard on Rodrigues Island, 4,653km (2891 miles) away. An eruptive column rose more than 25km (16 miles) above sea level, showering ash and pumice over the region and blotting out sunlight for two days. Pyroclastic flows travelled up to 40km (25 miles) across the sea and destroyed coastal communities in

STROMBOLIAN DISPLAY
Krakatau typically exhibits fairly mild Strombolian or Vulcanian eruptions. These produce basaltic lava flows and fountains or ash and lava bombs.

Sumatra. Situated in one of the busiest shipping lanes in the world, Krakatau's eruption was seen and logged by several vessels. The caldera collapse displaced enormous volumes of seawater and generated a series of tidal waves, or tsunamis, that devastated coastal settlements and killed more than 36,000 people. Two-thirds of the island was consumed in the collapse. Four decades later, a post-collapse cone, known as Anak Krakatau, appeared within the caldera.

EARLY COLONIZER
As Krakatau's vegetation has been repeatedly eliminated, species like this Morning Glory provide excellent models of tropical vegetation succession.

CHILD OF KRAKATAU
Anak Krakatau ("Child of Krakatau") has frequently erupted since 1928. Since its formation, it has been rising out of the sea and continues to grow with increased activity.

ASIA *southeast*

Rabaul

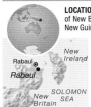

LOCATION On the east end of New Britain Island, Papua New Guinea

HEIGHT 688m (2,257ft)

TYPE Caldera

LAST ERUPTION 2014

Two major caldera-forming eruptions, about 7,100 and 1,400 years ago, formed the 14-km- (9-mile-) wide Rabaul caldera and Blanche Bay, a broad, sheltered harbour for the city of Rabaul. The outer flanks of the volcano are made up of pyroclastic-flow deposits. Three small volcanoes lie outside the caldera to the north and east. Several pyroclastic cones on the caldera floor have explosively erupted in recorded history, including the Vulcan cone, formed during a large eruption in 1878. The most recent period of unrest began in 1971 with frequent seismic activity. During 1984, up to several hundred tremors a day were recorded, and part

BILLOWING ASH CLOUD
During activity in 2000, Rabaul's Tavurvur cone erupted a cloud of ash and debris that travelled more than 1.5km (1 mile) into the atmosphere.

of the harbour was elevated by over 1m (3ft). In 1994, the powerful and simultaneous eruption of the Vulcan and Tavurvur cones forced the temporary abandonment of Rabaul, the largest city in New Britain. Between 1995 and 2003, the volcano remained active with intermittent ash emissions.

AUSTRALASIA *New Zealand*

Ruapehu

LOCATION About 40km (25 miles) southwest of Lake Taupo on the North Island of New Zealand

HEIGHT 2,797m (9,176ft)

TYPE Stratovolcano

LAST ERUPTION 2011

Ruapehu is the tallest mountain on New Zealand's North Island, and one of its most active volcanoes. A 100-cubic-km (24-cubic-mile) volcanic complex, it is surrounded by a ring of volcanic debris produced by a series of eruptions between about 22,600 and 10,000 years ago. The active vent near its summit contains a crater lake that formed about 3,000 years ago. There have been about 50 eruptions since the mid-19th century. Explosive phreatic eruptions from this crater lake have frequently produced lahars, one of which destroyed a railway bridge and train in 1953, killing 151 people.

EXPLOSIVE OUTPUT
During an eruption in 1996, the snow-covered Ruapehu released huge clouds of ash that covered surrounding ski fields and disrupted air traffic.

ANTARCTICA *central*

Mount Erebus

LOCATION On Ross Island in the Ross Sea, just off the Scott Coast, Victoria Land

HEIGHT 3,794m (12,447ft)

TYPE Stratovolcano

LAST ERUPTION 2022

Mount Erebus is the most southerly volcano on Earth to have been active in recorded history. Overlooking McMurdo Sound, it is one of three major volcanoes on Antarctica's Ross Island. Its summit has been modified by successive generations of caldera formation. A summit plateau at about 3,200m (2,000ft) marks the rim of the youngest caldera, within which the modern cone was constructed. The

JAMES CLARK ROSS

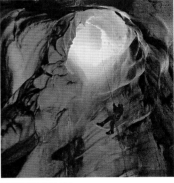

In 1831 the British naval explorer Sir James Clark Ross (1800–62) discovered the magnetic north pole while on expedition in the Arctic. From 1839 to 1843, he headed an expedition to the Antarctic, in the course of which he discovered the sea and island that now bear his name. Ross named the volcano after his ship.

elliptical crater at the summit, which is 600m (2,000ft) wide and 110m (360ft) deep, is unusual in having a small but persistent lava lake on its floor. The lava is also of unusual alkaline composition. High temperature anomalies associated with the lava lake can even be detected from space by satellites. The glacier-clad volcano was erupting when first sighted by Captain James Ross in 1841 (see panel, above), and Strombolian activity was also seen in December 1912, when

the volcano was visited by members of Scott's Antarctic expedition. More recently, fluxes in the emission of sulphur dioxide from the volcano have been measured as part of a global programme to try to correlate such gas emissions to the eruptive process. Continuous lava-lake activity with minor explosions, punctuated by occasional larger Strombolian explosions that eject bombs onto the crater rim, have been documented since 1972.

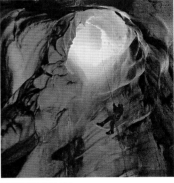

INSIDE EREBUS ICE
An abseiler descends into an ice cave near the crater of Mount Erebus. These caves were formed by volcanic heat rising to the surface from below.

MASSIVE PROFILE
With a bulk of 1,700 cubic km (400 cubic miles), Mount Erebus is one of the world's most impressive volcanoes.

Igneous Intrusions

34–35 The Earth's structure

40–41 The crust

45 Ores

62–65 Rocks

66–75 Igneous and metamorphic rocks

154–57 Volcanoes

Deserts 278–81

An igneous intrusion forms when magma – molten or partially molten rock – invades cracks in existing rock or stagnates within the Earth's crust and solidifies to form new rock. Magma is generated in the crust and upper mantle, and because it is hot and contains water and gases it migrates upwards. A small proportion erupts from volcanoes at the surface (see p.155). However, most magma cools and crystallizes beneath the surface to create igneous intrusions. Because they are formed underground, these rock formations, which include batholiths, dykes, and sills, are seen as features of the landscape only after erosion has stripped away the rocks that cover them.

Intruding Magma

Igneous intrusions form when cooling, crystallization, and loss of gases cause a once fluid and buoyant magma to solidify. Most rocks that form intrusions originate as either granitic or basaltic magma. Granitic magmas have a low density, and small volumes (a few kilometres across) rise slowly through the Earth's upper crust, pushing existing rock (termed country rock) aside. However, very large granite masses, which are perhaps 1,000 times larger, are formed when magma digests and incorporates heated country rock. This surrounding rock begins to melt, and elements of it are assimilated into the magma, which then sets. Such processes generally take place deep within the Earth's crust, and granite intrusions are typically found in areas of active mountain building. Basaltic magmas are less dense and more fluid. They inject into cracks and zones of weakness in country rock, and they are most commonly seen as dykes or sills (see opposite page). These intrusions have the same composition as extruded basaltic lava, but contain larger crystals. The crystals within basaltic magma can settle at different rates to form distinctive layers of rock that have different abundances of minerals.

INTRUSIVE CONTACT
Here, dark, sedimentary country rock has been assimilated into an intruding magma, which has crystallized to form this pink granite.

PEGMATITE
The very large crystals in this pegmatite (see p.67) in Utah, USA, were formed by an intruding granite magma that was rich in water and gases.

LAYERED INTRUSION
These igneous rocks with differing compositions were created in layers as dark, heavy crystals settled and cooled in still molten magma.

CRYSTAL STRUCTURE

As all igneous rocks are formed by cooling of magma, their primary characteristic is a mosaic of interlocking crystals. The tight jigsaw pattern of crystal boundaries gives igneous rocks their high strength. Crystal size depends largely on the rate of cooling. Intrusive magmas lose heat slowly into the surrounding rocks, so they have larger crystals than extrusive rocks, which cool rapidly in air or water. The minerals of igneous rocks are mostly silicates (see pp.56–61); in this microscopic view of granite, the feldspars and quartz are grey, while the biotite appears as several different strong colours.

Intrusive Rock Bodies

Different types of igneous intrusions can be named according to their shape, which is related to the type of rock that forms them. Granitic magmas are usually viscous, so they tend to form huge batholiths that have an area of at least 100 square kilometres (40 square miles). Basaltic magmas are more fluid, so they flow into thin gaps and cool vertically as dykes, which run through the rock strata, or they set horizontally as sills along bedding planes (the boundaries between layers of sedimentary rock). Because rock cools more slowly underground than it does on the surface, the rocks that form igneous intrusions have larger crystals than those formed when lava solidifies at the surface. Those that form dykes are medium-grained, and are usually dolerite (see p.67), whereas those that form larger intrusive bodies, such as batholiths and stocks, are coarse-grained gabbro or granite (see pp.66–67). Ring dykes, which are shaped like vertical tubes, are formed as magma fills cracks created by its upward pressure. Cylindrical plugs are formed from magma that has cooled inside volcanic vents. Laccoliths are large, lens-shaped intrusions that are usually formed of gabbro. Lopoliths are lenses and sheets that are also primarily gabbro, and are found in basins that have sagged under the weight of the intrusion.

SIERRA NEVADA, USA
Almost the entire Sierra Nevada range is a massive batholith, which was formed by multiple phases of granitic magma intrusion. This cooled beneath the surface of the Earth before being exposed by subsequent erosion.

TYPES OF IGNEOUS INTRUSION
Most igneous intrusions are classified by their size and shape. The intrusions shown here range from relatively small horizontal sills and vertical dykes, to conical ring dykes, volcanic plugs, lens-shaped laccoliths, and huge batholiths, which can be more than 100km (60 miles) in length.

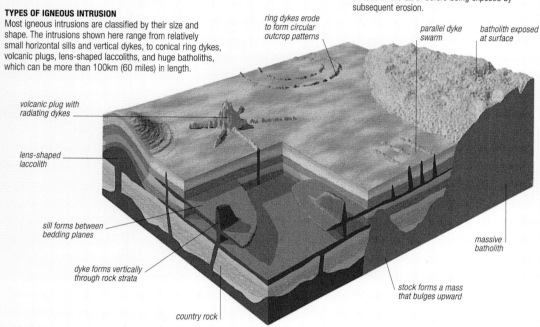

ring dykes erode to form circular outcrop patterns

parallel dyke swarm

batholith exposed at surface

volcanic plug with radiating dykes

lens-shaped laccolith

sill forms between bedding planes

dyke forms vertically through rock strata

country rock

massive batholith

stock forms a mass that bulges upward

LAND

EXPOSED DYKE
The typically hard rock of an igneous intrusion forms the core of this rocky crag in Colorado, USA. Debris eroded from the margins forms the more gentle slopes, which contrast with the sharp profile of the intrusion.

DYKE WALL
This dyke wall in Colorado, USA, has been formed by the erosion of the surrounding country rock, leaving strong igneous rock protruding from the ground.

ECONOMIC IMPORTANCE
Intrusive igneous rocks are so strong that they are widely quarried to make hard rock aggregate, which is required in huge quantities for use in the manufacture of concrete and to build roads. Some types of granite are also highly valued as large blocks of cut rock – known as dimension stone. Strong, unfractured, and light in colour, granite is sawn and polished to make thin slabs of glossy stone for use as cladding on buildings. The granite in this Sardinian quarry is cut out by powerful water jets and traditional wire saws, without using explosives.

SILL AND DYKE

The pale, banded, metamorphosed sedimentary rocks in this hillside exposure in Oregon, USA, are cut by two small intrusions of darker dolerite. One is a dyke that has filled a tension fissure cutting across the beds (bottom); the other is a sill that has injected along a single bedding plane (top).

IGNEOUS INTRUSION PROFILES

The pages that follow contain profiles of some of the world's igneous intrusions. Each profile begins with the following summary information:

AGE When the igneous intrusion was formed

TYPE Batholith, dyke, dyke swarm, ring dyke, sill, plug, stock, layered intrusion

ROCK The main rock type that makes up the intrusion

LAND

Skaergaard

LOCATION In the coastal mountains of eastern Greenland, just to the north of the Arctic Circle

AGE 60 million years

TYPE Layered intrusion

COMPOSITION Gabbro

Dissected by deep, glaciated valleys and fiords, and exposed in the soil-free mountains of Greenland, the Skaergaard intrusion is the world's most intensively studied large igneous body. Where it is exposed at the surface, the intrusion is about 10km (6 miles) across. It is notable for its vast sequence of layered rocks, which were formed by differentiation of a mass of basaltic magma. Crystals of dense, silicate minerals sank through the liquid remnants of the original magma, but were interrupted by convection currents as they did so. When this happened, the magma composition changed, with bands of rock forming from different minerals as the layers accumulated and cooled progressively. The dominant gabbro layers near the base of the intrusion contrast with the small layer of granophyre (see p.66), which was the last rock to crystallize at the end of the batholith's molten life.

CLEAN ROCK
The Arctic conditions lead to frost shattering (see p.90), which leaves bare outcrops of igneous rock. This makes Skaergaard the perfect place to study magmas and rocks.

COLOUR-CODED
Colour changes identify the mineral layers within the Skaergaard gabbro, and lumps of coarse, pegmatitic material are remnants of olivine-rich layers that fragmented before they solidified.

Mackenzie Dykes

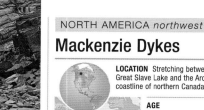

LOCATION Stretching between Great Slave Lake and the Arctic coastline of northern Canada

AGE 1,260 million years

TYPE Dyke swarm

COMPOSITION Dolerite

Hundreds of parallel intrusions form the largest dyke swarm in the world, which is exposed in the ice-scoured lowlands of the Canadian Shield. The dykes formed when magma intruded into fissures that opened along a zone of tension above an active mantle plume. Flood basalts formed where huge volumes of magma spilled onto the surface, and layered-gabbro intrusions have been detected at depth. The result is a trinity of forms that record the formation of new crust in the Earth's distant past.

MACKENZIE DYKE SWARM

Palisades Sill

LOCATION Formed along the New Jersey side of the Hudson River, northward from New York City, USA

AGE 190 million years

TYPE Sill

COMPOSITION Dolerite

For 80km (50 miles), the Hudson River flows beneath the dark wall of the Palisades Sill. Nearly horizontal, this dramatic outcrop is a sheet of intrusive dolerite. The sill is about 330m (1,100ft) thick, and intruded into now-eroded Triassic sandstones at shallow depth. Where the base of the sill is exposed, there is about 15m (50ft) of fine-grained dolerite that

HUDSON RIVER VALLEY
At the top of the sill is a colonnade of dolerite columns. Trees grow on the weathered fragments of these structures, which have fallen to the base of the cliff.

GEORGE WASHINGTON BRIDGE
Anchored within the diabase that forms the Palisades Sill, this suspension bridge spans the Hudson River from New Jersey to Manhattan in New York City.

chilled rapidly against the contact rock. Above this is a 5-m- (18-ft-) thick layer that is predominantly olivine (see p.56). This was the first mineral to crystallize, and its dense crystals sank through magma that was still 95 per cent liquid. The Palisades Sill is a clear example of the differentiation of magma, with rocks of varying compositions created from the same source.

Devil's Tower

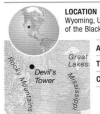

LOCATION In the Great Plains of Wyoming, USA, just northwest of the Black Hills of Dakota

AGE 40 million years

TYPE Plug

COMPOSITION Phonolite

The Devil's Tower originated as a volcanic vent, but it owes its spectacular appearance to erosion, which has removed the surrounding sedimentary rock to expose igneous rocks that cooled and solidified underground. The original volcano lost its gas pressure by exploding to the surface through a cover of sediment, but much of the magma remained in an underground plug that was nearly 300m (1,000ft) in diameter. As it cooled, the magma contracted and fractured, forming polygonal joints. When the cooling fronts moved inward, these fractures evolved into remarkably uniform columns. (A similar phenomenon occurs when mud dries to form polygonal cracks.) The tower reaches a height of 264m (867ft). Horizontal columns also intruded from the edge of the plug, but never extended far. Most of this fringe of horizontal columns has been eroded, though where it is not masked by debris, some can be seen in the flared base of the tower. The phonolite that forms the Devil's Tower is derived from continental crust; it contains less silica than the other rhyolitic magmas, and is distinguished by its small crystals of dark green aegirine, a pyroxene mineral that is rich in sodium.

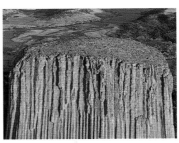

SUMMIT OF THE TOWER
The domed peak of the intrusion exposes the tops of the columns that form the mass of the Devil's Tower.

ROCK CLIMBING

The splendid columnar joints of the Devil's Tower intrusion provide perfect routes for the more serious breed of climber who enjoys long crack lines in very strong rock. However, natural cooling is never perfect, and some columns do merge or split, so climbers have to choose which joints they pursue with care.

WYOMING'S WONDER
The striking structure of the Devil's Tower has distinguished it as a sacred place for some Indigenous nations. It appeared in the films *Close Encounters of the Third Kind* and *Paul*.

ABOVE THE CLOUDS
On a cool morning, the Devil's Tower rises through a layer of cloud to mimic the situation about 15 million years ago, when much less of the intrusion was projecting above the surrounding land.

LAND

THE WINGED ROCK
Though the Navajo name suggests that Ship Rock flew to its present site, the geological wings are the remarkable dykes that now stand high as walls across the desert of New Mexico.

Ship Rock

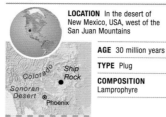

LOCATION In the desert of New Mexico, USA, west of the San Juan Mountains

AGE 30 million years

TYPE Plug

COMPOSITION Lamprophyre

Though it now stands 500m (1,700ft) above the surrounding plains, Ship Rock formed underground when magma cooled within the feeder pipes of a volcano. The lavas and pyroclastic deposits of the volcano have since disappeared, and subsequent erosion of the underlying soft shales has steadily lowered the surface of the surrounding plains. The resistant rock of the plug is far less worn, and stands high above the desert. The upper part of Ship Rock is a lava breccia which was created when rock was shattered by explosive eruptions within the volcano's vent. When this happened, the rocks that constitute today's summit were probably

less than 1,000m (3,300ft) below the Earth's surface. The lamprophyre that forms the intrusion is similar to dolerite, except that it contains more mica and less feldspar. It also contains small masses (xenoliths) of peridotite and eclogite. This suggests that the magma had a source deep within the upper mantle. In addition to feeding the main conduit, magma also ascended into radiating fissures, where it cooled to form the dykes that distinguish Ship Rock. These were exposed as the adjacent surface eroded, resulting in spectacular dyke walls. The rock ribs are about 3m (10ft) thick and they stand 20m (65ft) high, and the longest dyke stretches for nearly 3km (2 miles).

VOLCANIC REMNANTS
The rock exposures in the desert around Ship Rock are the volcanic remnants of intrusions that were once buried deep beneath the surface of the Earth.

El Capitan and Half Dome

LOCATION Within Yosemite National Park, Sierra Nevada, California, USA

AGE 82 million years

TYPE Batholith

COMPOSITION Granite and granodiorite

EL CAPITAN
Towering above Lower Yosemite Valley, this rock wall is a result of erosion, which has shaped this part of the huge Sierra Nevada batholith.

Known as the Incomparable Valley, Yosemite cuts deep into the largest and strongest of the Sierra Nevada granitic intrusions, and is bounded by the huge, steep vertical cliffs that make it so famous. Like all batholiths, the Sierra Nevada is not formed from a single type of granite. It consists of a series of intrusions that were emplaced over 50 million years during the Cretaceous Period, when the region sat above an active convergent plate boundary. Two of the batholith's most distinctive outcrops are El Capitan and Half Dome. The composition of the intrusions varies, as different magmas invaded different areas of country rock. El Capitan – Yosemite's highest rock wall – is formed of a true granite, but Half Dome – the area's steepest wall – is formed of granodiorite, which contains less alkali feldspar. Both are massive rocks with few weaknesses, as is typical of the largest batholiths.

HALF DOME
Yosemite's most famous landmark is the rounded end of a granite ridge between two deep valleys. One side of Half Dome was broken off at a major joint during glacial sculpting of the region.

Chuquicamata

LOCATION In the Atacama Desert of northern Chile, between the Andes and the Pacific Ocean

AGE 33 million years

TYPE Stock

COMPOSITION Granodiorite

DESERT COPPER MINE

The porphyritic granodiorite intrusion at Chuquicamata is associated with the largest of several porphyry copper ore bodies that are found throughout the western mountains of North and South America. Metal sulphides are dispersed throughout the stock, and half a million tonnes of copper are mined every year from over 4 billion tonnes of ore. Mining has created an open pit that is over 4km (2½ miles) long and 810m (2,650ft) deep.

SOUTH AMERICA *east*

Sugar Loaf

LOCATION Overlooking the entrance to Guanabara Bay, Rio de Janeiro, Brazil

AGE 800 million years

TYPE Batholith

COMPOSITION Granite

TOURISM

Rock that is almost white, bold curves, and towering cliffs are the characteristic features of mountains cut into the granites that form large batholiths. The resulting landscapes are often tourist attractions. Few are finer than Rio de Janeiro's two dramatic granite peaks, the Sugar Loaf and Corcovado, which can be ascended by cable car or tram.

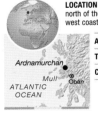

The Sugar Loaf is solid granite, and is actually a tiny part of a huge batholith that stretches along the Brazilian coast on either side of Guanabara Bay. It was formed by partial melting within a belt of metamorphic gneisses (see p.74) at the centre of a plate convergence zone that was active about 800 million years ago in late Precambrian times. Since then, erosion has left stumps of granite that form the magnificent peaks enclosing Rio de Janeiro. Sugar Loaf is not the highest promontory in the area as it rises to only 395m (1,295ft), but its rounded profile makes it so distinctive. The rounded outline is typical of exposed masses of strong, homogenous, joint-free granite that were once at the core of a massive batholith. When intrusive granite crystallizes, it is confined by horizontal and vertical stress, due to weight of the rock above it. Once exposed, these stresses are lost, and the rock relaxes, opening curved joints as it does so. Erosion strips away these shells of rock, each time exposing a smoothly curved surface.

BRAZIL'S GRANITE ICON
The smooth, rounded structures of the Sugar Loaf and the peninsula that extends to it are both formed from the huge mass of granite at the core of a batholith.

EUROPE *north*

Ardnamurchan

LOCATION A peninsula to the north of the Isle of Mull on the west coast of Scotland

AGE 60 million years

TYPE Ring dyke complex

COMPOSITION Gabbro

A complex of gabbroic ring dykes forms the Point of Ardnamurchan, which is about 8km (5 miles) across and is the most westerly point of the British mainland. A decline in pressure over a large magma chamber caused conical fractures to form. Magma intruded into these fissures to form dykes. Subsequent cone-shaped sheet intrusions then formed when magma pressure increased. Above these deep intrusions, huge volcanic craters erupted lavas and pyroclastic deposits.

POINT OF ARDNAMURCHAN

EUROPE *north*

Castle Rock

LOCATION Beneath the castle in the centre of the city of Edinburgh, Scotland

AGE 325 million years

TYPE Plug

COMPOSITION Dolerite

Edinburgh's great crag of black dolerite, which is now crowned by a castle and walled courtyard, is the exposed feeder pipe of an ancient volcano. The entire volcanic edifice has long been lost to erosion, and its only remnant is the basaltic magma that solidified deep beneath the eruption crater. The magma cooled more slowly than the extruded, air-cooled basaltic lavas and so formed large crystals, which are typical of dolerite. About one million years ago, during the last ice age, ice sheets swept over Castle Rock; however, the tough dolerite plug remained in place. Ice flowed around the crag, and deposited till debris (see p.95) in its unscoured lee. This resulted in the formation of the long, eastern rampart that now supports the High Street of Edinburgh's Old Town. The modern landform is a classic example of a feature known as a crag-and-tail.

EDINBURGH'S ROYAL ROCK
The hard black dolerite of Castle Rock's intrusive plug forms a crag far steeper than the ancient volcano that once existed here.

EUROPE *north*

Fair Head Sill

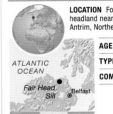

LOCATION Forming a coastal headland near Ballycastle, County Antrim, Northern Ireland

AGE 58 million years

TYPE Sill

COMPOSITION Dolerite

FAIR HEAD SILL

The great prow of Fair Head is the exposed edge of an 80m- (260ft-) thick dolerite sill, which intruded into Carboniferous sandstones as lavas poured across the rest of the region. The sill is scored by columnar cooling fractures, and is notable for its huge sandstone xenoliths.

LAND

EUROPE *north*

Whin Sill

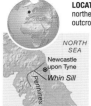

LOCATION Underlying England's northern Pennine hills, with outcrops at the edges

AGE 295 million years

TYPE Sill

COMPOSITION Dolerite

Whin is an Old English quarryman's word for hard black stone. This aptly describes the dolerite that forms Whin Sill. In the late Carboniferous Period, basaltic magma rose from beneath the Earth's crust and spread to intrude along shallow bedding planes in sedimentary rocks. The intrusion is complex, as it cuts across the bedding to form sheets at multiple levels and

COLONNADED ESCARPMENT
Rough columnar jointing in the dolerite adds texture to the long escarpment cliffs wherever Whin Sill emerges above the surface.

HADRIAN'S WALL

An escarpment that extends right across the Pennines marks the northern edge of Whin Sill. It was used as a natural line of defence in 122 CE, when the Roman emperor Hadrian built a wall to keep the Scots out of the Roman province of Britannia.

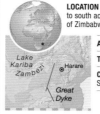

appears to have been fed by more than one magma chamber. Whin Sill is therefore not a single sill, but is actually a collection of cross-cutting sills and flat, lens-shaped intrusions. Nevertheless, at many outcrops it is a uniform and unbroken sheet of dolerite that is approximately 30m (100ft) thick.

AFRICA *south*

Great Dyke

LOCATION Stretching from north to south across almost the whole of Zimbabwe

AGE 2,600 million years

TYPE Layered intrusion

COMPOSITION Serpentinite

Called a dyke because it is long and thin at the surface, most of this intrusion is about 5km (3 miles) wide, and so is too large to be a true dyke. It consists mainly of peridotite (see p.68) and dunite, which have altered to become serpentinite (see p.71), though its upper layers consist of gabbro. Within the intrusion, crystal settling has created chromite bands that are the world's largest source of chromium ore. The mineral bands show a synclinal structure (see p.136), suggesting that the igneous body is actually a lopolith. However, its base is hidden, and its margins are both major faults, so Great Dyke may be best classified as a layered intrusion.

LINEAR INTRUSION
This satellite image shows the long, thin range of wooded hills that marks the line of the Great Dyke across the plains of Zimbabwe.

AFRICA *west*

Aïr Mountains

LOCATION In the northern part of Niger, Africa, within the southern Sahara Desert

AGE 410 million years

TYPE Ring dyke complexes

COMPOSITION Granite

MINING URANIUM

Granitic magmas contain traces of uranium. When erosion exposed the Aïr dykes, mountain streams carried the uranium as soluble oxides into sedimentary basins. There, low-oxygen conditions in nearly stagnant lakes caused precipitation of the uranium. Today, it is mined from the exposed basins in Niger's desert, but its original source was the Aïr intrusions.

The major ring dyke complexes of the Aïr Mountains were formed where three continental plates collided. Each complex is about 60km (37 miles) in diameter and contains one or more ring dykes that are 20–200m (65–650ft) thick. Before the plates welded together to form the huge basement block of northern Africa, their boundaries were subduction zones (see p.88–89), melting rock and generating magma. Around the junction, granitic magmas reached the surface, creating huge explosive volcanoes, whose lavas and pyroclastic deposits have since eroded. Further eruptive phases caused declines in the underground magma pressure,

resulting in large caldera collapses. Beneath the surface, gigantic blocks of rock subsided inside huge conical ring fractures. Then more magma rose through these new fractures. Some of the new magma fed younger volcanoes, but the magma that solidified within the fissures created ring dykes. Erosion has brought the modern surface down to a level that cuts through the ancient intrusive rings.

DESERT DYKES, AÏR MASSIF
Barren rock crags within a sea of sand dunes are the eroded stumps of old dykes and plugs in the Sahara's Aïr Mountains.

ANTARCTICA *north*

Salvesen Mountains

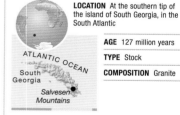

LOCATION At the southern tip of the island of South Georgia, in the South Atlantic

AGE 127 million years

TYPE Stock

COMPOSITION Granite

The roots of the island of South Georgia are intrusive igneous rocks, though they form only a small part of the modern outcrops. Stocks of white Cretaceous granite are intruded into Jurassic basaltic lavas and dolerite dykes – both of which are black, so they create striking colour contrasts in exposures across the Salvesen Mountains, the most jagged of the island's ice-bound mountain ranges. Both the granite and the basalt are remnants of rising magma formed on a divergent plate boundary where the southern Atlantic opened. The main ranges of South Georgia, famously crossed by Ernest Shackleton in 1916, are less rugged and precipitous than the Salvesen Mountains, as they are made of folded sandstones. These were formed from sand deposition, the sediment for which was derived from erosion of the igneous rocks and rifting continental blocks.

Hot Springs and Geysers

39	Heat transfer
47	Sulphur
106–07	Water
154–77	Volcanoes
Geothermal energy	194–95
Ocean tectonics	386–89

Hot springs, spouting geysers, fumaroles, boiling mud pots, and many mineral deposits are all evidence of the Earth's internal heat. They all depend upon the underground heating of water by hot rocks. Once hot water and steam have been generated, they are forced to the surface, often charged with dissolved minerals and gas. Hot springs produce warm water steadily, while geysers release intermittent bursts, often with copious amounts of steam. Fumaroles are vents where only steam and other gases escape.

WHITE DOME GEYSER
Hot underground rock turns groundwater to steam and boiling water, which spurts into the air from this geyser in Yellowstone, USA.

Geothermal Systems

The flow of internally generated heat up to the Earth's surface is almost constant over the globe, but in some places the heat flow is much higher and produces hot springs and associated geothermal phenomena. These zones of high heat-flow tend to occur at the margins of the Earth's tectonic plates and are normally associated with volcanism. Many active volcanoes are underlain by reservoirs of partially molten rock, centres of heat at abnormally shallow depths. Any groundwater coming into contact with these hot rocks is heated, but because it is under pressure at depth the boiling point is raised above normal and the water becomes superheated. Heated water tends to rise towards the surface, losing pressure as it does so. Flashing over into steam may then propel the water upwards with such force that it gushes forth above ground as a geyser.

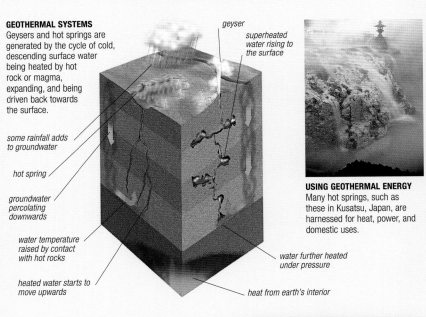

GEOTHERMAL SYSTEMS
Geysers and hot springs are generated by the cycle of cold, descending surface water being heated by hot rock or magma, expanding, and being driven back towards the surface.

geyser

superheated water rising to the surface

some rainfall adds to groundwater

hot spring

groundwater percolating downwards

water temperature raised by contact with hot rocks

heated water starts to move upwards

water further heated under pressure

heat from earth's interior

USING GEOTHERMAL ENERGY
Many hot springs, such as these in Kusatsu, Japan, are harnessed for heat, power, and domestic uses.

Water Deposits

Much of the water that feeds hot springs and geysers originates as rainwater, infiltrating below ground through joints or cracks to the local water-table. Where groundwater is heated by a geothermal source, it rises, drawing in additional cold groundwater towards the heat source as it does so. While underground, the heated water dissolves soluble minerals from the rocks, but, on reaching the surface, evaporation and cooling cause some of these dissolved minerals to be re-precipitated. The resulting deposits are dominated by silica, calcium, sodium, and potassium carbonates and chlorides, but include other minerals and metals, such as copper and lead, depending upon the local rocks.

SULPHUR CRYSTALS
Hot gases expelled from volcanic fumaroles often leave deposits, such as these sulphur crystals on Vulcano, in the Aeolian Islands.

BUBBLING CAULDRON
The constantly boiling waters of Punchbowl Spring, Yellowstone, USA, are contained within a raised rim of carbonate deposits (travertine), about 75cm (2½ ft) high, which is shaped like a punch-bowl.

HOT SPRING AND GEYSER PROFILES

The pages that follow contain profiles of a selection of the world's hot springs and geysers. Each profile begins with the following summary information:

TYPE Hot spring, mud pool, geyser, or fumarole

FREQUENCY Interval between periods of activity

LAND

LAND

Yellowstone

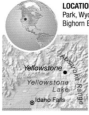

LOCATION In Yellowstone National Park, Wyoming, USA, between the Bighorn Basin and Snake River Plain

TYPE Geysers, hot springs

FREQUENCY Old Faithful 45–110 mins, Steamboat Geyser irregular

Yellowstone displays some of the most spectacular hydrothermal activity in the world, especially from its dramatic geysers (about 200 in number), of which Old Faithful is the most famous. There are also thousands of fumaroles, hot springs, and boiling mud pools. This thermal activity is a reminder that Yellowstone is the site of one of the largest explosive volcanic eruptions in recent history, which led to the collapse of an ancient volcano, about 600,000 years ago, to form a giant caldera. The magma chamber that fed the original volcano is still present, about 6km (3¾ miles) below Yellowstone, and is the heat source for the present geothermal activity. Plentiful rainfall keeps its hydrothermal system well charged with water. The geysers form where the underground circulation of hot water is restricted. Under pressure at depth, water is superheated without boiling. As it rises, pressure decreases, allowing steam to form, which further accelerates upward movement. Water and steam then gush out at the surface as geysers. Gushing ceases when

the hotwater reservoir is exhausted. The period between gushes depends on the time it takes for the underground system to recharge. Famously, Old Faithful produces columns of steam and water 30–55m (100–180ft) high every 67 minutes on average. At Yellowstone, the water temperature at 200m (660ft) depth is 200°C (392°F). This hot underground water not only supplies geysers but also springs. The rising hot water dissolves soluble minerals from the surrounding rocks,

TERRACED WATERFALLS AND POOLS
Each day, evaporation deposits two tonnes of carbonate mineral at Mammoth Hot Springs, which enlarges the ever-changing Minerva Terrace.

ERUPTING GEYSER
Castle Geyser is one of Yellowstone's 200 or so active geysers, which are a constant reminder that the region is still geologically active.

but, on surfacing, evaporation and cooling reverse the process, and mineral deposits form, such as those at Minerva Terrace. At Mammoth Hot Springs, terraces of limy flowstone (see p.241) contain beautiful, variably coloured pools at different temperatures – their colours produced by species of temperature-sensitive algae. Boiling mud pools are normally grey, but in Yellowstone's Fountain Paint Pot these too are multicoloured, and up to 20m (66ft) in diameter.

SURVIVING THE WINTER
For many generations, American Bison have trekked to Yellowstone's hot springs and pools in winter, to benefit from the warmer temperatures and to feed on the plants that grow there.

THE LODGEPOLE PINE
The dominant tree of Yellowstone is the Lodgepole Pine, so called because its straight trunk is ideal for use as tepee poles. Its needle-shaped leaves point downwards so the snow slips off them in winter.

LIFE AT HIGH TEMPERATURES

Prismatic Pool hot spring (shown right) is host to aquatic algae and other micro-organisms that can survive high temperatures. Investigation of their tolerance of extreme conditions is providing insights into how life may have originated in the extremely harsh environments of the early Earth, and whether life can exist in the even more difficult conditions found on other

planets. Thermophiles (heat-loving organisms) are among the most primitive forms of life. Over 4 billion years ago, the Earth's surface was hot, and the atmosphere charged with carbon dioxide, nitrogen, and methane. Similar conditions are found in hot springs today, making them natural laboratories for studying life's origins.

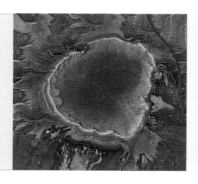

OLD FAITHFUL
After the famous geyser jets a spout of hot water around 50m (165ft) into the air every 45–110 minutes, water and steam droplets hang in the air for about four minutes.

Valley of Ten Thousand Smokes

LOCATION Near Katmai, in the Aleutian Range at the northeast end of the Alaskan peninsula, USA

TYPE Fumaroles

FREQUENCY Extinct

In 1912, about 22km (13 miles) of the Ukak valley was filled by the largest pyroclastic flow of modern times as it poured from vents near Mount Katmai (see p.158). The area is called the Valley of Ten Thousand Smokes because, for 15 years afterwards, vaporized water from the stream and wet sediments continued to escape the surface through small holes and cracks.

THE SMOKING VALLEY

Fly Geyser

LOCATION On the flat, high plains adjacent to the Black Rock Desert, northwest Nevada, USA

TYPE Geyser

FREQUENCY Continuous

This geyser has three outlets and is surrounded by hot-water pools contained by terraces of travertine (calcium carbonate). Over the years, the geyser has built up three rocky pinnacles of mineral deposits that continuously spout water. Some of the pools have been enlarged, allowing them to cool enough for bathing. Lying at an altitude of 1,300m (4,270ft), the Fly Geyser region receives an annual rainfall of less than 300mm (11¾ in). Nevertheless this, combined with run-off from the surrounding mountains, is sufficient to feed the local aquifer. The adjacent Black Rock Desert is the dry bed of a lake that covered about 22,420 square km (8,660 square miles) during the last ice age. Its sediments are floored by Pleistocene basaltic lavas, and the presence of the geyser shows that there must still be an associated magma chamber or intrusion at shallow depth, acting as a heat source.

THREE-VENTED GEYSER
Travertine terraces surround Fly Geyser's continuously spouting trio of cones, which jet from the ancient dry lake floor of the Black Rock Desert.

Soda Springs

LOCATION At the head of the Bear River valley, southeastern Idaho, USA

TYPE Geyser

FREQUENCY Hourly

It is claimed that Soda Springs is the only geyser in the world to be made by human activity. On 30 November 1937, a well was drilled for hot water to supply a swimming pool. At about 100m (330ft) down, the drill hit a reservoir containing a pressurized mixture of carbon dioxide and hot water. The drill-hole provided an escape route, sending a geyser 50m (165ft) into the air. The borehole was capped and fitted with a valve, which restricts activity to once an hour. This part of the Bear River valley contains numerous other hot springs, such as Steamboat Springs, so called because of the steampipe-like noise that the hot water makes as it gushes forth. In the 19th century, migrants travelling the Oregon Trail stopped here to bathe and to wash their clothes.

MANMADE GEYSER
Idaho's Soda Springs geyser is perhaps the only man-made geyser in the world, capped in such a way that it erupts every hour, on the hour.

El Tatio geysers

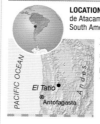

LOCATION Northeast of San Pedro de Atacama in the Chilean Andes, South America

TYPE Geysers, hot springs, and fumaroles

FREQUENCY Variable

At an altitude of 4,321m (14,176ft), the El Tatio geysers, hot springs, and fumaroles are the highest in the world. The area of thermal activity covers 10 square km (4 square miles) of a rift-valley floor. Water surfaces at temperatures of about 86°C (187°F) – the boiling point for water at this altitude. The water reservoir is within volcanic rocks capped by impermeable layers; faults conduct the hot water to the surface. The exact heat source is unknown, but is likely to be magma or an igneous intrusion beneath the valley.

FUMAROLES AT EL TATIO

Geysir

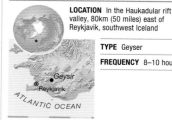

LOCATION In the Haukadular rift valley, 80km (50 miles) east of Reykjavik, southwest Iceland

TYPE Geyser

FREQUENCY 8–10 hours

The term geyser is derived from this feature, first described in 1294. Although there has been no volcanic activity here for about 10,000 years, temperatures below ground may reach 240°C (464°F). In 1915, Geysir became dormant, but earthquake activity in 1935 reactivated it for a time. After earthquakes in June 2000, Geysir currently erupts every 8–10 hours.

WATER BUBBLING IN GREAT GEYSIR

ATLANTIC *north*

Strokkur Geyser

LOCATION In the Haukadalur rift valley, 80km (50 miles) east of Reykjavik, southwest Iceland

TYPE Geyser

FREQUENCY 8 mins

Strokkur, whose name means "churn" in Icelandic, is one of the most famous geysers in Iceland. It became active following an earthquake in 1789, and erupted regularly until 1896, when another earthquake blocked its flow. In 1963, the conduit was cleaned out by local people, and it has been active ever since. The eruption of the geysers in this area is driven by rising hot and pressurized water flashing into steam. Water at a depth of 23m (75ft) is at a temperature of about 120°C (248°F), but the weight of the water in the conduit above prevents it from boiling. At 16m (52ft), the temperature of the water may rise above boiling point

WELLING UP
The interval between Strokkur's eruptions depends on the time its reservoir channels take to refill and raise their temperature back to the critical point.

and raise the water in the conduit slightly, setting off a chain reaction. The pressure decrease allows more water to boil and flash into steam, which in turn drives the water upwards with increasing velocity to erupt at the surface. When the water in the upper part of the conduit has erupted, an action lasting just a few minutes, the steam phase is exhausted and the cycle begins again.

REGULAR CYCLE
Strokkur's eight-minute repeat cycle owes its regularity to human intervention. In 1963, the conduit was cleaned to ensure regular eruptions.

With nearby Geysir (see opposite), Strokkur is one of Iceland's leading tourist attractions. There are a number of hot springs in the vicinity, as well as steam vents and colourful algal deposits.

EUROPE *west*

Chaudes-Aigues

LOCATION In the town of Chaudes-Aigues, in the Auvergne region of the Massif Central, France

TYPE Hot springs

FREQUENCY Continuous

The name Chaudes-Aigues literally means "hot waters". The inhabitants of this small medieval town still enjoy central heating developed from the world's first geothermal system, established in the 14th century. First used by the Romans in the time of Nero, over 30 hot springs provide the hottest spring waters in Europe, at 82°C (180°F). Today, Chaudes-Aigues is a spa town, famed for the curative powers of its hot, mineral-enriched spring waters. The hot springs are probably related to volcanic heat below the Chaîne des Puys (see p.163) and the famous spring waters of Volvic, even though Chaudes-Aigues lies 100km (60 miles) to the south. The Chaîne des Puys includes about 90 cinder cones and maars (explosion craters formed without basaltic lava flows), many aligned along northeast–southwest fissures. The last volcanic activity occurred about 6,000 years ago.

EUROPE *south*

Lardarello

LOCATION In the central region of the Colline Metallifere (Metalliferous Hills), Tuscany, central Italy

TYPE Hot springs

FREQUENCY Continuous

One of the world's largest geothermal fields and one of the oldest centres of geothermal power production, Lardarello has been generating electrical power since 1904. The heat source is located some distance below the surface in metamorphic rocks. Above, permeable limestones covered by impermeable shales and clays are folded into a dome, forming a trap for hot water and steam. Faults through the arch of the dome channel water to the surface, where they form hot springs.

USING STEAM TO PRODUCE ELECTRICITY

EUROPE *south*

Solfatara

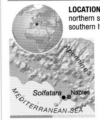

LOCATION West of Vesuvius, on the northern shore of the Bay of Naples, southern Italy

TYPE Hot springs, fumaroles

FREQUENCY Continuous

Solfatara is a crater lying within an ancient volcanic caldera called the Phlegrean Fields. About 40 hot springs and numerous fumaroles venting volcanic gases litter the caldera. Those of Solfatara's crater were believed by the Romans to cure a host of medical conditions, but are now the site of intense scientific research into the frequency and intensity of earthquakes, and changes in gas composition. The Bay of Naples, including the port of Pozzuoli, has risen and subsided by several metres over the last few millennia, accompanied by countless earth tremors (15,000 were detected between 1982 and 1984 alone). These movements affect the aquifer below Solfatara and water levels in the various wells of the region. The present Solfatara well was built in the early 19th century to extract alum (potassium aluminium sulphate) from the spring water. Deep magma chambers heat the overlying hydrothermal system, whose temperature varies from 200° to 250°C (390° to 480°F) if there is a rapid inflow of seasonal rain.

STEAMING FUMAROLES
Bocca Grande ("Large Mouth"), shown here, is the largest fumarole at Solfatara. It was described by Virgil as the entrance to Hell.

EXTREME ENVIRONMENT
These hot springs are produced by Ethiopia's
Dallol volcano, which lies 45m (175ft) below
sea-level. Biologists are interested in finding
out how microbial life can survive in such
extreme environments.

LAND

EUROPE *south*

Lipari

LOCATION One of the Aeolian Islands, in the Mediterranean Sea, 35km (22 miles) off northern Sicily

TYPE Fumaroles

FREQUENCY Continuous

Lipari is one of the Aeolian Islands, an active volcanic arc to the north of Sicily. The volcanoes are the result of subduction of the African plate beneath the southern edge of the Eurasian plate. This stimulates partial melting that feeds the volcanoes at the surface. In the past, volcanic activity on Lipari has produced highly viscous lavas, often with a glassy crust. Today, deep pockets of magma at depth produce vigorous fumarolic activity at the surface, and monitoring shows a wide range of temperatures over time, from about 200°C (392°F) to as much as 650°C (1,202°F). This is associated with increasing gas flow from depth and thermal expansion, which opens new conduit cracks. There is concern that volcanic activity is moving north from the nearby island of Vulcano and could cause an eruption near Lipari.

AFRICA *east*

Bogoria

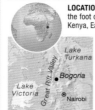

LOCATION On Lake Bogoria, at the foot of the Laikipia Escarpment, Kenya, East Africa

TYPE Hot springs and geysers

FREQUENCY Continuous

Hydrothermal features are rare in Africa but are present on the western and southern sides of Lake Bogoria. Here, groundwater is heated at depth by igneous intrusions buried below the valley's sediments and volcanic infill. As it rises, the hot water dissolves soluble salts (especially sodium carbonate) from the highly alkaline volcanic rocks. The hot springs feed soda lakes, which are home to alkali-tolerant algae and brine shrimps, upon which flamingos feed.

LAKE BOGORIA GEYSER AND HOT SPRINGS

ASIA *northeast*

Nagano

LOCATION In the central Japan Alps, northwest of Tokyo, Honshu island, Japan

TYPE Hot springs and fumaroles

FREQUENCY Continuous

The hot springs and fumaroles of the resort of Yamanouchi near Nagano are famous for their unusual inhabitants – Japanese macaque monkeys. The fumaroles expel sulphur-dioxide-enriched steam, and the hot springs, with their mineral-enriched waters, feed a series of pools. In the 1960s, the macaques discovered the benefits of immersing themselves in the hot water, and soon it became part of their behavioural culture, passed on from generation to generation. Today, they bathe in two specially excavated pools; people bathe elsewhere.

JAPANESE MACAQUES AT NAGANO
About 250 Japanese macaques use the hot springs at Nagano, where snow may persist for four months of the year.

ASIA *northeast*

Beppu

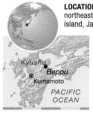

LOCATION In Beppu city, on the northeastern shore of Kyushu island, Japan

TYPE Geyser, hot springs, mud pools, and fumaroles

FREQUENCY 20–25 mins

The active volcanism of the Japanese islands means that the country is also rich in hot springs, which are now used for a variety of domestic, social, and industrial purposes. Beppu city is Japan's largest spa resort, with many fumaroles and mudpools, 2,849 springs, and the Tatsumaki Jigoku geyser. The volume of hot water that gushes from this hydrothermal system exceeds 136,000 kilolitres (36 million gallons) each day, giving it the second highest output in the world. Beppu has a number of hot springs, or jigokus

COMMUNITY BATHING

The curative and restorative properties of Japan's hot springs have been appreciated for hundreds of years, and bathing in the hot springs is something of a national recreation. Beppu's hot baths range from domestic tubs in private houses, to free municipal facilities for poorer citizens, to luxurious private baths where the rich can relax, meet friends, and gossip.

(a Buddhist word meaning "burning hell"), including: Blood Pond Jigoku, where red mud emerges with the boiling water; Ocean Jigoku, with its blue water; and Tatsumaki Jigoku geyser, which spouts water 20m (66ft) every 25 minutes. The springs are situated around two faults, Kannawa to the north and Asamigawa to the south. Their heat source is associated with 100,000-year-old buried lava domes which previously generated volcanism in the area. The abundant hot water, which surfaces at temperatures of 40–100°C (104–212°F), has been harnessed for domestic and industrial use, processing food, breeding fish, and producing geothermal power.

BEPPU'S STEAM POWER
These chimneys are not releasing pollutants, but excess steam from the geothermal energy plant, which powers local industry and heats homes.

HOT SPRING AT BEPPU
Blood Pond Jigoku is so called because the water is charged with red mud. The red colour is caused by iron oxide.

Waimangu

LOCATION In Waimangu Valley, near Rotorua, on the North Island of New Zealand

TYPE Fumaroles

FREQUENCY Continuous

The volcanic belt of North Island, New Zealand, was the site of several very large explosive eruptions in the past 10,000 years. Its volcanoes, near the southwest end of the Pacific's Ring of Fire (see p.157), are associated with subduction of the Pacific plate. Waimangu Valley is a bleak area of raw volcanic craters, boiling pools, and occasional geysers. It lies 10km (6 miles) southwest of Lake Tarawera and close to Lake Rotomahana. One geyser, which appeared in 1902 and lasted until 1905, spouted water 450m (1,475ft) into the air at intervals of between 5 and 30 hours and was the highest geyser plume ever recorded. Above the Waimangu Valley, there used to be scattered geysers and a series of tiered pink and white terraces, known as the White Terraces, which contained water-filled silica deposits. Maoris had traditionally used the site for bathing and cooking, and Waimangu was later

INFERNO CRATER

One of the attractions of Waimangu Reserve is this water-filled crater, which is a reminder of the 1886 eruption of Mount Tarawera.

regarded as the Yellowstone of its day, but early on the morning of 10 June 1886 all of these features were obliterated by a violent eruption of nearby Mount Tarawera. Within a couple of hours, a massive volume of basaltic scoria was blasted from a fissure 17km (11 miles) long. Fragments of the scoria fell over a huge area, some landing 30 km (18 miles) away. The eruption also buried three

WARBRICK TERRACE

The mineral deposits and hot-water pools of Warbrick Terrace are a reminder of the world-famous White Terrace, destroyed in 1886.

Maori villages, killing 155 people. One of these villages has now been re-excavated, in a similar way to Pompeii (see p.170), while the cliffs above Lake Rotomahana still steam perpetually. The cratered fissure is still visible as a chasm on Mount Tarawera.

Rotorua

LOCATION In the town of Rotorua, near Lake Rotorua, on the North Island of New Zealand

TYPE Geysers, hot springs, and fumaroles

FREQUENCY Virtually continuous

Rotorua is a modern lakeshore resort which lies in an extremely fertile region of volcanic hills and lakes around Lake Rotorua. Settled since the 14th century by the Arawa group of Maori tribes, the area lies in the north of the Taupo Volcanic Zone, which encompasses active geysers, fumaroles, hot springs, and boiling mud pools. This hydrothermal

activity is all that remains of one of the most violent volcanic eruptions of the last 2,000 years. Around 180 CE, extensive pyroclastic flows from the Taupo Volcanic Zone covered the area, devastating all life. The event passed without human record as the Maoris had not yet colonized the island. Today, Geyser Flat has seven active geysers; the most spectacular is Prince of Wales Feathers, a triple geyser that spouts to a height of 12m (40ft) and is always followed by an eruption of the Pohutu geyser, which rises to 30m (100 ft). The regular sequential timing of the geysers suggests that they are linked. They also have associated boiling mud pools, formed by steam leaking from underground geothermal reservoirs. Lady Knox geyser lies 24km (15 miles) to the southeast. It is artificially encouraged to spout to a considerable height by a daily dose of soap suds.

INDICATOR GEYSER

The Prince of Wales Feathers geyser resembles an emblem on the prince's coat of arms. It was named following a royal visit in the early 1900s.

GEOTHERMAL POWER

Wairakei near Rotorua was the second geothermal field to be developed after Lardarello (see p.189). Water heated to boiling point by an igneous intrusion rises through a permeable volcanic ash aquifer. Wairakei is liquid-dominated and benefits from high pressure. Since the geothermal aquifer has been punctured by boreholes, the rising water crosses its flash-point and turns to steam on its way to the surface.

WAI-U-TAPU MUD POOLS

A mixture of black sulphide with white silica and kaolin clay gives these mud pools their distinctive colour, while rising gas generates the bubbles.

CHAMPAGNE POOL

The hot waters of Champagne Pool are bordered with carbonate deposits and give off clouds of sulphurous gas, which fill the air with a rotten-egg smell.

WINTER WARMTH
Bathers enjoy a refreshing dip at the Svartsengi geothermal power plant in Iceland, which is one of the world's leading producers of such energy.

GEOTHERMAL ENERGY

The Earth's internal heat represents a vast and practically inexhaustible source of energy, yet such geothermal energy is exploited only to a limited extent. This may change if economically viable methods are found of harnessing it. Geothermal energy is mainly produced by natural radioactive decay of uranium, thorium, and potassium. Unlike energy from fossil fuels, it also generates little carbon dioxide, so does not contribute significantly to global warming (see pp.458–59).

Reservoirs in Rocks

The Earth's core is at a temperature of about 4,000°C (7,200°F). Heat constantly flows through the mantle and crust (see p.35) and escapes through the atmosphere to cold outer space. Over most of the planet's surface, such heat flow cannot be felt, let alone put to work. However, in areas where the heat energy of the Earth has been concentrated, such as active volcanic zones, geothermal energy can be exploited. Here, intense heat can boil groundwater that is close to the surface, creating underground reservoirs of superheated water or steam. If the water escapes upwards, it can emerge in hot springs, geysers, or fumaroles.

Hot springs were used to heat houses in Roman times, about two millennia ago, and the practice still continues in Iceland and Japan. This simple method of tapping geothermal energy does have some disadvantages. The water from hot springs often gives off sulphurous gases, and it can only be used locally, because its temperature soon falls if it is piped far from its source.

A much more flexible way of exploiting geothermal energy is to harness it for the generation of electricity. The world's first such geothermal power plant opened at Larderello, Italy, in 1904, and installations now operate in more than 20 countries worldwide. Most of these power plants tap into existing reservoirs of steam or water, which drive the turbines. The largest plant – at the Geysers in northern California, USA – generates more than 900 megawatts (MW) of electrical power, which is sufficient to supply about half a million homes.

NORTHERN HEAT
Although it is not far from the Arctic Circle, tomato plants can be grown in this geothermally heated greenhouse in Hveragerdhi, Iceland.

Low-Level Heat

Even in regions in which the Earth's heat flow is not concentrated, geothermal heat can be used as a way of saving energy. Through a closed system of underground pipes, a heat pump can transfer geothermal heat between a house and the ground beneath it, but it does not create the heat. During the winter, the ground temperature is often higher than the air in unheated rooms. As a result, water in the pipes collects heat from the ground and releases it indoors. In the summer, the situation is reversed. The pump acts as an air-conditioner, because the water absorbs heat as it flows indoors, and releases it back into the ground.

This way of exploiting geothermal energy works best in areas with continental climates, in which there is a large difference between winter and summer temperatures. In such conditions, a geothermal heat pump can cut electricity or fuel use by up to 50 per cent – saving money as well as reducing carbon dioxide emissions.

Untapped Resources

At present, geothermal sources meet only a tiny fraction of the world's total energy requirements, and most of this energy comes from regions in which there is geothermally heated water underground. However, much more widespread than these hydrothermal reservoirs are hot, dry rocks. If ways can be found to extract their heat, geothermal energy production could soar.

Several major engineering problems have to be overcome before this goal is realized. One is that zones of hot, dry rock are typically 2km (1¼ miles) or more underground. Once a borehole has been drilled to this depth, the heat energy has to be collected and brought to the surface. In experimental wells, water has been used to fracture the rock, creating fissures that fill with steam. The steam then travels to the surface, where its energy can be utilized. Enhanced Geothermal Systems (EGS) are currently being tested in a range of countries, and one station in the Netherlands is already operational.

LEADING GEOTHERMAL PRODUCERS

Geothermal energy currently provides for less than 0.02 per cent of the world's annual energy needs. With the exception of Iceland, most producers are located in the volcanically active Pacific Rim.

Country	Production (MW)	Global proportion (%)
USA	3,794	23.5
Indonesia	2,356	14.6
Philippines	1,935	11.9
Turkey	1,682	10.4
New Zealand	1,037	6.4
Mexico	963	5.9
Kenya	944	5.8
Italy	944	5.8
Iceland	754	4.6
Japan	621	3.8

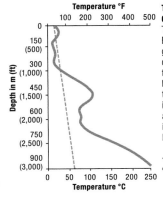

TEMPERATURE GRADIENTS
The temperature of the Earth's crust increases gradually with greater depth in most parts of the world. In regions of high geothermal activity, the temperature gradient is much more variable, and it can rise sharply in areas containing heated groundwater.

-- **Normal gradient**
— **Typical gradient in geothermally active region**

LAND

SHAPING THE SURFACE
Meandering across the Kenyan
countryside, the present channel of the
River Mara bypasses land scarred by old
meanders and ox-bow lakes.

RIVERS AND LAKES

Rivers flow and lakes form in all but the world's hottest, coldest, and driest places. They hold a small, but vital, proportion of the Earth's surface fresh water: lakes hold about half of one per cent, and river channels contain much less – about one-fortieth of one per cent. This tiny fraction is, however, disproportionally significant, because rivers shape the land. Rivers are one of the most powerful erosive forces on Earth – across the continents, the dominant landscape is one of hill slopes and river valleys. Given time, rivers can wear down mountains and carry them to the sea. Powered by the force of gravity, the world's rivers deliver about 20 billion tonnes of land surface to the oceans every year. Many rivers are old in geological terms, and have increased in size with age, forming systems that drain continents. Some have hollowed out vast caves and run largely underground. By comparison, most lakes are very young, temporary accumulations of water that will inevitably shrink and disappear.

Rivers

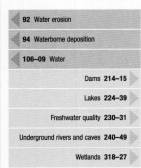

92	Water erosion
94	Waterborne deposition
106–09	Water
Dams	214–15
Lakes	224–39
Freshwater quality	230–31
Underground rivers and caves	240–49
Wetlands	318–27

Rain or melted snow that does not evaporate flows downhill across the land surface. This surface flow is channelled by small irregularities into tiny rivulets that merge and flow down gullies into streams. These accumulate water and increase in size along their courses, cutting valleys and receiving the inflow of smaller tributaries and seepage from ground water. The nature and form of a river depends on its rate of descent, the amount of water and the steadiness of its flow, and the underlying geology. These factors dictate the rates of erosion and deposition and, consequently, the degree to which the river modifies the landscape.

Drainage Basins

A river and its tributary streams form a system that collects all the run-off (surface flow) within its drainage basin. The mountains, or other high ground, that separate the drainage basins of different river systems are called the watershed. A notable example is the Continental Divide along the Rocky Mountains in North America, which separates eastward- and westward-flowing rivers. River systems form different patterns of drainage on the landscape, depending on the rock type, slope of the land, and movements of the Earth's crust. The most common pattern is dendritic, where tributary streams join a main river like the branches of a tree. The points where rivers join together are called confluences. In addition to its tributaries, groundwater seeping into the river channel may account for up to 30 per cent of a river's total volume. Most of the water carried in a river eventually flows into the sea.

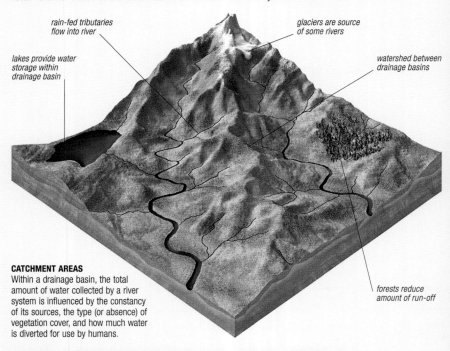

rain-fed tributaries flow into river

glaciers are source of some rivers

lakes provide water storage within drainage basin

watershed between drainage basins

forests reduce amount of run-off

CATCHMENT AREAS
Within a drainage basin, the total amount of water collected by a river system is influenced by the constancy of its sources, the type (or absence) of vegetation cover, and how much water is diverted for use by humans.

HEAD-WATERS TUMBLE
High in the mountains of Jotunheimen National Park in Norway, meltwater flows over bare rock on its way to becoming a river. Head-waters such as these need not necessarily flow through permanent channels.

LAND

Erosion and Transport

The size and amount of material carried by rivers varies enormously. Even the slowest, smoothest flow of water will wash away small particles of rock. The turbulent flow of a river, with churning currents and sudden changes in velocity, has much greater power. The flow of water erodes the banks and bed by friction, suction, and solution. This flow also carries with it large quantities of solid material – the products of weathering or the river's own processes. It is this material, known as a river's sediment load, that usually accounts for most of its erosive force, through the abrasion of sand, gravel, pebbles, and boulders rolling and bouncing along the river-bed. Most solid material is, however, transported in suspension as fine particles of clay and silt.

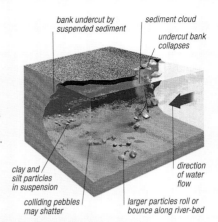

bank undercut by suspended sediment

sediment cloud

undercut bank collapses

clay and silt particles in suspension

colliding pebbles may shatter

direction of water flow

larger particles roll or bounce along river-bed

POTHOLES
Potholes, such as these in South Africa, form if hard boulders are "trapped" in depressions of softer rock and then swirled around by turbulent currents.

RIVER EROSION
Small particles are washed away, suspended in the water flow. Larger particles become concentrated on the river-bed where the current is strongest, and multiple impacts gradually reduce their size.

EROSION BY UNDERCUTTING
This section of canyon wall in Zion National Park, USA, has been undercut by the river – not by the gentle flow pictured, but by winter floodwaters crashing through the gorge.

Deposition

When the sediment-carrying capacity of a river decreases, deposition occurs. This can happen for a number of reasons – for example, where the current lessens as the river bends, widens, flattens, or floods it banks. Erosion and deposition are closely interlinked processes and may take place simultaneously on opposite banks of the same river bend. The most prominent features produced by deposition are point bars along river-banks, and riffles, sand bars, and islands in midstream. The floodplains and deltas seen in the lower courses of rivers are depositional features on a much grander scale.

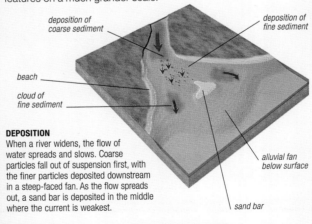

deposition of coarse sediment

deposition of fine sediment

beach

cloud of fine sediment

alluvial fan below surface

sand bar

DEPOSITION
When a river widens, the flow of water spreads and slows. Coarse particles fall out of suspension first, with the finer particles deposited downstream in a steep-faced fan. As the flow spreads out, a sand bar is deposited in the middle where the current is weakest.

SANDBANKS
Point bars are deposited along a river's banks below the surface, where the water flow is slowest.

EROSION AND DEPOSITION
Where the current is strongest, a river's bank is undercut; where the water moves more slowly on the other side, deposition occurs.

DRIED-UP RIVER
In summer, when water levels fall and evaporation increases, rivers such as the Yellow River shrink, revealing beds of dried mud.

YELLOWSTONE FALLS
The upper courses of many rivers – such as the Yellowstone River in Wyoming, USA (shown here) – include dramatic waterfalls, rapids, and gorges.

LAND

Upper Courses

In most rivers, the upper course is characterized by steep descent and the inflow of many small tributaries. The river is usually narrow and fast-flowing, and it erodes a distinctive V-shaped valley that zigzags between interlocking spurs. The rate of flow along the upper course may be highly variable, depending on seasonal factors such as snow-melt or heavy rainfall, and most mountain rivers are subject to regular flooding. Most of the valley-cutting takes place when the river is in flood. Flooding also maintains small floodplains where the river crosses flatter ground. When crossing a high, sloping plateau, a river may cut a deep gorge and become incised, particularly if the plateau is being uplifted.

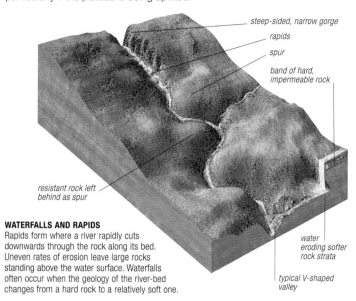

steep-sided, narrow gorge

rapids

spur

band of hard, impermeable rock

resistant rock left behind as spur

water eroding softer rock strata

typical V-shaped valley

WATERFALLS AND RAPIDS
Rapids form where a river rapidly cuts downwards through the rock along its bed. Uneven rates of erosion leave large rocks standing above the water surface. Waterfalls often occur when the geology of the river-bed changes from a hard rock to a relatively soft one.

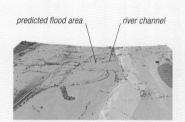

RAPID DESCENT
Rates of erosion are very high in steep mountain valleys where rivers race downhill.

WATERFALL
At the base of most waterfalls, such as this one in California, USA, the waters erode a plunge pool, which is enlarged by the scouring action of rock fragments from the cliff-face.

PREDICTING FLOODS

In the UK, the Environment Agency uses radar sensors carried by aircraft to determine precisely the height of land along the margins of rivers that are liable to flood. Using this information, scientists run computer simulations that predict the effects on the land of a river rising to a particular height. Areas that are more vulnerable to flooding can be identified and the flood defences planned accordingly.

predicted flood area

river channel

FLOOD RISK
This computer-generated image shows the effects of a 1.5-m- (5-ft-) high rise in the level of a river in the north of England.

FERTILE RIVER VALLEY
This combination of cone-shaped hills and a flat valley floor is typical of rivers flowing through limestone valleys in the tropics and subtropics.

LAND

Lower Courses

Rivers generally flow down gentler gradients in their lower courses compared to their upper reaches. Few such rivers follow one straight channel: some braid into numerous channels separated by temporary islands; others meander in great looping curves. A meandering river may cut rapidly downwards, forming an incised meander, or it may wander from side to side, creating a wide, flat valley with steep sides. Meanders that are cut off as the river shifts its course form distinctive ox-bow lakes. Deposition rates are high along lower courses and are often linked to seasonal flooding. Sediments deposited by floodwaters may build up into deep beds. Flooding also produces natural levees (raised banks) along stretches of riverbank.

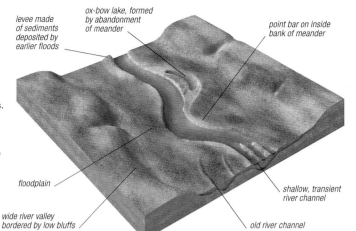

levee made of sediments deposited by earlier floods

ox-bow lake, formed by abandonment of meander

point bar on inside bank of meander

floodplain

wide river valley bordered by low bluffs

shallow, transient river channel

old river channel

FLOODPLAINS
A floodplain landscape (left) is subject to constant change as the river shifts its meandering course, leaving behind levees, ox-bow lakes, and abandoned channels. Because floodplains are flat and regularly covered with rich alluvial deposits, they are attractive both for agriculture and human settlement. Floodplains are some of the most densely populated areas in the world.

BRAIDING ON LOWER COURSE
This pattern of braided channels on a river in Denali National Park, Alaska, USA, will remain stable until the flow of the river changes.

CLEARING WATERWAYS
Rivers that are used for navigation often need to be dredged to clear them of sediment and vegetation. Some large river highways have to be dredged constantly in order to keep important navigation channels open. Here, a dredger is clearing a deepwater channel in the bed of the Mississippi – a river particularly prone to such problems, as it carries half a billion tonnes of sediment to the sea each year.

Deltas

Many rivers flow into the sea through a delta – a wide, flat region where the river branches into numerous distributary channels, and where new land is being created as the river deposits any material it is still carrying in suspension. The smallest particles may be carried a considerable distance out to sea, as fresh water slowly mixes with seawater. A typical delta takes the form of a broad fan that slopes gently seawards. Its shape depends on the balance between the river's flow, the tide, currents, and wave action. In some cases, the river's flow or coastal currents may be too strong for a delta to form. Some rivers form deltas when they enter lakes, but these lacustrine deltas, as they are known, have a steeper slope because water mixing takes place more rapidly.

BIRD'S-FOOT DELTA
Well-defined, elongate distributaries are characteristic of bird's-foot deltas, such as the Mississippi delta shown above.

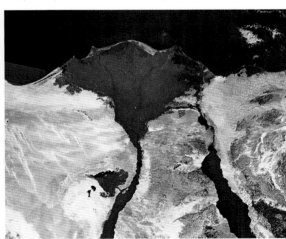

FAN DELTA
In this satellite image, the fertile Nile delta contrasts with the surrounding desert sands. Constant wave action keeps the delta's coastline smooth and distinct.

RIVER PROFILES

The pages that follow contain profiles of the world's main rivers. Each profile begins with the following summary information:

LENGTH Distance from source to mouth

VERTICAL FALL Height descended by river from source to mouth

BASIN AREA Area drained by the main river and all of its tributaries

MAIN TRIBUTARIES Subsidiary rivers that flow into the main river

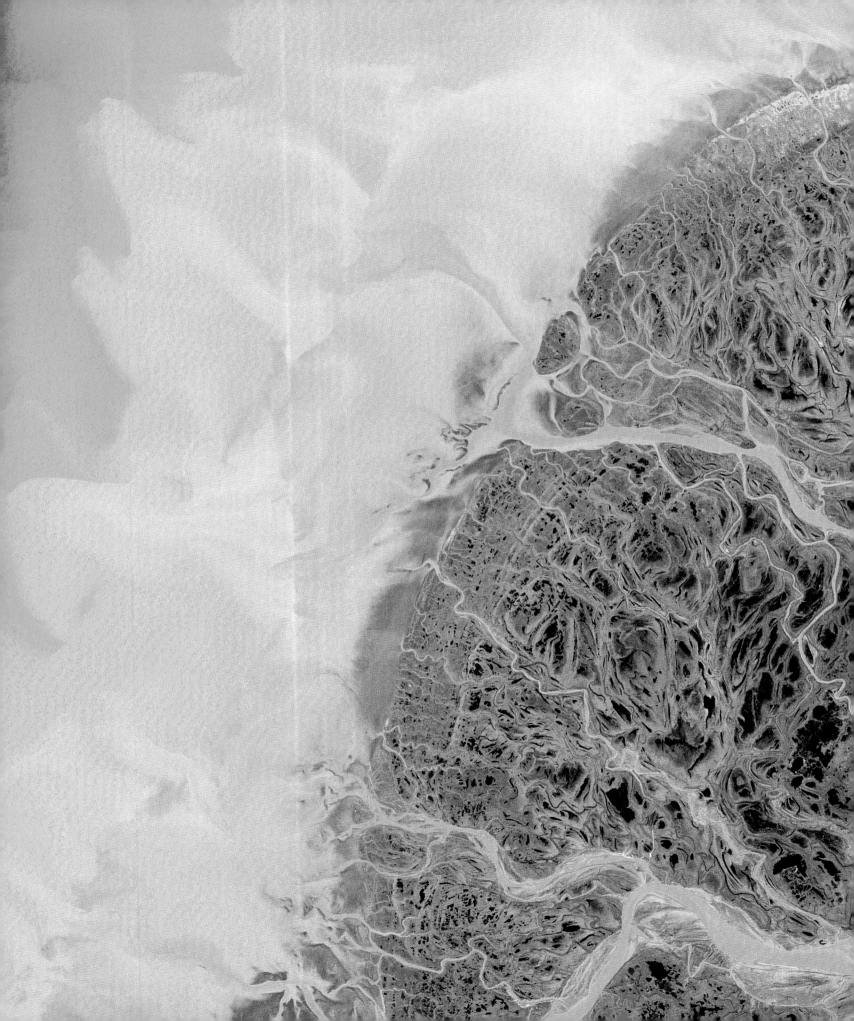

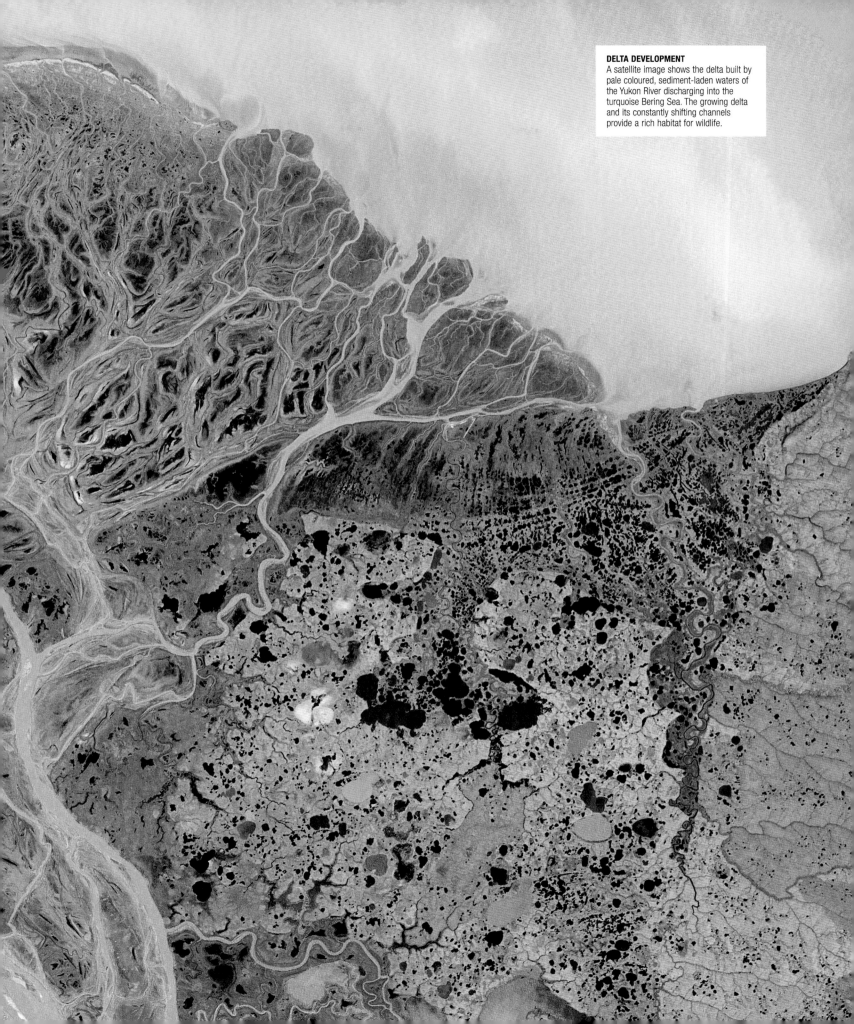

DELTA DEVELOPMENT
A satellite image shows the delta built by pale coloured, sediment-laden waters of the Yukon River discharging into the turquoise Bering Sea. The growing delta and its constantly shifting channels provide a rich habitat for wildlife.

LAND

Mackenzie

LOCATION In northeast Canada, flowing from Great Slave Lake into the Beaufort Sea in the Arctic Ocean

LENGTH
1,705km (1,060 miles)

VERTICAL FALL
156m (512ft)

BASIN AREA
1.8 million square km (700,000 square miles)

MAIN TRIBUTARIES
Liard, Great Bear, Peel

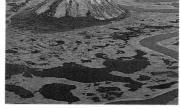

PINGO ON MACKENZIE RIVER DELTA
Isolated in Canada's Northwest Territories, the Mackenzie delta surrounds a pingo of sediment pushed up by the growth of a buried block of ice during the seven-month-long winter.

The basin of the Mackenzie River is the largest in Canada and the second largest in North America. From the western end of Great Slave Lake, the river flows northwest, mainly over ground subject to permafrost, and into the sea through a seasonally marshy delta. In winter, the river freezes along

CARIBOU
The Mackenzie lowlands, which are covered with a mosaic of forests, swamps, shrubland, and grassy tundra, are home to herds of Caribou.

its entire length. For the most part it is a broad river, 1.5–6.5km (1–4 miles) wide, which has steep gravel banks and frequently braids into channels. Mackenzie waters leaving Great Slave Lake are clear, any sediment having been deposited in the lake. The waters of the main tributary, the Liard, are very muddy, and the two flows do not completely mix for some 320km (200 miles) downstream of their confluence. Midway along its course, the Mackenzie meets Great Bear River, which drains Great Bear Lake.

FROZEN RIVER IN WINTER
Ice usually forms in mid-October, and parts of the river remain frozen until early June. Tributaries often thaw before the main river, causing flooding.

Hudson

LOCATION In New York State, USA, and flowing into the Atlantic Ocean at New York City

LENGTH
507km (315 miles)

VERTICAL FALL
1,371m (4,323ft)

BASIN AREA
34,628 square km (13,370 square miles)

MAIN TRIBUTARIES
Mohawk

From its source in Lake Tear of the Clouds, a post-glacial lake in the Adirondack Mountains, the Hudson flows southeast to its confluence with the Mohawk near Albany, then turns southwards for the last 240km (150 miles) of its

course. The lower Hudson Valley (and that whole section of eastern coastline) was drowned by the sea at the end of the last ice age, leaving a submarine canyon about 180km (110 miles) long. Tidal influence extends as far as Albany, which has a 1.4-m- (4½-ft-) high tide. A prominent feature of the lower valley is the Palisades (see p.180), a series of steep cliffs where the river cuts into hard rock. Since its exploration by Henry Hudson in 1609, the river has been commercially important. Nineteenth-century improvements, notably the Erie Canal, connected the Hudson with the Great Lakes, and the rapidly extending western "frontier". It remains a major river highway.

HUDSON PASSING MANHATTAN ISLAND
The shipping piers seen here along Manhattan's Lower West Side underline the Hudson's importance as a transport route.

St. Lawrence

LOCATION Flows from Lake Ontario to the Atlantic Ocean; forms part of the Canada–USA border

LENGTH
1,244km (760 miles)

VERTICAL FALL
75m (246ft)

BASIN AREA
1.3 million square km (500,000 square miles)

MAIN TRIBUTARIES
Ottawa, St. Maurice

The St. Lawrence River is the outflow for the whole of the Great Lakes system. It is a very young watercourse that began flowing only about 7,000 years ago, as a result of post-glacial uplift in the Great Lakes basin. The wide plains of the St. Lawrence valley formed before the river began to flow, when the area was inundated by a shallow sea. The upper part of its course is wide and for the first 184km (114 miles) forms the border between Canada and the USA. The river then narrows through a series of rapids towards Montreal, where it widens again. Lake St. Pierre, midway between Montreal and Quebec City, marks the start of the St. Lawrence estuary. Downstream of Quebec City, the lower estuary extends from Île d'Orléans to the island of Anticosti, where the river enters the

FROZEN FALLS IN WINTER
Water plunging over the Montmorency Falls on the Montmorency River in Quebec freezes as it falls, creating curtains of frozen spray.

Atlantic at the Gulf of St. Lawrence. Extensive engineering during the 20th century has transformed the river into the St. Lawrence Seaway.

NORTH AMERICA *central*

Mississippi

LOCATION In the USA, flowing from near the Canadian border in Minnesota to the Gulf of Mexico

LENGTH 3,780km (2,350 miles)

VERTICAL FALL 450m (1,475ft)

BASIN AREA 3.2 million square km (1.2 million square miles)

MAIN TRIBUTARIES Missouri, Ohio, Arkansas, Tennessee

Together with its many tributaries, the Mississippi River forms a vast and ancient river system. Its drainage basin extends over almost the whole of the United States from the Rocky Mountains to the Appalachians. The western tributaries drain the Great Plains, while the eastern ones

SLOUGHS ON MISSISSIPPI RIVER
This upstream view of the river in northern Iowa shows narrow, dead-end sloughs (backwaters) between a network of sand bars, which have been stabilized by vegetation.

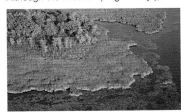

drain the Appalachian Plateau. The source of the Mississippi is Lake Itasca in northern Minnesota, which the river leaves as a clear 3-m- (10-ft-) wide stream that winds eastwards through extensive marshes, fed by other lakes and streams. It then turns southwards, skirting the higher ground around the western end of Lake Superior. Downstream of Minneapolis, the Mississippi begins to collect the upper-course tributaries that give rise to its name, which means "Father of Waters" in the native Algonquian language. The influx of the Missouri River near St. Louis brings huge quantities of sediment that change the character of the river. Downstream, at Cairo, Illinois, the inflow of the Ohio River marks the original beginning of the Mississippi delta, which has been

FLOOD CONTROL
Levees for protection against floods were first built on the Mississippi in the early 18th century. Long stretches are now ramparted in this way. The variable waters of the Ohio cause most floods, but in 1993, heavy rains over the northern Great Plains caused devastating floods upstream of Cairo. With little warning, East Dubuque and other towns on the upper Mississippi and lower Missouri were inundated by a so-called "100-year" event.

inching southwards for the last 100 million years. Flowing through a gap between the Cumberland and Ozark plateaus, the Mississippi follows a broad valley cut through the alluvial deposits by meltwater at the end of the last ice age. Although the Missouri ("Big Muddy"),

MINNESOTA MARSHES
The low-lying and fast-eroding shoreline of Lake Minnetonka reveals the delicate balance between land and water in the wetlands of Minnesota.

MILK RIVER MEETING THE MISSOURI
The Milk River gathers some of its distinctively pale sediment as it loops through southern Canada. It meets the Missouri in Montana.

4,120km (2,560 miles) long, is considered the Mississippi's main tributary, the Ohio – length 1,570km (975 miles) – contributes the greater volume of water. Only along its lower course, south of Cairo, does the Mississippi achieve the full majesty of "Ol' Man River", more than 1,600m (1 mile) wide and carrying about half a billion tonnes of sediment to the sea each year. The delta has shifted over an area of about 320 square km (200 square miles) during the last 6,000 years, and forms a branching extension into the Gulf of Mexico, southeast of New Orleans. It is currently shrinking, due to a combination of natural and human-induced factors.

LAND

Colorado

LOCATION Mainly within the USA, flowing from the Rocky Mountains to the Gulf of California

LENGTH
2,333km (1,450 miles)

VERTICAL FALL
4,320m (14,200ft)

BASIN AREA
632,000 square km
(244,000 square miles)

MAIN TRIBUTARIES
Green, Little Colorado, Gila

The Colorado, which drains the arid southwestern quarter of the USA, is the world's most formidable canyon-cutter. In response to the tilting uplift of the Colorado Plateau, the river has cut more than 1,600km (1,000 miles) of deep canyons, the most spectacular stretch of which is the 350-km- (220-mile-) long Grand Canyon in northern Arizona. The Colorado River rises from snowmelt high in the Rocky Mountains in northern Colorado, on the westward slopes of the Continental Divide. It flows southwest across the Colorado Plateau into Utah, where the confluence with the Green River in the Canyonlands region brings waters from the northernmost reaches of its drainage

CACTI BY THE COLORADO
The climate of the Colorado Plateau is hot and arid. Only cacti and low desert shrubs can grow – there are no trees, even at the river's edge.

basin in Wyoming. Downstream in northern Arizona, the main trunk of the Colorado's branching canyon system – the Grand Canyon – reaches 29km (18 miles) in width and cuts down through layers of sedimentary rock that record 2 billion years of geological history. Below the canyon, after exiting the plateau, the course turns southwards, forming the California–Arizona state line. Along its lower course the Colorado is a slow, meandering river, laden with fine silt and subject to flooding. The last 128km (80 miles) run through Mexico to a shrinking delta. The Colorado River is the most managed river in the world. Large-scale engineering began in the northern part of the drainage basin in the 1920s, with great tunnels cut beneath the Continental Divide to divert some of the river's water eastwards to the Great Plains. Subsequent attention has focused on the lower course. The 221-m- (726-ft-) high Hoover Dam created Lake Mead in 1935, and provides both electricity and irrigation water. Downstream, considerable amounts of water are diverted to the cities of Las Vegas, Phoenix, Tucson, San Diego, and Los Angeles. The Imperial Dam diverts most of the water that remains along the All-American Canal to California's Imperial Valley. Although the flow at the mouth of the Colorado is now much reduced, it has in the past effected great changes upon the coastline. The Gulf of California once extended much farther north until the Colorado changed its course

THE GRAND CANYON
Created by the river, the magnificent rock cliffs of the Grand Canyon are preserved by the aridity of the desert climate. At the base of the canyon, the river has cut down to 1.7 billion-year-old rocks.

in prehistoric times. Sediment deposited on the river's banks dammed the Gulf, creating the current shoreline and the great Salton Sink, a stretch of former sea bed that extends inland some 480km (300 miles). In 1905, a breached irrigation channel allowed the Colorado's waters to flood into the sink, creating the 112-km- (70-mile-) long Salton Sea.

COLORADO RIVER DELTA
This aerial view of the delta in Mexico shows some of the many distributary channels created when the river had a much greater flow.

REGULATING THE COLORADO

The 216-m- (710-ft-) high Glen Canyon Dam is located in Arizona about 175km (110 miles) upstream of the Grand Canyon. It was constructed in the 1960s to store water from winter rains to generate electricity later in the year, when the river's natural flow diminishes. By restricting the flow of the Colorado, the dam has had an alarming effect on the environment along riverbanks in the Grand Canyon. Instead of being scoured clean by regular winter floods, parts of the canyon are now starting to fill with sediment. In the 1990s, attempts were made to redress the balance by releasing artificial floods from the dam.

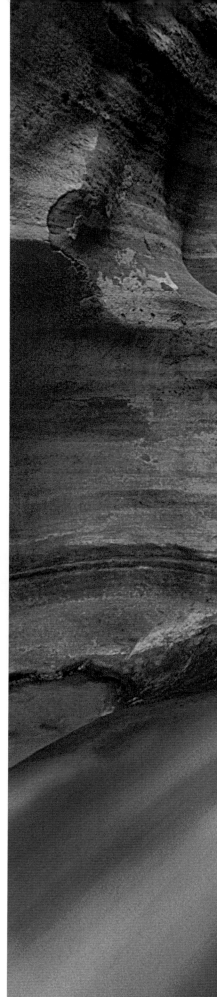

CUTTING POWER OF THE COLORADO
The erosive power of the Colorado River is clearly evident in this picture, which shows a deep channel cut through the colourful sedimentary rocks of the Grand Canyon.

LAND

Rio Grande

RIO GRANDE, TEXAS
Upstream of the Big Bend canyons, with the Chisos Mountains rising in the distance, the Rio Grande provides water to the arid landscape of the high plains.

LOCATION Flows from the Rocky Mountains to the Gulf of Mexico; is part of the USA–Mexico border

LENGTH
3,034km (1,885 miles)

VERTICAL FALL
3,650m (12,000ft)

BASIN AREA
445,000 square km (172,000 square miles)

MAIN TRIBUTARIES
Chama, Conchos, Pecos, Salado

Rising high in the San Juan Mountains of southern Colorado, the fifth longest river in North America flows through New Mexico into Texas. From El Paso on the high plains to Brownsville on the delta, the Rio Grande defines the border – 2,000km (1,250 miles) long – between Texas and Mexico. Fed by springs below the San Juan snowfield, its upper course flows eastwards along pine-forested canyons, then turns south into New Mexico, where the river has cut the Rio Grande gorge and White Rock Canyon. The middle course flows through basin-and-range terrain, skirting the San Andres Mountains, before descending through three great canyons, over 450m (1,500ft) deep, onto the Gulf Coastal Plain. In the USA, these canyons form Big Bend National Park. With much of its water being diverted for irrigation, the Rio Grande winds sluggishly across the coastal plain to the Gulf of Mexico. So much of the river's water is diverted that during long, dry periods surface flow at the delta disappears.

BIG BEND
The sheer walls of Santa Elena Canyon in Texas are the result of the river's great erosive power.

River Plate

LOCATION On the eastern coast of South America, with Uruguay to the north and Argentina to the south

LENGTH
290km (180 miles)

VERTICAL FALL Nil

BASIN AREA
4.2 million square km (1.6 million square miles)

MAIN TRIBUTARIES
Paraná, Uruguay

DEVIL'S THROAT AT IGUAÇU FALLS
Upstream from its confluence with the Paraná, the Iguaçu River plunges over a great, horseshoe-shaped waterfall into a steep-sided canyon called Devil's Gorge.

Not strictly a river, nor a marine gulf, the River Plate is a huge, funnel-shaped estuary formed by the mouth of the River Uruguay coinciding with the delta of the River Paraná. Extending over 35,000 square km (13,500 square miles), and 219km (136 miles) wide where it meets the Atlantic, the Plate receives the sediment load of three great rivers – the Uruguay, the Paraná, and the Paraguay (the major tributary of the Paraná) – that have a combined length of nearly 8,000km (5,000 miles). As a result, the Plate is choked by sediment that accumulates into great shoals, where the water depth can be reduced to as little as 1.5m (5ft). Constant dredging is necessary to maintain deep-water channels to the ports of Montevideo and Buenos Aires. The catchment area of the

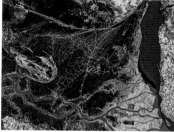

PARANÁ RIVER DELTA
This false-colour satellite image shows an area of marshland to the north of Buenos Aires, Argentina, where the Paraná and Uruguay rivers meet.

Plate extends over parts of Argentina, Bolivia, Brazil, Paraguay, and Uruguay. The River Paraná flows some 4,000km (2,485 miles) from the highlands of central Brazil to the Plate. Along its upper course, the most notable feature is the Guaira Falls, which carries several times the volume of Niagara Falls. Downstream, the River Iguaçu brings water still churning from a drop of 82m (270ft) over the spectacular Iguaçu Falls. The River Paraguay runs 2,550km (1,585 miles) from its source in the Mato Grosso region of Brazil, through the Pantanal wetlands, to its confluence with the Paraná. The river has a shallow gradient – most of its catchment area lies below 200m (650ft) – and it is subject to seasonal flooding. The River Uruguay rises on the western slopes of southern Brazil's coastal range and loops inland along a course of about 1,000km (1,600miles) to the Plate estuary.

Orinoco

LOCATION Flows through Venezuela from the Guiana Highlands to the Atlantic Ocean

LENGTH
2,151km (1,337 miles)

VERTICAL FALL
1,074m (3,523ft)

BASIN AREA
948,000 square km (366,000 square miles)

MAIN TRIBUTARIES
Guaviare, Meta, Arauca, Apure, Caroni

The second longest river in South America, the Orinoco flows in a great arc through Venezuela. Its drainage basin covers about 80 per cent of that country and about 25 per cent of Colombia. The upper reaches contain two remarkable features: Angel Falls, the world's highest uninterrupted waterfall, cascades down Auyan Tebui (Devil's Mountain); and Casiquiare Channel, 355km (221 miles) long, diverts some of the Orinoco's waters southwards into the Rio Negro, a tributary of the Amazon. On the east edge of the Guiana Highlands, the Orinoco flows through a series of boulder-strewn rapids (Region de los Raudales) before turning northeastwards. The lower course is marked by numerous swamps (notably the Llanos wetlands, see p.324) before branching into a delta some 436km (270 miles) from the Atlantic coast. The main channel discharges at Boca Grande ("Big Mouth"). The rainfall that feeds the Orinoco is very seasonal, and flooding along the lower course is frequent and extensive. At the end of the dry season in April, the river at Ciudad Bolivar may be 30m (100ft) lower than at the height of the rains, in July.

JIMMIE ANGEL

Angel Falls is named after American aviator Jimmie Angel (1899–1956), who first sighted the falls from the air in 1933. Four years later, he landed on top of Devil's Mountain, but damaged his plane in the process and had to trek back through rainforest to the nearest settlement. Angel, who achieved near-legendary status in Venezuela, died in Panama from injuries received in a flying accident.

ANGEL FALLS, VENEZUELA
Before joining the Orinoco (via the Caroni), the River Churun plunges 979m (3,212ft) over Angel Falls at Canaima, Venezuela.

WATTLED JAÇANA
Also known as the Lily Trotter, this bird has large, spidery feet that work like snowshoes, spreading its weight and enabling it to walk on floating vegetation.

elongated toe

ORINOCO DELTA
The delta covers about 36,500 square km (14,000 square miles) and is home to the critically endangered Orinoco Crocodile. Many of the islands are used for cattle ranching and growing cacao.

LAND

SOUTH AMERICA *north*

Amazon

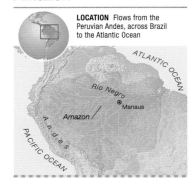

LOCATION Flows from the Peruvian Andes, across Brazil to the Atlantic Ocean

LENGTH 6,430km (3,990 miles)

VERTICAL FALL 5,500m (18,000ft)

BASIN AREA 7.1 million square km (2.7 million square miles)

MAIN TRIBUTARIES Juruá, Madeira, Negro

The Amazon is the world's greatest river, measured both by the size of its basin and by the volume of water it discharges. On a world scale, the Amazon delivers about 20 per cent of all the water reaching the sea from rivers. Its nearest rivals, the Congo and Yellow rivers, each deliver only about 4 per cent of the total. Some 16km (10 miles) wide at Manaus, while still 1,600km (1,000 miles) from the sea, it is truly a giant river. The Amazon flows from mountainsides and glacier-fed lakes high in the Andes of southern Peru, within about 160km (100 miles) of the Pacific Ocean. The upper course, known as the River Marañón, flows north, and then turns east, descending into the Amazon Basin – a vast depression sinking under the weight of material eroded from surrounding highlands. Here, it meanders for more than 3,200km (2,000 miles) from Iquitos to Belém. Marshy areas along the banks are liable to seasonal floods, but most of the basin is covered by dry-footed equatorial

AMAZON HEADWATERS
Many of the Amazon's Andean tributaries freeze in winter, as seen here in the Bolivia Cordillera Real.

GIANT RIVER OTTER
This highly endangered mammal lives in small family groups along quiet stretches of the river throughout the Amazon Basin.

BRAIDED CHANNELS NEAR MANAUS
Along its middle course, the river frequently braids between strings of low islands that constantly shift as the river channel alters following seasonal rains.

OVERGROWN RIVER
The river's placid backwaters are covered in mats of vegetation, such as these Giant Water Lilies, which provide a valuable habitat for many animals.

rainforest that experiences daily convectional thunderstorms. Along its middle and lower course, the Amazon receives many tributaries from the Guiana Highlands to the north, and the Brazilian Central Plateau in the south. These sources – Andean glaciers, daily rains, and the numerous tributaries – contribute to the 770 billion litres (170 billion gallons) of water that the Amazon pours into the Atlantic every hour. The flow of the river is so strong that there is no delta; instead it discharges through a wide, mangrove-fringed estuary that straddles the equator. Despite its size, the Amazon delivers less sediment to the ocean than the Mississippi, which has only one-tenth of the Amazon's flow.

DEFORESTATION

The Amazon Basin cradles the greatest expanse of rainforest on Earth – an irreplaceable resource that is being destroyed at an alarming rate. The Amazonian forests (see p.307) face a number of threats – logging, large-scale agriculture, and the slash-and-burn subsistence farming of internal migrants drawn to the Amazonian frontier. Stripped of cover, and deprived of the long-term release of the nutrients locked within the trees, Amazonian soils deteriorate rapidly into barren wastes, and add to the river's sediment load. Despite widespread concern and publicity, the deforestation continues, and some experts predict a catastrophic collapse of the entire forest ecosystem within the next 30–50 years.

LAND

LAND

Thames

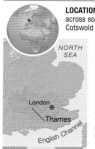

LOCATION In Britain, flowing across southern England from the Cotswold Hills to the North Sea

LENGTH
336km (210 miles)

VERTICAL FALL
108m (356ft)

BASIN AREA
13,600 square km
(5,200 square miles)

MAIN TRIBUTARIES
Colne, Kennet, Wey

The longest river in Britain, the Thames arises from springs in the Cotswolds, a ridge of limestone hills in south-central England. The upper course winds eastwards along the northern edge of the Chiltern Hills until it breaks through at the Goring gap near Oxford. Here, where it is known locally as the Isis, the river is about

THE THAMES BARRIER, LONDON
The Thames Barrier is a series of huge steel gates that can be raised to protect the city against flooding at high tide.

30m (100ft) wide. The lower course flows southeast along the wide, shallow Thames valley, which is characterized by gravel terraces and clays deposited by successive pulses of post-glacial meltwater. The river is tidal for the lower 145km (90 miles) of its length, and the tidal range at London, the lowest bridging point, is about 7m (23ft). Although London's inhabitants have a comfortable relationship with the Thames, the river in fact poses a considerable threat. Heavy rains combined with exceptionally high tides have flooded the city on many occasions. Below London, the Thames widens into an estuary that is about 8km (5 miles) wide, where it discharges into the North Sea.

Tagus

LOCATION Flowing across the Iberian Peninsula from eastern Spain to the Atlantic coast of Portugal

LENGTH
1,007km (626 miles)

VERTICAL FALL
1,590m (5,216ft)

BASIN AREA
81,600 square km
(31,505 square miles)

MAIN TRIBUTARIES
Gallo, Zezere

Rising in the Sierra de Albarracin about 140km (90 miles) east of Madrid, the Tagus is one of several rivers that drain the Meseta Plateau. Initially fast-flowing, through limestone gorges, the river slows

and widens as it flows westwards, past the city of Toledo, through increasingly arid landscapes. For a short distance the Tagus forms part of the border between Spain and Portugal, and the final 275km (170 miles) of its course flows through Portugal. The estuary at Lisbon is one of the world's finest natural harbours. The Tagus has been widely harnessed both for power and irrigation by both countries. Upstream of the border section, a dam at Alcántara in Spain has created an artificial lake, large by European standards, that extends back nearly 160km (100 miles) along the river's middle course.

TAGUS RIVER PASSING TOLEDO, SPAIN
The oldest parts of this city, founded before the time of the Romans, are perched high on a rocky outcrop that overlooks a bend in the river's gorge.

Severn

LOCATION In Britain, flowing from Wales through England to the Bristol Channel and the Irish Sea

LENGTH
290km (180 miles)

VERTICAL FALL
600m (2,000ft)

BASIN AREA
11,266 square km
(4,350 square miles)

MAIN TRIBUTARIES
Avon, Stour, Teme, Usk

Rising on Plynlimon mountain in western Wales, the upper course of the Severn descends 455m (1,500ft) in its first 24km (15 miles) into the Vale of Powys. After cutting the Ironbridge Gorge through hills on the English–Welsh border, the river follows a semicircular course through central England. One of its middle-course tributaries, the Warwickshire Avon, flows through William Shakespeare's home town of Stratford-upon-Avon. South of Gloucester the river widens and meanders to an estuary that discharges into the Bristol Channel below Newport. Locally, the lower course of the Severn is noted for its tidal bore, a wave that moves at speeds of about 25kph (15mph), reaching a height of nearly 3m (10ft).

Loire

LOCATION Flows from the Massif Central to the Atlantic coast south of the Brittany peninsula in France

LENGTH
1,020km (634 miles)

VERTICAL FALL
14,850m (4,500ft)

BASIN AREA
117,000 square km
(45,000 square miles)

MAIN TRIBUTARIES
Allier, Maine, Vienne

The longest river in France, the Loire rises among volcanic peaks in the Cevennes mountains at the south-east edge of the Massif Central. Its source is only about 160km (100 miles) from

Seine

LOCATION In northern France, flowing from the Burgundy region to the English Channel at Le Havre

LENGTH
780km (480 miles)

VERTICAL FALL
471m (1,545ft)

BASIN AREA
83,000 square km
(32,000 square miles)

MAIN TRIBUTARIES
Aube, Marne, Oise

The Seine defines what is often called the Paris Basin. Upstream of the city, the river and its tributaries have cut down through limestone to create a series of fertile plateaus that form the region known as the Île de France. Downstream of Paris, the Seine flows through chalk hills, and below Rouen follows a series of wide meanders, before widening into its estuary.

ALEXANDRE III BRIDGE, PARIS

CHATEAU OF SULLY-SUR-LOIRE
Tree-lined banks and chateaux make the Loire valley one of the most picturesque places in France.

the Mediterranean Sea and the mouth of the Rhône. Initially flowing north along a series of faulted depressions in the massif, the Loire then meanders through central France along a shallow, steep-sided valley. The river pivots west near the city of Orléans, passing Tours and Angers. Below Nantes, the Loire widens to an estuary that discharges into the Bay of Biscay at St. Nazaire. The river's flow is very irregular, and parts of the middle course, which is studded by strings of sandy islands, run almost dry in summer. In autumn and spring, heavy rains can cause flooding, especially in the lower reaches.

EUROPE *northwest*

Rhine

LOCATION Flowing from the Swiss Alps to the North Sea coast of the Netherlands

LENGTH
1,320km (820 miles)

VERTICAL FALL
2,339m (7,690ft)

BASIN AREA
220,000 square km (85,000 square miles)

MAIN TRIBUTARIES
Main, Moselle, Neckar

INDUSTRIAL RIVERS

The Rhine is heavily used by commercial vessels, such as these ships passing the Bayer chemical plant in Leverkusen, Germany. Most of the commercial traffic is concentrated in the industrialized section between Cologne and the sea. Despite strict controls, both over shipping and disposal of factory waste, the water is heavily polluted and contains high levels of microplastics.

The most commercially important river of western Europe, the Rhine rises high in central Switzerland. Its headwaters, the Vorderrhein and the Hinterrhein, are fed by snowmelt that reaches a maximum in June and July. The upper Rhine flows north into Lake Constance (see p.233) through a small delta. From the western end of the lake, the river continues westwards, plunging over the 30-m- (100-ft-) high Rhine Falls in northern Switzerland, and along the southern edge of the Black Forest mountains, before swinging north near Basel into the two-stage Rhine Valley. The upper part of the valley is a broad and ancient rift valley (see p.127) about 32km (20 miles) wide – with the Vosges mountains to the west and the Black Forest to the east – the sides of which contain the river's floodplain. Just below Basel, for about 160km (100 miles), the Rhine forms part of the border between Germany and France. Downstream, the German section of river has been straightened during the last two centuries by cutting through its many meanders. After passing Mainz, the Rhine enters the lower valley, a narrow, twisting gorge, 145km (90 miles) long, cut between the Hunsrück and Eifel hills on the western side, and the Taunus and Westerwald on the eastern side. South of Bonn, the valley widens onto a plain and the river's lower course passes the cities of Cologne, Düsseldorf, and Duisburg. The Rhine estuary begins below Emmerlich near the German–Dutch border, and the main stream enters the North Sea at Rotterdam. The River Meuse follows a parallel course to the sea and adds its waters to their combined estuary.

THE RHINE NEAR KAUB
The snaking course of the narrow lower valley is overlooked by hundreds of castles, some of which date back many centuries.

EUROPE *west*

Rhône

LOCATION Flowing from the Swiss Alps through France to the Mediterranean Sea

LENGTH
813km (505 miles)

VERTICAL FALL
1,800m (6,000ft)

BASIN AREA
96,000 square km (37,000 square miles)

MAIN TRIBUTARIES
Ardèche, Arve, Isère, Saône

The Rhône is the only major European river to flow into the Mediterranean Sea. Descending from the Rhône glacier in southern Switzerland, its upper course

CHAMOIS
This agile antelope grazes along the upper reaches of the Rhône in the French and Swiss Alps. In winter, it descends to lower levels to find food.

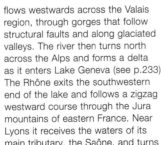

flows westwards across the Valais region, through gorges that follow structural faults and along glaciated valleys. The river then turns north across the Alps and forms a delta as it enters Lake Geneva (see p.233). The Rhône exits the southwestern end of the lake and follows a zigzag westward course through the Jura mountains of eastern France. Near Lyons it receives the waters of its main tributary, the Saône, and turns southwards. Its lower course follows a trough between the Alps and the Massif Central. Receiving tributaries on both banks, the Rhône proceeds through a series of gorges cut through rocky outcrops, separated by sediment-filled basins with terraced sides that are often inundated by seasonal flooding. As it approaches the sea, the river braids over an increasingly marshy landscape. The Rhône delta starts near Arles, where the river diverts into two main channels that flow across the Camargue wetland (see p.325) and into the sea west of Marseilles.

HEADWATERS OF THE RHÔNE
The ice-cold waters of the river cross the Vallée de la Clarée near Briançon in the French Alps.

RHÔNE AT AVIGNON
Along its lower reaches, the Rhône can present a placid, tranquil image that is quite at odds with its wild and racing Alpine origins.

EUROPE *central*

Oder

LOCATION Flowing from the Czech Republic to the Baltic Sea in Poland

LENGTH
886km (550 miles)

VERTICAL FALL
633m (2,080ft)

BASIN AREA
119,000 square km (46,000 square miles)

MAIN TRIBUTARIES
Neisse, Warta

The Oder rises in the mountainous northeast of the Czech Republic, but nine-tenths of its drainage basin lies within Poland, and a stretch of its lower course forms part of the Polish–German border. The river exits the mountains through a depression known as the Moravian Gate. It is a very wide river, up to 9.5km (6 miles) across in places. It is also very shallow, with an average depth of about 90cm (3ft). The river freezes for about one month in winter. As it approaches the sea, the main channel braids into several parallel streams. Below the city of Szczecin, it enters a lagoon and then flows into the Bay of Pomerania.

LAND

DAMS

Dams are some of the largest engineered structures on Earth. They supply about a sixth of the world's electricity, and they significantly reduce the risk of floods and droughts. However, in the industrialized world the construction of very large dams is almost at a standstill, largely because suitable sites have already been exploited. In developing countries, such as India and China, it continues apace.

Dams and Drainage

Earth-filled dams date back to the time of ancient Egypt, but concrete ones are a much more recent innovation. Dam-building became a matter of national prestige in the 20th century. The USA, for example, built a series of massive dams in the Colorado River basin (see p.206), and the former Soviet Union dammed its major rivers and supervised the building of the Aswan High Dam, a rock-filled barrage straddling the River Nile in Egypt (see p.217). These huge projects were followed by others in Africa and South America. One of them – the Itaipú Dam, on the border between Brazil and Paraguay – houses currently the third largest hydroelectric plant in the world.

Electricity production is the chief reason for building mega-dams but they also smooth out seasonal variations in water flow, preventing potentially dangerous floods. In China, this is one of the reasons behind the immense Three Gorges project, which regulates the River Yangtze (see p.220). It is hoped that the dam will stop disastrous floods, which over the centuries have claimed many lives.

SHAPED FOR STRENGTH
The Hoover Dam spans the lower reaches of the Colorado River. At 221m (726ft) high, it is the tallest dam in the USA.

DIVIDED BY DAMS
The erection of dams shown on the map above has limited the movements of the rare Indus River Dolphin. The dolphins are now confined to isolated pockets between dams, in which they interbreed and so jeopardize their survival.

plants need to become established on exposed soils. Instead, the silt accumulates in reservoirs, which reduces their working life.

For freshwater wildlife, large dams have mixed effects. Reservoirs create new habitats for animals, but dams form barriers that can be difficult to pass. This is a problem for migratory fish, and also for other species that normally keep on the move. Riverside plants decline when sediment levels drop, although reservoirs often make ideal homes for floating water plants. Unfortunately, these floating plants include invasive species such as water hyacinth, a problematic weed in warm parts of the world.

Gains and Losses

Dams make existing water supplies easier to exploit. In the American southwest, many farms depend on water from the Colorado Basin reservoirs, as do rapidly growing cities such as Phoenix and Las Vegas. In this region, dams attract plenty of environmental criticism, but current lifestyles there would be unsustainable without the water that dams provide.

In other parts of the world, small-scale dams are vital for rural life. In southern India, for example, thousands of hand-built reservoirs – which are known locally as tanks – capture rainwater from the annual monsoon. Without them, agriculture would be impossible during the dry season. But dams can also displace people, and threaten their livelihoods. By the time the Three Gorges project was completed in 2012, an estimated 1.3 million people had lost their homes and their farmland was submerged below a lake 600km (373 miles) long.

LARGEST HYDROELECTRIC DAMS

The world's hydroelectric dams have a combined capacity of about 700,000 megawatts (MW), but they exploit only about a quarter of this energy, owing to practical and environmental considerations.

Name	Location	Max power (MW)
Three Gorges Dam	China	22,500
Baihetan	China	16,000
Itaipú	Brazil–Paraguay	14,000
Xiluodu	China	13,860
Belo Monte	Brazil	11,233
Guri	Venezuela	10,235
Wudongde	China	10,200
Tucuruí	Brazil	8,370
Grand Coulee	USA	6,809
Xiangjiaba	China	6,448

Ecological Effects

The Three Gorges project epitomizes the environmental conundrum that dams can pose. The power they generate is renewable and theoretically pollution-free. These are major advantages for a country such as China, which still relies heavily on coal. But with extremely large dams, these benefits are offset by some serious drawbacks. By blocking sediment flow, dams deprive floodplains of fertile silt, which wetland

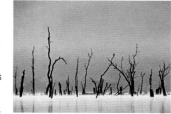

FLOODED FOREST
Rotting vegetation in newly flooded reservoirs releases methane – a gas that adds to the greenhouse effect.

LIQUID RESERVES
The Glen Canyon Dam on the Colorado River, USA, generates hydroelectric power. The water it traps is also used for drinking and irrigation.

LAND

Danube

LOCATION Flowing from southern Germany to the Black Sea coast in eastern Romania

LENGTH
2,860km (1,770 miles)

VERTICAL FALL
662m (2,174ft)

BASIN AREA
816,000 square km (315,000 square miles)

MAIN TRIBUTARIES
Drava, Sava, Tisza

Europe's second longest river, the Danube, rises in Germany from the confluence of the Brege and Brigach rivers. It flows eastwards into Austria, descending through the Hungarian

THE DANUBE PASSING BUDAPEST
The Chain Bridge, seen here in front of the Hungarian Parliament Building, joins what were once two separate cities – Buda and Pest.

Gate to the Little Alföld plain. After passing Budapest, the river flows southwards across the Great Hungarian Plain and between the Carpathian and Balkan mountains. Here the river has cut the deepest gorge in Europe, the Iron Gate, with sheer walls 800m (2,600ft) high. The Danube's lower course passes across the Wallachian Plain. As it nears the Black Sea, the river is deflected northwards by the ancient and eroded Dobruja Hills, before turning east once more. It discharges into the sea through three main distributaries, in a delta that has an area of about 5,000 square km (1,900 square miles).

FLOODING

During the summer of 2002, unusually heavy summer rains across central Europe caused the Danube to overtop its banks in many places. Millions of people living along its course were affected, and thousands of homes were inundated, as here in the Slovak capital, Bratislava. Downstream, the Hungarian capital, Budapest, was also affected. Upstream of Bratislava, in Austria, the Danube reached an all-time high as the region experienced the heaviest rainfall ever recorded. The town of Linz and part of Vienna were flooded, despite an excellent system of flood prevention.

Volga

LOCATION In Russia, flowing eastwards then southwards to the Caspian Sea

LENGTH
3,690km (2,293 miles)

VERTICAL FALL
258m (848ft)

BASIN AREA
1.4 million square km (525,000 square miles)

MAIN TRIBUTARIES
Oka, Kama

The Volga is Europe's longest river and one of Russia's most important transport routes. The river rises 228m (748ft) above sea level in the Valdai Hills about 400km (250 miles) northwest of Moscow, and its drainage basin extends eastwards to the Urals. Its upper course runs northeast through swamp forest, lakes, and reservoirs, then turns southeast along a narrow valley between the Danilov and Uglich highlands. Near the city of Nizhniy Novgorod, the Volga receives the waters of the Oka and increases considerably in size. The river continues its southeast course through damp forest until

it passes Kazan, where it turns southwards and flows through a progressively more arid landscape. Downstream of Kazan the river,

VOLGA DELTA
The delta (shown here in a satellite image) is lush and well-watered compared with the barren desert landscape that surrounds it.

swollen by the inflow of the Kama, flows between the Volga Hills and the Turgay Plateau. The most notable feature of its middle course is the Samara Bend – a remarkable loop 160km (100 miles) long, around a piece of high ground. Near Volgagrad the river splits in two. The Volga and its main distributary, the Akhtuba, follow a parallel course into the Caspian Depression where, downstream of Astrakhan, they discharge through a wide delta into the Caspian Sea (see p.238) about 30m (100ft) below sea level. The Volga freezes for about 100 days each year. Its importance for transport has been enhanced by the construction of canals, notably the Volga–Baltic Waterway completed during the 1960s. Numerous hydroelectric dams have also been built across its course.

VOLGA PASSING NIZHNIY NOVGOROD
Ice-free for about 200 days each year, the Volga has for centuries formed a river highway through the heart of European Russia.

Nile

LOCATION Flows north from Lake Victoria and the Ethiopian highlands to the Mediterranean coast in Egypt

LENGTH 6,648km (4,132 miles)

VERTICAL FALL 1,135m (3,725ft)

BASIN AREA 3.4 million square km (1.3 million square miles)

MAIN TRIBUTARIES White Nile, Blue Nile

ABU SIMBEL

The damming of the Nile to create Lake Nasser in the 1960s threatened to submerge several important relics of Egypt's ancient civilization, including the temple of Pharaoh Ramses II at Abu Simbel, built in the 13th century BCE. Engineers and archaeologists co-operated to carefully dismantle the temple, and reassemble it at a new site some 60m (200ft) above the waters of the lake. A smaller temple dedicated to Nefatari, Ramses' favourite wife, was moved to the same site.

The longest river in the world, the Nile is one of the few that are recognized to have more than one main source. The longest stream of the Nile – the White Nile – flows from the northern end of Lake Victoria (see p.235) as the Victoria Nile and over the Murchison Falls into the northern part of the Great Rift Valley (see pp.128–29). The river crosses the northern end of Lake Albert (exiting as the Albert Nile) and then descends through a series of gorges to the flat, marshy plain of the Sudd (see p.327). At Khartoum, the White Nile is joined by the waters of the Blue Nile, with its source at Lake Tana in the Ethiopian Highlands. From Khartoum to southern Egypt, the Nile passes through a series of six great cataracts and then enters a long, narrow valley between desert escarpments before it widens to a broad triangular delta north of Cairo. This delta region, and a long, narrow green ribbon along the Nile, make up

the 3 per cent of Egypt's land area that supports cultivation. Thanks to its two sources, the Nile provides both a constant source of water and an annual source of fertile silt. The White Nile, flowing from equatorial rain-fed lakes, ensures the Nile's constancy, but contributes only 16 per cent of its volume downstream of Khartoum. The Blue Nile contributes 84 per cent of the lower river's volume, and accounts for all of its variability. The regular annual flooding of the Nile, which delivers fine,

black, fertile silt across the floodplain, is caused by winter rains and snowmelt in the Ethiopian Highlands. The flood season begins in April at Khartoum and gets progressively later in the river's more northerly reaches – the high water of the floods at Cairo occurs in October. The completion of the Aswan High Dam in 1970 created a huge reservoir, Lake Nasser, which extends across the Egyptian–Sudanese border. The dam has significantly moderated the downstream flow of the Nile.

THE UPPER NILE VALLEY
As it flows through verdant forest-lined rapids in Uganda, the Victoria Nile is a foaming, turbulent, and fast-flowing river fed by heavy year-round rainfall across Lake Victoria's drainage area.

NILE CROCODILE
This dangerous, powerful reptile is found along the entire length of the Nile and White Nile, and also on rivers in central and western Africa.

muscular, vertically flattened tail

nostrils on top of snout

eyes on top of head

FERTILE NILE VALLEY
The banks of the Nile near Luxor are lined with fields. The farmed area extends only as far as the escarpment in the background.

THE NILE DELTA
This satellite image shows the Nile's enormous influence on Egypt. Most of the population live in the delta region, or scattered along the valley.

LAND

Congo

LOCATION Flows from the highlands of east Africa across the continent to the Atlantic Ocean

LENGTH
4,670km (2,900 miles)

VERTICAL FALL
1,760m (5,780ft)

BASIN AREA
3.5 million square km
(1.4 million square miles)

MAIN TRIBUTARIES
Kwa, Lualaba, Sangha, Ubangi

LIVING BETWEEN RIVER AND RAINFOREST
The Congo's banks are lined with villages like this. The villagers benefit from plentiful food and easy transport, but have to live with the risk of flooding.

The Congo is second only to the River Amazon in the volume of water it carries to the oceans and the third largest by discharge in the world. It is the second longest river in Africa, and sweeps in a great anticlockwise loop that drains most of central Africa. The Congo's headwaters are fed by rains on the western slopes of the East African Plateau near the southern end of Lake Tanganyika (see p.236) in Zambia. The upper course collects several smaller rivers and flows though rapids and small lakes before descending over the seven-stage Boyoma Falls upstream of Kisangani, and entering the Congo Basin – the vast, near-circular depression that extends over most of the river's catchment area. The river widens considerably over the lowlands, reaching widths of up to 13km (8 miles) in places, with strings of islands dividing the flow

UPPER CONGO
The upper reaches of the Congo are a hot, wet, sparsely populated and little-explored region of the African continent.

into several channels. Most of the basin is covered by tropical rainforest, and there are extensive marshes along the river margins and between the many tributary streams. This rain marsh, as it is known, is subject to flooding at all times of the year, especially in the western half of the basin. When tributaries that usually reach their highest level at different times happen to reach peak flow simultaneously, flooding can be extensive. The Congo leaves its basin flowing southwest through a narrow gorge, known as the Chenal or Corridor, that it has cut through the Bateke Plateau. It then widens and flows into the Malebo Pool,

a lake 27km (17 miles) wide. The city of Brazzaville is located on the northern shore, while Kinshasa stands on the southern shore. Upstream of the pool, the Congo is navigable for more than 1,600km (1,000 miles) of its length. Downstream of Malebo, a series of rapids and waterfalls, which long hindered European exploration of the river, block direct access to the sea. Over about 320km (200 miles), the river descends nearly 300m (1,000ft) and then continues along a gorge and through mangrove forests to the Atlantic. The flow of the Congo is so strong and constant that no delta has formed; instead the flow continues along a

submarine canyon, depositing its sediment in a great fan on the ocean floor. Many of the features on the River Congo (including the Malebo Pool, Boyoma Falls, and the town of Kisangani) were formerly named after British explorer Henry Stanley who, together with his French rival Pierre Brazza, was among the first Europeans to explore this region of Africa.

BOYOMA FALLS
Fish are a great natural bounty for the people of the Congo Basin. In swirling rapids like this one upstream of Kisangani, fish are harvested in conical traps hung from stout, wooden scaffolding.

AFRICA *west*

Niger

LOCATION Flows from Guinea through the southern Sahara to the Atlantic Ocean in Nigeria

LENGTH
4,180km (2,600 miles)

VERTICAL FALL
850m (2,800ft)

BASIN AREA
1.9 million square km
(730,000 square miles)

MAIN TRIBUTARIES
Bani, Benue, Kaduna

The major river of West Africa, the Niger rises in the Fonta Djallon highlands of northern Guinea. It flows northeast over the lowlands of the Sahel, then curves through the Sahara Desert before turning southeast to pass through Niger and Nigeria. The river flows through dense rainforest before reaching the sea, forming a delta in the Gulf of Guinea.

NIGER DELTA
The delta where the Niger meets the sea is dwarfed by its inland delta, but is still the largest in Africa and is mainly covered in dense mangrove forest.

AFRICA *southeast*

Zambezi

LOCATION Flowing from the highlands of central Africa to the Indian Ocean in Mozambique

LENGTH
3,540km (2,200 miles)

VERTICAL FALL
1,460m (4,800ft)

BASIN AREA
1.3 million square km
(500,000 square miles)

MAIN TRIBUTARIES
Kafue, Shire

Africa's fourth largest river rises as a small spring in the Kalene Hills of southwestern Zambia, and then flows eastwards, gaining water from many small tributaries. The Zambezi attains a width of about 1,700m (5,500ft) before plunging over the spectacular Victoria Falls, 108m (355ft) high. Downstream of the falls, where the river forms the border between Zambia and Zimbabwe, its deep, wide valley has been flooded by a lake 280km (175 miles) long, which formed following the construction of the Kariba Dam in 1959 to generate electricity for the region.

HIPPOPOTAMUS
Plant-eating hippopotamuses live in the shallows along the Zambezi's edge and can easily capsize canoes and other small craft.

VICTORIA FALLS IN ZAMBIA
More than double the height and width of Niagara Falls (see p.226), the local name for this magnificent waterfall, Mosi-oa-Tuya, translates as "the smoke that thunders".

LAND

ASIA *west*

Euphrates

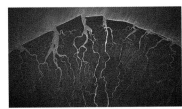

LOCATION Flows from the mountains of eastern Turkey to the Persian Gulf

LENGTH
2,800km (1,740 miles)

VERTICAL FALL
2,680m (8,800ft)

BASIN AREA
1.1 million square km
(430,000 square miles)

MAIN TRIBUTARIES
Tigris

Together with its main tributary and near twin, the Tigris (which is 1,890km/1,180 miles long), the Euphrates forms the largest river system in western Asia. Its headwaters originate in the highlands of eastern Turkey. The Tigris rises about 80km (50 miles) to the east, and collects much of its downstream waters from the Great and Little Zab rivers, which flow from the Zagros Mountains in Iran. For most of their length, the rivers follow nearly parallel courses to the sea, giving rise to the old name for Iraq – Mesopotamia ("land between two rivers"). As they

descend from the highlands, the two rivers follow deep, converging valleys that define a triangle of desert known as Al Jazirah ("the Island"). Near Baghdad they come within 80km (50 miles) of each other, only to diverge again. In their lower courses, the rivers' channels are branching and highly unstable. Their combined floodplain is dissected by numerous abandoned channels and irrigation canals. The two rivers meet at Basra and then follow a single course, known as the Shatt al 'Arab, through extensive reed-filled marshes to their outflow at the head of the Persian Gulf.

WARS OVER WATER

In the rain-starved Middle East, water is the most precious of all natural resources. The Ataturk Dam across the Euphrates is one of several that have been constructed in Turkey. These have greatly restricted the downstream flow of the river, and the governments of both Syria and Iraq have protested strongly about Turkey's "theft" of the water.

THE EUPHRATES VALLEY
Upstream of its confluence with the Tigris, the Euphrates flows along a high-sided valley, the floor of which has been levelled by successive floods and course changes.

LAND

Yangtze

LOCATION In China, flowing from the Himalayas to the East China Sea in the Pacific Ocean

LENGTH
6,300km (3,900 miles)

VERTICAL FALL
5,480m (18,000ft)

BASIN AREA
2 million square km
(760,000 square miles)

MAIN TRIBUTARIES
Han Shui, Ya-lung Chiang

The Yangtze is the longest river in Asia and the third longest in the world. It rises in the Kunlun Mountains, south of the Taklamakan desert, and flows

THE YANGTZE GORGES
At the top left of this image, the river can be seen snaking through the Yangtze Gorges before widening and looping across the downstream plain.

YANGTZE HEADWATERS
In the Tibet Autonomous Region of China, the upper course of the Yangtze River flows through icy mountain scenery that is almost barren of vegetation.

southeastwards across the Tibetan Plateau along narrow, steep valleys. Over the first 1,600km (1,000 miles), the Yangtze descends at an average rate of about 2m per kilometre (10ft per mile). The river then turns eastwards across Sichuan Province before flowing through the famous Yangtze Gorges. The three gorges have a combined length of about 90km (56 miles) and are no more than 180m (600ft) wide, with rock walls towering 350–600m (1,200–2000ft) above the river. It is here that the Three Gorges Dam, one of the largest in the world, was opened in 2012. Once across the dam, the Yangtze reaches depths of more than 150m (500ft), making it the deepest river in the world. The river exits the gorges onto a lake-strewn plain and loops south along a wide valley, exiting onto the southern edge of the North China Plain. Along parts of the lower course, the normal river level can be several metres above that of the plain, kept safely in check by natural and artificial levees. The waters of the Yangtze are essential to this densely

populated rice-growing area, where the summer drought can last for eight weeks. Conversely, when the rains finally arrive, the Yangtze frequently overtops its banks, flooding huge areas. After meandering northeastwards, the river divides into several distributaries and forms a delta. Shanghai is located between the outflow of the main southern distributaries. The Yangtze has always been China's most commercially important river, and is linked by the Grand Canal, 1,600km (1,000 miles) long, to the Yellow River and Beijing to the north.

FLOOD LEVEES

Constant maintenance is essential to keep the levees along the Yangtze in good condition and safely above the level of the river. Records reveal that there have been catastrophic floods at least 100 times in the last 2,000 years. The floods of 1932, 1954, and 2020 though not as high as some in the river's history, inundated the cities of Nanjing and Wuhan, killed about 300,000 people, and made 40 million homeless.

YANGTZE RIVER DOLPHIN
This crucially endangered mammal is found in the lower Yangtze. Overhunting and pollution have greatly reduced its numbers to probable extinction.

YANGTZE FISHERMEN
The crew of a traditional fishing boat cast their nets on the Yangtze as it flows along one of the many valleys it has cut through the undulating hills of Sichuan Province.

ASIA *east*

Yellow River

LOCATION In China, flowing across the north of the country from the Tibetan Plateau to the Yellow Sea

LENGTH
5,460km (3,390 miles)

VERTICAL FALL
4,500m (15,000ft)

BASIN AREA
1.9 million square km
(740,000 square miles)

MAIN TRIBUTARIES
Wei He, Fen He

JAPANESE CRANE
This endangered water bird is famous for its elegant and intricate courtship dance, and is considered to be a symbol of happiness and good luck.

Also called the Huang-He or Huang-Ho, the Yellow River is China's second longest river and the muddiest river in the world. It carries 30 times more silt per cubic metre of water than the River Nile, and it delivers as much sediment to the sea as the Ganges but transported in a much smaller overall flow. Rising on the Tibetan Plateau, the Yellow River follows an erratic course across northern China. Initially fast-flowing over barren upland, the river descends onto a high plateau covered by loess deposits (see p.77) that date from the last ice age and are

several hundred metres thick in places. It is here that the river collects its heavy load of sediment. The river continues southwards onto the North China Plain, then east to its current delta to the north of the Shandong peninsula. Between 206 BCE and 1949, some 1,092 major levee breaches resulted in catastrophic floods and the loss of millions of lives. The river's course has shifted some 26 times and by as much as 320km (200 miles). A flood control system has been put in place with embankments, reservoirs, and controlled flooding. The lower reaches are ramparted by levees, which in places lift the river bed several metres above the surrounding countryside.

YELLOW RIVER AT LAJIA
In Qinhai Province, western China, the Yellow River flows through narrow valleys and across small floodplains with fields of rape and barley.

HOKOU FALLS
The yellow colour of the river, caused by suspended particles of loess, shows as it falls between Shanxi and Shaanxi provinces in central China.

ASIA *northwest*

Ob'

LOCATION In Russia, flowing across Siberia to the Kara Sea in the Arctic Ocean

LENGTH
3,700km (2,300 miles)

VERTICAL FALL
Exact source unknown

BASIN AREA
3 million square km
(1.2 million square miles)

MAIN TRIBUTARIES
Chulym, Irtysh, Tom

The Ob' drains the sixth largest catchment area in the world, and some authorities combine it with the Irtysh to form a single river 5,400km (3,360 miles) long. The Ob' rises in the foothills of the Altai Mountains and flows northwards through progressively colder and drier forest landscapes, and then across open tundra, where it separates into two channels, the Greater Ob' and Lesser Ob'. The streams recombine before flowing into an elongated delta that has been partly submerged beneath the sea in the Gulf of Ob'. The river is frozen for up to 200 days per year.

ASIA *south*

Irrawaddy

LOCATION In Myanmar, flows from the Mishmi Hills to the Andaman Sea in the Indian Ocean

LENGTH
2,100km (1,300 miles)

VERTICAL FALL
Exact source unknown

BASIN AREA
411,000 square km
(158,000 square miles)

MAIN TRIBUTARIES
Chindwin

Rising from the meltwaters of east Himalayan glaciers, the Irrawaddy flows southwards to Mandalay and its confluence with the Chindwin. It then meanders through the dry central region of Myanmar, where it provides much-needed irrigation water. It follows a narrow valley east of the Arakan Yoma highlands, forming a triangular delta west of the Gulf of Thailand. Yangon is located on one of the numerous distributaries.

PANNING FOR GOLD AT PAGAN, MYANMAR
In northern Myanmar, the Irrawaddy and many of its tributaries contain placer deposits, carried down from the mountains when the rivers are in flood.

ASIA *south*

Mekong

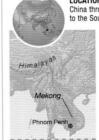

LOCATION Flows from western China through Laos and Cambodia to the South China Sea

LENGTH
4,000km (2,500 miles)

VERTICAL FALL
4,950m (16,250ft)

BASIN AREA
795,000 square km
(305,000 square miles)

MAIN TRIBUTARIES
Kang, Mun, Srepok

The longest river in Southeast Asia, the Mekong rises on the Tibetan Plateau, where it flows along deep valleys and is part of the Myanmar–Laotian border. Downstream it forms part of the Thai–Laos border, flowing through Cambodia before forming a delta near Hô Chi Minh City in Vietnam.

RAPIDS ON THE MEKONG RIVER
The border of Thailand and Laos sees the Mekong pass through spectacular rapids. Local people fish in swirling pools by the riverside.

ASIA *northwest*

Ganges

LOCATION Flows along the southern edge of the Himalayas to the Bay of Bengal in India

LENGTH
2,506km (1,557 miles)

VERTICAL FALL
3,030m (10,000ft)

BASIN AREA
1.6 million square km
(625,000 square miles)

MAIN TRIBUTARIES
Brahmaputra, Ghaghar,
Yamuna

HEADWATERS OF GANGES
The Bhagirathi Valley in Uttarakhand, India, is overlooked by the mountain glaciers that are the source of the Ganges.

GANGES AT VARANASI
The River Ganges is considered sacred by Hindus, and millions of pilgrims visit the city of Varanasi each year to immerse themselves in the holy waters.

FLOODING IN BANGLADESH

The Bengal Basin, within which the Ganges delta lies, covers 252,000 square kilometres (97,000 square miles) and is prone to seasonal flooding caused by monsoon rains. However, an even greater flood threat is posed by the tropical cyclones that sometimes sweep up the Bay of Bengal from the Indian Ocean. Storm surges inundate much greater areas than ordinary riverine floods and can extend over 75 per cent of Bangladesh's land area, most of which is less than 10m (33ft) above sea level. Despite the danger of flooding, the area has a population of about 90 million, and the rich, alluvial soils are intensively farmed. The floods of August 2007 and June 2022 affected almost the entire country. Homes were lost, crops and cattle were destroyed, and roads and bridges were washed away.

The River Ganges, sacred to the Hindu religion, drains about a quarter of India, and delivers more sand and silt to the sea than any other river in the world. Boosted at the head of its delta by the inflow of its largest tributary, the Brahmaputra, the Ganges discharges about 2 billion tonnes of sediment each year. Rising in the southern Himalayas, on the Indian side of the border with China, the Ganges is formed by the confluence of the Alaknanda and Bhagirathi rivers. The upper course flows westwards, then southwards, and descends onto the north Indian plain about 110km (70 miles) east of Delhi. The river's middle course meanders sluggishly eastwards along a wide floodplain covered with sandy alluvial deposits. Over a distance of more than 1,600km (1,000 miles), the river falls only about 180m (600ft). After passing Varanasi and Patna, the Ganges skirts the northern slopes of the Rajmahal Hills and turns southeast onto the Bengal lowlands. Here, the river divides into many distributaries, creating a network of interlacing channels that stretches 400km (250 miles) across the northern coast of the Bay of Bengal. The main western distributary is the Hooghly, which flows through Kolkata. Most of the Ganges delta is in Bangladesh, and Dhaka is situated at the northeastern corner. In total the Ganges delta covers an area of about 56,980 square km (22,000 square miles). The seaward fringes, known as the Sundarbans, are densely forested in places, and are one of the last refuges of the Bengal Tiger.

UPPER COURSE OF THE GANGES
In the mountainous north of India, every small area of riverside lowland, overlooked by Himalayan foothills, is intensely cultivated.

MOUTHS OF THE GANGES
This satellite view shows some of the river's thousands of silt-laden distributaries winding between forested areas of consolidated ground.

ASIA northwest

Indus

LOCATION Flows from the Tibetan Plateau across the Himalayas to the Arabian Sea

LENGTH
2,900km (1,800 miles)

VERTICAL FALL
4,848m (16,000ft)

BASIN AREA
1.2 million square km (450,000 square miles)

MAIN TRIBUTARIES
Chenab, Kabul, Jhelum, Sutlej

The greatest of the trans-Himalayan rivers, the Indus flows from headwaters fed by snowmelt high in southwest China. The upper course flows northwest through the disputed territory of Jammu and Kashmir, along a wide, faulted valley, between the foothills of the Karakoram and Zaskar mountains. The Indus then turns southwest into Pakistan, where it crosses the Himalayas south of the Hindu Kush,

INDUS RIVER IN KASHMIR
The river flows sedately past sandy terraces that correspond to higher levels in the past, when the river was swollen by snowmelt, and more turbulent.

INDUS RIVER AT SHARDU
In northern Kashmir the upper course of the river flows through a bleak landscape with the Karakoram Mountains in the background.

EARLY CIVILIZATION

The Indus – like the Nile, Euphrates, and Yellow rivers – gave rise to the first civilizations. Over 4,000 years ago, people of this region lived in cities of baked-brick houses, paved streets, and sewers. Some of them used animal-decorated seals like this one, written in an as yet undeciphered script.

flowing through gorges, the floors of which are more than 3,650m (12,000ft) below the surrounding peaks. After its confluence with the River Kabul between Peshawar and Islamabad, the Indus crosses the Potwar Plateau and descends onto the Punjab Plain, where it receives many tributaries flowing from the southern foothills of the Himalayas. Skirting the northern fringes of the Thar Desert, the last 480km (300 miles) of the river's course meander across a wide floodplain, and downstream of Hyderabad it splits into several distributaries, forming a delta known as the "Mouths of the Indus". The city of Karachi is located at the northern end of the delta.

AUSTRALASIA southeast

Murray

LOCATION In Australia, flows from the southern end of the Great Dividing Range to the Indian Ocean

LENGTH
2,590km (1,610 miles)

VERTICAL FALL
2,050m (6,750ft)

BASIN AREA
1.1 million square km (412,000 miles)

MAIN TRIBUTARIES
Darling, Murrumbidgee

The Murray, with its major tributary, the Darling, forms Australia's only major river system – one that drains the temperate southeastern corner of an otherwise mainly arid continent. Their combined catchment area contains

GALAH
The Galah is the most widespread cockatoo in Australia and is common around the River Murray.

more than 80 per cent of the country's irrigated farmland. From a source high in the Australian Alps, about midway between Canberra and Melbourne, the Murray flows northeast and then turns westwards, descending into the Murray Valley near the town of Wagga Wagga. The valley is flanked by the Riverina Plains of New South Wales to the north and the plains of northern Victoria to the south. Along its middle course the river

receives its main tributaries, which bring waters from the northern part of the catchment area. For the last 650km (400 miles) of its course, the Murray flows through South Australia along a series of deep gorges separated by shallow lakes. The river exits into the ocean through the Coorong Lagoon (see p.327), into Encounter Bay south of Adelaide. The level of the Murray is erratic, especially during droughts, and

the river dried up completely three times between 1800 and 1950. Recently, its flow has been increased by diverting water from the south-flowing Snowy River into its headwaters, as part of the Snowy Mountains hydroelectric scheme.

BIG BEND
High yellow cliffs border the River Murray at Big Bend, which is a 10-km- (6-mile-) long, 180-degree S-bend in the river's course.

LAND

Lakes

92–93	Erosion
94–95	Deposition
106	The water cycle
198–223	Rivers
Water quality	230–31
Underground rivers	240–49
Wetlands	318–27

A lake forms when surface water collects in a depression that has no immediate outlet. The water level rises until an outlet is formed, by which time a substantial amount of water may have accumulated. Lakes vary enormously in size and depth: some are no deeper than the average pond, while others are more than a kilometre deep and some have beds below sea-level. Some lakes are millions of years old, but most are much younger, and date back to the end of the last ice age. Not all lakes consist of fresh water. Some, mainly in arid regions, are saline, with levels of salinity varying from slight to six times that of seawater.

Inflow and Outflow

Lakes receive much of their inflow from surface water that may carry a considerable amount of sediment. Most lakes have a single outflow into a river system – those without a defined outflow are described as unchannelled. Some lakes in arid regions have no outflow at all – the inflow is balanced by evaporation. Geologically speaking, most lakes are short-lived, due to infilling by sediment and lowering of the water level through the down-cutting action of the outflow river.

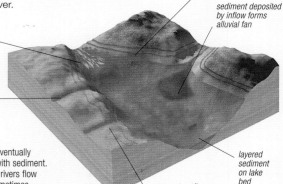

lake terrace marks position of past high-water level

sediment deposited by inflow forms alluvial fan

one of several lake inflows

former outflow at higher level

present outflow of lake

layered sediment on lake bed

LAKE FEATURES
Even the deepest lakes will eventually become shallow as they fill with sediment. Deposition is greatest where rivers flow into the lake. Infilling may sometimes give rise to wetlands.

Lake Formation

Some lakes, such as Baikal and Tanganyika (see p.236), formed in depressions along fault lines or in areas of rifting (see p.138). Others, such as the huge Caspian Sea (see p.238), were once part of an ocean, but have become cut off from the larger body of seawater. A few lakes have been formed by volcanic action, when lava dams a river or a caldera is formed. Most, however, are the result of glaciation, and this is reflected in their distribution. Most North American lakes, for example, are located in areas that 25,000 years ago were under a thick ice-sheet. Glaciers create lakes in two main ways: valley floors are hollowed and deepened by glacial erosion and later filled in by water; and the ridges of moraine deposited by glaciers (see p.255) form dams across valleys that trap water behind them.

FAULT LAKE
Loch Ness in Scotland fills a basin formed by an elongate, ancient structural fault, whose movement weakened the rocks that were further eroded by glaciations.

KETTLE LAKES
Kettle lakes, seen here in central Chile, form when large blocks of melting ice become surrounded by glacial debris. Such lakes have no outlet.

CALDERA LAKE
A near-circular caldera lake on the Kamchatka Peninsula in eastern Siberia is enclosed by the rim of an extinct volcano. The rounded outline is characteristic of this type of lake.

MORAINE LAKE
Peyto Lake, in Alberta, Canada, formed in a glaciated valley when meltwater became trapped by moraines deposited across the valley during successive stages of glaciation.

LAKE PROFILES

The pages that follow contain profiles of the world's main lakes. Each profile begins with the following summary information:

AREA Area covered by the lake

MAXIMUM DEPTH Deepest part of the lake

ELEVATION Height of lake above sea-level

OUTLET Main river flowing out of the lake

LAND

LAND

Great Bear Lake

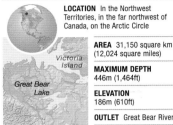

LOCATION In the Northwest Territories, in the far northwest of Canada, on the Arctic Circle

AREA 31,150 square km (12,024 square miles)

MAXIMUM DEPTH 446m (1,464ft)

ELEVATION 186m (610ft)

OUTLET Great Bear River

Great Bear Lake is the largest lake entirely within Canadian territory and the fourth largest in North America. Its northern shores lie within the Arctic Circle, and it is the largest inland body of water to exist at such a high latitude. The lake is irregular in shape and branches into three arms at the western end. The water is deepest near the eastern shoreline. The lake was formed by the scouring action of glaciers as they flowed from the North American ice-sheet down river valleys. Lakes formed in this way are often called piedmont lakes. When the glaciers melted at the end of the last ice age, their waters became trapped behind moraines, forming a huge lake that has

ECHO BAY ON THE NORTH SHORE
In summer, even the remotest northern bays are free of ice. In winter, the frozen lake and the shore are covered in snow.

since partially drained, leaving as remnants Great Bear Lake, Great Slave Lake (see below), and the numerous small lakes that lie between them. The lake is situated in gently rolling wilderness, with evergreen forest along its southern shores and tundra extending to the north. Its waters are often cited as an example of a pristine lake environment, although it has far fewer fish species than lakes of comparable size in more temperate regions, because its waters are much colder. The only settlement on its shores is the former

mining centre of Port Radium, named after the radioactive element, which was shipped from the mines here during World War II to make the first atomic bombs. Great Bear Lake receives inflow from the Whitefish River on its northwestern arm, and from other smaller rivers. The Great Bear River exits from the western end of the lake, flows westwards, and delivers its waters into the Mackenzie River about 480km (300 miles) upstream from the delta. Great Bear Lake and River both freeze over for about seven months each year.

BROWN BEAR
The Grizzly or Brown Bear is a top predator of forests around the lake's southern shores.

Great Slave Lake

LOCATION In the Northwest Territories of Canada, east of the Mackenzie Mountains

AREA 28,568 square km (11,027 square miles)

MAXIMUM DEPTH 614m (2,016ft)

ELEVATION 156m (512ft)

OUTLET Mackenzie River

Great Slave Lake is the second largest lake in Canada and the main reservoir for the Mackenzie River system. The lake is irregular in shape and has a rocky, mainly forested shore with wide bays and numerous islands. It is located along the boundary between the ancient rocks of the Canadian Shield and the newer strata underlying the Great Plains. At Great Slave Lake this division is marked by a large inlet that extends northwards from the northern shore. The lake's waters support a small commercial fishery. Other developments around its shores include the gold-mining town of Yellowknife, on the northern arm of the

lake, and several small tourist resorts. The main inflow is the Slave River, which enters the southern shore bringing water from Lake Athabasca and the Peace River. The Mackenzie River exits from the southwestern end of the lake and is the sole outflow. The lake freezes in winter and is ice-free for only four months of the year.

LAKE COASTLINE
The shores of Great Slave Lake are covered with dense coniferous forest, the tops of which stand out above a layer of morning fog that has formed on the lake's surface, enshrouding an offshore island.

Lake Winnipeg

LOCATION In south-central Canada, in the province of Manitoba southwest of Hudson Bay

AREA 23,750 square km (9,167 square miles)

MAXIMUM DEPTH 36m (118ft)

ELEVATION 217m (712ft)

OUTLET Nelson River

Lake Winnipeg is a shallow lake at the eastern edge of the Great Plains. It is the remnant of a much larger glacial lake that also included lakes Manitoba and Winnipegosis. Much of its water comes from the Saskatchewan River, which flows from the Canadian Rockies to the west. Another inflow, the Red River, drains the high plains. The Nelson River flows from the northeast end of the lake into Hudson Bay.

ROCKY SHORELINE

NORTH AMERICA *northeast*

Great Lakes

LOCATION On the border of the USA and Canada – Lake Michigan is entirely within the USA

AREA Superior 82,367 square km (51,159 square miles), **Michigan** 58,000 square km (36,024 square miles), **Huron** 59,570 square km (37,000 square miles), **Erie** 25,820 square km (16,037 square miles), **Ontario** 19,010 square km (11,807 square miles)

MAXIMUM DEPTH Superior 406m (1,333ft), **Michigan** 281m (923ft), **Huron** 228m (749ft), **Erie** 64m (210ft), **Ontario** 224m (735ft)

ELEVATION Superior 183m (601ft), **Michigan** 177m (580ft), **Huron** 176m (578ft), **Erie** 174m (571ft), **Ontario** 75m (246ft)

OUTLET St. Lawrence River

The Great Lakes form the largest system of freshwater lakes in the world, with a combined surface area of 244,000 square km (95,500 square miles). The system drains from west to east, discharging into the sea from Lake Ontario through the St. Lawrence River (see p.204). The gradient between the lakes is generally very shallow. Superior drains into Michigan through the St. Mary's River. Michigan has almost the same level as Huron, and the two are connected by the Straits of Mackinac, 6km (4 miles) wide. Huron drains through the St. Clair River, Lake St. Clair, and the Detroit River into Erie, which is drained in turn by

TIDES AND WAVES
Although the Great Lakes are unaffected by tides, the wind can produce waves that buffet the shores of Lake Superior like breakers against the seashore.

the Niagara River into Lake Ontario. The descent between Erie and Ontario is greater than between any of the others, and includes the spectacular Niagara Falls, 51m (167ft) high. The Great Lakes were formed during the last ice age, when successive episodes of glaciation eroded a landscape of mountains and river valleys. When the ice melted, Superior, Michigan, and Huron were all part of one huge lake (known as Lake Algonquin), which drained into the Ottawa River. Post-glacial uplift altered the drainage patterns, and the lakes began to take on their present shape about 10,000 years ago. The present shorelines became stable about 3,000 years ago, but the western end of Lake Erie is still sinking at about 2mm (1/10in) per year. Only the western and northern shores of Superior are rocky; the other shores are covered by thick glacial deposits, with concentric ridges of terminal moraine at the southern ends of Michigan and Huron. Numerous cities, including Buffalo, Chicago, Cleveland, Detroit, Milwaukee, and Toronto, are situated around the lakes.

NIAGARA FALLS
The Horseshoe Falls, 800m (2,625ft) wide, on the Canadian side of the Niagara River, are named after their distinctive shape.

CLIFFS CAUSED BY UPLIFT
These cliffs along the northern shore of Lake Superior were created when the land rebounded from the crushing weight of the glaciers that covered it during the last ice age.

URBAN SHORELINE
The cities around the Great Lakes withdraw huge amounts of water each day for municipal and industrial use. Most is returned to the lakes after appropriate treatment. Chicago, however, sends 10.9 billion litres (2.4 billion gallons) of water a day to the Mississippi. Combined with lower than average rainfall in recent years, this loss of water is beginning to affect the level of the lakes. The long-term effects of this fall remain unclear.

CLEARING A PATH
During the winter months, the Coast Guard uses powerful icebreakers to keep the Great Lakes shipping routes open – in this case Thunder Bay on Lake Superior.

LAND

Lake Seneca

LOCATION In the USA, in west central New York state south of Lake Ontario

AREA 174 square km (67 square miles)

MAXIMUM DEPTH 198m (650ft)

ELEVATION 135m (443ft)

OUTLET Seneca River

Lake Seneca is one of the famed Finger Lakes, a series of long, narrow lakes that lie roughly parallel with each other on the undulating uplands to the south of Lake Ontario. The lakes were formed more than a million years ago, during the last stages of the most recent ice age, when a series of shallow glaciated valleys running on a north–south axis became blocked at both ends by glacial moraines. Seneca is the second largest of the Finger Lakes and is by far the deepest, with parts of its bed lying 54.5m (180ft) below sea-level. The lake receives water from streams draining the surrounding hills, including Catherine Creek and the Keuka Lake outlet. Its outflow, the Seneca River, exits at the northeastern corner and, having received the waters of Lake Cayuga, flows into the Oswego River, which eventually discharges into Lake Ontario.

Lake Tahoe

LOCATION In the USA, in the Sierra Nevada mountains between Nevada and California

AREA 499 square km (193 square miles)

MAXIMUM DEPTH 500m (1,640ft)

ELEVATION 1,900m (6,200ft)

OUTLET Truckee River

The fifth deepest lake in the world, Lake Tahoe originated in a deep depression between two faults that subsequently became dammed by lava flow. Although fed by small streams all around its shores, the main inflow is at the southern end, through the Upper Truckee Marshes. The Truckee River flows out of the northwestern end of the lake into Nevada. Lake Tahoe is a popular tourist destination.

DEEP BLUE WATERS OF LAKE TAHOE

Lake Okeechobee

LOCATION In the USA, at the northern edge of the Everglades in central Florida

AREA 1,895 square km (732 square miles)

MAXIMUM DEPTH 4.5m (15ft)

ELEVATION 4m (13ft)

OUTLET Unchannelled

Lake Okeechobee is the main reservoir of fresh water feeding the Everglades' "river of grass" – although it no longer flows with enough force to keep the tide at bay in southern Florida. The lake lies in a shallow, sloping depression that was once part of the sea bed, and is dammed along its southern shore by an accumulation of peat. The lake receives water from the Kissimmee River and from Lake Istokpoga. There is no channelled outflow – the lake's waters flow south through the Everglades on a broad front. Since the 1950s, numerous canals have been constructed to divert the lake's water for agriculture, and to meet the increasing demand of resort towns along Florida's eastern coastline.

OKEECHOBEE PLANT LIFE
The marshy areas around the lake are colonized by a variety of aquatic plants, such as these waterlilies. Some are native; others are introduced.

Great Salt Lake

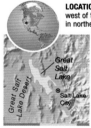

LOCATION In the USA, to the west of the Rocky Mountains, in northern Utah

AREA 2,500 square km (950 square miles)

MAXIMUM DEPTH 12m (39ft)

ELEVATION 1,280m (4,200ft)

OUTLET None

The Great Salt Lake is the largest inland body of salt water in the western hemisphere, and at 25 per cent salinity is about five times saltier than the sea. It is a remnant of the prehistoric Lake Bonneville, which covered most of the floor of the Great Basin (see p.282)

during the last ice age. The present lake is surrounded by great tracts of sand, salt flats, and salt marsh and receives most of its inflow from the Bear Creek and Weber rivers, which bring waters from the Wasatch Range along the eastern shore. The area of the lake is extremely variable – between 1964 and 1987, it more than doubled in size, reaching a maximum area of nearly 8,500 square km (3,300 square miles). A causeway constructed in the 1950s divides the lake into two parts. The smaller northern section is both shallower and saltier than the rest of the lake.

SALT LAKE SHALLOWS
Because the lake lies in a shallow basin, small changes in its level have large effects on its surface area, exposing or submerging large areas of its bed.

SALT-ENCRUSTED BOULDERS
Wind-driven waves splashing onto these lakeshore boulders have gradually covered the rocks with a thick, brittle coating of salt crystals.

SOUTH AMERICA *northwest*

Lake Titicaca

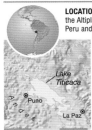

LOCATION In the central Andes, on the Altiplano plateau, bordered by Peru and Bolivia

AREA 8,772 square km (3,386 square miles)

MAXIMUM DEPTH 281m (923ft)

ELEVATION 3,812m (12,516ft)

OUTLET Desaguadero River

STRAITS OF TIQUINA
This view looking eastwards across the straits into the deeper section of the lake shows the bleak, rocky landscape that makes up much of its northern shoreline.

REED BOATS ON UROS ISLAND
Boats made from slim bundles of totora reeds are moored to an island made of larger bundles of reeds, forming a thick floating platform on which houses are built.

Lake Titicaca is the largest freshwater lake in South America, and the highest lake of any significant size in the world. The lake is situated in a structural depression at the northern end of the high Altiplano plateau that separates the Eastern and Western cordilleras of the Andes. Roughly oval in shape, the lake is 195km (120 miles) long and about 80km (50 miles) wide. It is divided into two unequal parts by a narrow strait known as the Tiquina. The lake bottom slopes down from west to east. The lake is widest at the northwest, and the deepest parts are near the northern end, where the shoreline is defined by the Bolivian Cordillera Real mountains. The southeastern part of the lake, below the Tiquina, is much shallower, with an average depth of about 9m (30ft). Titicaca receives inflow from across the northern part of the Altiplano and is fed by more than 20 rivers, the largest of which is the Ramis. Each year the level of the lake fluctuates by up to 4m (13ft), with the highest levels occurring after the summer rains between December and March. Evaporation in the arid mountain climate removes about 95 per cent of the lake's inflow. The sole outflow, the Desaguadero, exits from the shallow southern end of the lake and flows some 300km (185 miles) south to Lake Poopó, which has no outlet to the sea. The evidence of terraces made by former shorelines shows that Titicaca was much larger during the period immediately after the last ice age. Stretches of the present shoreline are covered by marshes and beds of totora reeds. These reeds are still used by some of the people living around the lakeshores, to make boats, houses, and floating islands in the traditional manner. Three fish species have evolved in the waters of the lake, which because of the high altitude contain about one-third less oxygen than fresh water at sea-level. These endemic species supported a small local fishery, but their population was devastated by the introduction of trout for commercial fisheries during the 1930s. The shipping route across Lake Titicaca is a vital link in the railway network that connects land-locked Bolivia with the sea.

VICUÑA
Adapted to high-altitude living, the Vicuña grazes on the grasslands of the Altiplano. Its domesticated cousin, the Alpaca, is raised for the fibres of its coat.

INCA SETTLEMENTS

The Inca people, who flourished until about 500 years ago, considered Lake Titicaca to be the sacred and ancestral home of their agriculture and their civilization. According to legend, their sun god, Manco Capac, sprang from the waters of the lake and founded the Inca dynasty at Cusco. The ruins of Inca buildings are still to be found around the lakeshores and on some of the islands, providing ample evidence of great skill at building with large blocks of unmortared stone.

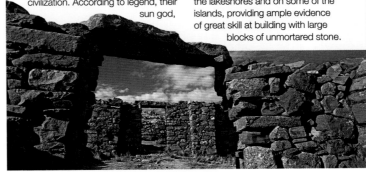

LAND

VITAL RESOURCE
Women in Burkina Faso have the important job of collecting drinking water from their village well. Water is often contaminated in this part of Africa, and waterborne diseases are widespread.

FRESHWATER QUALITY

Throughout human history, water has played a dual role: as a life-giving liquid and as a resource for waste disposal. Without careful management, these two uses can conflict, potentially dangerously. In the developed world, water pollution reached crisis proportions around the mid-twentieth century, but is now increasingly controlled. In developing countries, poor water quality remains a major health threat.

Natural Purification

In nature, pure fresh water is rare. Groundwater is normally purer than surface water, because it is filtered by porous rock. It contains dissolved minerals, but very little organic matter. Surface water also contains minerals, but, even when it looks clean, it often teems with living things and their dead remains. The vast majority of its living inhabitants are micro-organisms, such as bacteria. Most of the bacteria need oxygen to break down and purify organic matter in the water, and this process is most effective when the water contains high levels of dissolved air. Natural purification is also helped by water movement, becaues this stirs up sediment and dissolves more air. However, if water becomes overloaded with organic matter – for example, from human sources – the normal waste-disposal system may break down.

Eutrophication

In fresh water, two mineral nutrients – nitrogen and phosphorus – are essential for the growth of micro-organisms. These nutrients are often scarce, and the growth of micro-organisms stops once they have been used up. When untreated sewage and agricultural run-off enter the water, nitrogen and phosphorus levels can suddenly climb, and the micro-organism population booms. This creates more living matter, which uses up oxygen. As oxygen levels plunge, fish and other animals die, and the waterway fills with organic sludge. This process, which is known as eutrophication, is prevented by treating sewage before it is discharged, and by controlling fertilizer use.

ALGAL BLOOM
Lush growths of algae are a sign of eutrophication, which causes oxygen levels in water to fall.

Pollution

Untreated waste is also hazardous because it may contain disease-carrying micro-organisms. Normally, clean and waste water are kept apart during water treatment and waste management, but if this separation is breached – for example, through lack of sanitation – waterborne diseases can spread. To prevent this, water is periodically tested for indicator organisms, such as the bacterium *Escherichia coli*, a common resident of the human intestine. If large numbers of *E. coli* are detected in drinking water, immediate action is needed to trace their source.

Water pollution is caused not only by micro-organisms but also by industrial waste, which affects water quality in different ways. Synthetic organic chemicals can enter freshwater food chains,

WARNING SIGN
In this river, industrial pollution is very evident. Other kinds of water pollution can be much harder to detect.

and from there they can reach drinking water supplies. These chemicals include pesticides, herbicides, and pharmaceutical drugs, as well as substances that are utilized in manufacturing. Industrial waste can also contain inorganic pollutants, such as acids, zinc, and lead, many of which are known to be toxic. Safety limits for synthetic organics, which are relatively recent "inventions", are much more difficult to gauge, especially as thousands of new organics are synthesized every year.

Heat is another form of pollution that affects many urban rivers. One of the chief sources is industrial effluent, particularly water that is used to cool power stations. Heat affects the growth of plants and animals. It also reduces the amount of oxygen that water can carry – something that affects much aquatic life.

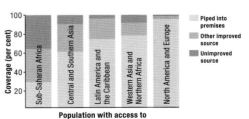

WASTE DISPOSAL
Water offers a cheap and easy way for industry to dispose of waste.

Future Trends

The increasing global population and the spread of industrialization mean that water pollution remain an acute problem in many parts of the world. According to a 2021 study, waste water adds around 5.6 million tonnes of nitrogen to coastal waters worldwide per year, contributing to eutrophication – a figure that is likely to increase. In developed countries, economic growth has helped to bring about improvements in water quality, but many developing countries are finding it difficult to follow this trend.

ACCESS TO CLEAN WATER
Within the developing world, campaigns have increased the number of people with access to clean, piped drinking water. But Sub-Saharan Africa, Southern Asia, and Central Asia still have major problems with water supply.

Chart: Population with access to driniking water facility, 2020

Coverage (per cent) — vertical axis from 20 to 100.

Categories: Sub-Saharan Africa; Central and Southern Asia; Latin America and the Caribbean; Western Asia and Northern Africa; North America and Europe.

Legend:
- Piped into premises
- Other improved source
- Unimproved source

LAND

Lake Nicaragua

LOCATION In the southwest of Nicaragua, only 15km (9 miles) from the Pacific coastline

AREA 8,150 square km (3,146 square miles)

MAXIMUM DEPTH 70m (230ft)

ELEVATION 32m (105ft)

OUTLET River San Juan

The largest body of fresh water in Central America, Lake Nicaragua is oval in shape and was once a wide bay on the ocean coastline. The present lake was formed when volcanic activity on

TOUCAN
The Chestnut Mandibled Toucan nests high in the forest canopy on hillsides around the lake.

the Pacific Rim (see p.157) produced a narrow ridge that extended from the Cordillera de Guanacaste mountain range northwards, cutting off the bay from the sea. This 19-km- (12-mile-) wide ridge now forms the lake's western shoreline. The bay's salt water has been gradually freshened by rain-fed streams flowing down the surrounding hills, combined with a steady outflow to the sea. This process has produced a unique combination of animal life, and Lake Nicaragua is the only lake in the world to have freshwater species of marine animals such as sharks and swordfish. The lake is dotted with hundreds of islands, some of which contain active volcanoes. Among the many rivers flowing into the lake is the River Tipitat, which drains Lake Managua, located to the northwest. Lake Managua was formed and freshened by the same processes that created Lake Nicaragua. The River San Juan flows from the southeastern corner of the lake through dense rainforest to the Caribbean Sea. Locally, the lake is famed for its apparent tides, which are in fact standing waves that oscillate across the lake surface – caused by winds blowing up the San Juan valley.

CONCEPCIÓN VOLCANO
This volcano on Ometepe Island is still active. Although it poses a threat to the island's inhabitants, its lower slopes are still intensively cultivated.

Lake Ladoga

LOCATION In the Karelia region of northwestern Russia, to the east of the Baltic Sea

AREA 17,675 square km (6,823 square miles)

MAXIMUM DEPTH 230m (755ft)

ELEVATION 5m (16ft)

OUTFLOW River Neva

The largest lake in Europe, Ladoga lies in a glacially modified depression. The water is deepest near the high, rocky cliffs of the northern shores. In the south, the lake is much shallower, with a low shoreline. Ice starts to form in December, and the entire lake freezes between January and May. Ladoga is fed by several rivers and two smaller lakes nearby. The River Neva flows from the southwestern corner of the lake and then through St. Petersburg about 40km (25 miles) to the west.

BATTERED SHORELINE
The bleak, rocky shores of Lake Ladoga are sometimes pounded by wind-driven waves up to 4.5m (15ft) high.

Loch Ness

LOCATION In Inverness, Scotland, in the northwestern part of the Scottish Highlands

AREA 56 square km (22 square miles)

MAXIMUM DEPTH 230m (755ft)

ELEVATION 16m (52ft)

OUTFLOW River Ness

Famous for the monster "Nessy" that is said to inhabit its depths, Loch Ness is a long, narrow lake at the northern end of the Great Glen, a valley excavated along an ancient fault, which bisects Scotland diagonally between the Grampian Mountains and the North West Highlands. Over time, movement on the ancient strike slip fault has displaced the rocks on either side by over 100km (60 miles). Erosion has excavated a valley 39km (22 miles) long and 1.5km (1 mile) wide in the weakened fault rocks. Subsequent glaciation has further deepened the lake to depths of over 100m (330ft) and filled it with glacial meltwater.

URQUHART CASTLE ON LOCH NESS

Lake Windermere

LOCATION In the Lake District of northwestern England, east of the Irish Sea

AREA 15 square km (6 square miles)

MAXIMUM DEPTH 64m (210ft)

ELEVATION 39m (128ft)

OUTFLOW River Leven

Windermere, the largest lake in England, formed at the end of the last ice age in a deep, glaciated valley that runs north–south through the Cumbrian Mountains. It is a long, narrow lake, deepest at the northern end. Fed by rain and snow from the surrounding highland, Windermere also receives waters from other nearby lakes, including Grasmere. Water flows out through the River Leven.

LAKE DISTRICT
Windermere is the largest of 17 major lakes that comprise England's Lake District.

Lough Derg

LOCATION In the southwest of the Republic of Ireland at the boundary of counties Tipperary, Galway, and Clare

AREA 118 square km (46 square miles)

MAXIMUM DEPTH 36m (118ft)

ELEVATION 34m (112ft)

OUTFLOW River Shannon

Lough Derg is located at the southern extremity of Ireland's central plain, about 30km (18 miles) upriver from the mouth of the River Shannon. The irregularly shaped lake lies in a glaciated section of the Shannon valley. The narrow southern arm, the deepest part of the lake, was formed during the last ice age, when ice was forced into the neck of the gorge through which the river now exits the lake. The northern area of the lake is shallow – with an average depth of 4m (13ft) – and surrounded by flat land and, in places, by large reed-beds.

LAND

EUROPE *central*

Lake Constance

LOCATION In the northern Alps, bordered by Austria, Germany, and Switzerland

AREA 541 square km (209 square miles)

MAXIMUM DEPTH 252m (827ft)

ELEVATION 395m (1,295ft)

OUTFLOW River Rhine

BATH-HOUSE ON LAKE CONSTANCE

Lake Constance occupies the lower part of the valley created by the Rhine Glacier during the last ice age. Measuring 64km (40 miles) long and 13.5km (8½ miles) wide, it is central Europe's second largest freshwater lake, and supplies drinking water to about 4.5 million people. The marshes along the lake margins support large numbers of waterfowl and shorebirds. The River Rhine enters the lake at the southeastern end, on the Swiss–Austrian border. At the western end of the lake, where the glacier spread out over low foothills, the lake splits into two arms. The Rhine exits through the southern arm at the German–Swiss border.

EUROPE *central*

Lake Geneva

LOCATION In the northwestern Alps, bordered by France and Switzerland

AREA 581 square km (224 square miles)

MAXIMUM DEPTH 310m (1,018 ft)

ELEVATION 372m (1,221ft)

OUTFLOW River Rhône

Lake Geneva is the largest of the Alpine lakes, forming a broad crescent, with the River Rhône entering from the southeastern tip and exiting from the southwestern tip. Although the lake is only 14km (8½ miles) across at its widest point, strong winds can cause violent storms, which are more characteristic of an inland sea. The northern shoreline is densely populated, the main cities being Geneva and Lausanne. The south-facing slopes along the lakeshore near Lausanne form the largest wine-producing area in Switzerland.

VITICULTURE ON THE NORTHERN SHORE

EUROPE *central*

Lake Como

LOCATION In Lombardy, northern Italy, near the southern edge of the Alps

AREA 146 square km (56 square miles)

MAXIMUM DEPTH 410m (1,346ft)

ELEVATION 198m (650ft)

OUTFLOW River Adda

ANCIENT RESORT
Como's spectacular mountain setting and mild climate have made it a popular resort for at least 2,000 years. Some towns around the lake date back to the Roman period.

Lake Como is the deepest of the Alpine lakes. It has three branches, each about 25km (15 miles) long and about 4km (2½ miles) wide. Its inverted Y-shape is defined by deep, faulted valleys that were produced during the formation of the Alps. The valley floors have been depressed and eroded by subsequent glaciations, which accounts for the lake's great depth – although it is an Alpine lake, parts of its bed are more than 200m (660ft) below the level of the Mediterranean Sea. Despite being surrounded by mountains – up to 2,400m (8,000ft) high at the northern end, descending to about 600m (2,000ft) in the south – the lake enjoys a mild Mediterranean climate and is a popular tourist destination. Lake Como receives most of its water from the snow-fed River Adda, which flows from the eastern Alps and enters the lake's northern arm. The lake also receives the inflow of numerous other rivers and streams. The Adda is the lake's sole outflow – it exits from the southeastern arm and is a tributary of the River Po.

LAKE COMO AND BELLAGIO
This view up the northern arm of the lake shows a high wedge of rock known as Bellagio (right centre). During the last ice age, this outcrop caused the lake-forming glacier to split into two arms.

LAND

Lake Chad

LOCATION On the southern fringes of the Sahara Desert, bordered by Cameroon, Chad, Niger, and Nigeria

AREA Highly variable

MAXIMUM DEPTH
10m (33ft)

ELEVATION
280m (920ft)

OUTFLOW Bahr el Ghazal

Lake Chad is a large, shallow, oval lake, with a highly variable surface area. The level of the lake fluctuates annually, according to patterns of rainfall in the western Sahel. Generally it covers between 10,000 and 25,000 square km (3,800 and 10,000 square miles). Changes in level have a major impact because the lake is shallow, averaging only about 4m (13ft). In 1970, when Lake Chad was at its maximum, it was the fourth largest lake in Africa, and since then it has shrunk dramatically. This is due to short-term drought and increased use of the lake's waters for

DRY LAKE BASIN
The Chad Basin has been drying out for centuries. During the drought of the 1970s and 1980s, the northern part of the lake almost dried out completely.

irrigation, but there is also a long-term downward trend. Lake Chad is all that remains of a much larger lake, which 10,000 years ago was about 50m (165ft) deeper. The present lake is

divided into two by a sandy ridge known as the Great Barrier, that runs east–west and is breached by a single channel. The northern part of the lake is the deepest, and has a shoreline of inlets, islets, and peninsulas, formed by dunes up to 30m (100ft) high. The southern part is shallower, and around the inflow of the River Chari the shoreline supports marshes. The Chari flows from the Ubangi Plateau

SHRINKING LAKE
The shallowing of the lake has made parts of it much harder to navigate and dramatically impacted the region's inhabitants, who rely on the lake for food and fresh water.

to the south and provides 95 per cent of the lake's inflow. The rain-fed water of the Chari also accounts for the lake's continuing freshness in a desert climate. If the level falls, it is the deeper, northern portion that is most likely to dry out completely, because the Great Barrier cuts it off from the Chari's inflow. When levels are at their highest, excess water flows westwards along the Bahr el Ghazal.

Lake Turkana

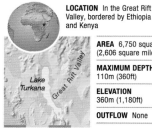

LOCATION In the Great Rift Valley, bordered by Ethiopia and Kenya

AREA 6,750 square km (2,606 square miles)

MAXIMUM DEPTH
110m (360ft)

ELEVATION
360m (1,180ft)

OUTFLOW None

Also known as Lake Rudolf, Turkana is a narrow lake, 265km (165 miles) long, running through the middle of the eastern arm of Africa's Great Rift Valley. It is the remnant of a larger lake that 5,000 years ago was three times as long, about 80m (260ft) deeper, and drained northeastwards into the White Nile. The present lake is deepest in the centre and at the southern end, where it is dammed by a lava flow. The lake has three islands, all of which are extinct volcanoes. The southern shores, once part of the lake bed,

consist of mud flats and dunes. The River Omo, flowing in at the northern end, brings water from the southern Ethiopian Highlands, and accounts for almost all the lake's inflow. With no outflow, Lake Turkana loses water mainly through evaporation in the hot, arid climate. The rate of evaporation is increased by strong winds, which often blow across the lake. Lake Turkana is subject to seasonal and periodic changes in level, but the overall trend is downwards. Since 1900, the level of the lake has dropped by about 15m (50ft).

VOLCANIC PEAK
The steep slopes of South Island, a large volcano (visible in the distance in this picture), descend to the deepest part of Lake Turkana.

NABUYATOM VOLCANIC CONE
The southern end of Lake Turkana is enclosed by a rugged volcanic landscape of old lava flows, and overlooked by extinct cinder cones.

FISHING AT TURKANA

The waters of Lake Turkana are rich in fish species, and fishing the lake has been the mainstay of some lakeside populations for thousands of years. The lake produces about 15,000 tonnes of fish each year, but stocks of the main food fishes – Nile Perch, tilapia, and tiger fish – are dwindling. To maintain the size of the catch, an increasing number of species are being harvested for food. Sport fishermen also use the lake.

AFRICA *east*

Lake Victoria

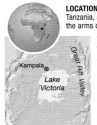

LOCATION Bordered by Kenya, Tanzania, and Uganda between the arms of the Great Rift Valley

AREA 59,900 square km (23,128 square miles)

MAXIMUM DEPTH 80m (262ft)

ELEVATION 1,134m (3,725ft)

OUTFLOW River Nile

Lake Victoria is the largest lake in Africa, the second largest freshwater lake in the world, and the main reservoir on the River Nile. The lake is roughly square in shape and is situated in the middle of the plateau that separates the western and eastern arms of the Great

Rift Valley. It formed about 2 million years ago, when tilting of the crust between the arms diverted surface water into a shallow depression on the plateau's surface. Lake Victoria has no major inflows – its level is maintained by year-round heavy rainfall across the lake's surface and run-off from the surrounding upland. Its shores are one of the most densely populated parts of Africa, and the cities of Kampala, Kisumu, and Mwanza are situated around the lake. Many of the lake's numerous islands are also inhabited. The northern shore has numerous long, thin peninsulas, between which there are areas of swamp that extend far inland and provide a refuge for the rare Sitatunga antelope. Lake Victoria is about 2 million years old – which is ancient by lake standards – and several hundred endemic species have evolved, mainly cichlid fishes.

The introduction of the Nile Perch (a food fish endemic to the River Nile and Lake Turkana) into the northern part of lake has proved an ecological disaster for local species. The perch is a voracious predator that has now spread to all parts of the lake, and the native fish population has been catastrophically reduced.

COLLECTING FRESH WATER
Lake Victoria's extensive shoreline provides ready access to water for millions of people.

TRANSPORT ROUTES
Steamships criss-cross the lake between cities, but life around the lakeshore mainly relies on even older forms of agriculture and transport.

AFRICA *east*

Lake Nakuru

LOCATION In western Kenya, in the eastern arm of the Great Rift Valley

AREA 40 square km (15 square miles)

MAXIMUM DEPTH 3m (10ft)

ELEVATION 1,760m (5,780ft)

OUTFLOW None

Lake Nakuru is a small, shallow, salty and alkaline lake – a remnant of a much larger lake that has shrunk to almost nothing over the last 10,000 years. At least once during the 20th century, Lake Nakuru has dried out completely. It is popular with tourists because of the hundreds of bird species that congregate there – particularly the spectacular flamingos.

FLAMINGOS FLOCKING ON THE LAKE

AFRICA *east*

Lake Manyara

LOCATION In Tanzania, southeast of the Serengeti Plain and northwest of the Masai Steppe

AREA 230 square km (85 square miles)

MAXIMUM DEPTH Not known

ELEVATION 945m (3,100ft)

OUTFLOW None

Lake Manyara lies at the southern end of the eastern arm of the Great Rift Valley, beneath the sheer rock walls of the valley's western escarpment. The lake has no perennial surface water sources, just a few occasional streams. Most of its inflow comes from beneath its shallow bed, having percolated down from the volcano-strewn Ngorongoro Highlands. This groundwater is saturated with minerals derived from volcanic rock. Despite there being extensive salt and soda deposits along areas of the lakeshore, its water is tolerated by catfish and bream, and by hippopotamuses. The Lake Manyara National Park, on the western shore, was established in the 1960s specifically for the protection of elephants, but is more famous for its tree-climbing lions. Visitors can also see more than 300 bird species.

AFRICA *east*

Lake Nyasa

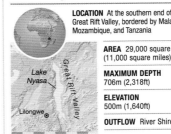

LOCATION At the southern end of the Great Rift Valley, bordered by Malawi, Mozambique, and Tanzania

AREA 29,000 square km (11,000 square miles)

MAXIMUM DEPTH 706m (2,318ft)

ELEVATION 500m (1,640ft)

OUTFLOW River Shire

Lake Nyasa, also known as Lake Malawi, is the most southerly of the Great Rift Valley lakes and was formed by the same rifting process that produced Lake Tanganyika (see p.236), which lies about 350km (220 miles) to the northwest. The long, narrow lake basin is delimited mainly by faults. Lake Nyasa is deepest in the north, where the bottom drops away sharply from the western shore. It becomes increasingly shallow towards the southern end, where its only outlet, the River Shire, flows sluggishly out of Monkey Bay to join the River Zambezi. In contrast, it receives water from no fewer than 14 rivers. Lake Nyasa has more fish species than any other lake in world, with perhaps as many as 1,000 different cichlid fishes living in its surface waters. These are very much threatened by pollution and overfishing. However, as with other deep tropical lakes, there is no vertical circulation of water. The deep waters lack oxygen and are saturated with noxious gases such as hydrogen sulphide, methane, and carbon dioxide, produced by decomposition of vegetation that has sunk to the bottom.

STILL WATERS
A placid, sand-fringed lagoon on the northern shore gives no clue that Lake Nyasa is the second deepest lake in Africa.

LAND

Lake Tanganyika

LOCATION At the southern end of the Great Rift Valley, between Burundi, Dem. Rep. of Congo, and Tanzania

AREA 32,000 square km (12,350 square miles)

MAXIMUM DEPTH 1,471m (4,800ft)

ELEVATION 773m (2,538ft)

OUTFLOW River Lukuga

The second deepest lake in the world, Tanganyika is a long, narrow lake extending 660km (410 miles) along a block-fault depression (graben) formed by the rifting process in East Africa about 15 million years ago (see pp.128–29). It has steep-sided, forested shorelines, with the Mitumba Mountains along the western shore towering about 2,000m (6,600ft) above the lake. The eastern shore is dominated by the highlands of the East African Plateau. There is lakeside lowland only where rivers enter or leave the lake. A subsurface ridge extends across the middle of the lake, dividing it into two extremely deep basins. The year-round heavy rain washes in large quantities of sediment that have accumulated in layers up to 5km (3 miles) deep. The main inflows are the River Ruzizi, which enters at the northern end, descending 600m (2,000ft) from Lake Kivu; the River Malagarasi on the eastern shore; and the River Kalambo, which enters on the southern tip, its waters having plunged over a 215-m- (705-ft-) high waterfall before reaching the lake. The single outflow, the Lukuga, exits midway along the western shore and adds its flow to the headwaters of the River Congo. Sometimes the outflow becomes blocked by floating vegetation, causing local flooding.

LEMON CICHLID
This species of cichlid, found only in the southern part of the lake, is one of dozens of species of cichlid that are unique to Lake Tanganyika.

EASTERN SHORE
The densely forested slopes of the Mohale Mountains in Tanzania descend smoothly to the water's edge – there are no beaches or lakeshore lowlands.

HENRY MORTON STANLEY

The British-born American journalist and explorer Henry Morton Stanley (1841–1904) famously "found" the "missing" British missionary-explorer David Livingstone in 1871 at Ujiji on the eastern shore of Lake Tanganyika. They then explored the northern end of the lake, looking for the source of the Nile. In 1874, Stanley organized and led a further expedition that crossed Africa coast to coast, from west to east.

Lake Baikal

LOCATION In Russia, south of the Central Siberian Plateau, near Mongolia

AREA 31,500 square km (12,160 square miles)

MAXIMUM DEPTH 1,741m (5,716ft)

ELEVATION 456m (1,497ft)

OUTFLOW River Angara

Baikal is the deepest lake in the world, containing about 20 per cent of the world's surface fresh water. The lake is 636km (395 miles) long and about 50km (30 miles) wide. It formed in a deep and ancient block fault some 25 million years ago, which also makes it the oldest lake in the world. The fault appears to be still active, as the lake is widening by about 2.5cm (1in) per year. On its western side, Baikal is lined by mountains; the eastern shoreline has gentler slopes. The deepest parts are in the centre and near the southern end. Layers of sediment on the lake bed are up to 7km (4 miles) thick in places. The continual descent of cold water keeps the bottom levels well oxygenated, and animal life is found at all depths. The lake receives inflow from many rivers, mainly along its eastern shore, and these include the Barguzin and the Selenga, which forms a delta as it enters the lake. Baikal also receives water below its surface from hot, freshwater vents in the lake bed. Irkutsk is situated near the southern end of the lake on its single outflow, the River Angara. The lake is ice-free for about six months each year. It was declared a UNESCO World Heritage Site in 1996.

STORM CLOUDS OVER LAKE
During the winter months, the surrounding mountains channel fierce storms along Lake Baikal.

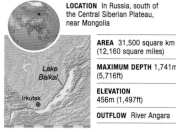

BAIKAL SEAL
The world's only freshwater seal is just one of the 1,500 or so plant and animal species that are unique to the lake.

FISHING

Lake Baikal provides an important source of food to the people living on its shores, despite remaining frozen much of the time. Local fishermen employ traditional methods to fish through a hole cut in the winter ice. The major food fish is the Omul, but 61 species of fish occupy the lake, and more than half of them are only found here.

Aral Sea

LOCATION North of the Karakum Desert, bordered by Kazakhstan and Uzbekistan

AREA Around 3,300 square km (1,270 square miles)

MAXIMUM DEPTH 58m (190ft)

ELEVATION 40m (121ft)

OUTFLOW None

Once the world's fourth largest inland body of water with an area of about 68,000 square km (26,300 square miles), the Aral Sea is now drying up and has been reduced to less than a tenth of its former size – making it one of the most conspicuous examples of human-made

RECEDING SEA
The Aral Sea was once busy with ships, the fishing industry alone employing more than 60,000 people. Now rusting hulks lay stranded as monuments to the vanishing sea.

desertification. Fifty years ago, it was a wide, shallow, saline sea – healthy and full of life. In the mid-20th century, its fisheries accounted for 3 per cent of the total catch of the former Soviet Union, but now the fish stocks have declined because the water salinity has increased. The Aral Sea was cut off from the oceans during the last ice age but maintained its size in the arid Central Asian climate because of the inflow of two mighty rivers – the Syr Darya, flowing down from the Tien Shan Mountains into the northern end, and the Amu Darya, bringing water from the Pamirs and entering along the southern shore, where it formed a large delta. In the 1930s, increasingly large

ARAL SEA SHORELINE
The volume of water entering the Aral Sea from the Syr Darya River has shrunk considerably because of the demand for water for crop irrigation.

amounts of water were diverted from these rivers to provide irrigation water for vast fields of cotton. While cotton production soared, the lake began to shrink, slowly at first and then alarmingly quickly. By 2007, it was reduced to four smaller lakes, and by 2009, the southeastern lake had disappeared, to be occasionally replenished in subsequent years. In 2005, a dam was built in an effort to replenish the North Aral Sea, and by 2008, its water level rose by 24m (79ft), and the salinity fell, allowing fish stocks to revive. The shrinking of the Aral Sea has reduced the humidity of the local climate, so that summers are now hotter and winters are colder. Millions of tonnes of dust and mineral salts have been blown away by the wind, adversely affecting the health of 5 million people who have to breathe the dust-laden air.

WATER PIPELINES
Toxic dust blowing from the desert created by the evaporation of the Aral Sea has contaminated water supplies over a wide area. Muynoq is located near the Amu Darya delta, once the region's greatest source of fresh water. It was a thriving port, but is now 56km (35 miles) from the coast of the Aral Sea. Today, its inhabitants must rely on water brought from distant reservoirs along pipelines such as the one shown below.

LAND

Caspian Sea

LOCATION On the borders of Azerbaijan, Iran, Kazakhstan, Russia, and Turkmenistan

AREA 371,000 square km (143,000 square miles)

MAXIMUM DEPTH 950m (3,120ft)

ELEVATION 30m (100ft) below sea-level

OUTFLOW None

The world's largest inland body of water, the saline Caspian Sea occupies an ancient, tilted depression in the Earth's surface. The water is deepest at the southern end, where the Elburz Mountains line its shores. Cut off from the Mediterranean Sea by falling water levels during the last ice age, the Caspian has no outflow, and owes its continuing existence to the Ural and Volga rivers, which supply much of its inflow. The eastern shore is defined by the cliffs of the arid Ustyurt Plateau, and the Karakum Desert lies to the southeast. The water's salinity varies from 1 per cent in the north, near the Volga inflow, to about 20 per cent in the shallow Kara Bogay Bay on the eastern shore, which is almost totally cut off from mixing with the rest of the Caspian's waters. (Ocean salinity

OIL EXTRACTION

The Caspian Sea's oil reserves equal those of the USA, and the region is being transformed by the oil industry, whose activities pose a threat to water quality. Oil extraction has expanded from Baku in Azerbaijan on the Caspian's southwest shoreline to include newly developed oil and gas fields in Kazakhstan at the northern end of the Caspian.

averages about 3.5 per cent.) For thousands of years, the water level of the Caspian Sea has experienced dramatic and unpredictable fluctuations. For example, the water rose 2.5m (8ft) between 1978 and 1995.

STERLET
This smaller relative of the caviar-producing sturgeon is one of the Caspian's most valuable food fish.

CASPIAN COASTLINE
Winter snows cover the foothills of Iran's Elburz Mountains and the banks of the Darya-ye Sefid river, where it flows into the sea's southern coastline.

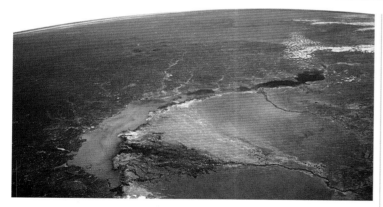

Lake Balkhash

LOCATION In southeastern Kazakhstan, south of the Central Kazakh Uplands

AREA 18,200 square km (7,025 square miles)

MAXIMUM DEPTH 25m (82ft)

ELEVATION 340m (1,115ft)

OUTFLOW None

Balkhash is a long, narrow, curved lake, situated in an east–west depression. The lake is divided into two parts by the Sarymsck Peninsula, which juts out from the southern shore. Water freshness in the wide and shallow western part of the lake

DIVIDED LAKE
The Sarymsck Peninsula, across the middle of Lake Balkhash, is visible near the centre of this space shuttle image.

has been maintained by inflow from the river Ili but increased water extraction from the river threatens this supply of freshwater. The lake narrows and deepens to the east, and there is little mixing between the two parts of the lake. The water in the eastern part has much higher levels of dissolved minerals because it receives only the inflow of minor rivers. The level of the lake fluctuates seasonally by about 3m (10ft), and during the last 100 years Balkhash's area has varied by up to 50 per cent – from 15,600 square km (6,000 square miles) to 23,400 square km (9,000 square miles). There are valuable copper deposits on the northern shore that are processed in the nearby town of Balkhash.

Sea of Galilee

LOCATION In northern Israel, close to the borders with Syria and Lebanon

AREA 166 square km (64 square miles)

MAXIMUM DEPTH 48m (157ft)

ELEVATION 212m (696ft) below sea-level

OUTFLOW River Jordan

Known also as Lake Tiberias, and renowned from the Bible, the Sea of Galilee is a small, shallow, pear-shaped lake situated in a deep, rifted depression that is an extension of Africa's Great Rift Valley. The River Jordan flows southwards along the rift and through the lake. To its north, around the inflow of the Jordan, is the narrow, fertile Gennesaret Plain, with the Golan Heights to the northeast in Syria. The lake's western shores lie beneath the sheer wall of the rift, and the southern shores are confined by a volcanic ridge. Inflow from the Jordan diminished until 2009 as water was diverted for irrigation upstream of the lake, leading to an increase in the lake's salinity. However, rising fresh water input since then has lowered the salinity.

SOUTHEASTERN SHORELINE
The Sea of Galilee is seen here from the southeastern shore. The wall of the rift is visible across the water to the west.

ASIA west

Dead Sea

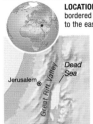

LOCATION North of the Red Sea, bordered to the west by Israel and to the east by Jordan

AREA 606 square km (234 square miles)

MAXIMUM DEPTH 330m (1,084ft)

ELEVATION 427m (1,401ft) below sea-level

OUTFLOW None

The Dead Sea lies in a depressed piece of continental crust (graben) that separates two tectonic plates, the Sinai Subplate and the Arabian Plate. To the north and south the basin is bordered by igneous intrusions called sills. Fed by the River Jordan, and having no outflow, the Dead Sea is the lowest lake in the world – its surface is the lowest point on the Earth's land surface, at 427m (1,401ft) below sea-level. It is also one of the saltiest lakes, with an average salinity of about 32 per cent. The Dead Sea achieved its greatest extent about 10,000 years ago, and it has been shrinking slowly ever since. It is currently about 50km (31 miles) long and up to 15km (9 miles) wide. Inflow from the River Jordan and the occasional

LAKESIDE SALT DEPOSITS
As the waters of the Dead Sea evaporate, they leave behind an unearthly landscape dotted with mounds of salt crystals and pools of extremely saline liquid.

rain shower are insufficient to balance the very high rates of evaporation. Recently the rate of shrinkage has increased because of the diminished flow of the Jordan. During the last quarter of the 20th century, the area of the lake declined by about 20 per cent, and the level of the lake is currently dropping at a rate of about 1m (39in) per year. The Lisan Peninsula, which once divided the lake into northern and southern parts, now forms the southern shoreline. What was the southern part of the lake now consists only of evaporation pans for salt and mineral production. In the deepest part of the lake, where the bottom waters are very dense and saline, there is little vertical mixing. The deepest water in the Dead Sea is many thousands of years older than the surface waters.

DEAD SEA COASTLINE
The paler sediments seen here on the barren shoreline of the Dead Sea indicate that the water level was once much higher.

ASIA southeast

Qinghai Hu

LOCATION In the Nan Mountains of north central China, east of the Qaidam Pendi basin

AREA 4,460 square km (1,772 square miles)

MAXIMUM DEPTH 38m (125ft)

ELEVATION 3,200m (10,500ft)

OUTFLOW None

Qinghai Hu is a shallow, slightly saline lake situated in a broad depression on a high plateau and ringed by mountains. It is the largest lake in China and is also known as Koko Nor, and alternatively as the Blue Lake because of its intense colour. The area of the lake varies seasonally, and approaches 6,000 square km (2,300 square miles) when water levels are at their highest. Terraces around the shores indicate much higher levels during the last ice age, when Qinghai Hu probably drained westwards into the Tarim Basin.

QINGHAI HU SHORELINE

ASIA southeast

Lake Toba

LOCATION In North Sumatra Province on the Indonesian island of Sumatra

AREA 1,100 square km (425 square miles)

MAXIMUM DEPTH 529m (1,737 ft)

ELEVATION 905m (2,970ft)

OUTFLOW River Asahan

Lake Toba is the largest caldera lake in the world and is situated high in the Barisan Mountains of northern Sumatra. The lake is oval in shape – 90km (56 miles) long by 30km (19 miles) wide and the cliffs of the enclosing caldera rim rise up to 1,200m (3,940ft) above the surface.

SAMOSIR ISLAND
Samosir, in the centre of Lake Toba, occupies about one-third of the lake. The rich volcanic soils on the island are intensively cultivated.

It is over 460m (1,518ft) deep. Lake Toba is situated on an active fault line that runs the length of Sumatra, and was formed as a result of four catastrophic explosive eruptions, the last of which took place about 74,000 years ago, which emptied and collapsed a subterranean chamber, producing a caldera that soon filled with water. Subsequent uplift produced the large island of Samosir, with an area of 640 square km (250 square miles), in the middle of the lake. At the northern end of the lake, water enters via the Sipiso-Piso waterfall. The River Asahan exits from the southern end of the lake and flows into the Indian Ocean.

AUSTRALASIA Australia

Kati Thanda-Lake Eyre

LOCATION In the state of South Australia, south of the Simpson Desert

AREA 9,700 square km (3,745 square miles)

MAXIMUM DEPTH 5.5m (18ft)

ELEVATION 16m (53ft) below sea-level

OUTFLOW None

A salt lake of variable size, Kati Thanda-Lake Eyre is the lowest point on the Australian continent. It receives water from ephemeral streams, fed by erratic rainfall over much of central Australia. Thought to have once dried out completely, the lake reached its present size following exceptionally heavy rains at the end of the 20th century.

DRY LAKE BED
This dried-out area of lake bed is covered by thick deposits of gypsum overlain by a thin crust of salt.

LAND

Underground Rivers and Caves

78	Limestone
78	Evaporites
90–91	Chemical weathering
92–93	Erosion by water
94–95	Deposition by water
110–11	Groundwater
198–223	Rivers

Beneath the surface in many of the Earth's limestone regions are large passages and caves. These are formed as limestone dissolves in slightly acidic water. As groundwater seeps downwards, tiny fractures gradually enlarge into open passages. When streams and rivers are swallowed by caves, erosion continues more rapidly. Underground rivers scour out huge waterfall shafts and dissolve great tunnels. Subsequently, chambers are modified by rock-fall from unsupported ceilings, and seepage water creates calcite deposits. Young, actively forming caves have noisy streams while older, inactive caves are silent and often decorated with strange formations and, in rare cases, prehistoric art.

Limestone in Solution

Limestone is rock mostly composed of calcium carbonate (see p.78) that dissolves in naturally acidic water. Rainwater absorbs carbon dioxide from the air and from microscopic soil organisms to become weak carbonic acid. As it seeps downwards, it can dissolve large amounts of limestone. If that water then drips out into a cave, it loses carbon dioxide into the air and deposits excess calcite (calcium carbonate) as stalactites and stalagmites (see opposite) in order to maintain its chemical equilibrium.

STRAW STALACTITE
Tiny crystals of calcite can be seen here growing inside a water drop at the end of a cave straw (miniature stalactite) just 5mm (1⁄5in) in diameter.

TRACING STREAMS

Underground streams can be tracked through large caves, but not where they are lost in narrow fissures. Then their courses are traced using fluorescent dyes, such as fluorescein. Even when diluted by many converging cave streams, this dye can be detected in minute quantities at a spring to prove the underground flow path from a distant sinkhole.

Underground Erosion

When streams pass underground, two main types of cave passage are formed – canyons and tubes. Underground streams and waterfalls above the water-table cut deep, twisting canyon passages, by removal of limestone in solution and sediment abrasion of their floors. Below the water-table, caves are full of slowly moving water, which dissolves limestone walls, floors, and ceilings equally, creating tubular tunnels. Many caves descend steeply to the water-table and then continue sideways, as flooded passages, until they intersect the surface, where the emerging water forms a spring (also known as a resurgence).

LIMESTONE LANDSCAPE

Sinking streams, bare rock pavements, and scattered sinkholes (or dolines) form the visible half of a limestone landscape. The unseen half is the network of cave passages, some with rivers, and others that are now dry and contain calcite deposits.

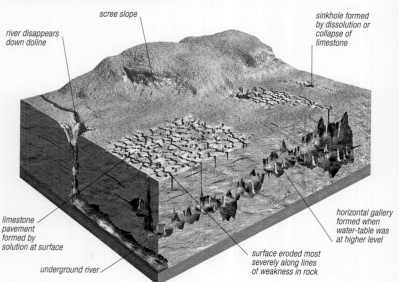

scree slope

river disappears down doline

sinkhole formed by dissolution or collapse of limestone

limestone pavement formed by solution at surface

underground river

surface eroded most severely along lines of weakness in rock

horizontal gallery formed when water-table was at higher level

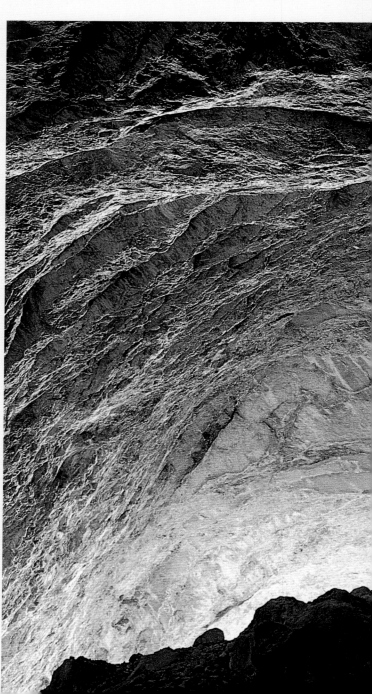

Cave Deposits

Sand, mud, calcite, gypsum, and guano are all common cave deposits, but the most widespread and beautiful are those made of calcite. Groundwater saturated with calcium carbonate deposits calcite in the form of stalactites where water drips from a cave roof. Stalagmites form where the water hits the cave floor. Banks of solid calcite on the walls and floors of caves are called flowstones. A rimstone pool forms on the surface of a flowstone deposit where water collects behind a wall of calcite. More calcite may be deposited as water flows over and down the wall. Most cave calcite contains traces of uranium; the radioactive decay of its atoms can be used to determine the age of deposition, and thus to date the cave.

ARAGONITE CRYSTALS
In Australia's Jenolan Caves, tiny crystals of aragonite (a form of calcium carbonate) grow under thin films of water derived from condensation or seepage.

CAVE PEARLS
These calcite spheres in Golconda Cavern, UK, are cave pearls, which grew by rolling around in a shallow pool disturbed by dripping water.

Karst Landscapes

Karst is defined as a limestone landscape with underground drainage. It is distinguished by extensive patches of bare rock, because the weathering of limestone produces little or no soil. Rock outcrops exposed to rainfall are fretted by grooves called karren, formed as weakly acidic rainwater runs across them. They start as narrow fissures on joints or meandering grooves, and evolve into deep gullies between pinnacles of remnant limestone. Sinkholes or dolines formed by collapses in the surface can merge to form large sunken regions, equivalent to valleys in other landscapes. As the sinkholes expand, the residual hills between them become cones, or can be undercut to form towers. The weathering of karst landforms relies on abundant rainfall and lush vegetation (to provide carbon dioxide through its decay), so the most mature karst and the largest caves form in wet, tropical areas.

LIMESTONE PAVEMENT
This pavement in England's Pennines was scraped clean by glaciers during the last ice age and then etched with deep fissures by post-glacial rainwater.

TOWER KARST SCENERY
Isolated limestone towers with vertical cliffs overlook the wide plain around Yangshuo, in Guangxi, southern China. They represent the ultimate mature karst landscape; tower karst is also known by its Chinese name, fenglin.

PINNACLE KARST
With razor edges honed by tropical rainfall, spectacular pinnacles and blades of limestone rise clear of the rainforest canopy on the karst hills of Gunung Mulu National Park, Sarawak, Malaysia.

LEVIATHAN CAVE
This cave, in Nevada, USA, was formed by a river that has now disappeared. The impressive arch shape is due to later collapse of the roof.

UNDERGROUND RIVERS AND CAVES PROFILES

The pages that follow contain profiles of some major underground rivers and caves. Each profile begins with the following summary information:

LENGTH Total length of all the passages within the cave system

DEPTH Vertical distance from the cave entrance to its deepest known point

NORTH AMERICA *west*

Carlsbad Cavern

LOCATION In the Guadalupe Mountains of southeastern New Mexico, USA

LENGTH 50km (31 miles)

DEPTH 316m (1,037ft)

Left high and dry in the mountains of the Chihuahua Desert, Carlsbad Cavern has no underground river and almost no dripping water, but is a gigantic relic of long-finished karst solutional processes. It is formed in Permian limestones that

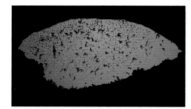

BATS AT THE CAVE ENTRANCE
Every evening, hundreds of thousands of Mexican Free-tailed Bats leave their daytime roost in Bat Cave to feed in the surrounding countryside.

PAPOOSE ROOM
This chamber gets its name from an Algonquian word for a small child. The cavity was originally created by sulphuric acid; the stalactites were later formed by carbonic acid.

STALAGMITE MIMICRY
This small stalagmite has grown a bulbous top with stalactite ribs hanging below, so that it is now always known as the "bashful elephant".

were invaded by fluids rising from the hydrocarbon reservoirs in the Delaware Basin, just to the east. These fluids were rich in corrosive sulphuric acid that dissolved the limestone, removing it in solution on an unusually large scale. The Big Room, which was formed about 4 million years ago, displays evidence of this very unusual origin. Although rarely noticed by visitors, there are thick banks of chalky white gypsum left by the chemical reaction between the sulphuric acid and the limestone. Only later was Carlsbad Cavern invaded by percolating rainwater, which both continued the excavation by solution, and also deposited the massive calcite stalactites and stalagmites. Just inside its gaping entrance, the spacious Bat Cave once held thick deposits of bat guano, most of which was mined out in the 1800s to use as fertilizer. It was the miners who first explored the extensive lower passages, which contain

many classic cave features. The descending Main Corridor is floored by huge limestone blocks, called breakdown, that have fallen from the faulted roof, totally changing its shape from the original solutional cave. Lower down in Carlsbad Cavern, the Boneyard is a complex of small, unmodified chambers and tunnels that show how the limestone was dissolved in random shapes when the cave was full of water. Beyond this, the famous Big Room is not so much a great chamber as a giant, twisting passage that is over 120m (400ft) wide in places. Most of Carlsbad Cavern is decorated with a profusion of calcite deposits, and the Big Room contains some very large stalagmites within its high, open space. The Big Room is the end of the cave section open to the public, but large tunnels descend beneath it, ending at the Lake of the Clouds, named after the huge bulbous stalactites that hang over a deep pool where the cave meets the present-day water-table.

DRIPSTONE DECORATIONS
Minor irregularities in the cave wall cause drips of water to deposit calcite, creating little flowstone terraces. Water dripping off the terraces deposits more calcite as curtains and fringes of stalactites.

PAINTED GROTTO
In this corner of Carlsbad Cavern's Big Room, calcite is coloured by natural pigments of iron oxides and hydroxides. The Big Room has an area of 33,210 square metres (357,480 square ft).

NORTH AMERICA *east*

Mammoth Cave

LOCATION In southwest Kentucky, USA, beneath the Chester Upland

LENGTH
563km (352 miles)

DEPTH 115m (377ft)

By far the longest cave system known in the world, Mammoth Cave has a huge network of passages on many levels. The nearly horizontal Carboniferous limestone in which it occurs carries water from sinkholes on the Pennyroyal Plateau to springs, along the Green River. There are many cave passages, each with a long, meandering canyon that passes into an elliptical tube beneath the water-table. Downcutting of the Green River has caused the water-table to fall in successive stages, leaving many large, dry, abandoned tubes. The highest level of caves

GYPSUM FLOWER

The curved crystals of gypsum flowers form where water carries calcium sulphate from the overlying sandstone down into the limestone and out into the caves.

contains sediments that are 2.3–3.5 million years old; these survive beneath the protective sandstone ridges of the Chester Upland. Most of the youngest caves drain through the ridges, where they connect with some of the older caves, while a few pass beneath the intervening Houchins and Doyel valleys. Except where shafts drop down through the limestone

ABANDONED TUBULAR CAVE
The many long, tubular tunnels of Mammoth Cave were formed below the water-table before it fell to a lower level.

beds, most caves follow the same bedding planes within the rock sequence. Consequently, there are numerous chance passage intersections, which have linked the caves into the one extraordinarily long system.

CENTRAL AMERICA

Windsor Caves

LOCATION Beneath the Cockpit Country of western Jamaica, close to the north coast

LENGTH 3km (2 miles)

DEPTH 20m (65ft)

The forest-covered hills in Jamaica's Cockpit Country are often likened to an upturned egg-box. They constitute a famous example of cone karst and are drained entirely underground. Windsor Great Cave is one of the few caves known in the karst; it is a high-level remnant of an ancient drainage system, now largely choked with calcite deposits. There is a small cave stream, but the main drainage lies unseen below the water-table at greater depths within the limestone.

WINDSOR GREAT CAVE

NORTH AMERICA *south*

Sistema Huautla

LOCATION In the Sierra Mazateca, part of the Sierra Madre del Sur, in Oaxaca, Mexico

LENGTH
56km (39 miles)

DEPTH 1,475m (4,840ft)

The limestone highlands of the Sierra Mazateca contain one of the world's finest collections of caves. Nita Nanta is the highest entrance to one of a dozen caves that converge in Sistema Huautla; each is a narrow stream canyon broken by a staircase of waterfall shafts. To the

south, the giant dolines of San Agustin drain into shafts that feed the underground river. The combined waters then drop down into a largely unexplored flooded zone. The main passage loops far below the water-table before emerging from underwater passages in Santo Domingo Canyon. As the canyon has cut deeper, it has triggered changes throughout the cave system. Fresh stream passages have been formed under now-abandoned tunnels. The oldest passages include chambers, modified by massive roof collapse, near the foot of the San Agustin shafts.

EXPLORING THE HUAUTLA CAVES
Cascading streamways and lashing waterfalls make exploration in Huautla both challenging and exciting for cavers.

SOUTH AMERICA *east*

Gruta do Janelão

LOCATION In the middle São Francisco valley of Minas Gerais, eastern Brazil

LENGTH 5km (3 miles)

DEPTH 150m (500ft)

Within the dense primary forest of the Peruaçú nature reserve, there is a broad, almost horizontal limestone bench, less than 200m (650ft) thick, which is broken by a series of wide, deep collapse dolines. These are windows into Janelão, a spectacularly large cave that winds beneath the bench. It is also a singularly beautiful cave. Gruta do Janelão means "cave of the windows", and it is named after the daylight that pours in to illuminate a cave passage about 90m (300ft) high and 60m (200ft) wide. A stream winds between great sandbanks on the cave floor, and terraces of ancient sediment rise high on the

walls, which suggest that a far larger river flowed through the cave in the distant past. Giant stalactites hang from the cave roof – one has been measured as 28m (92ft) long, and is perhaps the longest stalactite in the world. There are also massive banks of calcite flowstone on the walls and floors of the cave. The whole site is the product of karst processes that have been active for a very long time, but the exact age of the cave is not yet known.

CAVE OF THE WINDOWS
Daylight from holes in the roof (and the presence of a person sitting on a sandbank at the lower right in this photograph) make it possible to appreciate the sheer size of the cave.

GIANT CRYSTALS
Discovered in 2000, Mexico's Cave of Crystals preserves spectacular crystals up to 12m (39ft) long. Formed of gypsum, a calcium-sulphate mineral, they grew in hot, mineral-rich waters that filled the cave 500,000 years ago.

Ease Gill Caverns

LOCATION Beneath Leck and Casterton Falls in the northern Pennine karst, England

LENGTH 70.5km (44 miles)

DEPTH 211m (692ft)

Dozens of streams drain through this rambling series of caves, before resurfacing at Leck Beck Head. The present-day inlet passages are long, clean-washed stream canyons, interrupted by vertical waterfall shafts; some intersect with old, abandoned tunnels. The large, high-level tunnels, which date from the Pleistocene Epoch, have calcite decorations and are partly choked with in-washed sediments.

CAVE SHAFT
This deep, vertical shaft with its waterfall is created where a cave stream drops from one level to another. It forms a staircase-like descent through the limestone beneath Leck Fell in Yorkshire.

Lascaux

LOCATION In the side of the Vézère Valley within the Dordogne karst of western France

LENGTH 150m (500ft)

DEPTH 10m (30ft)

PRESERVATION

When the Lascaux cave was opened to visitors, the ancient paintings were damaged by new calcite deposition induced by environmental changes and were discoloured by algae from spores carried in through the enlarged entrance. Preservation and visitor access could not coexist. Visitors to Lascaux now see an exact replica of the cave and its paintings.

This tiny remnant of an ancient cave passage was long ago left high and dry in a limestone hillside. The passage was later truncated as erosion wore away the hillside, leaving an open entrance in the side of the Vézère Valley. Lascaux was found by people of the Solutrean and Magdalenian cultures in the Old Stone Age (or late Paleolithic), about 17,000

PALEOLITHIC CAVE ART
The perfect proportions of the animals within the ceiling frieze in Lascaux's Great Hall of the Bulls reflect the superb skills of bygone cave artists.

years ago. They walked into the open cave and well into its dark zone, where they painted on the upper walls and arched ceilings of the first five small chambers, using natural pigments of ochre, iron oxide, and charcoal. They produced beautiful figures of horses, aurochs, bulls, and stags, and even created narrative strips of hunting scenes. The cave was rediscovered in 1940, when local boys pursued their dog down a small hole in the rocks and soil that masked the entrance. Its wealth of Paleolithic art is regarded as the finest in the world.

Vercors

LOCATION In the limestone lower Alps of the Provence Alpes region, southeast France

LENGTH Numerous long caves

DEPTH 1,271m (4,170ft)

With an area of over 1,350 square km (521 square miles), Vercors is the largest karst area in Europe. Formed in Cretaceous limestone more than 400m (1,300ft) thick, it contains many long,

deep, and spectacular caves. Their passages are a mix of large, old, dry tunnels with calcite deposits, and narrow streamways with many fine waterfalls. The most famous cave is the very deep Gouffre Berger. Narrow shafts descend to the Grand Galerie, which is decorated with very large stalagmites. The tunnel follows the slope of the limestone beds for 3km (2 miles), to a flooded zone behind the resurgence in the Cuves de Sassenage.

GOURNIER CAVE LAKE, VERCORS
Just inside the resurgence entrance to the Gournier Cave, a placid lake lies beneath a profusion of calcite stalactites and flowstone cascades.

Grotte Casteret

LOCATION In the crest of the Spanish Pyrenees, within the Ordesa National Park, Spain

LENGTH 500m (1,600ft)

DEPTH 35m (115ft)

Grotte Casteret is a limestone cave in which water seeps in as a liquid and then freezes on entering the cave. Such ice caves, as they are known, are rare because they rely on delicately balanced microclimates. Though high in the Pyrenees, the ground is not frozen, so water can seep into the limestone. However, in winter cold air becomes trapped in the cave and remains there for most of the summer. Water freezes almost as soon as it drips into the cave, creating huge icicles and columns of glass-clear ice. The floor of the main chamber is a sheet of ice, which overflows in a frozen cascade to another ice floor in a lower chamber.

CURTAINS OF ICE
Glass-clear icicles drape the walls of Grotte Casteret wherever seepage water emerges from the limestone fissures.

ICE CAVE
This grand ice column in Casteret's main chamber is a rare and transient feature. If the cave warms, it will melt; if it cools, groundwater will be unable to enter and replenish the ice.

SINKHOLE ENTRANCE
The deep entrance dolines or sinks of this cave system are crossed by the River Reka before it finally sinks underground.

UNDERGROUND RIVER
The first part of Skocjanske Jame's river passage receives light filtering in from the cave's giant entrance.

EUROPE *south*

Skocjanske Jame

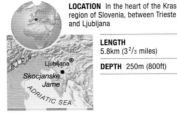

LOCATION In the heart of the Kras region of Slovenia, between Trieste and Ljubljana

LENGTH 5.8km (3²/₃ miles)

DEPTH 250m (800ft)

With a tourist pathway built high on the walls of its huge river passage, Skocjanske Jame is one of the most spectacular show caves in the world. Beside the village of Matuvan, the River Reka sinks through two giant dolines and runs beneath cliffs over 100m (330ft) high into the yawning cave entrance. There the underground river flows in a deep canyon cut down into a wider ancient passage, which now survives as high-level ledges adorned with calcite deposits. The abandoned gallery, which was the river's outlet before the canyon developed, is now blocked by massive calcite deposits and debris from a gigantic collapsed doline. The cave is the largest river sink in the Kras – the limestone region that gives its name to karst landscapes around the world.

EUROPE *southeast*

Optimisticheskaya

LOCATION Beneath the Dniester Valley, northwest of the Black Sea, in western Ukraine

LENGTH 165km (102 miles)

DEPTH 15m (50ft)

Limestone is not the only rock that can be dissolved to form caves. For more than a million years, groundwater has drained through a thin bed of gypsum lying beneath a limestone and clay cover in the area of the Ukraine called the Podil's'kyy flatlands. Water, flowing along joints and faults in the gypsum below the water-table, dissolved the walls to make each into a long, narrow fissure. These intersect with each other to create a seemingly endless underground maze, which has been left dry by the falling water-table. The Optimisticheskaya network is accessible only through a single entrance, dug through the floor of a doline in 1966 by local cavers. Removal of gypsum in solution continues today, creating another generation of caves to the south of Optimisticheskaya, where the gypsum still lies below the water-table.

AFRICA *south*

Cango Caves

LOCATION In the Swartberg Mountains to the east of Cape Town, South Africa

LENGTH 5.3km (3¹/₃ miles)

DEPTH 60m (200ft)

Isolated in a narrow outcrop of Precambrian limestones in the highlands near Oudtshoorn, the Cango Caves are remarkable for their wealth of calcite deposits. They are a remnant of a major channel below the water-table that was drained when nearby surface valleys were cut to lower levels. Subsequently, calcite deposition has occurred on a massive scale, leaving a spectacular

LARGE DRIPSTONE DEPOSIT

array of flowstone and large stalagmites. Bushmen left paintings on the walls of the entrance chamber thousands of years ago, but the entrance was lost, and rediscovered by local farmers only in 1780.

BOTHA'S HALL
The chambers of Cango Caves are sections of a larger main passage, separated from neighbouring sections by walls and columns of calcite.

ASIA *west*

Krubera

LOCATION High on the Arabika Massif of the Western Caucasus Mountains in Georgia

LENGTH 5km (3 miles)

DEPTH 1,710m (5,610ft)

Krubera was established as the world's deepest known cave following exploration in January 2001. Thick Mesozoic limestones form spectacular karst across the summits of the Western Caucasus, and explorations in the 1980s found a number of deep cave systems, including Krubera, draining the glaciated Ortobalangan Valley. Krubera is a steeply descending cave system with a single active streamway of narrow, meandering canyons, broken by nearly 40 major shafts and numerous smaller cascades. The bottom chamber is blocked by boulders that prevent access into the main river passage, which must exist to drain all the caves of the western Arabika. The proven resurgence of the cave water is at springs on the shore of the Black Sea, another 530m (1,740ft) below the level of the bottom of the known cave. In early 2003, the depth record went to Gouffre Mirolda in France, now known to be 1,733m (5,685ft) deep.

ASIA *southeast*

Nanxu Arch

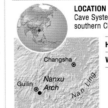

LOCATION Beside the Guanyuan Cave System, on the Li River in southern China

HEIGHT 60m (200ft)

WIDTH 40m (130ft)

The beautiful rock arch standing above Nanxu is the last surviving remnant of a large, old cave passage that once passed through the towering limestone hills in the karst of southeastern China. The arch's smooth profile is a result of surface weathering after the cave passage was exposed. The small river that formed the old cave now sinks just below the arch, into the younger Guanyuan Cave System.

LIMESTONE ARCH

LAND

ASIA *southeast*

Gunung Mulu

LOCATION In the rainforests of Sarawak, near the border with Brunei on the island of Borneo, Malaysia

LENGTH Over 320km (200 miles) in total

DEPTH 470m (1,542ft)

In 1978, British cavers discovered a series of gigantic caves with some of the largest passages in the world, all beneath the rainforests of the Gunung Mulu National Park. Deer Cave (Gua Payau) has a single passage over 100m (330ft) high and wide for its entire 1-km- (3,500-ft-) long route through a limestone ridge. Good Luck Cave (Lubang Nasib Bagus)

NESTING BIRDS

Thousands of swiftlet nests, made of moss and saliva, are glued to the walls of the Mulu caves, and swiftlets constantly fly through the darkness, navigating by echo location with distinctive clicks. Local people use swiftlet nests to make birds'-nest soup, but due to conservation measures in the cave the birds here are safe from all predators except voracious racer snakes, which are excellent climbers.

SKYLIGHT
Daylight filtering through the forest canopy beams down a hole in the roof of a gallery in the tourist section of Clearwater Cave.

has a long stream passage that leads into Sarawak Chamber – the world's largest known cave area, 700m (2,300ft) long and over 300m (1,000ft) wide beneath an arched limestone roof over 100m (330ft) high. Clearwater Cave (Gua Air Jernih) has over 110km (69 miles) of passages that have already been mapped. The lower levels have a major river draining through massive canyon passages, while the upper levels are huge, abandoned tubes decorated with calcite deposits. All these caves have been formed by the copious flows of rainwater draining from the sandstone mountain of Gunung Mulu (Mount Mulu) into the steep limestone ridges of Api and Benarat, and out onto the shale lowlands of the Melinau River. Wherever a river met the limestone, a cave formed and then matured into a large tunnel. As the ground surface lowered, the rivers created successive newer caves at lower levels and abandoned the older passages above. Dating of sediments in Clearwater Cave shows that the caves have evolved over a period of 2 million years. Throughout this period, surface erosion has progressively lowered the sandstone hills and shale lowlands. In contrast, erosion of the limestone has occurred largely underground, and minimal surface lowering has left the cavernous hills as ever-higher features of the changing landscape.

DEER CAVE, SARAWAK
The world's largest passage is lit by daylight from its southern entrance. The cave floor is covered by stream sediments and guano from the 3 million bats roosting overhead.

BATS LEAVING DEER CAVE
Every evening a seemingly endless procession of bats trails out of the cave entrance en route to feed on insects living in the nearby rainforest.

SARAWAK CHAMBER
Giant flashlights, held by cavers, light the largest cave chamber in the world, deep inside the mountain of Gunung Api.

ASIA *east*

Akiyoshi-do

LOCATION Beneath the Akiyoshi-dai Plateau, at the western tip of the island of Honshu, Japan

LENGTH
8.7km (5½ miles)

DEPTH 137m (450ft)

The rolling limestone plateau of Akiyoshi-dai contains a host of caves, the longest and most important being Akiyoshi-do. A stream enters the cave via two sinkholes and then passes through a short flooded section, before entering a passage with tall chambers that were created by successive collapses of the cave roof. The main passage is part of the lowest cave level, and its wide stream canyon has been carved into the floor of an original tubular cave that appears to be at least 100,000 years old. A layer of volcanic ash within the cave sediments is dated to about 34,000 years ago. The famous Hyakumai pools lie in the large stream passage that leads to the tall resurgence entrance in the foot of a cliff.

THE CASCADE OF HYAKUMAI-ZAI
Each terraced rimstone pool is retained by a calcite barrier that is deposited by the overflow of calcium-carbonate-saturated water.

AUSTRALASIA *Australia*

Koonalda Caves

LOCATION Beneath the Nullarbor Plain, close to the coast of South Australia

LENGTH 1,200m (4,000ft)

DEPTH 80m (260ft)

The flat, barren landscape of the Nullarbor Plain is broken by more than 130 large dolines, formed by surface collapse. Koonalda is one of these, where a very large cave passage is accessible below the debris. The Nullarbor Plain is made of Miocene limestone that has been drained by underground passages since it was uplifted about 14 million years ago. The caves formed as widely spaced tunnels, filled with slowly moving water at or below the water-table. During later phases of uplift, the water-table fell, leaving some of these passages dry. Further collapse of limestone ceilings has left high chambers, some of which have broken out at the surface, creating the collapse dolines that are the caves' only entrances. Koonalda has four chambers in its main passage, besides the one that forms its entrance. Between the collapse features, the stable sections of cave roof arch over long lakes of clear, brackish water. At the end of the third lake, the roof dips, and the cave continues as a submerged tunnel.

DOLINE ENTRANCE
The 40-m- (130-ft-) wide entrance of Koonalda Caves provided easy access for Aboriginal flint miners more than 20,000 years ago.

ASIA *southeast*

Ha Long Bay

LOCATION On the coast of Vietnam facing the Gulf of Tongking, close to the Chinese border

LENGTH Numerous short caves

DEPTH All less than 70m (230ft)

With its 1,969 limestone islands, Ha Long Bay provides one of the world's most distinctive and spectacular landscapes. Most of the islands are fringed by vertical cliffs, and many are just individual towers taller than they are wide. Since invasion by the sea, marine erosion has undercut the cliffs, so that the island profiles are even steeper than those of the inland towers well-known across the border in southern China. There are hundreds of caves within the islands of Ha Long Bay, but none is long or deep because they are limited by the sizes of the islands. Some caves are remnants of ancient systems, now decorated with calcite deposits. Other younger caves have natural tunnels at sea-level, which lead through the limestone hills into drowned dolines – hidden inland lakes that have no other access to the outside world.

HANG BO NAU
A vista of limestone hills and islands is framed by the mouth of the Bo Nau Cave, the remnant of an ancient passage on Island 268 in Ha Long Bay.

ASIA *southeast*

Tham Hinboun

LOCATION East of the Mekong River in the limestone mountains of Khammouan, Laos

LENGTH 12.4km (7¾ miles)

DEPTH 40m (130ft)

The Hinboun River flows through this cave for a distance of over 7km (4.5 miles), and is the world's longest navigable underground river. The cave originated with a switchback route, following fractures and bedding in the limestone. Subsequently, the high passages were eroded by retreating waterfalls, while the lower ones were removed by solution of their underwater ceilings. Today, the cave is an almost horizontal tunnel passing between two open karst basins.

AUSTRALASIA *Australia*

Jenolan Caves

LOCATION In the western flank of the Blue Mountains, in New South Wales, Australia

LENGTH 10.6km (6⅔ miles)

DEPTH 85m (280ft)

Australia's best-known tourist caves were discovered in around 1838 by escaped convict and bushranger James McKeown. They lie in a narrow outcrop of Silurian limestone that is crossed by the Jenolan River. Grand Arch, at just 120m (400ft) long, is all that remains of an ancient cave; but to either side the limestone contains a host of smaller cave passages with streams still flowing through the lower levels. Some Jenolan caves were formed in Permo-Carboniferous times before being completely filled with Permian carbonates and gravels. Later streams have re-excavated a few of the old caves, or cut new ones through the limestone.

JUBILEE CAVE
Delicate calcite crystals that grow in pools of calcium-carbonate-saturated water are a feature of this richly decorated part of the Jenolan Caves.

GRAND COLUMN
This large stalagmite in River Cave has developed an apron of stalactites and curtains below its overgrown top.

WALL OF ICE
As the Hubbard Glacier meets the sea in Disenchantment Bay, ice detaches from its 90m- (300ft-) high terminal wall. These chunks of ice form icebergs, which then float away into the Gulf of Alaska.

GLACIERS

About one-tenth of the Earth's land surface is covered in ice masses known as glaciers, which hold some 67 per cent of the world's fresh water. Glaciers are among the most beautiful natural phenomena, and they also have a profound effect on terrain, carving out landscapes and transporting vast amounts of rock across remote areas of the planet. Glaciers come in many different forms and with many different features. For example, some carry pinnacles of ice, streams of water, or even lakes on their surfaces. Others have volcanoes underneath them or long floodwater channels running through them. In recent years, glaciers have become an important focus of study. The melting of a large fraction of the Earth's glaciers over the coming decades, which is now inevitable due to climate change, will lead to a substantial rise in sea-level, although there is still uncertainty about the extent of the rise and how quickly it will occur.

Glaciers

29–31	The ice ages
93	Glacial erosion
95	Glacial deposition
107	Ice
109	The global water cycle
	Tundra **338–39**
	Ice-shelves and icebergs **398**

A glacier is an accumulation of many layers of snow, transformed over time and by their own weight into a mass of ice that persists from year to year. This ice mass slowly deforms and begins to move, mostly travelling downhill. A small glacier may be no bigger than a football field, but the largest, called ice-sheets, can be continent-sized. Glaciers are an important reserve of fresh water, and can act like time capsules, holding information – in the form of trapped air, chemicals, and dust – that tells us much about changing climatic conditions over hundreds of thousands of years.

Glacier Structure

Most glaciers except for the very largest descend from a mountain area and flow down a valley. The top of the glacier usually lies in a bowl-shaped area called a cirque. The ice below the cirque is often broken up by deep cracks called crevasses, formed in response to stress as the ice grinds over uneven bedrock. Over time, rocks fall onto the glacier or become attached to the moving ice and are plucked off the bedrock. They are deposited at the melting terminus (snout) of the glacier, leaving a debris pile called a terminal moraine, or at the sides (forming lateral moraines). Where glaciers join, rocks carried on their edges merge to form a surface stripe called a medial moraine. Broadly similar processes occur in larger glaciers, although these are less constrained by topography (see p.254).

UPSTREAM FEATURES
The Bagley Ice-field in Alaska, 8km (5 miles) wide, is typical of the upper areas of a glacier, which are composed of clean ice fractured by crevasses.

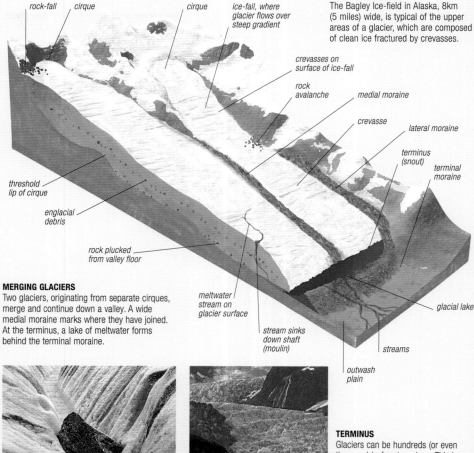

rock-fall   cirque

cirque

ice-fall, where glacier flows over steep gradient

crevasses on surface of ice-fall

rock avalanche

medial moraine

crevasse

lateral moraine

terminus (snout)

terminal moraine

threshold lip of cirque

englacial debris

rock plucked from valley floor

meltwater stream on glacier surface

glacial lake

stream sinks down shaft (moulin)

streams

outwash plain

MERGING GLACIERS
Two glaciers, originating from separate cirques, merge and continue down a valley. A wide medial moraine marks where they have joined. At the terminus, a lake of meltwater forms behind the terminal moraine.

MELTWATER STREAM
A glacier's surface may be broken up by streams of meltwater and studded with boulders.

TERMINUS
Glaciers can be hundreds (or even thousands) of metres deep. This is most apparent at a glacier's terminus, where an enormous cliff or mound of ice can be seen, as here in Harriman Fiord, Alaska.

MELTWATER CAVES
Ice caves, such as this one in Alaska's Muir Glacier, are formed by meltwater streams eroding vertical sinkholes, tunnels, and finally large chambers within the ice.

LAND

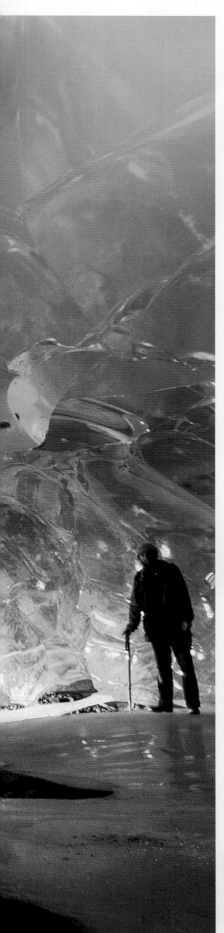

AIRBORNE SNOW

SNOW (85–90% AIR)

GRANULAR ICE (30–85% AIR)

FIRN (20–30% AIR)

BLUE ICE (LESS THAN 20% AIR, AS BUBBLES)

Glacier Ice

Newly fallen snow contains masses of air between individual snowflakes. As more settles onto a previously fall, the flakes beneath are compressed and deformed, by partial melting and refreezing, into smaller, rounder ice particles. These adhere to each other, initially forming small ice granules and then a material consisting of larger granules, known as firn. As more snow falls, further compression occurs, causing the firn to weld into solid, dense glacial ice.

BLUE ICE
Dense glacial ice owes its intense colour to its crystalline structure, which absorbs all except the shortest (bluest) wavelengths of visible light. Blue ice is stronger and contains less air than white ice.

STAGES OF FORMATION
As glacial ice is formed, there is a gradual decrease in the proportion of air to snow or ice. Depending on the location of the glacier, the process can take anything from 5 to 3,000 years.

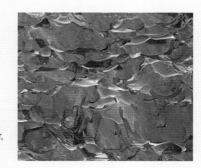

Movement

In the upper parts of a glacier, known as the accumulation area, there is a net gain of ice each year, as some of each winter's snowfall remains until the next winter. From here, ice flows downhill due to gravity. The mechanism of ice movement varies. In some glaciers (called warm-based), a thin layer of water forms under the glacier, allowing the ice to slide over the bedrock. In others (cold-based), the base of the glacier is frozen to the bedrock, so ice can move only by slow deformation. Eventually, the ice reaches the lower parts of the glacier, where it melts and evaporates or, in some glaciers, breaks off as icebergs. This lower zone of the glacier, where there is a net loss of ice over the year, is called the ablation area. The section between the accumulation and ablation areas, where ice gains and losses are in balance, is called the equilibrium zone. The difference between the glacier's overall ice gains and losses is called its mass balance.

CALVING ICE
Some glaciers end in lakes or the sea, where huge chunks of ice calve (break off) and float away as icebergs.

accumulation of ice from snowfall

accumulation area, above equilibrium zone

equilibrium zone

loss of ice by evaporation of surface meltwater

ablation area, below equilibrium zone

direction of movement of glacial ice

flow path of ice in glacier

snout

loss of ice in meltwater streams

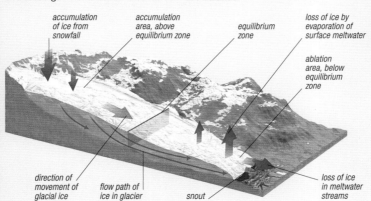

SHIFTING BALANCE
The position of the equilibrium zone along the length of the glacier varies from year to year depending on the climate. If ice accumulation lessens or ablation increases, the section of the glacier that is in balance will move up the glacier.

Advance and Retreat

Over a period of years, if more ice forms on a glacier than melts, the glacier must grow. It may widen, thicken, or advance (or any combination of the three). Conversely, if more ice is lost than forms, the glacier shrinks. The ice in its lower parts may thin and stagnate, or the snout may retreat – or both. Since 1900, most glaciers around the world have retreated, although within this time there have also been periods of advance, both by groups of glaciers and by individual glaciers (especially by surge-type glaciers with cyclical patterns of rapid advance and slow retreat). The vast majority of scientists believe the overall trend of retreat is due to global warming, which is, at least in part, due to human activity.

RETREATING GLACIER
The Rhône Glacier in Switzerland has retreated about 100m (330ft) per decade since 1880. These photographs of its terminus, taken from the same location, are from 2010 (above) and 2021 (left).

ADVANCING GLACIER
When a glacier advances, it simply brushes aside objects such as trees. Here, the Taku Glacier, which advanced more than 7km (4⅓ miles) between 1899 and 1989, pushes through a hundred-year-old forest near Juneau in Alaska.

Types of Glacier

Glaciers vary enormously in their size and form. The largest, called continental ice-sheets, and somewhat smaller glaciers called ice-caps, almost totally submerge the landscapes on which they lie. These glaciers are slightly dome-shaped. Driven by gravity, the ice in them flows outward from their centres towards their edges, where channels of ice flow, called ice-streams and outlet glaciers, are visible. Outlet glaciers are bounded by ice-free ground, while ice-streams are surrounded by slower-moving ice. The other main types of glaciers are found in areas where the ice flow is constrained by mountains. An ice-field is a huge expanse of ice at high altitude, partly hemmed in by mountains, that gives rise to numerous outlet glaciers. A cirque glacier is a small glacier mainly confined to a depression on the flank of a mountain. Where ice flows out of a cirque and extends far down a valley, the result is a valley glacier. This can be tens or even hundreds of kilometres long. When a valley or outlet glacier flows into a flat, lowland area, it may spread out into a wide expanse of ice called a piedmont lobe.

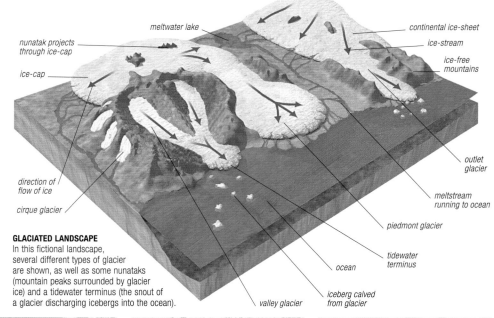

GLACIATED LANDSCAPE
In this fictional landscape, several different types of glacier are shown, as well as some nunataks (mountain peaks surrounded by glacier ice) and a tidewater terminus (the snout of a glacier discharging icebergs into the ocean).

ICE-SHEET MARGIN
Thin meltwater streams can be seen here on the surface of the Greenland Ice-sheet at one of the points where it meets the sea.

OUTLET GLACIERS
Several outlet glaciers from the Spitsbergen Ice-cap on Svalbard, Norway, flow through lines of nunataks towards the sea.

PIEDMONT GLACIER
This satellite image shows Alaska's Malaspina Glacier. This consists almost entirely of an enormous piedmont lobe, the world's largest.

VALLEY GLACIER
The Aletsch, a 22-km- (13½-mile-) long valley glacier, flows through the Swiss Alps. The lower part of the glacier is shown here.

CALIFORNIAN GLACIATED LANDSCAPE
McGee Creek in the Sierra Nevada, USA, is a U-shaped valley sculpted by a glacier during the last ice age. An array of cirques, horns, and arêtes can be seen among the mountains in the background. In the foreground, are distinct lateral moraines – ridges formed from rock debris deposited at the side of the glacier.

Glacial Erosion

Glaciers erode mountainous regions into characteristic shapes, some of which only become apparent when the glaciers eventually melt and disappear (as at McGee Creek, opposite below). Erosion occurs partly by the plucking of rocks from valleys and mountainsides as they freeze to glacial ice, and partly by the moving mass of ice and its rock load grinding or abrading the bedrock. At the head of a glacier, the ice typically carves out a cirque. Where two cirques back up against each other, ridges called arêtes are formed; three or more cirques can combine to produce a pyramidal peak called a horn. Downstream, a glacier moving through a valley will usually deepen and broaden the valley from a V-shape into a U-shaped profile. Another common feature is the hanging valley. This is seen where glacial erosion has occurred faster in the main valley than in a smaller tributary valley. When the glaciers melt, the tributary is left suspended high above the more deeply carved main valley. Landscapes like the one shown in the illustration on the right can be seen today in many mountainous regions of Europe and North America, such as parts of Norway, northern Britain, and the Rocky Mountains in Canada and the USA.

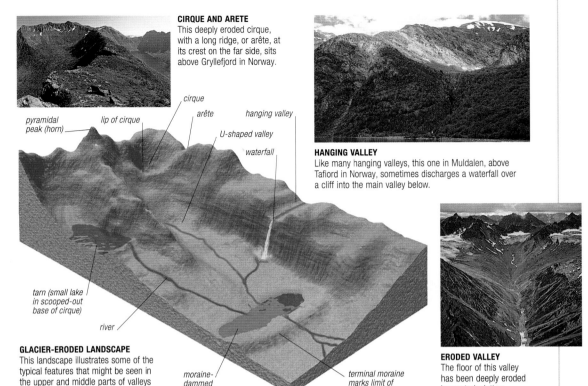

CIRQUE AND ARETE
This deeply eroded cirque, with a long ridge, or arête, at its crest on the far side, sits above Gryllefjord in Norway.

HANGING VALLEY
Like many hanging valleys, this one in Muldalen, above Tafiord in Norway, sometimes discharges a waterfall over a cliff into the main valley below.

GLACIER-ERODED LANDSCAPE
This landscape illustrates some of the typical features that might be seen in the upper and middle parts of valleys that were once filled with glaciers.

ERODED VALLEY
The floor of this valley has been deeply eroded by past glaciation.

Deposition

When glaciers melt, they deposit material on the landscape. Much of this matter is a mixture of particles, ranging in size from sand grains to boulders, called till. The deposits come in many forms. For example, a terminal moraine is a broad ridge of till marking the furthest point that a glacier advanced. An esker is a twisting ridge of sand and gravel that snakes its way across the landscape. It is formed from material deposited by a stream that once flowed under the glacier. A drumlin is a long, cigar-shaped mound of till that was smoothed in the direction of the glacier's flow. An erratic is a large, rounded boulder that a glacier carried far from its source in the mountains and eventually deposited at its terminus. When an ice block detaches from a melting glacier and is partially buried by deposits from meltwater streams, it melts and creates a depression in the layer of till called a kettle hole. If the depression fills with water, it becomes a kettle lake. When a meltwater channel becomes choked with coarse material, it causes the river to divide into a series of diverging and converging segments, known as braided streams.

KETTLE LAKE
This kettle lake lies in Dolma La Pass, one of the routes to Mount Kailas, the sacred mountain of Tibet, which is known locally as Ghang Rimpoche, or "jewel of snow".

ERRATIC
This erratic boulder in Yorkshire, UK, was transported by a glacier and then deposited on top of younger rock.

DEPOSITIONAL FEATURES
A melting glacier typically leaves several of the features seen here. Over large parts of lowland North America and Europe, deposits such as these can still be discerned in the landscape, providing clues to the region's glacial history.

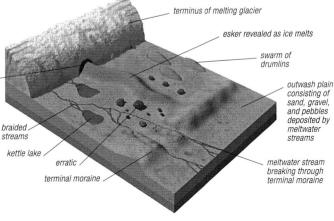

World Distribution

Glaciers can form only on areas of land where there is some snowfall and very low temperatures persist for much of the year. The only places that satisfy both these conditions are close to the poles or in high mountain areas. The world's largest glaciers are massive ice-sheets that cover Antarctica, while in high northern latitudes huge areas of glaciation exist over Greenland, parts of Iceland, Norway, and Alaska. Closer to the equator, the extensive high-altitude glaciers of the high Alps and Himalayas are diminishing, as are the smaller glaciers of the tropics, which are found in the northern Andes and on the highest mountains of Africa and Mexico.

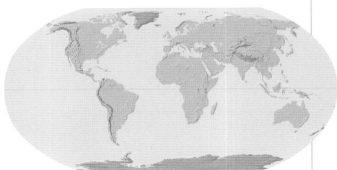

GLACIER PROFILES

The pages that follow contain profiles of a selection of the world's glaciers. Each profile begins with the following summary information:

TYPE Cirque glacier, valley glacier, piedmont glacier, outlet glacier, ice-cap, ice-field, or ice-sheet

TERMINUS Terrestrial snout, terrestrial lobe, tidewater (sea-calving) front, or lake-calving front (cirque, valley, piedmont, and outlet glaciers only)

AREA Surface area, excluding named tributaries (All glacier types)

LENGTH Distance from source to terminus (All glacier types except ice-caps and ice-sheets)

STATUS Advancing, retreating, stable, or uncertain (since 2000)

LAND

Trapridge Glacier

LOCATION On the eastern flank of Mt. Wood in the St. Elias mountains, Yukon, Canada

TYPE	Cirque glacier
TERMINUS	Terrestrial snout
AREA	4 square km (1½ square miles)
LENGTH	4km (2½miles)
STATUS	Stable

AFTER THE SURGE
Following its surge cycle, the Trapridge Glacier shows some surface mounds at its terminus because of the unevenness of the surface bed.

Trapridge is a surge-type glacier, in which periodically, over many years, a relatively rapid transfer of ice occurs from the upper to the lower parts, as the ice thickens and then surges forward. This is followed by a quiet phase, as ice builds up again in the upper glacier and thins lower down. Trapridge experienced a significant surge in the 1940s and another slower one between the late 1960s and 2005, when rates of flow in its lower parts peaked at 38m (123ft) per year. The glacier's flow was impeded by stagnant ice left from a previous surge, causing a conspicuous bulge that propagated downglacier. By 2005, the rate of ice flow had reduced and the bulge diminished.

Kennicott Glacier

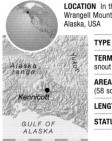

LOCATION In the southern Wrangell Mountains, southcentral Alaska, USA

TYPE	Valley glacier
TERMINUS	Terrestrial snout
AREA	150 square km (58 square miles)
LENGTH	42km (26 miles)
STATUS	Retreating

This glacier pours down from Alaska's Wrangell Mountains, passing the abandoned mining town of Kennicott near its terminus. Early in the 19th century, a wall of ice loomed above the town. But although its terminus has retreated less than 1km (^{3}/$_5$ miles) since 1860, the glacier has thinned considerably and much of its lower third is now covered in a thick layer of rock debris. In spring, side valleys of the main Kennicott Valley fill with lakes as the glacier dams the meltwater streams. As temperatures rise, the water escapes through channels beneath the ice and floods into the River Kennicott.

MEDIAL MORAINES
This view of the upper area of the Kennicott Glacier shows its well-developed medial moraines.

Black Rapids Glacier

LOCATION Occupies the Denali Fault valley in the east-central Alaska Range, southcentral Aaska, USA

TYPE	Valley glacier
TERMINUS	Terrestrial snout
AREA	135 square km (52 square miles)
LENGTH	39km (24 miles)
STATUS	Retreating

Sometimes called the Galloping Glacier, the Black Rapids is a surge-type glacier (see Trapridge, above). In 1936–37, the terminus advanced 6.5km (4 miles) in just three months. The photograph below reveals a typical feature of surge-type glaciers: dark, looped medial moraines on the surface.

LOOPED (OR S-SHAPED) MORAINES

Hubbard Glacier

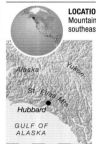

LOCATION In Canada's St. Elias Mountains, descending through southeastern Alaska, USA

TYPE	Valley glacier
TERMINUS	Tidewater (sea-calving) front
AREA	2,400 square km (925 square miles)
LENGTH	122km (76 miles)
STATUS	Stable

The Hubbard is the largest tidewater glacier in North America, discharging huge amounts of ice into Alaska's Disenchantment Bay. Over the course of the 20th century, the glacier advanced in fits and starts, and in 1986 it blocked off an arm of the bay called Russell Fiord. Over several months, there was a dramatic rise of water levels in the fiord, raising fears that it might become permanently blocked and overflow.

THE THREAT TO RUSSELL FIORD
This aerial view shows, at bottom right, the narrow mouth of the Russell Fiord, which flows into Disenchantment Bay, on the left.

However, the waters eventually broke through the ice dam in a torrential flood. In 2002, a similar sequence of events occurred, but since then the calving terminus of the glacier has remained more or less stationary.

THE VIEW FROM OSIER ISLAND
The terminus of the Hubbard Glacier in the Wrangell–St. Elias National Park is over 10km (6 miles) wide and 90m (300ft) high. Its base is punctuated with deep ice caves.

Bering Glacier

LOCATION In the Chugach Mountains, descending towards the Gulf of Alaska, Alaska, USA

TYPE Piedmont glacier

TERMINUS Lake-calving piedmont lobe

AREA 3,600 square km (1,400 square miles)

LENGTH 177km (110 miles)

STATUS Retreating

The Bering Glacier is both the longest glacier in North America and the largest by surface area. In places, its ice is 800m (2,600ft) thick. The upper region, called the Bagley Ice-field, is an ice basin about 90km (56 miles) long, almost completely filling several interconnecting valleys in

SURFACE TEXTURE
Parts of the lower and middle regions of the glacier have a highly fractured, patterned surface.

the eastern Chugach Mountains. This is where most of the Bering Glacier's ice forms. The ice-field gently slopes down towards the west from an altitude of about 2,100m (6,900ft). From its western end, a wide trunk of ice (which is an outlet glacier) turns southwest, feeding the lower area of the Bering, which consists of a piedmont lobe 42km (26 miles) wide. This lobe is fringed by the Vitus Lake, into which the glacier calves icebergs. From the melting edge of the glacier and the lake, streams flow a short distance into the Gulf of Alaska. In recent years the glacier has been releasing about 30 cubic km (7¼ cubic miles) of water annually into the North Pacific. The Bering Glacier has surged forward six times in the last 160 years – in 1995, it advanced about 750m (½ mile) in only two weeks. But over the long

ICE COLLAPSE
This view of the stagnating lower part of the Bering Glacier shows the vegetation growing on its surface and a crater-like area of ice collapse.

term, the glacier is in retreat – during the last century, it is estimated to have lost about 3 per cent of its surface area. Parts of the lower glacier are covered in rock debris due to melting, and in some areas the combination of a fine layer of rock particles and melting ice underneath has even allowed some vegetation to grow on its surface. The retreat of the glacier has revealed the remains of an ancient spruce forest that it overran about 1,500 years ago, as well as shells of marine organisms, indicating that part of the land that the glacier currently sits on was once under the ocean.

AERIAL MAPPING

Over the last 20 years, researchers in Alaska have used aircraft-borne laser devices to create precise contour maps of glacier surfaces such as the Bering Glacier's. As the aircraft flies over the ice, a laser beam is used to measure the distance from aircraft to glacier. Simultaneously, the aircraft's position is determined using GPS (the global positioning system). When the heights measured were compared with those given on old US Geological Survey charts, it was revealed that most of Alaska's glaciers have thinned in the past 50 years.

SNAKING TO THE SEA
Just a small part of the Bering Glacier's extensive terminus is seen here, with Vitus Lake in the foreground. The sinuous medial moraines are strikingly evident.

FRACTURED ICEFLOW
New Zealand's Franz Josef Glacier plunges hundreds of metres over an icefall, stretching its surface and breaking it into a mass of crevasses and pinnacles known as seracs. Like many of the world's glaciers, it is currently in retreat.

LAND

Worthington Glacier

LOCATION In the central part of the Chugach Mountains, southcentral Alaska, USA

TYPE Valley glacier

TERMINUS Terrestrial snout

AREA 7 square km (2¹/₂ square miles)

LENGTH 5.5km (3¹/₂ miles)

STATUS Retreating

A small valley glacier, the Worthington descends from 2,000m (6,500ft) up, and terminates close to a major highway. The changing shape of boreholes drilled through the glacier to the bedrock has revealed that at least 20 per cent of the glacier's movement is due to its internal deformation – the rest is due to slippage over the bedrock.

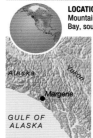

DOUBLE-ARMED TERMINUS

Columbia Glacier

LOCATION Descending from the Chugach Mountains to Prince William Sound, southcentral Alaska, USA

TYPE Valley glacier

TERMINUS Tidewater (sea-calving) front

AREA 885 square km (340 square miles)

LENGTH 45km (28 miles)

STATUS Retreating

One of the fastest-moving glaciers in North America, the Columbia is also one of the most rapidly retreating. Between the 1970s and 2001, the velocity of the ice near its terminus reached an astonishing 30m (100ft) per day, though it has since slowed. The glacier's fast forward movement has been cancelled out by the rapid calving of icebergs at its terminus. As a result, since 1982, the terminus has retreated about 24km (15 miles) and thinned considerably. Since 2011, it has split in two. The glacier's retreat cannot solely be attributed to global warming, as other

SURFACE DEBRIS
Parts of the lower glacier are covered in a thin layer of rock debris that has concentrated at the surface as the ice has melted.

SPLIT IN TWO
The Columbia Glacier has retreated so far that it has two branches; the West Branch (to the left in this picture) and the Main Branch.

nearby glaciers are not retreating as rapidly. Instead, they point to special factors related to the bedrock channel beneath the front of the glacier. This is very deep and slants backwards, and as a result, the front of the glacier is almost afloat on meltwater, explaining why it moves forward so quickly. The high ice velocity in turn increases the rate at which icebergs are calved, and causes ice to be lost faster than it can be replenished by the glacier's upper parts.

Margerie Glacier

LOCATION In the St. Elias Mountains, descending to Glacier Bay, southeastern Alaska, USA

TYPE Valley glacier

TERMINUS Tidewater (sea-calving) front

AREA 112 square km (43 square miles)

LENGTH 34km (21 miles)

STATUS Stable

The Margerie Glacier is one of several tidewater glaciers that originate from the eastern end of the St. Elias Mountains. They then descend to discharge ice into a large inlet of the Gulf of Alaska called Glacier Bay, which became a National Park in 1980 and a World Heritage Site in 1992. The terminus of the glacier is in Tarr Inlet at the northern end of the Bay. At one time, about 100 years ago, the Margerie was a mere tributary of a much larger glacier, the Grand Pacific Glacier, which filled the whole of Tarr Inlet. Earlier still, when the English naval officer Captain George Vancouver discovered Glacier Bay in 1794, the entire bay was filled with ice. Since then the Grand Pacific has shrunk and Margerie has become a calving glacier in its own right. Today, it is one of the most spectacular glaciers in the Bay, towering about 80m (265ft) above the sea surface. Unlike most other Alaskan glaciers, it has not markedly retreated in recent years.

WALL OF BLUE ICE
The Margerie Glacier regularly discharges large chunks of blue ice from its 1.5-km- (1-mile-) wide terminus into Glacier Bay, with thunderous cracking noises.

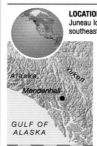

Mendenhall Glacier

LOCATION Descending from the Juneau Ice-field, Coast Mountains, southeastern Alaska, USA

TYPE Outlet glacier

TERMINUS Lake-calving front

AREA 75 square km (29 square miles)

LENGTH 22km (13¹/₂ miles)

STATUS Retreating

The Mendenhall is an outlet glacier from the Juneau Ice-field, a vast area of ice that straddles the border between Alaska and British Columbia. The glacier terminates in a lake 5km (3 miles) long, not far from the city of Juneau. This lake has existed only since 1929, gradually enlarging from meltwater as the Mendenhall Glacier has lost mass and retreated at rates of up to 200m (650ft) a year. The glacier has now almost pulled out of the lake, although once it can no longer calve icebergs, its rate of retreat should slow.

SURFACE MELTWATER
Streams, such as the one shown here, can develop where surface ice melts faster than it can be absorbed by the glacier.

NORTH AMERICA *northwest*

Malaspina Glacier

LOCATION On the coastal side of the St. Elias Mountains, southeastern Alaska, USA

TYPE Piedmont glacier

TERMINUS Terrestrial lobes

AREA 3,200 square km (1,240 square miles)

LENGTH 45km (28 miles)

STATUS Stable

Named after an Italian, Alessandro Malaspina, who explored this part of Alaska in 1791, the Malaspina Glacier is famous for containing the world's largest piedmont lobe. It is also the second largest glacier in North America. It consists of three lobes, each fed by different outlet glaciers that flow out of a large ice-field lying high up in the St. Elias Mountains

behind the Malaspina. The largest of these feeder glaciers is the Seward Glacier (seen in the image below as the blue ribbon of ice entering from the top). This feeder is itself 4km (2½ miles) wide at its lower end – it gives rise to the largest, central lobe of the Malaspina. To the west, the Agassiz Glacier (the blue stream at the top left of the image) feeds a smaller western lobe. To the east, the Hayden and Marvine glaciers give rise to a small eastern lobe.

ICE WORM
The tiny Ice Worm is one of very few species of animal known to be able to survive in glacier ice. This one is about 12mm (½ in) long.

Seismic surveys have shown that the Malaspina is up to 600m (2,000ft) thick, and that, roughly in the middle of the glacier, ice at its base is eroding the bedrock at a depth of about 300m (1,000ft) below sea-level. From there, the ice is actually forced upwards towards the lobe-shaped terminus of the glacier, where it melts and deposits the rock debris as till. All around the 100-km- (60-mile-) long terminus of the glacier is a region of stagnating ice, thickly overlain with rock debris. Surrounding this is a wide area of till studded with meltwater lakes and kettle holes. At its southern end, the glacier almost reaches sea-level at Sitkagi Bluffs on the Gulf of Alaska. There, a series of terminal moraines protects it from contact with the open ocean. If the sea-level were to rise sufficiently to connect the glacier to the ocean, the glacier would start calving icebergs and would probably retreat significantly. A remarkable feature of the Malaspina, and some other coastal glaciers in

ALTERNATING CURRENTS
Much of the Malaspina Glacier's surface is covered in a complex pattern of contorted medial moraines. These are an indication of surge behaviour – that is, patterns of alternating fast and slow ice flow.

Alaska, is the occurrence of Ice Worms. These small segmented worms are found in colonies of millions of individuals near the ice surface during the day; at night they burrow deep into the ice. The worms are thought to survive by feeding on green algae that grow near the surface of the glacier.

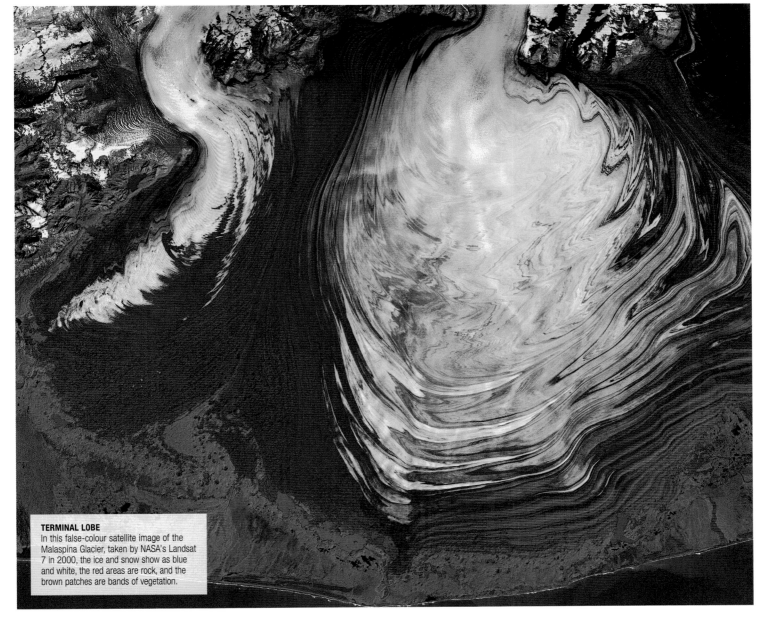

TERMINAL LOBE
In this false-colour satellite image of the Malaspina Glacier, taken by NASA's Landsat 7 in 2000, the ice and snow show as blue and white, the red areas are rock, and the brown patches are bands of vegetation.

ILULISSAT, NORTH GREENLAND
Icebergs from the Jakobshavn Glacier
float in the Illulisat Icefjord, Greenland.
The icebergs may stay here for years
before floating to the Atlantic.

ARCTIC

Greenland Ice-sheet

LOCATION Covering 80 per cent of the island of Greenland in the Arctic Circle

TYPE Ice-sheet

AREA 1.71 million square km (660,000 square miles)

STATUS Retreating

Greenland Ice-sheet

Jakobshavn Glacier

Despite its name, Greenland is mostly covered in a thick layer of ice. A single glacier, the Greenland Ice-sheet, makes up most of this ice, although smaller ice-caps occur at the edges. The main ice-sheet is the northern hemisphere's largest glacier, with an average thickness of about 1,500m (5,000ft) and a volume of 2.85 million cubic km (0.68 million cubic miles). Its surface is slightly domed, reaching about 3,290m (10,800ft) above sea-level, slightly to the southeast of the island's centre. A second, lower dome lies in the southeast corner. From these domes, ice moves slowly towards the edges of the ice-sheet. Around most of its periphery, the ice is constrained by coastal mountains, so there are few places where it meets the sea along a broad front. As a result, Greenland has no ice-shelves, but in many places, large outlet glaciers flow through valleys between the mountains and discharge enormous quantities of icebergs into the sea. One of the main outlets, Jakobshavn Glacier in west Greenland, is the world's fastest-flowing glacier. At its terminus, the ice flows at speeds of up to 2m (6.5ft) per hour, on average producing over 90 million tonnes of icebergs per day. The ice-sheet's high albedo (see p.447) keeps its summer

ICEBERG MAKER
Eqip Sermia Glacier is one of the biggest iceberg-calving glaciers in west Greenland. Its terminus is about 4km (2.4 miles) wide and in places rises 200m (660ft) above the sea. When chunks of ice fall off it, they can cause waves 10m (33ft) high.

BEACHED ICEBERGS
Most North Atlantic icebergs originate from the Greenland Ice-sheet. The icebergs enrich the sea around Greenland with fresh water and oxygen, attracting an abundance of fish.

WEATHER STATIONS
Since 1990, a network of AWS (automatic weather stations) have been established on the Greenland ice-sheet as part of the Greenland Climate Network project. The first station was "Swiss Camp", which was used as a field science and education site. By 2023, the network consisted of 31 AWS at 30 different sites, though these are now being replaced by new AWS operated by the Geological Survey of Denmark and Greenland. Each AWS records weather data, such as air temperature and wind speed, and transmits the data via satellite links to a research centre.

temperature far lower than it would be otherwise. Nevertheless, the ice-sheet has lost mass and volume every year since 1996 due to melting caused by global warming. At present, it is losing volume at a rate of about 150 cubic km (36 cubic miles) each year. While only a small fraction of it is expected to melt in the 21st century – leading to a sea-level rise by the year 2100 of around 28cm (11in) according to one recent study – it is now predicted that most or even all of the ice-sheet is destined to eventually melt under the present or likely near-future climatic conditions. If the ice-sheet was to melt completely, it is estimated that the world's oceans would rise by about 7m (23ft).

LAND

LAND

Athabasca Glacier

LOCATION Extending from the Columbia Ice-field, in the Rocky Mountains of Alberta, Canada

TYPE	Outlet glacier
TERMINUS	Lake-calving front
AREA	5 square km (2 square miles)
LENGTH	5km (3 miles)
STATUS	Retreating

EVIDENCE OF RETREAT
Piles of moraine surround the terminal lake of this glacier. High up on the horizon is the Columbia Ice-field.

The Athabasca is one of more than 20 outlet glaciers from the Columbia Ice-field, the largest body of ice in the Rocky Mountains, with a surface area of 350 square km (135 square miles). Remarkably, the water run-off from this ice-field flows into three different oceans: the Pacific, Atlantic, and Arctic. The top part of the Athabasca Glacier, which descends from the ice-field at an altitude of about 2,700m (8,860ft), consists of three huge ice-falls, which appear like a series of giant steps. Further down, the glacier descends gently to a terminal lake (in the image below, the lake is hidden by a terminal moraine). The depth of the ice over the whole glacier varies from about 90 to 300m (300 to 1,000ft). The Athabasca has been shrinking since about 1850 and is estimated to have lost more than 50 per cent of its mass since 1900. Currently, its rate of retreat is about 3m (10ft) a year, and it is losing depth at a rate of about 5m (16ft) a year.

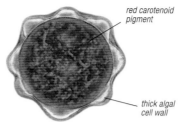

red carotenoid pigment

thick algal cell wall

WATERMELON SNOW
The patches of "pink snow" found on some glaciers, including the Athabasca, are actually colonies of bright red algae.

Kaskawulsh Glacier

LOCATION In the St. Elias Mountains, 35km (22 miles) south of Kluane Lake, Yukon, Canada

TYPE	Valley glacier
TERMINUS	Terrestrial snout
AREA	1,000 square km (390 square miles)
LENGTH	70km (43 miles)
STATUS	Retreating

Arguably the most spectacular glacier in Canada, the Kaskawulsh is notable for its sheer size and its impressive medial moraines (see below). These formed as a result of the merging of tributary glaciers into the main glacier trunk. The snout gives rise to the Slims River, which maintains the level of the Yukon's largest lake, Kluane Lake.

AERIAL VIEW OF KASKAWULSH GLACIER

Palisade Glacier

LOCATION In the southern part of the Sierra Nevada range, central California, USA

TYPE	Cirque glacier
TERMINUS	Lake-calving front
AREA	0.5 square km (⅕ square mile)
LENGTH	900m (3,000ft)
STATUS	Retreating

The Palisade is one of about 100 small glaciers in the Sierra Nevada that have survived since the last ice age. At its top is a horizontal crack, called a bergschrund, which formed when the glacier fractured and its main mass slipped down, leaving a wall of ice frozen to the rock.

THE HEAD OF THE PALISADE GLACIER

Pastoruri Glacier

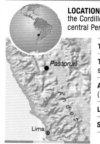

LOCATION In the southern part of the Cordillera Blanca, Andes, north-central Peru

TYPE	Cirque glacier
TERMINUS	Terrestrial snout
AREA	1.2 square km (½ square mile)
LENGTH	2km (1.2 miles)
STATUS	Retreating

The tropical regions of South America still harbour a few areas of glaciation, such as the Pastoruri Glacier in Peru. Occupying part of the flank of a 5,240-m- (17,200-ft-) high Andean peak, it has some steep, cliff-like edges and is usually covered in soft snow, with a few heavily crevassed areas. It has shrunk by more than half in the last 50 years.

PASTORURI GLACIER, LOOKING DOWNHILL

Southern Patagonian Ice-field

LOCATION In the Patagonian Andes, along the Chile–Argentina border, southwest of Patagonia

TYPE	Ice-field
AREA	10,000 square km (3,900 square miles)
LENGTH	330km (210 miles)
STATUS	Retreating

The Southern Patagonian Ice-field is the largest glaciated region in the southern hemisphere outside of Antarctica. It is the main remnant of a much larger area of glaciation (about 40 times bigger in area) that extended over much of the southern Andes about 10,000 years ago. Like all ice-fields, it is a single, extensive, mostly level-surfaced mass of ice that covers most of a mountain region, apart from the highest peaks and ridges. Monte Fitzroy at 3,405m (11,171ft) and volcano Lautaro at 3,623m (11,886ft) are two of the main peaks that the ice-field surrounds. At its widest point, the ice-field is about 50km (31 miles) across, and its average width is some 30km (19 miles). On its western side, nearly 50 significant outlet glaciers flow out of it, reaching sea-level in the rugged fiords on the coast of Chile. Many of these fiords are choked with ice. On the eastern side, which is mainly in Argentina, large outlet glaciers calve icebergs into extensive lakes at about 200–270m (650–900ft) above sea-level. Three of these lakes (Lago Martin, Lago Viedma, and Lago Argentino) can be seen in the image below. The lakes occupy valleys that were filled by large glaciers during the last ice age. Comparison of various aerial and satellite photographs of the ice-field taken since 1945 has shown that, with a few exceptions, most of its outlet glaciers have retreated over the past 50 years or so.

ICE-CAP REMNANT
This satellite photograph showing the vast extent of the Southern Patagonian Ice-field was taken looking towards the northeast.

LAKE-CALVING FRONT
Ice-block falls from this 60-m-
(200-ft-) high wall can produce
waves big enough to drown
people on the nearby shore.

Perito Moreno Glacier

LOCATION Extending from the
Southern Patagonian Ice-field,
southern Argentina

TYPE	Outlet glacier
TERMINUS	Lake-calving front
AREA	200 square km (80 square miles)
LENGTH	20km (12 miles)
STATUS	Stable

Perito Moreno is an impressive outlet
glacier from the Southern Patagonian
Ice-field (see opposite), descending
to the east of the ice-field, where it calves
icebergs into the largest lake in Argentina,
Lago Argentino. Unlike most outlet
glaciers in the region, its terminus has
scarcely changed its position in the last

GLACIER TERMINUS
When the Perito Moreno Glacier advances,
it forms a dam between Lago Argentino
(on the right in this photograph) and Brazo
Rico (on the left).

100 years. During that period, the glacier
has made 25 temporary advances spaced
anywhere from one to 16 years apart,
each time blocking off an arm of the
lake, called Brazo Rico. Whenever this
happens, a separate lake has been
created, with a rising water level that
floods a large area of grassland. After
six to 12 months, but always in summer,
the lake bursts through the ice dam,
discharging about a billion tonnes of water
in 24 hours and flooding nearby shores.

Vatnajökull Ice-cap

LOCATION In the southeast of
Iceland, covering about 8 per cent
of the island

Vatnajökull Ice-cap

TYPE	Ice-cap
AREA	7,900 square km (3,050 square miles)
STATUS	Retreating

BREIDAMERKURJÖKULL
This is the largest outlet glacier of the
Vatnajökull Ice-cap, seen here calving ice
into a small lake.

Vatnajökull, a classic ice-cap, is the second
biggest glacier by volume in Europe and
one of several ice-caps on Iceland. Roughly
ellipse-shaped, the glacier completely
covers the mountainous terrain it sits on
and is slightly domed. Its average thickness
is about 380m (1,250ft). Numerous outlet
glaciers, including surging glaciers, drain
into the sea from the ice-cap. Iceland lies in
a region of high tectonic activity, and there

are several active volcanoes and volcanic
fissures under the western parts of
Vatnajökull. In a few areas, heat from these
melts the base of the ice-cap, creating
subglacial lakes. In October 1996, a
volcanic fissure erupted a plume of ash and
steam some 3,000m (10,000ft) into the air.
Meltwater from the ice-cap filled the
subglacial caldera of the Grimsvotn volcano
(see p.163). When the caldera overflowed,
some 3,000 billion litres (65 billion gallons)
of water flooded out from beneath the
glacier (see panel below).

SUBGLACIAL ERUPTION
In 1998, an eruption of Grímsvötn blew
steam and ash to 10,000m (33,000ft)
above the ice-cap. An even bigger
eruption occurred in 2011.

FLOODING HAZARD

Glaciers can cause catastrophic
floods, and because the problem is
so common in Iceland, glaciologists
have adopted an Icelandic word,
jökulhlaup ("glacier flood"), to refer
to the phenomenon. Icelandic
jökulhlaups result from subglacial
volcanic activity, but in other parts

of the world, a more common
cause is the bursting of lakes
(which have either overfilled or
were previously ice-dammed), as
glaciers melt and disintegrate. The
Vatnajökull flood of 1996 (see above)
destroyed many roads and bridges
below the glacier.

LAND

Kongsvegen Glacier

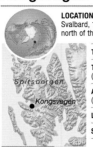

LOCATION In western Spitsbergen, Svalbard, 1,250km (775 miles) north of the Norwegian mainland

TYPE Outlet glacier

TERMINUS Tidewater (sea-calving) front

AREA 80 square km (31 square miles)

LENGTH 20km (12 miles)

STATUS Retreating

Svalbard is an archipelago under Norwegian jurisdiction in the Arctic Ocean. Its largest island, Spitsbergen, is about 80 per cent covered in glaciers, of which one of the best studied is the Kongsvegen. This is classified as a polythermal glacier, which means that there are large temperature variations within it – a cold surface layer of ice, 50–160m (160–520ft) thick, overlies

MERGED GLACIER TERMINUS
At its terminus, the Kongsvegen merges with another glacier, the Kronebreen. The merged glaciers calve icebergs into Kongsfjorden along a front 4km (2½ miles) wide.

a much warmer layer of ice at its base. Ice flow in polythermal glaciers tends to increase during the summer over the lower part of the glacier. This is thought to be caused by a build-up of meltwater beneath the glacier, lubricating its flow. In recent years, the terminus of Kongsvegen has been retreating. However, because it is a surge-type glacier (see p.253), this does not necessarily mean that the amount of ice in the glacier is falling. Indeed, detailed mass-balance studies conducted since 1987 have indicated that the glacier has actually slightly gained in ice mass overall.

DIRTY ICE CLIFF
This partially grounded cliff of ice, visible at one side of the glacier's terminus, contains medial moraines and englacial rock debris.

Jostedalsbreen Ice-field

LOCATION In southwestern Norway, west of the Jotunheimen Mountains and north of Sognafjord

TYPE Ice-field

AREA 400 square km (150 square miles)

LENGTH 60km (37 miles)

STATUS Retreating

Jostedalsbreen is the largest glacier on the European mainland, a remnant of a vast ice-sheet that covered the whole of Norway until about 10,000 years ago. It survives in southwest Norway primarily because of high regional snowfall rather than particularly cold temperatures, and this is reflected by a high rate of melting at the snouts of its 50-odd outlet glaciers.

West Svartisen Ice-cap

LOCATION In the coastal area of central Norway, at the latitude of the Arctic Circle

TYPE Ice-cap

AREA 200 square km (77 square miles)

STATUS Stable

The West Svartisen Ice-cap occupies a plateau near the Norwegian coast, with most of its surface lying above 1,100m (3,600ft). Like Jostedalsbreen, it owes its existence more to high snowfall than to low temperature. The ice-cap's surface is featureless and slightly domed, but at its edges are a number of outlet glaciers. One of these, Engabreen, is notable on several counts. First, it is the

ENGABREEN GLACIER
This arm of the West Svartisen Ice-cap descends to a lake that is just a few metres above sea-level.

lowest-lying glacial arm on the European mainland. Second, its mass balance has been studied since 1970, and until 1999 it showed a considerable gain of ice, and its terminus advanced. Between 1999 and 2015, the glacier retreated 350m (1,150ft), but the rate of retreat has since slowed. The third notable feature is the laboratory beneath its surface (see panel, right).

GLACIER LABORATORY

Tunnels have been drilled in the bedrock under the Engabreen Glacier as part of a hydroelectric power scheme. At the side of these tunnels, a laboratory has been excavated under the glacier, where the ice is about 200m (660ft) thick. This has allowed researchers to install instruments that continuously record the stress on the bedrock and the speed, temperature, and pressures within the ice as it slips across the rock. The results are helping scientists develop mathematical models for predicting glacier movement.

EUROPE west

Glacier d'Argentière

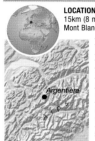

LOCATION In the Savoie Alps, 15km (8 miles) northeast of Mont Blanc, France

TYPE Valley glacier

TERMINUS Terrestrial snout

AREA 8 square km (3 square miles)

LENGTH 9km (6 miles)

STATUS Retreating

The Glacier d'Argentière is one of numerous glaciers that descend from the flanks of the Mont Blanc Massif, a spectacular mountain chain that lies on the borders of Switzerland, France, and Italy. Surrounded by some of the highest mountains in the Alps, the glacier descends from a wide ice accumulation area at an altitude of around 3,300m (11,000ft). Attempts have been made to tap the subglacial drainage of meltwater to provide hydroelectric power. To do this, tunnels were excavated under the glacier in the 1960s and a water intake put in place. The project was not entirely successful, due to changes in the course of the meltwater drainage, and the tunnels have now been closed. In 2005, a huge section near the snout of the glacier, containing some

NEW TERMINUS
Taken in 2009, this image shows the terminus of the main section of the Glacier d'Argentière.

ALPINE PURSUITS

The glaciers around the Mont Blanc Massif are heavily used by alpine sports enthusiasts. Cross-country skiing is possible on the upper parts of some glaciers – including the Glacier d'Argentière, which is so smooth that light aircraft can land on it – although care is needed to avoid the crevasses. Other popular pursuits include snow-boarding, ski-mountaineering, and ice-climbing. To facilitate this, night-time refuges have been set up in numerous, otherwise highly inhospitable, locations around the region.

400,000 cubic metres (14 million cubic ft) of ice, broke away from the rest of the glacier and disintegrated. Since then, the glacier has been in two parts. Of these, the main part terminates at a height of about 1,900m (6,200ft) at the top of a rock cliff, while lower down, a tongue of ice, covered in rock debris, continues down to a height of about 1,600m (5,200ft), which is a short but steep walk above the resort of Argentière. This lower part, which is shrinking, is occasionally replenished when seracs (blocks or pinnacles of ice) fall off the upper part.

EUROPE west

Mer de Glace

LOCATION In the Savoie Alps, 10km (6 miles) northeast of Mont Blanc, France

TYPE Valley glacier

TERMINUS Terrestrial snout

AREA 25 square km (10 square miles)

LENGTH 11km (7 miles)

STATUS Retreating

The Mer de Glace, or Sea of Ice, is so-called because alternating bands of light and dark ice at its surface have a wavelike appearance. It has retreated over 3km (2 miles) since the mid-19th century but is still the longest and largest glacier in France.

MER DE GLACE, HAUTE-SAVOIE, FRANCE

EUROPE central

Aletsch Glacier

LOCATION In the Bernese Alps, 25km (16 miles) south of the Brienzer See, southwest Switzerland

TYPE Valley glacier

TERMINUS Terrestrial snout

AREA 80 square km (31 square miles)

LENGTH 22km (13 1/2 miles)

STATUS Retreating

EUROPE central

Tschierva Glacier

LOCATION In the Rhaetian Alps, close to the border with Italy, southeastern Switzerland

TYPE Valley glacier

TERMINUS Terrestrial snout

AREA 5 square km (2 square miles)

LENGTH 3km (2 miles)

STATUS Retreating

The Tschierva is a small glacier that descends steeply from a 4,049m (13,284ft) peak, Piz Bernina, on the Swiss–Italian border. Formerly more than twice as long, the glacier is particularly notable for its large and clearly formed lateral moraines, which extend for about 2km (1 mile) downslope from the terrestrial snout of the glacier. About 40m (130ft) high, they indicate the extent to which the glacier advanced during the Little Ice Age in the 18th and 19th centuries (see p.455). Once, the area between the two moraines was completely filled with moving ice, but now it is just a hollow channel, about 700m (2,300ft) wide, with a floor made up of rock fragments and dust on top of a solid rock base.

The Aletsch is the longest and largest valley glacier in Europe. Its upper part consists of four distinct arms. These converge in an area to the south of the famous Jungfrau peak called the Konkordiaplatz ice plateau. From there, the glacier winds southwards then southwestwards towards the Rhône Valley. The Aletsch is up to 900m (2,950ft) deep, and the ice moves at speeds varying from about 200m (660ft) a year in the Konkordiaplatz to 10m (33ft) a year near the snout. The pattern of moraines below the snout indicates that the glacier has retreated about 6km (3 1/2 miles) since 1860.

LOWER ALETSCH GLACIER
Two dark medial moraines run nearly the entire length of the glacier below the Konkordiaplatz.

LAND

Rhône Glacier

LOCATION At the head of the Rhône Valley, in the eastern part of the Bernese Alps, Switzerland

TYPE Valley glacier

TERMINUS Terrestrial snout, with small lake

AREA 15 square km (6 square miles)

LENGTH 8km (5 miles)

STATUS Retreating

Recognized as the source of the River Rhône, the Rhône Glacier was once the largest in the Alps. Evidence from moraines shows that, 18,000 years ago, it reached as far as Lyons in France, more than 280km (175 miles) from its modern-day terminus. It has retreated about 1,150m (3,770ft) since 1900. About 1.5km (1 miles) wide on average, with a surface that is smooth in parts and deeply crevassed in others, the glacier was the subject of one of the earliest studies into glacial ice movement. In a classic investigation, a straight line of stakes driven into the glacier was observed between 1874 and

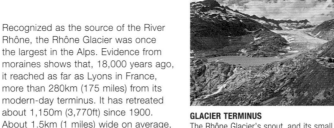

GLACIER TERMINUS
The Rhône Glacier's snout, and its small terminal lake, can be seen today only by climbing to a height of 2,250m (7,400ft).

1882. After a few years, the stakes had formed into a bow, proving that the ice in a glacier moves and that the movement is fastest in the middle.

Allalin Glacier

LOCATION In the Pennine Alps, southwestern Switzerland, close to the border with Italy

TYPE Cirque glacier

TERMINUS Terrestrial snout

AREA 8 square km (3 square miles)

LENGTH 6km (4 miles)

STATUS Retreating

Passing across the southeastern flanks of two alpine peaks – the Allalinhorn and Rimpfischhorn – the Allalin Glacier consists of a 4-km- (2$\frac{1}{2}$-mile-) long, gently sloping accumulation area and a steeply descending terminal tongue. Historically, the glacier has been a hazard to people living nearby. During the Little Ice Age (see p.455), it repeatedly advanced into the nearby Saas Valley, blocking the flow of the River Saaser Vispa and creating a lake called the Gletscherrandsee. Whenever the glacier retreated again, the lake would suddenly drain in a flood that devastated the valley. Since 1900, the glacier has retreated over 1.5km (1 mile). In 1965, a large piece of it broke off and cascaded into a dam construction site below, killing 88 people.

SURFACE TEXTURE
Deep crevasses and meltwater streams have carved a mosaic of rounded blocks on the lower part of the Rhône Glacier.

Kilimanjaro Ice-cap

LOCATION At the summit of Mount Kilimanjaro, northeastern Tanzania, close to the border with Kenya

TYPE Ice-cap

AREA 1.5 square km ($\frac{3}{5}$ square mile)

STATUS Retreating

One of the few remaining glaciers in Africa is at the top of Mount Kilimanjaro (see p.171). Despite the mountain's proximity to the equator, a large ice-cap once covered most of its summit. As recently as 1912, this covered more than 10 square km (4 square miles). Since then, over 85 per cent of it has been lost, leaving just a few fragmented areas of ice. These include a small ice-field, in two main parts, on the northern edge of the broad summit area. Even this ice is predicted to melt by about 2050. Climate change and greater regional aridity, with less new snowfall each year, are the undoubted causes.

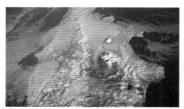

STEPPED ICE CLIFFS
Spectacular ice cliffs, typically about 25m (80ft) high, still exist at the edges of some ice fragments left on Kilimanjaro's summit.

Bogdanovich Glacier

LOCATION Within the Kliuchevskoi volcano group, Kamchatka Peninsula, eastern Russia

TYPE Valley glacier

TERMINUS Terrestrial snout

AREA 42 square km (16 square miles)

LENGTH 14km (9 miles)

STATUS Uncertain

The Bogdanovich Glacier forms part of a large glaciated area in Kamchatka, nestling in a valley surrounded by three huge volcanoes. It slopes gently down from a height of 2,800m (9,200ft) and is often covered in ash erupted from the Kliuchevskoi volcano (see p.173).

KAMCHATKA PENINSULA
The Bogdanovich Glacier is located in the glaciated area visible just below and to the right of centre of this satellite image of the Kamchatka Peninsula.

South Engilchek Glacier

LOCATION In the Tien Shan mountains, extending from western China into eastern Kyrgyzstan

TYPE Valley glacier

TERMINUS Terrestrial snout

AREA 150 square km (58 square miles)

LENGTH 60km (37 miles)

STATUS Retreating

One of almost 8,000 glaciers in Kyrgyzstan, the Engilchek Glacier has two branches – North and South – which wind their way through separate valleys in the Tien Shan mountains before converging. The South Engilchek is the sixth longest non-polar glacier in the world.

SOUTH ENGILCHEK GLACIER

ROUTE TO K2
Covered in huge chunks of rock, the lower half of the Baltoro is a challenge for trekkers passing through the magnificent Karakoram towards K2.

Baltoro Glacier

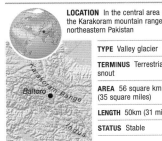

LOCATION In the central area of the Karakoram mountain range, northeastern Pakistan

TYPE	Valley glacier
TERMINUS	Terrestrial snout
AREA	56 square km (35 square miles)
LENGTH	50km (31 miles)
STATUS	Stable

Northern Pakistan, in and around the Karakoram mountain range, contains the greatest concentration of ice in Asia, with glaciers covering more than 13,000 square km (5,000 square miles). The Baltoro Glacier is one of the largest of these, occupying a long east–west valley running through the region. Between 2 and 3km (1 to 2 miles) wide for much of its length, the Baltoro is fed by over 30 tributary glaciers, which form ice-falls where they meet the main trunk glacier. The valley walls above the glacier vary from very steep to precipitous. Much of the lower area of the glacier is covered in thick rock debris. Emerging from its snout is the fast-flowing and dangerous Braldu River, a tributary of the Indus. Unlike most Asian glaciers, the Baltoro does not seem to be in fast retreat. Between 1990 and 1997, it actually advanced slightly, and some of its tributary glaciers have displayed surge behaviour. A trek along the Baltoro provides the only possible land access to the base of K2 (see pp.150–51), so the glacier has been well-trodden by mountaineers.

Durung Drung Glacier

LOCATION In the western part of the Zanskar Range of the Himalayas, Jammu and Kashmir, northern India

TYPE	Valley glacier
TERMINUS	Terrestrial snout
AREA	50 square km (19 square miles)
LENGTH	22km (14 miles)
STATUS	Retreating

One of the longest glaciers in the Himalayas, the Durung Drung is also one of the most accessible, from a road that crosses a pass, the Pensi-La, at 4,400m (14,435ft). Its surface consists principally of clean ice except for some thin rock debris at its snout and edges, and it has several tributary glaciers. From its deeply fissured snout arises the fast Stod or Doda river, a tributary of the Zanskar river.

Khumbu Glacier

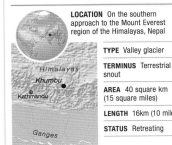

LOCATION On the southern approach to the Mount Everest region of the Himalayas, Nepal

TYPE	Valley glacier
TERMINUS	Terrestrial snout
AREA	40 square km (15 square miles)
LENGTH	16km (10 miles)
STATUS	Retreating

The Khumbu Glacier has two main segments. The upper segment consists of a steeply descending region of clean, deeply crevassed ice, the Khumbu Ice-fall. The world's highest glacier, this ice-fall is situated on the southwest flank of Mount Everest, immediately above the mountaineers' main base camp. At the bottom of the ice-fall, the glacier turns past the base camp to the south, and then descends down a straight, steep-sided valley, 9–10km (5–6 miles) long, towards the Himalayan foothills.

The glacier here has a relatively gentle downward slope and is covered in a thick layer of rock debris. Its snout has retreated a few kilometres since Everest was first climbed in 1953.

ICE FINS
These giant fin-like structures were squeezed up out of the side of the Khumbu Glacier as it was forced around a bend in its valley.

SERACS
As the East Rongbuk Glacier passed over irregular bedrock, it fractured to form isolated blocks of ice up to 25m (80ft) high.

East Rongbuk Glacier

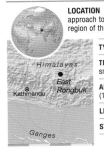

LOCATION On the northern approach to the Mount Everest region of the Himalayas, Tibet

TYPE	Valley glacier
TERMINUS	Terrestrial snout
AREA	25 square km (10 square miles)
LENGTH	14km (9 miles)
STATUS	Retreating

The East Rongbuk is part of a large, complex branching system of glaciers on the northern side of Mount Everest, and lies along one of the main routes for reaching the summit from Tibet. Its upper parts occupy several basins to the north of Everest's northeastern ridge. Ice flowing from these basins merges into a single stream that descends in a northwesterly direction. The East Rongbuk used to join lower down with the Middle Rongbuk glacier, but it now no longer does so, instead terminating at an altitude of about 5,600m (18,500ft). In common with several other glaciers in the region, the East Rongbuk has a large number of seracs, or ice pinnacles, on its surface. These can be up to 30m (100ft) high and are caused by successive fracturing of the ice as it passes over separate steep or irregular sections of bedrock. Each section produces a series of parallel cracks in the ice, and where these intersect, the pinnacles are formed.

LAND

RETREATING GLACIERS

Most of the world's glaciers have been melting and retreating for several decades, clearly due to the effects of global warming. Whatever the underlying causes of this warming (see pp.454–57), the undesirable results of glacier shrinkage are numerous – they range from acceleration of atmospheric warming due to reduced albedo (reflection of solar radiation back into space from ice), to sea-level rise and various other unwanted impacts on human life.

Thinning and retreating

Evidence that mountain glaciers are thinning and retreating worldwide comes from many different sources. These include ground observations, repeat aerial photography, and more recently satellite observations. Retreat has been occurring steadily since 1850, except for a slight slowing between 1950 and 1980 in some areas.

For some glaciers in the Americas, the cumulative losses in the past 50 years have been as high as 30,000–40,000kg per square metre, equivalent to an average of 30–40m (100–130ft) of thinning over the whole surface of the glacier. Glaciers in the Himalayas have suffered large losses, as shown by widespread, rapid growth of lakes on the glacier surfaces. Some glaciers in Norway advanced in the 1990s due to increased precipitation but since around 2000 the trend has reversed and most are now in retreat. In Africa, glaciers that once existed on the tops of Mount Kilimanjaro and Mount Kenya are expected to disappear entirely within the next 20 to 30 years.

GRINNELL GLACIER
The view of Grinnell Glacier in Montana, USA, taken in 1940 (left) contrasts with the same view in 2023 (below). The meltwater lake has increased enormously.

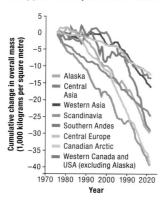

HEAVY LOSSES
This graph shows cumulative losses of ice in eight world regions since 1976. The greatest losses have been in central Europe and in western Canada and the USA.

Shrinking ice-sheets

Over the past 20 years or so, many outlet glaciers flowing from the Greenland and West Antarctic ice-sheets have increased their flow rates, thinned, and retreated. For example, the Kangerlussuaq Glacier – the largest outlet glacier on Greenland's east coast – experienced major retreats in 2004-2005 and 2016-2018, associated with average rates of ice flow near its terminus of up to 28m (90ft) a day compared to 15m (50ft) a day in the 1990s. Many other outlet glaciers such as the Helheim Glacier (Greenland's east coast) and Kjer Glacier (northwestern coast) have also retreated markedly. A 2022 study found that two large outlet glaciers that drain ice from the West Antarctic Ice Sheet (WAIS) – the Pine Island Glacier and Thwaites Glacier – have started retreating at a rate not seen in the last 5,500 years. Satellite data indicate that between 2002 and 2023, Antarctica as a whole has shed an average of 150 billion tonnes of ice per year, with the losses concentrated on the WAIS (parts of the East Antarctic Ice-Sheet (EAIS) have experienced some modest gains in ice mass).

Dangerous consequences

Glacier melting and retreat will have many unwelcome consequences for humans. These include sea-level rise (see p.456), reduced freshwater for irrigation, and loss of glacier runoff for hydropower. Melting of glaciers and ice-sheets currently contributes about 2mm (1/12in) a year to global sea-level rise. If it continues at this rate, this melting alone will cause another 15cm (6in) of sea-level rise by 2100. Combined with an additional rise caused by seawater expanding as it warms, this will have disastrous results for millions of people living in coastal areas and low-lying islands. Loss of irrigation water will badly hit some countries in South America and Central Asia, such as Peru and Kyrgyzstan. A further problem is that in many parts of the world, mountain lakes formed from glacier meltwater are steadily enlarging. Such lakes may overwhelm the terminal moraines that hold them, potentially causing disastrous outburst floods.

ABOUT TO BURST?
Tsho Rolpa is a glacial lake in Nepal that has been identified as being in danger of overwhelming its terminal moraine and causing a glacier outburst flood (GLOF).

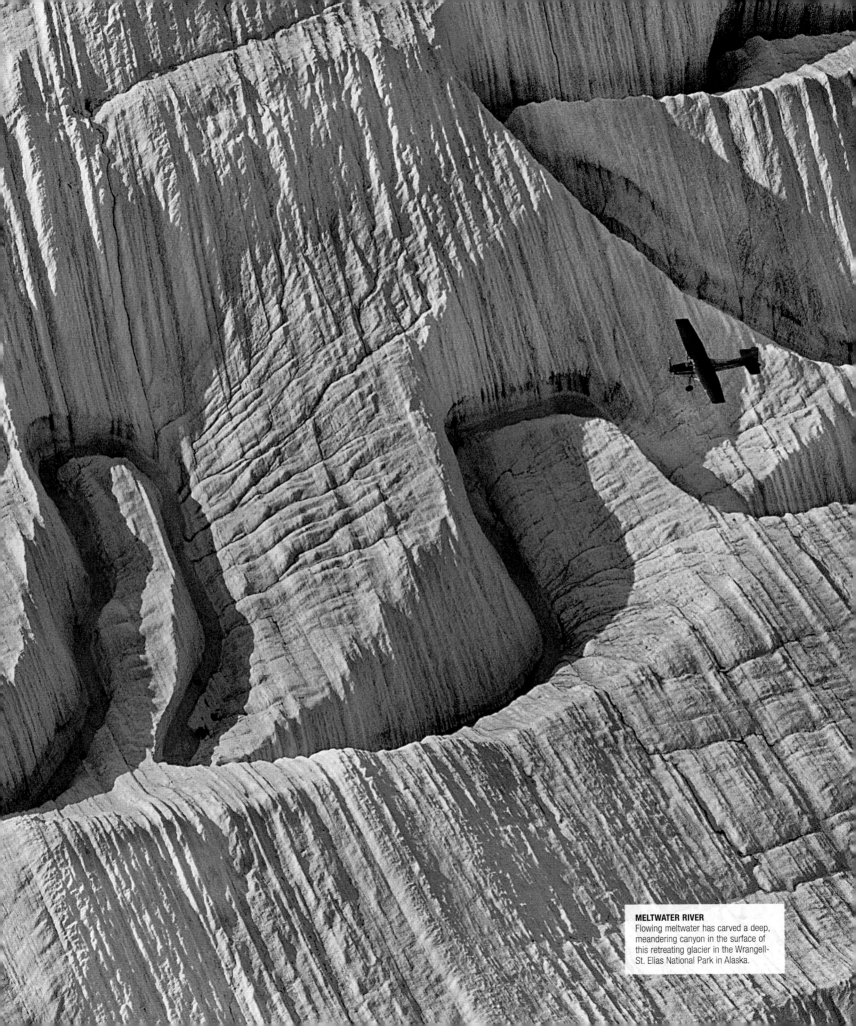

MELTWATER RIVER
Flowing meltwater has carved a deep, meandering canyon in the surface of this retreating glacier in the Wrangell-St. Elias National Park in Alaska.

ANTARCTICA

Antarctic Ice-sheet

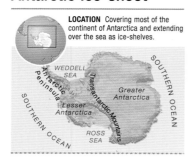

LOCATION Covering most of the continent of Antarctica and extending over the sea as ice-shelves.

TYPE Ice-sheet

AREA 14 million square km (5.4 million square miles)

STATUS Uncertain

The Antarctic Ice-sheet is by far the largest mass of ice on the Earth. Its volume is over 26.5 million cubic km (6.4 million cubic miles), and it holds over 60 per cent of the Earth's fresh water. It weighs so much that it depresses the Earth's crust by about 700-900m (2,300-3,000ft). The ice-

VEHICLE CONVOY
These vehicles are making their way across the ice-sheet close to the Adelie Coast of Greater Antarctica. The ridges of ice, called sastrugi, are caused by, and lie at right-angles to, the prevailing wind.

sheet consists of two parts, separated by the Transantarctic Mountains. The larger part, the East Antarctic Ice-sheet (EAIS), covers most of the landmass known as Greater Antarctica. It is over 4.5km (2¾ miles) thick in places, and its base is mainly above sea-level. The West Antarctic Ice-sheet (WAIS), over Lesser Antarctica, has a maximum thickness of 3.5km (2¼ miles), and its base lies mostly below sea-level. The rate of ice formation is slow (Antarctica is the driest continent,

ANTARCTIC ICE-SHEET

with snowfall averaging only a few centimetres per year), but once ice forms little of it melts or evaporates while it is still part of the ice-sheet. Instead, under the force of its own weight, the ice slowly deforms and moves towards the coasts, where numerous large outlet glaciers and ice-streams carry it down to the sea. The rate of ice movement varies from less than a metre a year in parts of the ice-sheet domes, to several hundred metres a year nearer the coasts. In many places the outlet glaciers extend over the sea as ice tongues or merge to form vast platforms of floating ice called ice-shelves. Increasing fragmentation of the shelves and calving of huge

icebergs causes ice streams on land to speed up. Many of the outlet glaciers draining Antarctica are immense. The largest, the Lambert, is 30-40km (20-25 miles) wide and over 400km (250 miles) long. Other notable features include the Byrd and Beardmore glaciers, which drain the EAIS into the Ross Ice-shelf, and the Rutford Ice-stream, which drains part of the WAIS into the Ronne Ice-shelf. The only ice-free areas in Antarctica are the peaks of the highest mountains and a few coastal areas. There are concerns that

ICEBERG OFF ANTARCTIC COAST
Antarctic icebergs form by breaking off ice-shelves or outlet glaciers flowing down from the Antarctic ice-sheet. This iceberg floating is highly eroded, indicating that it has been drifting for months.

ERNEST HENRY SHACKLETON

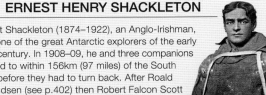

Ernest Shackleton (1874–1922), an Anglo-Irishman, was one of the great Antarctic explorers of the early 20th century. In 1908–09, he and three companions walked to within 156km (97 miles) of the South Pole before they had to turn back. After Roald Amundsen (see p.402) then Robert Falcon Scott reached the South Pole in 1911–12, he led another expedition to Antarctica in 1915, aiming to cross the continent. Unfortunately, his ship, *Endurance*, became locked in sea-ice and eventually sank. However, thanks to Shackleton's leadership, no lives were lost.

EMPEROR PENGUINS
The tallest and heaviest of all penguins, Emperors set up their breeding colonies on the newly formed sea-ice at the end of the Antarctic summer. The females lay a single egg then return to sea to feed, while the fasting males incubate the egg through the freezing winter. In 2023, diminished sea-ice formation, followed by early break up of the ice, caused a devastating loss of chicks in many Emperor colonies.

ICE-CORE ANALYSIS

Ice accumulates in annual layers in ice-sheets, somewhat like the rings of a tree-trunk. By drilling down into the Antarctic Ice-sheet, extracting a long cylinder, or core, of ice, and examining how the ice changes with depth, scientists are able to find out a great deal about the Earth's past, including fluctuations in climate. The longest ice-core drilled so far provides a record that goes back 800,000 years. Tiny, trapped air bubbles reveal changes in the atmosphere, including rising levels of carbon dioxide. Changes in atmospheric pollution are also apparent in the ice-cores.

the ice-sheet is shrinking due to global warming – the WAIS is thought to be particularly vulnerable as its base is below sea-level. If it totally melted, global sea-level would rise up by 3.6m (10ft). A 2022 study concluded that eventual total melting of the WAIS is likely if global temperatures rise 0.3°C (0.5°F) above current levels and remain raised, which now seems inevitable, though the melt could take up to several thousand years.

LAKE VOSTOK
The East Antarctic Ice-sheet overlies more than 360 lakes, the largest of which is Lake Vostok. Originally detected by ice-penetrating radar, the lake and any organisms in it are estimated to have been isolated for 15 million years.

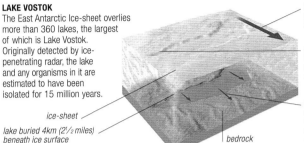

direction of movement of ice-sheet
outline of lake visible by satellite as flat area on ice surface
ice-sheet
lake buried 4km (2½ miles) beneath ice surface
bedrock
upper layers of lake water freeze to the ice-sheet and are carried beyond the lake edge

LAND

Franz Josef Glacier

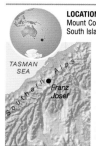

LOCATION To the northeast of Mount Cook, in the Southern Alps, South Island, New Zealand

TYPE Valley glacier

TERMINUS Terrestrial snout

AREA 20 square km (8 square miles)

LENGTH 10km (6 miles)

STATUS Retreating

Named in 1865, by a German geologist, after the Emperor of Austro–Hungary, the Franz Josef Glacier descends steeply from an altitude of 2,700m (8,860ft) in New Zealand's Southern Alps. The lower parts of the glacier used to flow through an evergreen forest near the coast. However, since 2008 the glacier's terminus has rapidly retreated, so by 2023 it only just reached below the tree-line. From the glacier's snout flows a fast and turbulent river, named the Waiho (Maori for "smoking water"; its name originates from the vapour rising from its ice-cold surface). The Franz Josef Glacier owes its existence primarily to a high rate of precipitation in the region – up to 30m (100ft) of snow falls on the glacier's ice-accumulation area every year. The glacier retreats and advances according to changes in regional climate change. Between 1983 and 2008, it advanced 1.5km (1 mile), but between 2008 and 2023, it retreated 2km (1¼ miles).

GLACIAL STRIAE
Scratches like these on rocks in the valley below the glacier were caused by boulders dragged by the ice.

GLACIER AND TEMPERATE RAINFOREST
The combination of a rainforest (foreground) with a glacier like the Franz Josef alomost in contact with it is unique to New Zealand.

Tasman Glacier

LOCATION Immediately east of Mount Cook, in the Southern Alps, South Island, New Zealand

TYPE Valley glacier

TERMINUS Lake-calving front

AREA 35 square km (14 square miles)

LENGTH 23km (14 miles)

STATUS Retreating

The Tasman is the largest glacier in New Zealand, flowing down a long valley to the east of Mount Cook (also known as Aoraki). It is up to 600m (2,000ft) thick in places and up to 2km (1 miles) wide. It originates in a large ice accumulation area at an altitude of 2,800m (9,200ft), beneath Mount Elie de Beaumont, and two-thirds of the way down it is replenished by the Hochstetter Ice-fall.

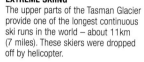

EXTREME SKIING
The upper parts of the Tasman Glacier provide one of the longest continuous ski runs in the world – about 11km (7 miles). These skiers were dropped off by helicopter.

This drains an enormous amphitheatre of ice beneath Mount Cook and Mount Tasman. The upper Tasman Glacier receives regular heavy snowfalls and is smooth and relatively crevasse-free. Further down are sections where it breaks up into a maze of crevasses, ice tunnels, and ice walls. The lower third of the glacier, which extends down to an altitude of about 815m (2,700ft), is covered in a 1-m- (3-ft-) thick layer of rock debris. This is the result of ice melting, which concentrates debris at the glacier's surface. At its bottom end, rapid melting since the 1960s has turned what were a few isolated craters and moulins in the centre of the glacier, and a small elongated lake along its eastern lateral moraine, into a large, melt-lake that extends for 6km (4 miles)

BRAIDED STREAMS
Meltwater streams flow down a 15-km- (9-mile-) long outwash plain below the glacier's terminus, converging as the Tasman River.

from the glacier's terminus. The terminus itself has over the last few decades been retreating at an accelerating rate, thus extending the meltwater lake, which is now expected to lengthen to a maximum of 16km (10 miles) over the next several decades. The glacier has also thinned dramatically since the beginning of the 20th century. Today, lateral moraine walls up to 140m (460ft) high lie on both sides of the glacier; around 1900, the ice was above those walls. The meltwater lake at the foot of the glacier is now up to 250m (820ft) deep and maintains a steady temperature close to 0.3–0.5°C (32.5–32.9°F), largely because icebergs are continually calved into it from the glacier's terminus – a process that started in about 1991. Some of the icebergs, which are full of rock debris, weigh millions of tonnes – a particularly large calving event, in which more than 30 million tons of ice were dropped into the lake, was triggered in February 2011 by the devastating Canterbury earthquake. Occasionally, calving takes place underwater, causing icebergs to pop up in the middle of the lake. For reasons not fully understood, silt suspended in the water does not settle out, and gives the lake a uniform grey colour. A demonstration of why some glaciers carry so much rock debris inside them and on their surfaces was given in 1991, when a large part of the summit of Mount Cook broke off. Millions of tonnes of rock cascaded down the Hochstetter Ice-fall and then over the Tasman Glacier. The rubble is now being carried slowly down the glacier and will eventually be dumped into the meltwater lake.

ICY LAKE
This lake sits on top of part of the disintegrating lower end of the Tasman Glacier and is rapidly extending up the glacier.

ICEBERG IN GLACIAL LAKE
This huge wall, full of rock debris, is part of an iceberg that was calved into the lake at the terminus of the Tasman Glacier. Aoraki, or Mount Cook, can be seen in the background.

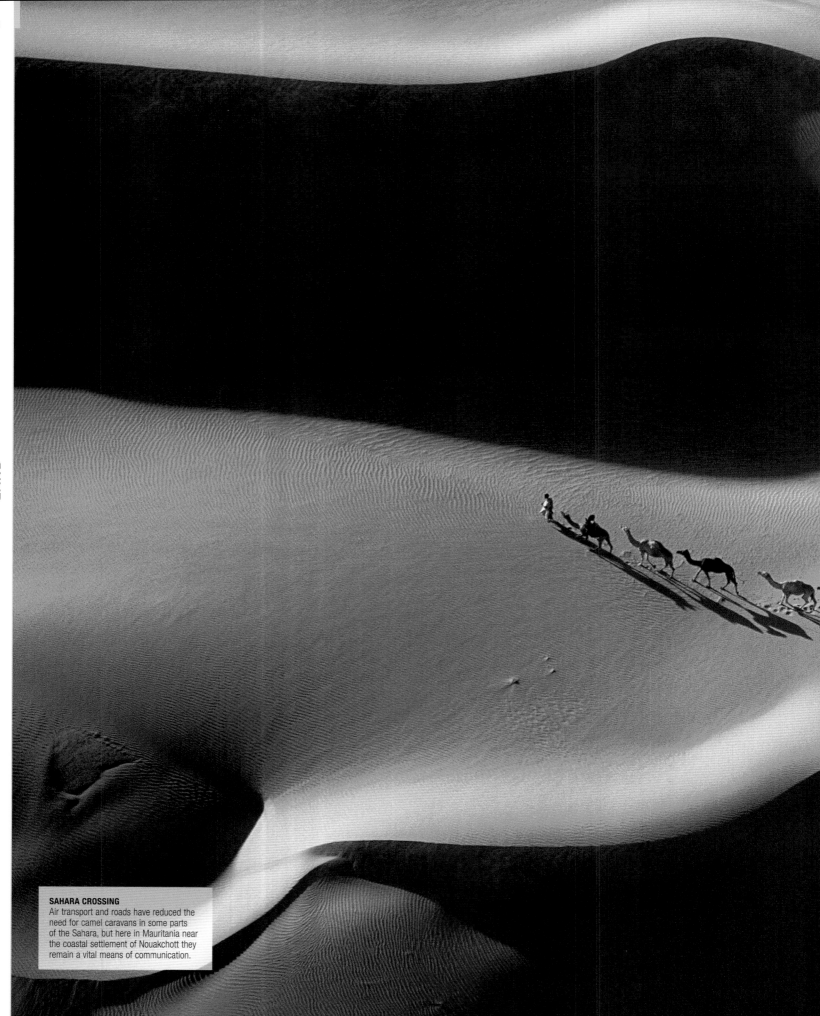

LAND

SAHARA CROSSING
Air transport and roads have reduced the
need for camel caravans in some parts
of the Sahara, but here in Mauritania near
the coastal settlement of Nouakchott they
remain a vital means of communication.

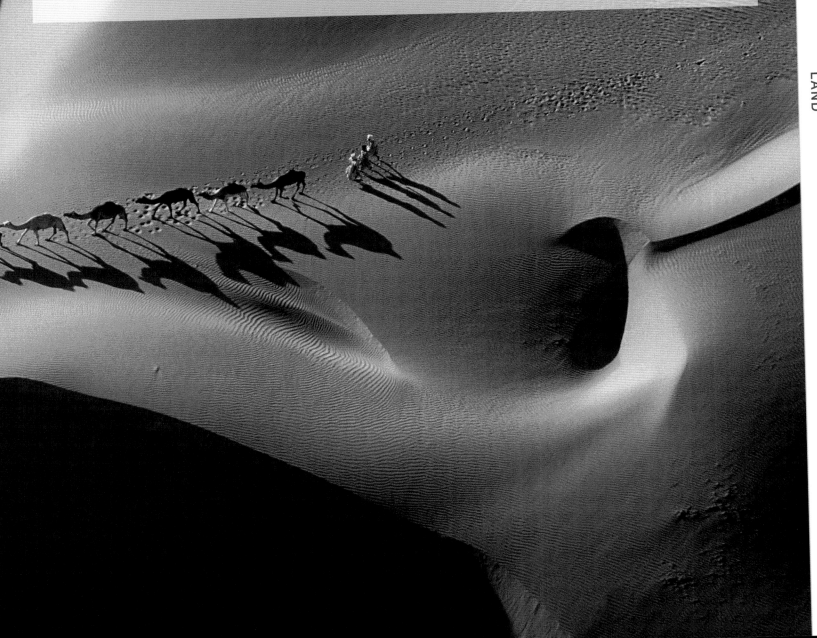

DESERTS

The term desert conjures up vistas of rolling, featureless sand-dunes and intense, dry heat. While these conditions are certainly found in many places, the world's deserts are far more varied in character, with some having rocky landscapes, some occupying high plateaus, and others dominated by open, salty lake beds. Nor are deserts hot all of the time. In some "cold deserts" the winters may be snowy or frosty, in strong contrast to the hot, dry summers. The temperature range between day and night may also be marked, with temperatures falling rapidly from dusk until dawn, and sometimes even dropping below freezing. Conditions for life in a desert are challenging, with water scarce most of the time, except for occasional downpours, which can lead to destructive flash floods. Wind-blown sand and grit is another hazard in many deserts, and it also acts as a potent agent of erosion, shaping rocks and forming dunes.

LAND

Deserts

90–91 Weathering

92–93 Erosion

94–95 Deposition

Desertification 290–91

Atmospheric circulation 450–51

Hot, dry climates 462

Deserts are among the most hostile places on Earth. They arise in regions that receive large amounts of sunshine, or where geographical features limit rainfall. High temperatures and dry winds quickly remove any moisture by evaporation. The dry ground becomes vulnerable to processes of erosion and weathering that shift sand and sculpt rock to produce beautiful but forbidding landscapes.

Types of Desert

Deserts have been classified in a number of different ways, but are usually defined according to their climate or their physical characteristics. Climatically, there are large differences between deserts. Subtropical deserts, such as the Sahara, have high temperatures all year and are described as "hot", whereas high-altitude or continental deserts, such as the Great Basin of North America and the deserts of Central Asia, experience cold winters and are described as "cold", although the summer temperatures may be high. Deserts may be just rocky or sandy, but several contain a mixture of surface features and textures. Where the occasional surface streams have no outlet, they evaporate, sometimes forming extensive deposits of salts. Many deserts are mainly or partly sandy, and the fine sand grains are easily moved and moulded by the wind, often into characteristic patterns or shaped dunes.

COLD DESERT
Cold deserts are found in temperate latitudes, on high continental plateaus, and, as above, at high altitude in the rain-shadow of the Andes.

ROCKY DESERT
Many deserts are set in rocky terrain, such as the Colorado Desert, where the rising sun is illuminating rock pinnacles and outcrops.

SEMI-ARID DESERT
In semi-arid deserts (below) the conditions are dry, but there is sufficient available moisture to sustain patches of vegetation.

Weathering and Erosion

Desert landscapes are created and maintained by erosion and weathering acting over long periods. Both processes break down soft rocks into smaller particles, and sculpt hard rock formations. Erosion is the process of reshaping by moving agents, such as wind-blown sand, which scours the surface of exposed rocks and chips off small particles (see opposite). In some deserts, such erosion sculpts rounded structures called yardangs from the bedrock. They rise above the desert sands like the upturned hulls of ships. Erosion also forms pediments – gently sloping surfaces, at the foot of rocky highland areas, that are still being broken down by erosion. Sudden storms, though rare, are another major cause of erosion, since deserts tend to have little vegetation to protect their surfaces. Dry streambeds, or wadis, suddenly fill with water, and the loose gravel and sand they contain is swept away, to be deposited downstream. Weathering happens within rocks themselves – temperature variations can cause desert rocks to crack under the stress of thermal expansion and contraction. Moisture such as dew on a rock surface can also penetrate cracks, then expand or freeze, breaking the rock open. If salt water comes into contact with rock, then the growth of salt crystals within it may also crack the rock apart.

DESERT SCREE
As weathering eats into exposed rocky outcrops, a scree of boulders and stones litters the desert floor below.

Landforms

The processes of erosion that happen in desert climates produce distinctive small- and large-scale landforms. Large landforms include inselbergs – isolated hills that stand out above the surrounding flat desert surface, and are all that remain of long-eroded mountain ranges. Australia's Uluru (also known as Ayers Rock) is a famous example. Mesas and buttes are towering platforms topped by flat, erosion-resistant surfaces. They mark the former ground level – their surroundings eroded after their hard cap was penetrated, perhaps by an ancient stream. Gravels of various grades are commonly found in deserts, often as alluvial fans at the feet of raised plateaus or spreading over the surface of rock pavements. Gravelly or rocky surfaces with a finer material beneath are called desert pavements. The term erg, originating in the Sahara, refers to an almost sea-like expanse of sand. Flat desert basins may contain lakes, which are typically dry for part of the year, or even for many years. These are known as playas, and the soils here are usually high in a mixture of salts, mostly of calcium and sodium. Some desert rocks are covered with a black or brown glossy coating called desert varnish, which usually contains large amounts of iron and manganese oxides. How this forms is not fully understood.

DESERT VARNISH
The shiny coating found on some desert rocks is called desert varnish. In some regions the varnish has been used as a canvas for primitive rock art.

DESERT PAVEMENT
These antelope footprints in the Kalahari Desert have revealed the fine sand beneath the gravelly surface.

MESAS AND BUTTES
A mesa is the steep-sided, eroded remnant of a plateau, like the ones seen here rising up above the flat plain of Monument Valley, USA. A butte is simply a small mesa.

butte

rock pillar

rock broken down into scree

SEA OF SAND
The Namib Desert in Africa is famous for the vast inland sea of majestic sand-dunes at Sossusvlei, which can reach heights of 200m (650ft).

Wind and Dunes

Dunes are large accumulations of sand, forming hills, ridges, or other features in a desert landscape. Half of all dunes are linear or seif dunes – roughly parallel ridges of sand that can extend for 20km (12 miles) or more. They are a particular feature of the Kalahari and Simpson deserts. Barchan dunes are curved, with two arms trailing downwind, and are often found around the edges of sand seas. Star dunes have three or more arms, usually irregularly arranged. They can reach heights of more than 300m (1,000ft), and are found mainly in the eastern Sahara, the Namib, and in the deserts of Central Asia. Parabolic dunes are also large, with long trailing arms, and are a notable feature of the Thar Desert and many coastal dune regions. Crescentic dunes form ridges, with slightly wavy edges. These tend to form in areas with a narrow range of wind directions, and where there is little vegetation.

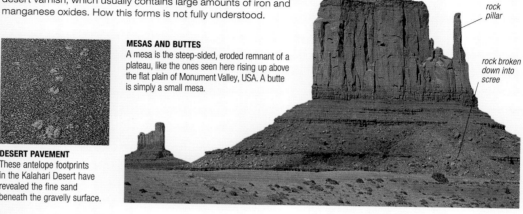

WIND-SCULPTED ROCKS
Wind is a powerful erosive force, especially if it contains sand particles. Here in Arizona, veins of calcite between bands of sandstone coloured with metal oxides emphasize the wave-like shapes of eroded sedimentary rock.

HOW DUNES FORM
Dune formation is a complex matter, still not fully understood, but in general dunes form where wind loses energy and drops the sand suspended within it. Wind-shifted sand soon becomes rippled, but over a longer period of time and with varied winds and textures of sand, distinctive dune shapes are created.

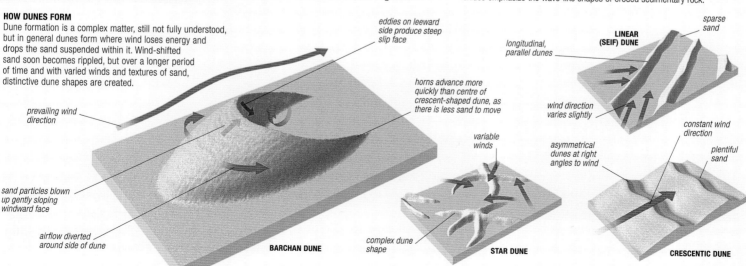

eddies on leeward side produce steep slip face

prevailing wind direction

horns advance more quickly than centre of crescent-shaped dune, as there is less sand to move

sand particles blown up gently sloping windward face

airflow diverted around side of dune

BARCHAN DUNE

complex dune shape

variable winds

STAR DUNE

longitudinal, parallel dunes

LINEAR (SEIF) DUNE

sparse sand

wind direction varies slightly

asymmetrical dunes at right angles to wind

constant wind direction

plentiful sand

CRESCENTIC DUNE

Water

Precipitation in deserts is scarce and unreliable. Average desert rainfall usually totals less than about 250mm (10in) a year, but in parts of the Sahara and coastal Chile, it is often exceedingly low – less than 1mm (¹⁄₁₆in) a year. Another key factor is that the dry, hot air soon removes water by evaporation, so that, even when the rains do fall, the water quickly turns to vapour, returning to the air before it can be used by animals or plants. In some deserts that experience cold winters, such as those of Central Asia, the main source of water is winter snow. Here again, availability of water is low, until the snows melt in the spring. Some coastal deserts, such as the Namib in southwest Africa and parts of the Atacama in Chile, receive moisture from fog, and this may be equivalent to as much as 130mm (5in) of rain annually. Dew is another important source of water, and if there is moisture in the air, it may form during the relatively cold, clear desert nights. In many deserts, such as the Thar on the India–Pakistan border, there are considerable stores of water underground, and these can be tapped by deep-rooted plants, and humans through drilling.

WADIS IN THE YEMEN
This low-orbit satellite image of the Arabian Peninsula reveals a vein-like network of dry wadis, which are prone to flash floods after seasonal rains.

WADI
A wadi is a steep-sided watercourse, such as a stream, that was formed in an earlier, wetter climatic phase and now flows only irregularly.

Animal Life

Although a desert may appear lifeless, night often reveals much activity from insects, reptiles, and small mammals in particular. These nocturnal animals avoid excessive water loss by sheltering from the heat of the day underground. Some species gain water from dew at night, or, in the case of the Darkling Beetle, from fog condensed on its body. Reptiles are perhaps the real desert specialists: they can withstand high temperatures and require little moisture, and when temperatures drop they become torpid until it warms up again. Snakes "swim" through sand by sinuous wriggling and burrow with ease, while many desert lizards use a tiptoe stance to raise themselves above the scorching surface. Large desert mammals, such as camels, are adapted for survival. Their woolly fur helps to slow down heating, and they can tolerate higher temperatures than other large mammals. They retain water by producing concentrated urine and dry dung, but can also lose large amounts of water without ill effect.

NAMAQUA SANDGROUSE
Male Namaqua Sandgrouses absorb water in their sponge-like breast feathers and carry it back to their chicks in the nest.

SIDE-WINDING ADDER
Almost invisible, this desert-dwelling snake lies buried in the sand, waiting for a suitable victim to approach.

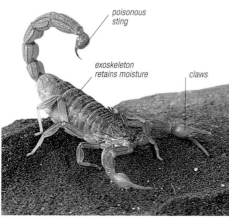

poisonous sting

exoskeleton retains moisture

claws

DESERT SCORPION
With a tough exoskeleton to help it retain moisture, this scorpion is well adapted for the dry desert climate.

DROMEDARY CAMEL
Over 90 per cent of the world's camels are Dromedaries. When dehydrated this desert specialist can drink 50 litres (11 gallons) of water in a few minutes.

LAND

Plant Life

Desert plants are generally of two main types: ephemeral annuals, which survive arid conditions as seeds, and perennials, which continue to survive and cope with periods of water scarcity. The annuals grow, flower, and set seed rapidly when conditions are more suitable, such as after rare rains, and some can complete their lifecycle in as little as two weeks. The flowers are often brightly coloured and large, perhaps because potential pollinators are scarce. The perennials have various strategies for enduring drought, such as sending down deep roots to tap underground moisture, storing water in swollen tissues, as many succulents and cacti do, or by developing underground organs such as bulbs and sending out stems and leaves only when conditions allow. Many have tough, reduced leaves to reduce water loss through transpiration, and spines to resist grazing. Cacti are typical of the American deserts, and, as well as being spiny, many species are waxy or woolly to reduce water loss further. In the African deserts, various species of euphorbia occupy the same ecological niche as cacti – they are also succulent, and often spiny. Cacti and other succulents typically have large, attractive flowers, mainly pollinated by insects, although some, including the Saguaro, are pollinated by bats. In Australian deserts, the spiky spinifex grasses can play an important role in stabilizing dunes.

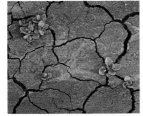

FOURWING SALTBUSH
The Fourwing Saltbush thrives in alkaline soils, and its leaves even taste salty. The first part of its name comes from the seed, which has four paper-like wings.

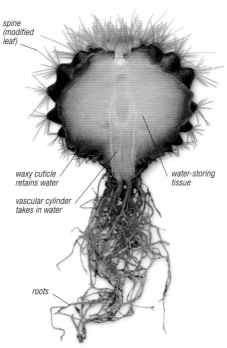

spine (modified leaf)

waxy cuticle retains water

vascular cylinder takes in water

water-storing tissue

roots

GOLDEN BARREL CACTUS
Native to the southwest USA and Mexico, this cactus sends its roots deep into the desert soil and stores water in its fleshy tissue, as revealed in cross-section. The outer cuticle helps to prevent water loss.

SONORAN DESERT IN DRY SEASON
Only the bright-red ripening fruits of the Prickly Pear Cactus stand out from the monotonous greys and greens of this parched desert landscape.

Desert distribution

The largest desert in the northern hemisphere is the Sahara. Further east, there are extensive deserts on the Arabian Peninsula, the Thar Desert in Pakistan and India, and the deserts of Central Asia, notably the Taklamakan and Gobi. In the USA, deserts are found mainly in California, Arizona, Nevada, and New Mexico, and in Mexico itself they occur in the northwest, especially in the states of Chihuahua and Sonora. The deserts of the southern hemisphere are mostly less extensive, occupying a narrow coastal belt in Chile and Peru, a swathe of southern Africa (mainly in Namibia, Botswana, and South Africa), and large patches of mainly central and western Australia.

DESERT PROFILES

The pages that follow contain profiles of the world's main deserts. Each profile begins with the following summary information:

TYPE Sandy, gravelly, stony, or rocky

AREA Surface area

RAINFALL Average annual rainfall

TEMPERATURE Summer maximum and winter minimum temperatures or average annual temperature

SONORAN DESERT IN BLOOM
With the rain, the desert is transformed by the sudden flowering of ephemeral annuals, which have to bloom and reproduce in a short period.

LAND

Great Basin Desert

LOCATION In the states of Oregon, Idaho, Nevada, Utah, Wyoming, Colorado, and California, USA

TYPE Sandy, gravelly

AREA 409,000 square km (158,000 square miles)

RAINFALL 250mm (10in)

TEMPERATURE
Max: 32°C (90°F)
Min: 4°C (40°F)

The Great Basin is the largest desert in the USA, lying mainly in Nevada and Utah, with minor extensions into neighbouring states. Sandwiched between the Sierra Nevada range to the west and the rest of the Rocky Mountains to the east, it lies to the north of the Mojave Desert. It is classed as a cold desert, due mainly to its northerly position, but also because of its high altitude, ranging between 900m and 1,980m (3,000ft and 6,500ft) above sea-level, although most of it is at 1,200m (4,000ft). The surface consists of broad valleys on a plateau, with remnants

prominent eyes with vertical pupils

COUCH'S SPADEFOOT TOAD
This toad can survive a drought by lying dormant underground, encased within a watertight cocoon made of shed skin.

hindfeet used to burrow 1m (3ft) or more into sandy soil

HARDY PLANTS
On the low hills of eastern Nevada, the dominant vegetation is the state flower, Sagebrush. This woody shrub has narrow, grey-green leaves and small yellow blooms.

of earlier lakes. The winters are cold, but the summers are hot and mostly dry. Because it lies in the rain-shadow of the Sierra Nevada, the Great Basin receives little rain, and about 60 per cent of the precipitation falls as winter snow. The gradual melting of the snow in spring seeps into the mostly fine-textured desert soils and only then is available to the plants. The Rocky Mountains also prevent most of the continental weather fronts from reaching the Great Basin, although there are occasional summer storms. Playas (temporary lakes) are a feature of the landscape, and their salty soils support vegetation dominated mainly by Saltbush. The vegetation tends to consist of shrubs, such as Sagebush, Blackbrush, a few cacti, and the bristlecone pine, which is one of the longest-living plants (up to 4,900 years). Among these plants live animals including jackrabbits, the Great Basin Rattlesnake, and the Desert Horned Lizard. In some areas, the concentration of sodium and calcium salts is so high that no plants can survive.

Mojave and Sonoran deserts

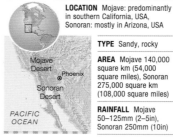

LOCATION Mojave: predominantly in southern California, USA, Sonoran: mostly in Arizona, USA

TYPE Sandy, rocky

AREA Mojave 140,000 square km (54,000 square miles), Sonoran 275,000 square km (108,000 square miles)

RAINFALL Mojave 50–125mm (2–5in), Sonoran 250mm (10in)

TEMPERATURE Mojave Max: 48°C (119°F) Min: -13°C (8°F), Sonoran Max: 48°C (119°F) Min: -13°C (119°F)

The Mojave lies in southern California, southeast of the Sierra Nevada range. This is a high desert, lying mostly at more than 600m (2,000ft) above sea-level. Death Valley in the north is an exception – at 86m (282ft) below sea-level, it is the lowest place in the USA, and it is also the hottest, reaching 57°C (134°F) in 1913. The plant life of the northern Mojave resembles that of the Great Basin, with a range of drought-tolerant shrubs (the extremely arid

DEATH VALLEY SALT FLATS
These strange polygons were created when high summer temperatures cracked the salt flat crust, allowing the water trapped below to evaporate and precipitate salt crystals.

JOSHUA TREE
Something of a symbol of the Mojave Desert, this large member of the yucca family thrives in cooler sites on the higher ground.

Death Valley has very little vegetation). The Sonoran Desert lies to the south of the Mojave, at lower altitude. This is a hot desert, but has winter rains that stimulate the growth and flowering of colourful annual plants. Cacti are also a prominent feature, and include species that grow to 12m (40ft) high. Trees with nitrogen-fixing bacteria in their roots make an important contribution to the fertility of the soil. Fauna of the region include the Desert Tortoise, Desert Scorpion, and a large lizard called the Chuckwalla.

VOLCANIC CRATER
Explosive volcanic activity during the Pleistocene Epoch created this huge maar crater in the southern part of the Sonoran Desert, north Mexico.

NUCLEAR STORE

Yucca Mountain, which lies within an active volcanic field in Nevada, has been earmarked as a suitable site for the storage of spent nuclear fuel and waste. Having conducted a series of scientific studies over two decades at a cost over $4 billion, experts concluded that it will be safe to store the materials, which will remain highly radioactive for thousands of years, sealed in canisters in tunnels about 200–300m (650–970ft) below the surface. However, other interest groups remain concerned about the area's tectonic stability.

AT THE END OF THE TUNNEL
A specially adapted machine emerges after boring an 8-km- (5-mile-) long investigative tunnel underneath Yucca Mountain.

NORTH AMERICA *south*

Chihuahuan Desert

LOCATION Between the Sierra Madre mountain ranges, Mexico, extending north into the USA

TYPE Sandy, stony

AREA 500,000 square km (194,000 square miles)

RAINFALL 400mm (150in)

TEMPERATURE
Max: 40°C (104°F)
Min: -30°C (-22°F)

Named after the Mexican province near its centre, the Chihuahuan is a high-altitude desert, lying mostly at 1,000–1,500m (3,240–4,860ft). It is the largest desert in North America, sandwiched

WESTERN DIAMONDBACK RATTLESNAKE
This highly venomous viper locates warm-blooded prey with the two heat-sensitive pits above its nostrils.

between the Sierra Madre ranges of Mexico, and extending north into New Mexico, Texas, and Arizona. Winters here are cool, with frosts occurring at night, but summers are very hot, and rain falls mainly during the summer, when it is least effective. In the lower sites, the soils are quite porous, with a calcareous base overlain by sand and gravel, and Creosote Bush and Tarbush grow here. On higher ground, such as the smooth-sided slopes known as bajadas, the thick-leaved, frost-hardy yuccas dominate. The bajadas have a rough surface of eroded rocks, boulders, and stones, and they also trap pockets of soil, in which plants are able to thrive. Higher still, where more water is available, the vegetation changes from desert grassland to a high-altitude scrub. These varied habitats allow the Chihuahuan to support a wide range of animal life, from tarantulas to bats. Desertification over the last few thousand years and grazing by large herds of cattle have led to the loss of the original desert grasslands.

TULAROSA BASIN
The shining white gypsum sand dunes of this beautiful valley at the northern end of the Chihuahuan Desert cover about 712 square km (275 square miles).

SOUTH AMERICA *south*

Patagonia Desert

LOCATION East of the Andes in the southern provinces of Chubut and Santa Cruz, Argentina

TYPE Gravelly

AREA 572,000 square km (221,000 square miles)

RAINFALL 100–260mm (4–10²⁄₅ in)

TEMPERATURE
Average: 5–13°C (41–55°F)

This cold semi-desert lies in the rain shadow of the Andes, and consists mainly of a series of dissected plateaus. Precipitation falls mostly in the long, bleak winters, with frosts likely any time between June and September. The region is subject to strong, steady, westerly winds, which severely restrict the vegetation and erode the soil.

BLEACHED BONES IN THE DESERT

Atacama Desert

LOCATION Along the coast of northern Chile, west of the Andes, between Arica and Vallenar

TYPE Rocky, salty

AREA 105,200 square km (40,600 square miles)

RAINFALL Less than 15mm ($\frac{3}{5}$ in)

TEMPERATURE
Max: 35°C (95°F)
Min: -4°C (25°F)

SATELLITE VIEW
Snow-capped volcanoes tower above the Atacama Desert in northern Chile. This is the driest place on Earth, as the dry lake beds at top right illustrate. On the left, the Atacama Fault runs from north to south.

The Atacama is the driest of all deserts, with some places having recorded no rain at all in living memory. It occupies a narrow coastal strip in northern Chile, averaging less than 160km (100 miles) in width, but extending about 960km (600 miles) south from the border with Peru, forming a huge strip of arid, hostile beach. Coastal fogs form in a few areas, providing moisture that allows the development of some plant communities, including cacti. On the eastern side, however, where the foothills of the Andes rise up, there is a small amount of rainfall in winter. Large deposits of salt are a feature of several parts of the Atacama, often in large salt basins, known as salars. When the salt flats flood after a storm they can support specialist invertebrates like the Brine Shrimp, whose eggs can survive several years of drought, and they are also home to large numbers of birds, including three species of flamingo. The Atacama has significant deposits of rare and valuable minerals, notably saltpetre (nitrates of potassium and sodium), which were formed in chemical reactions between the salty waters and volcanic debris. These are mined and extracted in a number of sites. The desert's dry climate, in which very few decomposing bacteria can exist, has preserved many features of archaeological interest, including Incan artefacts (500 years ago, this was the southern limit of their empire) and mummified remains from the Paleo-Indian civilization, thousands of years older.

SALT FORMATIONS
Many areas of the Atacama Desert are dominated by deposits of various salts, which sometimes accumulate into strange crystalline shapes.

VALLEY OF THE MOON
The Valle de Luna in northern Chile is a remarkable spectacle. Here salt, sand, and varied rock formations are set against the dramatic backdrop of the Andean peaks.

ASIA *west*

Arabian Peninsula

LOCATION Stretching from Syria in the north to Yemen and Oman in the south, east of the Red Sea

TYPE Sandy, gravelly

AREA 2.3 million square km (900,000 square miles)

RAINFALL 400mm (15in)

TEMPERATURE Max: 49°C (120°F) Min: 0°C (32°F)

In many ways, the deserts of the Arabian Peninsula are an extension of the Sahara (see pp.288–89), separated from it only by the Red Sea. The desert is very arid, with less than 100mm (4in) rainfall in some years, and has very hot summers and mild winters. It shares many features with the Sahara, such as large expanses of monotonous sand seas, which are difficult to traverse. Indeed, there are no permanent roads in this desert, making it impossible to drive over. The most famous expanse of sand is the Ar Rub 'al Khali, or Empty Quarter, in the south, which covers an area the same size as France. Mountain ranges run along much of the

THE EMPTY QUARTER

A vast area of sand dunes lies in the south of Saudi Arabia. This is the famous Ar Rub 'al Khali, or Empty Quarter, whose shifting sands buried the ancient city of Ubar.

Red Sea coast and also near the Gulf coast in Oman. Some gravel plains are found in the foothills, such as in the Hajar Mountains in the east. These ranges comprise some of the finest exposed rocks of the oceanic crust and contain fossils up to 300 million years old. The gravel plains often have a sparse cover of scrub, with species such as Ghaf, a leguminous tree that is an important source of fodder and firewood. In many places, the sands lie above oil-rich sedimentary rocks

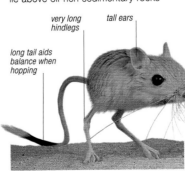

very long hindlegs

tall ears

long tail aids balance when hopping

JERBOA

Exclusively nocturnal, the Jerboa emerges from its burrow at night to feed on seeds, roots, and leaves. Its very long hind legs enable it to hop at speed over the desert sand – despite growing to only about 10cm (4in) long, it can leap as far as 3m (10ft).

on which the famous Gulf oilfields have been developed. Parts of the landscape bear evidence of less hostile conditions in the past. These include wetter climates, savannah vegetation, and river systems. Fossil remains of animals such as elephants, hippos, turtles, crocodiles, and freshwater molluscs have been recovered. Some parts of the desert have extensive salt flats, known locally as sabkha, and here the surface has a hard crystalline crust. This is often impervious and floods easily when the rare rains occur, creating temporary salty lakes. The Arabian Peninsula is used as a flight path by birds migrating between Asia and Africa, and the coastal lagoons, desert pools, and wadis provide vital refreshment. Recent human activity has caused the extinction of many animal species and the ecoregion is now listed as critical/endangered by the World Wildlife Fund.

WILFRED THESIGER

Born in Addis Ababa, Ethiopia, Wilfred Thesiger (1910–2003) studied in England before returning to Africa to work in the Sudan. An intrepid explorer, he was one of the first Europeans to visit the Ar Rub 'al Khali, where between 1945 and 1950 he lived among the local nomadic people, the Bedu. He described his experiences in the desert in his first book, *Arabian Sands*.

EXTREME WEATHER
Massive dust storms are common in Arizona, USA. Here, dust whipped up by severe winds is carried over 1,500m (5,000ft) into the air and blasted through the heart of the city of Phoenix.

ROCK AND SAND
Remnants of a former landscape, eroded
volcanic rocks rise above the shifting sands
near the Ahaggar Mountains in southeast
Algeria, at the heart of the Sahara Desert.

Sahara Desert

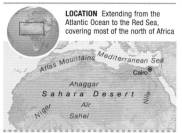

LOCATION Extending from the Atlantic Ocean to the Red Sea, covering most of the north of Africa

TYPE Sandy, gravelly, stony

AREA 9 million square km (3.5 million square miles)

RAINFALL 20–250mm (⅖–10in)

TEMPERATURE Average: 16–37°C (61–97°F)

The world's largest desert, the Sahara covers an area about the size of the USA. Although it contains many distinct geographical features, such as gravel surfaces (regs) and rocky floors (hammadas), its most famous features are the large ergs, or sand seas. Saharan sand tends to have a reddish colour, and the average thickness of the sand in the ergs ranges from 20m to over 100m (65–325ft). The ergs consist of dunes of different types. Some are true giants, and these megadunes may reach 300m (970ft) in height, with a crescentic, pyramidal, or linear shape. The Saharan landscape is moulded in large part by the wind. It creates the dunes and moves and shapes the sand, and also sculpts yardangs in the hammada areas, notably in Egypt west of the Nile, as well as in Libya and Chad. The overall net movement of the sands is more or less clockwise, from west to east along the Mediterranean side and the reverse in the Sahel region (the boundary between desert and

CAMEL CARAVAN
Dromedary camels have long been domesticated in the Sahara. They eat a wide variety of plants and, given the chance, will also feed on bones.

savanna) to the south, where it is driven mainly by the harmattan (northeasterly) winds. In the eastern, central, and southern Sahara there is an overall loss of sand from the ergs towards the Sahel, where the sand is generally accumulating and the desert extending, partly due to overgrazing. There are regular patterns to the prevailing winds, which can be strong and persist for days, carrying sand, soil, and dust and obscuring the landscape for miles around. Although vegetation is scant, oases, wadis, and other drainage channels support tough shrubs and palm trees. Several oases are found in moats, which are depressions, about 40–60m (130–195ft) deep, around inselbergs. Although there are low regions such as the Qattara Depression in northern Egypt, which is below sea-level, most of the Sahara occupies a tableland with an average altitude of 400–500m (1,300–1,600ft), and there are clusters of high ground in the centre, such as the massifs of Aïr, Ahaggar, and Tibesti. Rainfall is scant and intermittent, and some areas may see no rain at all for years

at a stretch. Storms develop from time to time, but these are localized, and can cause rapid erosion. Periodic droughts and famines have occurred every 40 years or so, especially across the southern Sahara – most recently in the Sahel in 2010 and 2012 – caused by changes in the surface temperatures of the North Atlantic Ocean. Animal species endemic to the Sahara show various adaptations to the intense heat. They include the Desert Horned Viper, which has strong, keeled scales that enable it to slip below the sand, where it lies protected from the heat of the day. At night, it waits in ambush for lizards, birds, and rodents. Another nocturnal desert specialist, the Fennec Fox, has furry soles that enable it to walk on soft, hot sand.

APE OR HOMINID?

In 2002, the Sahara yielded one of its most intriguing secrets: a fossil skull found in the Djurab Desert in Chad, which some scientists are claiming to be the oldest known human-like remains. The skull, thought to be between 6 and 7 million years old, combines both ape and early human (hominid) features, such as a chimp-like brain case and a prominent brow ridge. The new species has been named *Sahelanthropus tchadensis*, after the Sahel region of the southern Sahara where the fossil was discovered.

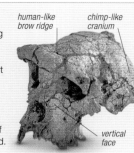

human-like brow ridge · chimp-like cranium · vertical face

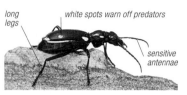

long legs · white spots warn off predators · sensitive antennae

DOMINO BEETLE
This ground beetle has unusually long legs, which raise its body clear of the hot desert sands when out hunting insects for food.

PLANT LIFE
Spiny shrubs dominate the vegetation in the rocky desert on the fringe of the Sahara in Morocco. Their thorns protect them from grazing animals.

DESERT ROSE
The mineral gypsum is often found in sand dunes, deposited by the evaporation of mineral-rich groundwater. The growth of the gypsum crystals is affected by the grains of sand, resulting in petal-like shapes clustered in "flowers".

NOMADS OF THE DESERT

The Sahara is inhabited by diverse groups, known as Tuareg, who share a common language and culture. The men wear a veil, or tagilmust, that covers almost their entire face. This is for spiritual reasons, but it also protects them from the desert sand and winds. The women wear a smaller veil, and the Tuareg refer to themselves as Kel Tagilmust, or "people of the veil". Traditionally nomadic or semi-nomadic herders of goats, sheep, cattle, and camels, for centuries, the

Tuareg also provided guides for the camel caravans of the trans-Saharan traders. The advent of roads and mechanized transport means that this source of external income has largely been lost, and today many Tuareg have been forced to settle in towns.

TUAREG OF THE SAHARA
The flowing blue dress of the Tuareg men absorb more heat than white, but this creates a cooling breeze through the robes.

LAND

DESERTIFICATION

Geological evidence – such as fossilized dunes and lake beds – shows that deserts change in the course of time. They shrink when the climate becomes more moist, and expand when it turns dry. Desertification, on the other hand, is the spread of desert-like conditions through the mismanagement of land. More than 40 per cent of the Earth's land surface is currently affected by this phenomenon, or at risk from it.

Living with Drought

In many parts of the world, rainfall follows a predictable pattern, with only minor variations from year to year. In arid regions, the climate is much more unreliable: just occasionally, heavy rain falls in sporadic storms, but years can then go by with little or no rain at all. Wild plants and animals are well adapted to these conditions, but, for humans, they are a major problem.

Where rainfall is scant and haphazard, shifting agriculture is a widespread way of life. By keeping on the move, nomadic herders (see p.344) can react to changing conditions, leading their animals to places where food can be found. In regions such as the Sahel – the belt of dry land south of the Sahara – this form of shifting exploitation has continued for several thousand years, but in recent decades several factors have put extra demands on arid land worldwide. One of these is a rise in the human population. With more people to feed, animal stocks tend to rise, until they strip away vegetation faster than it grows. This overexploitation leads to soil erosion, and at that point the desertification process sets in. Left unchecked, it can cause permanent changes in the landscape, and the original vegetation is unable to recover. Desertified landscapes sometimes resemble natural desert, but the remnants of soil – and the lack of balanced ecology – often reveal its recent origins.

The Sahel is particularly vulnerable to desertification, and so is much of southern Africa. But this is not solely an African problem, nor is it confined to the developing world. It currently threatens the southern USA, from California to western Texas, the southern fringes of Europe, as well as a wide expanse of south and Central Asia.

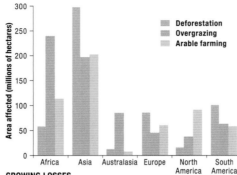

GROWING LOSSES
Desertification affects all the world's inhabited continents, but typically areas near existing deserts. In Africa, the main cause is overgrazing, while in Asia deforestation occupies first place.

Causes and Effects

Like global warming (see p.458), desertification has been the subject of intense debate among scientists. Most accept that overexploitation is a contributory factor, but opinions differ about whether it is the sole one. Some researchers think that the main trigger is local climate change, while others believe that human factors are largely to blame. The evidence is not easy to assess, because – just as with global warming – long-term trends can easily be masked by short-term change. Despite these differences, there is no disagreement about the effects: desertification can put land permanently out of action for farming.

According to the United Nations, more than 250 million people are directly affected by desertification, and 2 billion more are potentially at risk. As well as causing food shortages, desertification gives rise to other health problems, and it can displace entire populations, turning people into environmental refugees.

Combating Desertification

Despite headlines about "rolling back deserts", naturally expanding deserts cannot be stopped. Desertification, however, can be slowed and even halted by adopting measures designed to protect plant cover. Reducing livestock and planting drought-resistant trees and shrubs help to hold the soil in place, and, in regions where wood is used for fuel, they relieve the pressure on trees that grow in the wild. Paradoxically, improving the water supply does not necessarily relieve the problem. Extra wells can allow livestock numbers to increase beyond the land's carrying capacity, while excessive irrigation may increase salinity of the soil. In the Fertile Crescent, from Lebanon through to Iran, where agriculture began in the Middle East, irrigation in early recorded history damaged arid land so much that it remains unproductive today.

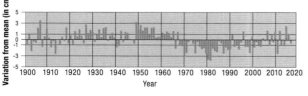

STORM WITHOUT RAIN
Caught by a sudden dust-storm, a group of women huddle together in Rajasthan, northwest India. Desertification is a major threat in this part of Asia (see graph).

RAINFALL IN THE SAHEL
Records from 1900 onwards show that rainfall in the Sahel is extremely variable. Since 1967, the annual departure from the long-term average has switched from being largely positive to almost entirely negative. This trend towards drier times has aggravated desertification.

DESERT FRONTIER
A giant sand-dune threatens to engulf a palm plantation in Morocco, on the edge of the Sahara Desert. Since the end of the last ice age, the Sahara has advanced and receded several times.

AFRICA *south*

Kalahari Desert

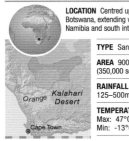

LOCATION Centred upon southern Botswana, extending west into Namibia and south into South Africa

TYPE Sandy

AREA 900,000 square km (350,000 square miles)

RAINFALL 125–500mm (5–20in)

TEMPERATURE
Max: 47°C (117°F)
Min: -13°C (10°F)

Ranging in altitude from about 900m (3,000ft) to 1,200m (4,000ft), the Kalahari Desert lies between the Orange River in the south and the Zambezi River in the north. Dominated by sandy ridges, the desert is also studded with lake beds and salt pans, which are usually dry. The soil is mostly fine-grained sand – bright red in some areas, and grey in others – and the dune systems are largely fixed and inactive. Parts of the Kalahari, notably the southwest, are known for their extensive linear dunes, many of which are partly vegetated. The vegetation is characterized by thorny scrub, with acacia trees and spiny shrubs, and the southern Kalahari, in particular, has a park-like landscape. Large herds of game, such as Springbok, Gemsbok, and Wildebeest, roam the Kalahari, moving long distances in search of occasional flushes of fresh vegetation. One of the commonest small mammals is the Meerkat, a highly social mongoose. Expert diggers, Meerkats are well adapted to the desert, creating elaborate communal burrows in the sand. The salt pans occasionally flood after heavy rains, attracting large numbers of birds, including both the Greater and Lesser Flamingo. The Kalahari is also home to the nomadic San people, about 40,000 in number.

AFRICAN WHITE-BACKED VULTURE
This impressive scavenger is a common sight in the Kalahari, either soaring overhead on its large, broad wings or feeding at a carcass.

SANDSTORM IN THE KALAHARI
Thorn trees provide the Gemsbok (centre) and Springbok with some shelter from sandstorms, which occur frequently in dry, sandy areas.

AFRICA *southwest*

Namib Desert

LOCATION On the Atlantic coast of Namibia, extending into southern Angola in the north

TYPE Gravelly, sandy

AREA 160,000 square km (62,000 square miles)

RAINFALL 15–100mm (⅗–4in)

TEMPERATURE
Max: 25°C (77°F)
Min: 10°C (50°F)

A long, narrow strip mostly less than 160km (100 miles) wide, the Namib stretches some 1,300km (800 miles) along the coast of southwest Africa. This is a dramatic coastline, with high dunes bordering the sea in some places.

GRANT'S GOLDEN MOLE
This unusual burrowing insectivore is found only in the Namib Desert and nearby areas. Covered in long, silky fur, it feeds on insects, such as termites and beetles, and larger prey like geckos, as above.

It receives very little direct precipitation – a mere 15mm (⅗in) a year near the coast – but coastal fogs provide much more moisture in the form of condensation. The dune systems are mostly active, with star dunes accounting for 10 per cent of all their number. Another unusual geological feature is the presence of rounded rock yardangs. Further inland, the Namib rises across a plain to an escarpment towards the east. The flora includes a large number of lichens (some brightly coloured), which are able to absorb moisture from the humid air when coastal fogs roll inland. A surprising number of animals survive in this hostile environment, ranging from small rodents to African Elephants.

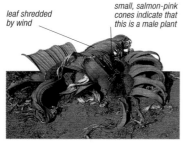

leaf shredded by wind

small, salmon-pink cones indicate that this is a male plant

WELWITSCHIA MIRABILIS
A speciality of the northern Namib Desert, this plant produces just two leaves during its entire lifespan of several hundred years.

SKELETON COAST
The surging dark blue ocean contrasts strikingly with the seemingly lifeless expanse of sand dunes in this dramatic coastline.

ASIA *southwest*

Thar Desert

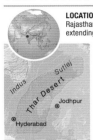

LOCATION In the province of Rajasthan in northwest India, extending into southeast Pakistan

TYPE Rocky, sandy

AREA 238,000 square km (91,900 square miles)

RAINFALL 100–500mm (4–20in)

TEMPERATURE
Max: 50°C (108°F)
Min: 3°C (37°F)

needle-like leaves

plumes of small flowers

TAMARISK
This heath-like small shrub is widely distributed in the Thar and many other desert regions in Asia. Its small, scale-like leaves help it conserve water.

The Thar (or Great Indian) Desert is centred on the province of Rajasthan, extending eastwards to the Aravali Hills. The landscape is mixed, with areas of sandy or gravel plains, broken rocks, and dune systems. Although most of the dunes are fixed and inactive, there are some regions of shifting sands. Parabolic dunes are a special feature of this desert. U-shaped, with fronts 10–70m (33–230ft) high and trailing,

SAND SCULPTURES
Undulating patterns have formed along the ridge of this linear dune in the Thar Desert. Such sculpting is the result of complex interactions of sand and wind.

partly vegetated, parallel arms up to 2km (1¼ miles) long, such dunes are unusual inland and are caused by unusually consistent prevailing winds. There is a marked absence of oases, but some areas, especially in the west, have been irrigated by the River Sutlej. Vegetation is generally sparse, and tends to be dominated by tough grasses and shrubs or small trees, such as tamarisks and acacias. The Thar is one of the last haunts of the highly endangered bird, the Indian Bustard.

ASIA *west*

Karakum Desert

LOCATION Occupying most of Turkmenistan, east of the Caspian Sea, south of the Aral Sea

TYPE Sandy

AREA 350,000 square km (140,000 square miles)

RAINFALL 100–200mm (4–8in)

TEMPERATURE
Average: -14–32°C (7–90°F)

Karakum means "black sands" in the language of Turkmenistan. It lies to the east of the southern end of the Caspian Sea, and west of the Amu Darya River. The landscape here is of cracked clay surfaces and crescentic dune systems. There is normally little or no rain between May and October, and dust-storms are a frequent occurrence. Crops, such as cotton, are grown in oases and in areas irrigated by the Karakum Canal, which takes water about 800km (500 miles) from the Amu Darya River west across the desert to Ashgabat. Black Saxaul trees are part of the dominant vegetation here, their very narrow leaves helping to reduce water loss by transpiration. The Sand Boa snake is also especially adapted to life in the desert: it has a flat head with eyes that face upwards, enabling it to keep watch for prey even when submerged in the sand.

DESERT GRAZING
The edge of the Karakum Desert has enough grass in spring to support grazing animals, although the grass quickly becomes parched.

ASIA *west*

Taklamakan Desert

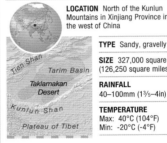

LOCATION North of the Kunlun Mountains in Xinjiang Province in the west of China

TYPE Sandy, gravelly

SIZE 327,000 square km (126,250 square miles)

RAINFALL 40–100mm (1³/₅–4in)

TEMPERATURE
Max: 40°C (104°F)
Min: -20°C (-4°F)

The Taklamakan is the second largest shifting sand desert in the world. Surrounded on three sides by snow-capped mountains, it lies between 500m and 2,000m (1,600ft and 6,600ft) and is internally drained by the Tarim river system, which flows into the Lop lowlands, depositing gypsum and salty sand there. The numerous rivers that enter the desert are fed by snowmelt from the mountains and are fringed with reeds, tamarisks, and even groves of poplars. However, away from the rivers, the plant life is sparse, with drought-resistant shrubs, such as Saxaul, helping to stabilize the dunes.

DESERT MOUNTAINS
An inhospitable mixture of stony plains and areas of shifting sand-dunes, this desert is aptly named: Taklamakan means "place of no return".

LAND

Gobi Desert

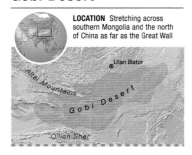

LOCATION Stretching across southern Mongolia and the north of China as far as the Great Wall

TYPE Stony, gravelly, sandy

AREA 1.3 million square km (500,000 square miles)

RAINFALL 10–250mm (²/₅–10in)

TEMPERATURE Max: 17°C (63°F) Min: -40°C (-40°F)

The largest desert in Asia, the Gobi takes its name from the Mongolian word for desert. To the north, it grades into steppe, and stretches to the Altai Mountains in the west, while its southern limit is set by ranges such as the Qilian Shan, forming part of the northern rim of the Plateau of Tibet. The landscape ranges from rocky massifs to wide valleys and plains, but very little of the Gobi is covered by sand. The central part consists of stony pavement, with sparse vegetation. It is extremely arid in the west, with no surface groundwater. Most rain falls during the summer, and the annual rainfall gradually diminishes across the region from about 250mm (10in) in the east to 10mm (²/₅in) in the west. The flora of the Gobi lacks the spring ephemeral annuals of many other deserts, probably because winter and spring here are very dry. One of the most ecologically important plants of the Gobi is Saxaul, a hardy, almost leafless woody shrub, which is resistant to drought and grows as scattered clumps in the desert. Saxaul grows to 2–4m (6½–13ft) in height, and it can survive even where the soil is sandy and unstable, helping to prevent further erosion. The shrub forms groves or woods on some of the adjacent mountain slopes, although in many

LIVING WITH THE GOBI

Mongols are the indigenous semi-nomadic people of the Gobi Desert region. Mongol tribes once herded their grazing animals (sheep, goats, cattle, and camels) over a wide area of the desert and steppe country, moving camp several times a year to avoid using up the desert's limited resources. They depended almost entirely on these animals, which yielded milk, sour cream, and yoghurt, as well as meat, skins, and wool. The Mongols also domesticated wild

BEASTS OF BURDEN
Camels are the favoured means of transport for the Mongols, and they are also sheared for their thick wool.

horses, using them to round up their flocks. The herdsmen and their families live in a type of lightweight, easily transportable tent, called a *ger*, which has a wooden frame and traditionally is covered in animal skins and woollen felt.

GRAVEL DESERT
Gravel covers large areas of the Gobi Desert. Most of the central regions, such as this area in China, are so arid that vegetation is extremely sparse and grazing animals must travel long distances to find adequate supplies of food.

places it is threatened by being overused for firewood. The sandy areas of the Gobi are home to the Long-eared Desert Hedgehog and small rodents such as the Dwarf or Desert Hamster. Several large mammal species are also found here. The last remaining wild Bactrian Camels roam the desert, mainly protected in a

reserve in the Southern Altai area (below left), and wild asses still live in the Gobi (below right), as do small herds of Blacktailed and Mongolian Gazelles. The world's only desert-dwelling bear, the Mazaalai (or Gobi Bear) is also found in the Altai region, but it is critically endangered. Reptiles endemic to the desert include the Gobi Gecko and the Tatar Sand Boa. There is increasing desertification of the Gobi, especially along its southern edge, with grasslands being overwhelmed by dust storms as a result of increasing human activity, including deforestation and grazing.

ASIATIC WILD ASS
The Khulan is a subspecies of Asiatic wild ass, which is the most horse-like of asses. It lives in the desert and desert steppes of Mongolia, particularly in the Gobi, in herds of up to 500 animals.

BACTRIAN CAMELS
Bactrian (or two-humped) Camels are native to the deserts of Central Asia and have long been domesticated, although about 1,000 still roam wild in a protected area of the Gobi Desert.

GOBI GURVANSAIKHAN
The Gobi Gurvansaikhan ("Three Beauties") National Park protects a large area of desert in southern Mongolia, seen here with the eastern Altai Mountains in the background.

Great Sandy Desert

LOCATION In the north of Western Australia, extending as far as the Indian Ocean in the northwest

TYPE Gravelly, sandy

SIZE 340,000 square km (130,000 square miles)

RAINFALL 300–400mm (10–12in)

TEMPERATURE Max: 42°C (108°F) Min: 25°C (77°F)

More than 40 per cent of Australia's land is classed as desert, and the Great Sandy Desert is one of the largest, occupying a vast, almost uninhabited area towards the north of Western Australia, southwest of the Kimberley Plateau. This is the only Australian desert to extend to the coast, forming the long sweep of Eighty Mile Beach where it meets the Indian Ocean. The sands here, as in most of Australia's deserts, are bright red, due to a coating of iron oxides on the sand grains. Rainfall is relatively high, with most rains arriving in the form of occasional thunderstorms. Daytime temperatures are in the range 38–42°C (100–108°F), and the short winter is warm, with temperatures averaging 25–30°C (77–86°F). The heat means that much of the moisture is lost by evaporation and is therefore not available to plants, with spinifex grasses among the few able to survive. Reserves of oil have been found beneath the sands and gravels of this desert, but these have yet to be exploited on any large commercial scale.

GREATER BILBY
Also known as the Rabbit-eared Bandicoot, this highly endangered marsupial digs a burrow up to 3m (10ft) long and shelters there during the day, emerging at night to feed on seeds, fruit, and insects.

INTRODUCED SPECIES

Dingoes were introduced to Australia from Southeast Asia some 3,500 years ago. They are regarded as a pest because they attack livestock, and they have been excluded from parts of Australia by the erection of fences.

Along with more recently introduced species such as foxes and domestic cats, Dingoes pose a severe threat to small native marsupials such as the Greater Bilby. Another alien pest, the European Rabbit, has displaced native fauna and destroyed vegetation, permanently changing the landscape in arid areas.

DINGOES
Now readily associated with Australian fauna, the Dingo is probably descended from the Indian Wolf.

SHIFTING DUNES
Although many desert dune systems are stable, some, like these in the Great Sandy, constantly change shape due to the wind.

Tanami Desert

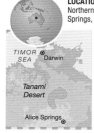

LOCATION In the centre of the Northern Territory, north of Alice Springs, Australia

TYPE Rocky, sandy

SIZE 37,500 square km (14,500 square miles)

RAINFALL 200–400mm (8–16in)

TEMPERATURE Max: 42°C (108°F) Min: 25°C (77°F)

Lying to the east of the Great Sandy Desert, the Tanami Desert contains a large area of red sandy plains, with intermittent ranges of hills. The plant life here is a combination of tough spinifex grasses and small shrubs, such as Saltbush, as well as occasional Spiny Acacias. A surprising feature of the Tanami, and the rest of Australia's arid interior, is the presence of Dromedary Camels, which were brought over by the early colonizers of the outback as pack animals. They became redundant with the arrival of motor transport, and today's feral herds are their descendants. The Tanami also has the largest population of the endangered Rufous Hare Wallaby. In the early 20th century, the discovery of gold led to the establishment of small mining communities, but the harsh, dry climate made economic extraction difficult.

Simpson Desert

LOCATION In the southern Northern Territory, western Queensland, and northern South Australia

TYPE Sandy

SIZE 145,000 square km (56,000 square miles)

RAINFALL 150mm (6in)

TEMPERATURE Max: 50°C (122°F) Min: 0°C (32°F)

Located in central Australia, the Simpson Desert is the world's largest sand dune desert, famous for its parallel red dunes, that grow to 38m (125ft) high and up to 120km (75 miles) long. Summer daytime temperatures average 35–40°C (95–104°F), and the winters are mild (18–23°C/64–74°F) with occasional night frosts. The rainfall is low and irregular, usually from thunderstorms, and the dominant vegetation is spinifex grasses and shrubs.

ULURU
Rising 348m (1,142ft) above the desert floor, Uluru is a vast mass of 300-million-year-old sandstone.

CHAMBERS PILLAR
This sandstone column helped early explorers navigate the Simpson Desert, and the rock still bears their carved inscriptions.

EASTERN GIBSON DESERT
Rock-strewn hills, spinifex grass, and mulgas are characteristic of the eastern Gibson Desert.

Gibson Desert

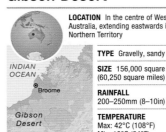

LOCATION In the centre of Western Australia, extending eastwards into Northern Territory

TYPE Gravelly, sandy

SIZE 156,000 square km (60,250 square miles)

RAINFALL 200–250mm (8–10in)

TEMPERATURE
Max: 42°C (108°F)
Min: 18°C (64°F)

The Gibson Desert lies between the Great Sandy and Great Victoria deserts. It takes its name from the Australian explorer Alfred Gibson, who attempted to cross it in 1874, but perished while searching for water. The landscape is mainly one of undulating beds of gravel, with areas of red sandhills, known as buckshot plains. Some parts are

ALEXANDRA'S PARROT
This nomadic, slender bird is unusual among parrots in being adapted to survive in the desert. Its diet consists mainly of small seeds gathered from the ground and tree blossoms.

rocky, and in others inselbergs protrude from flat plains. In the west, lies Lake Disappointment, a large salt lake that attracts great numbers of waterfowl. Daytime temperatures in the Gibson average 33–42°C (91–108°F) in summer, and 18–23°C (64–73°F) in winter. The rains are very unreliable in this region, coming mostly from the occasional thunderstorm, of which there are some 20 to 30 each year, and the vegetation is patchy. Alongside the usual spinifex grasses are mulgas (species of acacia, or wattle), which lend a green tint to the landscape, and colourful flowers such as the Yellow Coneflower appear after the rains. Red Kangaroos are a common sight.

slender build

long, narrow wings

tail feathers graduated in length

Great Victoria Desert

LOCATION In the southeast of Western Australia, extending east to the centre of South Australia

TYPE Sandy, gravelly

SIZE 338,500 square km (150,000 square miles)

RAINFALL 150–250mm (6–10in)

TEMPERATURE
Max: 40°C (104°F)
Min: 18°C (64°F)

Protected as a World Wildlife Fund Ecoregion, this desert borders Western Australia and South Australia, north of the Nullabor Plain. Mainly a sandy desert with extensive dune fields, it also includes gibber plains, on which the soil is covered by closely packed pebbles that are often glazed by iron oxides. They usually lack vegetation, although ephemeral annual species may appear after rains. In addition to the more familiar types of sand dunes, another kind, known as lunette dunes, also

occurs here. Lunette dunes are a special feature of the Australian deserts, especially in the south. They are crescent-shaped, consist mostly of clay, and are formed at the downwind edges of temporary lakes from dry, wind-blown lake-floor sediments. Although the summers are hot, winter frosts are not infrequent in parts of the Great Victoria Desert, especially on the higher ground. But the winters are short, with

temperatures rising quickly again by late September. Many parts of this desert have sufficient moisture to support open woodland, with eucalypts, common mulgas, and an understorey of grass hummocks. The Great Victoria Desert is famous

STURT'S DESERT PEA
The floral emblem of South Australia, this striking plant is able to survive the extremes of the inland desert climate.

for its reptiles, which number over 100 species, with geckos and skinks being particularly diverse. Many of the reptiles are active during the day, such as the bizarre Thorny Devil, whose camouflaged pattern makes it hard to spot against the stones and sand of the desert. Equally striking are the Parentie and Gould's Goanna, both of which can exceed 1.8m (6ft) in length. Little farming takes place here, although there are several Aboriginal reserves and some sheep and cattle ranches around the more fertile edges.

THORNY DEVIL
Despite its fearsome appearance, the desert-dwelling Thorny Devil is harmless to all except ants, and relies on its sharp spines for protection.

SAND DUNES AFTER RAIN
Although arid, the Great Victoria Desert can support scattered bushes along temporary watercourses, as here.

LAND

CLOUDFOREST
Epiphytes such as Spanish Moss clothe
the damp branches and trunks of trees
in this Costa Rican cloudforest, providing
niches for a vast number of species.

FORESTS

Forests cover 26 per cent of the Earth's habitable and non-habitable land surface and provide some of the richest habitats, offering an abundance of sites for plant germination as well as food and shelter for a wide range of woodland animals. Different types of forest are found in different regions of the world, from the dark boreal forest (taiga), which extends to the edge of the Arctic tundra, to the lush mixed broadleaf forests of temperate North America, Europe, and Asia, and the dense rainforests of the humid tropics. Forests also play a vital role in the global water and carbon cycles: in improving air quality, and in preventing soil loss through erosion. They were central to early human evolution, and even today the rainforests, in particular, are still home to many native peoples. Now, many of the world's natural forests are threatened with destruction by climate change and as land is cleared for housing, roads, timber, or agriculture. Some forests survive only in inaccessible areas or in enclaves on protected sites.

LAND

Forests

23 Plant life

109 The global water cycle

112 Kingdoms of life

115 The carbon cycle

Deforestation 308–09

Forests are categorized in various ways, usually according to the dominant type of tree. An important distinction is that between broadleaved and coniferous (usually needle-leaved) trees. Most coniferous trees are evergreen, which means that they retain a full complement of leaves throughout the year, but many broadleaved trees are also evergreen, and a few coniferous trees (such as larch) are deciduous. Deciduous trees shed all their leaves during the part of the year that is unfavourable for growth. This is the cold winter in temperate latitudes, but in the seasonal tropics, it may be the regular dry season.

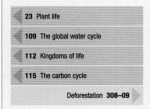

cluster of small flowers

compound leaf with five distinct leaflets

Forest Vegetation

Forest vegetation is dominated by the major tree species that form a high canopy. Conifers and their allies bear naked seeds. They have needle-like, scale-like, or blunt linear leaves, usually with a resinous coating that helps them survive cold winters. Broadleaves bear ovules in an ovary that develop into seeds only after fertilization. They have a variety of larger leaf shapes to better collect sunlight to drive photosynthesis (see p.112). Coniferous forests have less ground vegetation than broadleaved forests, since they are darker, and the resinous shed leaves produce an acid soil. In deciduous forests, the shed leaves form a nutrient-rich humus that fosters many more ground-cover plants. These are often adapted to grow and flower quickly in the spring, taking advantage of the brighter conditions before the trees above unfold their foliage.

BROADLEAVED FLOWER AND SEED
Many broadleaved trees, like Horse Chestnut, have showy flowers to attract pollinators such as insects, and some produce large, edible seeds.

SEEDS
spiky husk splits when it hits ground, releasing nut

shiny brown seed (or nut)

FLOWERS AND LEAVES

CONIFEROUS CONES
The "fruit" of most conifers, such as this Norway Spruce, is a cone, consisting of woody scales that turn brown when ripe.

LEAF TYPES
Broadleaves offer a large, flat surface area for light-gathering and gaseous exchange. Narrow conifer leaves reduce water loss by transpiration.

CLUSTER OF CONIFER LEAVES

BROADLEAF

Tropical Forests

Growth is rapid and lush in many tropical forests, and their complex structure offers niches for a multitude of animals and plants other than trees. Mature tropical rainforest is a multi-layered habitat. The canopy is composed of trees whose crowns almost meet. Beneath it lies a more broken understorey of young trees, and an often-dense ground cover of saplings and undergrowth. Tall trees whose crowns stick out above the canopy are known as emergents. In many tropical rainforests, the moisture levels remain high year-round, whereas in others the rainfall is seasonal, as in the forests of Southeast Asia, where the monsoon brings sudden high levels of rain. Some tropical forests are dry, such as those of eastern South America.

EMERGENTS

CANOPY

UNDERSTOREY

GROUND LAYER

RAINFOREST STRUCTURE
Distinct layers can be seen in a mature rainforest. The mid-level canopy is overtopped by lone emergent giants, and there is a loose understorey beneath.

RAINFOREST CANOPY
The dense canopy of almost interlocking crowns is clearly visible in this Madagascan tropical forest.

Temperate Forests

Temperate regions contain a wide variety of forest types. Many temperate forests are broadleaved deciduous, especially where the winters are hard, as in continental Europe, eastern North America, and many parts of eastern Asia. In such forests, there is a huge contrast between the forest in summer, at the height of the growing season, and in winter, when all the leaves have been shed and have accumulated on the forest floor. In China, Japan, and Korea, evergreen broadleaved trees are more common in the temperate forests, and sometimes form mixed stands with conifers. However, as these parts of the world have long been settled, many of the original forests have been destroyed, at least in part, and their rich, fertile soils used for agriculture. The remaining temperate broadleaved forests are nearly all managed in one way or another, although some are protected and are being allowed to revert to a more natural state.

yellow pigment from chloroplast remains

green colour from chlorophyll

SUMMER AND AUTUMN
Deciduous forests change dramatically with the onset of autumn, when the trees lose all their leaves in a short time.

CHANGING LEAVES
As they die, some leaves change colour from green to yellow, orange, or red, following the breakdown of the chloroplasts that contain their chlorophyll.

Coniferous Forests

There are several different kinds of coniferous forest, but the most extensive, and the best known, are the boreal forests of northern North America and northern Eurasia (taiga). These are dominated mainly by evergreen coniferous trees, such as species of spruce, pine, and fir, although larches, which are deciduous, are a prominent feature of some areas. Coniferous forests are also commonly found in many mountain regions, where they tend to develop above the highest levels of broadleaved woodland, with spruces and pines being the main trees present. An unusual type of coniferous forest is found in the high-rainfall coastal zone of the Pacific Northwest of North America. This temperate rainforest boasts several very tall species, such as Douglas Fir and Coast Redwood, while California's coniferous forest is home to the famous Giant Sequoia.

SUMMER AND WINTER
In northern or mountain coniferous forests, the trees retain a full covering of needles through the harshest winters. In fact, individual leaves are shed and replaced, but not all at the same time.

GREEN AND GOLD
Larch is an unusual conifer in that it is deciduous. Here, in France, its beautiful autumn colours stand out against the dark green of the evergreen conifers.

Forest Zones

Forests are the final development of mature vegetation on average soils with an adequate water supply. Under natural conditions, forests may persist for very long periods, and individual trees often grow to be several hundred years old. This, then, is the situation in natural forests, but few of today's forests are virgin, in the sense that they are completely uninfluenced by people or their activities. If forest is cleared and the vegetation allowed to recover, ecological succession may be observed, with a slow reversion to what is known as climax forest. A snapshot of succession may also be seen, for example, in the transition from the marshy vegetation around a lake, through wet scrub to mature damp woodland. Changes in forest type are also clearly visible with a change in altitude, ascending in a series of zones from lowland to alpine levels on a mountainside (see right). In temperate forest regions, such as in the European Alps, mixed deciduous forest tends to be found in lowland sites, and the mixture of trees changes through submontane and hill forest, with an increase in coniferous trees at montane levels. At about 2,000m (6,600ft), the climatic limit of tree growth is reached (the tree-line) and alpine grassland and scrub begin to dominate. In the tropics, the pattern of trees is different, with a zone of bamboo often found between the montane forest and scrub bands.

ALPINE ZONATION

In this European alpine landscape, the lowest level of the landscape features mixed forest, with trees such as Hornbeam and oaks dominant. This grades into montane forest, then a final forest zone of mainly coniferous species such as Silver Fir, spruce, and larch. Higher still, high winds, frost, snow, and ice discourage even coniferous trees.

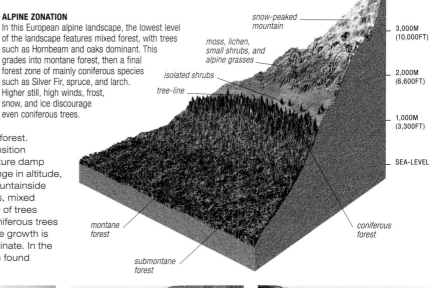

snow-peaked mountain

moss, lichen, small shrubs, and alpine grasses

isolated shrubs

tree-line

3,000M (10,000FT)

2,000M (6,600FT)

1,000M (3,300FT)

SEA-LEVEL

montane forest

coniferous forest

submontane forest

ISOLATED SHRUBS
Only hardy, specialist plants are able to survive above the tree-line.

NEEDLE-LEAF FOREST
Coniferous trees best tolerate the rigours of high altitude, and often form the tree-line.

MONTANE FOREST
Above 1,000m (3,300ft), the mix changes to species such as beech and scattered conifers.

SUBMONTANE FOREST
Lowland and hilly sites with good soils support a rich, mixed forest of mainly deciduous broadleaved trees.

BLUEBELL WOOD IN SPRING
Many deciduous woods, such as this one in the UK, have a ground flora of species such as the Bluebell that flower early, before the trees are in full leaf.

Animal and Plant Life

Although cold and challenging for much of the year, the boreal forests can still support animal life, especially during the summer, when they are home to seed -eating birds such as crossbills, and Spruce Grouse and Capercaillie, which feed partly on conifer needles. Many birds migrate to less harsh climates in the winter, but the mammals, which include bears and voles, either remain active beneath the snow or hibernate. In broadleaved temperate forests, conditions are more equable, and deciduous trees avoid the stresses of winter by shedding their leaves. This leaf litter is a bonus for the ground flora and fauna, and the soil of these forests is alive with invertebrates such as worms, spiders, ants, and beetles. Many woodland birds, such as warblers and tits, feed on caterpillars and other insects up in the canopy. Foxes, Badgers, and deer are among the larger mammals at home in these forests. Richest of all in plant and animal life are the tropical rainforests, where researchers are still discovering new species. Many kinds of epiphytes (plants that grow on the trees without being parasitic) thrive here, including bromeliads and orchids.

GILL POLYPORE
This bracket fungus is widespread in northern temperate forests, found mostly on the trunks of hardwood trees.

FOREST SIZES

The technology of satellite imaging has had an enormous impact on ecological monitoring. The comparison of satellite photographs with existing ground surveys allows detailed interpretation and analysis of such images, and enables the accurate mapping of the size of existing forests. The image below, taken from the International Space Station, for example, shows the forest cover in the Black Hills of South Dakota, USA. It is vital to build up such snapshots of vegetation cover if the effects of factors such as deforestation are to be accurately monitored. Satellite imaging is also used to track forest fires, enabling their paths to be predicted and control measures mobilized.

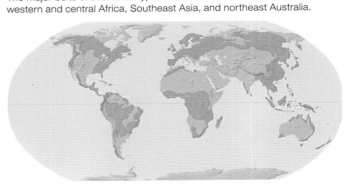

WEAVER ANTS
Ants are very common insects in tropical forests. Here, worker Weaver Ants in the Malaysian rainforest co-operate to cut leaves, from which they fashion their nest. The tree benefits as the ants deter other plant-eaters from browsing on the foliage.

MILTONIA ORCHID
Many plants in moist tropical forests, such as this orchid from Brazil, grow on the trunks and branches of trees.

GREAT SPOTTED WOODPECKER
Woodpeckers use their powerful bills to drill and probe the bark of trees for grubs and bugs. The stiff tail feathers act as a sturdy prop.

forward-facing eyes permit stereoscopic vision to judge distance

long tail aids balance

RUFFED LEMUR
This inhabitant of the rainforests of Madagascar feeds mainly on fruit, but will sometimes eat leaves, nectar, and seeds. It lives in the upper forest, and builds a nest of leaves in a tree hole or fork.

Forest Regeneration

Although forests often take many decades, or even centuries, to establish themselves, death and regeneration are part of the natural dynamic of forest ecosystems. Some trees are long-lived, growing for hundreds of years, but they do die and fall eventually, creating gaps in the forest into which more sunlight can penetrate. This opening up of the canopy encourages the often rapid germination of plants on the fertile forest floor, and among these will be seedlings or suckers of trees belonging to the same species as the fallen tree. In this way, natural forests tend to display a mosaic of patches of trees of differing ages as the forest regenerates in an uneven fashion. As well as individual trees succumbing to old age or disease, large stands of trees can be destroyed by fire or strong winds. The composition of the regenerating forest may well differ from that of the surrounding mature forest, as some species of tree, such as Douglas Fir, are better able to thrive in the more open conditions. Other, more shade-tolerant species such as Western Hemlock come to the fore as the forest returns gradually to a taller and denser and therefore darker state.

GAP IN A FOREST
Gap-regeneration is a natural phenomenon in woods and forests. When an old tree dies or is felled by strong wind, as here in New Zealand's Fiordland, it decays on the forest floor, gradually returning nutrients to the soil.

Forest Distribution

In general, coniferous forests are found at higher latitudes and altitudes than temperate forests. The vast boreal forest belt, for example, lies north of about 50°N. In temperate regions, the forest climate is one with an annual rainfall of 500–1,500mm (20–60in). Temperate forests are found in western and eastern North America, southern South America, western and central Europe, eastern Asia, and Australia and New Zealand. Tropical rainforest requires warm conditions and an annual rainfall of 2,000–3,000mm (80–120in). The major belts of this forest type are in Central and South America, western and central Africa, Southeast Asia, and northeast Australia.

FIRE AND REGENERATION
Fire (1) is a major regenerative factor in some habitats, such as this coniferous forest. A rich seed-bank in the ash-enriched soil (2) provides new growth, as germination follows the devastation (3).

FORESTS PROFILES

The pages that follow contain profiles of the world's main forested regions. Each profile begins with the following summary information:

TYPE Tropical rainforest, tropical dry forest, temperate rainforest, deciduous temperate forest, evergreen temperate forest, or boreal forest

AREA Surface area

LAND

North American boreal forest

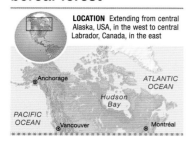

LOCATION Extending from central Alaska, USA, in the west to central Labrador, Canada, in the east

TYPE Boreal forest

AREA 6 million square km (2.3 million square miles)

ALASKAN FOREST
The deep northern boreal forests are broken up in some places by rivers, streams, and boggy lakes, as here in Alaska.

Boreal forest is one of the largest global biomes, covering more than 16 million square km (over 6 million square miles). It dominates much of northern North America, covering one-third of Canada's landmass. It is a counterpart to the Eurasian boreal forest or taiga (see p.312). The climatic conditions under which the boreal forest develops are generally cool and humid, with very cold winters, lasting from seven to nine months. Snow covers the ground for much of the year, often into the summer months. Coniferous trees, such as Black and White spruce, are well adapted to survive such rigorous conditions, and they dominate the North American boreal forest. The drooping branches shrug off the weight of heavy snow, and the waxy needles allow good control of water loss. Although conditions may be wet and boggy in summer, for much of the year water is

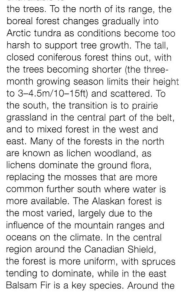

QUAKING ASPENS
The slender white trunks of Quaking Aspens add to the beauty of the boreal scenery here, in British Columbia.

frozen and therefore largely unavailable to the trees. To the north of its range, the boreal forest changes gradually into Arctic tundra as conditions become too harsh to support tree growth. The tall, closed coniferous forest thins out, with the trees becoming shorter (the three-month growing season limits their height to 3–4.5m/10–15ft) and scattered. To the south, the transition is to prairie grassland in the central part of the belt, and to mixed forest in the west and east. Many of the forests in the north are known as lichen woodland, as lichens dominate the ground flora, replacing the mosses that are more common further south where water is more available. The Alaskan forest is the most varied, largely due to the influence of the mountain ranges and oceans on the climate. In the central region around the Canadian Shield, the forest is more uniform, with spruces tending to dominate, while in the east Balsam Fir is a key species. Around the

Great Lakes, the boreal forest is mixed, with deciduous species including Sugar Maple and Yellow Birch. Quaking Aspens thrive on poor soil and are often found at montane levels. Mammals of the forest include the Canada Lynx and its prey, the Snowshoe Hare (trapping records have revealed their population cycles to be closely linked), and the Northern Flying Squirrel. Great Grey Owls are found here and also in the Eurasian boreal forest, and feed on voles, mice, and small birds.

BLACK SPRUCE
Here, in the northern boreal forest of Alaska, scattered Black Spruces tower above the willow scrub. This is the dominant tree in the open woodlands at the forest–tundra edge, from Alaska, across Canada to Newfoundland.

CANADIAN BOREAL FOREST
Mount Robson Provincial Park in British Columbia protects over 2,170 square km (840 square miles) of unspoilt boreal forest close to Canada's highest peak.

NORTH AMERICA *west*

Pacific Northwest rainforest

LOCATION Along the Pacific Coast from the Gulf of Alaska to northern California, USA, and Canada

TYPE Temperate rainforest

AREA 1.2 million square km (480,000 square miles)

These remarkable forests, the only temperate rainforests in the northern hemisphere, occupy a relatively narrow strip along North America's northern Pacific Coast. The rainfall is high (800–3,000mm/31–118in, mostly between October and March) and is supplemented in some areas by coastal fogs. These conditions are ideal to support the

HIGH RAINFOREST
Washington State's famous temperate rainforest receives a staggering 3,350mm (132in) of rain a year, resulting in lush vegetation, including many epiphytes.

WESTERN HEMLOCK
These shade-tolerant trees thrive in low light conditions, enabling them to slowly replace pioneer tree species, such as the Douglas Fir.

growth of some of the giants of the plant kingdom, notably conifers such as the Redwood, which can grow to over 110m (355ft) in height, Douglas Fir, Sitka Spruce, and Western Hemlock. Douglas Firs dominate in the Coast Ranges, Olympic Mountains, and northern Cascade Range, up to about 1,000m (3,300ft). Redwood forests are a feature of northern California and southern Oregon.

HABITAT LOSS

The Northern Spotted Owl is found only in the natural coniferous forests of the Pacific Northwest, where its habitat is under threat from logging activities. Each pair of owls requires about 11 square km (4 square miles) of unspoilt old-growth forest or heavily wooded canyons for successful breeding. The US Forest Service's best estimate is that between 3,000 and 5,200 Northern Spotted Owls remain on federal lands. A mere 30 pairs remain in British Columbia.

NORTH AMERICA *west*

California coniferous forest

LOCATION Along the western slopes of the Sierra Nevada, central California, USA

TYPE Evergreen temperate forest

AREA 43,600 square km (16,800 square miles)

These forests are famous for being the home of the world's largest tree species (in terms of bulk), the Giant Sequoia, or Big Tree. This is a high-altitude community, found mainly between 1,400m and 2,300m (4,540ft and 7,450ft). There are about 75 separate groves where the Giant Sequoias grow, often associated with White Fir, Sugar Pine, and Incense Cedar. Although not quite as tall as the Redwood or Douglas Fir (see above), Giant Sequoias may grow for over 2,000 years and reach almost 100m (325ft) in height. Fire is an important ecological factor in these coniferous forests. The cones of Giant Sequoias require the heat of forest fires before they can open to shed their seeds. Fire also burns the soil surface, exposing the mineral-rich layers

SURVIVING THE WINTER
The thick, spongy bark of the Giant Sequoia, seen here in Sequoia National Park, protects the tree from the winter cold as well as from the effects of fire.

YOSEMITE
Equally famous for its meadows, mountains, and waterfalls, Yosemite National Park also boasts three groves of ancient Giant Sequoias.

GENERAL SHERMAN
This, the largest living Giant Sequoia, is about 2,200 years old. It is 84m (275ft) tall and 9m (30ft) across at its base.

lying beneath, which are essential for seedling growth. The fully grown trees are protected by their spongy bark, up to 45cm (17½in) thick, which gives them an asbestos-like insulation from fire damage. Other trees in the region include Douglas Fir and Ponderosa and Lodgepole pines, with Engelmann Spruce and Bristlecone Pine at subalpine levels. The latter species is very slow-growing and includes the world's oldest-known living plant specimen – referred to as the Old Man, or Methuselah, it is nearly 4,770 years old. Mammals of these forests include Black Bears and Mountain Lions (Cougars).

LAND

East North American deciduous forest

LOCATION From the Great Lakes of southern Canada to the east coast area of the USA and south to Florida

TYPE Deciduous temperate forest

AREA 2.5 million square km (970,000 square miles)

Only scattered remnants of these once extensive forests now survive, but some of the finest are centred on the Appalachians (see p.140), whose foothills still retain considerable forest cover. On the western edge of the

SPRING IN THE APPALACHIANS
The ridges and valleys of the Great Smoky Mountains in the Appalachians, North Carolina, are clad with several types of oak and hickory, American Chestnut and Beech, and Basswood.

region, the deciduous forests are dominated by oaks and hickories, gradually giving way to prairie grassland in the drier interior. These forests are famous for their autumn colours, particularly in the north, where the leaves of the Sugar Maple may turn bright red.

AUTUMN COLOURS
In autumn, over 100 deciduous tree species cover the foothills of the Appalachians in a colourful display.

Northwest South American rainforest

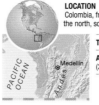

LOCATION On the west coast of Colombia, from the Panama border in the north, south towards Ecuador

TYPE Tropical rainforest

AREA 5,600 square km (2,160 square miles)

Of the many kinds of forest in South America, that found on the northwest coast is one of the richest, with many unique species, including 100 endemic birds. This Colombian forest, known locally as Chocó, certainly deserves

the designation of rainforest, as the region has one of the highest rainfalls in the world, at 11,770mm (465in) a year. Tall *Cecropia* trees are a feature of these forests, which are also rich in tree ferns, orchids, aroids, and bromeliads. Another prominent woody plant is the purple-flowered Glory Bush, but some of the most majestic, at higher altitudes, are the podocarps, conifers that can live for over 1,000 years. The rainforest is under serious threat from logging and mining operations, with a quarter having been lost already. Eight National Parks now protect some areas.

TROPICAL CLOUDFOREST
Lush undergrowth, rich in tree ferns, is typical of the damp cloudforest of La Planada, one of Colombia's best-known reserves.

Central American rainforest

LOCATION From the Yucatan Peninsula of Mexico, south through Belize to Costa Rica and Panama

TYPE Tropical rainforest

AREA 520,000 square km (200,000 square miles)

The narrow land bridge connecting Mexico with South America contains important pockets of diverse tropical forest. Central America has some of the most rapid rates of deforestation in the

prominent red eyes

adhesive pad at end of digit

RED-EYED TREEFROG
Like all treefrogs, the Red-eyed Treefrog has widely spread toes with adhesive pads to help it clamber in the foliage of the Central American rainforests.

world, and pressure on the remaining forest is high. Many sites are protected within reserves, such as the Kuna Yala reserve and the Smithsonian Tropical Research Institute on Barro Colorado Island, both in Panama, and La Tigra National Park in Honduras. Costa Rica, which lies in the middle of the land bridge, contains some of the richest rainforests, with species from the north and south intermingling with endemics.

The Corcovado National Park in Costa Rica is a vital nature reserve, conserving the most important stretch of lowland tropical rainforest on the Pacific Coast of Central America. The park has over 1,500 species of higher plants, 124 mammals, 117 reptiles and amphibians, about 6,000 insect species, and 375 birds, including the rare Harpy Eagle. The habitats range from coastal mangroves, through swamp forests,

to epiphyte-rich lowland rainforest, upland rainforest on higher ground, and finally, at the highest altitudes, to cloudforest, rich in tree ferns. Some of the trees of these forests, such as Cashews, reach heights of 50m (160ft), with some emergents, such as Vantanea, towering to 65m (210ft). Many of the plants are useful to humans, yielding oils, medicines, fibres, edible fruits and nuts, and even natural chewing gum from the Chicle tree. Relatives of the domestic avocado provide a reserve of genetic diversity useful to commercial plant breeders.

DANGLERS AND CLIMBERS
The almost constant moisture in cloudforests, such as this one in Costa Rica (right), allows a luxuriant growth of climbers and epiphytes to develop on the trees.

LOWLAND TROPICAL RAINFOREST
Tall emergent trees rise up above the canopy of this tropical rainforest lying on the bank of the River General, in Puntarenas, Costa Rica.

Amazon Rainforest

LOCATION Mainly in Brazil, Peru, and Bolivia, stretching from the Andes to the Atlantic Ocean

TYPE Tropical rainforest

AREA 6 million square km (2.3 million square miles)

The Amazon Rainforest is the largest area of tropical forest in the world. It occupies much of the basin of the River Amazon (see pp.210–11), which extends across most of northern Brazil, much of Bolivia, and also eastern Colombia and eastern Peru. This tropical lowland is a mosaic of different rainforest types, varying according to soil composition and the water regime. Most of the basin receives an annual rainfall of about 2,000mm (79in), but in

some places this rises to over 8,000mm (315in). Generally, the wetter the forest, the more diverse the flora, and the upper Amazon rainforests are as rich as those of the Colombian Chocó (see Northwest South American rainforest, left). The number of endemic plant species is estimated at 13,700 – nearly 80 per cent of the forest's flora. The rainforest has a complex structure, with distinct tree layers and a large number of lianas and vines, offering an array of sites for colonization by epiphytes, such as bromeliads, and various crevices and hollows that are occupied by many kinds of insects, frogs, and other animals. Some parts of the basin experience seasonal flooding, which transforms the dry forest floor into a vast lake. Such floods, which may last from 2 to 10 months, affect 100,000 square km (38,600 square miles) of forest, and the water can be as deep as 20m (65ft). In Brazil, the

SWAMP FOREST
The Tambopata-Candamo Reserve in Peru protects swamp forest, shown here, and drier habitats on higher ground. It is home to 1,200 species of butterfly.

swamp forests are called igapó forests. Plant diversity studies have revealed there to be as many as 300 woody species in just a single hectare of the Amazon Rainforest, including the much-prized Big-leaf Mahogany. Over 2,000 plant species are used by local people as a source of food, medicine (see panel, right), and other useful products. One of the most famous is the Brazil Nut, and others include natural rubber and the Açaí Palm, which provides palm-hearts to a thriving industry. Although tropical rainforest still covers much of the Amazon Basin, an estimated 745,289 square km (287,758 square miles) have been lost since 1970, mostly due to human activity, especially logging. Severe drought, which hit the region in 2005 and 2010, also poses a significant threat to the rainforest.

BUTTRESSES
The bases of many tropical trees extend outwards as distinct flanges. This gives them more stability in the often shallow, wet soil.

BROMELIADS
The fleshy, spirally arranged leaves of bromeliads form reservoirs that trap water, creating tiny ponds used by many animals, notably insects and frogs, as a home high up in the rainforest canopy.

SOUTHERN TWO-TOED SLOTH
Sloths spend almost all their lives up in the forest canopy, moving extremely slowly, hanging from the branches of trees, and feeding on leaves and fruit.

MEDICINAL PLANT

One of the most important traditional medicinal plants of the Amazon region is Pau d'Arco (shown here). Its inner bark is used for a remarkable range of ailments, including cancer, fungal infection, diabetes, dysentery, malaria, allergies, baldness, and acne; it is also used as a painkiller. Pau d'Arco is growing in importance as a herbal medicine in Europe and the USA.

LAND

ISOLATED RAINFALL
High temperatures and abundant rainfall in the Amazon Rainforest lead to high humidity. Showers here cast isolated shadows over just a small part of the forest's vast expanse.

DEFORESTATION

Humans have had a greater impact on forests than on any other land habitat. Forest clearance began with the development of agriculture, 10,000 years ago, and since then nearly a quarter of the world's tree cover has disappeared. Deforestation is now occurring most rapidly in the tropics, where it poses a major threat to plant and animal life and to the atmosphere, but the temperate world has actually lost far more of its natural forest cover – the difference is that it happened longer ago.

Temperate Deforestation

Trees are the natural vegetation across most of the temperate world, apart from arid areas. Deciduous forest once covered much of Europe, eastern Asia, and eastern North America, while coniferous trees grew on mountains and in the far north. That coniferous forest is still largely intact, but almost all deciduous forest has been cut down. In Europe, clearance was gradual, but in North America it was more abrupt. When European colonists arrived in the early 1600s, forests stretched from the Atlantic coast to the Mississippi. By the mid-1800s, most of this had gone.

Forest cover is now generally stable. In North America, the deciduous forest has partially regenerated, because land fell into disuse when cereal farming moved westwards, onto the open plains. But in most parts of the temperate world, remaining forests are managed as a resource. This is particularly true in the coniferous zone, where an area is totally cleared of trees and then replanted, a practice known as clear-cutting. As a result, natural "old growth" forest – with its varied wildlife – is becoming increasingly rare.

In the American northwest, conservationists have waged a long campaign to save the virgin coniferous forest that remains. In western Australia, a similar struggle centres on the region's unique eucalypt forests, which contain some of the continent's tallest and oldest trees.

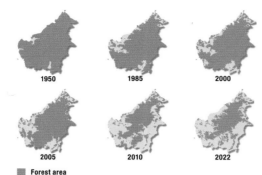

1950 **1985** **2000**

2005 **2010** **2022**

■ Forest area

BORNEO TREE LOSS
Borneo's rainforests are 130 million years old, the oldest in the world. However, between 2002 and 2019, there has been a 45 per cent loss of tree cover (31 per cent loss of primary forest).

Global demand for timber is a major pressure, too. In Borneo in the 1980s and 1990s, up to 240 cubic metres of wood were being taken from every hectare (3,400 cubic feet per acre). This compares with 23 cubic metres (812 cubic feet) in the Amazon. Rainforest soils are generally poor, so the cleared ground is often unproductive when used for agriculture. Despite many initiatives to protect it, tropical rainforest continues to disappear. At least 64 per cent of its original area has been cleared, and a further quarter may vanish by 2050.

Mangrove Forests

As well as rainforests, the tropics are home to mangroves – trees that grow in the zone between high and low tides. Mangrove forests develop on muddy shores, and, depending on local topography, they can be narrow and ribbon-like or they can reach far inland. They stabilize the coast, and they are also important breeding grounds for molluscs, fish, and wetland birds.

Mangrove wood is used for timber and charcoal, and in some parts of the tropics it is an important and sustainable resource. In recent years, the world's mangrove forests have suffered large-scale clearance, mainly to make way for aquaculture (see p.427). Mangrove removal has a direct impact on fisheries, and it also weakens the coast's natural defences against hurricanes and typhoons. However, commercial pressure means that these unique ecosystems are disappearing even faster than forests on land.

HISTORY IN THE HILLS
Scotland's Southern Uplands were once covered with natural forest. Centuries of deforestation, followed by recent replanting, have created the landscape that exists today.

Tropical Deforestation

In populous parts of the tropics, such as India and southern China, most of the original tree cover was cleared several thousand years ago, and remaining trees are now heavily exploited as a source of fuel. But in the humid equatorial belt, deforestation began in earnest only in the mid-20th century, and it is now happening at an unprecedented rate, especially in rainforests.

In some regions, the principal cause is agricultural. In Borneo, for instance, much of the continuing loss of forest cover can now be blamed on agriculture, urbanization, and infrastructure.

COASTAL RETREAT
Scarlet Ibises use mangroves as overnight roosts and as a place to raise their young, from Venezuela southwards to Brazil.

LAND

TREE HARVEST
In Canada, trees are usually removed by completely clearing an area. Although efficient, such clear-cutting harms wildlife and accelerates erosion.

ATLANTIC DRY FOREST
The Itacarambi Reserve protects a tropical dry forest near the Atlantic Coast of eastern Brazil.

SOUTH AMERICA *central*

South American dry forest

LOCATION Southern and eastern Brazil, Bolivia, western Paraguay, and northern Argentina

TYPE Tropical dry forest

AREA 4 million square km (1.5 million square miles)

There are two main areas of dry forest in South America: the chaco region of western Paraguay and nearby Argentina

and Bolivia; and the cerrado and caatinga regions of Brazil. The chaco and caatinga are the driest, with an average annual rainfall of less than 800mm (31in) and a dry season of five to eight months, while the cerrado has annual rainfall of 800–1,800mm (31–71in) and a dry season of three to four months. Many of the trees of these forests belong to the pea family and reach heights of up to about 30m (97ft). Another unusual feature is that they contain large numbers of cacti, particularly in the chaco. The original area occupied by these tropical dry forests has been much reduced, and they are classed as the most threatened of all tropical vegetation. Clearing of land for cattle-ranching is one of the main causes of habitat loss.

SOUTH AMERICA *south*

Southern Andean temperate rainforest

LOCATION On the western slopes of the Andes in central and southern Chile and some parts of Argentina

TYPE Temperate rainforest

AREA 76,000 square km (29,340 square miles)

High rainfall combined with sea mists produces wet conditions that are ideal

MONKEY PUZZLES
These coniferous trees are found only in the northern areas of the temperate rainforest, as here in Malacahuello, Chile.

for these temperate rainforests, which are found from sea-level right up to the tree-line of the Andes. In general, the lower altitude forests are broadleaved evergreens, with conifers dominating higher up, mixing with deciduous forest towards the tree-line. Trees include Lengas, podocarps, the famous Monkey Puzzle, and the Patagonian Cypress, which can live for up to 3,600 years and is valued for its timber. Consequently, intensive logging has made this one of the most imperilled ecosystems on Earth.

MOSS-COVERED LENGAS
These mosses and lichens clothe the trunks of Lengas (which belong to the beech family) in the Patagonian Andes.

EUROPE *southwest*

Mediterranean evergreen forest

LOCATION In the Mediterranean region, now mostly confined to Spain and Portugal

TYPE Evergreen temperate forest

AREA 625,000 square km (240,000 square miles)

Evergreen forests once covered large areas of the Mediterranean, from Portugal and Spain, to Turkey and parts of the Middle East, and also extending along the North African coast. Today, only scattered forests remain, and most of these have been significantly altered by people. In Portugal and Spain especially, managed evergreen woods, dominated by Cork and Holm

SPANISH IMPERIAL EAGLE
This highly endangered raptor is found only in central and southern Spain, where it hunts small mammals and reptiles in woodland and open country. There are over 530 breeding pairs only remaining in the wild.

SCATTERED OAKS
The Monfragüe National Park in Extremadura, Spain, protects fine expanses of dehesa. The trees here are mainly Cork and Holm oaks.

oaks, can be found in many areas. Dehesa is the Spanish term for this traditionally managed habitat, which is called montado in Portugal. It is derived from the original forest and is a park-like landscape of oaks, with the land also used for crops or as pasture for sheep, goats, and pigs. The trees reduce water loss from the soil, and their acorns also provide forage for pigs.

HARVESTING CORK

Cork is probably best known for its use as a stopper for wine bottles. However, it has a wide range of other uses too, from mats to floats, and as an insulating material and sound absorber. The bark of the Cork Oak grows as two distinct layers, and it is the outer layer that is harvested, leaving the inner bark to regenerate more cork. The bark is stripped from the trunks of the oaks every nine years, usually during July and August.

EUROPE *north*

European mixed forest

LOCATION Much of lowland and submontane Europe, from the British Isles to western Russia

TYPE Deciduous temperate and evergreen temperate forests

AREA 4 million square km (1.6 million square miles)

Mixed, mainly broadleaved deciduous forest forms the natural vegetation of much of the lowland and hill country in Europe, and is particularly well developed in central Europe. The main tree species are Common and Sessile oaks, beech, lime, Hornbeam, ash, elm, birch, and alder, with the exact mix depending on the soil type, drainage, and microclimate. Particular types of European mixed forest can be

NEW FOREST
This historic English hunting forest has been managed for centuries and now consists of an intricate mix of woods (as above) and heath.

BLACK FOREST
Conifers such as spruce and Silver Fir mingle with stands of beech in the hills and valleys of Germany's Black Forest. Most of the forest here is secondary, and much is planted.

recognized according to which trees are dominant, such as oak/birch, oak/ash, oak/Hornbeam, lime, or beech. In some areas, for instance on well-drained, sandy soils, pines mingle with deciduous trees or sometimes form pure stands on their own. However, Europe has been settled for a very long time and much original forest has long been cleared, and that which remains is far from its original condition. Beech woods can be found in many parts of Europe, especially central Europe, and some of the best of these remain at submontane levels of the major mountain ranges, such as the Alps,

FLY AGARIC
Common in European woods, especially under birch and oak, this toadstool is highly poisonous, but is unlikely to be mistaken for any other kind.

Pyrenees, Apennines, and Carpathians. In Slovakia, for example, they form about 30 per cent of the forest. At montane levels, spruce and fir often grow together with beech. Some of the forests that are closest to their natural condition, such as the Bavarian Forest straddling the German–Czech border and Bialowieska Forest in eastern Poland and adjacent Belarus, display a mosaic of woodland types within a single region, responding mainly to variations in local soil conditions. The natural processes of death and regeneration can continue unaltered by people in such places, with trees attaining heights of 40m (130ft) and living for 400 years before succumbing to decay or wind-felling. Dead wood is an important habitat for a range of woodland invertebrates, which in turn sustain a diverse population of birds, such as owls and woodpeckers. The mammals include wood mice, Wild Boar, Fallow and Red deer, and, in the case of the Bialowieska Forest, Grey Wolves and European Bison. This old forest also contains over a quarter of Poland's flora (about 550 species).

BIALOWIESKA FOREST
Wild Garlic (or Ramsons) carpets the floor of this ancient forest in eastern Poland, much of which has never been felled.

ARDENNES
The hilly Ardennes region, which straddles the border between France and southern Belgium, is well forested. Beech, oak, and spruce are the major tree species here, mainly in managed forests.

COPPICING

Traditional management of many European woods involved harvesting shoots and poles from selected trees, such as ash and hazel, a process known as coppicing. The stumps then resprout from the base, producing another crop of wood, which can be used to make fences, as below. A few trees are left to grow to their full height. Coppicing results in an open forest floor, allowing a diverse woodland flora to flourish.

LAND

Eurasian boreal forest

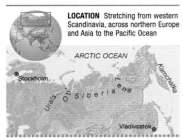

LOCATION Stretching from western Scandinavia, across northern Europe and Asia to the Pacific Ocean

TYPE Boreal forest

AREA 8.75 million square km (3.4 million square miles)

Part of the world's second largest biome and one of the least disturbed by human activity, the Eurasian boreal forest, or taiga, extends over 10,000km (6,210 miles) from Scandinavia to the east coast of Russia. It is broadest in Siberia, where it stretches from 49°N to 72°N. In Europe, the main tree species is Norway Spruce, although Scots Pine dominates in some parts of the boreal forest here, especially in the more oceanic west. Further east, Siberian Spruce, Siberian Fir, and Siberian Stone Pine also occur. Although evergreen conifers cover large stretches of the boreal forest, deciduous trees flourish in some areas, often mixed with the conifers. These trees

WOLVERINE
This powerful member of the weasel family is found in both boreal forest belts. It takes a wide range of food, from eggs and birds, to lemmings and berries, and can even kill reindeer.

BOREAL FOREST, FINLAND
The autumn colours of birch, Finland's national tree, brighten the landscape of the boreal forest in Lapland. Birch rapidly colonizes disturbed soil, and often grows near rivers and lakes.

SIBERIAN BIRCH FOREST
Silver Birch is especially prevalent at the interface between the taiga and tundra. Birch forest is less dense than coniferous forest, and the light reaching the forest floor encourages a thick layer of herbs to develop.

include larch, birch, alder, and rowan, with larch prevalent in a vast expanse of the Yakutiya region of north-central Russia. Towards its northern limit, the taiga starts to thin out, with trees more widely spaced or restricted to sheltered sites such as valleys, and eventually the trees give way to tundra vegetation. To the south, the taiga grades into mixed woodland or, further east, into steppe habitats. In Russia, the northern taiga is interspersed with expanses of bogs and marshland, where impeded drainage produces conditions too wet for forest. Dwarf shrubs such as Bilberry, Cranberry, and Crowberry flourish here. The trees of the boreal forest survive very harsh conditions, with winter temperatures in Siberia plunging to -60°C (-76°F). These mainly dense, dark forests tend to have a rather sparse ground flora but abundant mosses and lichens. The taiga still has stable populations of mammals such as Sable and Eurasian Brown Bears, and typical birds include Capercaillie and Siberian Jays.

NENETS

The Nenet people, who number about 45,000, live in the northern boreal forests and tundra of Siberia in Russia, where they herd Reindeer, on which they depend for food, clothing, tools, and transport. Each summer, they move north with the Reindeer herds to fresh grazing lands in the tundra as it becomes free of ice, and then return to the protection of the taiga during the autumn and long winter. The Nenet have perfected the art of exploiting the harsh environments of the Siberian boreal forests and adjacent tundra. Their diet of Reindeer meat is supplemented by sea and river fish, seals, and wildfowl. However, in the face of increased links with modern society, the Nenet are now struggling to retain their traditional lifestyle.

TAIGA
These mammal tracks are preserved in snow beneath a stand of tall conifers in the Russian boreal forest. In many areas, snow covers the ground for several months.

AFRICA *central*

Central African rainforest

LOCATION From the coasts of Cameroon, Equatorial Guinea, and Gabon to Uganda and Burundi

TYPE Tropical rainforest

AREA 1.9 million square km (1.2 million square miles)

The rainforests of central Africa account for more than 80 per cent of the continent's rainforest, but 2 per cent of this has been removed by human acitvity since the 1990s. The forest is centred on the basin of the Congo River (see p.218). The rainforests of the Democratic Republic of Congo (DRC), in particular, are the richest in the whole of Africa, with about 11,000 plant species, including relatives of impatiens and begonias, and over 400 species of mammal, such as the Bonobo and African Forest Elephant. The trees include valuable timber species such as Sapele, Iroko, and African Mahogany.

MOUNTAIN GORILLAS
The mountain forests of Uganda, Rwanda, and Burundi are home to the world-famous Mountain Gorillas, whose dense fur keeps them warm even at 4,000m (13,200ft).

COASTAL RAINFOREST
Mayombe Forest Reserve in the Congo and DRC (shown here) protects an important tract of tropical rainforest. West Africa's forests are being lost at a faster rate than any others.

AFRICA *southeast*

Madagascan rainforest

LOCATION On the Masoala Peninsula in the northeast, and on hills in the east, Madagascar

TYPE Tropical rainforest

AREA 38,000 square km (14,670 square miles)

leaf-shaped body

prehensile tail

fused toes for gripping branches

PARSON'S CHAMELEON
This chameleon lives mainly in trees, is slow-moving, and so relies on camouflage to avoid predators.

Most of Madagascar's rainforests lie on the eastern side of the island and are now restricted to a narrow, broken strip following the hills just inland from the coast. Some of the best-preserved of the lowland forests are on the Masoala Peninsula in the northeast. The canopy of the forest is low (about 25m/81ft) and, in addition to hardwood trees such as ebony, there are many palms, as well as epiphytes including orchids and ferns, with 80 per cent of the plant species being endemic. However, due to illegal logging, the forest was listed in 2010 as a World Heritage Site in danger. Of the 300 species of reptiles identified, 95 per cent are endemic, including two-thirds of the world's chameleons. Madagascar is renowned for its lemurs, which are found only there and on some nearby islands. Those species that inhabit the rainforest include the Ruffed Lemur (see p.303) and the nocturnal Aye-aye. Over 500 species of Madagascar's plants have medicinal uses, most famously the Rosy Periwinkle, which is now used worldwide for treating some types of childhood leukaemia.

MASOALA NATIONAL PARK
This reserve, covering about 2,300 square km (888 square miles), protects Madagascar's largest remaining rainforest, as well as nearby coral reefs.

ASIA *east*

Asian mountain mixed forest

LOCATION In eastern China, from Shaanxi Province to the Pacific coast and in southern Japan

TYPE Deciduous temperate and evergreen temperate forests

AREA 1.9 million square km (760,000 square miles)

The mixed forests of Asia tend to be restricted to mountain sites, partly because much of the lowland has been cultivated. In China's Sichuan Province, the forests consist of a mixture of needle-

WUHUA LAKE
This forest-fringed lake lies in Sichuan Province. Giant Pandas are still found in these bamboo-rich mountain forests.

BAMBOO FOREST
Bamboo forests are found in several parts of China, notably in the east. In some areas, the bamboo is cropped to provide a wide range of products, from poles to food and medicine.

leaved evergreen trees such as hemlock and spruce, with many broadleaved deciduous trees such as birch, maple, and cherry. In the Qin Ling Mountains, the forests are largely of oaks, pines, and firs. A notable feature of these forests is the abundance of bamboo in the understorey. In Japan, the mountain forests contain Japanese Beech and Cedar, while at higher altitudes conifers such as Nikko Fir dominate, along with Erman's Birch.

LOWLAND RAINFOREST IN BORNEO
Occasional tall, emergent trees rise above the
general level of the canopy in this rainforest,
while the gap created by the river encourages
rapid growth of young trees and climbers.

Indonesian rainforest

LOCATION From the island of Sumatra in the west to Papua New Guinea in the east

PACIFIC OCEAN

Borneo

Sumatra

Jakarta

New Guinea

INDIAN OCEAN

TYPE Tropical rainforest

AREA 1.43 million square km (540,540 square miles)

Indonesia's extensive rainforests form about 10 per cent of the world's forest, but some 1,000 square km (386 square miles) were being removed each year by logging, until 2010, when a moratorium on new logging was imposed. With more than 500 mammal species, around 1,500 bird species, 1,000 reptiles and amphibians, and 20,000 plant species, they are also some of the richest habitats in existence. In lowland areas, a mixed rainforest usually develops below about 950m (3,100ft). Typically this has three main layers of trees, of up to 100 different species. Many belong to the family Dipterocarpaceae, which is virtually absent from the tropical forests outside Asia. About 18m (60ft) tall, with trunks that are branch-free for most of their length, many trees rise into the canopy, a few overtopping it as emergents. Many species flower during the same season, having been free of flowers for several years. The ground

vegetation is made up of young trees, shrubs, palms, and ferns. The palms are more diverse than anywhere else in the world, with 975 species, but the largest family are the orchids, numbering 6,500 species. Some plants that grow here are remarkable, such as the Rafflesia (see below). Swamp forests are widespread in Indonesia, occurring where the ground is waterlogged for much of the year, such as close to rivers that are prone to flooding or on wet, peaty ground. Many of the coasts have mangrove forests in the tidal zones, and these show a transition towards inland rainforest. Above about 950m (3,100ft), montane forest develops. Here the trees are mostly less than 15m (50ft) tall and have smaller, more leathery leaves. Such habitats are rich in epiphytes, including mosses, ferns, orchids, and lichens. Cloud and mist often form, and it is this aerial moisture that supports the epiphytes.

ORANGUTANS
The orangutan, one of our closest relatives, uses its long arms and large hands to help it clamber through the tangled tropical forest.

LAND

petals resembling rotting flesh attract insects

centre of flower releases carrion-like scent

RAFFLESIA ARNOLDII
This species produces a single flower that is larger than that of any other plant. The flower of one individual weighed 11kg (23lb) and measured 1m (3ft) in diameter.

MANGROVE FORESTS
Thriving in marine or brackish mud around many tropical coasts, some species of mangrove produce "knees" (pneumatophores), special aerial growths that bring air into the root tissues.

RAINFOREST FIRES

Fires are a major threat, even to moist rainforest, and Indonesia has suffered badly, both from natural fires and from those started deliberately as an aid to logging operations or for clearing trees to make way for alternative land-use. Once a forest fire has taken hold, particularly in dry conditions, such as those of the droughts in 1997 and 2003, the proximity of the trees allows it to spread rapidly, destroying not only the trees themselves but virtually all the associated wildlife.

LAND

Northeast Asian mixed forest

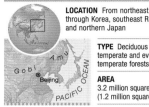

LOCATION From northeast China, through Korea, southeast Russia, and northern Japan

TYPE Deciduous temperate and evergreen temperate forests

AREA 3.2 million square km (1.2 million square miles)

A typical example of this type of mixed forest is that found on the foothills of the Sikhote-Alin' Range in Ussuriland, southeast Russia. It is mostly composed of Cedar Pine, Black Fir, and local species of spruce, ash, linden, maple, and walnut. Lush undergrowth includes plants such as wild vines and ginseng. Mammals include Musk Deer and, in some parts, the rare Siberian Tiger.

LOWLAND FOREST, USSURILAND, RUSSIA

Southeast Asian rainforest

LOCATION In Myanmar, Laos, Thailand, peninsular Malaysia, Cambodia, and Vietnam

TYPE Tropical rainforest

AREA 1.7 million square km (655,000 square miles)

Like those of Indonesia (see p.315), the rainforests of mainland Southeast Asia are dominated by dipterocarp trees, which account for 40 per cent of the understorey and about 80 per cent of emergents. The tallest trees reach heights of 40–50m (130–162ft). The valuable timber tree Teak is native to Myanmar, Thailand, and Laos, where it is usully found mixed with other trees. But logging has greatly reduced the natural stands, and Teak plantations in various Asian countries, Africa, and the Americas now help to meet the incessant demand. In peninsular Malaysia, the rainforests contain large

MONTANE RAINFOREST
The forests surrounding Thailand's highest peak, Doi Inthanon, range from tropical to a more temperate cloudforest towards the summit.

RAJAH BROOKE'S BIRDWING
Both sexes of this large butterfly feed on flowers, and the males (shown here) often drink from wet mud on the forest floor.

elongated forewing typical of birdwing butterflies

jet-black body

distinctive contrast of green on black base

numbers of rattans. These palms are harvested for thatching, matting, and basketry materials, and are a source of dyes and medicines. Many familiar tropical fruits originate from the Southeast Asian forests, such as bananas and mangoes, and also durian.

LOWLAND RAINFOREST
These tropical trees in southern Thailand have typically straight trunks, branching mainly towards the crown.

New Zealand mixed forest

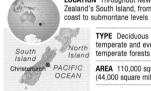

LOCATION Throughout New Zealand's South Island, from the coast to submontane levels

TYPE Deciduous temperate and evergreen temperate forests

AREA 110,000 square km (44,000 square miles)

Temperate forests still cover a large area of New Zealand's South Island. Common trees include Southern and Silver beech and members of the podocarp family, long-lived conifers that were widespread in the Mesozoic Era but which today are mostly confined to the southern hemisphere. Deciduous trees such as Tawa and Kamahi often dominate in the forest canopy, and these tend to be overshadowed by tall podocarps, such as Kahikatea, Rimu (or Red Pine), and Totara.

COASTAL FOREST
New Zealand has large areas of well-preserved coastal lowland forests, as here in Catlins, in the far south of South Island.

Australian eucalypt forest

LOCATION In northern, eastern, southwestern, and southeastern Australia, including Tasmania

TYPE Evergreen temperate forest

AREA 1.01 million square km (389,900 square miles)

Eucalypt forest is scattered through various regions of Australia, but not in the interior, which is too arid for tree

JOSEPH BANKS

Englishman Joseph Banks (1743–1820) was one of the great botanical explorers and collectors. Aged 25, he joined James Cook's expedition to the South Pacific on the *Endeavour*. In 1770, they landed in eastern Australia, where Banks amassed a vast collection of botanical specimens then unknown to Europeans. Appropriately, Captain Cook named their landing place Botany Bay.

growth. Australia has about 450 species of eucalypt, nearly all of which are evergreen. Different species dominate in different areas: Karri in the north and southwest; Sydney Blue Gum in the east; and Australian Mountain Ash in the southeast. Fires are frequent during dry weather, and most gum trees have fire-resistant bark. They can also regenerate rapidly from seed or by sending up suckers. Koalas spend nearly all their lives up in the trees in the forests of east Australia, feeding on gum leaves at night. One of the smallest possums, the Honey Possum, is only found in the forests of the southwest. The eucalypt forest is being threatened by plant dieback disease, alien species, and human activity, with 15,000 square km (580 square miles) cleared between 2000 and 2004.

BLUE MOUNTAINS
The Blue Mountains, to the west of Sydney, get their name from the bluish haze created by sunlight filtering through a mist of eucalypt oil released by the gum trees covering their slopes.

OLD-GROWTH RED GUM TREES
Many species of eucalypts shed old bark as they grow, revealing lighter new bark underneath, as seen in these old red gums in Victoria.

FAST-FLOWING STREAM
This stream is in the Daintree rainforest, North Queensland. The forests in this area contain 41 per cent of Australia's freshwater fish species.

Northeast Australian rainforest

LOCATION In the coastal ranges of northeast Queensland, from Cape York south to the Connors Range

TYPE Tropical rainforest

AREA 33,200 square km (12,800 square miles)

The largest expanse of Australia's remaining tropical rainforest is found between Cooktown and Townsville, centred on Cairns. The forests are best developed on the eastern slopes of the coastal ranges, where the average annual rainfall is highest, at more than 1,500mm (59in). The canopy is uneven, varying from 20m to 40m (65ft to 130ft) in height, and the undergrowth features clambering vines and epiphytes, as well as Walking Stick Palms, ginger, and aroids. The trees include valuable timber species such as Queensland Maple, kauri pines, and Red Cedar (though logging is now illegal in much of the forest), and strangler figs. At the highest altitudes, over 1,000m (3,300ft), the trees are shorter, and these cloudforests have a dense growth of ferns and mosses. Australia's geographical isolation means that it has a high number of ancient and unique species, and Queensland's tropical rainforests account for much of the country's biodiversity. About 20 per cent of Australia's cycads (see p.23) are found here, as is Idiot Fruit, a primitive flowering plant that originated 120 million years ago. Two species of tree kangaroo, Lumholtz's and Bennett's, live in the rainforest, along with many other endemic mammals, birds, reptiles, and frogs.

LUMHOLTZ'S TREE KANGAROO
This nocturnal marsupial is a highly agile climber, using its cushioned, rough-soled feet to cling to branches and its long tail as a counterbalance.

SATIN BOWERBIRD
To attract the female, the glossy black male builds an elaborate bower on the rainforest floor, which it decorates with any blue object it can carry in its beak.

DAINTREE RAINFOREST
This rainforest, a major part of the Wet Tropics World Heritage Site created in 1988, clothes the coastal hills near Cairns and stretches right down to the sea.

ABOVE THE CANOPY

Skyrail is a cableway near Cairns that runs for 7.5km (4⅔ miles) from the base of the MacAllister Range to Kuranda, a village high up in the rainforest. There are 114 gondolas, from which tourists can gaze down at the forest and its inhabitants with the minimum of disturbance. There are also two stops where the cableway descends through the canopy and people can get out and walk through the forest on boardwalks.

LAND

LAND

SEASONAL FLOODING
Waterlilies bloom near the mouth of the
River Limmen, which lies in the northeast
of Australia's Northern Territory. This
river has extensive areas of seasonally
inundated floodplains along its banks.

WETLANDS

Wetlands are found all over the world, in places where drainage of water is impeded or in areas adjacent to the floodplains, deltas, and estuaries of large rivers. A wetland is any waterlogged or flooded terrestrial area that has a covering of water plants, either rooted or free-floating. The water is often just 1m (3ft) deep. Not all wetlands are flooded all the year round – for example, some wetlands occur in areas that are seasonally flooded by rivers; others form where spring meltwater cannot drain away because of underlying permafrost. Wetlands are highly important ecosystems, supporting a wide variety of specially adapted plants and animals, most notably vast numbers of water birds. They may act as reservoirs of water for human populations, and are also important as a source of fish and for recreational activities. Many of them are increasingly threatened by encroachment – with large areas being drained for housing or agricultural use – and also degraded by different types of pollution.

Wetlands

◄ 106–107 Salt water and fresh water

109 The global water cycle and the local water cycle

198–223 Rivers

224–39 Lakes

Tundra 338–39 ►

Wetlands can be divided into freshwater wetlands, where the water is mainly derived from rainfall, and saltwater wetlands, which are generally coastal and influenced by sea water. Inland saltwater wetlands occur where evaporation has concentrated salts in the surface layers of the soil. Many freshwater wetlands occur at lake and river margins, especially along the floodplains of lowland rivers, and soil saturation can be highly seasonal.

Freshwater Wetlands

The main types of freshwater wetland are fens, bogs, swamps, and marshes. Fens and bogs, known collectively as mires, are habitats rich in wet, peaty soils. Bogs receive water from rainfall alone, and form acid peat. Fens receive groundwater as well as rainfall – they are generally wetter, and form neutral or alkaline peat. Mires are most widespread in temperate regions with high rainfall, as in the north and west of North America and Europe. They also occur at high altitudes if conditions are cool and damp. There are many gradations between fens and bogs, with mixed mires or intermediate mires occurring in some areas. The terms swamp and marsh classify wetlands by the plants that grow in them – a swamp is a wetland forest, while a marsh is a wet grassland.

WICKEN FEN
This wetland reserve in Cambridgeshire protects a vital remnant of the once widespread fens of eastern England.

PEAT CUTTING
Large areas of Ireland are covered in bogs, especially in the west, as here in County Clare. The peat is traditionally cut, dried, and used for fuel, but is a non-renewable resource.

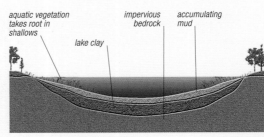

aquatic vegetation takes root in shallows · impervious bedrock · accumulating mud · lake clay

WETLAND SUCCESSION
Freshwater wetlands often occur on peat bogs, which form as lakes are infilled by sediment. In the first stage of this process, mud collects on an impervious lake bed.

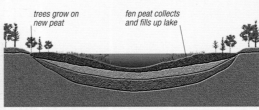

trees grow on new peat · fen peat collects and fills up lake

INTERMEDIATE STAGE
Fen peat develops above the lake mud as the partly decomposed remains of plants, such as sedges and rushes, gather under the influence of mineral-rich (alkaline) groundwater.

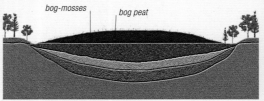

bog-mosses · bog peat

PEAT BOG
As fen peat fills the lake, the surface is isolated from alkaline groundwater, and slowly becomes more acidic. This encourages the growth of bog-mosses, whose remains gather as bog peat.

Saltwater Wetlands

Most saltwater wetlands occur at the interface of land and sea, where fresh and salt water mix. A range of habitats forms where large rivers with wide floodplains braid to create extensive deltas. These include brackish (slightly salty) and saltwater lagoons, and salt marshes with salt-tolerant vegetation covering the rich estuarine mud. In warm climates, salty lagoons are often managed by people as a method of salt production. In the tropics, mangrove swamps are often found in tidal creeks and gullies. Mangrove trees germinate fast in the mud, their stilt-like roots stabilizing the soil and allowing the plants to survive tidal sea-level changes. Saltwater wetlands also occur inland, frequently in arid regions, and even in some deserts, where intermittent flooding creates temporary water bodies.

SALT MARSH
Salt marshes develop in flat, coastal regions that are regularly flooded at high tide, such as here in the Chincoteague Inlet, USA.

FOSSILS IN PEAT

Bogs preserve material because wet peat contains little oxygen and so slows down the processes of decay. The peat bogs of northern Europe, particularly those in Denmark, have yielded some remarkably well-preserved human bodies, about 2,000 years old, still with their hair, skin, and fingernails intact. These include the remains of the man shown here, thought to have been the victim of ritual sacrifice.

LAND

Plant Life

Wetlands present a variety challenges and opportunities to plants. Many aquatic or marsh plants have light, spongy stems and leaves, adapted to transport atmospheric oxygen to the roots, allowing the plant to keep respiring. Some, such as Water Hyacinth and Water Fern, also have waxy or hairy leaves, which resist waterlogging. By contrast, salt-marsh plants often display features shared by desert plants, such as a thick cuticle and narrow leaves. Salt water is not easily absorbed by plants (as salt solutions tend to draw water out of plant tissues by osmosis), so many salt-marsh plants suffer effective drought conditions, at least some of the time. Some plants growing at the edges of creeks, ponds, or lagoons, such as reeds, are held above the water by woody stems and can survive short-term changes in water level.

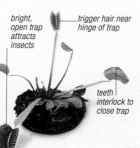

bright, open trap attracts insects

trigger hair near hinge of trap

teeth interlock to close trap

VENUS FLYTRAP
Insects caught in this plant's leaves provide essential nutrients that are rare in the soils of the eastern USA, where it grows.

SPHAGNUM MOSS
This fast-spreading inhabitant of bogs and fens grows half-submerged. Its decomposing remains form many types of peat.

MARSH MARIGOLD
The Marsh Marigold, which is common in temperate freshwater wetlands, produces brightly coloured flowers in spring and summer.

Animal Life

Fish and many aquatic invertebrates thrive in wetlands where there is sufficient water available, and in turn they provide food for a host of other animals, most notably a wide range of water birds. Herons and egrets stalk their prey in shallow water, ducks filter-feed at the surface or dive for food, while waders are adept at snatching invertebrates at or just below the surface of the wet mud. A large number of reptiles, including crocodiles, alligators, and freshwater turtles, are at home in wetlands. Grass snakes often swim in search of frogs, and tropical wetland species include the world's largest snake, the Anaconda. The mammals have produced many wetland specialists, especially among the rodents, such as beavers, Muskrats, and Capybaras. Larger herbivores that are at home in wetlands include the Asian Water Buffalo and hippopotamuses.

ATLANTIC MUDSKIPPER
Mudskippers are a unique fish that can climb out of water, using its fins to grasp.

JAGUAR
The Jaguar's preferred habitat is swamps or seasonally flooded forests. It is a strong swimmer, and its diet includes Capybaras, fish, and caimans.

SEASONAL GATHERING
Wetlands such as Botswana's Okavango Delta burst into life in the wet season, when herds of game such as these Burchell's Zebras gather.

WETLAND PROFILES

The pages that follow contain profiles of some of the world's main wetlands. Each profile begins with the following summary information:

TYPE Marsh, bog, fen, swamp, mire, delta, or lagoon

WATER Fresh, acid fresh, brackish, or salt

AREA Surface area

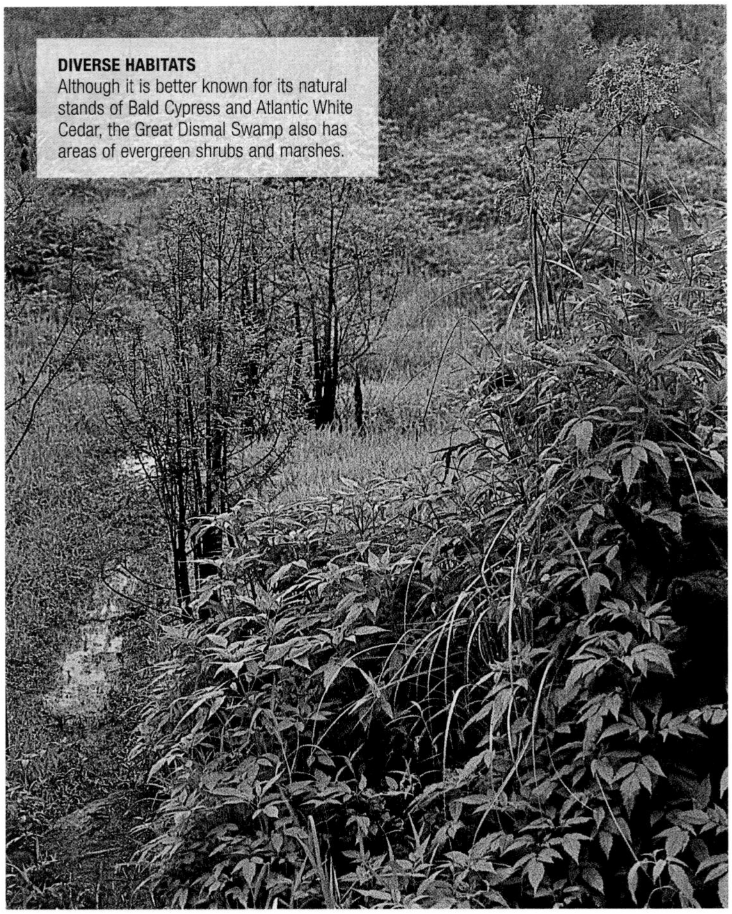

DIVERSE HABITATS
Although it is better known for its natural stands of Bald Cypress and Atlantic White Cedar, the Great Dismal Swamp also has areas of evergreen shrubs and marshes.

NORTH AMERICA *east*

Great Dismal Swamp

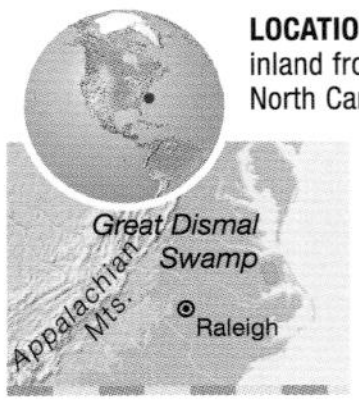

Great Dismal Swamp
Appalachian Mts.
Raleigh

LOCATION About 40km (25 miles) inland from the Atlantic Ocean, in North Carolina and Virginia, USA

TYPE Swamp

WATER Acid fresh

AREA 1,550 square km (600 square miles)

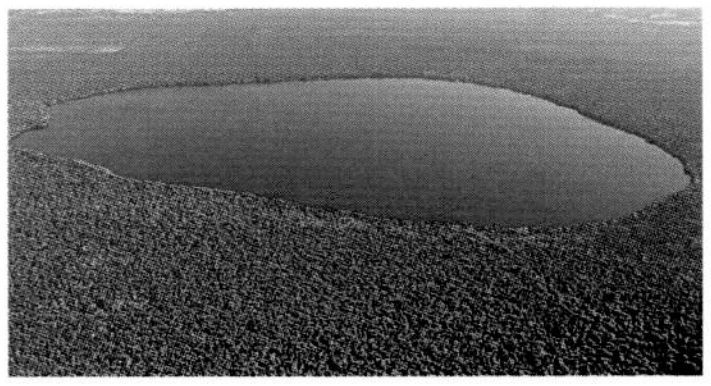

LAKE DRUMMOND
This lake, edged with Bald Cypress trees, was named after William Drummond, Governor of North Carolina in the 1660s – though it lies in Virginia.

Despite its discouraging name, the Great Dismal Swamp is a beautiful complex of wetland habitats in which tracts of wet forest are interspersed with scrub-shrub wetlands and areas of peat bog. The swamp is unusual in being located above sea-level – most swamps occur in natural depressions, but the Great Dismal has been rebounding slowly since the end of the last ice age. A 45-m- (150-ft-) deep layer of clay below the swamp prevents its waters seeping away. The variety of sites results in a rich diversity of plant and animal life. Forests of Bald Cypress, Atlantic White Cedar, and Black Gum (or Tupelo) dominate the marshy ground. Ground vegetation includes the rare semi-evergreen Log Fern. At the centre of the swamp lies Lake Drummond – the lake and swamp waters support catfish, Yellow Perch, Redfin Pickerel, and the partially sighted Swampfish. Almost 100 species of bird breed here, including Wood Duck and Barred Owls, and mammals include White-tailed Deer, Raccoons, Otters, Black Bears, and even Bobcats. About 250 years ago, the swamp was as large as 5,700 square km (2,200 square miles), but logging, agricultural ventures (George Washington was an early investor), and roads, canals, and ditches have severely reduced its size. About 430 square km (170 square miles) of the swamp is now protected in an area designated as a National Wildlife Refuge.

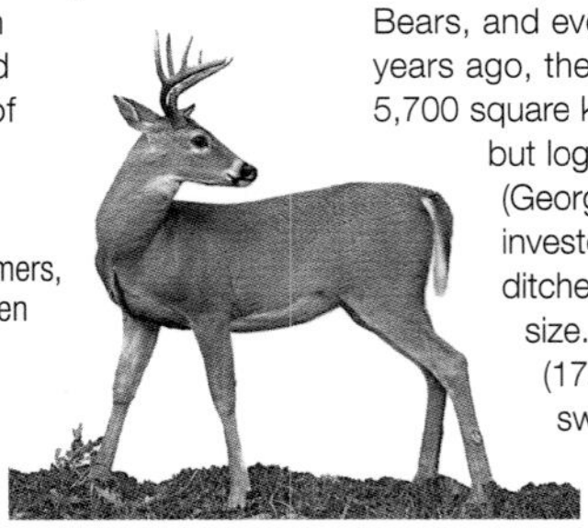

WHITE-TAILED DEER
Known to be good swimmers, White-tailed Deer are often seen in wetland areas, feeding on green plants, including aquatic species, in the summer months.

NORTH AMERICA *southeast*

Okefenokee Swamp

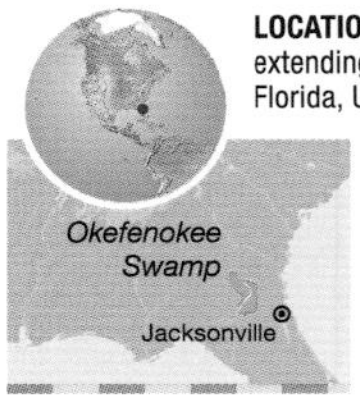

Okefenokee Swamp
Jacksonville

LOCATION In southeast Georgia, extending south into northern Florida, USA

TYPE Swamp, marsh

WATER Acid fresh

AREA 1,770 square km (685 square miles)

The Okefenokee Swamp is dominated by an enormous mire lying inside a large, concave depression that is drained by the Suwannee River. Peat deposits have accumulated over the centuries to the extent that the swamp's surface is now over 30m (100ft) above sea-level, and the peat is up to 4.5m (15ft) thick. In some areas the peat floats on waterlogged substrata, so that the whole surface quakes when trodden on. This is the origin of the swamp's name, from a Hitchiti language term meaning "land of the trembling earth". Mires of this type, known as quaking bogs, are rare. The swamp's waters are sediment-free, but tinted the colour of black tea by tannic acid released from decaying vegetation and peat. Although mires dominate the landscape, pines grow on some drier islands, especially towards the northern margin. Recent wildfires in 2007 and 2011 have burned significant areas of the swamp. Another distinctive feature are large areas of marsh called water prairies. Bladderworts, sundews, and pitcher-plants are all found in the swamp, compensating for the nutrient-poor soil with a carnivorous diet. Alligators thrive, as does the world's largest freshwater turtle, the Alligator Snapping Turtle. Snakes include the Florida Cottonmouth, which, unlike other water snakes, swims with its head held out of the water. Birds abound, with over 230 species recorded, including the rare Red-cockaded Woodpecker. The swamp is also home to a population of Florida Black Bears.

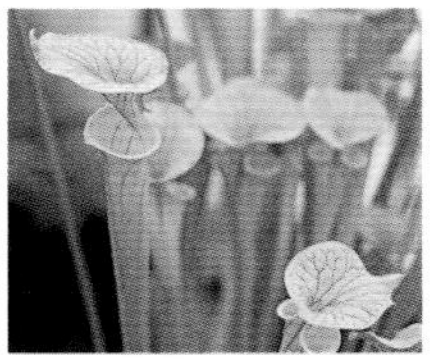

GOLDEN TRUMPET
This is one of three species of pitcher-plant found in the Okefenokee. The pitchers are highly modified leaves, in which insects are trapped and digested.

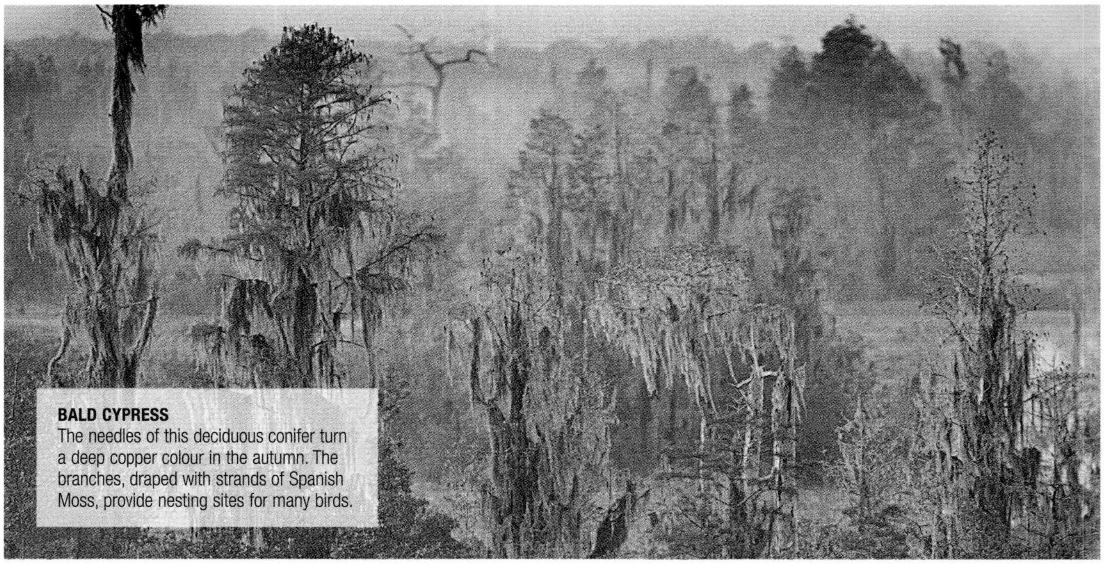

BALD CYPRESS
The needles of this deciduous conifer turn a deep copper colour in the autumn. The branches, draped with strands of Spanish Moss, provide nesting sites for many birds.

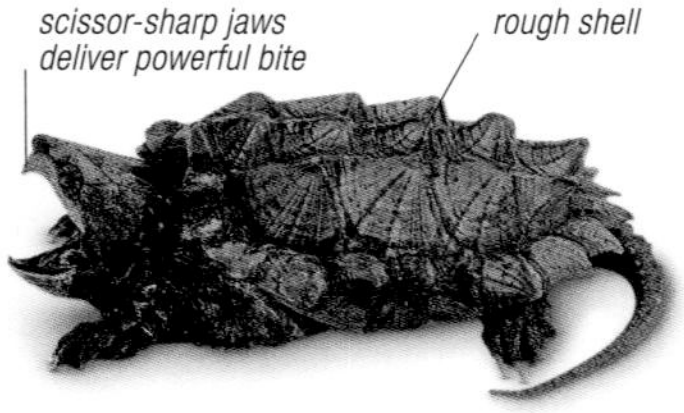

scissor-sharp jaws deliver powerful bite
rough shell

ALLIGATOR SNAPPING TURTLE
This fearsome-looking reptile lurks in the swamp's dark, tannic waters, ready to lunge at unwary prey, which it secures in its powerful jaws.

NORTH AMERICA *southeast*

Everglades

LOCATION Extending south from Lake Okeechobee to Florida Bay, Florida, USA

TYPE Swamp, marsh

WATER Fresh, salt

AREA 10,360 square km (4,000 square miles)

GULF OF MEXICO
Orlando
Miami
Everglades

SAWGRASS
This plant is actually a member of the sedge family, rather than a grass, and gets its name from the serrated edges to its long leaves.

MANGROVES
Mangroves line many of the narrow rivers and creeks near the coast of the Everglades. Here, seedlings emerge from the rich coastal mud, putting out stilt roots for support.

North America's only subtropical wetland, the Everglades, begins at Lake Okeechobee (see p.228). Water flowing from the lake seeps slowly through the low-lying land towards the Gulf of Mexico, creating a wetland wilderness with prairie-like expanses of sawgrass. This grass-dominated seepage area, known locally as the river of grass, is dependent on fire, usually ignited by lightning strikes, which burn back old grass that would otherwise impede the flow of water. Hardwoods are found on pockets of higher ground, mainly along limestone ridges, and these islands of trees are known as hammocks. The Everglades also has large areas of cypress marshes and, along the coast, mangrove swamps, salt marshes, and estuaries. Much of the original Everglades has been lost to development (see panel, right). Dykes, put in place to assist agriculture and

STREAM CHANNELS
The swamp is dissected by a multitude of channels that slowly carry the sluggish waters towards the Gulf of Mexico.

other development, have altered the flow of water so that much of it is now diverted to the coast instead of draining into the soil. Agricultural fertilizers have also encouraged the spread of cattail weeds, which overwhelm the sawgrass. A National Park in the south protects some of the ecosystem, which is home to birds such as Snail Kites, Purple Gallinules, and Anhingas. The Everglades is also famous for being the only place where crocodiles and alligators coexist.

WILDLIFE HAVEN
The Everglades has a large and varied population of birds. Here, a Blue Heron is seen in the waters of the Big Cypress National Reserve.

LAND RECLAMATION

The Everglades originally occupied most of the southern tip of the Florida Peninsula. However, southern Florida's warm climate has attracted a large human population. Many areas of land that were once part of the Everglades ecosystem have been drained and reclaimed for urban use. This drainage has gradually reduced the natural flow of water through the Everglades, with devastating effects on its ecology.

HOUSING DEVELOPMENT
These houses surround a water run-off pond, a drainage feature required by Florida law. The pond collects water from the housing development and helps to prevent possibly polluted water reaching the Everglades.

LAND

Pantanal

LOCATION In the Mato Grosso and Mato Grosso do Sul states of Brazil, extending into Bolivia and Paraguay

TYPE	Swamp, marsh
WATER	Fresh
AREA	129,500 square km (50,000 square miles)

VICTORIA WATER LILY
This magnificent aquatic lily has large floating leaves, about 1–2m (3–6ft) in diameter, which can support the weight of a child.

At around 13 times the size of the Everglades, the Pantanal is the world's largest and most biologically diverse freshwater wetland. However, it now has critical/endangered status because of wildlife poaching, the contamination of its waters by pesticides, and gold-mining

CAPYBARAS
The Capybara is at home on land and in water. Its ears, eyes, and nostrils are set on top of the head, so it can remain alert while swimming.

MARSH AND FLOODED FOREST
This aerial view shows the complex of inter-related habitats that make up the Pantanal, with vegetation-covered waterways cutting between forested embankments.

effluent. A low-lying floodplain with rich alluvial soil, the Panatanal occupies about one-third of the basin of the upper Paraguay River. It is a dynamic ecosystem governed by the annual flooding of the river, for which it acts as a sponge. Each year the waters rise by several metres, transforming a vast area of dry grasslands and forests into temporary swamps, islands, pools, and water channels. Floating aquatic plants, such as Water Hyacinth and Water Ferns, often carpet the water, in which piranhas, caimans, and anacondas are to be found. Caranda Palms grow on the flooded land, while on the higher ground the mixed forest includes fig trees and Acuri Palms. The Pantanal is home to the endangered Jaguar, the Capybara – the world's largest rodent – tapirs, and Capuchin and Howler monkeys. Birds include the rare Hyacinth Macaw, and there are thousands of species of butterfly.

Llanos wetlands

LOCATION Centred upon the River Orinoco and its tributaries in western Venezuela

TYPE	Swamp, marsh
WATER	Fresh
AREA	10,000 square km (3,860 square miles)

SCARLET IBIS
This brightly coloured bird breeds in treetop colonies, often over water, and feeds in flocks in lagoons, swamps, and estuaries.

The Llanos is a term used to describe the huge expanse of savanna plains that stretch across Colombia and Venezuela from the Andes to the Orinoco Delta. Within this region, the Llanos wetlands comprise large areas of swamps and seasonally flooded savanna around the River Orinoco (see p.209) and its tributaries. The heaviest rains fall in May, at the end of the dry season, causing the rivers to overflow and transform the surrounding forests and grasslands into temporary wetlands. These support about 70 species of water birds, including around 90 per cent of the world's population of the endangered Scarlet Ibis.

FLOODED LLANOS PLAIN
Floods during the wet season (May to October) create islands of forest surrounded by extensive pools and swamps, which gradually dry out after the rains cease.

Flow Country

LOCATION In the counties of Caithness and Sutherland in the far north of Scotland

TYPE	Bog
WATER	Acid fresh
AREA	4,000 square km (1,540 square miles)

The Flow Country is a landscape of huge, rolling peatlands in the extreme north of Scotland, and it contains some of the best-developed blanket bogs in the world. Blanket bogs are wet peat formations that hug the contours of the land and which

BOGBEAN
The white flowers of Bogbean stand out from the dark, peaty waters in which it grows.

grow very slowly, as little as 1mm (1⁄16in) a year, under the influence of a cool, wet climate. Sphagnum mosses, which ultimately create and sustain the bogs (see p.321), thrive, and ridges, hummocks, and small pools are rich in a variety of other plants.

DUBH LOCHANS
Small, dark pools, known locally as dubh lochans, break up this vast and otherwise even expanse of blanket bog.

EUROPE *central*

Biebrza Marshes

LOCATION Centred on the Biebrza River Valley, in Suwalki, Lomza, and Bialystok provinces, northeast Poland

TYPE Bog, fen, marsh

WATER Fresh

AREA 1,000 square km (390 square miles)

The Biebrza Marshes are a complex mixture of habitats, including river channels, lakes, extensive marshes, wooded areas on higher ground, and peat bogs that are some of the best preserved in the world. Many parts show a classic succession from riverside fen through raised bog, grading into wet woodland. These

GREAT SNIPE
This wader breeds on the Biebrza Marshes, where male birds gather in groups called leks to attract females by posturing and making croaking and bubbling calls.

varied habitats support a wide range of wildlife. The birds found here include many species of waders, gulls, terns, and both White and Black storks. Mammals include wetland specialists such as Elk, Muskrats, and Eurasian Beavers.

WATER MEADOWS
The water meadows of these marshes are regularly flooded, and the resulting rich alluvial soil supports a varied community of wetland plants.

EUROPE *central*

Hortobágy

LOCATION Centred on the upper reaches of the River Tisza on the Great Hungarian Plain, Hungary

TYPE Lagoon, marsh

WATER Fresh, salt

AREA 115 square km (44 square miles)

Hortobágy lies in the Hungarian steppe region, known as the Puszta, an area consisting mainly of semi-arid grassland. Regular flooding by the River Tisza has turned part of this otherwise dry landscape into a wetland encompassing streams, lakes, lagoons, salt marshes, and river-valley woodland. The region became Hungary's first National Park in

COMMON CRANES
Large flocks of these migratory birds often stop over at Hortobágy to feed on insects, frogs, and plants.

1973 and a World Heritage Site in 1999. Now recognized as one of the world's best wetland reserves, it hosts over 300 bird species. Regulation of the flow of the River Tisza has reduced flooding, and the park authorities now use canals to preserve some of the wetland habitats.

REED-BEDS
The lakes of Hortobágy are fringed by dense beds of reeds, which are used for shelter by ducks, herons, and other birds.

EUROPE *west*

Camargue

LOCATION Lying within the Rhône Delta, on the northwest shore of the Mediterranean Sea, France

TYPE Lagoon, marsh

WATER Fresh, salt

AREA 850 square km (330 square miles)

The Camargue is based on sediments deposited by the River Rhône (see p.213) as it braids to form a huge delta, just to the west of Marseille. This diverse wetland comprises salt marshes, saltwater and brackish lagoons, freshwater ponds, rivers, grazed flooded meadows, reed-beds, and dunes. The plants found here include salt-tolerant tamarisk bushes, Sand Crocuses, and the white-flowered Sea Daffodil, as well as many orchids. But, aside from its white horses and black bulls, it is for its birds that the Camargue is best known. The breeding birds include its famous colony of Greater Flamingos (now numbering about 10,000), as well as Black-winged Stilts and Avocets. Its location also gives it great significance as a refuelling stop for migrating birds flying

BRINE SHRIMP
These translucent crustaceans thrive in temporary salt lagoons, providing food for flamingos and other birds.

between Europe and Africa. Much of the original wetland has been drained for conversion to agriculture (see panel, right), and a series of dykes and canals control the flow of the water. This, along with pollution of the Rhône, has led to a deterioration of some of the habitats. In 1970, the entire Camargue was designated a Regional Natural Park, and some smaller areas have since been given greater protection as nature reserves.

RICE CULTIVATION

Large areas of the Camargue, especially in the north, have been drained and are used to grow cereals, fruit, and vegetables under irrigation. Rice, which has been cultivated here since the Middle Ages, in particular thrives in the region's rich alluvial soil. The small plots of land in which the rice is grown are flooded with water from the irrigation canals in spring, and the crop is harvested mechanically during September and October, once the rice-fields have been drained.

FERAL HORSES
The Camargue's famous white horses are actually grey when young, only becoming white when four or five years old. For centuries they have been used by local farmers and breeders to round up the black bulls that also roam the salt marshes.

LAND

Okavango Delta

LOCATION In the Ngamiland District of northern Botswana, extending into the northern Kalahari Desert

TYPE Delta, swamp

WATER Fresh

AREA 2,000 square km (770 square miles)

The lagoons, swamps, and savannas of the Okavango Delta form what is arguably the greatest wildlife wilderness in southern Africa. The river starts its life as the Cubango in Angola, to the northwest, but instead of flowing to the coast, it spreads out through the flat land of northern Botswana, eventually dissipating in the parched savannas and desert sands of the northern Kalahari. Each year, the rains cause the Okavango River to flood, generally peaking in May, and then the waters spread, covering more than 13,000 square km (5,000 square miles). In high-flood years, the waters even reach the Makgadikgadi Pans (via the Boteti River), a huge area of salt pans about 200km (125 miles) to the southeast, turning it into lagoons that provide a temporary haven for wildlife, most notably large flocks of flamingos. Although most of the floodwaters disappear into the dry atmosphere and arid, sandy soils, enough remains to form a network of clear streams, lagoons, and swamps, bringing a rapid greening of the landscape and a massive influx of herbivores, which in turn draw predators such as lions and leopards. The Okavango Delta's aquatic vegetation includes bladderworts (free-floating plants that trap and digest mosquito larvae), and dense beds of phragmites reeds and Papyrus. These reed-beds are home to the Sitatunga, a semi-aquatic antelope whose widely splayed hooves are an adaptation for walking on soft, muddy ground. Kingfishers, Pel's Fish-owl, and the African Fish-eagle are among over 500 species of bird found in the delta, along with about 160 species of reptiles, 40 species of amphibians, and 90 species of fish.

SWAMP IN RIVER DELTA
The waters of the Okavango River disperse through a series of braided channels and spill over into the delta, depositing vast amounts of sediment.

FLOODING IN THE KALAHARI
The annual flooding of the Okavango River inundates large areas of land in the northern Kalahari, creating a temporary wetland with islands of grass and palm trees.

AFRICAN ELEPHANTS
Although they live in a variety of habitats, African Elephants will usually seek fresh water to drink – once a day, if possible, or every few days. They also bathe to help keep their skin in good condition.

ECOTOURISM

Ecological tourism has developed rapidly in the Okavango Delta in recent years. One of the main tourist destinations is Botswana's Moremi Wildlife Reserve, in the northeast of the delta. The reserve, established in 1963, covers about 3,000 square km (1,160 square miles), and includes large islands of dry land as well as lagoons and swamps. Ecotourism has brought welcome employment for large numbers of local people, who are skilled at identifying and tracking the wildlife.

MOKORO SAFARI
These local guides are leading a party of ecotourists through the wildlife-rich waters of the delta in dug-out canoes, or mokoros.

AFRICA *central*

Sudd

LOCATION Centred on the upper reaches of the White Nile river, southern Sudan

TYPE Marsh

WATER Fresh

AREA 34,500 square km (13,300 square miles)

One of the world's largest inland wetlands, the marshes of the Sudd cover a vast area in the upper reaches of the White Nile (see p.217). This is a landscape of reed-beds and Papyrus, with areas of open water that are often choked by dense mats of floating Water Hyacinth. The rains fall mostly between April and September, keeping the marshes wet, and flooding of nearby grasslands and woods trebles the wetland's extent to over 100,000 square km (38,600 square miles). The Sudd is a haven for wildlife in an otherwise dry region, and it positively teems with birds, especially during the migration periods, when the number of species exceeds 400. The eastern edge of the Sudd is scarred by the Jonglei Canal, an ambitious and uncompleted project designed to divert water and to drain part of the marshes. Abandoned for financial and political reasons, this project has left a trough about 5m (16ft) deep, 75m (245ft) wide, and 360km (225 miles) long, which blocks the migrations of large mammals such as giraffes, elephants, and hippopotamuses.

SHOEBILL STORK
Also known as the Whale-headed Stork, this distinctive African bird is seldom seen far from water. It feeds mostly on lungfish, and also frogs and small mammals.

NILE LECHWE
Of all antelopes, Lechwe are the most at home in wetlands. They require a good supply of high-quality green vegetation, such as that found in the Sudd marshes.

AUSTRALASIA *New Zealand*

Waituna Lagoon

LOCATION In the Southland region, at the southern tip of South Island, New Zealand

TYPE Lagoon, bog

WATER Fresh, salt

AREA 35 square km (13 1/2 square miles)

Now part of the Awarua Wetland spanning 200 square km (77 square miles), Waituna is the southernmost of all the world's recognized wetlands, comprising areas of salt marsh and peatland, as well as the lagoon itself. It is unusual in that it contains areas of cushion bog, which have vegetation that is normally associated with upland conditions rather than coastal regions. The cushion plant *Donatia novae-zelandiae* grows here, as do gentians, sundews, Comb Sedge (*Oreobolus pectinatus*), Wire Rush, Tangle Fern, and Manuka. Insects are highly diverse, with over 80 species of moth having been recorded. Endemic birds include the Australasian Bittern, South Island Fernbird, and Variable Oystercatcher, while Royal Spoonbills and Grey Teal are regular visitors to the lagoon.

MANUKA
Some areas of the Waituna wetlands are dominated by Manuka, an evergreen shrub that yields a much-prized honey.

AUSTRALASIA *Australia*

Coorong

LOCATION At the mouth of the Murray River, southeast South Australia, Australia

TYPE Lagoon

WATER Brackish, salt

AREA 490 square km (190 square miles)

YOUNGHUSBAND PENINSULA
Sand dunes, partly covered by coastal scrub, stretch over 145km (90 miles) from the mouth of the Murray River, protecting the lagoons of the Coorong from the open ocean.

BLACK SWAN
Native to Australia, the Black Swan is exclusively vegetarian, feeding mainly on aquatic plants and sometimes grazing on land.

The Coorong lies at the mouth of the Murray River (see p.223), about 80km (50 miles) south of Adelaide. The river opens into Lake Alexandrina and then connects with the narrow channel of the Coorong, which lies behind the Younghusband Peninsula. The latter is a narrow spit on which an impressive array of sand dunes faces the Southern Ocean. These run parallel to a set of ancient dunes (dating from the Pleistocene Epoch) on the landward side. It is between these two rows of dunes that the saltwater lagoon of the Coorong stretches southwards for more than 100km (60 miles). Other habitats found here include temporary freshwater lakes and mudflats, particularly in the south. In 1940, barrages were built between the Coorong and Lake Alexandrina, preventing seawater from reaching the lake. This altered the hydrology of the area considerably, and reduced the natural flow of the Murray into the Coorong. Water extraction for irrigation projects upstream has also reduced the flow, and if it decreases much more in the future there is a risk that the gap connecting the Coorong to the sea will close completely. This would prevent fish and other animals from migrating between the Coorong and the sea. The Coorong is one of the best sites for viewing wildfowl in Australia.

Thousands of Black Swans, Cape Barren Geese, and countless waders flock to these waters, which are also home to the largest breeding colony of Australian Pelicans. In all, 335 species of bird have been recorded in the lagoon.

MUD CRABS
Mud crabs are a prominent feature of the Coorong. These crustaceans thrive on the mudflats and in the shallow, brackish waters of the coastal lagoons.

LAND

WILDEBEEST MIGRATION
Mass migrations to follow the best
grazing, such as that undertook by
the Blue Wildebeest in east Africa,
are common among large herbivores.

GRASSLANDS AND TUNDRA

Vast areas of the Earth's surface are covered by apparently unchanging expanses of natural grassland. In fact, this uniformity is often an illusion, for grasslands vary widely: they may be hilly or punctuated by rocky outcrops; they include both lush and semi-arid areas; and some, especially in the tropics, are dotted with trees, pools, and marshes. Despite experiencing harsh weather, grasslands are one of the world's most productive biomes. Tundra is another biome that covers a large part of the Earth's land, mainly in the Arctic. Conditions in this desolate, treeless terrain are even more extreme, with long, cold, perpetually dark winters followed in spring by an explosion of life.

23	Plant life
27	Spreading grasslands
Tundra 338–39	
Cereal cultivation 350	

Grasslands

Natural grasslands mostly develop over deep, fertile soils, in areas with seasonally variable precipitation and a growing season of 120–200 days. The rainfall must be too low to support forests, but too high for either scrublands or deserts to take hold. Fierce winds and large swings in temperature between day and night, and from one season to the next are typical of many grasslands. Their boundaries are ill-defined and ever-changing due to natural climatic fluctuations and changes, and to human influence, including livestock grazing, deliberate burning, and the conversion of land to arable or pastoral crops. There are two major types of grassland: temperate and tropical, which differ both in structure and in the variety of their plant and animal species.

Temperate Grasslands

Temperate grasslands are found mainly in continental interiors, far from coasts and their rain-bearing winds. Summers here are usually warm and dry, with occasional violent storms, but winters may be bitterly cold. Most rain falls in late spring and early summer. Wind has a major impact on temperate grasslands: it fans the flames of fires in late summer, and it dries the land by accelerating the process of evaporation. The vegetation is dominated by perennial grasses, which have extensive root systems that join up to form a tightly woven mat known as turf. Herbs, sedges, flowers, and specialized shrubs also occur, particularly in wetter grasslands such as the tall-grass prairies of North America. Trees are generally restricted to sheltered hollows, the edges of streams and drainage channels, or marginal areas where grassland grades into park-like open woodland.

PAMPAS GRASS
Many grass species grow long flower-heads in summer, which disperse clouds of pollen on the breeze. Those of the Pampas Grass, show here in Uruguay, are among the most spectacular.

BUFFALO GRASS
Native to the short-grass prairies of North America's Great Plains, this species features prominently in the diet of bison (also known as buffalo).

PRAIRIE CONEFLOWER
The leaves and flowers of this plant were used by Great Plains Indigenous peoples to make tea.

AUSTRALIAN RANGELAND
This temperate grassland, found in southeast Australia, is studded with eucalyptus and paperbark trees.

Tropical Grasslands

Tropical grasslands, or savannas, are typically more varied in character than temperate grasslands. There is normally a scattered cover of trees and bushes, which may be sparse in some places and dense enough to form thickets in others. Some savannas, such as those in parts of east Africa, are actually a form of open woodland, in which a ground cover of grass is always present. Many of the grasses in savannas grow very tall during periods of rain; those found in the monsoon grasslands of the Indian Himalayan foothills are the tallest on Earth. Unlike temperate grasslands, savannas are usually warm all year, but a more important distinction is the much higher rainfall, which is concentrated into distinct wet seasons. Bush fires play an even more significant role in savannas than in cooler grasslands.

KANGAROO GRASS
One of the main grasses of Australian savannas, this species grows to a height of 3m (10ft) or more.

DROPSEED GRASS
In Africa, the edible seeds of some dropseed grasses are eaten by indigenous peoples.

CERRADO LANDSCAPE
This South American savanna is dotted with small, gnarled trees and countless red termite mounds. Its open areas are mixed with scrub and patchy dry forest.

PRAIRIE LANDSCAPE
Temperate grasslands change with the seasons. In winter, the plants lie dormant, while in the summer the grasses turn ripe yellow, as here in North Dakota, USA.

Animal Life

With the exception of the African savannas with their herds of antelope, zebra, and other large herbivores, grasslands may at first seem to be devoid of wildlife. This impression is misleading, because many grassland species are nocturnal or spend much of their life underground. A host of animals, from voles and moles to earthworms, beetles, and other invertebrates, tunnel under the turf. In doing so, they aerate the soil and create homes for non-tunnelling species such as snakes and birds. Animals that remain above ground are often fast movers, since hiding places are at a premium in most grasslands; they include the Ostrich and gazelles. Other strategies for avoiding predators include camouflage, the use of warning calls or signals (one example being the white rumps of certain rabbits), and herd formation. Many herbivores live in herds because predators find it harder to single out victims, and there are more animals on the lookout.

PRAIRIE DOGS
A group of prairie dogs keep watch for danger, ready to bark the alarm and dive for cover.

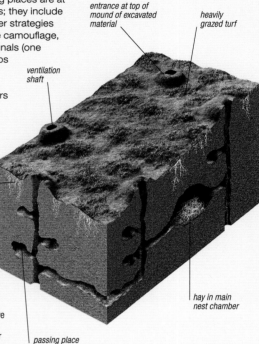

entrance at top of mound of excavated material

heavily grazed turf

ventilation shaft

grass roots bind surface layers together

hay in main nest chamber

passing place in vertical access tunnel

LIFE IN A PRAIRIE-DOG TOWN
Prairie dogs are highly sociable ground squirrels of the North American prairies. They live in burrow systems known as towns, which provide shelter from predators and harsh winter weather, and a safe place to raise young. Prairie dogs have powerful front limbs with thick claws for excavating and scraping soil. Their sustained grazing keeps the vegetation near their colony short.

GRASS BURNING

Whether started naturally or by people, fire plays a crucial role in the ecology of grasslands. It removes dead plant material, and the ash created boosts the fertility of the soil, promoting new grass growth. Burning also inhibits colonization by woody shrubs and trees. Here, Masai people are undertaking a controlled burn of old grass in the Serengeti.

BLACK-FOOTED FERRET
This endangered carnivore hunts prairie dogs, and lives in their abandoned burrows.

BURROWING OWL
Lack of trees forces this grassland owl to nest in burrows, often those of prairie dogs.

COMMON BROWN
Butterflies, such as this species from Australia, are important pollinators of many grassland flowering plants.

HAIRY ARMADILLO
When threatened, this armoured resident of arid South American grasslands rolls up into a tight ball.

Grassland Distribution

The largest temperate grasslands are the prairies, which occupy much of central North America, and the Asian steppes, which stretch in a band from easternmost Europe to northern China. Other temperate grasslands are the Pampas of southeast South America, the veld of eastern South Africa, and the rangeland of southeast Australia. The main savannas are in Brazil, in east and southern Africa and to the south of the Sahara, and in northern Australia. Smaller savannas exist in the Indian subcontinent.

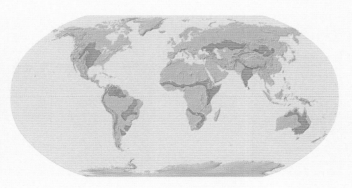

GRASSLAND PROFILES

The pages that follow contain profiles of the world's main grasslands. Each profile begins with the following summary information:

TYPE	Temperate or tropical
AREA	Surface area

LAND

BADLANDS NATIONAL PARK
Named after its strange rock
formations sculpted by wind and rain,
this reserve in South Dakota protects
the largest area of mixed-grass prairie
in the USA.

NORTH AMERICA *central*

Great Plains

LOCATION Occupying much of
North America between the Rocky
Mountains and River Mississippi

TYPE Temperate

AREA 3 million square km
(1.2 million square miles)

By far the largest expanse of grassland
in North America, the Great Plains stretch
from the southern Canadian provinces
of Alberta, Manitoba, and Saskatchewan,
through the USA's Midwest, and south
almost to northern Mexico. As recently

FOXTAIL GRASS
In summer, Foxtail Grass produces long, feathery
flower-heads. Later in the season, its seeds are
eaten by a wide variety of rodents and birds.

WESTERN MEADOWLARK
One of the most
characteristic prairie birds,
the meadowlark perches
on a fence post or large
bush to deliver its rich,
flute-like territorial song.

as the early 19th century, this immense
area was covered by grasslands known
as prairies, but most of the fertile land
is now under intensive agriculture. It is
especially suitable for growing cereals,
such as wheat and corn (maize), and has
become the main grain-producing region
of North America. Only one per cent of
the natural grassland survives, divided
into three main types: short-grass prairie,
mixed-grass prairie, and tall-grass prairie.
Short-grass prairie occurs mostly in the
west in the rain-shadow of the Rocky
Mountains, where the annual rainfall is
about 250mm (10in). Here, most grasses
are 20–50cm (8–19in) high. Tall-grass

**COMMON
SUNFLOWER**
This is the wild
ancestor of the
popular garden
plant. It flourishes in
the rich, dark soils
of the Great Plains.

prairie is found in the eastern districts
of the Great Plains, with an annual
rainfall of 650–1,000mm (26–39in).
Some of the grasses growing here,
such as Indian Grass, Big Bluestem,
and Cordgrass, may reach 1.5m (5ft).
In central areas, mixed-grass prairie
forms a zone of transition between
short- and tall-grass prairie. The
open prairie landscape used to
be preserved partly by grazing by
native herbivores, which included
Bison, Pronghorn, and prairie dogs,
but in places the steady decline of
these animals has led to changes
in the length and composition of
the turf. Before European colonization,

ROLLING PLAINS
Undulating hills and wide, grassy plains are
typical of many parts of the prairies, as here
in Nebraska, USA. The harsh winters and
strong winds prevent all but a handful of
trees from gaining a hold.

the prairies were also home to North
American tribes such as the
Lakota and Pawnee.

RETURN OF THE BISON

Until the early 1800s, up to 60 million
Bison (or Buffalo) roamed the
short-grass prairie in huge herds.
Hunting by settlers from the USA's east
coast reduced the Bison population to
just 2,000 by 1885. Today, there are
more than 350,000, mainly in National
Parks and cattle ranches. There are
plans to create a "buffalo commons",
allowing herds to move freely over
the prairies of Montana, the Dakotas,
Nebraska, Kansas, and Oklahoma.

SOUTH AMERICA *north and east*

South American tropical grasslands

LOCATION In eastern Colombia and Venezuela, south-central Brazil and Paraguay, and northern Argentina

TYPE Tropical

AREA 2.7 million square km (1 million square miles)

MANED WOLF
Neither wolf nor fox, this member of the dog family lives in Brazil, Paraguay, and northern Argentina. Its stilt-like legs help it to move through long grass, and to leap on rodent prey.

South America has two enormous savannas: the Llanos and the Cerrado. Lying either side of the River Orinoco in Venezuela and eastern Colombia, the Llanos is a wilderness of rough grassland interspersed with marshy areas (see p.324) and light woodland. Various species of tussocky *Trachypogon* grass are dominant, while typical animals include Savanna Rabbits and Carpenter Ants. Today, the Llanos is a major

BRAZILIAN CERRADO
Open grassland shades into scrub and palm forest in the Cerrado, creating a marginal habitat rich in species.

cattle-ranching region. The Cerrado occupies much of Brazil and Paraguay, covering an area the size of western Europe. It borders the Amazon Basin (see pp.211 and 307) to the north, with the Pantanal wetland (see p.325) to the west and the Atlantic forests in the east. As many as five per cent of the world's animal species live in this savanna, including the Giant Anteater and tall, flightless birds called rheas. The

Cerrado's ecological richness is due in part to its mosaic of habitats, ranging from grasslands to dry forests (see p.310). But agriculture is rapidly encroaching on the Cerrado; less than three per cent of it is protected. To the south, in Argentina's arid chaco region, there are sandy, desert-like grasslands.

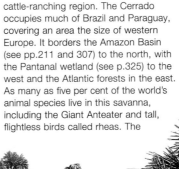

SOUTH AMERICA *southeast*

Pampas

LOCATION In northern Argentina and Uruguay, extending from the Andes foothills to the Atlantic Ocean

TYPE Temperate

AREA 700,000 square km (270,000 square miles)

Named after a Quechuan word meaning "flat surface", the Pampas is a huge plain in the southeastern corner of South America. It is centred on the lowlands of the River Plate (see p.208). Apart from a few low hills, most of the

region is just above sea-level. Deposits of volcanic ash provide a rich soil. The eastern Pampas, which includes much of Uruguay and the Argentinian province of Buenos Aires, enjoys a year-round mild climate, with cool summers and an annual rainfall of up to 1,200mm (47in). Snow and frosts are very rare. The original vegetation of coarse grasses, including the famous Pampas grass (see p.330), has greatly reduced, as ranching developed and other non-native grasses more suitable to cattle and horses were introduced. Clusters of trees cling to the damper hollows or drainage channels. Animals of the Pampas include the endangered Pampas Deer, Geoffroy's Cat, and the Argentine Tortoise. Further west, in the Argentine provinces of La Pampa and Córdoba, the landscape is different. Here, the rainfall is half that of the eastern Pampas, and the sandier soil supports a semi-arid grassland made up

PAMPAS NEAR BUENOS AIRES
Much of the Pampas has been converted to fertile agricultural land or is used for settlement. It is the most densely populated part of Argentina.

of feather grasses and other species. Towards the Andes, the land rises and becomes increasingly dry, eventually turning into semi-desert.

CATTLE RANCHING

Millions of cattle are raised on the Pampas, and gauchos, or cowboys, still follow the herds on horseback. To some extent, their livestock have replaced the ecosystem's natural grazers and fulfil a similar function. However, the mix of native grasses has largely been replaced by a smaller number of introduced fodder species.

PATAGONIAN PAMPAS
Blasted by an icy, dry wind known as the pampero, this desolate plateau in southern Argentina provides a contrast to the lush Pampas further north.

AFRICA *east*

Serengeti Plains

LOCATION In northwest Tanzania east of Lake Victoria, extending north into southwest Kenya

TYPE Tropical

AREA 23,000 square km (8,900 square miles)

RESTING LIONESSES
A group of lionesses rests on an anthill, which offers a good vantage point over their territory. The pride's mature male will probably not be far away, but plays little role in hunting. The lionesses provide most of the food, moving as a pack to ambush their victims.

HUGO VAN LAWICK

One of the earliest and most famous wildlife film-makers, Hugo van Lawick (1937–2002) was a Dutch naturalist who first visited Africa in his twenties. His many documentaries for TV and cinema did much to increase public awareness of east African conservation issues. In 1964, he married the primatologist Jane Goodall. Together, they studied three generations of wild chimps.

Lying close to the equator in east Africa, the Serengeti Plains support an exceptionally varied collection of plants and animals, all adapted to the complex intermingling of grassland and open woodland. The main grassland areas lie in the southeast, with wooded savanna typified by flat-topped acacia trees dominating in the north and west. In

NGORONGORO CRATER
The world's largest unbroken caldera, this huge relic of past volcanic activity looms over the eastern Serengeti. Its walls are over 600m (1,970ft) high.

some places, the undulating plains are interrupted by rocky outcrops known as kopjes, composed of hard Precambrian rocks. The Serengeti's reddish soil is derived mainly from volcanic ash. Relatively low in organic materials compared with temperate grasslands, it is replenished with vital nutrients by frequent dry-season bush fires, either natural or human-made. Billowing clouds of smoke may darken the horizon for days at a time. Another major factor shaping the landscape is grazing pressure. The Serengeti holds the largest populations of grazing and

browsing mammals in the whole of Africa. During the two rainy seasons, from November to December and from March to May, these herbivores are widely dispersed over the plains, but when the savanna is yellow and parched, they migrate in large herds in search of fresh grass and drinking water.

SERENGETI IN THE WET SEASON
With the onset of the rains, thorny shrubs and acacia trees are flushed with new leaf growth. Acacia woodland provides valuable shade and browsing for many of the Serengeti's animals.

These migrations include 1.7 million Blue Wildebeest; 260,000 zebras; and 40,000 Thomson's Gazelles. Such an abundance of big game provides enough food for many predators and scavengers, including big cats, hyenas, jackals, crocodiles, and

BLUE WILDEBEEST AND ZEBRA
Although these herbivores graze side by side in large herds, they eat different foods. The zebras can consume drier, tougher grasses, while the wildebeest harvest softer grasses ignored by the zebras.

six species of vulture. Equally important are the countless dung beetles that roll away 75 per cent of all the dung dropped in the Serengeti, burying it in nests for their larvae to eat. Both the Serengeti National Park and the Ngorongoro Preservation Area are World Heritage Sites, and ecotourism raises funds for conservation projects. Despite this protected status, wildlife in the region is at risk from poaching, serious overgrazing (caused by excessive numbers of cattle and linked to climate change), and diseases spread by domestic dogs and livestock.

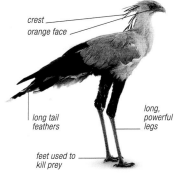

crest
orange face

long tail feathers

long, powerful legs

feet used to kill prey

SECRETARY BIRD
This distinctive relative of the birds of prey stalks slowly through long grass, watching the ground for its prey – mainly insects, snakes, lizards, and small mammals.

MASAI

The Masai live in the Serengeti and the neighbouring Masai-Mara. Unlike some other African nomadic peoples, they do not rely on hunting, and so historically Masai land has tended to be rich in wildlife. Instead, they herd cattle, goats, and sheep, and collect fruits and other plant products from the wild. Their main traditional food is fermented cow's milk; meat is usually eaten only during ceremonies.

LAND

Central Asian steppe grasslands

LOCATION Extending from Ukraine, east through southern Russia and Kazakhstan, to Mongolia and China

TYPE Temperate

AREA 2.5 million square km (975,000 square miles)

The steppes of Central Asia stretch almost a third of the way around the world and include some of the least populated areas on Earth. In the west, the steppes reach the northern shores of the Black Sea, while they extend east in an irregular band almost to the mountains of China's northeastern provinces on the Pacific. This immense grassland zone is bordered to the north by the Eurasian boreal forest (see p.312), and to the south by the Central Asian deserts (see pp.293–95). Dramatic swings in temperature are a defining feature of the steppes; on the grassy upland plateaus of Mongolia, it may reach 40°C (104°F) in summer and plunge to -20°C (-4°F) in winter.

The annual rainfall of 200–600mm (8–23½in) is sporadic, with prolonged droughts and sudden thunderstorms. Thick snow blankets the ground in winter. When the snow melts in spring, the steppes are briefly ablaze with a profusion of flowers, including irises, hyacinths, crocuses, and tulips. During the short summer, feather and fescue grasses and sedges dominate the turf. Apart from a few stunted trees confined to the sheltered valleys, the steppes are bereft of tall vegetation, although in northern China there is a more park-like landscape, with a scattering of trees such as elms. Animals of the steppes are adapted to the extreme climate in various ways. The Saiga Antelope has a bulbous, trunk-like nose that warms and moistens the cold, dry air it breathes, while rodents such as the Steppe Lemming, susliks, and gerbils dig burrows for shelter.

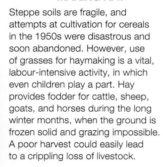

MAKING HAY

Steppe soils are fragile, and attempts at cultivation for cereals in the 1950s were disastrous and soon abandoned. However, use of grasses for haymaking is a vital, labour-intensive activity, in which even children play a part. Hay provides fodder for cattle, sheep, goats, and horses during the long winter months, when the ground is frozen solid and grazing impossible. A poor harvest could easily lead to a crippling loss of livestock.

SAVING A SPECIES

Built like a heavy pony, Przewalski's Horse is the only truly wild horse. At home on steppes and in semi-deserts, it once ranged from the Urals to northern China. It was extinct in the wild by 1969, but more than 1,000 survived in zoos. Small herds bred in captivity are now being reintroduced into the wild at suitable sites.

SOUTHERN RUSSIAN STEPPES
Many of the steppes in Russia, Ukraine, and Kazakhstan have been converted to semi-natural grasslands due to heavy grazing by cattle.

NORTHERN MONGOLIAN STEPPES
The mountains of southern Siberia rise above this high-altitude plateau near Ulan-Uul. A sparse covering of dusty turf is often all that distinguishes the high steppes from semi-desert.

ASIA south

Indian savanna grasslands

LOCATION Within a zone stretching northwards from south-central India to the Himalayan foothills

TYPE Tropical

AREA 35,000 square km (13,515 square miles)

ELEPHANT GRASS
In the grasslands of Nepal, this relative of sugar cane grows up to 8m (26ft) high. It provides the Indian Rhinoceros with both shelter and the bulk of its diet.

Forming a discontinuous area and with few large expanses remaining, Indian savanna grasslands occur mainly in inland regions, around the fringes of deserts, and in hilly country. Those Terrai-Duar grasslands remaining after the encroachment of agriculture represent one of the most threatened grassland ecoregions in the world. They form a narrow strip along the southern margin of the Himalayas, from Nepal east to Bhutan and India. The hot, humid summers and monsoon rains enable tall grasses, including giant reeds and canes, to grow. These "grass forests" have the highest densities of tigers in India. Further south, the much lower rainfall of the Deccan Plateau gives rise to coarser, semi-arid pastures with a scattering of trees such as Sandalwood. This is the haunt of numerous reptiles, among them the highly venomous Indian Cobra.

DEW-COVERED GRASSES
Morning dew clings to grasses in the Corbett National Park in the Indian Himalayan foothills. Torrential monsoon downpours create swampy conditions here, in which annual grasses thrive.

GRASSLAND PLATEAU
The interior uplands of southern India are known as the Deccan Plateau. Pockets of wooded savanna form a patchwork with dry forest, farmland, and villages in this highly populated region.

AUSTRALASIA Australia

Australian savanna grasslands

LOCATION Stretching from the far north of Western Australia, through Northern Territory, into Queensland

TYPE Tropical

AREA 1.2 million square km (463,500 square miles)

Australia's savanna grasslands are one of the largest, most unique, and unspoiled grassland ecoregions in the world, but they have been threatened by invasive non-native weeds and pests.

RED KANGAROO
The largest of the kangaroos, this marsupial is common in wooded savannas. Its numbers double in years of high rainfall.

short forelegs
large, muscular hindlimbs
very long, strong tail

AUSTRALIAN BAOBAB
Although mainly found in Africa, and particularly Madagascar, one species of baobab is endemic to northwest Australia.

They form a wide band between the hot interior deserts and northern coastal forests, with the Great Dividing Range to the east. It is hot and humid during the wet season (December to March), but little or no rain falls during the cooler dry season (May to August). Wooded savanna is mainly found in the wetter north and east, but gum trees, also known as eucalypts, are present in most areas. One group of uniquely Australian grasses are the spinifex, which bristle with spear-like leaves that minimize water loss, allowing them to thrive in the desert too.

TOUGH GRASSES
Scattered clumps of hardy grasses such as spinifex typify these grasslands, as here in the Barkly Tableland, Northern Territory.

LAND

Tundra

◀ 29–31 The ice ages

◀ 90 Physical weathering

◀ 250–75 Glaciers

◀ 338–39 Grasslands

The term tundra is derived from a Lappish word meaning "barren land". It is an apt description of the low-growing vegetation found in the Arctic, beyond the northern limit for tree growth. The windswept landscape here is frozen for much of the year, but its surface layers melt in spring. Summers are short and intense, with 24-hour daylight, while the Sun barely rises during the long, ferociously cold winters. Patches of tundra-like habitat also exist above the tree-line on some high mountains.

Permafrost Zones

Land that has been frozen for at least two years is classified as permafrost, and it has a major influence on the landscape and ecosystem of the tundra. A zone of permanent permafrost is found at the highest latitudes, spreading out from the North Pole. Immediately to the south is the zone of semi-permanent permafrost. Here, the topmost layer usually thaws out to a depth of a few centimetres each summer, and this so-called active layer supports plant and animal life. Further south still lies the sporadic permafrost zone, in which the active layer is deeper and where the surface freezes less often; however, the underlying frozen soil prevents meltwater from draining away.

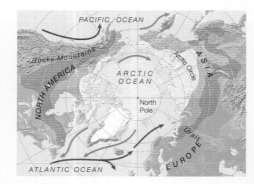

ARCTIC PERMAFROST ZONES
Three zones of permafrost surround the Arctic, arranged from north to south, determined by latitude and the influence of ocean currents.

☐ Permanent permafrost
☐ Semi-permanent permafrost
☐ Sporadic permafrost
→ Warmer ocean current
→ Colder ocean current

Periglacial Landforms

Periglacial landforms are those features that develop under the action of hard frosts, often in permafrost conditions. A classic periglacial feature is the polygonal patterning that appears in the surface of tundra soils, formed by the repeated freezing and thawing of water seeping along surface cracks. Rapid thawing of the active surface layers may give rise to a creeping movement of the soil known as skin flow. The accumulation of ice trapped between the surface layers of the soil and its deeper frozen layer may result in disruption of the surface, splitting and buckling it, or heaving it up into the shape of a small hill or mound. Rounded ice-heave hills of this type are called pingos, while irregularly shaped masses of ice are referred to as ice wedges. Other landforms are created by the partial thawing and refreezing of the permafrost, especially in sedimentary rocks. Where the bedrock is exposed, it may be weathered into angled fragments, to produce areas of coarse rubble called blockfields.

SPLIT-HILL PINGO
A pingo is a blister-like mound that develops as the ice at its centre expands. The pressure of the expanding ice core may cause its surface to tear apart, as seen here.

ICE LENS
When compacted ice trapped in soil develops convex surfaces, it is called an ice lens.

FOSSIL ICE WEDGE
Blocks of ice sometimes leave their traces as fossils, indicating cold conditions in an earlier age.

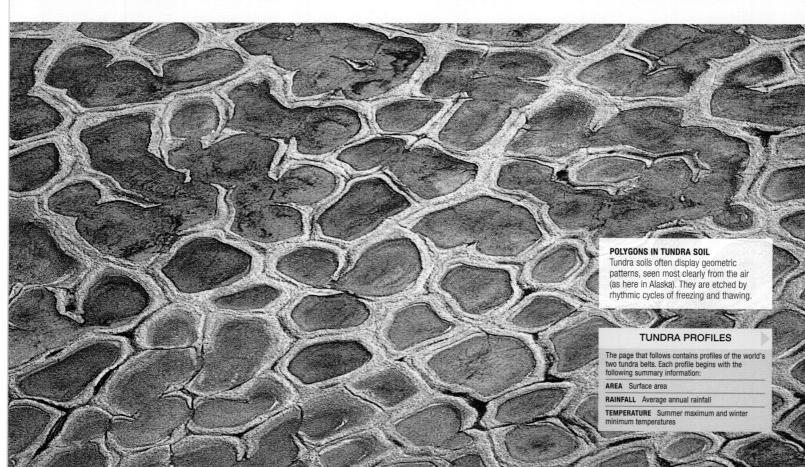

POLYGONS IN TUNDRA SOIL
Tundra soils often display geometric patterns, seen most clearly from the air (as here in Alaska). They are etched by rhythmic cycles of freezing and thawing.

TUNDRA PROFILES

The page that follows contains profiles of the world's two tundra belts. Each profile begins with the following summary information:

AREA Surface area

RAINFALL Average annual rainfall

TEMPERATURE Summer maximum and winter minimum temperatures

North American tundra

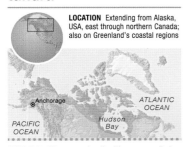

LOCATION Extending from Alaska, USA, east through northern Canada; also on Greenland's coastal regions

Anchorage

PACIFIC OCEAN

Hudson Bay

ATLANTIC OCEAN

AREA 5.3 million square km (2 million square miles)

RAINFALL 50–200mm (2–8in)

TEMPERATURE -60°C–24°C (-76°F–75°F)

Tundra is one of North America's biggest ecoregions and covers a large swathe of land north of the boreal forest (see p.304) and a narrow coastal strip around Greenland. Flat landscapes predominate although some periglacial landforms break up the monotony (see polygons and split-hill pingo on opposite page), as do a few rugged mountain ranges. In spring, the ice and

ARCTIC FOX
In winter, this tundra specialist has a white coat with long outer hairs and thick underfur for superb insulation; its summer coat is chocolate brown and much thinner.

snow melt to reveal a thick mat of lichens, mosses, sedges, and specialist Arctic flowers such as saxifrages. Where the soil is deeper, there are Crowberries, Bilberries, and

DENALI NATIONAL PARK
Named after the Native Athapaskan word for Mount McKinley, this huge reserve in southern Alaska (seen here in autumn) conserves alpine tundra and glaciers.

other ground-hugging shrubs. Dwarf willows and birches grow in southern areas. The tundra briefly teems with invertebrates, especially mosquitoes, beetles, and other insects, which provide food for migrant geese, ducks, and wading birds. Like all tundra, it is threatened by global warming, resulting in the release of large volumes of the greenhouse gas methane and the acceleration of the warming process.

DWARF WILLOW
This miniature tree creeps along the ground, growing less than 7.5cm (3in) high to avoid the drying effect of the relentless tundra wind.

Eurasian tundra

LOCATION From Iceland in the west, eastward through northern Scandinavia, Russia, and Siberia

ARCTIC OCEAN

Stockholm

Urals *Ob* *Siberia* *Lena* *Kamchatka*

AREA 3.3 million square km (1.3 million square miles)

RAINFALL 200–300mm (8–12in)

TEMPERATURE -60°C–25°C (-76°F–77°F)

The Eurasian tundra features a range of landscapes, from the soggy plains of Siberia, crossed by numerous rivers, to the various archipelagos scattered through the southern Arctic Ocean, such as Franz Josef Land and Novaya Zemlya. In central Siberia, winters are particularly extreme, while the

ARCTIC POPPIES IN BLOOM
These poppies are one of the northernmost plants in the world. Like many tundra flowers, they concentrate warmth by turning to track the course of the Sun.

ARCTIC WILLOW
The fluffy seed heads of this Arctic Willow have exploded to release hundreds of frost-resistant seeds, which will be spread by the wind to germinate next year.

SIBERIAN TUNDRA IN EARLY SPRING
As temperatures rise here in spring, meltwater floods the ground. Repelled by the permafrost, it forms millions of small lakes, pools, and bogs.

Scandinavian tundra is the warmest, with an average temperature of -8°C (18°F) in winter. Most of the plants are small, long-lived perennials. Mosses, lichens, sedges, and rushes dominate in the northern tundra, with

plant diversity increasing steadily from north to south. As in the North American tundra, the short but prolific 90-day growing season draws migratory animals to the region. More than 200 million birds arrive to breed,

including sandpipers, ducks, and geese. Reindeer (known as Caribou in North America) spend the winter in the taiga zone, then migrate north to calve and to graze lichens on the open tundra.

FUTURE HARVEST
Working in steeply terraced fields, a girl tends a rice crop in southern China. Terracing is one solution to the global need for increased agricultural land.

AGRICULTURAL AREAS

When farming first began, about 10,000 years ago, it was a fringe activity carried out by small numbers of people, and it had little impact on the wider world. Since then, agriculture has developed beyond recognition. It now supports almost all of the human race, and approximately 35 per cent of the world's land surface is used for farming of some kind. During its remarkably short history, agriculture has transformed the world's landscapes, creating some of the greatest physical and ecological changes that the Earth has ever seen. Today, over 1 billion people are actively involved in farming. The land that they work varies enormously, from steep rice terraces to table-flat prairies, and from intensive mixed farmland to arid rangeland, where domestic animals live much like their relatives in the wild. Taken together, the world's farmland is a testament to human ingenuity – and perseverance – in the constant drive to produce food.

LAND

Agricultural Areas

◀ 110–11   Water resources

◀ 290–91   Desertificationt

Urban areas 352–53 ▶

Industrial areas 372–79 ▶

Fishing 426–27 ▶

Many different factors affect the way that land is farmed. Some are connected with the local soil and climate, but others relate to economics, or to local conditions. In many parts of the world, the predominant form of agriculture is arable farming, which is any kind of farming that involves growing crops. Where conditions are less suitable for crops, arable farming is often replaced by pastoralism, or raising livestock. Arable farming and pastoralism are the two principal types of agriculture, but they embrace an enormous variety of farming practices. Together, these have created the agricultural landscapes that exist today.

Shifting and Settled Agriculture

In shifting or nomadic agriculture, farmers keep on the move and generally do not own the land that they use. This kind of agriculture was probably the first to be practised, but it is relatively unproductive and cannot support dense populations. Today, it is restricted to regions where the climate is harsh or where the soil is not sufficiently fertile to support full-time settlement. In most other parts of the world, land is farmed on a settled or sedentary basis. Settled farmland is generally divided up by boundaries indicating ownership – a feature that has become the hallmark of agricultural l`and as a whole. In some regions of the world, such as northern Europe, these boundaries can date back hundreds of years.

FIELD SIZES
These fields, in Burgundy, France, are larger than their predecessors a century ago, because farm machinery now needs more space.

MOVING ON
Reindeer herding is one of the few forms of shifting agriculture practised outside the tropics. The herds keep on the move in search of lichens, rather than grass.

Scale and Intensity

The world's earliest farmers worked on a subsistence basis, which means that they produced only enough food for themselves and their immediate dependants. By contrast, many of today's farms – particularly in the developed world – are run as commercial enterprises. They specialize in particular crops or animals, and often raise them on a substantial scale. Subsistence farms are typically small: in Southeast Asia, for example, some cover less than 1 hectare (2.5 acres). At the other extreme, some commercial farms in Australia or North America are over 100,000 times that size. From an economic standpoint, farms also differ in the intensity with which the land is worked. In intensive farming, a large amount of labour or investment is put into a relatively small amount of land, and the yields are often high. In extensive farming, inputs are more thinly spread, but, because the amount of land is greater, overall production is much higher. Most subsistence farms are run on an intensive basis, while commercial farms can be intensive or extensive.

MECHANIZED HARVESTS
Modern agriculture relies on mechanization and fossil fuels. For some crops, the energy used in fuel is greater than the energy produced as food.

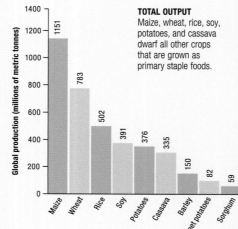

TOTAL OUTPUT
Maize, wheat, rice, soy, potatoes, and cassava dwarf all other crops that are grown as primary staple foods.

Global production (millions of metric tonnes)

Crop	Production
Maize	1151
Wheat	783
Rice	502
Soy	391
Potatoes	376
Cassava	335
Barley	150
Sweet potatoes	82
Sorghum	59

TOTAL COMMITMENT
The American Great Plains are one of the world's most important cereal-growing regions. Here, agriculture has replaced the original grassland habitat.

Farming and Landscape

Shifting agriculture leaves few permanent marks on the landscape, whereas settled farming can bring about huge and far-reaching changes. In the Near East, for example, Mesopotamian farmers dug elaborate irrigation channels over 6,000 years ago, and in mountainous regions, from the Andes to Southeast Asia, farmers stabilized the ground by building terraces – many of which are still in use. Over time, farming can create an entirely artificial landscape that is maintained by human intervention. In northern Europe, for example, farmers cut down large expanses of forest in Neolithic times. The result is a patchwork of fields and woodland that looks semi-natural, but which is almost entirely artificial.

CHANGE OF USE
The slopes of this extinct volcano in Madagascar have been intensively farmed, which has led to erosion and gully formation on its flanks.

NORMAN BORLAUG

US agriculturalist Norman Borlaug (1914–2009) was a leading figure in the Green Revolution – an international crop-breeding programme that helped to boost dramatically yields of wheat and rice from the 1960s onwards. In recent years, the Green Revolution has been criticized for its reliance on agrochemicals, even though it has transformed food security in what were once famine-prone regions of the world.

Farming and Soil

In nature, soil forms very slowly, yet it erodes slowly as well. When land is cultivated, this natural balance is disrupted, because erosion rates increase. One of the most catastrophic examples of this effect occurred during the 1930s, on North America's Great Plains. Here, the rapid expansion of cereal farming, coupled with overgrazing, made the soil vulnerable to wind erosion. During the Dust Bowl years, millions of tonnes of it simply blew away. Today's farmers are generally more alert to the danger of soil erosion, but it remains one of the major challenges facing global agriculture. The problem is worst where the climate is dry, the soil loose, and the population pressure high. In the loess plateau of northern China, these factors combine to produce the world's highest erosion rates. Soil here is lost at the rate of up to 5 tonnes per hectare (2 tonnes per acre) per year, much of it into the Yellow River.

GULLY EROSION
This deep gully in the African tropics is caused by loss of plant cover through overgrazing, allowing heavy rain to wash away the soil.

RAISING THE DUST
Plants play a key role in preventing soil erosion. In dry areas, such as this farmland in Oregon, USA, ploughing is potentially risky, because it exposes the soil to the wind.

Agriculture and Wildlife

Farming changes natural habitats and replaces their original plants and animals with ones that live under human control. Although there are at least 2 million wild species on the Earth, the list of domesticated species includes less than 30 kinds of animals and about 200 kinds of food-producing plants. In farmland, these are given all the resources that they need to grow, while their wild competitors are ploughed up, cut down, poisoned, or fenced out. The result, after 10,000 years of farming, is a world where ecosystems are simplified. Wildlife fares worst where only one crop is grown (see Monoculture, p.345) and where high levels of agrochemicals are used. Mixed farmland and rangeland are better for wildlife, because they have more in common with the original habitats.

DRAWN TO WATER
Dromedaries, seen here drinking at a well near Timbuktu, Mali, are no longer found wild in their original habitat.

AGRICULTURE PROFILES

The pages that follow contain profiles of agricultural types. Each profile begins with the following summary information:

AREA	Approximate extent (where known)
LAND USE	Shifting or sedentary
ECONOMIC TYPE	Extensive or intensive
AGRICULTURAL TYPE	Arable, pastoral, or mixed
LEADING PRODUCERS	Main producer countries, listed in order of output

LAND

SUBSISTENCE
Hunting and gathering

DISTRIBUTION In the tropics and subtropics (chiefly in the southern hemisphere); also in high latitudes, including polar regions (northern hemisphere only)

AREA 17,500,000 square km (6,750,000 square miles)

LAND USE Shifting   **ECONOMIC TYPE** Extensive

AGRICULTURAL TYPE Not applicable

LEADING PRODUCERS Not known

By far the oldest form of land use, hunting and gathering stretches back to the beginning of human history. Until about 10,000 years ago, it supported the world's entire human population, but its importance waned rapidly once arable farming began. True hunter-gatherers are now very rare. They are restricted to remote regions such as central Amazonia and parts of the Arctic, but their self-sufficient lifestyle is being eroded by

COMMUNITY ON THE MOVE
Among the world's last hunter-gatherers, only a minority of San people, from southwest Africa, still follows their traditional nomadic way of life.

contact with the outside world. Hunter-gatherers usually live in small nomadic bands and rely entirely on food that can be caught or collected in the wild. In the tropics and subtropics, this typically consists of a range of animal food, together with fruit, roots, and seeds. At high latitudes, hunting becomes increasingly important, and meat can make up almost all the diet. To survive, hunter-gatherers rely on a detailed knowledge of their environment and of the seasonal changes that affect the food supply.

FOREST MARKSMAN
A hunter aims a blowpipe in the lowland rainforest of eastern Ecuador. Such traditional weapons are handmade, using locally available materials.

SUBSISTENCE AND COMMERCIAL
Nomadic herding

DISTRIBUTION In mid-latitudes in Africa, the Middle East, and Central Asia, on low ground and at altitude; in high latitudes in Eurasia. Extremely rare in the Americas

AREA 26,000,000 square km (10,000,000 square miles)

LAND USE Shifting   **ECONOMIC TYPE** Extensive

AGRICULTURAL TYPE Pastoral

LEADING PRODUCERS Not known

Also known as nomadic pastoralism, this form of agriculture is an effective use of land in arid or cold climates, where the soil is unsuitable for growing crops. Animals raised include sheep, goats, cattle, Yaks, and Reindeer, and they can wander freely while being watched by their owners. In dry areas such as the Sahel, in central Africa, nomadic herders typically keep on the move, which helps them cope with

REINDEER ROUND-UP
The original nomadic lifestyle of Reindeer herding is gradually dying out in parts of the Arctic, where it has evolved into a commercial activity.

unpredictable rainfall. In mountainous regions, such as the Himalayas, herders lead their animals between winter and summer grazing areas, a practice known as transhumance. In general, nomadic herding is a subsistence activity, with meat, milk, and animal products supplying herders and their families with most of their food. Reindeer herding, however, is carried out on a commercial scale.

MIXED HERDS
Sheep and goats forage for food around a Bedouin camp in the Jordanian desert.

SUBSISTENCE
Slash-and-burn agriculture

DISTRIBUTION In tropical regions worldwide including the Amazon Basin, sub-Saharan Africa, and Southern and Southeast Asia; rare or absent in Australasia

AREA 20,000,000 square km (7,700,000 square miles)

LAND USE Shifting   **ECONOMIC TYPE** Extensive

AGRICULTURAL TYPE Arable

LEADING PRODUCERS Brazil, Democratic Republic of Congo, Indonesia

One in 25 of the world's population still relies on slash-and-burn for their livelihood, which has sustained them for thousands of years. In this type of cultivation, which was the original

form of agriculture in the tropics, temporary clearings are hacked out of forests or woodlands, and the felled vegetation is set on fire. Crops such as cassava and sweet potatoes are planted in the newly cleared ground, drawing on the ash that fertilizes the soil. Initially, yields are good, but the soil's fertility soon declines. After a handful of years, the farmer abandons the ground and moves on to clear a new plot elsewhere. Slash-and-burn looks destructive, but it can be efficient and environmentally benign. Yields are small, but the energy returns are high, because no inputs of

EMERGING CROP
Two months after the ground has been cleared, plants begin to appear, but soil nutrients will soon be exhausted.

agricultural chemicals or fossil fuels are involved. Once land has been abandoned, it gradually reverts to its natural state, and after several decades may be used again. Slash-and-burn can support much denser populations than nomadic herding. However, it has one major weakness: if the land is worked for too long, it can become permanently

infertile. This is an increasing problem in regions such as Southeast Asia, where land is in short supply.

CLEARING THE LAND
In Roraima State, Brazil, a farmer burns off vegetation. Without the help of tractors or bulldozers, farmers often leave trees to lie where they fall, and crops are planted between them.

LAND

Plantation agriculture

DISTRIBUTION Chiefly in the tropics, in the Americas, Africa, Asia, and Australasia, in lowlands and at altitude; common on tropical islands

AREA 8,000,000 square km (3,100,000 square miles)

LAND USE Sedentary   **ECONOMIC TYPE** Intensive

AGRICULTURAL TYPE Arable

LEADING PRODUCERS Malaysia, Brazil, Mexico, India, Cuba

Most of the large-scale commercial crops that are grown in warm climates are produced in plantations. These include tea, coffee, cocoa, bananas, palm oil,

and sugar, as well as fibre crops, such as sisal and cotton. Several minor crops, such as pineapples and nuts, are also grown in this way. Three features typify this kind of agriculture. Crops are cultivated on large estates and are almost always destined for export

CUTTING SUGAR CANE
Sugar cane is one of the world's most important plantation crops. Over 1 billion tonnes of cane are harvested each year; much of it is cut by hand.

MONOCULTURE

In some forms of agriculture – including most plantations – the same crop is grown on a large scale year after year. This type of cultivation, called monoculture, gives important economies of scale, but it also creates a giant reservoir for pests and diseases. These have to be tackled by using high levels of agrochemicals: non-organic cotton, for example, is one of the most sprayed crops in the world.

rather than for local use. Secondly, the crops are often bushes or trees, rather than plants that have to be sown and harvested each year. Finally, much plantation agriculture had its origins in colonial times, and many plantations are still owned by corporations based overseas. For local people, plantations provide a source of employment, and often of housing and schooling. However, like all monocultures (see panel, above), their long-term nature makes them economically inflexible – a problem if prices fall.

TEA HARVEST
Workers pick young shoots in a tea plantation in northern India. The tea bushes are interplanted with trees that provide timber and shade.

RUBBER TAPPING
Latex – the source of natural rubber – trickles from a tree. Once a key commodity, natural rubber has been largely replaced by synthetic substitutes.

Rice cultivation

DISTRIBUTION Chiefly in the tropics and subtropics, in southern and Southeast Asia; also in the Americas, Africa, southern Europe, and Australasia

AREA 1,650,000 square km (637,000 square miles)

LAND USE Sedentary   **ECONOMIC TYPE** Intensive

AGRICULTURAL TYPE Arable

LEADING PRODUCERS China, India, Indonesia, Bangladesh, Vietnam

Rice is the world's second most important cereal crop, with a history of cultivation dating back at least 5,000 years. It was originally domesticated in Asia, and this continent remains by far the largest producing region. Unlike maize and wheat, rice is used almost exclusively for human consumption. There are thousands of varieties of cultivated rice, and the crop can be grown in two different ways. Upland rice is planted in dry soil and raised much like other cereals. It accounts for less than 10 per cent of the global harvest. Lowland rice is typically sown in nurseries. After about four weeks, it is then transplanted into paddies, or flooded

fields, which have been ploughed and cleared of weeds. This method is highly labour-intensive, particularly on sloping land where terraces have to be built and maintained. It is also very productive, given sufficient water and warmth. Modern hybrid plants mature rapidly and can produce up to three crops each year. Rice cultivation spread to Europe in medieval times and from there to the Americas. In these regions, rice production is more mechanized, and pre-germinated seed is sometimes planted from the air. But this large-scale farming is still the exception. Most of the world's rice is grown on a small-scale or subsistence basis, and less than 5 per cent of the total crop finds its way into global trade.

BUFFALO POWER
In southern and Southeast Asia, Water Buffalo are important as draught animals in wet conditions. Their dung also helps to fertilize paddy fields.

METHANE AND RICE

Rice paddies are a major source of methane – a gas that is produced when organic matter is broken down in surroundings that are low in oxygen. Methane is an effective greenhouse gas, and is estimated to be responsible for up to 20 per cent of global warming. Other forms of agriculture also contribute. Grazing livestock produce large quantities of methane when they digest their food.

GREENHOUSE GAS
Flooded rice fields produce high levels of methane. It is generated when rice stalks and roots rot underwater, where oxygen cannot reach them.

LOWLAND RICE
Young rice plants grow quickly in the warmth of the tropics. In ideal conditions, rice can be ready for harvesting within 12 weeks of transplantation.

COTTON FIELDS IN SUDAN
NASA's Terra satellite captured the thermal emissions from an extraordinary mosaic of irrigated fields at El Gezira in Sudan, where the waters of the Nile are used for growing cotton in an otherwise semi-arid region.

Cattle ranching

DISTRIBUTION Principally in mid-latitudes (northern and southern hemispheres), in areas with moist or semi-arid continental climate

AREA 29,000,000 square km (11,200,000 square miles)

LAND USE Sedentary **ECONOMIC TYPE** Extensive

AGRICULTURAL TYPE Pastoral

LEADING PRODUCERS USA, China, Brazil, Argentina

The world's most important source of meat is cattle, which is the primary product of cattle ranching. Some breeds, however, are kept chiefly for their milk (see Dairy Farming, opposite). Cattle country is normally dry, open, and relatively unproductive, and large areas are required to support cattle herds. Historically, this kind of space has always been rare in Europe – except in a few regions such as central Spain – but European colonists found it in abundance when they reached South America, North America, and Australia. Here, they developed commercial cattle ranching on a large scale, and these regions still remain key producers of beef. In Africa, the spread of cattle ranching was limited by animal diseases such as rinderpest, which spreads to cattle from wild grazing mammals, such as antelopes, buffalo, and Giraffes. In the wake of deforestation, however, ranching is now expanding in tropical areas – a trend that is causing major environmental concerns, as is overgrazing (see panel, right). Cattle ranches themselves have changed with improvements in

technology. In early ranches, cattle roamed freely, under the supervision of cowboys (or gauchos). But in the 1870s, the invention of barbed wire introduced cheap and effective fencing, enabling herds to be more easily controlled. In North America particularly, this was followed by improvements in transport, which allowed animals to be carried to commercial feedlots, where they would be fattened up for sale. The feedlot system is now highly mechanized, and animals receive a computer-controlled diet designed to maximize returns to the farmer. In energy terms, cattle farming remains

CATTLE DRIVE
In North America, cattle fill the ecological role once occupied by American Bison (or Buffalo). Today, cattle outnumber bison by more than 5,000 times.

OVERGRAZING

In arid regions, vegetation is slow to recover from the effects of grazing. Wild grazers evolve a long-term balance with their food supply, but domesticated cattle can cause severe damage if they are stocked at higher densities than the land can support. The result – after several decades – is severe erosion, which can make the land valueless for farming. Rangeland erosion is a problem throughout the world, but North America is particularly badly affected.

EXTREME EROSION
At Mungo, Australia, sheep and rabbits have accelerated natural erosion, creating a scarred landscape.

NATURAL PARTNERS
Originally from North America, the horse has been introduced to cattle farms worldwide, but, in many regions, feral horses still live beyond human control.

an inefficient form of agriculture, because it takes more nutrients to produce a unit of beef than any other kind of animal food. However, it does enable farmers to exploit marginal land that would otherwise go unused.

Sheep farming

DISTRIBUTION In the subtropics and higher latitudes (northern and southern hemispheres), in regions with widely varying climates, on low ground and mountains

AREA 34,000,000 square km (13,000,000 square miles)

LAND USE Sedentary

ECONOMIC TYPE Extensive or intensive

AGRICULTURAL TYPE Pastoral

LEADING PRODUCERS China, India, Australia, Iran, New Zealand

Compared to cattle, sheep are hardy animals, with a high tolerance of heat, cold, and drought. Some are looked after by nomadic herders, but most of the world's 1 billion sheep belong to settled farmers and are raised in several different terrains. In northwest Europe – where many of today's commercial breeds were developed – sheep are nurtured on good lowland grazing and on moorland and hills. But in many other parts of the world, they are raised in open rangeland, which has a drier climate and less food. To make up for this, rangeland sheep roam over much bigger areas. In Australia, some sheep farms cover more than 1,000 square km (386 square miles), and, where grazing is poor, each sheep may need more than 20 hectares (49 acres) of land to survive. Sheep are very efficient foragers, and they can have a significant effect on the landscape. In regions with moist climates, they keep pastures closely cropped, which helps to suppress weeds and maintain the grass. In arid regions, however, they can strip away the plant cover, leading to soil erosion. Wherever they live, sheep are raised chiefly for their meat. In the past, wool was an equally important product, and breeds, such as the Merino from Spain, were developed for their luxuriant fleeces. Some of today's experimental breeds have relatively sparse wool, more like their relatives living in the wild.

SHEEP FARMING IN WINTER
Although sheep are raised in many climates, they cannot forage in deep snow and so need food supplements, such as hay, which is being provided here in northern England.

A CHANGE IN TRADITION
Sheep shearers clip a fleece so that it comes away as one piece. But with the invention of synthetic fibres, fleeces no longer command high prices.

COMMERCIAL

Dairy farming

DISTRIBUTION Restricted to temperate regions, in areas with a moist climate; on low ground, and on seasonal pastures in mountains

AREA 5,400,000 square km (2,100,000 square miles)

LAND USE Sedentary   **ECONOMIC TYPE** Intensive

AGRICULTURAL TYPE Pastoral

LEADING PRODUCERS USA, Russia, Germany, France, New Zealand

In dairy farming, animals are raised for their milk and for products that are derived from it, such as butter and cheese. The two most important milk-producing animals are cows and water buffalo, with much smaller quantities being provided by goats and sheep. However, of all dairy animals, cows are the only ones that support a complete

GRAZING COWS
Renowned worldwide for the richness of their milk, these Jersey cows are grazing on a farm in South Africa's Western Cape.

farming system based on milk alone. Milk is highly nutritious, and cows need a high-quality diet to maintain productivity. Such food is found in the rich pastures that typify the dairy-farming landscape. Because cows have to be milked twice a day, they need easy access to a farm. As a result,

SIMPLE DIET
Goat's milk is an important product in many parts of the developing world. Unlike cows, goats do not need high-quality grass.

fields tend to be small. In some parts of the world – such as New Zealand – the animals can be kept outside throughout the year, but in many dairy regions they are brought under cover and fed during the winter months. Milk is one of the most perishable of all agricultural products. At one time, fresh milk could be sold only for local consumption, and cities were supplied by dairy parlours housing what were known as urban cows. Today, refrigerated transport allows milk to be carried long distances before it is processed.

MILKING TIME
In modern milking parlours, milk is chilled to slow the growth of bacteria. Milk may also be flash-heated, or pasteurized, to extend its shelf life.

LAND

COMMERCIAL

Horticulture

DISTRIBUTION Widely scattered across tropical, subtropical, and temperate regions, often near centres of population; dependent on irrigation in dry areas

AREA 4,000,000 square km (1,500,000 square miles)

LAND USE Sedentary   **ECONOMIC TYPE** Intensive

AGRICULTURAL TYPE Arable

LEADING PRODUCERS USA, the Netherlands, Israel, Spain, Ecuador

POTATO HARVEST
A harvesting machine unearths potatoes in Idaho, USA. Potatoes are a versatile crop, equally suitable for subsistence growers or large farms.

Soft fruit, salad crops, and vegetables are raised on an intensive basis in commercial horticulture, also known as market gardening. The actual range of crops depends on local climate and conditions, and, in many cases, several different types are grown side by side. Landholdings are often small, but in some regions – such as southern Spain, the Netherlands, and the Imperial Valley in California, USA – they are more substantial, with major investment in glasshouses or irrigation. These centres of production include some of the most intensively worked arable land in the world (see panel, right). Traditionally, commercial horticulture is located close to towns and cities, because perishable crops must be delivered to market without delay. But with the rapid expansion of road and air freight, and improvements in refrigeration, the lines of supply grow longer each year. This has created new markets for countries such as Mali, Ecuador, and Thailand, which

GOING FOR GROWTH
The mild winter climate here in southern Texas, USA, gives farmers a head start over those further north.

FLOWERS UNDER COVER
Cut-flower cultivation (right) is a highly intensive form of horticulture that can be very lucrative, despite high costs.

are able to export fresh produce during winter in the northern hemisphere. Compared to other kinds of agriculture, commercial horticulture is very sensitive to changing tastes, and it frequently focuses on low-volume crops that command premium prices.

SALINIZATION

In intensive horticulture, and other kinds of agriculture, irrigation can cause salinization – a damaging build-up of salt in the soil. Salinization occurs when irrigation water mobilizes salts that are naturally present in the ground. Salt build-up reduces crop yields, and eventually may make land unfit for farming.

LAND

Cereal cultivation

DISTRIBUTION Worldwide; commercial production (apart from rice growing) takes place chiefly in temperate regions

AREA 7,390,000 square km (2,853,295 square miles)

LAND USE Sedentary

ECONOMIC TYPE Extensive or intensive

AGRICULTURAL TYPE Arable

LEADING PRODUCERS China, USA, India, Russia, France

The first plants to be domesticated were cereals, which are of supreme importance today. Cereals are grasses that are cultivated for their edible seeds (or grains). Together, just three cereals – maize, rice, and wheat – account for over half the world's food, a remarkable situation given the large number of potential cereals that grow in the wild.

JOHN DEERE

The American blacksmith and engineer John Deere (1804–88) invented the all-steel plough, which set off one of the most far-reaching changes in agriculture. Unlike wood-and-iron ploughs, Deere's plough could slice through densely matted soil, opening it up for arable farming. By 1867, 10,000 ploughs a year were sold by Deere's company.

GLOBAL HARVEST
Surrounded by a sea of harvested wheat, this Australian farm lies in an area where cereal growing dates back less than 60 years.

Cereals are cultivated all over the globe, but the main centres of production are in temperate regions, except for rice (see p.345), which is grown chiefly in the tropics. As crops, cereals have two main advantages: their grain contains high levels of nutrients; and it can be stored for extended periods as long as it is kept dry. This makes grain easy to transport and to trade – a form of commerce that dates back to the world's oldest civilizations. The cereal trade is now ensuring adequate supplies of food worldwide, and over 470 million tonnes of grain are traded each year. In some parts of the world, cereal farming is still at subsistence level, but in developed countries, it has become a high-technology business, conducted on an increasingly large scale. Improvements in crop varieties and the use of agrochemicals have led to an extraordinary growth in yields. In the United States, for example, average maize yields stood at less than

MAIZE IN FLOWER
Multi-purpose maize can be harvested while it is green as fodder for livestock, as well as when ripe for its nutritious grain.

grain enclosed by modified leaves or husks

MAIZE   **RYE**

ears contain two rows of grain

KEY CEREALS
Maize, wheat, and rice are the world's major cereals. In addition, rye and barley are important in temperate regions, while sorghum and millet are popular in the tropics.

individual grains

modern varieties have short awns

long bristles, or awns

BARLEY

hollow stem

WHEAT   **RICE**

2 tonnes per hectare (about 30 bushels per acre) in 1900. The figure is now six times higher. Productivity has also increased because more land is being ploughed (see panel, left), particularly in the American Midwest, the Pampas of South America (see p.333), and the wheat belt of Australia. For the "big three" cereals, however, the growth in yields is now levelling off, and vacant land is in short supply. Many agronomists therefore believe that genetically modified crops (see panel, above) are the key to future growth, despite environmental concerns.

GM CROPS

Many agriculturalists believe that genetically modified (GM) crops – particularly cereals – have a vital role to play in combating world hunger. Crops can be modified to produce higher yields and to resist pests. GM crops, however, remain controversial, because of worries about their possible effect on human health, on the environment, and on the economy of poorer nations.

WINNOWING GRAIN
An Egyptian separates wheat in the traditional way, by tossing it in the air. Winnowing removes the inedible husk that surrounds each grain.

Mixed farming

DISTRIBUTION Worldwide in the tropics and in temperate regions; absent in areas with cold or arid climates

AREA 54,000,000 square km (14,000,000 square miles)

LAND USE Sedentary

ECONOMIC TYPE Extensive or intensive

AGRICULTURAL TYPE Arable and pastoral

LEADING PRODUCERS China, India, USA, Russia, France

In agriculture, specialization can increase efficiency, but it can also create problems if crops or animals are struck by bad weather or disease. This is one of the reasons why mixed farming is so widespread throughout the world. In this form of agriculture, farmers raise a range of crops and livestock, rather than concentrating on a single product. In ecological terms, mixed farming is more balanced than specialized agriculture, because animal food is frequently grown on the farm itself, while animal manure helps to improve the fertility of the soil. Mixed farming is more variable than any other kind of agriculture. At one extreme, some mixed farms – particularly in developing countries – raise more than 20 different sources of food. In China, for example, these include cereals, vegetables, fruit, and winter root crops, as well as chickens, ducks, and pigs. Many Chinese farms also have fish-ponds, which create high-protein animal food from kitchen scraps and farmyard waste. In other parts of the developing world, mixed farms work in a similar way, although the crops and animals raised

PREPARING FOR WINTER
Haymaking allows farmers to turn surplus grass into winter feed. In many farms in eastern Poland, hay is still raked and piled by hand.

depend on local climate and traditions. In the Bolivian Andes, for example, the traditional staple, called *chuño*, is prepared by drying potatoes, oca, or *ysaño* – three tuber-bearing crops that originated there more than 1,000 years ago. As well as poultry, Bolivian livestock includes llamas and guinea-pigs, both of which were domesticated in that part of the world. In developed countries, mixed farms are usually larger, and their produce less varied. European and North American farms typically grow cereals and root crops, and most of the harvest is destined to feed livestock,

INTENSIVE FARMING
Chickens are the world's most numerous farm animals. With a population of about 15 billion birds, they outnumber people by over two to one.

which includes cattle, pigs, and poultry. Until the 1950s, it was usual for farms to include all these animals, but today's farms often concentrate on just one. This increasing specialization reaches a peak in pig and poultry farms, where hundreds or thousands of animals are raised in intensive conditions. Unlike traditional mixed farms, these industrial-style units rely solely on brought-in feed.

Mediterranean-type agriculture

DISTRIBUTION In the Mediterranean region and areas with similar climate in North and South America, southern Africa, and Australia

AREA 5,500,000 square km (2,100,000 square miles)

LAND USE Sedentary

ECONOMIC TYPE Extensive or intensive

AGRICULTURAL TYPE Arable and pastoral

LEADING PRODUCERS Spain, Italy, USA, South Africa, Australia

With its distinctive climate of wet winters followed by long, dry summers, a characteristic kind of agriculture developed in antiquity in the Mediterranean region and has since spread to several other parts of the

world. Although Mediterranean-type agriculture is based on a variety of plants, two of them – the olive tree and the grapevine – epitomize this kind of farming. Olives are long-lived trees, capable of fruiting for centuries. However, they cannot survive where winter temperatures drop below about -5°C (23°F). Grapevines are much more hardy, but, like olives, they need strong sunshine to produce a useful crop. Only in a Mediterranean climate do these two plants thrive side by side. Mediterranean summers are usually too warm and dry for cereals to continue growing, so cereal crops are often sown in the autumn and then harvested in late spring. Where farms are still run on traditional lines, the result is a patchwork landscape of olive groves, vineyards, and cereal

PRESSING THE GRAPES
Traditionally, grapes are pressed by trampling them with bare feet. In modern wineries, presses do this work, processing thousands of tonnes of fruit a day.

fields. When Spanish colonists crossed the Atlantic, this kind of agriculture travelled with them, and so grapevines are now grown extensively in Chile, Argentina, and California, together with many other Mediterranean-climate crops, such as citrus fruits, peaches,

GROWING GRAPES
Although grapevines originated from western Asia, today's vines are grafted onto rootstocks from North American plants, which are resistant to insect pests.

and figs. South Africa and Australia have also become important wine-exporting regions, as well as producers of soft fruit. But the spread of crops has not been only one way: tomatoes and sweet peppers were unknown in Europe until the early 1500s, when they arrived from the Americas. Five centuries later, these crops are seen as a quintessential part of the Mediterranean daily diet.

FAMILY LABOUR
Olives are harvested either while they are still green and semi-ripe, or when they are almost black and ripe.

LAND

HEIGHT ADVANTAGE
The skyscrapers of Chicago's central
business district project above the clouds.
They clearly delimit the city centre,
separating it from the rest of the city.

URBAN AREAS

Two hundred years ago, most people lived in small, rural communities, and there were few large cities. Only London is known to have had over a million inhabitants, and just a handful of cities had populations in excess of 100,000 people. Today, about half of the world's population lives in cities. This rapid and continuing growth in urban populations suggests that by the year 2050 seven out of every ten human beings will be city dwellers. Accommodating large numbers of people has led to accelerated urban development, both upwards in the form of taller buildings, and outwards as cities increase in area. Cities are growing so fast that in some parts of the world, such as New York and Tokyo, they are merging into one another, covering vast areas known as megalopolises. About three per cent of the Earth's land surface is already urban, and it is likely that this proportion will double over the next 20 years. Urban areas increasingly dominate economic and social life.

Urban Areas

31	Human migration
340–51	Agricultural areas
Population growth	362–63
Industrial areas	372–79

The human world is now essentially urban. This is not simply because half the world's population lives in cities. Cities dominate contemporary life. They exert such enormous power that even the most remote rural regions are usually bound up in supplying their need for food, raw materials, and energy. As a result they become focal points for transport, trade, and technological advances. Cities are the centres of cultural life, with their many museums, theatres, and other attractions. Social changes that might take centuries in the countryside can sweep through urban areas in just a few years. Cities' impact on the environment can be equally dramatic. Urban development is not only burying vast areas under concrete; it is also having a profound effect on the world's climate – at a local level, by creating heat islands, and at a global level, by filling the atmosphere with the by-products of its massive energy consumption.

Location

Most cities start life as small settlements, and grow only because their location gives them some advantage over other places. In the past, cities such as London and Paris grew up at natural crossing places over major rivers. A few developed at strategic defence sites. Others, such as Johannesburg, were sited near sources of valuable minerals. In the developed world, many older urban areas, such as Pittsburgh in the USA and the cities of the Ruhr valley, Germany, grew up where there were ready sources of coal and iron. Now, as the world becomes increasingly dominated by global trade, local resources are less important, and access to markets is the key factor. Big cities operate globally, not locally. Most of the world's major cities, such as New York and Tokyo, are close to navigable waterways. Forty per cent are actually located on tidal estuaries or directly on sea coasts, and are major centres of road, rail, and air transport.

NIGHT LIGHTS
The clusters of cities in the USA, Europe, and Japan stand out in this composite satellite image of the Earth at night.

PORT OF HONG KONG
Inlets around the coast of Hong Kong Island make ideal harbours, such as this one at Aberdeen. They also provide shelter for local craft during typhoons.

Internal Zones

The concentration of some activities in particular areas gives most cities a distinct structure. In the 1920s, it was suggested that cities were arranged in rings around the central business district, with first a transition zone of industry and poor housing, then a ring of workers' houses, a ring of suburbs, and finally an affluent commuter belt. In 1939, it was argued that cities developed not in rings, but in wedge-shaped sectors along transport routes. Another later idea

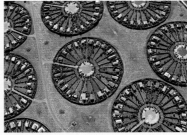

A PLANNED CITY
Most European cities grew haphazardly. Few have a planned structure like this circular housing development near Copenhagen, Denmark.

was that cities develop around not just one centre or nucleus, but several. Elements of all these models can be identified in most older western cities, but cities are changing, with increasing decentralization and the growth of "edge cities" on urban fringes. Analysis of cities in other parts of the world has led to a further series of models, emphasizing, for example, the central mosque and street markets for Islamic cities, and wealthy central areas and poor fringes for Latin American cities. New cities, such as Brasília in Brazil, are planned from the outset.

LAND

Population Trends

Urban populations are growing everywhere in the world. After 2008, for the first time in history, more people were living in urban areas around the world than in the countryside. By 2030, another 1.5 billion will be added to the world's urban population. In the more developed countries, already highly urbanized, over 86 per cent of people will live in cities by 2050. Some of the largest cities in the developed world are already almost as big as they are ever likely to get, and Tokyo's population has probably peaked. The megacities of the future are in the developing world: by 2025, Delhi in India may be home to 28.6 million people; Mumbai, also in India, may be nearly as big. However, both of these are likely to be eclipsed by China's Pearl River delta, centred on the city of Guangzhou (formerly Canton). Already, 40 million people live in an area smaller than Los Angeles. Soon the whole region may merge to create the biggest single urban area the world has ever seen, with a population completely dwarfing Tokyo's. The population of Guangzhou alone soared from just 380,000 in 1980 to nearly 27 million in 2023.

MASS MOVEMENT
As urban populations soar, the pressure placed on mass transit systems will grow, with underground systems moving the greatest numbers of people.

CROWDED SHANTYTOWN
Shanty dwellings, such as these in Mexico City, arise when large numbers of people migrate into cities and have nowhere to live.

City Climates

A large city is nearly always a little warmer than the surrounding countryside. On warm, sunny days, the difference in temperature may be 5°C (10.5°F) or more. The reasons for this effect, called the urban heat island, are complex. One cause is waste heat given off from city buildings, trains, and cars, which can contribute up to half as much heat as the Sun. Another cause is the way concrete and brick in the city soak up extra heat from the Sun during the day, and then release it at night. Even pollutants can add to the warming by trapping heat near the ground. Pollutants can cause other changes to the climate, too. Dense fogs called smogs were common in many industrial cities in the last century as domestic fires, industrial furnaces, and steam trains poured out soot that encouraged the condensation of moisture in the air. Clean-air legislation, and changes in fuel use have cut the worst of these smogs. However, in the 1980s, a new kind of smog, called photochemical smog, began to appear in sunny cities with heavy traffic, such as Los Angeles. This smog is a result of the way emissions from car exhausts react with sunlight to produce a thick, brown haze.

CITY SMOG
This policeman on traffic duty in Kuala Lumpur wears a face mask to protect himself from the fumes.

HEAT ISLANDS
Large urban areas develop their own microclimates. This diagram shows the temperature gradient at night for a city in a temperate region. The buildings in the city centre radiate more heat than the surrounding low-density housing and parkland, keeping it warmer.

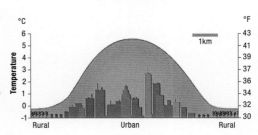

BRIGHT LIGHTS
The centres of many cities are brightly illuminated at night. Here, Times Square in New York displays a dazzling array of neon lights.

URBAN AREA PROFILES

The following pages contain profiles of selected major cities. Each profile begins with the following information:

AREA Area covered by the city and its suburbs

POPULATION People living in the city and its suburbs

FOUNDED Year in which city was founded

LAND

New York

LOCATION At the mouth of the Hudson River, New York State, eastern seaboard of the USA

AREA 11,344 square km (4,380 square miles)

POPULATION 21.4 million

FOUNDED 1609

The largest city in North America, New York is also one of the world's pre-eminent financial, commercial, and cultural centres. Situated on the east coast of the USA, about midway between Washington DC and Boston, New York lies on the same latitude as Rome and Istanbul. However, it experiences a very different climate, with bitterly cold winters, often with heavy snow, and hot summers. The city's origins date back to 1609, when European colonists set up a trading post on Manhattan Island at the mouth of the Hudson River. Sheltered from the Atlantic by Long Island, and with access to the interior of the country via the Hudson River, Manhattan provided the ideal gateway to North America. Because it has the best harbour on the eastern seaboard, Manhattan attracted many European traders, and first developed as the Dutch settlement of New Amsterdam. It was a flourishing port long before it was captured by the British in 1664 and renamed New York. It is estimated that more than half of the people and goods that have ever entered the USA came in through the docks of New York. Between 1892 and 1954, some 17 million immigrants landed at Ellis Island, another island in the mouth of the Hudson River, close to Manhattan and Liberty Island, where the Statue of Liberty now stands. Today, the point of arrival for many people in North America is New York's JFK airport. Until the 1950s, New York was by far the world's busiest seaport and, although it has declined in importance since the opening of the St. Lawrence Seaway in 1959, it

BROOKLYN BRIDGE
Spanning the East River, Brooklyn Bridge was the first suspension bridge to be made of steel. This view of Manhattan is taken from its walkway.

remains the second largest seaport in the USA. As the city expanded, it spread over adjacent islands and also onto the mainland, giving rise to Brooklyn, the Bronx, Queens, and Staten Island. Each district was connected to the other by ferries and bridges, and later by tunnels and rail links. In 1898, Manhattan and the four

ROCKS IN THE PARK
Despite the deposition of 10 million cartloads of stone and earth to create Central Park in 1858, outcrops of the hard rock on which Manhattan's skyscrapers are built are still visible.

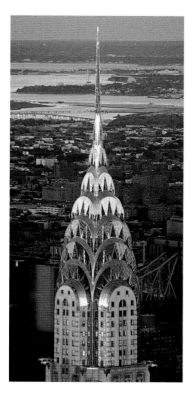

CHRYSLER BUILDING
Completed in 1930, the stainless-steel art deco tower of the 77-floor Chrysler Building is one of the best-known landmarks of central Manhattan.

other boroughs united to form Greater New York, a city second only to London in size. It also adjoins other urban areas, such as New Jersey, from which it draws some of its workforce. By 1930, New York was the largest city in the world, the solid bedrock of Manhattan Island providing a firm foundation for some of the world's tallest buildings, including the Empire State Building, 381m (1,250ft) high, which was completed in 1931. For a long time, New York was the world's biggest industrial centre, but competition from countries with cheaper land and labour had a devastating effect. Over three-quarters of

a million factory jobs disappeared between 1960 and 1990. However, as both shipping and industry dwindled, the city reinvented itself as a centre for stock-market traders, currency traders, and investment bankers. At the beginning of the 21st century, New York remains one of the world's most dynamic, prosperous, and influential cities, and its skyline, with its towering skyscrapers, remains one of the great icons of modern urban life.

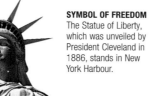

SYMBOL OF FREEDOM
The Statue of Liberty, which was unveiled by President Cleveland in 1886, stands in New York Harbour.

WASTE DISPOSAL

New York generates more waste than any other city in the world. It is estimated that New York generates nearly a quarter of a tonne of rubbish every second, and almost 24,000 tonnes every day. Until recently, the rubbish was burned and dumped beyond the city limits in 11 landfill sites. Incineration causes pollution, and people living close to landfill sites are starting to object to the mounting tide of garbage. The giant landfill site at Fresh Kills on Staten Island, which closed in 2001, brought some relief, but the problem refuses to go away.

MANHATTAN ISLAND
Skyscrapers can only be built where there are solid foundations. Most of the tallest buildings in Manhattan are built on granite to the south and centre of the island.

LAND

LIGHTING UP THE NIGHT SKY
Many of the world's cities are growing as people migrate to their peripheries in search of more living space. Urban energy use, as seen here in night-time Chicago, USA, is so great that it produces measurable warming of the air.

NORTH AMERICA *northeast*

Toronto

LOCATION On the northern shores of Lake Ontario, Ontario Province, Canada

AREA 2,344 square km (905 square miles)

POPULATION 6.7 million

FOUNDED 1793

Toronto is the commercial and financial heart of Canada. The city owes its success to its situation on the shores of Lake Ontario, giving it direct access to Atlantic shipping via the St. Lawrence Seaway and to the major US industrial centres across the Great Lakes. The business area, close to the lake, is dominated by many tall buildings, including the CN Tower, which at 553m (1,814ft), is the world's second tallest free-standing structure.

CN TOWER DOMINATES THE SKYLINE

NORTH AMERICA *north*

Chicago

LOCATION On the southwestern shores of Lake Michigan, northeast Illinois State, USA

AREA 7,006 square km (2,705 square miles)

POPULATION 9.05 million

FOUNDED 1816

Located in the centre of the North American continent, Chicago's position on the shores of the Great Lakes makes it a natural meeting point for traffic of all types. However, the flat terrain of the Midwest and the city's site beside Lake Michigan make its climate unpredictable,

ELEVATED RAIL ROUTE
Also known as the "El", the Chicago Transit Authority train runs across the city on elevated tracks; here it is crossing the Chicago River.

and wind storms are common. Chicago is a major industrial area, with massive chemical and steel-making empires. The harbour, created in 1834, allowed ships to deliver their cargoes directly onto land instead of anchoring offshore to unload. Lake Michigan has no natural harbours.

FRANK LLOYD WRIGHT

Twentieth-century Chicago was the centre for a new school of architecture, led by Frank Lloyd Wright (1867–1959). His Prairie Style is based on low-level houses, constructed from mass-produced materials. There are over 100 Wright-designed buildings, including the famous Robie house, in the Chicago area.

The channel that was cut through the offshore sand bar gave ships access to the Chicago River and shelter from storms. The docks now handle 82 million tonnes of cargo every year. Chicago's proximity to America's farming heartland also makes it a major focus of the food industry; its famous stockyards used to process 18 million head of livestock a year. At times, the city has gone into serious economic decline, yet it remains one of the world's most industrious and dynamic cities.

RIVER CROSSING
A series of drawbridges, dwarfed by the surrounding skyscrapers, span the river in downtown Chicago.

NORTH AMERICA *west*

Los Angeles

LOCATION On the plain of the southwestern Pacific coast, California State, USA

AREA 6,351 square km (2,452 square miles)

POPULATION 15.2 million

FOUNDED 1781

Sprawling across the Californian coastal plain, Los Angeles comprises 80 different towns and is the world's largest urban area. Although only the seventh largest city by population, its cars and freeway system have made its inhabitants uniquely mobile, enabling Los Angeles to spread over a wide area. It is bounded to the east by mountain chains running north–south, and is separated from the San Fernando valley to the north by the Hollywood Hills and Santa Ana

Mountains. Founded by Spanish missionaries in the 18th century, Los Angeles is now North America's leading industrial and technological centre. However, the city has a serious problem with pollution. The smogs, for which Los Angeles is famous, are caused by the city's warm, dry climate. Onshore sea breezes trap the pollutants against the surrounding mountains, where temperature inversions (cool air below warm air) may prevent their dispersal.

WATER RIGHTS

Los Angeles is warm, sunny, and very dry, with less than 380mm (15in) of rain a year falling in the south of the city. Providing fresh water is a major problem, and supplies were being maintained from increasingly distant sources. Today, water reclamation and conservation facilities help to avoid conflict with farmers and Indigenous peoples, who also need the water.

LOS ANGELES SMOG AT SUNSET
During the day, pollution levels in downtown Los Angeles rise as people drive to and from work. In the evening, smog may develop, as here, that could last until morning.

HOLLYWOOD
The Hollywood sign, erected in 1923, originally advertised a housing development; nowadays it is synonymous with the film industry.

NORTH AMERICA

Mexico City

LOCATION On the dry lake bed of Texcoco, at the southern end of Mexico's high central plateau

AREA 2,530 square km (977 square miles)

POPULATION 21.8 million

FOUNDED 1325–50

Although it is 2,238m (7,385ft) above sea-level and lies within an active earthquake zone, Mexico City is not only the capital and commercial centre of Mexico, it is also one of the three largest cities in the world, along with Tokyo and New York. Due to its altitude, the climate is mild and spring-like for most of the year, unlike other parts of Mexico. Surrounded by mountains, the city is built in a lake basin with no outlet channel. Drainage of Lake Texcoco, and other lakes, has lowered the water table in the area and caused buildings in the city to subside. Mexico City also has an air-pollution problem due to a combination of climatic conditions, rapid urban growth, and sulphur dioxide emissions from the nearby volcano Popocatépetl (see p.161).

ALAMEDA PARK
Despite overcrowding, Mexico City has many beautiful areas, including this park, which is right in the heart of the commercial centre.

CENTRAL LIBRARY
Designed by Juan O'Gorman, the Central Library is a mixture of modern and traditional styles. It is an important repository of Spanish literature.

SOUTH AMERICA *east*

Rio de Janeiro

LOCATION On Guanabara Bay on the Atlantic coast, Rio de Janeiro State, Brazil

AREA 2,020 square km (780 square miles)

POPULATION 12.5 million

FOUNDED 1536

Sitting on a wide bay, with dazzling beaches and steeply domed mountains that include the Sugar Loaf (see p.183), Rio de Janeiro is one of the most dramatically sited cities in the world.

It was founded by Portuguese explorers in the 16th century, who called it Rio de Janeiro ("River of January") because they thought Guanabara Bay was the mouth of a great river. Early economic activity centred around growing sugar cane and whaling, but there was more significant growth when gold and diamonds were discovered in the neighbouring district of Minas Gerais. Until 1960, when the government transferred to Brasília, Rio was the capital of Brazil. It is now the second largest city after São Paulo, and one of South America's key centres of finance, trade, and transport. The mountains around Rio make it a crowded city, where the contrast between wealth and poverty is sharply apparent. Sleek beach-front suburbs are backed by shanty towns, called favelas, built on slopes that are too steep for normal housing. Despite these contrasts, Rio is a vibrant city, with a busy port that handles a huge proportion of Brazil's trade.

IPANEMA BEACH
Every weekend, thousands of people head for the famous Ipanema beach to sunbathe, swim, and play volleyball.

FAVELA
Built on a steep hillside in the Catumbi district, the wooden shacks of this favela are supported on stilts.

OVERLOOKING THE CITY
The giant statue of Christ the Redeemer is one of the landmarks of Rio de Janeiro. It stands on top of the domed peak of Corcovado, which is 710m (2,330ft) high.

POPULATION GROWTH

At the beginning of the 20th century, there were about 1.5 billion people on the Earth. Since then, the world's population has increased fivefold to a current total of more than 8 billion people, and, despite recent falls in the growth rate in some countries, the overall total is still rising fast. For the planet as a whole, the size of the human population is of crucial significance, because sheer numbers lie behind many of the environmental problems that affect the world today.

Continuing Growth

The current surge in human population began in the mid-1700s, with the start of the Industrial Age. At that time, the annual growth rate stood at less than 0.2 per cent – a modest figure that meant the population doubled only over several centuries. From this low starting point, growth gradually accelerated, launching population levels onto an exponential path. By the early 1980s, annual growth reached a peak of nearly 2 per cent, bringing doubling time down to just 35 years. This record growth fuelled doomsday predictions of a world swamped by humankind. In the last two decades, the situation has subtly changed. Humanity continues to multiply, but growth rates have eased back slightly from their 1980s peak. This fall has been due to an unexpectedly sharp drop in fertility rates – the average number of children born to each woman during reproductive life. In developed countries, total fertility rates have now fallen below 2.1 children, which is the figure needed to keep the population stable (assuming no net immigration). In some developed countries, such as Spain, birth rates fell as populations aged, but the arrival of young immigrants has helped lift birth rates again. In the world's developing nations, the picture is more complex. Some of these countries are already emerging from the same shift, known as demographic transition, as that seen in the industrialized world. China, for example, has a total fertility rate of about 1.3, which is well below replacement levels, as a result of stringent legal controls on family size. But in many other developing nations, fertility levels are still high enough to maintain rapid growth. In sub-Saharan Africa,

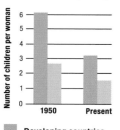

DECLINING FERTILITY
In the 1950s, developed and developing countries all had fertility rates well above replacement levels (2.1). Developed countries are now below replacement, while in developing countries the rate has almost halved.

(Bar chart: Number of children per woman, comparing 1950 and Present for Developing countries and Developed countries)

- Developing countries
- Developed countries

CONTINUING GROWTH
Africa is the world's fastest-growing region. With a fertility rate of over 6, Nigeria is likely to more than double its current population by 2050.

AGE PROFILES
The age profile of developed nations, such as Italy, is often narrow at the base, reflecting sustained low fertility rates. By contrast, Nigeria's indicates much higher fertility rates, coupled with lower life expectancy.

(Age profile chart: Age groups 0–9 through 80+, Population (millions), comparing ITALY and NIGERIA)

most countries have fertility rates above 4.0, while some are greater than 7.0. These differences are evident when the age profiles of national populations are compared (see above). In a typical developed country, there is a roughly even mix of people of different ages, tapering off from the age of 60. In developing countries, there is a greater bias towards younger age groups, and, as today's children move up the profile into childbearing age, there will be rapid population growth for decades to come.

Uncertain Future

According to recent UN estimates, the human population may reach a plateau of 11 billion by 2100, but a more pessimistic set of assumptions foresees continued rapid growth. In either event, humanity's impact on the Earth seems certain to increase, raising questions about whether such huge numbers can be sustained. Some economists argue that improvements in biotechnology will enable the human race to feed itself, and that similar "technological fixes" will be found for problems such as global warming (see pp.458–59). But most ecologists are not convinced and foresee widespread deprivation and environmental destruction, fuelled by humanity's growing need for raw materials, water, and food.

LOOKING AHEAD
Human population growth has gathered pace over the last 250 years – in this graph, three UN projections demonstrate how fertility could affect future population growth. The medium projection assumes fertility stabilizes at replacement levels – about 2.1 children per woman. The high projection assumes 2.8 children, and the low projection 1.6.

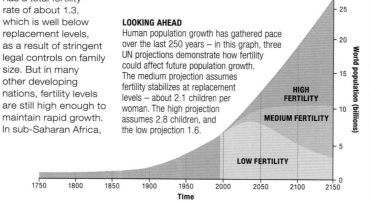

(Graph: World population (billions) over Time from 1750 to 2150, showing HIGH FERTILITY, MEDIUM FERTILITY, and LOW FERTILITY)

MASS MOVEMENT
Commuters leave after a day's work at a steel mill in Baotou, China, the world's most populous country – and also one of the most polluted, after rapid industrialization.

LAND

SOUTH AMERICA *east*

São Paulo

LOCATION On a plateau 50km (30 miles) from the Atlantic coast, São Paulo State, Brazil

AREA 3,649 square km (1,409 square miles)

POPULATION 23.08 million

FOUNDED 1554

Situated on the Tietê River, in the southeast corner of the Brazilian plateau, São Paulo is the biggest city in the southern hemisphere and one of the four largest in the world. It is also one of the fastest-growing cities, with an influx of about 190,000 people each year. The region's climate and soil are ideal for growing coffee, but there are also deposits of iron ore, which led to its industrial development. São Paulo is a major transport hub, with links to the port of Santos, 95km (60 miles) to the east.

CITY OBELISK

SOUTH AMERICA *southeast*

Buenos Aires

LOCATION On the shores of the River Plate, Buenos Aires State, Argentina

AREA 3,437 square km (1,327 square miles)

POPULATION 16.7 million

FOUNDED 1580

As capital of Argentina, Buenos Aires is the second largest city in South America and its leading port. Located on the River Plate (see p.208), 240km (150 miles) from the Atlantic, the shore is gently shelving, so a channel was cut to give access to larger ships. Exports, such as grain and meat, have made Buenos Aires a focal point for rail and road transport. It is also the main industrial centre of Argentina.

AVENIDA 9 DE JULIO
Buenos Aires has many tree-lined boulevards, including Avenida 9 de Julio, the widest street in the world, which is 130m (425ft) across.

EUROPE *central*

Berlin

LOCATION At the confluence of the Spree and Havel rivers, central Brandenburg Region, Germany

AREA 1,368 square km (528 square miles)

POPULATION 4.01 million

FOUNDED before 1244

An amalgamation of towns around a historic centre, Berlin is Germany's capital and also its largest city. It is situated in the northeast of Germany, on flat land surrounding two rivers, the Spree and the Havel, which meet in the district of Spandau. Its position in the centre of Europe (on a line between Vienna and Copenhagen), its recent history, and its excellent transport routes

TIERGARTEN IN WINTER
Berlin's famous Tiergarten park was destroyed in World War II, but has now been replanted and includes one of the best zoos in Europe.

make it the obvious gateway to central and eastern Europe. The centre of Berlin is linked with the districts of Potsdam, Spandau, Charlottenburg, and other local areas by a well-established network of canals. For 40 years, it was divided by the Berlin Wall, which isolated the western half of the city within communist East Germany. In 1989, the wall was demolished, and the following year Germany reunified. Berlin soon regained its status as the capital of a reunited Germany, and is once again one of the leading cities in Europe.

REICHSTAG BUILDING
After reunification, Berlin's parliament building was restored, with the addition of this striking glass dome designed by the British architect Norman Foster.

EUROPE *east*

Moscow

LOCATION On the banks of the Moskva River in the European plain of western Russia

AREA 7,295 square km (6,154 square miles)

POPULATION 17.3 million

FOUNDED Before 1147

ST. BASIL'S CATHEDRAL
With its ten colourful onion domes, St. Basil's Cathedral in Moscow's Red Square is like no other cathedral in the world.

Situated far inland, Moscow has a strategically important location on the Moskva (Moscow) River. The river links the city to seas in the north and south via the Moscow Canal and the Volga River waterways; it also connects the Baltic region to the north with the Caspian Sea to the southeast. Although Moscow shares the same latitude as Copenhagen and Glasgow, it lacks the

SOVIET ACHIEVEMENTS
This monument, celebrating industry and farming, was a centrepiece of the Soviet Exhibition of Economic Achievement. Now the building is used as a showroom for goods from around the world.

ameliorating marine influence of the sea. Its climate is continental, with very cold winters and hot, humid summers. Despite being frozen for five months of the year, the Moskva River has been an important routeway since medieval times, and the Kremlin is built close to its banks. The Moscow Canal was built by forced labour in the Soviet era, and the city's industrial and economic might stems from that time. Moscow is the largest city in Europe, considerably bigger even than Paris. The city is a transport hub for a vast hinterland that covers all of Russia and the countries around it. Moscow is not only Russia's rail- and air-traffic centre; it is also one of its largest ports. In 1712, the country's capital was moved to St. Petersburg, and Moscow went into decline until the Communists came to power in 1917. They reinstated Moscow as capital and enforced a programme of industrialization. Since the collapse of the Soviet Union in 1991, many factories have closed as their products failed to compete on the world market. Though Moscow's financial and service sectors boomed in post-Soviet times, the impact of Russia's war on Ukraine and isolation from the West remains to be seen.

made of stainless steel

APARTMENT BLOCKS IN THE SUBURBS
Many Muscovites live in small, two-room flats in giant apartment blocks, which were erected on the outskirts of Moscow in the 1950s.

EUROPE *west*

London

LOCATION On the lower reaches of the River Thames in southeastern England, UK

AREA 1,738 square km (671 square miles)

POPULATION 11.2 million

FOUNDED Before 43 CE

THE SHARD
Standing 309.6m (1,016ft) tall, the Shard near London Bridge opened in 2013 and is the tallest building in Western Europe.

PARLIAMENT BUILDINGS
The Houses of Parliament and Big Ben were built in 1860 after a fire destroyed an earlier building.

London, one of the world's oldest and most historic cities, is the capital of the United Kingdom. Although on the same latitude as Warsaw, it is influenced by the Gulf Stream, a warm oceanic current, and so has a mild, temperate climate. The city was established by the Romans over 2,000 years ago on a site now occupied by the city's financial district. During the 19th century, as the hub of the British Empire, it became the world's largest city and encompassed many small towns, including Highgate and Wimbledon. Today, despite the highly detrimental impact of the UK leaving the European Union, it remains one of the financial capitals of the world, and one of its most cosmopolitan

cities. London is at risk of flooding from high surge tides, entering the Thames estuary. The threat is increasing because southeastern England is slowly sinking, while tides are rising at a rate of 60cm (2ft) per century. To protect the city, the Thames Barrier, which was opened in 1984, was built at Woolwich.

GREEN BELTS

In the 1950s, the expansion of the London built-up area was deliberately halted by City planners, who set aside a broad band of countryside around the capital called the Green Belt. Strict planning controls limited building within this area. So effective have these controls been that even today, there is a green ring of open land around the capital.

EUROPE *west*

Paris

LOCATION On the River Seine in the Ile de France region, northern France

AREA 2,853 square km (1,102 square miles)

POPULATION 11.06 million

FOUNDED Before 50 BCE

Paris has been one of the world's key cities for hundreds of years. Situated in the Paris Basin, an area of flat valleys and plateaus about 400km (250 miles) across, it is divided by the Seine into the well-known Right and Left Banks. As the Seine passes through Paris, it flows around two islands, the boat-shaped Ile de la Cité, where Notre Dame is

EIFFEL TOWER
The instantly identifiable Eiffel Tower was built for the Great Exhibition in 1889. At 319m (1,051ft) tall, it dwarfs the buildings around it.

located, and Ile de St-Louis, which was once swampy pasture land but is now an elegant residential area. Few parts of the old city exceed 100m (30ft) above sea-level; Montmartre, the artists' quarter and site of the Sacré Coeur, is one of the higher parts, being located on a hill called the Butte. Paris is not only the political and cultural capital of France, but also the country's financial and commercial heart. Although the Paris region lacks any real resources, the city's position as capital made it France's dominant industrial centre up until World War II. Since then, many heavy industries have moved out. New high-tech industries have developed in the suburbs, but Paris is now more a centre of commerce and one of the world's leading tourist destinations.

PARIS SHOP
Renowned for its food, Paris not only has good restaurants, but also many small shops, such as this greengrocer, which offer a wide range of foodstuffs.

LA GRANDE ARCHE
This modern triumphal arch in the centre of the new business district is also an office block.

LAND

EUROPE *south*

Rome

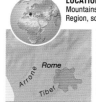

LOCATION West of the Abruzzi Mountains, on the River Tiber, Lazio Region, southern Italy

AREA 1,145 square km (422 square miles)

POPULATION 3.1 million

FOUNDED Before 753 BCE

Centre of the great Ancient Roman Empire and home of the Catholic church for almost 2,000 years, Rome is Italy's capital and largest city. Situated 28km (17 miles) inland from the west coast of southern Italy, Rome originated as a shallow crossing point on the Tiber and later became a market place. This low-lying site meant that, until 1870 when embankments were built to contain it, the Tiber flooded regularly, and inundated parts of the city. Rome is also famous for its seven hills. The Palatine Hill, above the Forum, has been settled since Roman times, as have the Capitoline and Quirinal hills. Later expansion of the city incorporated most of the remainder. Millions of tourists come to Rome each year to marvel at its treasures; they provide the bulk of the city's income. The city is also a major administrative and financial centre.

THE SPANISH STEPS
A favourite place for both tourists and locals to congregate, the steps were built to link the Trinite dei Monti church with the Piazza di Spagna.

ST. PETER'S BASILICA
The magnificent Basilica of St. Peter's, with its great dome designed by Michelangelo and Bramante, is one of the greatest architectural treasures of Rome.

PRESERVING THE PAST

Rome's long history poses many problems to city planners. There are ancient remains buried everywhere under the city centre, and most new buildings present another hard choice between building or preserving the past. Construction work to extend Rome's inadequate subway has been constantly halted as archaeologists battle to record ancient relics before they are destroyed.

EUROPE *south*

Istanbul

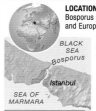

LOCATION On either side of the Bosporus straits, straddling Asian and European Turkey

AREA 1,471 square km (568 square miles)

POPULATION 16 million

FOUNDED About 650 BCE

Known previously as Byzantium and Constantinople, Istanbul is the only city in the world to stand on two continents, Asia and Europe. This position has made it a unique meeting point of East and West for over 2,000 years. With the finest natural harbour in the region, it is a major trade hub and controls shipping passing into the Black Sea from the Sea of Marmara.

VIEW OF CITY FROM BEYOGLU DISTRICT
Old European Istanbul, called Stambul, is visible from the modern part across a narrow channel of the Bosporus, the Golden Horn.

AFRICA *north*

Cairo

LOCATION On the banks of the Nile at the head of the delta, between Upper and Lower Egypt

AREA 2,010 square km (776 square miles)

POPULATION 20.2 million

FOUNDED 3011 BCE

Situated about 200km (125 miles) south of the Mediterranean Sea, Cairo is Egypt's capital and Africa's largest city. Its origins date back 6,000 years, when the Ancient Egyptians founded their capital Memphis here. The reasons for their choice of site still hold true today. Cairo is located in one of the few areas of Egypt to have fertile soil and plentiful water. Vast expanses of desert stretch away from the city fringes on all sides except the north, where the lush delta region is watered by the Nile (see p.217). The Muqattam Hills border the eastern side. Cairo receives less than 25mm (1in) of rain a year, and gets all its water from the river. However, the annual flood cycle of the Nile ended in 1971 with completion of the second Aswan Dam. Estimates of Cairo's size today vary, but

THE PYRAMIDS AT GIZA
The Giza plateau was the royal burial ground for Memphis, but today it is being extensively developed to accommodate the needs of tourists.

its population seems likely to be more than 20 million, and it is growing by the year. Rapid growth has created problems such as widespread poverty, traffic congestion, and pollution. To combat congestion, the Egyptian government is building a new capital 45km (28 miles) to the east, currently called the New Administrative Capital (NAC). To the west, modern skyscrapers rise along the well-irrigated Nile shoreline, while further east there are ancient medieval streets filled with noisy souks (bazaars). Beyond these to the southwest, rise the Great Pyramids that play a key role in the city's tourist industry.

MOSQUE AND CITADEL
Dominating the skyline of Islamic Cairo is the heavily fortified complex of Saladin's Citadel and the Mosque of Mohammed Ali.

AFRICA *west*

Lagos

LOCATION On the southwest coast of Nigeria and four nearby islands, Lagos State, Nigeria

AREA 1,966 square km (759 square miles)

POPULATION 16.6 million

FOUNDED 1472

The city of Lagos is situated on the Bight of Benin, an arm of the Atlantic Ocean. It covers several islands, linked to each other by bridges and landfills, as well as parts of the mainland, such as Apapa, the main port area. Estimates of Lagos's population range from 13 to 26.6 million. It is Nigeria's commercial and industrial hub, and it may now be Africa's biggest city.

MARKET IN LAGOS

AFRICA *south*

Johannesburg

LOCATION On the Witwatersrand ridges, Guateng Province, South Africa

AREA 4,040 square km (1,560 square miles)

POPULATION 14.5 million

FOUNDED 1886

Johannesburg is South Africa's largest city and the largest in Africa south of the Equator. It is built on the highest part of the interior plateau known as the Highveld, near the spot where gold was discovered in 1886. The city grew from a mining camp into a modern city, which is combined with its neighbouring city Pretoria to the north to create a metropolitan area in the old gold-mining region with a population of over 14 million. Although no gold mining is carried out in Johannesburg today, white spoil heaps are littered around the city. Soweto is situated to the south-west, sprawling across some 100 square km (39 square miles) of bare terrain.

SHANTY TOWN
Most of Johannesburg's population live far from the centre, and far from their places of work, in shanty towns on the city's outskirts.

OVERVIEW OF CITY
The high-rise buildings of the central business district quickly give way to low-level urban sprawl.

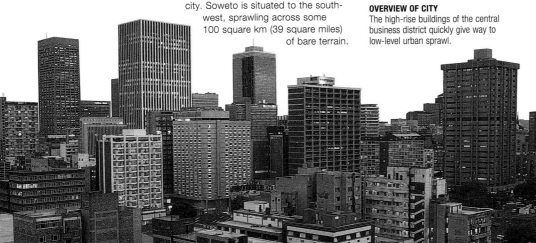

ASIA *west*

Tehran-Karaj

LOCATION On the southern slopes of the Elburz Mountains, Tehran Province, northern Iran

AREA 1,704 square km (658 square miles)

POPULATION 14.1 million

FOUNDED 1220

The capital of Iran, Tehran-Karaj is one of the Middle East's largest cities. It lies 2,100m (6,930ft) above sea-level at the foot of the highest peak in the Elburz Mountains, Mount Damavand. Although there are iron-ore deposits nearby, the city's rapid growth over the last 80 years is due to oil. Tehran-Karaj today shows little sign of its ancient history despite its proximity to Rayy, a city built about 2,500 years ago.

TEHRAN-KARAJ AND THE ELBURZ MOUNTAINS

ASIA *southwest*

Karachi

LOCATION On the shores of the Arabian Sea, Sindh Province, southern Pakistan

AREA 1,124 square km (434 square miles)

POPULATION 15.7 million

FOUNDED 1728

Located on the northwestern edge of the Indus river delta, on the shores of the Arabian Sea, Karachi is a major transport hub. It is also the largest city and main seaport of Pakistan. Its modern docks, situated partly on what was once the island of Kaimari, are among the busiest in Asia, handling its own produce, and much of that from landlocked Afghanistan. The city also has a large airport and is a focus for major railroads and highways. Being on the edge of the Sind desert, Karachi has a hot, dry climate, with barely 200mm (8in) of rainfall a year. This creates problems in providing year-round fresh water for the city's inhabitants. In 1947, Karachi was a small city of less than half a million people, but internal migration, and an influx of refugees after independence, greatly increased its size. Today, it is overcrowded, with areas of extreme poverty and pollution problems.

BUSTLING STREETS
The central market, or Sadir, is one of several bazaars in Karachi's crowded centre.

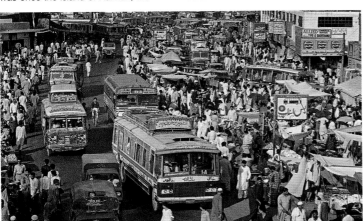

ASIA *south*

Dhaka

LOCATION By the Buriganga River, in the south-centre of Bangladesh

AREA 619 square km (239 square miles)

POPULATION 19.1 million

FOUNDED Over 1,000 years ago

The capital and largest city of Bangladesh, Dhaka may be the most densely populated big city in the world. A population of nearly 20 million crowd into a very tiny area –

LALBAGH FORT
This unfinished 17th century Mughal fort complex in Old Dhaka is a key tourist attraction. Of all the buildings inside the fort, Bibi Pari's Mausoleum is the most important.

NARAYANGANJ BOAT MARKET
With the onset of the monsoon, little handmade wooden boats line up for sale at Bangladesh's boat markets.

that's over 30,000 per square kilometre. It is a third more crowded than even Mumbai and eight times as dense as Los Angeles central, the USA's most packed centre. Most people in Dhaka live in high-rise blocks, or in the shanties on the city fringes, and the average income is very low. Yet it is a major cultural, economic, and scientific hub and South Asia's third city in GDP, driven in part by its huge garment industry.

LAND

ASIA southwest

Delhi

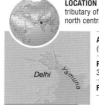

LOCATION On the Yamuna River, a tributary of the Ganges, Delhi State, north central India

AREA 2,344 square km (905 square miles)

POPULATION 32.2 million

FOUNDED 1st century BCE

India's capital, Delhi, is now combined with neighbouring cities in the National Capital Region (NCR) to form the third largest metropolitan area in the world. There has always been a city on the site since the 1st century BCE. Because of its strategic position at the focal point of an important north–south and east–west trade route, early rulers fortified their city with an impressive array of defences, such as the Red Fort of Shah Jahan,

built in 1639. Less than a century ago, Delhi had just a quarter of a million inhabitants; now it is 130 times as big. The upsurge started with the influx of refugees in the wake of independence, but it has continued unabated ever since. The city stands pivotally between the Indus and Ganges valleys, making it a transport hub for all northern India. Five national highways converge here, and it is a key railway junction. Old Delhi, and many of the ancient capitals, are situated along the banks of the Yamuna River. New Delhi, with its grandiose, colonial government buildings, has a powerful administrative role but is also a key financial, industrial, and information-technology centre.

TRAFFIC CONGESTION IN OLD DELHI
The sheer volume of traffic in Old Delhi's streets, including pedicabs and bullock carts, creates some of the world's worst traffic congestion.

THE RED FORT
With its 2km (1¼ miles) of massive red sandstone walls, the magnificent Red Fort is the largest of Old Delhi's monuments.

EDWIN LUTYENS

New Delhi was the grand capital of India, started by the British in 1911. Its chief architect was Sir Edwin Lutyens (1869–1944), who was inspired by Christopher Wren's plans for London and Pierre L'Enfant's for Washington DC. The Rashtrapati Bhavan (President's House, 1913–30), a mix of classical structure with Indian decoration, is but one of his designs.

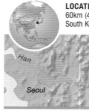

CULTURAL MIX
Delhi is a cosmopolitan city with Hindi, Punjabi, Urdu, and English all widely spoken. There are also many different cuisines.

ASIA southwest

Mumbai

LOCATION On seven islands off the Konkan coast, Maharashtra State, southwestern India

AREA 976 square km (377 square miles)

POPULATION 24.9 million

FOUNDED 1530s

Bounded to the east by the Western Ghats and to the west by the Arabian Sea, Mumbai (formerly Bombay) is situated on a promontory at the northern end of the Konkan coast. Because of its position, it receives the full force of the southwest monsoon, and in July rainfall averages 661mm (26in). Established by the Portuguese in the 16th century on the site of an

ancient settlement, Mumbai is India's second largest and most prosperous city. It has an excellent natural harbour that was improved in the 19th century by land reclamation, joining the seven islands and linking them with the mainland. In the late 19th century, Mumbai's proximity to the Suez Canal, across the Arabian Sea, gave access to the British market for its cotton goods. Although still important, the cotton textile industry is now shrinking; however, the city's industrial base is becoming more diversified and includes "Bollywood", the world's largest film industry. Mumbai remains India's busiest port, but it is also the centre of India's economy and its commercial and financial hub.

CONTRASTING ARCHITECTURE
Many of Mumbai's older buildings, dating from the 19th century, are built in a uniquely Indian and florid style of Gothic architecture.

HIGH–RISE RESIDENTIAL BLOCKS
Many of Seoul's inhabitants live in the huge area of high-rise apartments along the south bank of the Han River.

ASIA east

Seoul

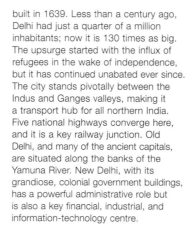

LOCATION On the Han River, 60km (40 miles) from the Yellow Sea, South Korea

AREA 2,769 square km (1,069 square miles)

POPULATION 23 million

FOUNDED 1394

BUSY STREET AT NIGHT
Seoul has the same traffic-congestion problems as other major cities, despite a network of highways.

Located in a natural basin and surrounded by mountains, such as the Bukhan Range to the north, Seoul is considered one of the most attractive capital cities in the world. The Han River, which runs through the city, divides it into two distinct areas. Most people live in high-rise apartments on the south side of the river but work in the city's business district

to the north. This causes massive congestion on bridges over the river as people commute to work. Founded in 1394 as the capital of the kingdom of Choson, Seoul is a city with a striking mix of ancient and modern. It grew rapidly in the 1960s and is South Korea's main financial, commercial, and industrial centre, as well as being a major aviation hub for northeast Asia.

ASIA *east*

Tokyo

LOCATION On the shores of Tokyo Bay, in the southeast of Honshu Island, Japan

AREA 8,231 square km (3,178 square miles)

POPULATION 37.7 million

FOUNDED 1457

Formerly known as Edo, Tokyo is Japan's capital city and the largest city in the world, still dwarfing even the Jakarta metropolis. Tokyo proper is smaller than this, but it merges with the cities of Yokohama, Kawasaki, and Chiba to form the largest concentration of people the world has ever seen, with nearly 40 million inhabitants. The ancient city of Edo was situated on land reclaimed from the Sumida estuary, but Tokyo spreads far over

TOKYO CITY HALL
Among the many striking, new buildings in Tokyo is the Tokyo Metropolitan Government Building, designed by architect Kenzo Tange.

TOKYO AT NIGHT
With more bright lights and neon than any other city, the skyline of Tokyo at night is a spectacular sight.

land reclaimed both from the river and the sea. Lying at the same latitude as Washington DC, it has a similar climate, with hot, humid summers. Twice in the last century, Tokyo was almost annihilated. Because the city lies in one of the world's most geologically active areas, where our tectonic plates converge (see p.89), earth tremors are common.

FLYOVERS IN TOKYO
In such a crowded city, an efficient transport system is essential. Many people travel large distances from out of town to offices in the city.

In 1923, the Kanto earthquake killed 100,000 people and laid the city centre flat. Then in World War II, it was bombed by the American airforce. However, within a few years the city's economy was booming, and has been ever since. Tokyo's economy is gigantic, with over 800,000 businesses and the biggest labour force in the world. The service sector is focused mainly on the districts of Maronouchi and Nihonbashi near the old Imperial Palace, while manufacturing is centred on the shores of Tokyo Bay, with easy access to the port of Yokohama.

OVERCROWDING

Tokyo is one of the world's most overcrowded cities. With space at a premium, land prices and rents are among the highest in the world. A striking response to the lack of space has been Tokyo's capsule hotels with their rows of sleeping units, called pods, which are no bigger than a bed.

ASIA *east*

Beijing

LOCATION On the North China Plain, in an arc of the Yen Mountains, China

AREA 4,284 square km (1,654 square miles)

POPULATION 18.5 million

FOUNDED 3000 BCE

Set at the northern corner of the North China Plain, on the alluvial deposits between the Yongding and Chaobai rivers, Beijing is built in a massive curve in the Yen Mountains. Lying at about the same latitude as Greece, it has warm summers, but bitter winds from the north bring icy cold winters. As the winds are dry, they rarely bring snow but they do carry fine yellow loess (see

p.77). Beijing has been a political and cultural focus for longer than perhaps any other city in the world. In 1929, fossil hominids, collectively called Peking Man, were discovered to the southwest of the city. The site has been settled by modern humans for 5,000 years and has been a key trading centre for at least 2,000 years. Beijing has been China's capital ever since the Mongol emperor Kublai Khan chose it for his headquarters in 1272. The Great Wall passes to the north of the city, and once protected it from invasion. The Grand Canal to the south was historically

important as it brought prosperity, being the only north–south route for transport and communication. Under Communist rule, Beijing has not always been sympathetic to its past. In the drive to modernize, many older buildings were demolished, especially in the run-up to the 2008 Beijing Olympics. Yet several old buildings survive, including the Forbidden City, once home of the Chinese emperors but now a major tourist attraction. Beijing is China's third largest city after Guangzhou and Shanghai and a leading industrial centre.

OLD BEIJING
Parts of old Beijing are characterized by high-density, single-story square buildings. Many are now being replaced.

WALLS OF THE FORBIDDEN CITY
Once separating the emperor from his people, the walls of the Forbidden City today enclose a series of museums.

THIRD RING ROAD AT TWILIGHT
Beijing is a sprawling city with an extensive transport network. Roads are busy, but many people still commute on their bicycles.

LAND

Hong Kong

LOCATION East of the Pearl River on the south coast of China, by the South China Sea

Hong Kong

Lantau Island

AREA 290 square km (112 square miles)

POPULATION 7.4 million

FOUNDED About 2,000 years ago

Once one of the world's busiest ports, Hong Kong has been eclipsed by Ningbo-Zhoushan, Shanghai, and the Pearl River delta. Its buildings are crowded onto the northern shores of Hong Kong Island and spread some distance up the slopes of Victoria Peak and other mountains that surround the bay. Hong Kong faces Kowloon, on the mainland of China, across Victoria Harbour, one of the most spectacular stretches of natural deep water in the world. The climate is humid and monsoonal, winters are cool and dry, and summers frequently bring typhoons. Hong Kong was just a fishing village until 1842, when the British took it over; they turned it into the key trading hub of eastern Asia. It was returned to Chinese rule in 1997.

ACROSS THE HARBOUR
Across Victoria Harbour from Hong Kong Island lies Kowloon, to which it is linked by ferries, three road tunnels, and a subway.

HAPPY VALLEY RACECOURSE
A relic of British colonial days, Happy Valley racecourse was first established on reclaimed marshland in 1846.

Shanghai

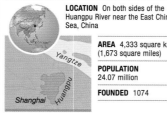

LOCATION On both sides of the Huangpu River near the East China Sea, China

Yangtze

Shanghai *Huangpu*

AREA 4,333 square km (1,673 square miles)

POPULATION 24.07 million

FOUNDED 1074

China's second biggest city, Shanghai is one of the world's leading seaports. Sited on a bend of the Huangpu River, it spreads across the delta of the Chang Jiang River and is criss-crossed by many canals and waterways. Its mild, maritime climate is due to its position in the centre of China's coastline, near the East China Sea. The city originally developed to make clothing from local cotton. When Britain forced China to open to western trade in the 19th century, Shanghai prospered. By the 1930s, it was a leading seaport. Now it is a major industrial city, with shipbuilding plants, iron-and-steel and chemical works, and textile factories.

COLONIAL BUILDINGS ON SHORELINE
Known before 1949 as the Bund, Zhong Shan road is a famous avenue along the Shanghai waterfront with grand, European-style mansions.

TAI CHI AT ZHONG SHAN
Where once Europeans arrived, now Chinese citizens practice their tai chi.

Guangzhou

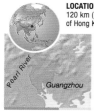

LOCATION On the Pearl River, about 120 km (75 miles) north-northwest of Hong Kong, in southern China

Pearl River

Guangzhou

AREA 4,535 square km (1,751 square miles)

POPULATION 26.9 million

FOUNDED 3rd century BCE

Guangzhou's history dates back to the Qin dynasty in the 3rd century BCE. However, it is only in the last few decades that it has really developed and become the focus of China's staggering economic progress. Industrial growth in Guangzhou has been among the most rapid in history, and the city has become the hub of a massive urban agglomeration, making it home to over 85 million people. Many predict that it will even outstrip Tokyo in the not too distant future.

CANTON TOWER
A symbol of Guangzhou's growing confidence, the 600-m- (2,000-ft-) high Canton Observation Tower is the fifth tallest freestanding structure in the world.

Manila

LOCATION On the eastern shore of Manila Bay, Luzon Island, Philippines

Manila
Manila Bay

Laguna de Bay

AREA 1,911 square km (738 square miles)

POPULATION 24.9 million

FOUNDED 1574

SMOKEY MOUNTAIN RUBBISH DUMP
Manila's rubbish dumps are the only means of survival for many of the city's poorest people, who scour them for food and scraps to sell. This dump was levelled in the 1990s.

Lying on the swampy delta of the Pasig River, Manila is the capital of the Philippines and its largest city. It was first settled by Spanish colonists in the 1600s because of its superb natural harbour. Sandwiched between the Pasig River and the Bataan mountains, the city is sheltered from extreme weather and has a warm, tropical climate. Metro Manila, as the Manila urban area is called, has grown rapidly since independence in 1947. Manila's setting once earned it the name Pearl of the Orient, but today it has poor housing and traffic congestion, though it is much cleaner than it was at the beginning of the century.

MANILA CATHEDRAL
The Catholic cathedral is one of the more elegant remnants of Manila's colonial past. It has been rebuilt seven times due to damage by earthquakes.

Kuala Lumpur

LOCATION On the junction of the Kelang and Gombak rivers, West Malaysia State, Malaysia

AREA 2,163 square km (835 square miles)

POPULATION 8.9 million

FOUNDED 1857

Kuala Lumpur is one of the world's youngest capitals. A century and a half ago, it was just a camp set up by Chinese tin-miners in the middle of the Malaysian jungle. The name Kuala Lumpur comes from the Malay for "Muddy River Mouth", a reference to the muddy river-bed from which tin was scooped by giant dredgers. Due to its strategic position at the confluence of the Kelang and Gombak rivers, Kuala Lumpur grew rapidly as a tin centre, despite its humid, often malarial climate. When Malaysia became independent in 1963, Kuala Lumpur became the capital and was perfectly placed to benefit from Malaysia's booming economy.

PETRONAS TWIN TOWERS
One of the world's tallest buildings at 452m (1,483ft), the Petronas Twin Towers were designed by Cesar Peli and finished in 1996.

Singapore

LOCATION On the island of Singapore, off the Malay Peninsula, Republic of Singapore

AREA 523 square km (202 square miles)

POPULATION 5.9 million

FOUNDED 100 CE

The city state of Singapore is one of the world's smallest, yet most successful, countries. Lying off the coast of Malaysia, about 137 km (82 miles) north of the Equator, it comprises 60 islands of which the largest by far is Singapore Island. Low-lying with a humid, tropical climate and abundant rainfall, Singapore Island was once blanketed by lush jungle and mangrove swamps, but it is now all but entirely covered by urban sprawl. Known originally as Temasek, the city got its current name, which means "Lion City" in Sanskrit, from the many tigers that once lived in the region. Singapore's success story began in 1819, when the British colonial administrator Sir Stamford Raffles gained possession of Singapore for the British East India company. The city's strategic location between the Indian Ocean and South China Sea, and its huge deepwater harbour, rapidly made it one of the the two busiest ports in the

world along with Hong Kong (see opposite page), and it is still among the busiest. On any day, there are at least 600 ships in port, and a ship sails in or out of the harbour every ten minutes. In 2019, Singapore docks handled 626.1 million tonnes of shipping, over half of which were container ships. Singapore's most remarkable period of growth has been since its independence in 1965. The economy has boomed and includes thriving electronics and oil industries. The people of Singapore are now among the richest in the world.

ORCHARD ROAD DISTRICT
This is Singapore's main shopping street and site of some of its most expensive hotels and shops.

Jakarta

LOCATION At the mouth of the Ciliwung River, Java Island, Indonesia

AREA 3,546 square km (1,369 square miles)

POPULATION 33.7 million

FOUNDED About 400 CE

Situated on swampy plains at the mouth of the Ciliwung River, Jakarta is a steamy, tropical city where humidity typically rises above 85 per cent. It suffers from both frequent flooding and

a shortage of drinking water. The area was first settled around 400 CE by Hindu people, but got its current name from Muslims who took over in 1527. Dutch traders captured Jakarta in 1619, and for 300 years it was their trading centre in Southeast Asia. The harbour still handles much of Indonesia's foreign trade, and its factories produce a wide range of goods. After Indonesia's independence in 1949, the capital grew rapidly and continues to do so and is now the world's second largest urban metropolis.

ISTIQLAL MOSQUE
Jakarta is a predominantly Muslim city, and the Istiqlal mosque is the largest mosque in Southeast Asia and one of the largest in the world.

Sydney

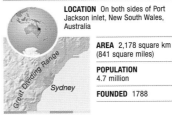

LOCATION On both sides of Port Jackson inlet, New South Wales, Australia

AREA 2,178 square km (841 square miles)

POPULATION 4.7 million

FOUNDED 1788

Built on low hills around a magnificent natural harbour, Sydney is Australia's oldest and largest city. Its waterside location and temperate, humid climate, with warm summers, attracted indigenous

Australians tens of thousands of years ago, but in 1788 it was taken over by the British. By 1800 it was a successful commercial port, first trading in whale products, then in wool and wheat grown in Australia's interior. Sydney gradually spread north across the harbour, as well as east to include the beaches for which it is famous. The city has an excellent transport system, including the Sydney Harbour Bridge, which was completed in 1932. Sydney is the financial and industrial hub of the South Pacific and remains the main gateway to Australia.

SYDNEY HARBOUR BRIDGE
Built in 1932, Sydney Harbour Bridge links downtown Sydney with the suburbs on the northern shore with a giant steel span.

MOVING IRON
Heavy machinery unload large amounts of
imported iron ore at the Lianyungang port
in Jiangsu Province, China.

INDUSTRIAL AREAS

Two and a half centuries ago, one of the most profound changes ever to affect human lifestyles began to spread across the world from northwest Europe: the Industrial Revolution. Up until that time, most people lived in the country and worked the land. The Industrial Revolution brought the first large factories, spawned the first major urban centres, and transformed landscapes. Vast areas were turned over, not only to the factories and mills, but also to their support networks: homes for workers; transport networks to supply the factories and distribute their products; and mines and quarries to feed them with raw materials. From its European roots, the tidal wave of industrialization has gradually gathered pace across the world from Europe to the USA, Canada, Japan, and Russia. It has now spread right across the world, with most countries having considerable industry. The scale of industry in China now dwarfs the rest of the world.

Industrial Areas

45	Ore deposits
80–81	Fossil fuels
340–55	Agricultural areas
352–71	Urban areas

The first great industrial areas were located close to major sources of raw materials and power. They spread out across the Ruhr coalfield in Germany, over the ironfields of Pittsburgh, USA, and many other places where coal and iron were nearby. Heavy industries still tend to develop near their source materials, especially in the developing world. Yet, as power became easier to access, transport of goods and workers more efficient, and components lighter and easier to move, modern industry has shifted away from traditional industrial regions. Although most high-tech industries are based in metropolitan areas, these businesses do have more flexibility in choosing their locations.

Recent Development

Europe and North America have been industrialized for so long that new developments have little impact. The most dramatic growth is now occurring in the world's most recently industrialized countries, such as Malaysia, Korea, and Taiwan. Over the last 50 years, these countries have changed rapidly from exporting raw materials to processing them. They invested first in heavy industries and then in high-tech products. They have also expanded by encouraging global companies to locate within their boundaries. Industrial development has now slowed in many Asian countries, except for China. The area along the southern coast from Shenzhen to Zhuhai, and on the Yangtze around Pudong, is currently undergoing the most rapid industrial development the world has ever seen.

OIL REFINERY
The world's largest single industrial plants are oil refineries. They cover vast areas with pipelines and storage tanks.

Landscape Impact

No human activity has a more dramatic effect on the landscape than industry. It covers vast areas with concrete, brickwork, and steel, obliterating natural features. The most striking impacts are the huge holes in the ground made by open-cast mines and quarries, such as the Bingham Quarry in Utah, USA, where over 7 square km (2¾ square miles) of the Earth's surface has been blasted away. The hills created by spoil dumps can be equally dramatic – for example, the mountains of white china clay in Cornwall, UK, and the gigantic New Cornelia Tailings in Arizona, USA.

LANDFILL SITES
Industry generates a huge amount of waste, which must be disposed of. One option is to create landfill sites, such as this one in Hong Kong. As it is a densely populated area, the site will create much-needed space for building.

Types of Industry

Industry is divided into three types: primary, secondary, and tertiary. Primary industry uses raw materials from the land and sea: mines and quarries extract coal, oil, copper, and many other materials; while fishing, farming, and forestry exploit living resources. In the world's most advanced economies, primary industry is not as important as it once was, but it still accounts for a large share of the wealth of developing nations. Secondary industry takes raw materials and manufactures them into new products, such as cars and furniture. This type of industry was the economic backbone of the world's oldest industrial nations. Now, much of their wealth comes from tertiary industries. These do not produce goods, but instead offer a range of services, from transport, retail sales, and customer-service operations run from call centres to health care, education, and leisure.

INDUSTRIAL LANDSCAPE
The Ruhr Valley in Germany has been industrial for over 150 years. Huge power plants like this one are needed to satisfy the demand for energy.

LIGHTER AND CLEANER
Modern, high-technology manufacturing industries (such as circuit-board assembly seen here in China) often have less impact on the immediate environment than traditional industries.

PRIMARY

Power generation

DISTRIBUTION The world leaders in power generation are the USA, China, Japan, Russia, Canada, Germany, France, and India

Electrical power is generated where there is demand for it, rather than close to the natural resources, so centres of power generation generally coincide with the major centres of population and wealth, that is, in North America, Europe, China, and Japan. More than 70 per cent of the world's electricity is generated for less than a fifth of its population. Nearly

SOLAR POWER
Solar energy is seen here being utilized in Australia, but it can also be used in colder parts of the world.

every American and European has access to electricity, while in many developing countries only a lucky few have an electrical supply. Electricity is generated in power stations and fed out to its users across a network of cables called a grid. In some places, the cables are buried underground, but in others electricity is carried by overhead cables hung between pylons that stretch across the landscape for mile after mile. Nearly two-thirds of the world's electricity is thermal power. This is generated by burning fuels to heat water, which creates steam to drive turbines that turn electrical generators. In the developed world, the majority is still generated by burning fossil fuels – coal, oil, and natural gas. In developing countries,

POWER STATION
The majority of the world's electricity is generated by burning fossil fuels in power stations, such as this one in Wakefield, England.

a small but significant proportion comes from burning biomass – that is, plant and animal matter converted into fuel – but in the developed world the use of biomass is negligible. The two other major sources of power are nuclear and hydroelectric. Over the last 50 years, both these sources have been extensively developed so that they generate 10 and 16 per cent each of the world's electricity. Alternative energy (from sources other than fossil fuels and nuclear power) and renewable sources, such as windmills, solar panels, geothermal power stations, and tidal and wave power stations, now provide more than 10 per cent of electrical energy, and the proportion is growing.

WIND POWER
These wind turbines in California, USA, generate power, but their aesthetic value is still hotly debated.

DISPOSING OF NUCLEAR WASTE

Nuclear power generation creates radioactive waste, and there remains fierce debate over how to dispose of it safely. Gaseous waste is vented into the air, and in some places low-level radioactive liquid is pumped into the sea. The most dangerous waste is buried in sealed containers. At Hanford, Washington, USA (shown here), low-level waste was dumped in canisters in open trenches over a vast area. The site is now undergoing a massive clean-up operation, costing $2 billion.

PRIMARY

Fossil fuels

DISTRIBUTION Saudi Arabia, the USA, Russia, Iran, and China are the main producers of crude oil; China, the USA, and India are the main producers of coal

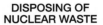

Burning fossil fuels – coal, oil, and natural gas – provides 75 per cent of the world's energy. It also produces 75 per cent of the world's greenhouse gas emissions, driving climate change (see pp.454-457) and causing health-harming pollution. Fossil fuels form in the ground from the remains of

UNDERSEA EXPLORATION
Demand for gas (and oil) has led to offshore exploration for resources under the sea bed, as here in the Gulf of Mexico.

organisms that lived millions of years ago. Most coal is derived from plants that grew in tropical swamps during the Carboniferous Period, over 300 million years ago. Some of the world's biggest coal reserves are in Siberia, Russia. Major reserves are also found in northern Europe, the Damodar Valley in India, and the Appalachians and the Midwest in the USA. Oil is made mainly of the remains of tiny organisms that lived in warm seas. As it is a liquid, oil does not settle in seams, but seeps up through rock until it meets an impervious layer. It accumulates under this layer, forming an underground reservoir. Oil prospectors look for suitable structures where oil might collect, and use seismic surveys to confirm its presence. The biggest reserves are in the Middle East. Saudi Arabia is the leading producer, supplying about 12 per cent of the

world's oil. Major reserves are also found around the Caspian Sea, in Venezuela, and Texas, USA. Oil is often found in areas that are either distant from world markets, or in hostile environments such as the Arctic (Alaska), tropical rainforests (Nigeria and Indonesia), or under stormy seas (the North Sea). So oil exploration, extraction, and transport is expensive, and spillages can cause environmental catastrophes. Natural gas forms in a similar way to oil, and is also found in underground reservoirs. Because it burns less dirtily

and can be piped direct to homes, it has become the fastest-growing energy source. It now provides almost a quarter of the world's energy needs. Natural gas is often found close to oilfields, so there are major reserves in the Middle East, but the biggest reserves are around the southern end of the Caspian Sea. With oil reserves being used up, attention has turned to unconventional sources such as oil shale and tar sands, which hold a lot of thick oil. However, they are expensive and environmentally hazardous to extract.

COAL MINING
In some developing countries, such as Vietnam, extracting coal is highly labour-intensive and poorly paid.

OPEN-CAST MINING
This open-cast mine in Germany is extracting lignite, a type of coal, using huge machines and few people.

Metals

DISTRIBUTION Leading iron ore producers are China, Australia, Brazil, and Russia; the main producers of aluminium include Australia, Guinea, Jamaica, and India

Only a few metals, such as gold (see p.46), are found in a pure state. Most are impure ores, mixed with other minerals, which must be refined to produce the pure metal. Some metals are then mixed to make alloys. Steel, for example, is a mixture of iron and carbon. Humans first used metals about 8,000 years ago, and they began to extract metal from ore about 6,500 years ago. Until recently, metals were usually mined and processed in the same area. Iron production, for example, was sited in places near outcrops of iron ore that also had an abundant

SHIP BUILDING
One of the largest demands for steel comes from the shipbuilding industry, which utilizes millions of tonnes of steel each year.

PROSPECTING FOR GOLD
Gold miners in French Guiana search for alluvial gold in sedimentary rocks. It can be prospected in this way because it exists naturally in a pure form.

supply of wood for fuel. With the Industrial Revolution, the need for coal for large-scale smelting shifted the iron industry to sites where iron ore was associated with coalfields, such as the Ruhr in Germany. By the late 20th century, the rising demand for metals in the developed world had outstripped supply. This meant looking further afield for metallic ores, which often separated the processes of extraction and refinement. Ships now carry huge quantities of ore from countries such as Brazil to processing plants on distant coasts near to the major markets. Iron ore is one of the most abundant metal ores in the Earth's crust, second only to aluminium. Earth's most important iron ore deposits are found in sedimentary

rocks. The USA once led the way in steel production, but it has been hit by rising costs, and China is now the world's leading producer, followed by the European Union, Japan, and India. Aluminium is extracted mainly from bauxite, an ore which occurs in feldspars (see p.60) and other silicate minerals, which break down in tropical conditions to form a surface crust. Leading producers include Australia and Guinea, though large quantities are also mined in southeastern Europe and Russia. Base metals, such as copper and lead, are mined all around the world, but rare and precious metals come from fewer places. Demand for very rare elements, such as neodymium and terbium, for use in electronics has soared – but 85 per cent comes from China alone, and a further 10 per cent from Australia.

OPEN-CAST COPPER MINING
The huge demand for copper has led to excavation of open-cast mines, such as the Bingham copper mine, Utah, USA, which is the largest human-made hole in the world.

PRODUCING NICKEL
Ores, such as chalcopyrite and pentlandite, are refined to produce nickel, as here at a nickel-smelting plant in Noril'sk, Russia.

WATER POLLUTION

Mining metals can pollute nearby rivers and streams and may even cause the water to change colour – as here, near Telluride in Colorado, USA. Today this area is well known as a ski resort but in the past zinc and copper were mined in the surrounding mountains. This has damaged some of the alpine ecosystems and contaminated streams. In places, the water does not meet the standards for agriculture or for drinking water.

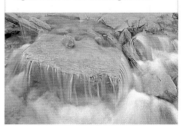

PRIMARY

Non-metallic minerals

LOCATION Non-metallic minerals are distributed, extracted, and processed throughout the world, although some mineral ores are rare and highly localized

The world's commonest minerals are non-metallic. They are used to make bricks, ceramics, insulation, fertilizers, and many other chemicals, including the salt that is an essential part of our diet. Sediments are a rich source of building materials, such as clay, sand, gravel and stone. Soft, dull clays, deposited on sea beds long ago, were among the first

SULPHUR HILLS
Huge piles of quarried sulphur await shipment on the dock of the Vancouver Harbour Sulphur Mine, British Columbia.

minerals ever used. Clay bricks 10,000 years old have been found under the ancient city of Jericho, Israel. Today, clays are among the most widely used, and widely distributed, of all mineral sources: kaolinite (see p.60) is used for paper-making, and pottery; montmorillonite for lubricating oil drills and cleaning wool; illite for bricks; and vermiculite (see p.60) for electrical insulation. Rare minerals such as gemstones tend to form under extreme conditions linked to volcanic activity. So most gems are found in, or near, mineral veins –

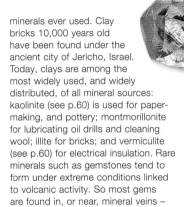

DIAMOND CRYSTAL
Uncut diamonds, such as this one, can vary in colour and in translucence, varying from transparent to opaque.

these are cracks in the Earth's surface where hot, mineral-rich fluids have risen from deep in the Earth's crust. Diamonds (see p.47), for example, formed in the crust over 3 billion years ago, and are brought to the surface by volcanic eruptions. They are found in rocks called kimberlites (see p.68), named after the area in South Africa that produces more diamonds than all other sources combined.

CEMENT PLANT
Large, dusty cement works, such as this one in Staffordshire, England, produce cement by heating crushed limestone and clay.

EXTRACTION OF PHOSPHATES
Morocco possesses three-quarters of the world's phosphate reserves. This open-cast phosphate mine is near Casablanca.

SECONDARY

Chemicals

DISTRIBUTION The chemical industry is centred on economically advanced countries such as the USA and Japan; Germany is the main producer in Europe

The chemical industry is one of the world's most diverse, with products ranging from pesticides to perfumes. The sources of raw materials are very diverse but the products can be divided into two main kinds: organic (carbon-based chemicals, such as plastics) and inorganic (all non-carbon-based

chemicals, most of which are used for fertilizers). Initially, the organic chemical industry was centred on coalfields, as its raw materials were obtained from coaltar. Today, most organic chemicals are oil-based, and chemical plants are sited near oil refineries or close to the markets for the products, such as cities. Inorganic chemicals include phosphate (see above) and potash, which are dug out in vast quantities and used to produce fertilizers. Since chemical plants may have a negative environmental effect, they are often sited in old industrial areas where their impact is less marked, or in developing countries, where labour is cheaper.

PETROCHEMICALS PLANT
Petrochemicals, which are used mainly for producing plastics, are made in huge plants, covering thousands of hectares.

TERTIARY

Transport

DISTRIBUTION The USA is the world leader in road, rail, and air transport, followed by India, Brazil, China, and Russia; Chicago and Atlanta have the busiest airports

All countries have transport networks, but their development, extent, and efficiency is highly variable. At a local and regional level, road and rail links are important, whereas rail, air, and ship are commonly used for countrywide and international movement of people and freight. The USA has the most extensive rail network, followed by China,

PASSENGER TRAFFIC
Over 2,000 flights a day take off from the world's busiest airports, such as Chicago, USA, which handles more than two planes every minute.

although China has the world's largest network of high-speed rail with nearly 40,500 km (25,000 miles) of lines.

CONTAINER TRAFFIC
Much of the world's freight is transported in containers. Victoria Harbour in Hong Kong, shown here, is the world's second busiest container port.

LAND

Textiles

DISTRIBUTION China produces most cotton, followed by India and the USA; Australia and China lead the way in wool production; China produces the most clothes

WOOL USED FOR CARPET WEAVING
The principal source of wool used in wool-weave carpets is from Australia. The wool is dyed into a multitude of different shades.

WORKERS AT SILK FACTORY IN CHINA
The Chinese discovered silk 3,700 years ago. Today, China produces more silk than anywhere else, manufacturing it in factories, such as this one at Hotan, China.

The raw materials used to make textiles are either natural or synthetic. Natural fibres include linen, wool, cotton, and silk, of which wool and cotton are by far the most important. Australia has 78 million sheep and is the leading wool producer. Cotton grows in subtropical areas with a long growing season and enough dry weather for the crop to be harvested. It is still the main raw material of the textile industry, but its dominance is being gradually eroded by synthetic fibres, such

HARVESTING COTTON
Over 95 per cent of cotton in the USA is harvested mechanically, as shown here in Mississippi. The machines are called spindle pickers or strippers.

as acrylic, nylon, and polyester, which are not woven but mechanically, thermally, or chemically bonded together. (Four-fifths of the fibres processed in American mills 60 years ago were cotton; now it accounts for just over a third.) Meanwhile, the weaving industry, and many American and European clothing companies, have moved to developing countries where labour is cheap. Guangdong Province in southern China is now the centre of the world's textile and clothing industry.

Food processing

DISTRIBUTION The world's leading food processors are the USA, Germany, and the UK; China produces and consumes the most food

Food processing involves all stages of production from harvesting to marketing. Whether it simply needs to be sorted and washed, or requires elaborate preparation, almost every type of food is processed, or prepared in some way, to keep it fresh, lengthen its shelf-life, or make it look more appealing. Many foods deteriorate en route to markets, so many processing plants are located close to the growing regions. The biggest food-processing centres tend to be in large cities, such as Chicago, USA, which have good transport links with the farming areas. The cities provide labour for the plants and distribution centres for the processed food.

SALMON PROCESSING IN RUSSIA

High technology

DISTRIBUTION World leaders in advanced technology include Silicon Valley and Boston in the USA, and various eastern countries, such as Japan and Korea

A key development of the last 50 years is the explosive growth of entirely new technologies, such as semi-conductors, computer software, and robotics. The combination of low energy and raw materials needs, and a lightweight product, means these high-technology industries are free to locate almost anywhere. They are often sited in attractive places, near universities, in order to attract the necessary skilled workers. The most famous of these sites are Silicon Valley in California and Route 128 in Boston, USA, but similar centres have appeared all around the world, in places such as Singapore, Seoul-Inch'on in South Korea, and T'aipei-Hsinchu in Taiwan.

SILICON VALLEY
One of the world's leading centres for high-technology industry is Silicon Valley in California, USA.

CONSTRUCTING CIRCUIT BOARDS
Basic components of high-tech equipment are made in developing countries, where labour is cheap, as here in Guadalajara, Mexico.

SERVICED BY SATELLITE
Modern telecommunication is a high-tech industry. Today, it can provide communication services from almost anywhere on Earth. This satellite dish is in rural Bavaria, Germany.

RECYCLING

High-tech industries are generally considered clean, but equipment, such as computer monitors, which rapidly becomes obsolete, is hard to dispose of. As a result, high-tech waste has joined the rising mountain of industrial and household waste created each year by modern society. The damage to the environment caused by disposing of this waste, and also of making the materials in the first place, has spurred efforts to recycle materials. However, as yet, only a tiny fraction of our waste is recycled.

LAND

TERTIARY
Services

DISTRIBUTION Service industries are associated with urban and industrial areas; world leaders include the USA, Europe, and India

Without exception, the industrial focus in developed countries has shifted away from the traditional manufacturing industries and towards service industries, such as banking and retailing. In countries such as the USA and the UK, the service sector is now by far the biggest employer, and the wealth of cities, including New York and London, is in part based upon the financial services they provide. Many developing countries are following the same path. Indeed, the proportion of the workforce involved in the service sector is often taken as a measure of their development. Because these industries provide a service, they

STOCK EXCHANGE
The financial service sector, which includes stock markets, creates jobs and wealth in major cities around the world.

SHOPPING CENTRES
The Bluewater shopping complex in the UK is built on London's outskirts and therefore needs good road access and large car parks.

are located where the service is needed, typically in major urban and industrial centres. The growth of service industries is one of the most significant factors in the expansion of major megalopolises (see pp.353–55) across the world, as more and more people and businesses gather to provide services for each other. Services are normally divided into three kinds. Consumer services are provided direct to the consumer by, for example, shops, high-street banks, and estate agents. Producer services are directed towards other businesses and include finance, warehousing, consultancy, and advertising. Finally, there are public services, such as transport, education, and health care.

TERTIARY
Tourism

DISTRIBUTION Leading tourist destinations are France, Spain, the USA, Italy, China, and the UK; leading earners from tourism are the USA, Italy, France, and Spain

Leisure and tourism is the world's fastest-growing industry, accounting for 1 in 5 new jobs. Spending on tourism is one of the main ways money moves between countries – over 8 per cent of all the world's export earnings are from tourism. The biggest earners from this worldwide travel boom are North America and Europe. The biggest spenders are Americans, Germans, Japanese, and the British. In recent years, tourists from these countries have ventured further afield, and tourism is now the mainstay of the economy of countries such as Thailand, to which millions of visitors travel each year to enjoy tropical beaches. One of the fastest-growing tourist destinations is China. Before 1980, it was hard for the Chinese to travel around their own country, or for foreign tourists to visit. As travel restrictions were lifted, so tourism has developed rapidly, but France, Spain, and the USA remain the world's top tourist destinations. In each tourist area, visitors tend to concentrate at a few top attractions, such as the Great Wall in China or Disneyland in Florida, USA, and this places huge pressures on local facilities. Often the landscape is

SEEING POLAR BEARS IN THE WILD
Churchill, Canada, is one of the only places where tourists can see Polar Bears in their natural habitat, making it a popular "safari" destination.

CORAL REEF
Exploring coral reefs and relaxing on white sandy beaches are major attractions of tropical beach resorts, such as this one in French Polynesia.

DAMAGE TO NATURE
Geysers and hot springs, such as those seen here in Rotorua, New Zealand, are very popular, but the influx of visitors can be damaging to these fragile natural features. The damage is sometimes deliberate, as souvenir hunters and vandals attack them. More often, the problem is due to sheer numbers of people and cars. At many sites, the damage is caused by facilities built to cope with tourists, such as car parks.

completely transformed by the construction of transport facilities and hotels. In the 1950s, most villages on southern Spain's Costa del Sol, for instance, were small fishing villages, but from the 1960s onwards, the availability of cheap flights began to draw huge numbers of visitors from northern Europe to its warm beaches. By the 1980s, 7.5 million visitors a year were travelling to the Costa del Sol, and the villages were transformed by giant hotels and multi-lane highways. Ironically, all this tourist development has begun to deter tourists, and visitors may be further put off by increasing heatwaves.

OCEAN

STORMY SEAS
The Norwegian Sea, which has excellent
fishing grounds, is a marginal sea of
the Atlantic Ocean, the world's second
largest ocean.

OCEANS AND SEAS

What clearly makes the Earth unique in the Solar System is the vast expanse of ocean water that dominates its outer surface. Oceans and seas cover over two-thirds of the Earth's surface area. In fact, the volume of water contained in the oceans and seas is so great that if the Earth's surface was converted to a smooth sphere without topography, it would be entirely covered by a layer of seawater about 2,500m (8,200ft) deep. The floor beneath this great body of water includes such features as the most extensive mountain range on the planet, the deepest trench, and the largest structure built by living organisms. Life first evolved in the oceans and, today, they support a vast diversity of species, ranging from microscopic organisms to the biggest animal in the world, the Blue Whale. The oceans are also the driving force that powers and modifies the world's climate, transporting huge quantities of solar-derived energy across the globe in the process.

OCEAN

Ocean Water

17 Oceans and continents

106–09 Water

115 The carbon cycle

Coastal pollution 438–39

The oceans of the world contain about 1.35 billion cubic kilometres (a third of a billion cubic miles) of seawater. This vast quantity of water is not uniform but varies in several physical attributes, including temperature, salinity, pressure, and level of illumination. The variation in these properties is mainly vertical, dividing the oceans into layers, each of which has specific attributes. However, there is also horizontal variation – between tropical and temperate latitudes, for example – and seasonal changes as well.

Pressure

The units by which oceanographers measure pressure are called bars (one bar is equal to 100,000 Newtons per square metre). At sea-level, all objects are subject to atmospheric pressure, which averages about one bar. Underwater, pressure increases at the rate of about one bar for every 10m (33ft) increase in depth, due to the weight of the overlying water. This means that at 90m (300ft), the total pressure is 10 bars or 10 times the surface pressure, while at 4km (2½ miles), the pressure is over 400 times that at the surface. The huge pressures pose a considerable challenge to human exploration of the deep ocean, and animals living there exhibit special adaptations.

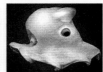

COPING WITH PRESSURE
The Dumbo Octopus lives as deep as 1,500m (5,000ft), compared with a maximum of 200m (650ft) for the Common Octopus. Its adaptations include a semi-gelatinous body, webbed arms, and fins above the eyes.

DUMBO OCTOPUS

COMMON OCTOPUS

Light

Light cannot penetrate far through ocean water. However, different colours, or wavelengths, of light penetrate to varying extents. White light, such as sunlight, contains a mixture of wavelengths, ranging from red and orange (long wavelength) at one end of the spectrum to blue and violet (short) at the other. Ocean water strongly absorbs the longer wavelengths, and the amount of very short wavelength (violet) light in sunlight is small. As a result, only a little light in the blue area of the spectrum reaches much beyond a depth of 45m (150ft), and divers who descend into this zone see everything as a blue-grey colour. Below a depth of about 200m (650ft), even this blue light has been absorbed, and so the ocean is almost completely dark. Because they rely on light to photosynthesize, phytoplankton are restricted to the upper layers of the oceans, and this in turn affects the distribution of other marine organisms. A few organisms that live at depth are able to generate their own light.

LIGHT PENETRATION
Visible light in the red/orange area of the light spectrum is absorbed in the top 10–15m (33–50ft) of even the clearest seawater. Most other colours of the spectrum are absorbed in the next 30m (100ft), so only a little blue light reaches deeper than 45m (150ft).

Red Orange Yellow Green Blue Violet

Depth (m)
0
30
60
90

WITH FLASH   **WITHOUT FLASH**

TRUE COLOURS
Even at 20m (65ft) depth, features such as coral reefs appear blue-grey under ambient light conditions. Only by illuminating the reef with a torch or photographic flash are the real colours of reef organisms revealed.

 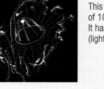

COLOBONEMA
This small jellyfish lives at depths of 100–800m (330–2,650ft). It has bioluminescent (light-producing) tentacles.

Salinity

The salinity, or saltiness, of the oceans averages about 35 grams of salt per litre (1,000 grams) of water, often expressed as 35ppt (parts per thousand). This concentration varies, being highest at the surface of warm seas, due to water evaporation from the surface, and lowest in polar oceans near the mouths of rivers, where there is a high rate of freshwater inflow. The salts in the oceans came originally from minerals that dissolved from rocks into rain-water and were then carried in rivers to the sea. The dissolved salts exist in the form of ions (charged particles), the most important being sodium, chloride, magnesium, and sulphate ions, although there are many others in smaller quantities.

HALITE (ROCK SALT) CRYSTAL
Analysis of seawater trapped in ancient salt crystals has shown that its ion content has varied a little over millions of years.

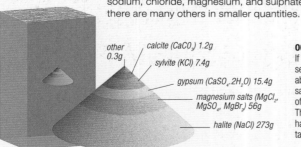

other 0.3g
calcite ($CaCO_3$) 1.2g
sylvite (KCl) 7.4g
gypsum ($CaSO_4.2H_2O$) 15.4g
magnesium salts ($MgCl_2$, $MgSO_4$, $MgBr_2$) 56g
halite (NaCl) 273g

OCEAN WATER SALTS
If 10 litres (17½ pints) of seawater were evaporated, about 353g (12½oz) of salts would be obtained, of the types shown left. The largest component is halite (sodium chloride, or table salt).

Gulf of Mexico — USA

Mexico

South America

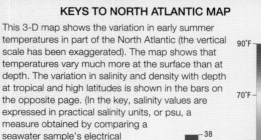

KEYS TO NORTH ATLANTIC MAP

This 3-D map shows the variation in early summer temperatures in part of the North Atlantic (the vertical scale has been exaggerated). The map shows that temperatures vary much more at the surface than at depth. The variation in salinity and density with depth at tropical and high latitudes is shown in the bars on the opposite page. (In the key, salinity values are expressed in practical salinity units, or psu, a measure obtained by comparing a seawater sample's electrical properties to that of a standard salt solution. The psu equates roughly to ppt, or parts per thousand.)

Density (kg/m³)
1,050
1,040
1,030
1,020

Salinity (psu)
38
37
36
35
34

Temperature
90°F — 32°C
— 30°C
70°F — 20°C
50°F — 10°C
30°F — 0°C

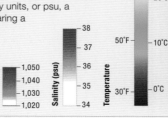

OCEAN

Sound

Sound travels further and faster in seawater than in air – its speed in seawater is about 1,500m (5,000ft) per second, more than four times faster than in air. The velocity is decreased by a decrease in temperature and increased by an increase in pressure (depth) of the water. The effects of vertical variations in temperature and pressure on sound velocity mean that in most ocean areas there is a layer of minimum sound velocity at about 1,000m (3,300ft) depth. This layer is called the SOFAR (Sound Fixing and Ranging) channel. It is exploited for long-distance communication underwater by people using hydro-acoustic and sonar listening devices, and, it is believed, by whales and dolphins.

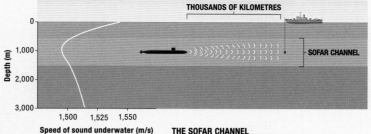

THOUSANDS OF KILOMETRES

SOFAR CHANNEL

Depth (m)

Speed of sound underwater (m/s)

THE SOFAR CHANNEL
Low-frequency sounds generated in the SOFAR channel are "trapped" in it, by refraction or bending of the sound waves towards the centre of the channel. As a result, sounds can travel for long distances in this ocean layer, a phenomenon that has been exploited by submarines.

CETACEANS AT RISK

There is growing evidence that human use of sound underwater, such as military use of sonar for submarine detection, can damage the sensitive hearing apparatus of whales and dolphins, which forms part of their own sonar-based navigation system. This can cause the animals to become disoriented, often with fatal results.

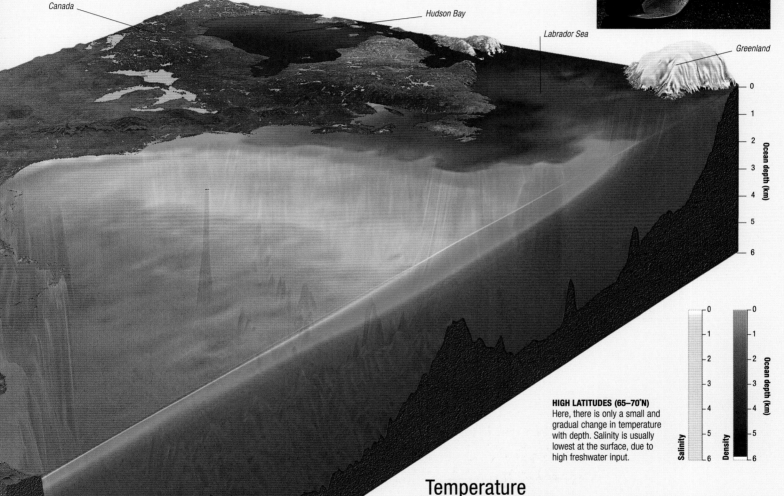

Canada

Hudson Bay

Labrador Sea

Greenland

Ocean depth (km)

HIGH LATITUDES (65–70°N)
Here, there is only a small and gradual change in temperature with depth. Salinity is usually lowest at the surface, due to high freshwater input.

Salinity

Density

Ocean depth (km)

Temperature

Temperature varies significantly over the surface of the oceans, but there is little variation at depth. In tropical and subtropical regions, intense solar heating warms the upper ocean to produce a broad band of water with temperatures higher than 25°C (77°F). Below the surface, from a depth of about 300m (1,000ft) down to about 1,000m (3,300ft), the temperature steeply declines to about 8–10°C (46–50°F). This region of steep decline is called a thermocline. Below 1,000m (3,300ft), the temperature decreases more gradually to a uniform 2°C (35.6°F) throughout the deep oceans. In mid-latitudes (40–50°N and S), the temperature structure of the ocean is similar to that in the tropics, except that the seasonal variation in surface temperature is much more marked, from about 17°C (63°F) in summer to 10°C (50°F) in winter, and there is a more gradual thermocline. In high latitudes and polar oceans, surface temperatures are in the range 0–5°C (32–41°F), although the temperatures in some parts and at some times may drop below 0°C (seawater freezes at a few degrees below zero, the exact temperature depending on its salinity). Below the surface, there is a gradual change to the uniform deep-water temperature of 2°C (35.6°F).

TROPICAL LATITUDES (0–22°N)
Here, the ocean surface is warm and very saline due to high water evaporation. There is a sharp decline in temperature and salinity, and rise in density, down through the first 1,000m (3,500ft).

Salinity

Density

Ocean depth (km)

OCEAN

Ocean Tectonics

40–41	The crust
86–87	Tectonic plates
88–89	Plate boundaries
136–53	Mountains
157	Volcano distribution

The sea floor lies at depths ranging from zero to more than 11,000m (36,000ft) below the surface of the oceans. In many places, this surface is a flat, monotonous expanse – the abyssal plains. Elsewhere, it has been heavily shaped by a combination of tectonic and volcanic activity to form features such as massive mountain ranges, seamounts, deep trenches, and basins. The ocean crust is comparatively young – no part is greater than 200 million years old – because it is continuously reworked by the plate-tectonic processes. It is covered by layers of much younger sediment, shaped by deep-ocean processes.

Mid-ocean Ridges

Mid-ocean ridges are the largest features of the sea floor. In fact, they are the largest single geological feature on the Earth's surface. They consist of an interconnected chain of mountains that twists and branches for about 65,000km (40,000 miles) over the floor of the world's oceans. These mountains occur at divergent plate boundaries (see p.88), where new oceanic lithosphere is created from magma that wells up from Earth's mantle, then cools and hardens. Overall, the ridges rise several thousand metres above the general level of the sea floor. In some places, volcanoes rising from the ridges break the surface of the sea; Iceland, in the North Atlantic, is an example of this. Two different types of ridge are recognized: fast- and slow-spreading. Slow-spreading ridges, such as the Mid-Atlantic Ridge, form new oceanic lithosphere at a rate of only about 2–5cm (1–2in) a year. They have deep depressions (rift valleys), which are 10–20km (6–12 mile) wide, running down their centres. Fast-spreading ridges, which include the East Pacific Rise, spread at 10–20cm (4–8in) a year, and lack rift valleys.

MID-ATLANTIC RIDGE
A section of the Mid-Atlantic ridge reaches the sea surface in Iceland, where it is visible as a series of deep cracks running through an area known as the Thingvellir Rift valley.

Hydrothermal Vents

Located on or near the mid-ocean ridges, at an average depth of about 2,100m (7,000ft), are some spectacular features called hydrothermal vents. These vents are like hot springs (see p.201), continuously spewing out copious amounts of mineral-rich seawater at temperatures as high as 400°C (750°F). Some of the vents have tall chimneys, formed from dissolved minerals that precipitate out when the hot vent water meets cold deep-ocean water. Black smokers are vents in which a dark cloud of sulphides precipitates out of the vent water as it exits the chimney. The minerals dissolved in the water, and bacteria that thrive on these substances, support a diverse community of organisms, including giant tubeworms and clams.

LIFE AT VENTS
These blind Rift Shrimps and the Vent Crab are found in large numbers around mid-Atlantic hydrothermal vents.

SUBMERSIBLES

Manned submersible vehicles have played an important part in unlocking the secrets of the sea floor. In 1977, for example, scientists aboard the submersible *Alvin* discovered the previously unknown phenomenon of hydrothermal vents in the Pacific Ocean. *Alvin* had a maximum depth of 4,500m (14,800ft). Recently upgraded, *Alvin* was certified for operating down to 6,500m (21,000ft).

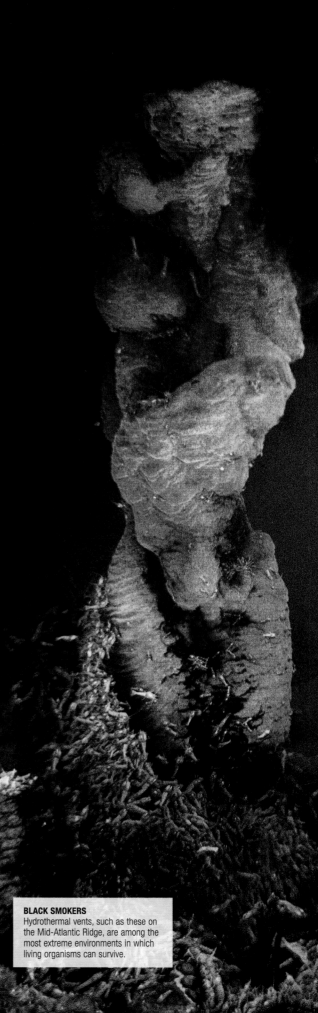

BLACK SMOKERS
Hydrothermal vents, such as these on the Mid-Atlantic Ridge, are among the most extreme environments in which living organisms can survive.

Submarine Volcanoes

Volcanic activity occurs under the sea surface in a number of different settings, including on or near mid-ocean ridges, above hotspots beneath the Earth's crust, and within volcanic island arcs (see p.157). It can take hundreds of thousands of years for a submarine volcano to grow to the point where its summit reaches the surface. Many become extinct before this happens and end up as seamounts that then gradually erode away. When one does reach the surface, a spectacular eruption can result. Examples are the sudden surface appearance of a volcano called Kick'em Jenny in the Caribbean in 1939, and the eruptions that built up the island of Surtsey, south of Iceland, in 1963. In most cases where volcanic islands are formed by such eruptions, they tend not to last long, as they are quickly eroded by wave action – a number of submarine volcanoes in the southwest Pacific are known to have created whole series of short-lived islands. In a few cases, if the underlying volcanic source is strong, a submarine volcano may grow into a large, long-lived volcanic island.

FUKUTOKU-OKANOBA
This submarine volcano in the west Pacific has erupted at, or just under, the sea surface on several occasions, most recently in 2010. The eruptions sometimes leave rafts of hot, steaming pumice floating on the surface, as here.

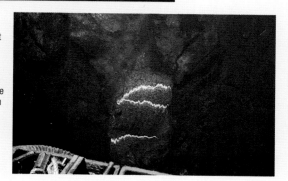

WEST MATA
Located near Tonga, this volcano's summit currently lies about 1,150m (3,800ft) underwater. In 2009, a robotic submersible filmed it erupting huge amounts of lava, seen here with surface splits that reveal the hot magma inside.

Island Chains and Seamounts

Some submarine volcanoes are erupted onto the sea floor at some distance from any plate boundaries, above what are known as hotspots. Many scientists think that these hotspots lie at the top of mantle plumes – streams of magma rising up from deep in the mantle (see p.39). As oceanic lithosphere passes over a hotspot, magma erupting up through it creates a subsea volcano that will grow as long as it remains over the hotspot. If the plate moves in a straight line, and the hotspot stays in the same place, periodic eruptions may form a line of volcanic features called a hotspot chain. Depending on the persistence of a hotspot, the strength of the magma eruptions, and the speed of plate movement, different types of hotspot chains may develop. They range from chains of substantial volcanic islands, such as the Hawaiian Islands, to linear chains of seamounts, the conical remains of extinct volcanoes that lie below the ocean surface. Guyots, slightly different phenomena with flat tops, are volcanoes that broke above the surface before later subsiding. The Hawaiian Island chain, together with a line of seamounts that extends to its northwest (the Emperor Seamounts), constitute a single hotspot chain that stretches for 6,000km (3,700 miles) across the Pacific.

HAWAIIAN ISLANDS
This satellite photograph shows several of the Hawaiian Islands, including Hawaii at bottom right and Oahu at top left. Hawaii is composed of five merged volcanoes, only two of them still active.

OCEAN

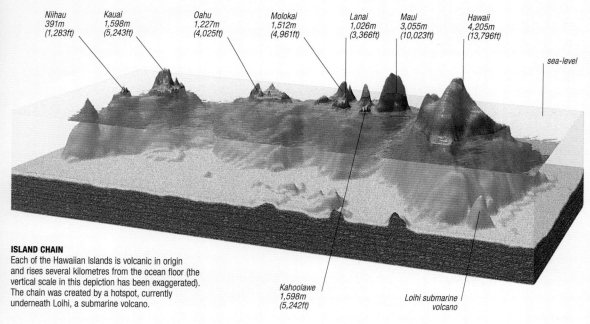

| Niihau 391m (1,283ft) | Kauai 1,598m (5,243ft) | Oahu 1,227m (4,025ft) | Molokai 1,512m (4,961ft) | Lanai 1,026m (3,366ft) | Maui 3,055m (10,023ft) | Hawaii 4,205m (13,796ft) |

sea-level

ISLAND CHAIN
Each of the Hawaiian Islands is volcanic in origin and rises several kilometres from the ocean floor (the vertical scale in this depiction has been exaggerated). The chain was created by a hotspot, currently underneath Loihi, a submarine volcano.

Kahoolawe 1,598m (5,242ft)

Loihi submarine volcano

Ocean Trenches and Island Arcs

The deepest parts of the ocean are found in arc-shaped depressions in the ocean floor called deep-sea trenches. These trenches form at the convergent plate boundaries called subduction zones where one plate descends beneath another (see p.89). Where oceanic lithospere is subducted beneath oceanic lithosphere, the result is the development of an arc of volcanic islands above the overriding plate and parallel to the trench. One example of a trench–island arc combination is the Java–Sunda Trench and Indonesian Island Arc on the edge of the Indian Ocean. Most of the world's deep-sea trenches descend to depths of over 6,000m (20,000ft). The very deepest, the Mariana Trench, plunges to 11,034m (36,201ft). Despite the total darkness, high pressure, and near-freezing temperatures, a surprising variety of life-forms are able to live in the trenches, including some limpets and clams.

ANAK KRAKATAU
This small island is part of the Indonesian Island Arc, a classic volcanic island arc lying on the northeastern aedge of the Indian Ocean.

AMPHIPOD
Shrimp-like animals such as this small crustacean have been found in even the deepest ocean trenches.

OCEAN FLOOR CONVERGENCE
Where oceanic lithosphere subducts under a neighbouring plate, escaping water lowers the melting point of the mantle rock above. Magma forms and rises to the surface, eventually creating a gentle arc of volcanic islands.

continental crust
volcanic island arc
deep-sea trench
magma chamber
subducting lithosphere releases water
asthenosphere
oceanic crust

wave movement

Wave origination
At the sea surface, a series of high-energy waves is triggered. These move off in two directions.

block of sea floor suddenly shoots up several metres, pushing up the column of water above it

powerful shockwaves spread out in all directions

Sea floor rupture
A massive rupture rips across the sea floor. At the same time, a huge block of seabed is suddenly thrust upwards.

rupture line may be hundreds of kilometres long

epicentre – the spot on the sea floor above the point in the Earth's interior where the rupture started

in the open sea, tsunami waves are evenly spaced with a long wavelength

Tsunami Generation

Subduction zones, where oceanic lithosphere descends beneath a neighbouring plate (either oceanic lithosphere or continental lithosphere), are common sites for earthquakes. A powerful earthquake occurring at a subduction zone, particularly if it involves a rupture that causes a large section of sea floor suddenly to be thrust upwards or to sink, may trigger a tsunami – a series of high-energy waves that spread out from the ocean area above the disrupted sea floor. Not all tsunamis are caused by large earthquakes at subduction zones – other possible causes include cataclysmic volcanic eruptions close to the sea, mass downslope movements of sediment underwater, big landslides into the sea, or impacts of large objects from space. However, subduction zone earthquakes are the most common cause of the largest, most serious tsunamis, such as the Indian Ocean Tsunami of 2004 and the 2011 tsunami that devastated part of Japan.

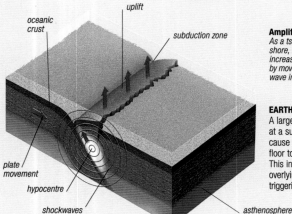

uplift
oceanic crust
subduction zone
plate movement
hypocentre
shockwaves

Amplification
As a tsunami wave approaches shore, it slows and its height increases. These changes are caused by movements of seawater under the wave interacting with the sea floor.

EARTHQUAKE TRIGGER
A large earthquake occurring at a subduction zone may cause a massive slab of sea floor to be thrust upwards. This in turn pushes up the overlying column of water, triggering a tsunami.

passage of each wave is accompanied by a circular movement of seawater under it

asthenosphere

tsunami wave grows higher as the sea floor slopes upwards

Tsunami Propagation and Impact

The series of high-energy waves that comprise a tsunami spreads out in all directions if it is caused by a disturbance at a single point. The waves travel in two opposite, directions if – as is often the case – the cause is a linear rupture in the sea floor. In the open ocean, these waves travel at about 800kph (500mph) and have a very long wavelength – with the distance between wave crests as much as 200km (125 miles). The waves can propagate thousands of kilometres across oceans, only gradually losing energy. The initial signs of a tsunami about to hit shore depend on whether a wave trough or crest arrives first. If a trough arrives first, the sea recedes dramatically, whereas in the case of a crest, the incoming wave appears offshore. As the wave nears the shore and interacts with the sea bottom, it slows to less than 80kph (50mph), and its height increases. Usually, it does not break like a normal wave but surges forwards powerfully. A large tsunami can lift and smash boats, buildings, and vehicles. After about 15 minutes, the water pulls back strongly, drawing vast amounts of debris out to sea. The cycle can repeat several times over an hour or so.

INDIAN OCEAN TSUNAMI
The Indian Ocean tsunami of 26 December 2004 left a colossal amount of destruction on coasts across the region, including here at Banda Aceh in Indonesia.

WARNING SYSTEMS

Tsunami warning systems are in place in the Pacific, Atlantic, and Indian oceans. These systems have three main components: a network of seismometers for detecting tremors that might trigger tsunamis; a means of detecting tsunamis in the open ocean – usually pressure sensors on the sea floor linked to surface transmitters; and procedures for quickly sending alerts to affected populations. This map, created moments after the 2011 Japanese tsunami was detected, shows predicted wave heights across the Pacific; red indicates the greatest height.

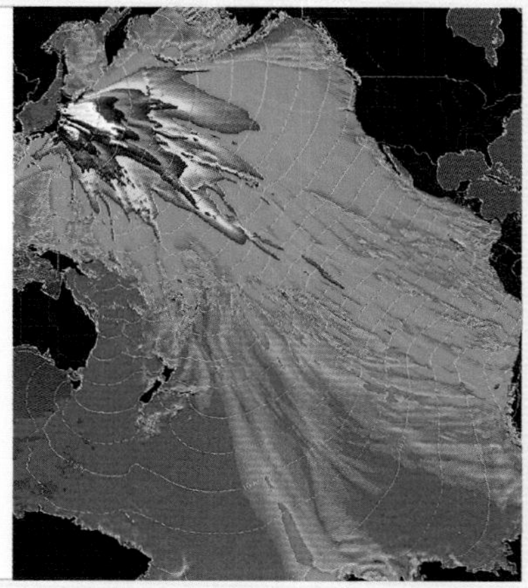

World Distribution

This map shows the distribution of mid-ocean ridges, hydrothermal vents, deep-sea trenches (marking convergent boundaries associated with subduction zones), and major zones in which seamounts are found. Subduction zones, the sites of the types of earthquakes that most commonly cause destructive tsunamis, are found mainly around the Pacific.

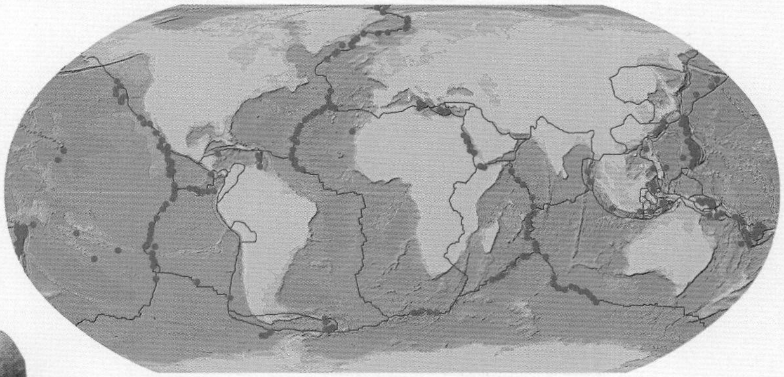

TSUNAMI CHARACTERISTICS
Out at sea, tsunami waves have a low amplitude (height) and may pass unnoticed. Once it reaches the shore it surges forwards, causing massive destruction. The first wave is usually followed by others, typically at intervals of about 20 minutes.

WAVES OF DESTRUCTION
Tsunamis generated by offshore earthquakes and displacement of the seabed have enormous destructive power. Here, the tsunami of March 2011 sweeps across the low-lying coastal fields and towns of northeastern Japan.

Shelves and Plains

40–41 The crust
86–87 Tectonic plates
88–89 Plate boundaries
230–31 Freshwater quality
426–27 Fishing

Although hidden from view, the ocean floor is swept by violent currents, silent storms, and cascading waterfalls. It is the scene of massive submarine slides and speeding turbidity currents that flow for a thousand miles. There is an irregular topography of steep scarps, mounded drifts, giant sediment waves, and imposing canyons, as well as some of the largest, flattest plains on the planet. And everywhere there is a constant gentle rain of sediment from above. The ocean bottom can be divided into three realms – continental margins, abyssal floors, and mid-ocean ridges (p.88).

Continental Margins

The continental margins consist of three parts: the continental shelf, slope, and rise. The continental shelves are areas that were dry land during the last ice age and have since been flooded. They slope gently away from modern-day shorelines to a depth of about 200m (660ft). Past the edge of the continental shelf, called the shelf break, is the more steeply descending continental slope, and beyond that the continental rise, which stretches down to the abyssal plain. Many continental shelf areas are rich in deposits of oil and natural gas and also methane hydrate. The latter, a solid but volatile substance, is potentially a highly valuable energy source, but the question of how to extract it safely from the sea floor has yet to be answered.

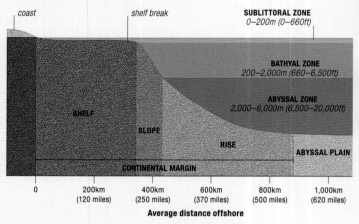

coast — shelf break

SUBLITTORAL ZONE
0–200m (0–660ft)

BATHYAL ZONE
200–2,000m (660–6,500ft)

ABYSSAL ZONE
2,000–6,000m (6,500–20,000ft)

SHELF
SLOPE
RISE
ABYSSAL PLAIN
CONTINENTAL MARGIN

| 0 | 200km (120 miles) | 400km (250 miles) | 600km (370 miles) | 800km (500 miles) | 1,000km (620 miles) |

Average distance offshore

LOBSTERS
Lobsters inhabit the sublittoral zone. Some species migrate from place to place, often in groups, as here.

BRITTLESTAR
Brittlestars, relatives of starfish, inhabit the bathyal and other depth zones, often in huge numbers.

ANGLERFISH
Anglerfish swim in the bathyal and abyssal zones. Females have a luminescent organ that attracts prey.

CONTINENTAL MARGIN
This is a typical continental margin profile. Each part (shelf, slope, and rise) is associated with a depth zone, and each zone (sublittoral, bathyal, and abyssal) is associated with different types of organisms.

LIGHT TRAPS
The Sea Mouse is an unusual type of floor-dwelling worm that lives in the sublittoral and bathyal zones at depths of up to 2,000m (6,600ft). The animal has numerous iridescent spines on its surface that are so efficient in handling light they have attracted the attention of experts in fibre optics. The Sea Mouse itself is thought to use the spines for defence, their colour acting as a warning to predators.

WRECK OF RMS RHONE
The world's continental shelves are littered with shipwrecks, such as this one in the Caribbean Sea. The ship sank in a hurricane in 1867.

Submarine Canyons

At many locations on the continental margins, the continental slope has been eroded into deep, V-shaped valleys, called submarine canyons. The upper part of these canyons is often close to the point on the continental shelf where a large river runs into the sea. The canyons are excavated by a combination of sediment slumping down the slope and turbidity currents, which are large-scale movements of water and silt, like underwater avalanches, triggered by earthquakes or floods. Once they have formed, the canyons continue to act as passageways for sediment, which flows through them and finally comes to rest on the abyssal plain in a wide fan shape. The largest of these fans are associated with major sediment-carrying rivers, such as the Ganges and Amazon.

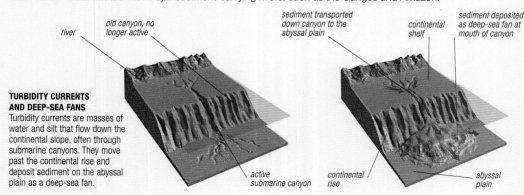

river

old canyon, no longer active

sediment transported down canyon to the abyssal plain

continental shelf

sediment deposited as deep-sea fan at mouth of canyon

TURBIDITY CURRENTS AND DEEP-SEA FANS
Turbidity currents are masses of water and silt that flow down the continental slope, often through submarine canyons. They move past the continental rise and deposit sediment on the abyssal plain as a deep-sea fan.

active submarine canyon

continental rise

abyssal plain

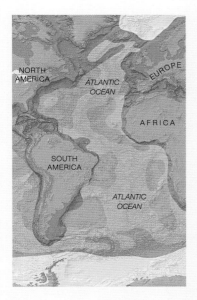

Ocean Sediments

Most of the ocean floor is covered in layers of sediment. In some places, these are over 5km (3 miles) thick and have taken more than 100 million years to accumulate. On the continental shelf, most sediments are gravels, sands, and muds that are derived from the erosion of rocks on land and transported by rivers to the sea. Deep-sea sediments come from many different sources. Some are fine silts that have migrated down from the continental shelf through mechanisms such as turbidity currents. Others derive from sand and dust that has been blown by winds off the continents, or from icebergs that carry rock particles away from glaciers in polar regions and then drop the particles as they melt. Authigenic sediments are precipitates of chemicals, such as iron oxide, from seawater, in forms such as manganese nodules (see below). Another important type of sediment is biogenic ooze, the skeletal remains of microscopic organisms that once lived in the ocean. There are two major varieties: calcareous oozes consisting of calcium carbonate; and siliceous oozes consisting of silica. Oozes composed of the shells of different foraminiferan species (see p.454) have been used to obtain information about ocean surface and floor temperatures and climatic conditions in the past, going back over 100 million years.

▨	Clay	▨	Glacigenic ooze
▨	Calcareous ooze	▨	Siliceous ooze
▨	Sand and mud		

SEA-BED SEDIMENTS
This map shows the distribution of the different sediments found on the floor of the Atlantic Ocean. Oozes are composed of the shells of countless millions of tiny organisms, such as those shown at far right.

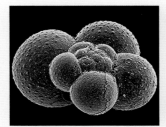

GLOBIGERINA
This is a foraminiferan, a simple organism with a calcium-carbonate shell.

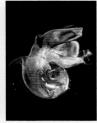

PTEROPOD
Pteropods are tiny molluscs with shells of calcium carbonate.

RADIOLARIAN
Radiolarians are single-celled, floating, animal-like organisms with silica shells.

DIATOMS
Diatoms are delicate, photosynthesizing life-forms with silica shells.

Abyssal Plains

The abyssal plains begin where the continental margins end. They occupy extensive areas of the seafloor at depths of 4,000–6,000m (13,000–20,000ft) and are the deepest parts of the oceans apart from the deep-sea trenches. They are the flattest, and some of the most featureless, places on Earth, with a gradient of just 1:1,000. Despite the pitch blackness of the water, the huge pressure, and the freezing cold, many different animals live on the abyssal plains, including several species of worms, shrimps, brittlestars, sea cucumbers, and some extraordinary fish.

MANGANESE NODULES
These potato-sized lumps litter wide areas of the abyssal plains. They contain many valuable metals, but so far no satisfactory way has been devised for mining them.

World Distribution

Areas of continental shelf (light purple on the map below) exist around the edges of all continents but vary greatly in their width, from just a few kilometres off the west coast of much of the Americas to up to 900km (560 miles) off the coast of Siberia, Russia. The narrowest shelves are found at convergent plate margins. Abyssal plains (dark purple on the map) are most common in the Atlantic and Indian oceans, but are relatively rare in the Pacific.

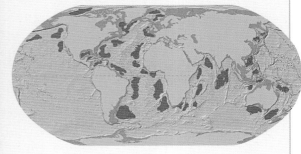

OCEAN

Circulation and Currents

384–85 Ocean water

392–93 Shelves and plains

Tides and waves 432–33

Atmospheric circulation 450–53

Global warming 458–59

Seawater is in constant motion, and not simply due to waves and tides. Throughout the oceans, there is a constant circulation of water, both across the surface and more slowly in the ocean depths. Many interlinked processes play a part in causing and maintaining these currents. They include solar heating of the atmosphere and the surface of the oceans, the winds that this heating generates, the Earth's rotation, various processes that affect the temperature, salinity, and density of surface waters, as well as the position of continents and the shape of oceans.

Surface Circulation

The large-scale pattern of surface movement in the oceans is called the surface circulation. It is driven by winds, but modified by the Coriolis effect (see p.450), which results from the Earth's spin. In the northern hemisphere, the Coriolis effect deflects wind-driven surface movements of water (known as wind drag) slightly to the right (as shown below), and in the southern hemisphere to the left. However, through an accentuation of the Coriolis effect called Ekman transport, the average water motion in the top few hundred metres is almost at right-angles to the wind direction. The overall effect of the predominant winds and Ekman transport on the surface of the oceans is a pattern of large-scale circular movements of water, called gyres, which rotate clockwise in the northern hemisphere and anticlockwise in the southern. Specific components of these gyres are called boundary currents. The boundary currents on the eastern side of oceans are generally weak, cold, and move towards the Equator. In contrast, those on the western sides of oceans tend to be strong, warm, and move away from the Equator.

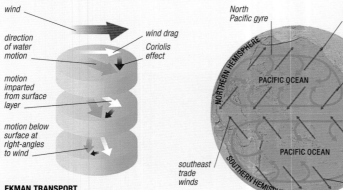

wind
direction of water motion
wind drag
Coriolis effect
motion imparted from surface layer
motion below surface at right-angles to wind

EKMAN TRANSPORT
At the ocean surface, the Coriolis effect deflects water motion slightly from the direction of wind drag. This motion produces a drag in the next layer down, which is also deflected, and so on. On average, the water is pushed at 90° to the wind direction.

North Pacific gyre
westerly winds
northeast trade winds
NORTHERN HEMISPHERE
PACIFIC OCEAN
Equator
PACIFIC OCEAN
South Pacific gyre
southeast trade winds
SOUTHERN HEMISPHERE
westerly winds

PACIFIC GYRE FORMATION
Ekman transport curves water to the right of the predominant winds in the northern hemisphere, and to the left in the southern hemisphere, producing two gyres, or circular movements, of surface water.

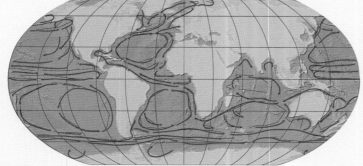

SURFACE CURRENTS
This map shows the overall pattern of surface currents. In the Atlantic, Pacific, and Indian oceans, much of the circulation is in the form of circular gyres. These consist predominantly of currents carrying warm water away from the Equator and colder water towards it.

→ Warm ocean currents
→ Cold ocean currents

Local Currents

Local currents are movements of water that are the result of localized interaction between tidal forces and the shapes of coastlines. Tides cause regular variations in sea-level, which are most noticeable around coasts and are more pronounced in some locations than others. These vertical changes in water level can occur only through horizontal movement of water, into and out of bays, for example. It is these movements that bring about local or tidal currents, and they can be especially significant where large volumes of water are funnelled through narrow channels, especially around promontories, between islands, and up estuaries. Because they are tide-related, these currents continuously vary in their strength and direction hour by hour over the daily tidal cycles, and also over monthly cycles (see p.432). At particular locations, they can cause phenomena such as tidal bores and whirlpools.

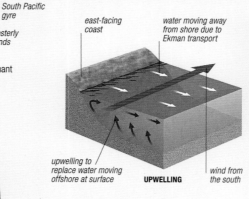

BASS STRAIT
Bass Strait, between Tasmania and the mainland Australian State of Victoria, is famous for its strong currents, which are partly wind-driven and partly tidal.

Vertical Transport

Vertical transport refers to the movement of water from the surface to depth (downwelling) or vice versa (upwelling). One cause is an increase in the density of surface water, through cooling or an increase in its salinity. A prime example is sinking of water under sea-ice in polar regions. As the water cools, it is made more saline when salt is rejected from seawater as it freezes (see p.399). Other causes include wind patterns that cause water masses to converge in an area (where they are forced down) or to diverge (where water must upwell to replace the diverging water). Where a wind blows parallel to a coast, this can also cause upwelling or downwelling. Because deep waters are rich in nutrients (derived from the breakdown of the remains of organisms that have sunk from the surface), upwelling has important biological effects as it brings these nutrients to the surface, which helps the growth of plankton.

east-facing coast
water moving away from shore due to Ekman transport
upwelling to replace water moving offshore at surface
UPWELLING
wind from the south

PLANKTON BLOOM
Growth spurts of plankton, seen here colour-enhanced in red, often occur in upwelling zones as nutrients are brought to the surface.

COASTAL UPWELLING AND DOWNWELLING
In the northern hemisphere, Ekman transport produces water motion to the right of the wind direction. When a southerly wind blows up an east-facing coast, this forces water away from the coast (above) and causes upwelling near the coast. When the wind blows from the north (right), water is pushed towards the coast and then sinks.

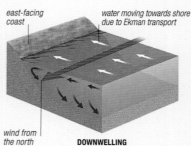

east-facing coast
water moving towards shore due to Ekman transport
wind from the north
DOWNWELLING

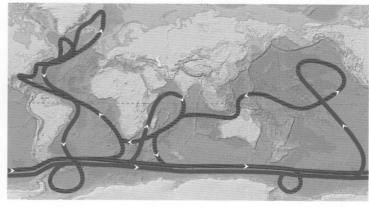

NARUTO WHIRLPOOL
Spectacular whirlpools caused by strong tidal currents develop several times a day in the narrow channel that links the Sea of Japan to the Pacific Ocean.

Deep-Water Currents

The world's deep-water currents are set in motion by differences and changes in the density of water masses, principally by the downwelling of dense, cold, salty water in polar and subpolar regions. Once this water reaches a depth level of equal density, it spreads out, often over long distances. In other parts of the world, slightly less dense water rises to the surface as it mixes with and absorbs heat from water masses above it. This type of circulation is called thermohaline, "thermo" referring to temperature and "haline" to saltiness. Deep-water currents move slowly, no more than a few metres a day. Once a body of water sinks, it can spend hundreds of years away from the surface until it rises again to become part of the surface circulation.

**THERMOHALINE
CONVEYOR BELT**
This circulation is driven by the cooling and sinking of water masses to great depth in the North Atlantic. The cold water circulates down through the whole of the Atlantic and penetrates into the Indian and Pacific oceans, before returning as warm upper ocean currents to the South Atlantic.

━ **Warm surface current**
━ **Cold, salty, deep-ocean current**

OCEAN

Reefs

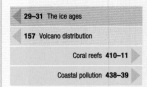

29–31   The ice ages

157   Volcano distribution

Coral reefs   410–11

Coastal pollution   438–39

There are two types of coral reefs: cold-water and warm-water. Cold-water coral reefs are found in the deeper, darker parts of oceans. Warm-water reefs are built from the remains of small marine organisms and are found in many shallow seas. They are mainly limestone, consisting of the build-up of cemented skeletons and shell fragments of animals that once inhabited the reef. On the surface of this limestone is a thin skin of living organisms. When these organisms die, their hard parts remain as part of the reef structure, helping the reef to grow in size over time. Coral reefs support a highly diverse fauna and flora, and protect shorelines from erosion. They are vulnerable to changes in environmental conditions such as water temperature, salinity, and pollution.

BORA BORA
Bora Bora lies in the south-central Pacific Ocean. It consists of a volcanic island surrounded by a lagoon and a necklace of submerged reefs and elongated islands that constitute a barrier reef.

How Coral Reefs Form

The main reef-forming organisms, known as hard or stony corals, belong to a group of animals called cnidarians. Individual animals are called coral polyps. As it grows, a coral polyp secretes limestone, building on the rock underneath. Some reef-building corals consist of single large polyps, but the majority live as colonies that, as they grow, create community skeletons in a variety of shapes. An important contributor to the life of coral polyps is the presence within their tissues of unicellular organisms called zooxanthellae. Although polyps use their tentacles to capture plankton for food, most of their nutrition comes from these zooxanthellae. The zooxanthellae use photosynthesis to convert the carbon dioxide produced by the polyps' respiration into the nutrients that polyps need for growth and so for production of limestone. They also provide oxygen for the polyps. Other organisms that add their skeletal remains to reefs include molluscs and echinoderms. Further contributions are made by boring and grazing organisms. These fragment some of the coral skeletons into sand, which fills the gaps in the skeletons. Algae and encrusting bryozoans cement the coral fragments and other debris into a solid reef.

HARD CORAL POLYP ANATOMY
The polyp's gut cavity sits within a fluted, cup-like limestone exoskeleton. Food is taken in, and some waste products are expelled, through the mouth. The tissue around the gut cavity lays down limestone, which helps to build the reef.

tentacle

mouth

gut cavity

limestone exoskeleton

connecting tissue between adjacent polyps

POLYP OF A HARD CORAL
This reef-building species, called a plate or mushroom coral, consists of a single large polyp with many green, pink-tipped tentacles. Often mistaken for anemones, mushroom corals inhabit shallow reefs.

Fringing Reefs

Fringing reefs form a fringe of coral around a tropical island or along part of the shore of a large landmass, with little or no lagoon between the reef and shore. They are the most common type of reef worldwide. Fringing reefs consist of several zones that are characterized by their depth, structure, and coral communities. The reef crest is the part the waves break over, marked by a line of breaking surf. In front of the reef crest is an area called the buttress zone, where there are spurs of coral growing out into the sea separated by grooves. This is the region of most active coral growth. Shoreward of the reef crest is the reef flat, a shallow, flat expanse of limestone, sand, and coral fragments that may become partly exposed at low tide.

CORAL SPUR
In this area on the surface of a spur at the front of a fringing reef, several colonies of stony coral are growing in about a metre of water.

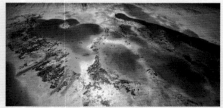

REEFS IN PALAU
This aerial view shows a large area of fringing reefs (the light coloured areas) around islands and exposed coral flats in the Republic of Palau in the western Pacific.

Dangers and Stresses

Many different types of stress can damage and destroy coral reefs. Some damage is due to human activity (see p.411), and a significant threat is posed by rising sea-levels due to global warming (see p.459). Natural disturbances include various coral diseases, tropical storms, and an increase in predation by animals such as parrotfish, snails, and the Crown of Thorns Starfish. The latter has been a particular concern since the 1960s. One theory for its emergence is that there has been a loss of some of its natural predators. Another is that the addition of nutrients from human use of the coastal zone has increased the availability of planktonic food for young starfish.

CROWN OF THORNS STARFISH
The large Crown of Thorns Starfish, which eats coral polyps, has become a blight on coral reefs throughout the Indo-Pacific region.

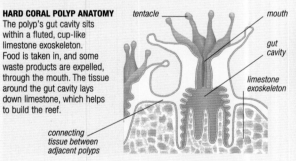

Barrier Reefs

A barrier reef is a coral reef that parallels the shore but is separated from it by a sizeable lagoon. Large and continuous barrier reefs occur in association with continental land-masses. The largest in the world are the Great Barrier Reef off the coast of Queensland, Australia, and the Belize Barrier Reef in the Caribbean. Smaller barrier reefs can also be seen around sinking volcanic islands. Barrier reefs contain the same zones as fringing reefs, but with some additions. In front of the buttress zone, there is frequently a steeply descending wall called a reef face, or drop-off, which is usually rich in coral formations. The main bulk of the reef behind the reef crest is often several kilometres wide and may include patch reefs separated by submerged areas.

REEF FACE
A diver swims around the top of the reef face on part of Australia's Great Barrier Reef.

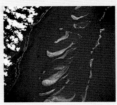

GREAT BARRIER REEF
This satellite image of part of the Great Barrier Reef shows that it is an assemblage of reefs rather than a single structure. .

Coral Reef Distribution

Warm-water reefs cover about 600,000 square km (230,000 square miles) of the world's marine areas. Reef-building corals can grow only in shallow, clear water, where there is plenty of sunlight, water temperatures of at least 18°C (64°F) but preferably 25–29°C (77–84°F), average salinity of 36ppt, and not too much wave action, turbidity, or silting. The right conditions for reef growth are found mainly in the tropical areas of the Pacific, Indian, and Atlantic oceans, and predominantly in the western parts where waters are warmer than in the eastern parts. Coral reefs are absent from any coastal areas where there is a large amount of sedimentation from river run-off. There are no reefs near the mouths of the large rivers of east Asia, for example.

Atolls

An atoll is a ring of coral reefs, or low-lying islands made of coral, that encloses a shallow central lagoon. Atolls are often elliptical in shape, but many are irregularly shaped. The process by which they form was first explained in the 1840s by the English naturalist Charles Darwin. Darwin proposed that fringing reefs can transform into barrier reefs and barrier reefs into atolls. He theorized that these transitions result from the upward growth of coral on the edge of a gradually sinking volcano, and that the ring-like appearance of an atoll with a central lagoon result from the total submergence of the summit of the volcano. Other scientists have contributed to Darwin's theory, especially on the importance of temperature restraints in reef formation and the contribution of wave patterns to atoll shapes. Today, it is realized that sea-level rise may contribute to atoll formation as much as volcanic subsidence.

MALDIVES ATOLL
This example of a small atoll is in the North Male' island group in the Indian Ocean.

JAMES DWIGHT DANA

From his observations on various coastlines in the South Pacific, the American geologist James Dana (1813–95) found evidence to support Charles Darwin's ideas on how atolls form (see right). Dana also developed a theory explaining how the shapes of atolls are related to wind and wave patterns, which affect the supply of nutrients to reef-building organisms.

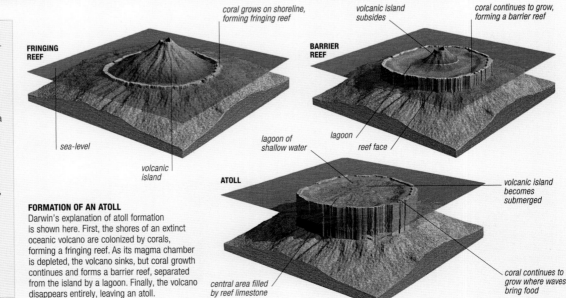

FRINGING REEF
coral grows on shoreline, forming fringing reef
sea-level
volcanic island

BARRIER REEF
volcanic island subsides
coral continues to grow, forming a barrier reef
lagoon
reef face
lagoon of shallow water

ATOLL
volcanic island becomes submerged
coral continues to grow where waves bring food
central area filled by reef limestone

FORMATION OF AN ATOLL
Darwin's explanation of atoll formation is shown here. First, the shores of an extinct oceanic volcano are colonized by corals, forming a fringing reef. As its magma chamber is depleted, the volcano sinks, but coral growth continues and forms a barrier reef, separated from the island by a lagoon. Finally, the volcano disappears entirely, leaving an atoll.

OCEAN

OCEAN

Polar Oceans

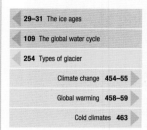

29–31   The ice ages

109   The global water cycle

254   Types of glacier

Climate change   454–55

Global warming   458–59

Cold climates   463

The polar oceans comprise the Arctic Ocean in the northern hemisphere and the Southern Ocean in the south. They differ from other oceans in several respects, not least the vast amounts of ice that float on them in various forms, although parts of the Atlantic and Pacific oceans are also covered with ice for some of the year. This ice coverage has helped stabilize world climates, insulating polar oceans from solar radiation in summer and preventing heat loss in winter. However, this balance is threatened by global climate change. The polar oceans are less stratified by temperature and density than other oceans and have different water circulation patterns.

Ice-Shelves

Ice-shelves are extensions of ice-sheets and other glaciers over the sea. Although there are a few in the Canadian Arctic, most of the world's ice-shelves, including two enormous ones, the Ross and Ronne-Filchner ice-shelves, exist around the coasts of Antarctica (see p.273). Because they originate from snow, ice-shelves contain fresh water. At their seaward edges, they are marked by cliffs that are up to 60m (200ft) high and may extend up to 900m (3,000ft) underwater. Underneath the ice-shelves, there is some circulation of seawater. Water drawn in from beyond the edge of the shelf freezes underneath the ice, rejecting salt as it does so. This creates a layer of cold, salty, dense water that sinks and flows away from the continental shelf area beneath the ice to deeper water. Since the early 1990s, some of the smaller Antarctic ice-shelves have completely broken up (see p.455). This has raised concerns for the future of the ice-sheets covering Antarctica, with implications for possible accelerated sea-level rise, as ice-shelves appear to provide a barrier to ice flow off the ice-sheets.

BRUNT ICE-SHELF, ANTARCTICA
The edge of an ice-shelf is usually marked by ice cliffs. Here the cliffs are fronted by sea-ice on which some Emperor Penguins are gathered.

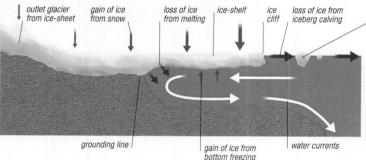

outlet glacier from ice-sheet | gain of ice from snow | loss of ice from melting | ice-shelf | ice cliff | loss of ice from iceberg calving | iceberg

grounding line | gain of ice from bottom freezing | water currents

ICE-SHELF GAINS AND LOSSES
Ice-shelves gain ice from glaciers flowing into them, from seawater freezing to their undersurface, and from new snowfalls. They lose ice mainly by icebergs calving (breaking off) from their front edges.

Icebergs

Icebergs are pieces of ice that have broken off the edges of ice-shelves or from the fronts of glaciers where they reach the sea. They range from tabular icebergs, hundreds of square kilometres in area, to yacht-sized chunks of ice (called bergy bits), and car-sized ones (called growlers), which are just as dangerous to shipping. In the Southern Ocean, most icebergs have broken off ice-shelves around Antarctica. They initially drift westward in a coastal current around the continent and many become trapped in bays. A few drift several hundred kilometres from Antarctica, some reaching as far north as 40°S. In the northern hemisphere, most icebergs have originated from glaciers flowing off Greenland, Ellesmere Island in the Canadian Arctic, or Alaska. A few reach as far south as 45°N. Because of their glacial origin, icebergs contain fresh water. The average density of icebergs is about 87 per cent of the average density of seawater. As a result, an iceberg typically floats with about 87 per cent of its volume underwater.

PINNACLE

ICEBERG TYPES AND SHAPES
Icebergs come in various regular shapes, including pinnacle, pyramid, wedge, and blocky, but there are also many irregularly shaped icebergs. Striated icebergs contain layers of rock debris from the glacier from which they were calved.

IRREGULAR

BLOCKY

STRIATED

DRY DOCK

PANCAKE ICE
These scientists in Antarctica are investigating the properties of pancake ice, circular pieces of sea-ice up to several metres in diameter and 10cm (4in) thick.

Circulation in Polar Oceans

The circulation of water in polar oceans is influenced by factors such as changes in the surface temperature and salinity of the water caused by the formation and melting of sea-ice. Both polar oceans are sources of cold, dense bodies of water that sink towards the ocean floor and then continue towards the Equator. These flows are key drivers of the Earth's deep-water circulation. In the Arctic, there is a constant flow of cold water and sea-ice across the North Pole from Siberia towards Greenland in a current called the Transpolar Drift. To the east and southwest of Greenland, this cold water dips under warmer Atlantic water. Further north, the Atlantic water plunges beneath less saline Arctic water then moves through the Arctic Basin. An equivalent circulation operates in the Southern Ocean (see p.428).

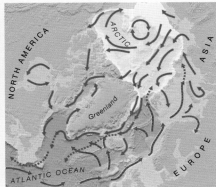

← Warm surface current	← Cold surface current
◄▪▪ Warm sinking current	◄▪▪ Cold sinking current

ARCTIC CURRENTS
This map shows the basic pattern of surface currents and downwellings in the Arctic and North Atlantic oceans. Observations of ice drift have indicated the surface flows. One of the most pronounced currents is the southerly movement of cold water to the east of Greenland.

Sea-Ice

Unlike icebergs and ice-shelves, sea-ice is seawater that has frozen. It contains little salt as most is rejected as it forms, and it floats on the sea because it is less dense than seawater. About 7 per cent of the world's oceans are covered in sea-ice, nearly all of it in polar regions. New sea-ice forms in distinct stages. Early on, a viscous, slushy layer of ice crystals, called grease ice, develops at the sea surface. The ice thickens and hardens, and is broken up by wave action into platelets. As these collide the result is a mosaic of platter shapes with curling edges, called pancake ice. Later stages include first-year ice, which is more than 30cm (12in) thick, and multi-year ice, which has survived at least two summers' melt and can be 3–10m (10–35ft) thick. Fast ice is sea-ice that is securely frozen to the shore or an ice-shelf, while pack ice is any other area of sea-ice.

GREASE ICE
A Crabeater Seal surfaces through grease ice – an early stage in sea-ice formation – in the Weddell Sea.

ICE LEAD
A lead is a narrow fracture or passageway through sea-ice, which makes the ice navigable by surface vessels and some marine mammals.

POLYNYA
Polynyas are ice-free areas within a region of sea-ice. Upwelling of warm water in a localized area is one way they are formed.

Oceans of the World

It is generally accepted today that the Earth has five oceans, although the Southern Ocean was officially recognized only in 2000. Three long-recognized oceans, the Atlantic, Indian, and Pacific, extend northward from the Southern Ocean, separating the world's major landmasses. The fifth and smallest ocean, the Arctic Ocean, caps the north polar region. Around the edges of these oceans are smaller regions called seas, bays, and gulfs. Some of the seas, such as the Sargasso Sea, are only vaguely defined, but others, such as the Mediterranean Sea and the Caribbean Sea, are almost completely surrounded by land or by arcs of islands.

MEDITERRANEAN SEA
An offshoot of the Atlantic Ocean, the Mediterranean Sea was the centre of the classical world. This coast is on the volcanic island of Santorini, in the Cyclades Islands, Greece.

OCEAN

ATLANTIC OCEAN
A major area for marine commerce and fishing, the Atlantic is known for its storms, though lighthouses on its coasts, such as this one at Minots Ledge, USA, help to protect shipping.

CARIBBEAN SEA
The warm waters of the Caribbean Sea support large numbers of marine species, such as these Bottlenose Dolphins.

INDIAN OCEAN
The Indian Ocean is littered with clusters of coral islands, the best known being the Seychelles and the Maldives. The Maldive Islands lie to the southwest of Sri Lanka and consist of more than a thousand islands, of which two are shown here. Most of them are uninhabited.

YELLOW SEAHORSE
Seahorses are fish that swim upright and have prehensile tails. This species is common in the Indian Ocean.

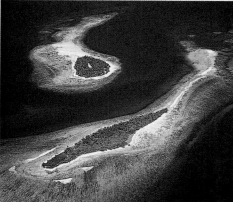

OCEANS AND SEAS
This map shows the range of depths in the world's oceans and seas. Surrounding all the landmasses are shallow areas of continental shelf (the lightest shade of blue). Beyond the shelves, the ocean basins fall into two main depth zones: a deeper, mostly flat region at about 5,000m (16,400ft) and a more elevated, rugged zone at 1,000–5,000m (3,300–16,400ft), lying mainly around the mid-ocean ridge areas.

- 0–500m
- 500–1,000m
- 1,000–2,500m
- 2,500–5,000m
- over 5,000m

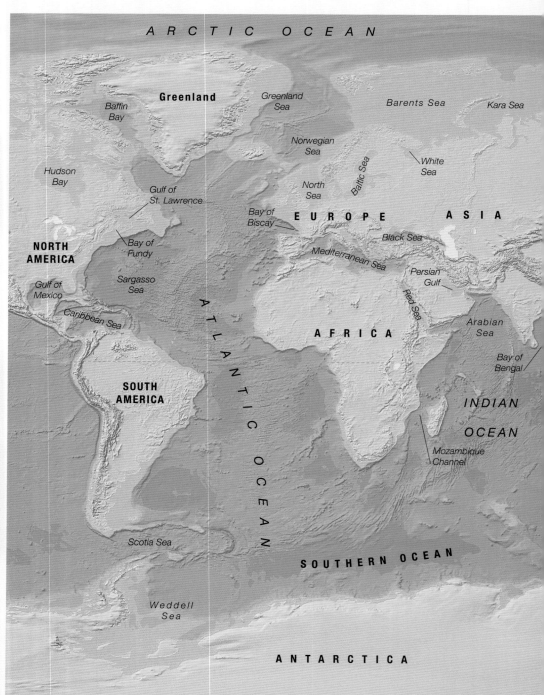

ARCTIC OCEAN

Greenland

Baffin Bay

Greenland Sea

Barents Sea

Kara Sea

Norwegian Sea

White Sea

Hudson Bay

Baltic Sea

North Sea

EUROPE

ASIA

Gulf of St. Lawrence

Bay of Biscay

Black Sea

NORTH AMERICA

Bay of Fundy

Mediterranean Sea

Persian Gulf

Gulf of Mexico

Sargasso Sea

Red Sea

Arabian Sea

Caribbean Sea

AFRICA

ATLANTIC OCEAN

Bay of Bengal

SOUTH AMERICA

INDIAN OCEAN

Mozambique Channel

Scotia Sea

SOUTHERN OCEAN

Weddell Sea

ANTARCTICA

ARCTIC OCEAN
The Arctic Ocean consists of a deep basin surrounded by a ring of seas, which partly cover continental shelves. The volcanic island of Jan Mayen, seen here, lies on the edge of one of these marginal seas, the Greenland Sea.

WALRUS
Walruses are insulated from the Arctic cold by a thick, subcutaneous layer of blubber. When sunbathing, their heavily creased skin flushes deep pink.

FERDINAND MAGELLAN

The Portuguese explorer Ferdinand Magellan (1480–1521) led the first expedition to circumnavigate the globe, though he died during the voyage. Magellan embarked on the enterprise in 1519 hoping to discover a new route to the Spice Islands (now the Maluku Islands or Moluccas) by finding a passage from the Atlantic into the Pacific. In November 1520, he discovered such a passage near the tip of South America, today called the Straits of Magellan.

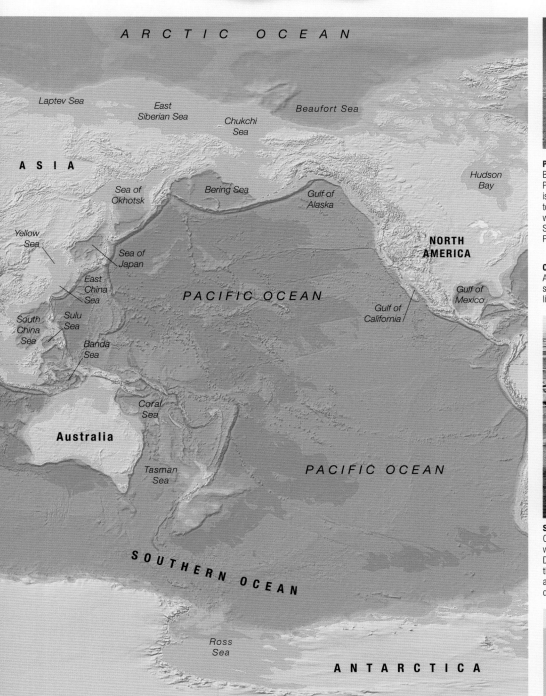

PACIFIC OCEAN
By far the largest of the oceans, the Pacific is dotted with thousands of islands, many formed from (or on) the tops of undersea volcanoes, and fringed with coral reefs. This little island in the Solomon Islands group in the western Pacific is a typical example.

CRIMSON ANEMONE
Anemones are found worldwide, in all seas and at various depths. This species lives in the northeastern Pacific.

SOUTHERN OCEAN
Covered in ice for much of the year, the Southern Ocean is whipped by high winds in winter, producing mountainous seas. Despite the severe conditions, an amazing diversity of wildlife thrives there. It is home to some of the world's great whales and large numbers of birds, such as these Chinstrap Penguins drifting on sea-ice in the Weddell Sea.

OCEAN AND SEA PROFILES ▶

The pages that follow contain profiles of the world's oceans and major seas, bays, and gulfs. Each profile begins with the following summary information:

AREA Surface area

MAXIMUM DEPTH Depth of the deepest part of the ocean, sea, or gulf

INFLOWS Major inflows including oceans, basins, seas, rivers, fiords, and glaciers

OCEAN

ARCTIC OCEAN

Arctic Ocean

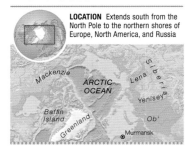

LOCATION Extends south from the North Pole to the northern shores of Europe, North America, and Russia

AREA 14 million square km (5.4 million square miles)

MAXIMUM DEPTH 4,665m (15,305ft)

INFLOWS Atlantic Ocean, Pacific Ocean; rivers Mackenzie, Ob', Yenisey, Lena, Kolyma

The Arctic Ocean is the smallest of the world's oceans. It is also the least studied, partly because about a third of its surface is permanently covered in ice. The ocean comprises a central region around the North Pole and several marginal seas. Most of the Arctic is in continuous darkness from November to February and continuous daylight from May to August. The floor of the Arctic Ocean has two distinct parts: an extensive depression (the Arctic Basin) in the centre, with an average depth of about 4,000m (13,500ft), and an area of continental shelf around the edges. The central basin is divided by a long submarine ridge, the Lomonosov Ridge, into two large sub-basins. These are called the Amerasia Basin (on the North American side) and the Eurasia Basin (on the European side). The continental shelf area is much wider on the Eurasian

ROALD AMUNDSEN

The Norwegian Roald Amundsen (1872–1928) is best known for leading the first expedition to reach the South Pole, but his accomplishments in the Arctic were equally illustrious. Between 1903 and 1906, on the sloop *Gjoa*, he led the first expedition to negotiate the Northwest Passage, a long sought-after sea route through the Canadian Arctic Archipelago, linking the Atlantic Ocean to the Pacific. As part of the expedition, Amundsen also established the position (at the time) of the North Magnetic Pole.

COPEPOD
Small crustaceans called copepods are a key component of the Arctic food chain. They are herbivores, feeding on phytoplankton, and are themselves eaten by fish, squid, sea birds, seals, and some whales.

than the Amerasian side. Many islands and archipelagoes sit on the shelf area. Sea-ice covers the Arctic Ocean more or less permanently above a latitude of about 75°N. In most of the rest of the ocean, sea-ice occurs in winter but retreats in summer. The permanent ice-cap is composed of pack ice – pieces of sea-ice that may be more than 10m (33ft) thick. Wind and currents (see p.394) keep the pack ice in continuous motion, forming cracks and open pools. Since the 1960s, the Arctic ice-cap has thinned considerably, causing loss of

habitat for some of the fauna, especially polar bears. It has also allowed increased navigation, bringing with it consequent economic exploitation and sovereignty issues. Melting of the ice-cap in itself would not affect sea-levels (because it is already afloat) but would strip the ocean of its reflective cover, allowing the water to absorb and retain more solar heat. Warmer waters would, in turn, speed melting. In addition to sea-ice, the Arctic Ocean contains thousands of icebergs and larger bodies called ice islands. These have broken off glaciers and ice-shelves, mainly around Greenland and the Canadian Arctic. The main input of water into the Arctic Ocean comes from the Atlantic Ocean (see p.399). A second input of water enters from the Pacific Ocean through the shallow Bering Strait.

BOWHEAD WHALE BREACHING
The Bowhead Whale is found only in Arctic and subarctic waters. Its broad back has no dorsal fin, which may make it easier for it to swim beneath the polar ice, to which it stays close all year round.

TEMPORARY NORTH POLE
Although the North Pole is in a geographically fixed position, signs stuck in the ice to mark its position can be accurate only temporarily because the ice drifts a few kilometres per day.

ICE-BREAKERS

Ice-breakers are ships that combine powerful engines, heavy displacement (weight) for their size, and reinforced hulls. A modern ice-breaker can break through sea-ice that is 1.8m (6ft) thick at a steady 6kph (3¾ mph), or through 6.5m (21ft) of ice by ramming. The bow of an ice-breaker is shaped so that when the ship is driven forward, momentum causes it to ride up on top of the ice, which is crushed by the weight of the ship. Today, ice-breakers are involved in the supply of remote settlements, creation of navigation channels for commercial ships, scientific exploration, and tourism.

ICE FACTORY
Disko Bay in northern Greenland, 300km (185 miles) north of the Arctic Circle, contains one of the most productive ice floes in the northern hemisphere, averaging 20 million tons of ice per day. Icebergs are towed from here into harbours to be chipped into ice cubes.

OCEAN

ARCTIC OCEAN *south*

Barents Sea

LOCATION North of Norway and European Russia, and south of Svalbard and Franz Josef Land

AREA
1.4 million square km (542,000 square miles)

MAXIMUM DEPTH
600m (2,000ft)

INFLOWS Norwegian Sea (Atlantic Ocean), Arctic Basin

Of all the Arctic seas, the Barents Sea is unique in that a substantial part of it is ice-free all year round. A predominantly shallow sea, it covers an area of continental shelf cut by several trenches. Ice encroaches into northern parts of the sea during the winter, and

UNDERWATER LIFE
The floor of the Barents Sea is rich in invertebrates, including feather stars, anemones, sea cucumbers, and starfish, as shown here.

a cold current flows from the north all year round. The combination of mixing warm and cold currents and changes in surface density over the year causes frequent upwellings of nutrient-rich water, and this leads to strong plankton blooms in spring and summer. The plankton support large populations of fish, which in turn sustain other animals and a productive fishing industry.

SOUTHERN COAST
The Barents Sea coast of northern Norway and Russia contains many shallow bays and inlets, though in the west it is deeply indented with fiords.

ARCTIC OCEAN *south*

White Sea

LOCATION Forms an inlet adjacent to the Barents Sea in the northern part of European Russia

AREA 90,000 square km (35,000 square miles)

MAXIMUM DEPTH
340m (1,115ft)

INFLOWS Barents Sea; rivers Onega, Northern Dvina

The White Sea is an almost completely landlocked body of water with an average depth of 60m (200ft). Its floor, which consists of continental shelf, is broken up by various troughs and ridges. The White Sea connects to the Barents Sea through a narrow strait known as the Gorlo ("throat") and to the Baltic Sea via canals and Lake Onega to its southwest. The sea can be navigated year-round, albeit with the help of ice-breakers in winter. It is on a key route linking the economically active parts of northwestern Russia with the rest of the world.

HARP SEAL PUP
The White Sea is the breeding ground for a large population of Harp Seals, believed to be some 1.5 to 2 million animals strong.

ARCTIC OCEAN *south*

Kara Sea

LOCATION Bounded by the islands of Novaya Zemlya in the west, and Severnaya Zemlya in the east

AREA 880,000 square km (340,000 square miles)

MAXIMUM DEPTH
620m (2,035ft)

INFLOWS Barents Sea, Laptev Sea; rivers Ob', Yenisey

The Kara Sea formed as a result of deglaciation after the last ice age. Most of the sea sits over continental shelf with depths of less than 200m (655ft). The sea floor contains many broad terraces that step down from the southeast (where it is very shallow) to the north and west. Two deep troughs cut through the shelf. A huge volume of fresh water pours into the

DEEP INLETS
Several deep inlets of the Kara Sea cut into the mainland. Those visible at the foot of this satellite image are estuaries at the mouths of the rivers Ob' and Yenisey.

Kara Sea during the summer from rivers along the Siberian coast. Because of this, surface salinity varies considerably, from about 34ppt throughout most of the sea in winter down to 10–12ppt near the mouths of the rivers Ob' (see p.233) and Yenisey in early summer. From December to June, most of the sea is ice-covered. In contrast, between July and October, over 90 per cent is ice-free, although surface temperatures never rise above 4–5°C (39–41°F). During the summer, the Kara Sea is an important route for transport of goods into and out of western Siberia, including oil and natural gas since large deposits were discovered in the Ob'-Yenisey region. The sea is also an important fishing ground, but there are concerns about the threat to marine life posed by the dumping of nuclear waste in the recent past.

ARCTIC OCEAN *south*

Laptev Sea

LOCATION Bounded by Severnaya Zemlya in the west and the New Siberian Islands in the east

AREA 714,000 square km (276,000 square miles)

MAXIMUM DEPTH
2,980m (9,775ft)

INFLOWS Kara Sea, Arctic Basin; rivers Lena, Khatanga

Of the various Arctic seas, the Laptev Sea is one of the coldest and most inhospitable. It has been called the ice factory of the Arctic, as it produces most of the ice transported by the Transpolar Drift, a current that carries water and sea-ice from the coasts of Siberia and Alaska across the North Pole to the Fram Strait, east of Greenland. Almost all of the Laptev Sea is ice-covered from October to May. In winter, there are gales and blizzards; in the summer, frequent fogs. Much of the sea covers a shallow continental shelf, which breaks off abruptly to depths of 2,000m (6,600ft) or more in the far north. The sea is influenced by large amounts of run-off from large Siberian rivers, several of which form extensive deltas.

TIKSI HARBOUR
Tiksi is the main port on the Laptev Sea coast but is active only from June to October. For the rest of the year, the harbour is frozen over. Timber is the main cargo that goes through the port.

BROKEN, DRIFTING SEA-ICE (TOP RIGHT)

ARCTIC OCEAN *south*

East Siberian Sea

LOCATION Bounded by the New Siberian Islands in the west and Wrangel Island in the east

AREA 936,000 square km (361,000 square miles)

MAXIMUM DEPTH 155m (510ft)

INFLOWS Laptev Sea; rivers Kolyma, Indigirka, Alezeya

The East Siberian Sea is the shallowest of the Arctic seas. Its average depth is just 10–20m (35–65ft) in its western and central parts and 30–40m (100– 130ft)

in the east. Freshwater input from the Kolyma River runs as high as 10 million litres (2.6 million gallons) per second in the spring. The salinity of the sea ranges from almost zero near the river-mouth to 33ppt in the north. It is ice-covered for most of the year, and navigation through it is possible only in August and September.

ARCTIC OCEAN *south*

Chukchi Sea

LOCATION Lying to the northeast of eastern Siberia and to the northwest of northern Alaska

AREA 582,000 square km (225,000 square miles)

MAXIMUM DEPTH 110m (360ft)

INFLOWS Bering Sea (Pacific Ocean), East Siberian Sea, Arctic Basin

The Chukchi Sea is a mostly shallow sea situated between Alaska, Siberia, and the central Arctic Basin. The sea is fed by low-salinity waters from the

Pacific Ocean via the narrow, shallow Bering Strait. This water mixes with colder and more saline water in the Chukchi Sea; the mixed water then moves into the Arctic Basin. In the northeast, a submarine channel feeds some of the mixed water into the Beaufort Sea. The Chukchi Sea is totally iced over between December and May. It supports large populations of walrus and several species of seal.

A SUMMER FOGBOW

ARCTIC OCEAN *south*

Baffin Bay

LOCATION Southwest of northern Greenland and northeast of Baffin Island in the Canadian Arctic

AREA 689,000 square km (266,000 square miles)

MAXIMUM DEPTH 2,100m (6,900ft)

INFLOWS Arctic Basin, Labrador Sea (Atlantic Ocean); glaciers of West Greenland

Separated from the Arctic Basin by Ellesmere Island and parts of the Canadian Archipelago, Baffin Bay is a mostly deep-water Arctic sea. From

ICEBERGS IN DAVIS STRAIT
Many icebergs calved from glaciers in western Greenland drift across Baffin Bay. From there, the Baffin Current moves them into the Labrador Sea.

mid-August to October, the bay is largely free of sea-ice, but in winter an ice cover forms and moves southeast under the prevailing winds. Combined with Arctic pack ice entering through the northern sounds, and icebergs that have broken off glaciers in northwestern Greenland, this makes the bay unnavigable for much of the year. A significant feature is the North Water polynya, an area of open water where thick ice cover would be expected. Its presence may be related to the warming effect of the West Greenland Current, which enters the bay from the south.

ARCTIC OCEAN *south*

Beaufort Sea

LOCATION North of northern Alaska and northwestern Canada, bounded in the east by the Amundsen Gulf

AREA 476,000 square km (184,000 square miles)

MAXIMUM DEPTH 4,680m (15,350ft)

INFLOWS Chukchi Sea, Arctic Basin; rivers Mackenzie, Colville

The Beaufort Sea has a shallow coastal area and a deep offshore area, the Beaufort Deep. Ice covers the sea for most of the year, but it is navigable near the coast from August to October. The Mackenzie River deposits billions of tons of fresh water and 15 million tons of sediment into the sea each year.

MELTING SEA-ICE

ARCTIC OCEAN *south*

Greenland Sea

LOCATION Bounded by the Arctic Basin to the north, Svalbard to the northeast, and Iceland to the south

AREA 1.2 million square km (465,000 square miles)

MAXIMUM DEPTH 4,800m (16,000ft)

INFLOWS Arctic Basin, Norwegian Sea (Atlantic Ocean)

Sandwiched between the Arctic Basin to the north and the Atlantic Ocean to the south, the Greenland Sea is an important exchange region for different bodies of water moving in the global system of ocean circulation. A constant current of warm and salty Atlantic water flows north through the eastern part of the sea into the Arctic Basin. Simultaneously, enormous amounts of shallow fresh water and ice move south via the Fram Strait between Greenland and Svalbard and down the western part of the sea as the East Greenland Current. As they pass

through, these waters are modified by cooling (Atlantic water) and melting (Arctic ice). In each case, the water subsequently sinks to the sea-floor when it meets the less dense water masses. From November to July, ice covers most of the sea and includes drifting Arctic pack ice, locally formed sea-ice, and icebergs. Surface water temperatures range from -1°C (30°F) in the north in February to 6°C (43°F) in the south in August. Moderately dense populations of a variety of plankton thrive in the Greenland Sea during spring and summer.

BLACK-LEGGED KITTIWAKES
Seen here perched on an iceberg in the western part of the Greenland Sea, Black-legged Kittiwakes are one of the main gull species that breed on the Greenland coast.

NORTH ATLANTIC STORM
Cyclonic storms and rough seas are
common in the North Atlantic above the
latitude of 30°N. These wave-battered
rocks are at Porspoder on the
Brittany coast in France.

Atlantic Ocean

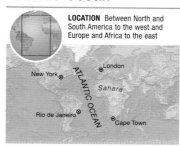

LOCATION Between North and South America to the west and Europe and Africa to the east

AREA 77 million square km (29.7 million square miles)

MAXIMUM DEPTH 8,605m (28,230ft)

INFLOWS Arctic Ocean, Southern Ocean; rivers St. Lawrence, Mississippi, Orinoco, Amazon, Paraná Congo, Niger, Loire, Rhine; Mediterranean Sea

The Atlantic Ocean is the second largest of the world's oceans, and it has several tributary seas. An immense mountain range extends from north to south along the Atlantic sea floor, covering almost a third of its area. Known as the Mid-Atlantic Ridge, this began to form about 180 million years ago, and during the Mesozoic Era sea-floor spreading on either side of the ridge opened up the Atlantic by separating the Americas from Europe and Africa. Rising from the ridge system and its flanks are several islands, including Iceland, the Azores, and Ascension Island. On either side of the Mid-Atlantic Ridge lie basins that are smooth and flat in some areas, but disrupted in others. Some large volcanoes are found singly or in rows in the Atlantic basins. These rise to form sea mounts and, in some places, islands (such as Bermuda). The surface waters of the Atlantic Ocean move in two large gyres, a clockwise one in the North

ISOLATED ISLAND
Tristan da Cunha in the South Atlantic is the most remote inhabited island in the world. A volcano rising from 3,700m (12,140ft) below sea-level, it is almost entirely surrounded by high cliffs.

ATLANTIC COD
Cod is an important commercial fish species of the North Atlantic, but overfishing has severely depleted stocks in several regions.

Atlantic and an anticlockwise one in the South Atlantic. In addition, there is a distinct deep-water circulation. Dense cold water formed at the poles sinks to the bottom and moves slowly at depth towards the Equator. These water masses converge and mix in the South Atlantic, where a portion of the water wells up, bringing nutrients to the surface. Prevailing weather in the Atlantic Ocean varies with latitude. The higher latitudes are characterized by variable westerly winds, changeable weather, and frequent storms. Closer to the Equator, there are belts of higher pressure systems, and constant easterly trade winds. This region has few intense storms, though in late summer/early autumn, the low latitudes of the North Atlantic are affected by hurricanes, which arise in the mid- and eastern Atlantic and move west and northwest.

BASKING SHARK
Basking sharks range throughout Atlantic waters. They swim with their mouths wide open, trapping small crustaceans in their comb-like gill rakers.

ROCKY SHORELINE
The Atlantic Ocean has many rocky coasts, such as this shoreline of eroded pink granite, on Mount Desert Island, Maine, USA. The glacier-sculpted island is the home of Acadia National Park.

LIGHTHOUSES

The Atlantic Ocean has been the most heavily travelled of the world's oceans for hundreds of years, and many of its coasts are prone to storms and fog. To safeguard the large numbers of passing ships, thousands of lighthouses have been constructed on Atlantic coasts since the 1700s. Today, over 200 lighthouses are in active use on the eastern US seaboard alone. Lighthouses are constructed at important points on a coastline: at entrances to estuaries or harbours, and on isolated or sunken rocks or shoals. Today, most lighthouses are automated – the days of permanent lighthouse-keepers are almost gone.

OCEAN

Hudson Bay

LOCATION A large inlet from the Atlantic Ocean, extending into east and central Canada

AREA 819,000 square km (316,000 square miles)

MAXIMUM DEPTH 270m (900ft)

INFLOWS Rivers Albany, Churchill, Moose, Nelson, Severn, Grande Rivière de la Baleine

Hudson Bay is a large, shallow body of water, averaging about 128m (420ft) in depth, which occupies a depression in the vast ancient rock mass known as the Canadian Shield. A huge area of Canada, and even small parts of Minnesota and North Dakota in the USA, drains into the bay. Because of this river input, its surface waters are of low salinity. The main input of sea water comes from the northwest (through a northern part of the bay called Foxe Basin), while the main outflow is in the northeast through Hudson Strait, which connects Hudson Bay to the Atlantic. The eastern coast of the bay is rocky, with some high cliffs, but its other shores are marshy and low-lying. Hudson Bay contains a great quantity of dissolved nutrients, which allows considerable plankton growth in spring and summer. Small, shrimplike crustaceans occupy the open waters, and many invertebrates, such as molluscs and worms, live on the sea floor. Fish include Atlantic Cod and Greenland Halibut, as well as salmon migrating to the rivers to spawn. The shores of the bay are sparsely settled, principally by Indigenous nations and Inuit, who support themselves mainly by fishing and hunting sea mammals.

COLD WATERS
Ice clogs up much of Hudson Bay from November until June, and surface water temperatures never rise above 9°C (49°F) even in the summer months.

HENRY HUDSON

Hudson Bay is named after Henry Hudson (c.1570–1611), an English explorer who sailed into the bay in August 1610 while trying to find the Northwest Passage. After being trapped there by ice for 10 months, mutineers cast Hudson adrift in a small boat.

BELUGA
A substantial number of Belugas, or White Whales, enter the bay during the summer months, where they feed on bottom-dwelling fish and invertebrates.

Gulf of St. Lawrence

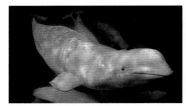

LOCATION At the mouth of the St. Lawrence River in Canada, extending to western Newfoundland

AREA 155,000 square km (60,000 square miles)

MAXIMUM DEPTH 2,300m (7,550ft)

INFLOWS Atlantic Ocean; St. Lawrence River

The Gulf of St. Lawrence is a cold-water sea that acts as a gateway to the interior of North America. About 14 million litres (3 million gallons) of fresh water are discharged every second into the gulf from the St. Lawrence River. The main inflows and outflows of water from the gulf, as well as most maritime traffic, pass through Cabot Strait, which separates Newfoundland from Nova Scotia to the south. Ice floes are a prominent feature from February to April and are a hazard to ships. The deep waters of the gulf contain good fishing grounds.

STRAIT OF BELLE ISLE
This northern outlet from the gulf separates the island of Newfoundland (shown here) in the south from Labrador on the Canadian mainland in the north.

Bay of Fundy

LOCATION Between the Canadian provinces of Nova Scotia and New Brunswick

AREA 9,300 square km (3,600 square miles)

MAXIMUM DEPTH 365m (1,200ft)

INFLOWS Atlantic Ocean; St. John River

The Bay of Fundy is best known for its fast-running tides, which produce changes in water level as great as 17m (55ft), the highest in the world. High cliffs channel the bay's waters until they separate into two narrow arms at its northeastern end. In these, the tidal fluctuations are magnified by the narrowness and shape of the bay. The tidal surge in the northern arm, Chignecto Bay, produces a bore up to 1.8m (6ft) in height up

FLOWERPOT ROCKS
Near Hopewell Cape, the tides have carved away the bases of cliffs, producing sandstone pillars with trees on top, known locally as the Flowerpot Rocks.

COASTAL SALT MARSH
Salt marshes, some cut through by rivers, are numerous along the 1,300-km- (800-mile-) long coastline of the Bay of Fundy.

the Petitcodiac River. Over several decades, Passamaquoddy Bay in the northwest of the Bay of Fundy has been the focus of studies into the feasibility of harnessing its hydroelectric potential through damming or some other method. Environmental considerations, engineering difficulties, and the huge costs involved have so far impeded any development.

ATLANTIC OCEAN *west*

Sargasso Sea

LOCATION In the western part of the North Atlantic Ocean, southeast of Bermuda

AREA
5.2 million square km
(2 million square miles)

MAXIMUM DEPTH
7,000m (23,000ft)

INFLOWS None

LIFE-SUPPORT SYSTEM
Eels, baby turtles, crabs, and fish such as these juvenile Filefish thrive in Sargassum.

The Sargasso Sea is a large, clockwise-circulating region of the North Atlantic, created by three currents around its edge: the south-moving Canaries Current to the east; the westward-moving North Equatorial Current to its south; and the Gulf Stream to its west

and north. Due to the Coriolis effect (see pp.394 and 450), the movement of these currents around the perimeter of the sea forces ocean water towards its centre. As a result, the middle of the Sargasso Sea is about 1m (3ft) above the level of water along the eastern US seaboard. In summer, an excess of evaporation over rainfall results in a thick lens of warm salty water forming in the middle of the sea. This inhibits the upwelling of nutrient-rich, colder water from the ocean bottom, and this lack of nutrients results in a sparsity of plankton. However, the mats of Sargassum seaweed support a complex web of animal life.

EELS
All river-dwelling eels in Europe and eastern USA are born in the Sargasso Sea. Adult eels migrate there to mate when they are 10–15 years old.

SEAWEED MATS
The Sargasso Sea is so named because of the extensive mats of a yellow-brown seaweed, Sargassum, that float on the surface thanks to tiny gas-filled bladders.

ATLANTIC OCEAN *west*

Gulf of Mexico

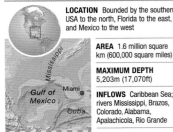

LOCATION Bounded by the southern USA to the north, Florida to the east, and Mexico to the west

AREA 1.6 million square km (600,000 square miles)

MAXIMUM DEPTH
5,203m (17,070ft)

INFLOWS Caribbean Sea; rivers Mississippi, Brazos, Colorado, Alabama, Apalachicola, Rio Grande

WATERSPOUT
Tornadoes over water, called waterspouts, are common in the gulf. High humidity and temperatures in the range 30–35°C (85–95°F) are predisposing factors.

The Gulf of Mexico is an almost landlocked body of water that is edged by a hydrocarbon-rich area of continental shelf 40–320km (25–200 miles) wide (see panel, below). Fringing the continental shelf is a coastal zone of mangrove swamps, tidal marshes, beaches, lagoons, and estuaries. A current of warm water enters the gulf

from the neighbouring Caribbean Sea via the Yucatán Channel in the southeast. This exits again via the Straits of Florida into the Atlantic Ocean to the east, forming the Gulf Stream (see p.452). The Gulf of Mexico supports a large fishing industry, focused especially on shrimp but also on fish such as Red Snapper. A recent threat to the industry has come from over-growth of plankton in northern parts of the gulf, depleting oxygen levels in the water. The cause is thought to be heavy run-off of nitrate fertilizers from the land.

ROSEATE SPOONBILLS
These waders inhabit the gulf's coastal zone. They feed by sweeping their bills through the water and snapping them shut when they feel a shrimp or fish.

DRILLING FOR OIL

The continental shelf regions of the Gulf of Mexico contain large reservoirs of oil and gas. These resources have been explored and developed extensively since the 1940s, with numerous offshore wells drilled. They supply a substantial proportion of Mexico's and the USA's domestic energy needs.

FLORIDA KEYS
The Keys are a string of low-lying islands that project into the Gulf of Mexico from the tip of Florida. They are chiefly composed of ancient coral reefs underlain by thick limestone.

LIVING REEF
Like a forest on land, this coral reef at the Vatu-i-Ra Passage, Fiji, has a complex structure that provides homes and food for a wealth of animals.

THREATS TO CORAL REEFS

Warm-water coral reefs are extremely rich marine habitats and by far the largest structures created by living things. However, because they grow slowly and have stringent environmental requirements, they are easily harmed by human activity. At least 20 per cent of the world's coral reefs have been so badly damaged that they are effectively dead. Without urgent action, about 40 per cent of the reefs may disappear in the next 30 years – a prospect that conservationists are fighting hard to prevent.

Reefs in Danger

Corals are soft-bodied animals that protect themselves by building chalky skeletons. The skeletons last far longer than the animals themselves, and they build up to form reefs. The surface of a reef is like a living skin. It contains millions of coral animals, or polyps, and even larger numbers of microscopic algae called zooxanthellae. These live inside the polyps, and they help to nourish their hosts by harnessing the energy in sunlight (see pp.396–97).

In a healthy reef, some branching corals grow more than 20cm (8in) a year. Their upward spread is limited only by the sea itself, because few corals can stand more than an hour of exposure at low tide. This growth offsets the damage caused by storms and by coral-eating animals, so the reef expands or maintains its size. But if anything harms either the polyps or their zooxanthellae, this delicate balance is upset. Growth is no longer maintained, and the surface of the reef begins to erode. If the coral does not recover, the entire ecosystem begins to decline, because corals provide reef animals with food and habitats.

DEAD ZONE
Staghorn corals are the fastest-growing coral and among the most easily damaged. They are seen here littering the sea bed in the Caribbean.

Causes of Decline

Coral reefs have always been used as a local resource, and they can be seriously harmed by overfishing or by removing the coral itself for personal or commercial gain. They can also be affected by pollution from an area far away.

Coral requires clear, nutrient-poor water, because this favours their zooxanthellae and keeps competing free-living algae in check. Conversely, anything that makes seawater more turbid or more fertile can suppress their growth. In many parts of the tropics, farming and logging have increased soil erosion, producing sediment-laden run-off, which finds its way into the sea. This shades the coral, robbing it of the light that its zooxanthellae partners need to survive.

Reefs are also affected by coastal and tourist development, because sewage raises water nutrient levels, causing overgrowths of algae. Sewage treatment can prevent this, but adequate funds are often lacking. Even in developed countries, many coastal communities still discharge raw sewage into the sea.

STRANGERS ON THE SHORE
In coastal areas with coral reefs, tourism has to be carefully managed to avoid creating environmental damage.

Raised Sea Temperatures

Despite these threats, local reef damage can be reversed. In protected areas, some reefs have been quick to recover when pollution and fishing are controlled. But coral reefs are also threatened by global warming (see pp.458–59) – a worldwide phenomenon that is far harder to control.

When sea temperatures are raised, coral reefs are harmed. Coral polyps are put under stress, and so they respond by expelling their algae – in a process called coral bleaching. Bleaching reverses itself if the water then cools, but, if raised temperatures continue for a prolonged period, the coral polyps die.

Human addition of carbon dioxide to the air also turns the oceans more acidic. This could severely impair corals' ability to grow their skeletons.

SCATTERED REMAINS
Fragments of old coral litter a beach at Puako, Hawaii, and contrast with the black volcanic sand.

BLEACHING EVENTS
This graph shows the incidence of coral bleaching across the world over three decades. There has been a general upward trend and a year of catastrophic bleaching in 1998.

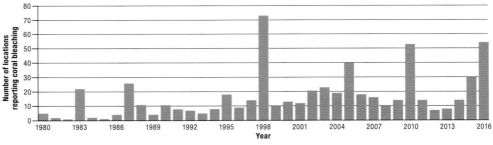

OCEAN

OCEAN

ATLANTIC OCEAN *west*

Caribbean Sea

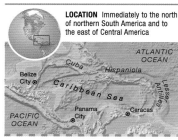

LOCATION Immediately to the north of northern South America and to the east of Central America

AREA 2.75 million square km (1.1 million square miles)

MAXIMUM DEPTH 7,685m (25,215ft)

INFLOWS Atlantic Ocean; rivers Magdalena, Coco, Patuca, Motagua

A popular resort area, the Caribbean Sea is best known for its warm tropical climate, turquoise waters, and beautiful white beaches. Beneath the surface, the

BEACH IN CUBA
One of the Caribbean's main assets is its beaches, which are important for many reasons, not just tourism. They protect coasts from wave action, espdecially during hurricanes, and provide boat landing sites and material for construction. In recent decades, beach erosion has become a major concern.

sea occupies five submarine basins. From southeast to northwest, these are the Grenada, Venezuelan, Colombian, Cayman, and Yucatán basins. The islands of the West Indies, which fall into three main groups, lie in a curve that extends mainly around the northern and eastern perimeter of the sea. The Lesser Antilles are a string of about 20 small volcanic islands that form an arc across its eastern boundary. The four large islands of Cuba, Hispaniola, Jamaica,

and Puerto Rico in the northern part of the sea are known as the Greater Antilles. The Windward Passage, a major shipping route between the USA and the Panama Canal, passes between Cuba and Hispaniola. In the southern part of the Caribbean Sea, near the coast of Venezuela, there is a separate group of three islands: Aruba, Curaçao, and Bonaire. The main surface current in the Caribbean Sea is an extension of the North Equatorial Current. Surface water enters the sea in the southeast, through channels between the Lesser Antilles, flows in a northwesterly direction, and exits through the Yucatán Channel in the northwest (between Cuba and Mexico's Yucatán Peninsula) into the Gulf of Mexico. Trade winds from the northeast dominate the region, and tropical storms occur in late summer and autumn in the northern Caribbean, sometimes reaching hurricane velocity. Most of the islands, as well as some of the

mainland coasts, are fringed with coral reefs that support large numbers of fish, and invertebrates including spiny lobsters and conches. These reefs have, however, suffered considerable damage over the past 30 years through a combination of storms, coral disease, increased tourism, seawater temperature rise, and coastal development.

MANATEE CONSERVATION

The Caribbean is home to a subspecies of manatee that was once common but is now endangered. The West Indian Manatee is threatened by loss of habitat, hunting, entanglement in nets, and collisions with boats. In Belize, one of its remaining strongholds, a project is underway to track the mammal, educate local people about its conservation, and reduce destruction of its habitat.

PANAMA CANAL

Much of the shipping in the Caribbean is en route to or from the Panama Canal, which links the Atlantic to the Pacific. A huge feat of engineering, the 64-km- (40-mile-) long canal opened in 1914 after decades of construction. Over 30,000 workers died during the project from malaria and yellow fever.

SOFT AND STONY CORALS

The Caribbean Sea contains both soft corals, such as the Sea Fan in the background of this picture, and stony corals, such as the purple-coloured coral (called Blue Crust Coral) growing in front of it.

GREAT BLUE HOLE
Situated on Lighthouse Reef, east of Belize, this sinkhole (see p.241) through submerged limestone is 125m (410ft) deep. It formed in the last ice age, when sea levels were lower.

OCEAN

Norwegian Sea

LOCATION To the west of Norway, north of the North Sea; bordered by the Greenland Sea

AREA
1.4 million square km (534,000 square miles)

MAXIMUM DEPTH
3,970m (13,020ft)

INFLOWS Central North Atlantic Ocean; numerous Norwegian fiords

The Norwegian Sea is the most northerly of the marginal seas of the Atlantic Ocean. Although cut through by the Arctic Circle, it is kept free of ice by the warm Norway Current, a branch of the North Atlantic Current. Strong tidal currents on two areas of the Norwegian coast cause notable whirlpools: the Lofoten Maelstrom near the tip of the Lofoten Islands; and the Saltstraumen near Bodø.

SALTSTRAUMEN WHIRLPOOL

North Sea

LOCATION Extending from the east coast of Great Britain towards the mainland of northwestern Europe

AREA 570,000 square km (220,000 square miles)

MAXIMUM DEPTH
700m (2,300ft)

INFLOWS Central North Atlantic Ocean; rivers Elbe, Weser, Ems, Rhine, Scheldt, Thames, Humber

About 1.6 million years ago, much of the North Sea's southern half was part of the European mainland, and in the intervening period ice-sheets advanced

RECLAIMED LAND
Areas of reclaimed land in the Netherlands, called polders, are kept safe from the sea by dykes.

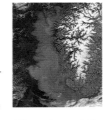

PLANKTON BLOOM
A large bloom of plankton fogs the waters in a 160-km- (100-mile-) wide band off the coast of southern Norway.

and retreated across the region several times. Even today, on the sea floor off northern England, there are large banks of glacial moraine (see p.252), such as the Norfolk and Dogger banks. The main inflow of water is the warm North Atlantic Current from the west. Colder, less saline waters enter from the Baltic Sea via the Skagerrak channel between Norway and Denmark. The constant mixing of cold and warm waters leads to a rich supply of nutrients. As a result, the North Sea has productive fisheries, although stocks have declined markedly in the past 30 years. Since the 1950s, the North Sea has become a major site of offshore oil and gas production.

Bay of Biscay

LOCATION In an indentation of the western European coastline, north of Spain and west of France

AREA
223,000 square km (86,000 square miles)

MAXIMUM DEPTH
4,735m (15,525ft)

INFLOWS Rivers Loire, Dordogne, Garonne, Adour

The Bay of Biscay lies partly over an area of continental shelf and partly over an abyssal plain. A number of islands dot the French area of the shelf. Navigation in the bay is often difficult because of strong northwesterly winds that cause rough seas. The coasts vary from rocky cliffs to areas of lagoon and marsh. Fishing and oyster culture are important industries here.

COAST OF BELLE ILE, FRANCE

Black Sea

LOCATION At the southeastern extremity of Europe and the western extremity of Asia

AREA 422,000 square km (163,000 square miles)

MAXIMUM DEPTH
2,200m (7,200ft)

INFLOWS Mediterranean Sea, Sea of Azov; rivers Danube, Dniester, Dnieper, Kizil Irmak

An almost completely landlocked body of water, the Black Sea mostly occupies a deep, broad submarine basin separating Europe from Asia. It connects to the Mediterranean Sea in the southwest by means of two narrow channels – the Bosporus and the Dardanelles – and the intervening Sea of Marmara. A separate enclosed sea, the Sea of Azov, lies to the northeast.

BLACK SEA COAST OF TURKEY

Baltic Sea

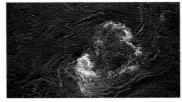

LOCATION Separating the Scandinavian Peninsula from the rest of continental Europe

AREA
386,000 square km (149,000 square miles)

MAXIMUM DEPTH
449m (1,473ft)

INFLOWS Rivers Vistula, Oder, Western Dvina

One of the world's largest areas of brackish (moderately salty) water, the Baltic Sea is a remnant of a huge water-covered region that formed about 8,000 years ago as the ice-sheet that covered Scandinavia retreated towards the Arctic. Today, the Baltic is a semi-enclosed sea, which receives freshwater drainage from a large area of northern Europe. The Baltic Sea links to the

CRETACEOUS LIMESTONE
Cliffs of limestone are a common feature in the Baltic, such as on the east coast of the Danish island of Møn, shown here.

North Sea on its western side, and there is a constant outflow of low-salinity water and, at depth, an inflow of saltier water. Winters are long and cold, summers short and cool, and storms are frequent. Pack ice accumulates in the Gulf of Bothnia in the north in winter and early spring, when navigation is suspended. With coasts on nine different countries, the Baltic Sea is of great commercial importance to northern Europe.

WIND FARMING

The Baltic Sea has about 19 to 20 offshore wind farms, including the world's first, built at Vindeby, Denmark, in 1991, and the Bockstigen wind farm off the Swedish island of Gotland (below). These projects show that offshore wind farming is economically viable for producing renewable energy, and more farms are planned. The main objection to such farms is that they are a potential hazard to birds.

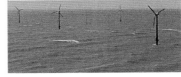

ALAND, FINLAND
The low-lying Åland Islands rise from a narrow strait between Finland and Sweden. They mark the entrance to a northern arm of the Baltic called the Gulf of Bothnia.

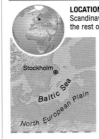

OCEAN

Mediterranean Sea

LOCATION To the south of Europe and western Turkey and to the north of North Africa

AREA 2.5 million square km (970,000 square miles)

MAXIMUM DEPTH 4,900m (16,000ft)

INFLOWS Atlantic Ocean; Black Sea; rivers Nile, Rhône, Po, Ebro

OCTOPUS ON SEA FLOOR
Although Mediterranean marine life has been depleted over the past 40 years through human activity, some species, such as octopuses, remain common.

The world's largest inland sea, the Mediterranean occupies an elongated depression between Europe and Africa that originally formed tens of millions of years ago. About 6 million years ago, tectonic activity closed the Straits of Gibraltar, cutting the Mediterranean off from the Atlantic Ocean to the west. About a million years later, further tectonic activity opened the Straits, and thick salt deposits were formed over a period of around 1 million years. Finally, a rise in sea-level allowed Atlantic waters to flood back in. Today, the Mediterranean Sea remains linked to the Atlantic Ocean via the Straits of Gibraltar in the west, a channel just

13km (8 miles) wide and 320m (1,050ft) deep. It also connects to the Red Sea via the Suez Canal in the southeast (see p.418) and to the Black Sea in the northeast. Underwater, the Mediterranean sea floor is composed of a series of moderately deep basins separated by sills (see p.179). One of the main sills runs between the island of Sicily and Cape Bon in Tunisia, dividing the sea into a western half and a larger eastern half. The Mediterranean Sea also contains many smaller regions lying over shallower water, such as the Adriatic Sea to the east of Italy and the Gulf of Gabes off the eastern coast of Tunisia. Large islands in the Mediterranean include Crete and Cyprus in the eastern half, Sardinia, Corsica, and Majorca in the western half, and Sicily at the toe of Italy. The Aegean Sea, between Greece and Turkey, is dotted with the many islands of the Greek Archipelago. Although a large amount of fresh water flows into the Mediterranean Sea from rivers, the total amount is only about a third of the water lost by evaporation. As a result, the sea is highly saline. There is also a continuous inflow of surface waters of

JACQUES COUSTEAU

As the inventor of SCUBA, and through his films and books about the underwater world, the Frenchman Jacques Cousteau (1910–97) made a huge contribution to subaquatic exploration in the 20th century and to public awareness of the marine environment. The Mediterranean Sea was the site of both Cousteau's original experiments with SCUBA and some important discoveries, such as the unearthing in 1957 of an ancient Greek wine freighter, buried beneath the seafloor off Marseilles.

STRAITS OF GIBRALTAR
Two series of solitons, or large pulses of water travelling in the open sea, can be seen above, propagating from the Straits of Gibraltar at the top of this picture. They are triggered by inflowing Atlantic water accelerated by its passage through the straits.

ALEXANDRIA
One of the busiest Mediterranean ports is Alexandria, located on the edge of the Nile Delta. A former capital of Egypt, Alexandria now handles over 80 per cent of the country's exports and imports.

normal salinity into the Mediterranean from the Atlantic via the Straits of Gibraltar, and to a lesser extent, from the Black Sea. These inflows are partly offset by outflows of saltier water at depth. The dominant current in the Mediterranean Sea is eastward, flowing along the north of the African coast, a continuation of the current that brings seawater in from the

Atlantic. This current is most powerful in summer, when evaporation is at its strongest. Tides are generally weak throughout the whole region, two exceptions being the Gulf of Gabes and the Adriatic Sea. The Mediterranean Sea has long been important as a route for trade, (see panel, right) and remains so today, particularly since the opening of

the Suez Canal. Marine mammals, fish, sponges, corals, and other invertebrates were once plentiful. However, a rise in population and the expansion of industry and tourism over the past 40 years have led to severe pollution of many of the coastal areas. This, together with overfishing, has severely depleted some fish stocks and other marine life.

SEA FLOOR TREASURE
Trading has been conducted by sea in the Mediterranean for over 3,000 years. As a result of navigational errors, storms, and wars, the present-day sea floor is littered with ancient wrecks, providing many sites of archaeological interest. Here, a diver brings a basket from the surface to collect the remains of bowls, bottles, and cups from a wreck that sank off the coast of Turkey in the 11th century CE. The oldest known shipwreck in the Mediterranean (at Ulu Burun, also off the coast of Turkey) dates from the 14th century BCE.

LAVA FORMATIONS
Along with some other Aegean Islands, Lesbos has seen considerable volcanic activity in the past, as evidenced by these lava formations on its coast.

OCEAN

INDIAN OCEAN

Indian Ocean

LOCATION Between Africa and Australia, extending from southern Asia to the Southern Ocean

AREA 69 million square km (26.5 million square miles)

MAXIMUM DEPTH 7,258m (23,812ft)

INFLOWS Rivers Ganges, Indus, Tigris, Euphrates, Zambezi, Limpopo, Murray

BANDED SEA SNAKE
These snakes are common in shallow lagoons in the eastern part of the Indian Ocean. Their tails act like rudders, making them highly efficient swimmers.

SCULPTED ROCKS
These curious granitic rock formations, on the coast of the island of La Digue in the Seychelles, have been carved by a combination of wind, rain, and seawater erosion.

The Indian Ocean is one of the youngest of the world's ocean basins and the most complex in terms of the structure of its floor. Both the Indian and Southern Oceans have formed over the past 120 million years following the break-up of the ancient continent of Gondwana. The Indian Ocean narrows towards the north, where the Indian Peninsula divides it into two large bodies of water, the Arabian Sea on the west and the Bay of Bengal on the east. The most striking feature of the ocean floor is a massive mid-ocean ridge, shaped like an upside-down Y, which occupies about a fifth of the ocean's total floor area (see p.401). One arm runs to the south of Africa and joins the Mid-Atlantic Ridge. The other runs to the south of Australia,

STILT FISHING
This method of fishing is still practised today in parts of Sri Lanka and Thailand. The fishermen and -women perch on poles in shallow water to avoid scaring away the fish.

where it eventually joins the Pacific–Antarctic Ridge. The two ridge arms join roughly in the centre of the ocean and continue north as the Mid-Indian Ridge. Creation of new oceanic crust at the various ridges is continuing to separate Africa, Antarctica, and Australia from each other. Along an extensive arc at its northeast margin, the Indian Plate is subducting beneath the southeast edge of the Eurasian Plate to form the Indonesian volcanic arcs of Sumatra and Java. A continental shelf extends to an average distance of about 120km (75 miles) from

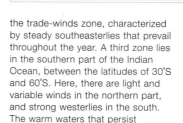

MACARONI PENGUINS
The Indian Ocean extends as far south as 60°S. On a remote southern island, Kerguelen, there are many colonies of mammals and birds, including this colourful species of penguin.

the landmasses around the ocean. In places, the shelf and continental slope are penetrated by deep canyons carved by outflows from the Ganges, Indus, and Zambezi rivers, which have also deposited vast amounts of sediment onto the abyssal plains. The northern part of the Indian Ocean has a monsoon climate, characterized by rain-bearing winds that change direction twice yearly (see p.469). These changes are accompanied by twice-yearly reversals in the pattern of ocean currents, something that distinguishes the northern part of the Indian Ocean from all other oceans. The monsoon zone is also prone to destructive tropical storms that form over the open ocean and head in a westward direction. South of the monsoon zone lies

the trade-winds zone, characterized by steady southeasterlies that prevail throughout the year. A third zone lies in the southern part of the Indian Ocean, between the latitudes of 30°S and 60°S. Here, there are light and variable winds in the northern part, and strong westerlies in the south. The warm waters that persist

throughout much of the Indian Ocean provide an ideal environment for a rich variety of marine life. Some 4,000 species of fish live near the shores, many unique to this ocean. Coral reefs are abundant around all islands in the tropical areas, as well as along the western coasts of Thailand and Myanmar and the east coast of Africa.

ZHENG HE
One of the first explorers to travel widely across the Indian Ocean was a Chinese Admiral, Zheng He (1371–1435). Accompanied by a huge fleet of ships, he visited most parts of the region, including India, Arabia, and east Africa, during 7 separate voyages in 1405–1433. His objectives included trading in goods, gathering knowledge about the region, and extending the sphere of Chinese political influence.

INDIAN OCEAN *northwest*

Red Sea

LOCATION Between the northeastern corner of Africa and the Arabian Peninsula

AREA 450,000 square km (175,000 square miles)

MAXIMUM DEPTH 3,040m (9,975ft)

INFLOWS Arabian Sea (via the Gulf of Aden)

One of the warmest of the world's seas, the Red Sea occupies a depression formed by the Great Rift Valley (see p.128), the primary branch of the East African Ridge System, which has developed over the past 50 million years. Hydrothermal vents on the sea floor are evidence of continuing tectonic activity. The Sinai Peninsula splits the northern end of the sea into two arms: the shallow Gulf of Suez on the west; and the deeper Gulf of Aqaba on the east. Hardly any rain falls on the Red Sea and no rivers flow into it, but water evaporates at a high rate owing to the strong solar radiation. As a result, the Red Sea is highly saline. To make up for evaporation losses, surface water flows in from the Gulf of Aden in the south and continues as a northward-moving surface current. The Red Sea contains a rich marine life based on extensive coral reefs. Since the 1970s, this marine life has attracted increasing numbers of recreational scuba-divers, whose activities have caused great damage to the reefs.

SEA ROSE
This curious flower-like object, commonly seen attached to rocks at shallow depths, is actually a ribbon of eggs that was laid by a nudibranch, or sea slug.

SUEZ CANAL

First opened to shipping in 1869, the Suez Canal connects the Gulf of Suez at the northern end of the Red Sea to the Mediterranean Sea. One of the busiest waterways in the world, the 163-km- (101-mile-) long canal provides a short cut for maritime traffic travelling between ports in North America and Europe and those in the Middle East, southern Asia, East Africa, and Oceania. The canal has no locks because the Gulf of Suez has roughly the same sea-level as the Mediterranean Sea.

GULF OF SUEZ
The turquoise waters of the Red Sea are completely surrounded by desert and, behind a coastal plain, mountains that rise as high as 2,000m (6,500ft).

ANEMONE GARDENS
The thousands of animal species found on Red Sea reefs include many Giant Anemones. The clownfish swimming close by secrete a mucus that protects them from the stinging tentacles.

INDIAN OCEAN *northwest*

Persian Gulf

LOCATION Forms an inlet of the Arabian Sea, between Iran and the Arabian Peninsula

AREA 241,000 square km (93,000 square miles)

MAXIMUM DEPTH 110m (360ft)

INFLOWS Rivers Tigris, Euphrates, Karun

Sometimes called the Arabian Gulf or just the Gulf, the Persian Gulf is best known for the large deposits of oil and natural gas that lie around its shores and beneath its thin floor. The region contains approximately half of the world's proven oil reserves and a third of its proven natural gas reserves. A warm and salty sea, the gulf receives a large inflow of both water and sediments from the Shatt al 'Arab, a river formed from the confluence of the Tigris and Euphrates rivers in the northwest (in Iraq). Its eastern shore is mountainous but its western shore is low-lying, with many islands, lagoons, and tidal flats. After oil, fishing is the most important industry in the Persian Gulf.

DESALINATION

The Gulf region receives little rainfall so water is scarce, though there are limited groundwater reserves. Over 850 desalination plants have been built on the western side of the Persian Gulf. Together they produce over 22,600 million litres (5,900 million gallons) of desalinated water per day. They return warm, concentrated seawater back into the Persian Gulf.

STRAIT OF HORMUZ
Entrance to the Persian Gulf (left) is gained from the Gulf of Oman (right) via the Strait of Hormuz.

INDIAN OCEAN *north*

Arabian Sea

LOCATION To the east of the Arabian Peninsula and to the southwest of the Indian Peninsula

AREA 3.9 million square km (1.5 million square miles)

MAXIMUM DEPTH 5,803m (19,038ft)

INFLOWS Rivers Indus, Narmada

The Arabian Sea occupies the western part of the northern Indian Ocean. Two northern extensions, the Gulf of Aden and Gulf of Oman, connect it to the Red Sea and Persian Gulf respectively. Various island groups lie around its margins, including the Seychelles to the southwest and the Maldives to the southeast. The sea supports an extensive fishing industry.

FISHERMEN ON DHOWS

INDIAN OCEAN *west*

Mozambique Channel

LOCATION Between Mozambique on the African mainland to the west and Madagascar to the east

AREA 1 million square km (386,000 square miles)

MAXIMUM DEPTH 3,000m (9,800ft)

INFLOWS River Zambezi, Rio Lúrio

An important shipping route for the countries of east and southern Africa, the 1,000-km- (620-mile-) long Mozambique Channel is fed by the waters of the Zambezi and all the major rivers of Madagascar. For many decades, it was believed that a continuous warm current, the Mozambique Current, passed down through the channel.

THE COELACANTH

The Mozambique Channel is one of just two places in the world with an identified population of Coelacanths, primitive fish that were believed to have been extinct for 65 million years until, in 1938, one was caught off the east coast of South Africa. Coelacanths live at depths greater than 150m (500ft), where they feed on small bottom-dwelling fish and cuttlefish.

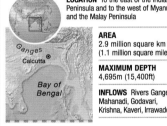

Recently it has been shown that no such current exists. Rather, water movements in the area are dominated by a series of anticlockwise eddies that reach to the channel bottom.

INDIAN OCEAN *north*

Bay of Bengal

LOCATION To the east of the Indian Peninsula and to the west of Myanmar and the Malay Peninsula

AREA 2.9 million square km (1.1 million square miles)

MAXIMUM DEPTH 4,695m (15,400ft)

INFLOWS Rivers Ganges, Mahanadi, Godavari, Krishna, Kaveri, Irrawaddy

The Bay of Bengal occupies the northeastern corner of the Indian Ocean. Its smaller eastern part, the Andaman Sea, is separated from the rest of the bay by two island groups, the Andaman and Nicobar islands. The Bay of Bengal is bordered by a wide area of continental shelf cut through by canyons, and its floor is dominated by the world's largest submarine fan – the Bengal Fan. This distinctive feature at its northern extremity is the wide and thick fan of sediments from the River Ganges (see p.222). Intense tropical storms with high winds and torrential rain occur in spring and autumn. Combined with inflow from rivers during the summer monsoon, these markedly reduce the surface salinity of the bay's waters.

LIMESTONE ROCK OUTCROPS
On the Andaman Sea coast of Thailand, at Phang-Nga, there are many spectacular undercut limestone cliffs.

KO PANYEE, THAILAND
This picturesque fishing village rests precariously on wooden piles driven into the floor of the bay.

PIROGUE BEACHED ON NOSY BE
The island of Nosy Be lies just off mainland Madagascar, on the eastern side of the Mozambique Channel. Pirogues are boats of a simple construction, typically used by Madagascans for fishing.

OCEAN

Pacific Ocean

LOCATION Between Asia and Australia to the west and North and South America to the east

AREA 156 million square km (60 million square miles)

MAXIMUM DEPTH 10,924m (35,840ft)

INFLOWS Southern Ocean; rivers Yukon, Columbia, Amur, Yellow, Yangtze, Mekong

By far the largest ocean, the Pacific covers more than a third of the Earth's surface and holds more than half of its liquid water. It also contains the deepest point in the world's oceans, the Challenger Deep in the Mariana Trench. About three-quarters of the Pacific Ocean lies over a single tectonic plate, the Pacific Plate. Marking the boundary between this and the adjacent oceanic Nazca Plate in the southeastern Pacific is a large mid-ocean ridge, the East Pacific Rise. Associated with plate movements away from this ridge, and occurring all around the Pacific Ocean's Ring of Fire (see p.157), are various deep-sea trenches marking subduction zones with abundant volcanic and earthquake activity. The eastern part of the Pacific Ocean has high mountain ranges on the continental side for most of its

PACIFIC BREAKERS
The Pacific has the world's largest fetch, the area over which wind can interact with surface water to create waves. This results in some immense breakers, like these ones in Hawaii.

RED TIDE
A red tide is seawater discoloured by a bloom of certain types of plankton, some of which produce toxins. Red tides occur in many parts of the Pacific, including the coasts of California (seen here).

length. It has a narrow continental shelf, few large rivers discharging into it, and a regular coastline, except for fiords in the extreme north and south. In contrast, the western Pacific has an irregular continental boundary, many large marginal seas, huge rivers flowing in off the Asian mainland, and some wide areas of continental shelf.

The surface waters over most of the Pacific Ocean move in two large gyres (see p.394). The main deep-water circulation is a northward spread of cold water at depth. This reaches as far north as Japan and then moves to the east. The main winds of the Pacific are twin belts of westerlies (blowing from west to east) between latitudes of 30° and 60°, both north and south of the Equator. Between the westerlies are the more predictable trade winds, which blow towards the Equator. Typhoons affect two large areas of the western Pacific in late summer and early autumn. There are numerous islands in the Pacific Ocean, many of them part of volcanic island arcs. Under the Nazca Plate off Ecuador is a volcanic hot spot that created the Galápagos Islands. Over millions of years, a remarkable collection of animals and plants reached this isolated spot from South America and then diversified on the different islands (see p.114). Charles Darwin's study of this unique fauna and flora in the 1830s influenced the formation of his ideas about evolution.

MARINE IGUANA
One of many unusual animals found in the Galápagos Islands, the Marine Iguana is the only lizard that forages underwater for food.

GIANT KELP FOREST
The northeastern coasts of the Pacific are covered in forests of Giant Kelp. Among the world's fastest-growing plants, they thrive along rocky reefs.

ARTIFICIAL ISLANDS

Faced with a severe shortage of suitable land for building, Japan has turned to its Pacific coastal areas as potential construction sites, and created several artificial islands. There are now several in Tokyo Harbour and two near the port city of Kobe. Artificial islands are usually built by enclosing an area of shallow water with a seawall and then filling it in. Similar projects are now planned in other parts of the world, including the Baltic and Mediterranean seas.

KANSAI AIRPORT
This airport was constructed on an island built some 5km (3 miles) off the coast of Osaka, Japan, between 1987 and 1993.

ISLANDS IN PALAU, MICRONESIA
Consisting of coral structures built on top of a submarine ridge, these are typical islands of the west Pacific, except for the way they have been uplifted and eroded into humped shapes.

OCEAN

Bering Sea

LOCATION Between western Alaska, USA, and the northeastern extremity of the Asian landmass

AREA
2.3 million square km (890,000 square miles)

MAXIMUM DEPTH
4,097m (13,442ft)

INFLOWS Central North Pacific Ocean; rivers Yukon, Anadyr'

The Bering Sea is the most northerly marginal sea of the Pacific Ocean, connecting to the Arctic Ocean in the north via the Bering Strait. In the south, the sea is fringed by an arc of volcanic islands, the Aleutians. In the southwest, the sea floor is separated by ridges into three basins. The Bering Sea has a severe climate, and in winter its northern part is covered in ice. It has a high biological productivity, however, with explosions of plankton growth occurring twice each year. The plankton supports large numbers of fish, mammals such as Sea Otters, and on the continental shelf area in the northeast, invertebrates such as molluscs, starfish, and sponges.

SEA-ICE BREAKING UP IN SPRING

Gulf of Alaska

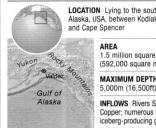

LOCATION Lying to the south of Alaska, USA, between Kodiak Island and Cape Spencer

AREA
1.5 million square km (592,000 square miles)

MAXIMUM DEPTH
5,000m (16,500ft)

INFLOWS Rivers Susitna, Copper; numerous large iceberg-producing glaciers

The Gulf of Alaska is a roughly triangular-shaped body of water, the apex of which is in Prince William Sound near the port of Valdez, which is North America's northernmost ice-free harbour. Two sides of the triangle extend from there to Cape Spencer (near the entrance to Glacier Bay) in the southeast and to Kodiak Island and the Alaskan Peninsula to the southwest. A warm current, the Alaska Current, moves through the gulf up the southeastern coast of Alaska, and a continuation, the Alaska Stream, runs out past the Alaskan Peninsula towards the central North Pacific. The southern part of the gulf lies over a deep abyssal plain, and in the southwest this plunges into the eastern part of the Aleutian Trench, to the southeast of Kodiak Island. There are several deep fiords and inlets along the Alaskan Gulf Coast. Many large glaciers, such as the Hubbard Glacier (see p.256), discharge icebergs into these inlets, from where they slowly float out to sea.

BREACHING ORCA
Orcas, or Killer Whales, are a common sight in the Gulf of Alaska, preying upon seals, sea lions, fish, squid, and occasionally sea birds and Sea Otters.

The *Exxon Valdez* disaster of 1989, caused when an oil tanker went aground in Prince William Sound, was one of the worst-ever oil spills in terms of damage to the environment. Fourteen years later, toxic chemicals were still seeping into the Gulf of Alaska from contaminated beaches.

Gulf of California

LOCATION Between the west coast of mainland Mexico and the peninsula of Baja California

AREA 160,000 square km (62,000 square miles)

MAXIMUM DEPTH
3,050m (10,000ft)

INFLOWS Rivers Fuerte, Sonora, Yaqui, Colorado

Also known as the Sea of Cortés, the Gulf of California is a 1,200-km- (750-mile-) long inlet from the Pacific Ocean into the west coast of Mexico. It lies over the boundary between the Pacific and North American plates and is thought to have formed by sea-floor spreading over the past 6 million years. The Gulf of California consists of two distinct regions. The northern part is mostly shallow – due to the accumulation of silt

BAHIA DE LOS ANGELES
This area of the Gulf of California, surrounded by desert, has been compared to the Galápagos Islands because of its diverse marine life.

BALL OF JACK FISH
Various types of jack are among the 800 species of fish found in the gulf. The balling behavior seen here is thought to be a defence against predators.

transported by the Colorado River (see p.206) over thousands of years – and has a large tidal range. The southern part is much deeper and less affected by tides, and it includes the Guaymas Basin, which has many hydrothermal vents. The large amount of sunlight in the region, combined with the nutrients provided by upwelling, allow high plankton growth, which supports some rich fishing grounds. Unfortunately, the gulf's environment is now threatened by overfishing, unregulated tourism development, and insufficient freshwater flow (most water from the Colorado River is now diverted for residential and agricultural use).

PACIFIC OCEAN *northwest*

Sea of Okhotsk

LOCATION Off the far eastern coast of Russia, to the west of the Kamchatka Peninsula

AREA
1.6 million square km (611,000 square miles)

MAXIMUM DEPTH
3,372m (11,063ft)

INFLOWS Sea of Japan; rivers Amur, Uda, Okhota, Penzhina

The Sea of Okhotsk is a large, cold sea that freezes over during the winter months and is frequently covered with fog. Except where it laps the northern-most Japanese island of Hokkaido in the south, it is surrounded by Russian territory: the Kamchatka Peninsula in the east; the Russian mainland and the island of

THE WORLD'S LARGEST EAGLE
The coasts of the Sea of Okhotsk are a breeding ground for Steller's Sea Eagle.

Sakhalin in the west; and the Kurile Island arc in the southeast. The floor of the sea slopes gently from north to south and then plunges to depths greater than 3,000m (10,000ft) to the northwest of the Kuriles. Tides and currents in the sea are strong, with a tidal range reaching 13m (43ft) in some locations. The Sea of Okhotsk is exceptionally important to Russia as a fisheries resource. Oil and gas deposits have also been found on the continental shelf area in the northern part of the sea.

ISLAND IN THE KURILES
The volcanic origin of Ushishur Island is clear in this view of its lake-filled central crater.

PACIFIC OCEAN *northwest*

Sea of Japan

LOCATION To the northwest of Japan and to the east of the Korean Peninsula and southeastern Siberia

AREA 978,000 square km (377,600 square miles)

MAXIMUM DEPTH
3,742m (12,276ft)

INFLOWS East China Sea; rivers Tumen, Ishikari, Shinano, Agano, Mogami, Teshio

The Sea of Japan is almost completely enclosed by mainland Asia and the islands of Japan. A current of warm water (the Tsushima Current) flows into the sea from the East China Sea to the south. Outflow is mainly into the Sea of Okhotsk in the north, via the La Pérouse Strait, and into the Pacific Ocean to the east through the

FISHING FOR SQUID

During summer and autumn, large fishing fleets set off into the Sea of Japan to fish for squid. Bright halogen lamps, specially water-proofed, are lowered underwater or floated on the surface to attract the squid, which feed at night. The lights are so brilliant that the fleets can be seen from space.

Tsugaru Strait. The northern part of the sea freezes in winter. Occasional upwellings of nutrient-rich water encourage strong plankton blooms.

RUGGED COASTLINE
The coasts around the Sea of Japan are mostly rocky, as here on the island of Honshu.

PACIFIC OCEAN *northwest*

Yellow Sea

LOCATION To the southwest of the Korean Peninsula, extending to the coast of east-central China

AREA 530,000 square km (205,000 square miles)

MAXIMUM DEPTH
103m (338ft)

INFLOWS Rivers Yellow, Yangtze, Liao He, Luan He, Yalu, Han

The Yellow Sea is a shallow, flat-bottomed, and partly enclosed inlet of the Pacific Ocean. Its main bulk lies to the southwest of the southern part of the Korean Peninsula (see p.132). Further north, there are two extensions

of the sea: Korea Bay in the northeast; and the Gulf of Bo Hai in the northwest, into which flows the waters of the Yellow River. The Yellow Sea derives its name from the colour of the silt-laden water that is discharged into it from the Yellow River, the Yangtze, and other major rivers. Most of the Yellow Sea floor is continental shelf that only 10,000 years ago was dry land or the estuary of the Yellow River. There are many sand shoals near the coast of China, and several small islands lie off the coast of Korea. A warm current flows from the southeast into the middle of the Yellow Sea, and there are southward-flowing currents along the coasts. Tides in the sea are considerable, especially along the west coast of the Korean Peninsula, with a maximum tidal range of 8m (27ft). The surface temperature of the sea varies remarkably, from close to freezing in the north in winter to as high as 28°C (82°F) in some of the shallower parts during the summer. The Yellow Sea is particularly rich in bottom-dwelling fish, and this is exploited by fleets of trawlers from China, Japan, and North and South Korea.

GREEN TURTLE
The Yellow Sea is the northern limit of the range of the Green Turtle, which lives near coasts and grazes on algae and sea grass.

SILT-LADEN WATERS
In this satellite image of the Yellow Sea, the Yellow River can be seen discharging its load of silt at top left, as can the Yangtze at bottom left.

OCEAN

East China Sea

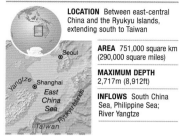

LOCATION Between east-central China and the Ryukyu Islands, extending south to Taiwan

AREA 751,000 square km (290,000 square miles)

MAXIMUM DEPTH 2,717m (8,912ft)

INFLOWS South China Sea, Philippine Sea; River Yangtze

The East China Sea is a shallow sea with an average depth of just 349m (1,145ft). It is separated from the Yellow Sea (see p.423) to the north by an imaginary line that runs from Cheju Island, off the tip of the Korean Peninsula, to the mouth of the River Yangtze in mainland China. To the south, it connects to the South China Sea (see below) through a shallow strait, the Taiwan Strait, between mainland China and the island of Taiwan. Much of the western part of the sea covers an area of continental shelf. Its deepest section, the Okinawa Trough, stretches for several hundred kilometres along the western side of the Ryukyu Islands in the east. A warm current (a branch of the Kuroshio Current) enters the East China Sea to the east of Taiwan and moves north, with branches penetrating into the Yellow Sea and into the Sea of Japan (see p.423) to the northeast. During summer, warm, moist winds from the western Pacific bring heavy rain to the East China Sea, sometimes accompanied by typhoons (hurricanes). In winter, cold, dry winds blow across the sea from the Asian mainland. Fishing is one of the major sources of income in the region, with the main catches being anchovy, tuna, shrimp, mackerel, and the eel-like Cutlass Fish. The East China Sea serves as the main shipping route from the South China Sea to ports in Japan and elsewhere in the North Pacific.

TYPHOON
Hurricanes in the western Pacific are known as typhoons. Many form to the east of the Philippines and then move northwest over the East China Sea.

MANTA RAY
Manta Rays are most often encountered near the surface of lagoons. These gigantic fish, measuring up to 7m (23ft) across, swim by flapping their large pectoral fins.

HYOPCHAE BEACH ON CHEJU
This idyllic beach is on the west coast of Cheju Island in the northern part of the East China Sea. Offshore are some of the world's most northerly coral reefs.

South China Sea

LOCATION Bounded by southern China, eastern Southeast Asia, the Philippines, and Indonesia

AREA 3.7 million square km (1.4 million square miles)

MAXIMUM DEPTH 5,016m (16,457ft)

INFLOWS Rivers Xi Jiang, Mekong, Red, Tha Chin, Chao Phraya

ANG THONG NATIONAL PARK
Lying in the Gulf of Thailand, this marine park consists of 42 forest-topped islands that were created by sea-level rise.

WHALE SHARK
The world's largest fish at over 12 tonnes, these plankton feeders are commonly sighted in some areas of the South China Sea, such as in western parts of the Philippines.

The largest marginal sea of the western Pacific, the South China Sea stretches for over 2,700km (1,700 miles) around the southeast edge of the Asian mainland. Two offshoots of the sea, the Gulf of Tongking and the Gulf of Thailand, indent Southeast Asia. The northeastern part of the sea, close to the Philippines, lies over a deep, diamond-shaped submarine basin, of which the most southerly part is called the South China Basin. North of this lies a broad area of continental shelf that extends for up to 240km (150 miles) from the coast of China and Vietnam and includes the Taiwan Strait and Gulf of Tongking. To the south, off southern Vietnam, the shelf narrows and then connects with the Sunda Shelf, one of the largest areas of continental shelf in the world, which underlies the whole of the southern part of the South China Sea. This shelf is dotted with many small islands and island groups. The weather in the South China Sea is marked by monsoons and summer typhoons. Annual rainfall is as high as 4,000mm (160in) on parts of the Sunda Shelf. Both fishing and shipping are of high economic importance. Fish caught in the South China Sea supply as much as 50 per cent of the animal protein consumed by the people of Southeast Asia. The major ports and commercial centres around the sea include Hong Kong, Manila, Hô Chi Minh City, Singapore, and Bangkok.

TEK SING PORCELAIN

A famous shipwreck in the South China Sea is that of the *Tek Sing*, an unusually large Chinese junk that ran aground on a coral reef about 500km (300 miles) southeast of Singapore in 1822, while en route to Java. The ship was carrying about 350,000 pieces of Chinese porcelain, much of which has been salvaged by divers and subsequently auctioned, following discovery of the wreck site in 1999.

PACIFIC OCEAN *west*

Sulu Sea

LOCATION Between northeastern Borneo (Malaysia) and the central islands of the Philippines

AREA 260,000 square km (100,000 square miles)

MAXIMUM DEPTH 5,600m (18,400ft)

INFLOWS Celebes Sea

The Sulu Sea is one of several warm seas around Borneo. It connects in the northwest to the South China Sea via passages on either side of the island of Palawan, part of the Philippines. In the southeast, the Sulu Sea is fringed by the Sulu Archipelago, comprising over 4,000 small coral and volcanic islands. Pearl fishing is often carried out from stilt houses built on top of the reefs.

LIMESTONE CLIFF, PALAWAN

PACIFIC OCEAN *west*

Banda Sea

LOCATION Bounded by Sulawesi to the northwest and the central and southern Maluku Islands to the east

AREA 470,000 square km (180,000 square miles)

MAXIMUM DEPTH 7,440m (24,410ft)

INFLOWS Molucca Sea, Arafura Sea

The Banda Sea is a warm sea that laps the shores of the Maluku Islands or Moluccas, formerly the Spice Islands, of Indonesia. It connects to numerous other marginal seas, including the Molucca Sea in the north, and the Arafura Sea in the southeast, via straits between various island groups. The people of these islands are still involved in the cultivation of spices such as cloves and nutmeg, as well as fruit, coconuts, and tapioca.

NEIJALAKKA ISLAND, MALUKU, INDONESIA

PACIFIC OCEAN *southwest*

Tasman Sea

LOCATION Lying between New Zealand and the southeastern coast of Australia

AREA 2.3 million square km (900,000 square miles)

MAXIMUM DEPTH 5,945m (19,500ft)

INFLOWS Southern Ocean; Coral Sea

The most southerly of the marginal seas of the Pacific, the Tasman Sea lies in the belt of westerly winds known as the roaring forties and is noted for its storminess. The sea merges with the Coral Sea (see below) in the north, and in the southwest connects to the Indian Ocean via the Bass Strait (see p.394) between Tasmania and Australia. Its southern part overlies a broad, deep basin, the Tasman Basin, while the northeastern part covers an elongated trough, the New Caledonia Depression. These two deep areas are separated by a ridge called the Lord Howe Rise. Lord Howe Island lies near the boundary of this rise and the Tasman Basin. A dominant current in the western part of the sea is the warm, southward-moving East Australia Current. During winter and spring, colder water moves up from the south. The Tasman Sea's resources include oil deposits in the Gippsland Basin at the eastern end of Bass Strait.

VOLCANIC COAST
Lord Howe Island is the eroded remnant of a 7-million-year-old volcano located in the northern part of the Tasman Sea. The world's most southerly coral reef surrounds the island.

PACIFIC OCEAN *southwest*

Coral Sea

LOCATION To the southeast of Papua New Guinea, east of northeast Australia and west of Vanuatu

AREA 4.8 million square km (1.8 million square miles)

MAXIMUM DEPTH 9,165m (30,070ft)

INFLOWS West Central Pacific Ocean; rivers Fly, Purari, Kikori

The Coral Sea is a large marginal sea that extends from northeastern Australia some 1,300km (800 miles) east to the island chain of Vanuatu (formerly the New Hebrides). A major feature is the Great Barrier Reef, the world's largest coral reef, which stretches for 2,010km (1,250 miles) along the east coast of Queensland. However, the reef occupies only a small area of the whole Coral Sea. To its east, and north of latitude 20°S, there are many individual reefs and small islands, known collectively as the Coral Sea Islands Territory, which rise from an area called the Coral Sea Plateau. To the northeast of the Coral Sea Plateau lies a large, deep basin, the Coral Sea Basin, and there are other deep basins much further east towards Vanuatu. Tectonic plate boundaries lie at the northern and eastern margins of the sea, where the Indian and Australian plates are being forced beneath adjoining plates. These boundaries are marked by deep trenches, notably the Vanuatu Trench. The main current in the Coral Sea is the South Equatorial Current, which flows from the east. As it approaches Queensland, it splits into two branches, one of which moves down the east coast of Australia as the East Australia Current while the other deflects north towards Papua New Guinea and then back east across the northern part of the Coral Sea. The sea has a subtropical climate and is subject to typhoons, especially from January to April. The Coral Sea's marine life is highly diverse and includes hundreds of species of stony (reef-building) corals, over 4,000 types of mollusc, and thousands of species of fish.

CLOSED ANEMONE
This giant anemone, with attendant clownfish, is one of several thousand species of cnidarians (corals, sea anemones, jellyfish, and hydroids) that are found in the Coral Sea.

HARDY REEF
Located about 100km (60 miles) off the coast of Queensland, near the Whitsundays, this is a typical section of the Great Barrier Reef.

FINAL JOURNEY
A Norwegian fishing boat takes aboard
a catch of herring. In the North Sea,
industrial fishing of this species has
brought about a collapse in stocks.

FISHING

Fish is the only major food that is still largely gathered from the wild. It accounts for about 10 per cent of all the protein eaten by humans, which makes it an essential resource. The world's fish stocks, however, are falling rapidly, and, despite improvements in technology, the global catch is actually falling. As fishing nations seek to manage dwindling supplies, they are frequently torn by the conflict between short-term exploitation and long-term gain.

Food from The Sea

The worldwide harvest of sea fish has expanded from about 5 million tonnes per year in 1900 to about 90 million tonnes in 2012. This huge rise has been brought about by a revolution in fishing vessels and equipment, including factory trawlers, which can catch and process several hundred tonnes of fish a day. Since the early 1990s, however, the total catch has levelled out, and even dropped, indicating that the world's fisheries cannot expand any more.

Concern about overharvesting has led many coastal nations to introduce fishing quotas, as well as strict controls over the mesh sizes of nets. Setting quotas is a difficult exercise, because fish species differ in their breeding biology, and also in the rate at which they grow. A catch that is sustainable for one species may drive another towards commercial extinction – the point where it becomes too rare to be worth catching by professional fishermen.

READY FOR MARKET
A Senegalese woman checks fish that are drying in the sunshine. In recent years, West Africa's fishermen have faced intense competition from vast industrial trawlers.

Boom and Bust

Fish populations often undergo natural oscillations, which overfishing can disrupt. With the European Herring, for example, historical records show a 50-year cycle going back several centuries. For most of this time, fishing had little effect, but, during the 1960s, the catch expanded greatly, and herring stocks went into a sudden decline. They have still not fully recovered. The fate of the Peruvian Anchoveta was even more dramatic. Found off the west coast of South America, this finger-sized species once provided 30 per cent of the world's entire fish catch. By the mid-1980s, its fishery had all but collapsed.

Although overfishing causes great ecological damage, it is not this problem but economic factors that are more likely to bring it under control. Faced with unpredictable catches and growing competition, many developing countries are turning to fish-farming as a way of meeting their food needs.

EASTERN STYLE
In Kerala, India, a fisherman adjusts his net. This kind of fishing yields a sustainable harvest, because enough adult fish survive to reproduce.

Aquaculture

Practised for centuries in Southeast Asia, fish-farming or aquaculture can be an efficient method of producing high-protein food. Unlike mammals, fish are cold-blooded, so most of their feed contributes to their growth rather than to keeping their bodies warm. China currently leads the world in this form of farming, producing more than 25 million tonnes of fish and crustaceans (such as shrimps) each year.

On a small scale, fish-farming dovetails well with other kinds of agriculture, because fish and other aquatic animals can be fed on a wide range of organic waste. However, large-scale fish-farming does have an environmental price. Organic waste escapes into the water, and so do preventive treatments such as antibiotics, which often damage habitats that are used by fish in the wild. By 2020, almost 50 per cent of the world's entire fish production, nearly 88 million tonnes, came from fish farms.

SHORT-LIVED RETURNS
The Peruvian Anchoveta fishery began in the 1950s, as a source of fish-meal and fish-oil. After the boom years of the late 1960s, the fish population crashed when overfishing reinforced the natural effects of El Niño (see p.453).

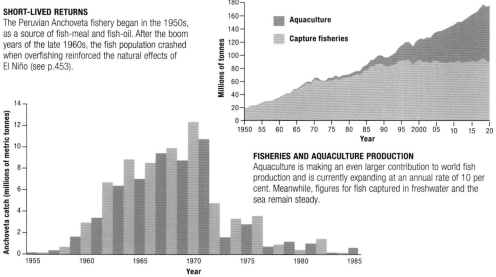

FISHERIES AND AQUACULTURE PRODUCTION
Aquaculture is making an even larger contribution to world fish production and is currently expanding at an annual rate of 10 per cent. Meanwhile, figures for fish captured in freshwater and the sea remain steady.

OCEAN

Southern Ocean

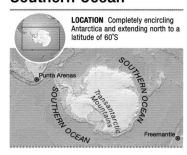

LOCATION Completely encircling Antarctica and extending north to a latitude of 60°S

AREA 20 million square km (7.8 million square miles)

MAXIMUM DEPTH 7,235m (23,735ft)

INFLOWS Summer melting of sea-ice and icebergs calved from Antarctic ice-shelves

KRILL
These shrimp-like animals are a key part of the food chain in the Southern Ocean. The biomass of Antarctic krill is thought to be larger than that of Earth's human population.

LIGHT-MANTLED SOOTY ALBATROSS
Several types of albatross inhabit the Southern Ocean. Breeding pairs of the species shown here nest on islands such as South Georgia and the Macquarie Islands.

The fourth largest of the world's five oceans, the Southern Ocean has no natural boundary – it is defined as the sea lying south of the 60°S latitude. Over most of its extent, it has a depth of 4,000–5,000m (13,000–16,500ft), its floor consisting of a series of basins separated by submarine ridges. The Antarctic continental shelf is unusually deep, as the whole of Antarctica is depressed by the weight of ice pushing down on it. The flow of currents in this ocean is complex. Close to Antarctica, highly saline cold water forms as salt is rejected from seawater freezing beneath the ice-shelves. This cold salty water sinks, then moves north at depth. Further from Antarctica, there is a northerly movement of cold but less saline surface water. This sinks beneath warmer water in an area called the Atlantic Convergence, which extends all around Antarctica at latitudes between 50°S and 60°S. To replace the various bodies of cold water

TABULAR ICEBERGS
Numerous tabular icebergs (see p.398) are present in the Southern Ocean, especially in areas close to the Antarctic landmass. After breaking off ice-shelves, they initially drift westward in Antarctica's coastal current. The largest are hundreds of square kilometres in area.

forming, sinking, and moving north, there is a continuous upwelling of water, called Circumpolar Deep Water, into intermediate areas of the Southern Ocean, bringing nutrient-rich waters to the surface from depth. The world's largest ocean current, the Antarctic Circumpolar Current, lies in the same area as the Atlantic Convergence. It is driven by westerly winds, which move water perpetually eastward. In winter, much of the Southern Ocean freezes over, the overall area of sea-ice around Antarctica growing seven- or eightfold. This ocean has the strongest average winds on Earth. The combination of high winds, large

HITCHING A LIFT
One recently identified threat to Southern Ocean ecosystems comes from the increasing amount of rubbish washing up on its shores. Marine biologists have established that by acting as rafts, plastic bottles and other floating objects are responsible for accelerating the spread of small marine organisms to new locations around the world, where they may displace native species. The Southern Ocean is particularly vulnerable because of its isolation.

waves, and ice makes ship navigation hazardous. The abundant planktonic life in the upwelling zone provides the basis for a complex food web. The most important organism in the food chain is krill, a crustacean that eats sea-ice algae and phytoplankton, and in turn provides food for fish, whales, seals, and birds. Fisheries focus on harvesting krill and bottom-dwelling fish called toothfish.

OCEAN

SOUTHERN OCEAN *south*

Weddell Sea

LOCATION To the east of the Antarctic Peninsula and west of Coats Land, Antarctica

AREA
2.8 million square km
(1.1 million square miles)

MAXIMUM DEPTH
3,000m (10,000ft)

INFLOWS Icebergs calved from the Ronne-Filchner Ice-shelf

YOUNG WEDDELL SEAL
Weddell Seals live around the pack ice of the entire Southern Ocean. They swim beneath the ice and can break through to the surface to create breathing holes.

The Weddell Sea is a large bay in the coast of Antarctica, forming part of the region of the Southern Ocean that lies south of the Atlantic. Ice and water moves around the sea in a clockwise direction. Overlying the southwestern portion of the Weddell Sea is the Ronne-Filchner Ice-shelf. A high

MIDNIGHT SUN
The southern part of the Weddell Sea has continuous daylight from November to February.

proportion of the cold, dense water that lies at the bottom of the world's oceans was originally formed beneath this ice-shelf. On its west side, along the coast of the Antarctic Peninsula, the sea overlies a 240-km- (150-mile-) wide area of continental shelf, which also extends for about 480km (300 miles) beyond the edge of the Ronne-Filchner Ice-shelf. The part of the Weddell Sea beyond the ice-shelf is more or less permanently covered in thick pack ice, which has greatly restricted exploration.

SASTRUGI
A research scientist surveys the thick pack ice at Cape Norvegia at the eastern end of the Weddell Sea, which has had its surface formed into ridges (sastrugi) by wind action.

SOUTHERN OCEAN *south*

Ross Sea

LOCATION To the east of Victoria Land and to the west of Marie Byrd Land, Antarctica

AREA 960,000 square km (370,000 square miles)

MAXIMUM DEPTH
2,500m (8,300ft)

INFLOWS Icebergs calved from the Ross Ice-shelf

ICEFISH
The ghostly-looking Icefish, in common with other Antarctic fish species, contains a substance in its tissues that prevents it from freezing.

The Ross Sea is an indentation into the coast of Antarctica, forming a part of the section of the Southern Ocean lying south of the western Pacific. Its southerly arm is overlain by the Ross Ice-shelf. The sea is generally shallow, its floor extending northward as a broad

continental shelf, before plunging into the depths of the Pacific–Antarctic Basin. Of all the seas around Antarctica, the Ross Sea is the least prone to sea-ice, being free of pack ice in summer, and so is the most accessible to shipping.

SEA FLOOR UNDER ICE
Under the sea-ice in shallow water near McMurdo Sound, the marine life is dominated by mobile organisms, such as starfish.

SOUTHERN OCEAN *north*

Scotia Sea

LOCATION Northeast of the Antarctic Peninsula and southeast of Tierra del Fuego, Argentina

AREA 900,000 square km (350,000 square miles)

MAXIMUM DEPTH
4,000m (13,000ft)

INFLOWS Southern Ocean to the west of Drake Passage

The Scotia Sea is an elongated body of water that stretches east from the 965-km- (600-mile-) wide channel, Drake Passage, that runs between the southern tip of South America (Cape Horn) and the northern tip of the Antarctic Peninsula. The Scotia Sea occupies its own small tectonic plate, sandwiched between the South American Plate to the north and

CAPE HORN
The area south of Cape Horn is notorious for its rough seas, caused partly by strong currents as water funnels through Drake Passage.

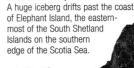

the Antarctica Plate to the south. Underwater, the sea consists of two adjoined deep basins that are almost completely surrounded by a ridge, the Scotia Ridge, from which rise various islands. These include South Georgia to the northeast, the South Sandwich Islands (a volcanic island arc) to the east, and the South Shetland Islands to the south.

LOOKOUT POINT, ELEPHANT ISLAND
A huge iceberg drifts past the coast of Elephant Island, the eastern-most of the South Shetland Islands on the southern edge of the Scotia Sea.

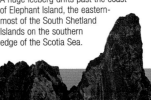

OCEAN

EROSIONAL COASTLINE
These cliffs pounded by Atlantic waves are near the village of Port, County Donegal, Ireland. They show features typical of long-term marine erosion.

COASTS

Coasts are the places where the Earth's oceans and seas meet the land. Within these regions can be found a great variety of different landforms, which are shaped by the interaction of breaking waves, the rise and fall of tides, the effect of currents, and slow changes in the levels of land and sea. Coasts can also be affected by land-based phenomena – such as glacier advance and retreat, lava flows, and the discharge of sediment from rivers – and weathering by wind, rain, or frost. Variations in these processes and factors over time and along the length of shorelines cause constant change, sometimes eroding coasts and sometimes building them up. Last but not least, the activities of people also influence coastlines – about half of the world's human population lives on or within 100km (60 miles) of a coast. These activities include the building of coast defences such as sea-walls and groynes, the reclaiming of land from the sea, and the creation of artificial islands.

Coasts and sea-level 434–35

Erosional and depositional coastlines 436–37

Tides and Waves

Tides and waves are two phenomena that are both highly noticeable in coastal areas and can have a profound effect on coastal landforms. Tides are regular rises and falls in sea-level that occur all over the oceans but are most obvious near coasts, where they can cause strong currents and are important to coastal life. Waves are wind-generated and are instrumental in erosional and depositional processes, including the formation and maintenance of beaches.

Tides

Tides are produced by gravitational interactions between the Moon, the Sun, and the Earth. The Moon's gravitational attraction is dominant, pulling the ocean water into a bulge on the side facing the Moon, while centrifugal force creates an equal bulge on the opposite side of the planet. As the Earth rotates, these bulges sweep over the oceans, causing peaks and troughs in sea-level, called high and low tide. The difference in height between high and low tide is known as the tidal range. Each ocean or sea has its own slightly different tidal pattern – two high and low tides a day (semi-diurnal), one per day (diurnal), or very little tidal change at all. In addition, there is a 28-day cycle, caused by the relative alignment of the Moon and Sun. The tidal range is most affected by the horizontal component of the gravitational force and by the shape of nearby land. In narrow estuaries, the tidal range can be very large and may cause strong surges, known as tidal bores.

INTERTIDAL ZONE
Many different types of organisms have adapted to living in the zone between high and low tides.

TIDAL BORE
Experiencing the Qianjiang Tidal Bore is an annual tradition for nearby residents, as seen here in 2011.

DAYS 0   7   14   21   28

Tidal height measured over one month at a specific location (west Atlantic)

SPRING   NEAP   SPRING   NEAP   SPRING

new moon

full moon

SPRING TIDES
Spring tides occur when the Sun, the Earth, and the Moon are aligned, at the times of new moon and full moon.

SPRING AND NEAP TIDES
Twice every 28 days (around days 0 and 14), the Sun and the Moon's gravitational pull combine to produce extra-large bulges in the Earth's oceans, leading to a high tidal range called spring tides. At other times (around days 7 and 21), the Sun partly cancels out the effects of the Moon, causing a smaller tidal range called neap tides.

last quarter moon

first quarter moon

NEAP TIDES
Neap tides occur at lunar first and last quarters, when the Sun's and Moon's pull on Earth are at right angles.

WAVE ENERGY
The intensity of wave energy release can have a profound effect on the form of a coast. These waves are breaking in a rough sea on the coast of Oregon, USA.

Waves

Ocean waves are caused by the action of wind blowing across the surface of the sea. Their size depends on the speed of the wind, the length of time the wind blows (wind duration), and the distance over which the wind blows (the fetch). A series of waves can be characterized by their wavelength (distance between wave crests), frequency (number of crests that pass each minute), and height (amplitude). In the area where the wind is blowing, the sea often has a chaotic form because the wind generates waves with many different frequencies that interfere with each other. As the waves travel away from the generation area, they become sorted by frequency into a steady pattern called a swell. Overall, waves transfer energy, not water, across the sea surface. When they reach a coast this energy is released in the form of breakers.

CHAOTIC AND REGULAR WAVES
Where they are generated, waves give the sea a chaotic form (above right). Far from the generation area, waves come onshore in a regular pattern (above left).

movement of individual particles of seawater

INTERMEDIATE WAVES   SHALLOW WATER WAVES   BREAKER ZONE   SURF ZONE   SWASH ZONE

WAVELENGTH

water movement occurs to depth of half the wavelength

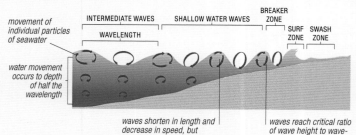

waves shorten in length and decrease in speed, but increase in height (amplitude)

waves reach critical ratio of wave height to wave-length and begin to break

MOTION IN WAVES
The particles of water in a choppy sea do not move forward with the waves. Instead, they gyrate in little circles or loops. Underwater, the particles move in smaller and smaller loops, until at a depth equal to half the distance between the wave crests, there is hardly any movement at all.

OCEAN

Beaches

A beach is a deposit of sedimentary particles, varying in size from mud to boulders, but mostly sand that occupies a coastal area between high and low tide. The particles are derived from erosion of the land, or fragmentation of marine structures such as corals or shells, brought to the beach by rivers and waves. Beach deposits never stay in the same place for long but are continually moved around – onshore, offshore, or along the shore – through the action of currents, waves, tides, and changes in land- or sea-level. The average intensity of wave action is important in determining the nature of a beach. Low-energy waves tend to build beaches, whereas high-energy waves tend to erode them. There are many zones on a beach. The nearshore zone extends from where waves break to the upper end of the swash zone (the area covered and uncovered by each wave surge). Further up most beaches is a berm, an accumulation of beach material with a flat top surface and a relatively steep seaward slope.

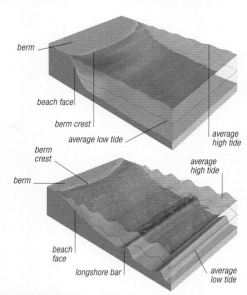

SANDY BEACH
Beaches of fine sand or coral fragments are associated with low-energy waves and a calm sea.

GRAVEL BEACH
Pebble or gravel beaches, such as this one in Dorset, England, tend to be associated with higher energy waves.

HUGE DUNES
Coastal dunes, such as these at Arcachon, France, are caused by wind blowing sand off the dry part of a beach.

STORM PROFILE
During stormy weather, higher energy waves erode the beach, flattening the beach face and carrying material offshore, where it is deposited as longshore bars. The result is a beach with a storm profile.

NORMAL PROFILE
Fair weather and a low-energy swell (regular waves) for several days results in accumulation of beach material, giving a steep beach face. This is sometimes called a swell profile.

BEACH PROTECTION

Groynes are structures built out from a beach into the sea with the object of slowing beach erosion. They are supposed to work by trapping sand particles that are being moved along the beach by wave action (longshore drift). Ultimately, groynes often prove ineffective as they starve other beach areas of sand. The beach shown below, in Ustronie Morskie on the Slovincian coast of Poland, reveals several past attempts to slow erosion.

Tsunamis

A tsunami is a powerful pulse of wave energy that can propagate for a long distance, and at high speed, through the surface of an ocean. A tsunami has a long wavelength but low amplitude, so that in the open ocean its passage is hardly noticed. However, when a tsunami reaches shallow water, its amplitude increases dramatically. Sometimes it forms a huge wave, as high as 30m (100ft), which can cause tremendous amounts of damage as it breaks on a shore. Tsunamis are usually set off by submarine earthquakes or the sudden slumping of large masses of sediment underwater, and they occur regularly in some parts of the world. In recent years, scientists have recognized the rarer phenomena of massive megatsunamis, which produce waves many times higher than those generated by normal tsunamis.

AFTERMATH
When a tsunami hits a shore, it can smash buildings and carry boats far inland. The damage shown here occurred on Okushiri Island, Japan, in 1993.

OCEAN

Coasts and Sea-Level

29–31 The ice ages

41 Isostasy

386–89 Ocean tectonics

392–93 Shelves and plains

Erosional and 436–37 depositional coastlines

Coastal pollution 438–39

Global warming 458–59

A coast is a strip of land in immediate contact with the sea, extending from the shoreline inland to the first major change in terrain. Coasts are dynamic rather than static. They quickly alter in response to natural forces such as wave action and tides, land-based processes, and sea-level change. These forces and processes constantly push and pull at coastlines, and, as a result, their shape and location changes. There are also many different types of coasts. The types and the processes that shape them are in turn determined by the various causes of sea-level change.

Sea-Level Change

Changes in sea-level are of two different types, regional and global. There are two important causes of a global change in sea-level – an increase or decrease in the extent of the world's ice-sheets and glaciers (see below) and an increase or decrease in the size of a spreading ridge. When these grow larger, sea-level drops because water becomes locked up in the ice; and when they melt, sea-level rises. Two other parameters that affect global sea-level are ocean temperature and the size of ocean basins (see right). Regional sea-level change occurs when a specific area of land rises or falls relative to the general sea-level. Two of the main causes are tectonic uplifting of land, which is common in regions where oceanic crust is being forced beneath continental crust, and glacial rebound, which is a slow rise of an area of land after an ice-sheet that once weighed it down has melted. Sea-levels have constantly fluctuated throughout the Earth's history, not least in the recent past. For example, about 15–18,000 years ago, during the last ice age, sea-levels were about 120m (400ft) lower than they are today.

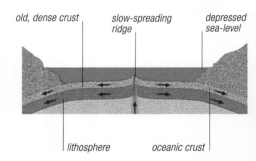

old, dense crust · slow-spreading ridge · depressed sea-level

lithosphere · oceanic crust

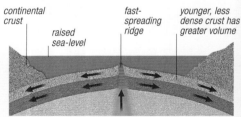

continental crust · fast-spreading ridge · younger, less dense crust has greater volume

raised sea-level

OCEAN BASIN CHANGE
A rise in sea-level can occur when new crust is produced at a fast-spreading ridge. The new crust swells because it is relatively hot and buoyant, pushing the ocean water upwards.

CHANGES IN SEA-LEVEL OVER GLACIAL CYCLES
When temperatures rise, glacial ice melts and flows into the sea and seawater expands, raising sea-levels. The rise is partially offset by depression of oceanic crust and, in some regions, by land rising following glacial rebound.

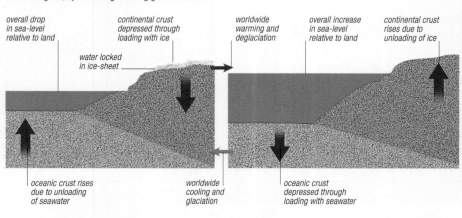

overall drop in sea-level relative to land · continental crust depressed through loading with ice · worldwide warming and deglaciation · overall increase in sea-level relative to land · continental crust rises due to unloading of ice

water locked in ice-sheet

oceanic crust rises due to unloading of seawater · worldwide cooling and glaciation · oceanic crust depressed through loading with seawater

Emergent Coasts

Emergent coastlines result from land being uplifted faster than sea-level has risen since the last ice age. The causes are either tectonic activity (for example, on the USA's Pacific Coast) or slow glacial rebound (such as in parts of Scandinavia and New England, USA). On emergent coasts, areas that were formerly sea floor may become exposed above the shoreline, and former beaches may become the tops of cliffs. Emergent coasts are typically rocky, but usually have a smooth shoreline.

RAISED BEACH
The tops of these cliffs in the Bay of Fundy, Canada, are former beaches that were raised by glacial rebound.

FIORD COAST
This coast, a type of drowned coast, is located in the Lofoten area of Norway. Fiords are narrow, steep-sided, and, unlike rias, penetrate far inland.

Types of Coasts

Coasts are either primary or secondary. Primary coasts are relatively young ones that have been created by terrestrial processes or by sea-level change. Examples of primary coasts are volcanic coasts, where lava has flowed to the sea and solidified, and deltas, where a river deposits sediment as it enters the sea. Coasts in which sea-level change has played a significant part include drowned and emergent coasts. Other types of primary coast are those dominated by glaciers or glacial moraines, and karst coasts, where a region of limestone has been eroded and later inundated by the sea. Secondary coasts are older ones that have been shaped by marine erosional and depositional processes (see pp.436–37) or the activities of marine organisms (see Reefs, pp.396–97).

DROWNED KARST
This coast of eroded limestone in Ha Long Bay, Vietnam, has been inundated by a rise in sea-level.

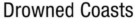

VOLCANIC COAST
This coast on the Snaefellsnes Peninsula, Iceland, was created by lava from a nearby volcano. Since the coast formed, it has started to undergo some wave erosion.

DELTA COAST
A delta, such as this one, the Wadi al-Mujib on the Dead Sea, Jordan, is the result of sediment deposited by a river.

Drowned Coasts

Drowned coasts (sometimes called submergent coasts) are formed as a result of a global or regional rise in sea-level. Several types are recognized. A ria coast is a river-eroded hilly landscape, with valleys running perpendicular to the coast, that has been partly drowned. The coast usually has deep bays or estuaries, intervening headlands, and many offshore islands. There are examples in southwest Ireland and Great Britain, northwest Spain, and the eastern USA (such as Chesapeake Bay, see p.102). A fiord coast, seen in Norway, Chile, and New Zealand's South Island, for example, develops when a glaciated mountain landscape becomes partly drowned, forming long deep bays with steep sides. A Dalmatian coast, named after the coast of Dalmatia, Croatia, is a series of valleys formed by folds running parallel to the coast that has been almost completely submerged.

COASTLINE PAST AND PRESENT
The eastern coast of North America has changed a lot in the past 15,000 years. Sea-level rise has drowned many ancient valleys.

---- **Eastern coastline of North America 15,000 years ago**

RIA COAST
Southwest Ireland, in the counties of Cork and Kerry, contains a series of elongated bays typical of a ria coast.

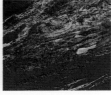

DALMATIAN COAST
This satellite photograph shows the classic drowned Dalmatian coast in the Adriatic Sea.

Erosional and Depositional Coastlines

92 Erosion by water

94 Waterborne deposition

392–93 Shelves and plains

432–33 Waves and tides

434–35 Coasts and sea-level

Erosional and depositional coastlines are two forms of secondary coast, which have been shaped primarily by marine processes. Erosional coasts occur where land is being eroded by the sea faster than sediment can be deposited. They are typically associated with emergent coasts (see p.434), where the land is being lifted by tectonic processes. Depositional coasts occur where sediment is being deposited on a coast faster than it is eroded. They are often associated with tectonically inactive continental margins, where the coast is slowly subsiding.

STACKS
Stacks – such as the Twelve Apostles off Victoria, Australia – represent an advanced stage in the erosion of a headland.

Processes of Erosion

Energy for the processes of coastal erosion is supplied mainly by waves and currents. As they beat against cliffs, waves compress air within cracks in rocks, and on re-expansion the air can shatter the rock. Mechanical erosion refers to waves hurling beach material against cliffs and abrading the rock. Where waves encounter headlands, a significant factor is refraction (bending) of the wave fronts, which causes their erosive energy to be directed towards the headlands and away from the bays. Waves arriving on a coast at an angle also generate a longshore drift (movement of sediment along a shore), which shifts material eroded from the headland towards adjacent bays. Groundwater that has emerged at the surface and surface-water run-off also gradually wear channels through the headland.

REFRACTING WAVE
The bending of the wave front towards the headland can be seen clearly here in this bay in southern Ireland.

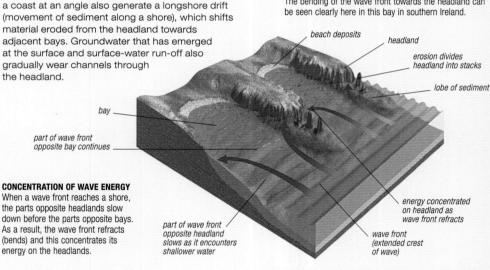

beach deposits

headland

erosion divides headland into stacks

lobe of sediment

bay

part of wave front opposite bay continues

CONCENTRATION OF WAVE ENERGY
When a wave front reaches a shore, the parts opposite headlands slow down before the parts opposite bays. As a result, the wave front refracts (bends) and this concentrates its energy on the headlands.

part of wave front opposite headland slows as it encounters shallower water

energy concentrated on headland as wave front refracts

wave front (extended crest of wave)

Erosional Landforms

At headlands subjected to wave erosion, various distinctive features develop in a classic sequence. First, sea caves form at the bases of the cliffs on the sides of the headlands, usually along faults and joints (planes of weakness in a rock, see p.122). Wave action gradually deepens and widens the caves until they penetrate right through the headland to form an arch. Next, the roof of the arch collapses to leave an isolated rock pillar called a stack. These are ultimately worn down to stumps. Along cliffs, wave action cuts into the base of the cliff. This produces a horizontal platform at sea-level called a wave-cut platform, which provides some protection from further erosion by dissipating wave energy. Tectonic uplift sometimes raises the land in short bursts, and it is then worn into a series of wave-cut platforms called marine terraces.

ARCH
This spectacular example of a wave-eroded arch is at Loch Ard Gorge (named after a ship that was wrecked there) on the coast of Victoria, Australia.

CAVES
The coast of Cornwall, England, is dotted with numerous erosional features, including these caves at Nanjizel, near Land's End.

Processes of Deposition

Like coastal erosion, coastal deposition is brought about primarily by wave action. However, whereas erosion occurs in areas of high wave energy, deposition happens in environments where most of the energy has been dissipated before it reaches the shore, allowing sand or other material in the water to settle out. The deposited sediment comes mostly from rivers, but some comes from eroded headlands or from offshore. An important mechanism of transport and deposition is that of longshore drift. When wind and waves strike a shore obliquely, the movement of surf (swash) propels water and sand up the shore at an angle, but backwash drags them back down at a right angle to the shore. The overall effect is to move water and sand gradually along the shore. Eventually, where the sediment-carrying water arrives at a lower energy environment, the sediment settles out and builds up to form features such as spits, sand bars, and barrier islands. Some depositional processes modify one type of landform into another. For example, the development of a beach may lead to a spit, which may eventually extend across the mouth of a bay to create a baymouth bar.

SPIT
This spit at Hurst Castle is one of the most prominent depositional features on the southern coast of England.

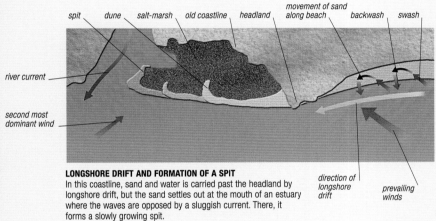

spit   dune   salt-marsh   old coastline   headland   movement of sand along beach   backwash   swash

river current

second most dominant wind

LONGSHORE DRIFT AND FORMATION OF A SPIT
In this coastline, sand and water is carried past the headland by longshore drift, but the sand settles out at the mouth of an estuary where the waves are opposed by a sluggish current. There, it forms a slowly growing spit.

direction of longshore drift   prevailing winds

Depositional Landforms

Depositional processes form a number of characteristic features around coasts. A spit is a sandy peninsula with one end attached to land. An offshore bar is an unattached sand bar (a general name for any sand deposit) that lies parallel to the coast. About 13 per cent of the world's coasts have offshore bars. Where a spit extends right across the mouth of a bay, or two spits have merged, the result is called a baymouth bar. A tombolo is a sand bar that connects an island to land or to another island. Barrier islands are sand bars that coalesce to form long, linear islands parallel to the coast. They are typical of the eastern and Gulf coasts of the USA, and often slowly migrate shorewards. Other types of depositional landform include salt-marshes, which often develop on the landward sides of spits and bars, and mudflats, which form from the deposition of fine silt in sheltered tidal water, particularly in estuaries.

TOMBOLOS
This example of a tombolo (a sand bar between two islands) is in Puget Sound, near Seattle, USA.

ALASKAN MUDFLAT
Large mudflats on the Alaskan coast are the result of glacial erosion of nearby mountains.

OCEAN

BEACHED
Cargoes that have leaked from wrecked ships are major sources of coastal pollution. Synthetic organic chemicals can have particularly damaging effects.

COASTAL POLLUTION

Biologically as well as physically, coasts are frontiers between land and open sea. They teem with life, but are also highly vulnerable to pollution from contaminated rivers as well as waste that has been discharged at sea and washed up by inshore currents. For humans, coastal pollution may be a minor nuisance only or a major economic threat, but for coastal wildlife its effects are far-reaching but often unseen.

Marine Litter

More than 8 million pieces of litter are estimated to enter the seas and oceans every day. Fifty years ago, wood was the main ingredient, but it has now been superseded by plastic. The lightness and durability of plastic enable it to reach some of the remotest coastlines on the Earth. During the 1990s, Chilean scientists identified 1,500 pieces of plastic waste on the shores of Livingston Island, in Antarctica, while in Ducie Island, in the South Pacific, a researcher found nearly 1,000 pieces of litter on a stretch of beach under 3km (2 miles) long. They included 189 buoys, 71 plastic bottles, 44 pieces of rope, and 29 lengths of plastic pipe. Ducie Island is uninhabited, and the nearest continental coast is more than 4,500km (2,800 miles) away.

Plastic is chemically inert, but it causes physical problems by not rotting. Marine animals, such as dolphins, seals, and turtles, become entangled in plastic debris, while seabirds often mistake plastic fragments for food. To counter this threat, an international convention forbids the disposal of plastics at sea – but, as yet, there is no equivalent agreement about litter that comes from land.

LIFE SENTENCE
This young Herring Gull has plastic tangled in its beak. Gulls are naturally curious, and their feeding habits put them at risk from waste.

Oil Spillage

In public imagination, few sights evoke the idea of pollution more vividly than coastal wildlife covered with oil. Birds are foremost among the victims, but casualties include all kinds of inshore animals, from seals to fish and molluscs. Oil clogs up fur, feathers, and gills, and it is toxic if it is swallowed. When the densest part of crude oil coalesces and sinks, it can remain in the sea-bed sediment for years. In a study off the coast of Massachusetts, USA, in 2000, oil contamination was found from a tanker accident that had occurred in 1969.

Large-scale oil spills generate intense media coverage and calls for emergency action. However,

CLEANING UP
Off the coast of South Africa, rescue workers treat penguins that are covered in oil. Rapid action is needed to treat oil-damaged birds.

SOURCES OF OIL POLLUTION AT SEA
Shipping accounts for nearly half of the oil waste found at sea, while roughly a third is from urban and industrial waste. However, more than 8 per cent of sea pollution comes from natural sources, such as oil seepages on the sea bed.

■ Shipping
■ Waste and run-off
■ Atmospheric pollution
■ Natural sources
■ Offshore oil production

scientists are divided on the best way to respond. Although detergents break up visible slicks, the dispersed oil can then enter marine food chains, where it may cause hidden environmental damage. As with all forms of pollution, prevention is better than cure, and here the record is more promising. Double-hulled tankers, together with better navigation systems, have helped cut major oil spills by four-fifths in the last 30 years.

Surprisingly, accidental spills actually account for less than 20 per cent of oil that finds its way into the seas. Coastal areas are also polluted by land-based discharges, and by the oil from small craft through to large ships. Badly maintained engines in pleasure boats probably create more oil pollution than all the world's oil spills combined. But, because this pollution comes from many small sources, it is largely invisible and thus very hard to control.

Sewage and Industrial Waste

Coastal waters are common dumping grounds for urban and industrial waste. Sewage is sometimes screened (in a process known as primary treatment), but in most coastal communities it is discharged without any treatment at all. In time, sewage decomposes, but industrial pollutants are often more persistent. Heavy metals – such as cadmium and lead – may remain in sea-bed sediment for a century or more.

OUT OF SIGHT
When waste is discharged on the shore, it often returns with each tide. Long underwater outfall pipes aid successful offshore dispersal of such waste.

Until relatively recently, coastal pollution was a local problem. Estuaries were often contaminated, but nature's recycling systems meant that there was little human impact farther out to sea. Today, the sheer volume of waste means that this is no longer the case. In New York and New Jersey in the USA, for example, nearly 10 million tonnes of sewage are dumped each year, creating a submarine desert that stretches many kilometres from the shore. However, because it affects the sea bed, and not the shore itself, this form of pollution often goes unnoticed.

ATMOSPHERE

DARK SKIES OVER THE DESERT
Some climates are strongly influenced by
physical features. The hot, dry climate of
Death Valley, USA, is due to its position in
the rain-shadow of the Sierra Nevada.

CLIMATE

The weather changes from one day to the next and is evident as variations in temperature, precipitation, wind, and clouds. Examine the weather over many years, however, and a pattern emerges: winters are cold, with snow and ice; summers are warm; and rainfall in June is about the same as in November. This pattern of weather, repeated over many years, constitutes the climate of a particular region. Obviously, different parts of the world have different climates. In equatorial regions, the weather is always warm and usually wet. Deserts are dry, and polar regions are cold. The different climate regions result partly from the Sun, which shines more strongly at the tropics than anywhere else, and partly from the way the atmosphere, and oceans, transfer the Sun's warmth away from the equator. If the Earth had no atmosphere or oceans, it could have no climates. The way in which the air and water produce climates is a complicated but fascinating story.

Structure of the Atmosphere

◀ 18 Oxygenating the atmosphere

Energy in the atmosphere **446–47** ▶

Atmospheric circulation **450–51** ▶

Global warming **458–59** ▶

Weather **450–51** ▶

The atmosphere forms distinct layers around the Earth. These layers are remarkably uniform in chemical composition, but their density decreases with altitude. Within each layer there is a constant temperature change with height that alters abruptly at its boundaries. The lowest layer, called the troposphere, is where life exists and weather occurs because of the heating effect of the Sun. The Sun's rays pass through the atmosphere and warm the Earth's surface, causing the air to move and water to evaporate or condense. This causes weather and creates the different climate regions. Harmful ultraviolet rays are filtered out by the ozone layer, a gas that forms a thin layer in the stratosphere.

Layers

The lowest layer in the atmosphere is the troposphere, where air moves vertically and horizontally, and is thoroughly mixed. Air temperature decreases with height. As warm air rises, it loses some of its energy and becomes cooler. Eventually, it reaches a level at which it cools no further. The air above it is no denser, and so it rises no further. This level marks a boundary, called the tropopause. It is the first of several such boundaries that divide our atmosphere into its distinct layers. It separates the troposphere from the stratosphere, which itself ends at the stratopause. Above lies the mesosphere, and the thermosphere. At the top of the thermosphere, the temperature rises to about 1,000°C (1,800°F) due to the absorption of ultraviolet radiation. The upper limit of the thermosphere, the thermopause, extends up to 1,000km (625 miles). This outermost layer gradually merges with the vacuum of the Sun's atmosphere (which we call space), so Earth's atmosphere has no distinct upper boundary.

AURORA
The Northern and Southern Lights, also known as the Aurora Borealis and Australis, appear in the thermosphere.

RADIATION FOG
The troposphere is warmed by the Sun during the day but at night it cools again by radiating heat. The cool early morning air may contain fog.

LIVING AT HIGH ALTITUDE

Air pressure decreases with altitude, and above about 4,000m (13,200ft) there is insufficient oxygen for people to sustain strenuous activity – unless they are acclimatized. These tea pickers in Darjeeling, India, have greater lung capacity and breathe more deeply than lowlanders. Their blood has a greater affinity for oxygen and can absorb it faster. These unique adaptations are genetic and are therefore passed on from one generation to the next. It allows the tea pickers to work on the upper slopes and carry heavy loads all day without becoming unduly breathless.

THERMOSPHERE
This layer extends up to the thermopause, as high as 1,000km (625 miles). The temperature in the lower part of this layer remains constant with height, but increases rapidly above 88km (55 miles).

MESOSPHERE
The temperature of the mesosphere remains constant with height through the lower mesosphere, but above about 56km (35 miles) it decreases with height to about -80°C (-112°F) at the mesopause.

STRATOSPHERE
Temperature in the stratosphere remains stable up to about 20km (12 miles), then increases, due to absorption of ultraviolet radiation. This layer's upper boundary, called the stratopause, is at about 48km (28 miles).

TROPOSPHERE
The air temperature falls as one climbs through the troposphere, at a rate of about 6.5°C (43.7°F) per 1,000m (3,280ft). At high altitudes, the temperature drops to about -30°C (-22°F) at the poles and -65°C (-85°F) at the Equator.

Height above sea-level

130km / 80 miles

120km / 74 miles

110km / 68 miles

100km / 62 miles

90km / 56 miles

80km / 50 miles

70km / 42 miles

60km / 36 miles

50km / 30 miles

40km / 24 miles

30km / 18 miles

20km / 12 miles

10km / 6 miles

Sea-level

LAYERS OF THE ATMOSPHERE
The atmosphere consists of layers, principally the troposphere, stratosphere, mesosphere, and thermosphere. Each is defined by the way temperature changes within its limits. The layers are separated by clear boundaries.

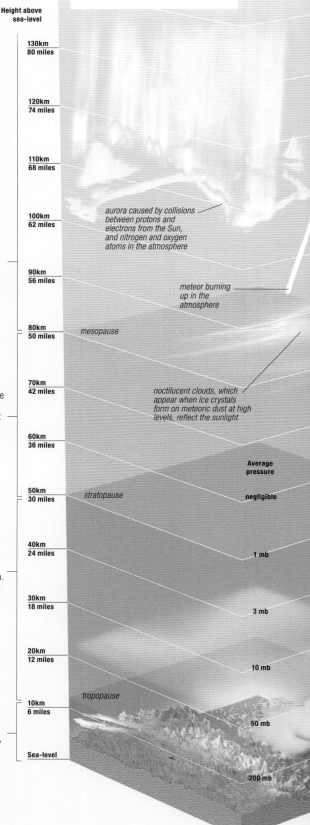

aurora caused by collisions between protons and electrons from the Sun, and nitrogen and oxygen atoms in the atmosphere

meteor burning up in the atmosphere

mesopause

noctilucent clouds, which appear when ice crystals form on meteoric dust at high levels, reflect the sunlight

Average pressure

stratopause — negligible

1 mb

3 mb

10 mb

tropopause

50 mb

200 mb

1000 mb

above this level the air is so thin that each molecule may have a different temperature

Average temperature

60°C
140°F

-10°C
14°F

-80°C
-112°F

-90°C
-130°F

-80°C
-112 °F

-50°C
-58°F

-30°C
-22°F

-10°C
-14°F

-20°C
-4°F

-40°C
-40 °F

thin layer of ozone gas absorbs harmful radiation from the Sun

-60°C
-80°F

-60°C
-80°F

all weather occurs in the lowest layer of the atmosphere

15°C
59°F

SAMPLING THE AIR

Balloons are used to sample the more inaccessible parts of the atmosphere. The Jimsphere seen here is an inflatable weather balloon, which is used to determine wind speed and direction prior to the launch of space vehicles, and missiles. Its shiny surface enhances radar signals, and the conical "spikes" on its surface help to stabilize its flight. Other types of weather balloons are released daily from weather stations all over the world.

Composition

Air consists mainly of three gases: nitrogen, oxygen, and argon, in constant proportions. It also contains water vapour, but in variable amounts. In addition to these principal constituents, there are minute quantities of 10 other gases. These are carbon dioxide, neon, helium, methane, krypton, hydrogen, nitrous oxide, carbon monoxide, xenon, and ozone. These form about 0.04 per cent of the atmosphere by volume. Air also contains even smaller amounts of gases that come from the surface, such as ammonia, nitrogen dioxide, hydrogen sulphide, and sulphur dioxide. Finally, there are dust and smoke particles.

NITROGEN
78.08%

OTHER GASES: ARGON 0.93%
CARBON DIOXIDE 0.037%
NEON 0.018%
HELIUM 0.005%

WATER VAPOUR VARIES FROM 0–4%

OXYGEN
20.95%

ATMOSPHERIC GASES
The major constituents of air are nitrogen, oxygen, and argon, which total 99.96 per cent by volume. The composition is the same in the troposphere and stratosphere.

Airborne Water

Water evaporates from the surfaces of oceans, lakes, rivers, and from wet ground. It also evaporates from cloud droplets and raindrops, many of which fail to reach the ground. Water vapour is a gas, and it is present in all air. The amount of water vapour varies widely. Desert air often contains almost no water vapour, but even the moist air of the humid tropics seldom contains more than about four per cent of water vapour by volume. The humidity of the air is a measure of the amount of water vapour it contains. However, the most often quoted measure is relative humidity, which is the amount of water vapour present expressed as a percentage of the amount of water vapour needed to saturate the air. This varies according to the amount of water vapour and the temperature of the air. Warm air can hold more water vapour than cold air.

NOCTILUCENT CLOUDS
This type of cloud is sometimes seen on summer nights in high latitudes. Unlike other clouds, it forms in the mesopause, and shines by reflecting light from the Sun.

RAINBOW
This rainbow in Idaho, USA, is produced when light is refracted, on entering raindrops, then reflected from their rear, and refracted again as it leaves.

THE SNOWLINE
At altitude, low temperatures stop snow from melting, even in summer. The lower limit of the permanent snow, seen here on Cotopaxi in Ecuador, is called the snowline.

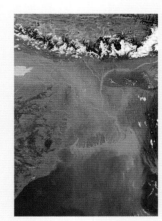

POLLUTION
Clouds of aerosol particles are seen here in the skies along the southern edge of the Himalayas in northern India, streaming south over Bangladesh. This type of industrial and urban pollution now affects much of southern Asia.

Atmospheric Particles

Air contains solid particles, called aerosols, that are so small they remain temporarily suspended. They fall so slowly that upcurrents repeatedly carry them aloft. Nevertheless, individual particles do not remain airborne for very long. Large particles return to the ground within minutes, and small particles, such as smoke, within a few hours. Some fall to the ground and some collide with surfaces as they are carried horizontally by the wind, but most are washed to the ground by rain and snow. Many particles are lifted into the air as a result of natural events. Volcanic eruptions eject vast quantities of ash and dust, while desert winds carry sand and dust over great distances – for example, Saharan dust occasionally falls in North America. Fires, ignited by lightning or humans, release smoke and ash that can be carried upwards by rising hot air. Even pollen and spores released from plants, fungi, and bacteria become airborne, along with the aerosols produced by various agricultural and industrial practices. Water vapour in the atmosphere condenses onto certain minute particles, including dust, smoke, and sulphate. Some aerosols reflect sunlight back towards space, partially offsetting the effects of global warming. These cloud condensation nuclei, as they are known, allow clouds to form.

ETNA ERUPTION
Mount Etna, Sicily, erupted in 2002. This satellite image shows a cloud of ash travelling southwards across the Mediterranean Sea.

VOLCANIC ASH
This is a magnified image of a particle of volcanic ash. Many tonnes of ash are produced during a volcanic eruption.

ATMOSPHERE

Energy in the Atmosphere

Atmospheric circulation 450–53

Climate change 454–57

Global warming 458–59

Climate regions 460–63

Weather 464–85

The Sun supplies the energy to produce our weather. The Sun's radiant heat is absorbed by the land and sea surface, and air is warmed from below by contact with it. Air movements transport this heat throughout the troposphere. Warmth from the Sun also provides energy for water to evaporate. Water vapour enters the air, and when it condenses to form clouds the water molecules release the latent heat that was absorbed when the water first evaporated. A huge amount of energy is involved in this process: an average summer thunderstorm releases as much energy as burning 7,000 tonnes of coal in less than one hour; and a tornado releases enough energy to light the streets of New York City for one night.

Solar Budget

The balance between the amount of solar energy received by the Earth's surface, its atmosphere, or the clouds, and the amount of energy reflected or radiated back into space, is known as the solar budget. The temperature at the visible surface of the Sun is about 5,800°C (10,472°F). It radiates its heat into space, but in all directions. The Earth, which is about 150 million km (93 million miles) away, receives only a tiny proportion of it. The amount of solar energy arriving as shortwave radiation at the topmost level of the atmosphere is known as the solar constant. Of the solar energy that reaches the top of the atmosphere, about 30 per cent is lost due to backscatter by air molecules, and reflection by clouds, land, and sea. Reflection will vary both with the amount of cloud cover and the nature of the Earth's surface (see opposite). About 70 per cent of all incoming radiation is absorbed, mostly by the land and sea, but also by the air and clouds. The surface radiates the energy it has absorbed as longwave radiation, maintaining the balance between heat gains and losses. About 45 per cent of the radiation reaching the Earth is visible as light.

SOLAR POWER

The Sun's energy can be used to heat homes and generate power. Solar collectors on the roof can heat water and solar cells can convert sunlight into electrical power. A solar furnace, seen here, uses hundreds of mirrors to focus light and heat onto a target in a central tower. It heats to more than 2,000°C (3,600°F), raising steam for power generation or for research.

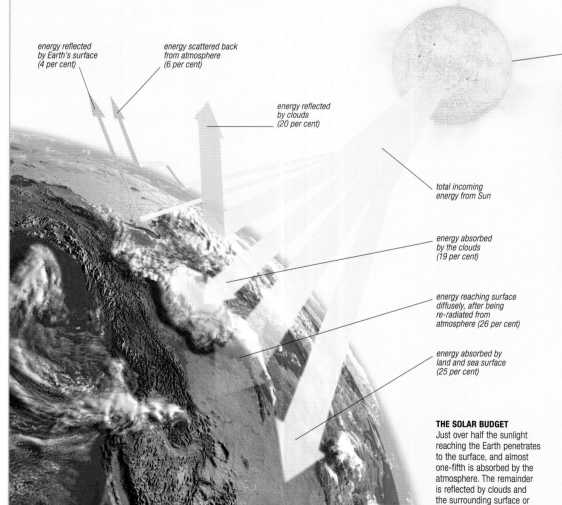

energy reflected by Earth's surface (4 per cent)

energy scattered back from atmosphere (6 per cent)

energy reflected by clouds (20 per cent)

Sun sends out shortwave radiation in all directions

total incoming energy from Sun

energy absorbed by the clouds (19 per cent)

energy reaching surface diffusely, after being re-radiated from atmosphere (26 per cent)

energy absorbed by land and sea surface (25 per cent)

SOLAR ACTIVITY

The solar flare, shown above, is produced by a sudden release of the Sun's energy. It creates a huge amount of radiated heat, which has a direct effect on our climate.

THE SOLAR BUDGET
Just over half the sunlight reaching the Earth penetrates to the surface, and almost one-fifth is absorbed by the atmosphere. The remainder is reflected by clouds and the surrounding surface or scattered back into space by air molecules.

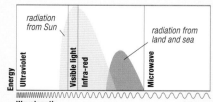

radiation from Sun

radiation from land and sea

Energy

Ultraviolet | Visible light | Infra-red | Microwave

Wavelength

WAVELENGTHS

Incoming solar radiation is emitted at shorter wavelength than that which is radiated back into space. Ultraviolet wavelengths are absorbed by oxygen and ozone, and the longer infra-red wavelengths are absorbed by carbon dioxide and water vapour.

Greenhouse Effect

Radiation from the Sun is most intense at short wavelengths. The atmosphere is almost completely transparent to this radiation, which passes through the air and is absorbed by the Earth's surface. When the land and sea are warm, they radiate the energy they have absorbed, but at longer wavelengths. Some of the longwave radiation is absorbed and re-radiated back to the Earth by atmospheric gases, including water vapour, carbon dioxide, methane, and ozone. This warms the gases, and although the Earth's radiation eventually escapes into space, it is retained long enough to warm the atmosphere. This is the greenhouse effect, and the gases causing it are called greenhouse gases. Without the natural greenhouse effect, temperatures would be 30°–40°C (54–72°F) cooler. However, since 1900, the concentration of carbon dioxide in the atmosphere has increased by about 30 per cent due to an increase in the burning of fossil fuels, atmospheric pollution, and deforestation (trees absorb carbon dioxide). Most scientists agree that the addition of carbon dioxide and other greenhouse gases to the air is changing the temperature balance, producing an enhanced greenhouse effect and increasing average temperatures over the Earth as a whole.

SUNFLOWERS
The Sun's energy is important to most living things. Green plants, such as these sunflowers, need sunlight to photosynthesize food. The flowers turn to follow the path of the Sun.

GLOBAL WARMING
The lower atmosphere gradually becomes warmer when longwave solar radiation emitted by the Earth's surface is trapped in the atmosphere by carbon dioxide, water vapour, methane, and other greenhouse gases.

some heat emitted by greenhouse gases escapes to space

some heat emitted by greenhouse gases heats surface

greenhouse gases absorb longwave radiation

escaping longwave radiation

diffused incoming radiation

solar radiation deflected back into space

incoming solar radiation

ATMOSPHERE

Absorption Variables

Different surfaces reflect different proportions of the radiation falling on them. This is because pale colours reflect light and heat, and dark colours absorb them. The percentage of the radiation that a surface reflects is known as the albedo of that surface. Freshly fallen snow is highly reflective. It has an albedo of 75–95 per cent, usually written as 0.75–0.95. Bright cloud is also very reflective. Its albedo is 0.70–0.90. So, surfaces with a high albedo absorb very little heat, while dark surfaces absorb solar radiation much better. A broad-leaved forest, with an albedo of 0.10–0.20, absorbs 80–90 per cent of the radiation falling on it. A coniferous forest absorbs even more. If the surface of a large area is changed, the albedo may be altered. Removing forest and replacing it with farm crops, for example, increases the albedo from about 0.10 to about 0.20, and so more sunlight is reflected. When grassland is replaced with concrete, the albedo increases from about 0.15 to 0.22. Changes of this kind can affect the local climate, because increased reflection reduces convection in the air, and this in turn affects the amount of cloud cover.

DARK CANOPY
This forest canopy in the Amazon Basin is dark and will absorb sunlight. This affects the air temperature in the forest and the rate at which water evaporates.

ICE AND SNOW
Most of Antarctica is covered by ice and snow. This reflects almost all of the sunlight, so the layers beneath the surface remain cold.

Changes in State

Water exists in three different forms (or phases). It occurs as liquid water, as solid ice, and as water vapour, which is an invisible gas. Most substances can exist in these three phases, but water is remarkable in that it can exist in all three states at temperatures commonly experienced at the Earth's surface. For example, in winter, the surface of a pond may partly freeze over. Here water is present as ice (solid) and water (liquid), and in the air above the pond, there is water vapour (gas). All three are present in the same place, at the same time. When ice melts or changes directly into water vapour, and when liquid water evaporates, energy is needed to break the bonds holding the water molecules together. The molecules absorb this energy, but it does not change the temperature of the ice or water. The energy is latent (or hidden) heat. The same amount of heat is released when water vapour condenses, or changes directly to ice, and when liquid water freezes into ice. The release of latent heat when water vapour condenses provides the energy that fuels the growth of huge storm clouds. These clouds usually have great vertical depth due to unstable atmospheric conditions. This instability occurs because latent heat is released from water vapour as it condenses from the air, making the rising air warmer than the surrounding air, so it continues to rise.

WATERSPOUT
This waterspout, off the Bahamas, is visible because low pressure inside the wind funnel causes water vapour to condense into droplets.

HEAVY RAINSTORM
Warm air rises and cools over the hills around Lake Shastina, California, USA, causing water vapour to condense and resulting in a rainstorm.

TROUBLED SKIES
High-altitude polar clouds – seen here over Lapland – indicate unusually low temperatures in the stratosphere. Ozone thins rapidly in these conditions.

THE OZONE LAYER

Ozone is a natural component of the atmosphere, and a gas that plays a crucial part in the survival of life on Earth (see p.445). In the stratosphere, the ozone layer screens out short-wave ultraviolet (UV) radiation – a form of solar energy that can damage or kill living cells. During the 20th century, the ozone layer was seriously thinned by atmospheric pollutants, but, following concerted international action in the late 1980s, the damage is gradually being reversed.

Ozone and Life

Toxic and highly reactive, ozone is a relatively rare form of oxygen that has three atoms in each of its molecules (O_3), rather than the normal two (O_2), which many organisms breathe. Most of the world's ozone is found in the ozone layer, which is 20–25km (12–16 miles) above the Earth's surface. Here, ozone is continually formed from O_2, and it reverts back into O_2 when it breaks down again.

In the ozone layer, ozone molecules absorb UV-B and UV-C radiation – short-wavelength forms of ultraviolet that carry high levels of energy. They re-emit this energy as heat. As a result, the ozone layer acts as a screen, preventing most of this radiation from reaching the ground. Any depletion of the ozone layer is potentially dangerous for life, because UV-B and UV-C radiation can disrupt organic molecules, producing cancerous changes in living cells.

Ozone Depletion

The ozone layer has been stable around the Earth for millions of years, but that situation ended in the late 1920s, when chemists synthesized the first chlorofluorocarbons (CFCs). These non-flammable gases turned out to have many uses: as solvents, aerosol propellants, and industrial cleaners, and also as coolants in refrigerators and air-conditioners. Unfortunately, CFCs can destroy ozone if they escape into the atmosphere. They can continue to do damage for decades or even centuries after they have been released, because they are stable and do not dissolve in rain.

During the early 1970s, two American chemists – Mario Molina and Sherwood Rowland – identified the theoretical threat of ozone depletion. Initially, reaction was sceptical, but, when in 1983 scientists at the British Antarctic Survey discovered an ozone "hole" over the South Pole, the abstract threat suddenly became real.

SOUTHERN EXPOSURE
The Antarctic ozone hole forms during the southern winter and reaches a maximum each spring. In this composite satellite map, the zone of greatest deficiency is shown in red.

International Action

The reality of ozone depletion prompted a rapid response. In 1987, two dozen industrialized countries signed the Montreal Protocol, which committed them to phase out CFC production by 1996. In the period leading up to this deadline, "ozone-friendly" products were developed to take their place. In 1990, a further agreement was drawn

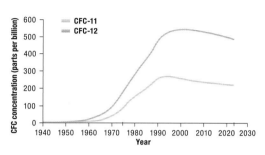

ATMOSPHERIC CFC CONCENTRATIONS
Since the 1950s, levels of CFC-11 and CFC-12 have risen steeply before flattening out in response to a production ban. The concentrations are highest in the northern hemisphere, where most of the world's production was based.

up to include most developing nations, with a ten-year extension to allow them to adopt ozone-friendly technology. As a result of these measures, atmospheric CFC levels have reached a plateau and may already be starting to fall.

How long the ozone layer takes to repair itself will depend on the residence time of different CFCs – the period that an average molecule stays in the atmosphere – and also on the amounts originally released. For example, CFC-11, which was widely used in refrigeration, aerosols, and plastic foams, has a residence time of 45 years, while that of CFC-115 is 500 years. Current data suggest that the ozone layer might be near normal by 2050, though scientists remain wary of newly synthesized substances that may also have ozone-depleting effects.

TASK FORCE
To prevent CFCs escaping into the atmosphere, old refrigerators and freezers here in southern England have their coolant removed before being scrapped.

Ground-Level Ozone

Ozone can be dangerous when it is near the ground. Such ozone is created by lightning, by any kind of machinery that creates electric sparks, and also by internal combustion engines. The exhaust gases of internal combustion engines include nitrogen oxides and organic compounds, and these can form ozone when they react – particularly in the presence of sunlight. In traffic-dense urban areas, such photochemical reactions can raise the air's ozone content to ten times its normal background level.

Ground-level ozone is toxic to plants, and it irritates the lining of the lungs, causing asthma and bronchitis. This form of pollution can be reduced by improvements in engine design, and by the use of catalytic converters in car exhausts.

ATMOSPHERE

ATMOSPHERE

Atmospheric Circulation

394–95 Circulation and currents

444–45 Structure of the atmosphere

446–47 Energy in the atmosphere

Air is constantly moving. Worldwide, this movement constitutes the general circulation of the atmosphere, transporting warmth from equatorial areas to high latitudes, and returning cooler air to the tropics. It comprises three sets of "cells". These cells produce wind systems called prevailing winds, which drive the surface waters of the ocean, creating currents. The winds are deflected in opposite directions north and south of the Equator by the Coriolis effect. In the upper troposphere, fast-moving jet streams form due to differences in temperature and pressure at the boundaries of air masses. They can increase the intensity and movement of low-pressure systems, resulting in climate cycles, or oscillations. El Niño is one such oscillation that affects weather patterns on a global scale. The behaviour of the atmosphere is unpredictable, so it is difficult to know when these climatic events will take place.

Coriolis Effect

Air flowing towards, or away, from the Equator invariably follows a curved path that swings it to the right in the northern hemisphere and to the left in the southern hemisphere. The reason for this was discovered in 1835 by Gustave-Gaspard de Coriolis (see panel, below). It used to be called the Coriolis force and is still abbreviated as CorF, but no force is involved, and nowadays it is known as the Coriolis effect. It occurs because the Earth is rotating anticlockwise about its axis, so as air moves across the surface, the surface itself is also moving beneath it but at a different speed. The magnitude of the Coriolis effect depends on the latitude, and the speed of the moving air. The Coriolis effect is greatest at the poles.

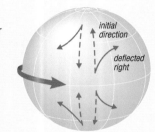

HOW CORIOLIS EFFECT WORKS
All points on the Earth's surface complete one revolution every 24 hours, so a point at the Equator travels further and faster than a point at higher latitude. This rotation causes the air to shift to the right north of the equator and to the left south of the equator.

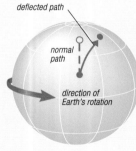

DEFLECTING AIR
Air moving over the Earth's surface travels at a different speed to the Earth's rotation, so when plotted on a map, it appears to be deflected.

STORM BUILDUP
These storm clouds, forming off the southeast coast of the USA, are in the northern hemisphere, and so rotate anticlockwise.

GUSTAVE-GASPARD DE CORIOLIS

The mathematician and mechanical engineer Gustave-Gaspard de Coriolis (1792–1843) was born in Paris, France. In 1816, he became a tutor at the École Polytechnique, where he carried out research on friction and hydraulics. He went on to discover the Coriolis effect, publishing his results in 1835. He died in Paris aged only 51.

Fluid Circulation

The air in the Earth's atmosphere moves in response to heating by the Sun. When a fluid is heated, its molecules absorb energy. This allows them to move faster, so they move further apart, and the fluid expands. Expansion makes it less dense, because a given volume contains fewer molecules than before. Denser fluid, which is heavier, then sinks beneath the less dense fluid, pushing it upwards. Air, a fluid, gets heated by contact with the Earth's surface that has been warmed by the Sun. As warm air rises, air pressure decreases, as there is a smaller amount of air above it, pressing down. This allows the air to expand, but this uses energy, so it also cools. When the air reaches a level where its density equals that of the air immediately above it, it stops rising. If the air then subsides, the pressure on it increases. It absorbs energy and its temperature rises. These changes give rise to the different cells of the Earth's circulation system.

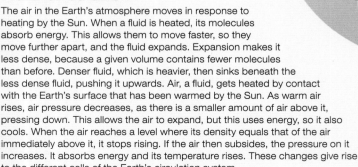

polar easterlies blowing away from high-pressure system over the North Pole

inner core of jet stream flows much faster than outer shell

polar-front jet stream

subtropical jet stream flows at about 30°N all year round

direction of Earth's rotation

cloud indicating position of intertropical convergence zone

cool air subsides at the South Pole, flowing towards mid-latitudes as the polar easterlies

HOW AIR MOVES IN CELLS

Atmospheric circulation in each hemisphere consists of three cells. At the Equator, warm air rises, moves north and south, and subsides at the tropics. The air that flows back to the Equator creates the Hadley cells. Cold air subsides over the poles, flows to mid-latitudes, where it meets air from Hadley cells, so rises, and returns to the poles, forming polar cells. Rising mid-latitude air divides, flowing to the poles and to the Equator, forming Ferrel cells.

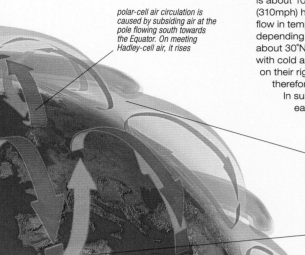

polar-cell air circulation is caused by subsiding air at the pole flowing south towards the Equator. On meeting Hadley-cell air, it rises

air in Ferrel cells rises to the tropopause and then divides, some air flowing to the pole and some towards the Equator

winds in mid-latitudes produced by Ferrel cells flow from the west

northeasterly trade winds deflected to right by Coriolis effect

air within Hadley cells rises moist at the Equator and subsides dry at the subtropics

southeasterly trade winds

roaring forties

INTERTROPICAL CONVERGENCE ZONE

In equatorial regions, the meeting of trade winds at the Equator is known as the intertropical convergence zone. Here, a band of cloud indicates its position.

Jets and Waves

Jet streams are narrow, winding ribbons of strong wind in the upper troposphere. They occur at the top of fronts, marking the boundary between air masses at different temperatures (see pp.466–67). As they are produced by a temperature difference, they are also known as thermal winds. Sometimes waves form in jet streams, and travel along with them. Wind speed at the centre of a jet stream is about 105kph (65mph), but speeds of 500kph (310mph) have been recorded. Polar-front jet streams flow in temperate latitudes, between 30 and 50°N and S, depending on the season. The subtropical jet stream is at about 30°N and S all year round. These jet streams blow with cold air on their left in the northern hemisphere, and on their right in the southern hemisphere. They blow therefore from west to east in both hemispheres. In summer, there is a jet stream that blows from east to west at about 20°N, crossing Asia, southern Arabia, and northeastern Africa.

STORM DAMAGE

During the night of 16 October 1987, there was a severe storm in the UK, caused by an unusual northwards shift of the jet stream.

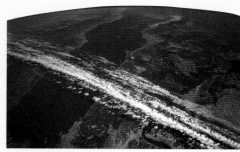

JET STREAM

These bands of cirrus cloud mark the position of the jet stream over the Red Sea and Egypt. The cloud forms in air that is lifted as it is drawn into the core of the jet stream.

Prevailing Winds

Although the wind may blow from any direction, when directions at a particular place are compared over a long period, it is usually found that the wind blows from one direction more than it blows from any other. This is the prevailing wind. The general circulation of the atmosphere produces belts of prevailing winds around the world. In the tropics, wind blowing towards the equator, as part of the Hadley cells, is deflected to the right, and forms the northeasterly and southeasterly trade winds. Air flowing away from the high pressure over the poles is deflected to the right and forms the polar easterlies. Between these, in middle latitudes, the Ferrel cells produce westerly winds. Over the world as a whole, the force of the easterly winds is precisely balanced by the force of the westerlies. If this were not so, the rotation of the Earth would either speed up or slow down. In sandy deserts, and in polar regions, prevailing winds produce dunes, and drifts, with characteristic shapes. Sand and snow grains are blown up a slope, and tumble down the far side. This creates high dunes and snow drifts, with sharp, sinuous crests, extending for great distances. Knowing about prevailing winds is important for locating structures such as airport runways. Sailors used to depend on prevailing winds to cross oceans and would avoid places, such as the doldrums, where there is little wind.

HIGH DUNES

The sinuous crest of these Namibian Desert dunes shows that the prevailing wind flows along the dune, sometimes a little to the left or to the right.

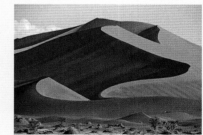

WIND SYSTEMS

Global wind belts result from the way air circulates. Seasonal differences in the northern hemisphere are due to large landmasses where pressure systems change from high in January to low in July.

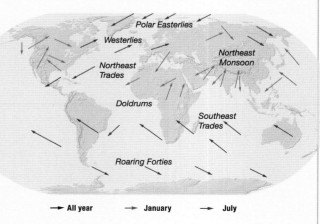

Polar Easterlies

Westerlies

Northeast Trades

Northeast Monsoon

Doldrums

Southeast Trades

Roaring Forties

→ **All year**   → **January**   → **July**

Moisture and the Oceans

Oceans cover about 70 per cent of the surface of the Earth, and they play a major part in the formation and regulation of climates. Because water warms and cools much more slowly than dry land, the oceans moderate temperatures, making summers cooler and winters warmer than they would be on a completely dry planet. Ocean currents (see p.394) also transport heat from the Equator into high latitudes. The North Pole has a much warmer climate than the South Pole because it lies at the centre of the Arctic Ocean, rather than the centre of the continent of Antarctica, and there is always liquid water beneath the sea-ice. Precipitation – rain, snow, hail, fog, dew, and frost – consists of water that came originally from the ocean, and that returns there through rivers and groundwater. This is an obvious contribution to climate, but it is not the only one. The condensation of water vapour in rising air releases latent heat (see p.447) that makes the air continue to rise. This is how storm clouds form. The energy of tornadoes and hurricanes is provided mainly by condensation. Clouds also shade the surface by reflecting incoming sunlight, which has a cooling effect. The climatic importance of the oceans is most evident when rains fail and food crops wither in a drought.

GULF STREAM
This satellite image of the east coast of the USA shows temperature differences in the ocean due to warm (orange and yellow) and cool (blue and green) currents. The warm current is the Gulf Stream, which transfers heat polewards.

ICEBERGS
These icebergs near Cuverville Island, Antarctica, provide information about ocean currents as they move north into warmer waters and then melt.

WATERSPOUT
When a tornado moves over water, it is called a waterspout. This one occurred off the coast of Spain and killed six people. The spray ring at the base consists of water whipped up from the surface.

Winds and Currents

Prevailing winds, blowing over the ocean, drive the surface water along particular paths to form ocean currents. The trade winds drive equatorial currents westwards in both hemispheres. As they approach land, they turn away from the Equator into higher latitudes. Here the Coriolis effect swings the currents further, until they enter the middle latitudes, where prevailing winds come from the west. They then form currents flowing eastwards, but the Coriolis effect continues to deflect them. Finally, they head back toward the Equator. This circular movement forms gyres, one in each major ocean, apart from the Southern Ocean. On the western sides of oceans, currents are narrow, deep, fast-flowing, and carry warm water; on the eastern side, they are wide, shallow, slow-moving, and carry cool water. In the Southern Ocean, there is no land to deflect the Antarctic Circumpolar Current. It is the only current to flow all the way around the world. Currents do not flow precisely in the direction of the wind. This is because friction between the surface layer of water and the layer below, combined with the Coriolis effect, deflect the current at about 45° to the right of the wind direction in the northern hemisphere, and to its left in the southern hemisphere.

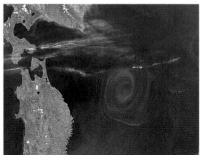

EDDY OFF JAPAN
The winds around pressure systems drive eddies (small currents), such as this one, visible as a green spiral below the white streaks of cloud.

TYPHOON
Typhoon Odessa (left) formed in the tropical trade-wind belt of the western Pacific. It moved northwards until it met the mid-latitude westerlies.

Variation and Oscillation

Climates change due to cyclical variations in the distribution of pressure over varying periods. An example of a climate cycle is the North Atlantic Oscillation (NAO), which results from pressure differences between the Azores high and Iceland low. When pressure is higher than average over the Azores, and lower over Iceland, the NAO index is said to be positive (high). At this time the jet stream flows strongly over the Atlantic, bringing mild, wet winters to Europe and warm, dry winters to the Mediterranean. When the pressure difference is small, the NAO index is said to be negative (low). This blocks the jet stream, so in winter cold air enters Europe from Asia, and rainfall increases in the Mediterranean region. The NAO is unpredictable, but tends to occur every couple of years.

HARSH WINTER IN CANADA
Tuktoyaktuk, northwest Canada, shown here experiences harsher than normal winters due to climatic oscillations.

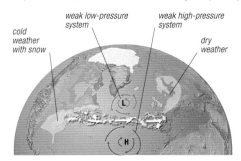

cold weather with snow
weak low-pressure system
weak high-pressure system
dry weather

NORTH ATLANTIC OSCILLATION (NEGATIVE)
A negative (low) North Atlantic Oscillation exists when pressure systems are weak, bringing mild winters to the northwestern Atlantic, cold weather with snow to northeastern North America, cold, dry winters to northern Europe, and wet weather to the Mediterranean.

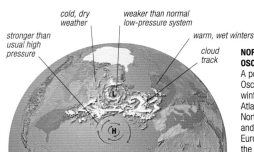

cold, dry weather
stronger than usual high pressure
weaker than normal low-pressure system
warm, wet winters
cloud track

NORTH ATLANTIC OSCILLATION (POSITIVE)
A positive (high) North Atlantic Oscillation index brings cold winters to the northwestern Atlantic and northeastern North America, mild winters and higher rainfall to northern Europe, and dry weather to the Mediterranean region.

El Niño

El Niño is a reversal in the normal flow of the South Equatorial Current in the Pacific that brings dramatic changes in the weather at intervals of between two and seven years. The effects of El Niño become evident in late December, and are associated with a change in the distribution of air pressure over the South Pacific. This change is known as the Southern Oscillation. The full cycle is called an El Niño–Southern Oscillation (ENSO) event and includes La Niña, the opposite of El Niño. Ordinarily, pressure is high over the eastern South Pacific and low in the west. This produces the trade winds that drive the South Equatorial Current, carrying warm water away from South America and toward Indonesia.

This pattern produces heavy rain over Indonesia and extremely arid conditions along the coast of Peru and northern Chile. During an ENSO event, the pressure difference weakens or reverses, as do the trade winds. Heavy rain falls over Peru and Chile, causing deserts to bloom, while Indonesia experiences drought, sometimes leading to serious forest fires.

FLOODING
Heavy rains caused by El Niño cause extensive damage. Here, a mudslide in California, USA, has flowed down a slope and covered a section of road, blocking it to traffic. Other risks include floods and landslides.

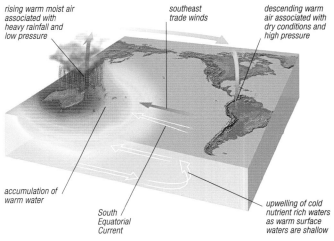

rising warm moist air associated with heavy rainfall and low pressure
southeast trade winds
descending warm air associated with dry conditions and high pressure
accumulation of warm water
South Equatorial Current
upwelling of cold nutrient rich waters as warm surface waters are shallow

NORMAL CLIMATIC CONDITIONS
Ordinarily, a low-pressure system over Australia draws the southeast trade winds across the eastern Pacific from a high-pressure system over South America. These winds drive the warm South Equatorial Current towards the coast of Australia. Off the coast of South America, the warm water layer is only 100m (300ft) deep, so upwellings of cold water bring nutrients to the surface.

EL NINO HUMAN IMPACT

El Niño brings drought to Indonesia and northeastern South America. Soils dry out and crack. This farmer must hope there is enough irrigation water to keep his crop alive, and that the rains will soon return. Indonesian farmers commonly clear the ground by burning off old, dead vegetation in preparation for planting. In El Niño years, the fires may blaze out of control, as they did during the unusually severe El Niño of 1997–98. When subsistence crops, such as rice and corn, fail, the government has to purchase the shortfall of food from elsewhere. In 1997, when over 105 million people in Indonesia were still dependent on agriculture to make a living, over five billion tonnes of rice were imported. The cost puts considerable additional strain on the country's economy.

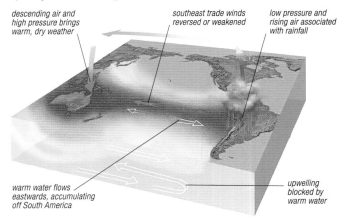

descending air and high pressure brings warm, dry weather
southeast trade winds reversed or weakened
low pressure and rising air associated with rainfall
warm water flows eastwards, accumulating off South America
upwelling blocked by warm water

EL NIÑO EFFECT
During El Niño, the pressure systems that normally develop over Australia and South America are much weaker or reversed, which is reflected by the flow of the trade winds and ocean currents. Warm water to a depth of about 152m (500ft) flows eastward, blocking the normal upwelling of nutrients along the west coast of the Americas, and devastating fish stocks.

ATMOSPHERE

Climate Change

16–31 The Earth's past

100 Meteorite impacts

252–53 Glaciers

446–47 Energy in the atmosphere

Global warming 458–59

The world's climates are changing constantly. In the past they have been both warmer and cooler than they are now, due to catastrophic events and cycles that have natural causes. At present, the average global temperature is increasing, but the rise is not spread evenly around the Earth and some regions are becoming cooler. Several factors are contributing to the present warming. Most climatologists agree that part of it is due to an enhanced greenhouse effect, which is caused by the release of certain gases when fossil fuels – coal, oil, and natural gas – are burned.

Past Evidence

Reliable written climate records are available for about the last 200 years, but until recently there were few weather stations in much of Africa, Asia, the Middle East, and South America, so scientists must reconstruct past climates from other types of evidence. Tree rings, for example, vary in width, depending on growing conditions, and details of climate over more than 8,000 years have been obtained from bristlecone pines in California, USA. Many beetles live only within a narrow temperature range and their wing cases, which are sometimes preserved in the soil, can provide evidence about the climate in which the insects lived. Plant pollen also preserves well, and can reveal the plant communities that inhabited a particular area and, thus, the climate at the time. Scientists also study cores drilled from sea-bed and lake-bed sediments, and from polar ice-sheets, in search of clues to water temperatures, rainfall, and the composition of the gases in the atmosphere.

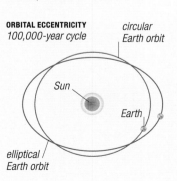

UNICELLULAR FORAMINIFERAN
This is a magnified image of the shell of a temperature-sensitive animal found in sedimentary rocks.

POLLEN GRAIN
Scientists use pollen to reconstruct past climates. It is extracted from cores drilled in sediment or peat.

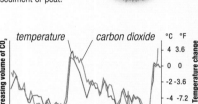

CARBON DIOXIDE VARIATION
From this graph, using data from Vostok Station in Antarctica, it can be seen that the amount of carbon dioxide in the atmosphere and temperature have varied together over the past two ice age cycles.

DESERTIFICATION
Deserts increase in area when climates become hotter and drier, as here in the Sahel, where the desert has encroached on a village.

Orbital Cycles

Fluctuations in the Earth's orbit around the Sun and rotation on its axis over time are reflected by cyclical changes in the climate. When these fluctuations coincide, temperatures fall sufficiently to trigger an ice age. The Earth's orbital path extends from almost circular to slightly elliptical over a cycle of about 100,000 years, altering the distance between the Earth and the Sun. The tilt of the Earth's axis varies over 42,000 years, altering the area directly exposed to the Sun. The rotational axis of the Earth wobbles over 25,800 years, causing the dates of the solstices and equinoxes to move.

ORBITAL VARIATIONS
The orbit and rotation of the Earth is not constant, but changes cyclically over time. Its orbit fluctuates from elliptical to circular (when the orbit is described as more or less eccentric), and it is also subject to tilt about its axis, and wobble through movement of the axis. Over time, these changes in the Earth's orbit affect temperature and can be correlated to specific events, such as ice ages.

MILUTIN MILANKOVICH

Serbian mathematician and climatologist Milutin Milankovich (1879–1958) was determined to test a suggested link between climate and the levels of solar radiation reaching the Earth. He identified three cyclical changes (see left) that might be relevant, and calculated the dates when these combined to minimize and maximize solar radiation over hundreds of thousands of years. The dates coincided with the ice ages.

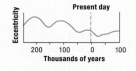

ORBITAL ECCENTRICITY
100,000-year cycle

circular Earth orbit

Sun

Earth

elliptical Earth orbit

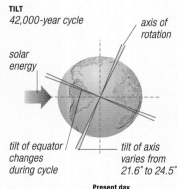

TILT
42,000-year cycle

axis of rotation

solar energy

tilt of equator changes during cycle

tilt of axis varies from 21.6° to 24.5°

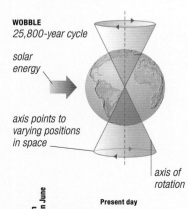

WOBBLE
25,800-year cycle

solar energy

axis points to varying positions in space

axis of rotation

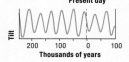

Eccentricity
Present day
200 100 0 100
Thousands of years

Tilt
Present day
200 100 0 100
Thousands of years

Earth–Sun distance in June
Present day
200 100 0 100
Thousands of years

Other Causes

The present climatic warming may be due, in part, to changes other than higher concentrations of greenhouse gases. It may be a natural trend as the Earth emerges from the Little Ice Age, a cold period that lasted from the 15th to the mid-19th century. Warming may also be caused by natural oscillations. The North Atlantic Oscillation (see p.453) has been mostly positive since 1990, bringing mild winters to Europe and possibly increasing the flow of warmer water into the Arctic Basin, melting the sea-ice in places. One of the hottest years on record (2016) may partly have been due to a strong El Niño event. Warming may not be a predominant, or even the main, cause of the shrinkage of glaciers in tropical Africa – factors such as reduced snowfall may be equally or more important. Changes in solar output may also affect climate. Throughout history, periods when few sunspots were recorded have coincided with cold conditions. A period of low sunspot activity called the Maunder Minimum was the coldest part of the Little Ice Age. An intensification of the solar wind, which happened during the 1990s, may go some way to explaining the warming that occurred in that decade – a stronger solar wind reduces cloudiness, allowing more radiation to enter and be absorbed by the atmosphere (see p.447). However, most climate scientists have not been convinced by these arguments. The overall consensus is that global warming and associated climate change are at least partly and in all likelihood predominantly the result of human activity and not just part of some natural cycle.

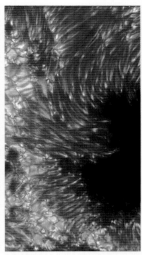

SOLAR ACTIVITY
Sunspots reflect variations in solar activity. They have two distinct regions: a dark centre, called the umbra, and a brighter, hotter surrounding area – the penumbra.

FOREST FIRES
Scientific evidence suggests that there is a link between climate and fire. Burning vegetation releases carbon dioxide into the atmosphere, and the smoke screens out solar radiation. Here, flames and smoke engulf Northfork in Wyoming, USA.

VOLCANIC ERUPTIONS
Clouds of fine dust and ash particles rise from Cerro Negro, a volcano in Nicaragua, Central America. Most of the particles will fall to the ground in a matter of hours, but any that enter the stratosphere will remain airborne much longer, shading the Earth's surface and slightly reducing the temperature.

The Current Trend

At present, average global temperature is rising. The temperature is recorded in three ways. Surface weather stations record temperatures at least twice each day – there are about 10,000 stations on land and 7,000 at sea. Weather balloons monitor air temperature at different altitudes. Finally, satellites monitor atmospheric temperature above about 2km (6,500ft). The most reliable data indicate that average global temperature rose by about 0.8°C (1.44°F) between the early 20th century and 2016. Since 1975, the increase has been at a higher rate of about 1.5°C (2.7°F) per century. When temperature data obtained with modern instruments is compared with paleoclimatic data (from past evidence such as tree rings), it is clear that not only has there been a sharp recent temperature rise, but also the Earth is warmer today than at any time in the past 1,500 years.

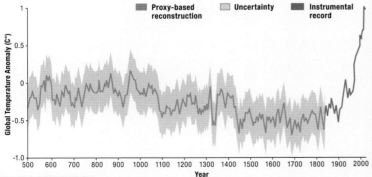

TEMPERATURE CHANGES
The graph shows estimated variations in global temperature going back 1,500 years using palaeoclimatic data up to 1850 and data from modern instruments thereafter. The shaded areas denote uncertainty in the historic data.

Changes in the Arctic

Since the 1970s, when it first became possible to study the phenomenon accurately using satellites, a highly significant decline has occurred in sea-ice coverage in the Arctic Ocean. This shrinkage has been faster than that predicted by climate models. The sea-ice coverage reaches a minimum in September each year, with an all-time record low on September 16, 2012. Other notable lows where the sea ice coverage plumbed were in 2007, 2016, and 2022. The loss of sea-ice provides evidence of global warming, separate from other ice-related evidence such as the retreat of glaciers (see p.270). One of the possible adverse effects of the loss of Arctic sea-ice may be increased warming due to reduced albedo (see p.447). However, it is likely that average plankton mass will increase in the Arctic Ocean, which could lock up more carbon.

METHANE RELEASE
A likely effect of higher temperatures in the Arctic is increased release of the greenhouse gas methane from thawing permafrost. Here, escaping methane has just been ignited.

SEA-ICE SHRINKAGE
Images of Arctic sea-ice coverage in 1980 (left) and 2012 (below), as measured by satellites, show the marked shrinkage in "core" ice coverage (the bright white areas) remaining each September.

Rising Sea-level

Since the late 19th century, most of the world's glaciers have retreated, and their melt water has contributed to a rise in global sea-level. In a warming planet, sea-level has also increased due to thermal expansion, because, above 4°C (39°F), warm water is less dense than cold water. Over the past 100 years, global sea level has risen at an average rate of about 1.5mm (1/16in) a year, but since 1993 the rise has been faster, at a rate of about 3mm (1/8in) per year. If melting were to start breaking up the West Antarctic Ice-sheet (see pp.272), the rise could accelerate considerably. Projections of future sea-level rise vary widely, because some conditions may accelerate the change, while others could dampen it. Climate models based on the current rate of increase in greenhouse gases indicate that global sea-level may rise by 20–50cm (8–20in) above its 2013 level by 2100.

ISLANDS AT RISK
Many small coral islands, such as this one in the Maldives, have a maximum height of less than 1m (3ft), making them highly vulnerable to sea-level rise.

RISING TREND
This graph traces changes in sea-level. It combines instrumental data from the past 160 years with a range of estimates for earlier times and the future.

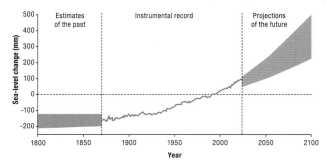

- Range of uncertainty in historical data
- Range of future projections

Extreme Weather

A predicted feature of climate change resulting from global warming includes more frequent occurrence of extreme weather events such as heat waves, droughts, and episodes of heavy rainfall causing major floods. Over the past few decades, this prediction has turned into reality. One study compared the incidence of extreme weather events from 1951 to 1980 and from 1981 to 2011. The researchers found that only about 0.2 per cent of the world's land area was hit by extreme weather in the former years, but about 10 per cent was affected in the latter period. Some examples of the types of events that most scientists think have been made more likely by global warming include the extended heat wave that hit Europe in 2003; Cyclone Nargis, which devastated Myanmar in 2008; the 2010 floods in Pakistan and Australia; and the 2012–2013 North American Drought, which was one of the costliest natural disasters in US history. Although no specific weather event can be said with complete certainty to be a direct effect of climate change, the trend of more frequent extreme weather events can be. The trend is so pronounced that it is no longer realistic to ascribe the majority of extreme weather events to chance or to natural climate cycles, such as El Niño and La Niña.

FLOODING
People use a boat to transport a motorcycle down a flooded street as Typhoon Ketsana passes through the central Vietnamese city of Hue in 2009.

DROUGHT
River levels dropped to an all-time low during a drought that affected the Amazon basin in 2010. Tree deaths caused the rainforest to release an estimated 8 billion tons of carbon dioxide.

STORMS
Hurricane Isaac, part of one of the most destructive Atlantic hurricane seasons of all time, batters Gulfport, Mississippi, USA, in August 2012.

LOSS OF HABITAT
The Arctic Ocean is now expected to be almost entirely ice-free for part of the year by about 2035. As sea-ice is used by polar bears when they hunt, its decline threatens their survival.

Changing Biomes and Habitats

Over time, climate change is likely to alter the distribution of natural biomes (see p.114) and habitats. As parts of the world become warmer, woodland will encroach into grassland, for example, while coniferous forests will spread into the tundra zone. In some areas, the movement may be to higher altitudes rather than to higher latitudes. The distributions of both plant and animal species will gradually alter. A study of some 2,000 species published in 2011 found that this has already been happening, with species migrating northwards at an average rate of 17.6km (11 miles) per decade and to higher elevations at an average of 12.2m (40ft) per decade. The extent of some biomes may become diminished rather than altered by side-effects of climate change, for example, mangrove swamps and other types of coastal wetland may shrink as sea-levels rise. For some plants and animals, particularly those that occupy expanding biomes, the changes will open up new opportunities, but for others survival will become more difficult. Where a habitat is gradually being lost, or where there is no higher ground to move to, or where changes are taking place too quickly for a species to adjust, losses and even extinctions are likely to occur.

CETTI'S WARBLER
The distribution of this small songbird, which first bred in the UK only in 1961, has been moving steadily north through England and Wales in recent years, probably as a result of milder winters and perhaps more subtle climatic shifts.

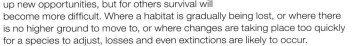

SHRINKING SWAMPS
Mangrove swamps, such as the Sundarbans in Bangladesh and northeastern India, are shrinking worldwide. This threatens the survival of animal inhabitants, such as tigers in the Sundarbans.

GLOBAL WARMING

Since the early 1800s, carbon dioxide levels in the atmosphere have risen by nearly 50 per cent, chiefly through the increased use of fossil fuels. This has accentuated the greenhouse effect (see p.447), making the Earth warm up, and during the coming century global warming is likely to accelerate still more. For the world's human population, this represents a huge and momentous change.

Changing Climates

Scientists generally agree that global warming is a phenomenon caused by human activity rather than the result of a natural climate cycle. Forecasting the effect of global warming is extremely difficult, because many variables are involved. Earth's dynamic systems, such as its atmosphere and oceans, also mean that warming will not be felt equally all over the world. According to current computer forecasts, areas at high latitudes will experience the most rapid warming over the next 50 years. This phenomenon can already be seen in the Antarctic Peninsula, where average summer temperatures have risen by about 3°C (5.4°F) since the 1970s. Paradoxically, warmer parts of the world may also experience some abnormally cold periods as weather patterns shift.

Agriculture and Health

Global warming and accompanying climate change can affect agriculture in a number of ways, such as through the direct influence of higher temperatures on crop yields, changes in the amount and pattern of rainfall, and the impact of a rising sea-level on land availability. Effects may include the expansion of subtropical deserts and reductions in crop yields in tropical regions, though agricultural yields are predicted to rise in many temperate regions. Another area likely to be affected is human health. Projected impacts range from excess deaths due to heat waves to alterations in the distribution of some infectious diseases. Climatic conditions strongly affect diseases transmitted by snails and insects, altering their geographic range. For example, the area of China where the snail-borne disease schistosomiasis occurs is expected to enlarge, and changes are also predicted in the distribution of mosquito-borne diseases such as malaria and dengue fever.

DISEASE VECTOR
Mosquitoes of the genus *Aedes* transmit dengue fever and several other infectious diseases. By altering their distribution, global warming is expected to expose several hundred million additional people to dengue.

Coastal Populations

The sea-level rise expected to occur this century as a result of global warming (see p.456) greatly increases the risk of flooding of coastal areas, and so is likely to affect the approximately 23 per cent of the world's population who live on or near a coast. Many of these people live on river deltas, which are particularly vulnerable to flooding. A serious consequence of flooding from the sea will be the contamination of both surface and underground freshwater supplies, worsening existing shortages. In some areas, extensive areas

CITY AT RISK
Severe flooding hits Dhaka, the capital of Bangladesh. It is feared that projected sea-level rise in the 21st century will displace millions of the low-lying city's population.

of farmland, especially rice paddies, will be lost to the sea. Some wealthier countries will be able to protect their coastal populations by building higher sea defences, but eventually this approach is likely to become impractical.

Tackling Global Warming

Possible responses to global warming include attempts to control greenhouse gas emissions and geoengineering solutions. Examples of emissions control include obtaining more energy from non-polluting, renewable resources (such as solar and wind power) and individuals reducing their energy use. Geoengineering solutions, which at present are mostly only in the research stage, fall into two main categories. First are solutions that concentrate on capturing carbon dioxide from the atmosphere, whether by artificial or natural means, and turning it into a form that can be permanently stored away. A second set of suggested approaches attempt to reduce solar heating of the atmosphere. These range from the relatively simple, such as putting solar reflective surfaces on the tops of buildings, to much more expensive and speculative ones, such as installing orbiting mirrors in space to reflect solar radiation.

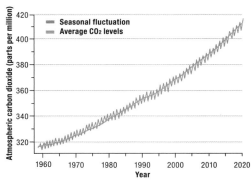

THE KEELING CURVE
Named after the American scientist who first plotted it, this graph shows the rise in atmospheric carbon dioxide – thought to be a major cause of global warming – since 1958.

ATMOSPHERE

Climate Regions

276–77 Deserts

298–17 Forests

318–19 Wetlands

330–37 Grasslands

338–39 Tundra

Weather 464–85

Scientists divide the world into regions according to their climates. The first system, devised by Aristotle, based the regions on the height of the Sun above the horizon. This method produced three belts, the torrid, temperate, and frigid zones, separated by the tropics of Cancer and Capricorn and the Arctic and Antarctic circles. Modern climate maps show many more regions, the distribution of which is determined only partly by latitude. The elevation above sea-level and distance from the ocean are just as important. Continental interiors have climates very different from those of ocean islands in the same latitude, and mountains have colder and wetter climates than those of the adjacent lowland plains.

World Climates

One of the most significant differences between climate regions is the amount of solar radiation they receive. Polar regions receive the least energy, so Arctic and Antarctic climates are characterized by extreme cold. With increasing distance from the poles, these conditions give way to temperate climates, with warm summers, cool or cold winters, and moderate precipitation. Exceptions are continental interiors, where distance from the ocean produces cool deserts. Subtropical deserts, such as the Sahara, are arid. Equatorial climates are hot and, as oceans cover most of the equatorial belt, they are wet. The map below (and the profiles of climate types on the pages that follow) are based on the classification devised by Wladimir Köppen.

- Hot climates with year-round rain
- Hot climates with monsoon rain
- Hot climates with seasonal rain
- Hot, dry climates
- Cool, temperate maritime climates
- Warm, temperate climates
- Cool, temperate continental climates
- Cold climates
- Mountain climates

Classification

Climate classifications arrange climates according to their main characteristics, and give each climate a short, simple name to identify it. Modern classifications are of two types: generic and genetic. Generic classifications are based mainly on aridity and temperature, and relate these to plant growth. The most widely used generic classifications were devised by the German Wladimir Köppen and the American Charles Thornthwaite. There are fewer genetic classifications. These are based on the physical causes of particular types of climate and are related to the general circulation of the atmosphere and its implications. The German Hermann Flohn and the American Arthur Strahler have proposed generic classifications.

VENICE
Winters in Venice are mild, even though the city is slightly further north than New York (see left). The Alps protect it from the cold, continental air masses of Europe.

NEW YORK
In winter, the climate of New York is influenced by cold, continental air masses and so has low temperatures and snow.

CLIMATIC BOUNDARY
The Sneffels Mountains in Colorado, USA, are visibly colder than the San Juan Forest in the foreground.

CLIMATE PROFILES

The pages that follow contain profiles of the world's main climate regions. Each profile begins with the following summary information:

ANNUAL REGIME FOR Name of example locality

TOTAL PRECIPITATION Total annual rainfall

TEMPERATURE OF WARMEST MONTH Average figure for the warmest month of the year

TEMPERATURE OF COLDEST MONTH Average figure for the coldest month of the year

TROPICS

Hot climates with year-round rain

DISTRIBUTION In equatorial regions, comprising the northern part of South America, southern Central America, and central Africa

ANNUAL REGIME FOR Belém, Brazil

TOTAL PRECIPITATION 2,439mm (96in)

TEMPERATURE OF WARMEST MONTH 32°C (90°F)

TEMPERATURE OF COLDEST MONTH 30°C (86°F)

KENYAN GIANT FIG
Fig trees, such as this one in Kakamega Forest, western Kenya, are typical plants of tropical rainforests. Some species of fig are parasites that use other trees for support and then strangle them.

HOT AND WET
Mist shrouds the lush and varied vegetation in this tropical rainforest in Limoncocha, Ecuador.

FOREST ANTELOPE
The Bongo is the largest, and most distinct, of the forest antelopes. It lives in the lowland, montane, and highland forests of tropical Africa.

This type of climate is hot – the temperature at sea-level rarely falls below 18°C (64.4°F) – and wet, with abundant rainfall distributed evenly through the year. The midday Sun is always overhead somewhere within the tropics, as it moves between the tropics of Cancer (June) and Capricorn (December) during its annual cycle. As a result, seasonal differences are small and there is little variation in day length. The climate supports luxuriant, evergreen rainforest in the lowlands, on mountains, giving way, with increasing height, to other types of forest culminating in cloud forest. It is estimated that half of the world's animal species live in these regions.

TROPICS

Hot climates with monsoon rain

DISTRIBUTION In Southeast Asia, Korea, southern Japan, and the northern tip of Australia; also in localized areas of west Africa, and northeastern South America

ANNUAL REGIME FOR Darjeeling, India

TOTAL PRECIPITATION 3,037mm (120in)

TEMPERATURE OF WARMEST MONTH 19°C (66°F)

TEMPERATURE OF COLDEST MONTH 8°C (46°F)

Monsoon climates have two seasons. In the Indian subcontinent, where the monsoon is most extreme (see p.469), the winter season is very dry. The prevailing winds blow outward, from land to sea, carrying dry air from a large area of high pressure over Central Asia. In spring, the humidity begins to rise, but there is no rain. The wind direction reverses as a low-pressure system develops over Tibet and the intertropical convergence zone shifts northwards. In mid-June, the weather changes rapidly. There are torrential rains which, by the end of July, have reached all of southern Asia, extending as far as Japan.

CAUGHT IN THE MONSOON RAINS
Here in Shimla, as in the rest of India, people eagerly await the onset of the monsoon rains to relieve the hot, humid conditions.

TROPICS

Hot climates with seasonal rain

DISTRIBUTION In the Sahel region of Africa, much of east Africa, coastal parts of southern Central America, and parts of South America

ANNUAL REGIME FOR Dodoma, Tanzania

TOTAL PRECIPITATION 570mm (22in)

TEMPERATURE OF WARMEST MONTH 31°C (88°F)

TEMPERATURE OF COLDEST MONTH 26°C (79°F)

Many parts of the tropics experience constant high temperatures, but have highly seasonal rainfall, with over 95 per cent of their annual total falling during the wet season. Regions with this type of climate, such as east Africa, lie between the intertropical convergence zone and the subtropical deserts. The intertropical convergence zone moves with the seasons. As it approaches in summer, it brings heavy rain and there are many violent thunderstorms. As the intertropical convergence zone retreats in winter, the dry season commences. It affects plants and animals in much the same way as the cold winters of higher latitudes. When there is no water, plants become dormant and large animals migrate. The wet season causes rapid regrowth of the plants, and the wildlife returns.

SERENGETI PLAINS
Animals graze near some acacias during the wet season in the Serengeti National Park, Tanzania.

ATMOSPHERE

MID-LATITUDES

Hot, dry climates

DISTRIBUTION In mid-latitude deserts and adjacent semi-arid regions in the subtropics of the northern and southern hemispheres

ANNUAL REGIME FOR I-n-Salah, Algeria

TOTAL PRECIPITATION 17mm (¾ in)

TEMPERATURE OF WARMEST MONTH 45°C (113°F)

TEMPERATURE OF COLDEST MONTH 21°C (70°F)

The Sahara is an example of a mid-latitude desert, with a hot, dry climate. It has areas that receive no rain for several years but then experience a violent, short-lived storm. This is because hot air rises over the equator and sinks over the subtropics, creating stable areas of high pressure and very dry conditions. In contrast, semi-arid regions, such as the Sahel, usually have a short rainy season. It occurs in summer when the intertropical convergence zone, during its seasonal migration, is closest, bringing the desert margins under the influence of the equatorial rain belt.

PINNACLES DESERT
Clouds over the semi-arid Pinnacles Desert, Nambung National Park, Western Australia, herald the onset of the brief rainy season.

MID-LATITUDES

Cool, temperate maritime climates

DISTRIBUTION In the mid-latitudes of North America, southern South America, the western margin of Europe, Tasmania, and New Zealand

ANNUAL REGIME FOR Cork, Ireland

TOTAL PRECIPITATION 1,049mm (41in)

TEMPERATURE OF WARMEST MONTH 20°C (68°F)

TEMPERATURE OF COLDEST MONTH 9°C (48°F)

Weather systems in the middle latitudes most often travel from west to east, driven by the prevailing winds and the overlying jet stream. Air reaching the western coasts of the continents has crossed the ocean, where it acquired large quantities of moisture. When the air rises to cross the coast, much of this moisture falls as precipitation. Frontal depressions develop between tropical and polar air, and they also cross the ocean, delivering more precipitation. This regime produces mild climates, with winter temperatures falling only slightly below freezing and summer temperatures that rise above 10°C (50°F), usually with only brief spells of hotter weather. These climates are moist, with rain distributed through the year, though often with slightly more rain in winter than in summer. Cool maritime climates provide good conditions for farming. The wetter western areas favour pasture and are used for dairy and beef production, with sheep on the higher ground. Fruit and vegetables are also grown. Further east, it is drier, and the climate is suitable for growing cereals.

COUNTY KERRY, IRELAND
Prevailing winds from the Atlantic Ocean rise to cross the rocky cliffs on the northern coast of the Dingle Peninsula, resulting in precipitation.

MID-LATITUDES

Warm, temperate climates

DISTRIBUTION In southeastern USA and California, parts of South America and Australia, countries around the Mediterranean, and Cape Province, South Africa

ANNUAL REGIME FOR Rome, Italy

TOTAL PRECIPITATION 730mm (29in)

TEMPERATURE OF WARMEST MONTH 30°C (86°F)

TEMPERATURE OF COLDEST MONTH 11°C (52°F)

BEE-EATER
This medium-sized Mediterranean bird feeds on insects caught in flight, including bees, hence its common name.

Regions with mild, wet winters and warm or hot, dry summers have a warm, temperate, or Mediterranean climate. Los Angeles, for example, has daytime temperatures, ranging from 18°C (64°F) in January to 28°C (82°F) in August. Rainfall averages 381mm (15in) a year, falling mainly between December and March, with no rain at all in August. The natural vegetation is adapted to the dry conditions. It comprises certain species of coniferous trees, such as the Aleppo Pine of the Mediterranean, as well as evergreen shrubs with tough, leathery leaves, such as heathers and broom. The taller shrubs and small trees, such as the olive, form a type of vegetation known as maquis, whereas vegetation consisting of smaller shrubs and herbs is known as garrigue or chaparral. Many plants, such as rosemary, produce aromatic oils and are cultivated as herbs. The climate is excellent for growing a wide variety of fruits, especially grapes.

MEDITERRANEAN VEGETATION
This view of a valley between Cogolin and Collobrières, southern France, shows the typical vegetation of the region.

HUNTER VALLEY
Warm, temperate climates provide ideal conditions for vineyards, such as these in the Hunter Valley, New South Wales, Australia.

MID-LATITUDES

Cool, temperate continental climates

DISTRIBUTION In the continental interiors of North America and Europe, parts of Sweden, eastern Asia, and northern Japan

ANNUAL REGIME FOR Minneapolis, Minnesota, USA

TOTAL PRECIPITATION 725mm (28½in)

TEMPERATURE OF WARMEST MONTH 28°C (82°F)

TEMPERATURE OF COLDEST MONTH -6°C (21°F)

Regions with continental climates experience much more extreme temperatures than those areas with maritime climates. This is because the land heats and cools more rapidly than the sea. Summers are warm or hot, and winters are cold and often severe. They are also influenced by high- and low-pressure weather systems. Precipitation is distributed evenly through the year. The central European climate, though more severe than that further to the west, is moderated by the fact that air reaching the region from the Atlantic is still relatively mild and moist. Vladivostok, on the Pacific coast, receives air that has crossed the entire landmass of Eurasia and so experiences more severe conditions. Regions with temperate continental climates are extensively cultivated and support large-scale cereal production.

SNOW-COVERED LANDSCAPE
Snow-covered corn stubble and grain silos are characteristic of grain-growing areas such as here in Ellsworth, Wisconsin, USA.

POLAR REGIONS

Cold climates

DISTRIBUTION In the Arctic and Antarctic circles, above 66.5° north and 66.5° south, including Greenland, and the extreme north of North America, and Eurasia

ANNUAL REGIME FOR Vostok Station, Antarctica

TOTAL PRECIPITATION 4.5mm (⅒in)

TEMPERATURE OF WARMEST MONTH -32°C (-26°F)

TEMPERATURE OF COLDEST MONTH -67°C (-89°F)

Latitudes above 66.5° are extremely cold and very dry. At Qaanaaq, on the coast of northern Greenland, average daytime temperatures in July reach 8°C (46°F). The temperature in February, the coldest month, averages -20°C (-4°F). Inland, temperatures are much lower. Most of the Arctic is covered by the Arctic Ocean, which is fed with water from further south. Although ice-covered for most of the year, there are breaks in the ice where sea and air meet, preventing the temperature falling as low as it would otherwise. Antarctica is colder because it is a continent and high above sea-level. Surface temperatures measured above the ice-sheet, about 4,500m (14,750ft) above sea-level, average -49°C (-56°F) at the South Pole. Stable high pressure over the Antarctic and Greenland ice-sheets produces winds that blow outwards, preventing milder, moister air from entering. The low temperatures mean that the air can hold very little moisture. Consequently, the climate is extremely arid and all precipitation falls as snow that remains on the ground and accumulates gradually over time.

GREENLAND ICE-SHEET
Air in contact with this huge ice-sheet is cooled, creating an area of high pressure. Winds blowing from it are both cold and extremely dry.

WORLDWIDE

Island and coastal climates

DISTRIBUTION On the coasts of continents, especially the west coasts of continents in temperate latitudes, and oceanic islands worldwide

ANNUAL REGIME FOR Honolulu, Hawaii

TOTAL PRECIPITATION 643mm (25in)

TEMPERATURE OF WARMEST MONTH 28°C (82°F)

TEMPERATURE OF COLDEST MONTH 24°C (75°F)

STEEP COASTAL CLIFFS
The Na Pali coast of Kauai, Hawaii, is strongly influenced by the sea, which brings plentiful rain.

Climates dominated by the proximity of the sea are cooler in summer and warmer in winter, compared with continental interiors in the same latitude. The maritime air also brings abundant precipitation, but there are exceptions. The very dry coastal Atacama and Namib Deserts occur in latitudes 15–30° south, near the edges of the subtropical high-pressure regions, where dry air is flowing away from the continental interior. Some oceanic islands are relatively dry, for the same reason. For example, Las Palmas in the Canary Islands has only 176mm (7in) of rain a year, with a temperature range from about 17°C (63°F) to 24°C (75°F). The Solomon Islands, at 6.2° south, have a tropical climate, with temperatures ranging from 29°C (84°F) in winter to 32°C (90°F) in summer, with 3,000mm (120in) of rain a year.

CANARY ISLANDS
This satellite image shows dust being blown from the African mainland by seasonal sirocco winds. They flow towards the Canary Islands, but are dry and carry little rain.

WORLDWIDE

Mountain climates

DISTRIBUTION Mountains, plateaus, and mountain ranges throughout the world, starting at about 600m (2,000ft) above sea-level

ANNUAL REGIME FOR Les Escaldes, Andorra

TOTAL PRECIPITATION 808mm (32in)

TEMPERATURE OF WARMEST MONTH 26°C (79°F)

TEMPERATURE OF COLDEST MONTH 6°C (43°F)

The temperature of the air decreases with height by about 6.5°C per kilometre (3.6°F per 1,000ft). This has two consequences. The first is that the temperature in mountain climates decreases with increasing elevation. The second is that mountain climates are wetter than the adjacent lowland climates. This is because air is forced to rise over mountains, the water vapour condenses to form clouds, and these deliver precipitation on the windward slopes. The lee side is often dry, because it is in the rain-shadow. Although temperature decreases with height everywhere, actual mountain climates vary according to latitude, because the temperature beginning at sea-level varies in different locations. Mountain vegetation changes with height in much the same way that it changes with latitude (see p.306).

SNOW-CAPPED MOUNTAINS
The snow-covered peaks of the La Sal Mountains, USA, indicate that temperatures are lower above the snow-line than below it, preserving the snow.

CHANGING WEATHER
Unstable atmospheric conditions can produce rapid changes in weather. This storm cloud in New Mexico, USA, will probably cause a violent thunderstorm and heavy rainfall.

WEATHER

"Weather" is a general term for the constantly changing atmospheric conditions that prevail at a particular place and time. It is a result of the physical interactions between sunshine, air, and water. Sunlight heats the air more in some places than others, leading to differences in air pressure, which in turn cause the air to move in the form of wind to eliminate the differences. The Sun's heat also causes water to evaporate, only to condense again to form clouds as the air carrying the water vapour rises and cools.

Changes in the weather can be rapid, as when a warm, sunny morning gives way to a cold, wet afternoon, or slow, as in the seasonal change from summer to winter. Over longer periods, weather is less changeable: there are warm summers and cool ones, and hard or mild winters, but over the years these seasonal variations cancel each other out. The average weather over many years makes up the climate for a particular place.

ATMOSPHERE

Air Masses and Weather Systems

394–95 Circulation and currents

450–53 Atmospheric circulation

Living with hurricanes 470–71

Precipitation and clouds 472–73

An air mass is a body of air with uniform characteristics. Different air masses meet and interact along boundaries called fronts. There is a close link between air masses and patterns of circulating air known as pressure systems: warm air masses are associated with low-pressure systems and cool air masses with high-pressure systems. Over the course of a year, the distribution of air masses and pressure systems changes. These shifts can have significant effect on the weather. For example, continental air is hot in summer, causing low pressure over the land, and bitterly cold in winter, resulting in high pressure.

Pressure Systems

Warm, rising air produces an area of low surface pressure (or cyclone). Dense, subsiding, cool air forms an area of high surface pressure called an anticyclone. The difference in pressure between the centre of a pressure system and the air outside is called a pressure gradient. Air flows along this gradient from areas of high to low pressure – the steeper the gradient, the stronger the wind. The moving air is deflected by the Coriolis effect (see p.450), so in the northern hemisphere air flows anticlockwise around cyclones and clockwise around anticyclones.

HIGH AND LOW PRESSURE
Air spirals into a low-pressure area (cyclone), then rises, perpetuating the low pressure below, but creating high pressure above, where air diverges. In high-pressure areas (anticyclone), the situation is reversed.

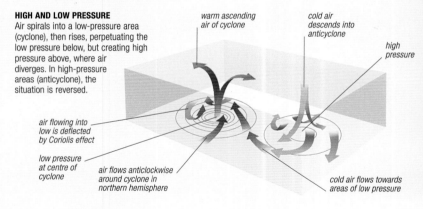

warm ascending air of cyclone

cold air descends into anticyclone

high pressure

air flowing into low is deflected by Coriolis effect

low pressure at centre of cyclone

air flows anticlockwise around cyclone in northern hemisphere

cold air flows towards areas of low pressure

Air Masses

An air mass is a volume of air that covers most of a continent or ocean and acquires characteristics from the surface below it. Air over a continent will be drier than air over the ocean, and it will be warmer in summer and cooler in winter, because of the different rates at which the land and sea warm and cool with the seasons. Once it has acquired these characteristics, the almost homogenous air constitutes an air mass. Air masses extend from the surface all the way up to the tropopause (see p.444). Air masses have names that describe their origins: those that form over land are continental, and those forming over the sea are maritime. Depending on the latitude in which they form, air masses may be arctic, polar, tropical, or equatorial. These names can then be combined to produce six types of air mass: continental arctic; continental polar; continental tropical; maritime arctic; maritime polar; maritime tropical; and maritime equatorial. There is no continental equatorial type, because there is no continental land mass in equatorial regions that is large enough to produce continental equatorial air. Arctic and polar air are essentially identical at surface level, but they differ in the upper atmosphere. When two air masses meet, they do not simply merge together, because they are at different temperatures. This means that the density of the air is different in the two air masses. Instead of mixing, the denser air moves beneath the less dense air, lifting it from the surface. The boundary between two air masses is known as a front, named for the air behind it: as a warm front passes, warm air replaces cool air; a cold front brings cooler air.

NORTH AMERICAN AIR MASSES
The weather at a particular place is usually influenced by several air masses. For example, North American weather is influenced by the six air masses shown below: cool polar air flows from the north, and warm tropical air from the south. In winter, tropical maritime air comes in from the Pacific Ocean.

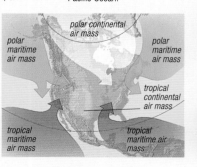

polar continental air mass

polar maritime air mass

polar maritime air mass

tropical continental air mass

tropical maritime air mass

tropical maritime air mass

DEEP CLOUD
When two air masses meet along a cold front, air is forced upwards and becomes unstable, resulting in the formation of deep cumulus clouds.

VILHELM BJERKNES

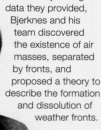

The Norwegian Vilhelm Bjerknes (1862–1951) is one of the founders of scientific meteorology. He established the Bergen Geophysical Institute in 1917, and set up a series of weather stations throughout Norway. Using the data they provided, Bjerknes and his team discovered the existence of air masses, separated by fronts, and proposed a theory to describe the formation and dissolution of weather fronts.

Fronts

On a weather map, the position of a front is indicated where it meets the ground, but fronts also extend upwards, often all the way to the tropopause, the boundary between the troposphere and the stratosphere (see p.440). Fronts are 100–200km (60–120 miles) wide, and they form along the edges of air masses. They also slope, so a frontal system, comprising a warm and cold front with warm air between them, is shaped rather like a bowl, but with uneven sides. A warm front slopes at about 0.5°–1°, and a cold front at about 2°. Cold fronts travel at an average speed of 35kph (22mph), and warm fronts at 24kph (15mph). At a cold front, cool air undercuts warm air, forcing it steeply upwards along the line of the front. Strong convection currents develop, leading to the formation of storm clouds and heavy rainfall. Cold fronts are often associated with low-pressure systems and unsettled conditions blowing away from a high-pressure region. At a warm front, warm air rises over cold air more gradually. Moisture condenses in the rising air, producing clouds and precipitation, the type depending on how rapidly the air rises. Because it moves more quickly, a cold front will eventually overtake a warm one, lifting it clear of the ground. This is known as an occlusion. A warm occlusion occurs when the cold front rises over the warm one, and a cold occlusion occurs when the cold front undercuts the warm front.

GRADUAL SLOPE OF A WARM FRONT
This warm front over Illinois, USA, has produced layered, stratus-type clouds. It meets the ground about 300km (186 miles) away.

STORM FRONT
A fast-moving cold front may "shovel up" warm air, producing huge storm clouds. Each cloud dissipates quickly, but another forms, creating a squall line, as seen here.

COLD FRONT
A cold front advances into the warm air ahead of it, forcing air to rise up the relatively steep frontal slope. This triggers the formation of cumulus-type clouds.

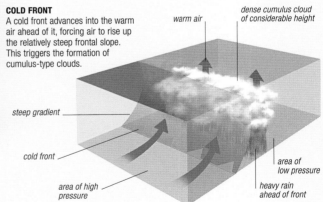

warm air

dense cumulus cloud of considerable height

steep gradient

cold front

area of high pressure

area of low pressure

heavy rain ahead of front

Jet Streams

Jet streams are narrow ribbons of fast-moving air that circle the Earth, close to the tropopause, at 100–500kph (60–310mph). They are produced by the difference in temperature between air masses. Because warm air is less dense than cold air, pressure decreases with height more rapidly in a column of warm air than in cold air. This difference in temperature is reflected in a difference in pressure that increases with height. It is this difference that generates the wind, which blows with the cold air on its left in the northern hemisphere and on its right in the southern hemisphere. This means the jet streams blow from west to east in both hemispheres. The exception is a jet stream that develops in summer over India and Africa, which flows in the opposite direction.

WARM FRONT
Air rises more slowly up the shallow slope of a warm front and produces stratus-type clouds. The clouds warn people of its approach as the front overhangs the region ahead of it.

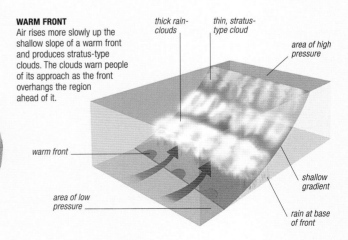

thick rain-clouds

thin, stratus-type cloud

area of high pressure

warm front

area of low pressure

shallow gradient

rain at base of front

JET-STREAM VARIATION
Jet streams do not blow in a constant direction. Shallow waves, known as Rossby waves, develop, which become bigger over about three weeks and then disappear as a steady flow resumes.

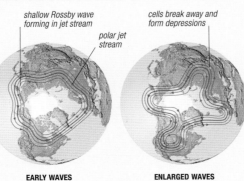

shallow Rossby wave forming in jet stream

polar jet stream

cells break away and form depressions

EARLY WAVES

ENLARGED WAVES

High- and Low-Pressure Weather

As cold air sinks, it becomes warmer, creating an area of high pressure at the surface and bringing settled weather with cloudless skies. It is usually dry and sunny in summer and cold and frosty in winter. However, if the descending air traps warmer air beneath it, fog may form. Warmth at the surface causes air to rise, producing low surface pressure. Moisture condenses in the rising air, producing both cloud and precipitation, the type depending on how rapidly the air rises. Winds blowing into low-pressure regions are generally stronger than the winds blowing away from high-pressure regions because the pressure gradient is usually steeper around low-pressure systems. The very high winds and heavy rain of a tropical cyclone are extremely destructive. In temperate areas, low-pressure cyclones bring less severe weather and are known as depressions.

HIGH-PRESSURE SYSTEMS
Clear skies, dry air, and fine weather indicate anticyclonic (high-pressure) systems, such as seen here in the Namib Desert. Air flowing outwards prevents moister air from entering.

LOW-PRESSURE SYSTEMS
Cyclonic (low-pressure) weather, as seen here in Brazil, is usually dull, with grey cloud, often accompanied by persistent, light rain.

STORMY WEATHER FRONT
A cold front is lifting warm, moist, unstable air, causing these storm clouds to form over Norfolk, England. The clouds will produce heavy rain.

CLEAR SKIES
This satellite image of South Australia shows anticyclonic conditions in summer, which bring high temperatures and almost no rain.

Mid-Latitude Depressions

Between polar and tropical regions, low-pressure areas, called mid-latitude depressions, travel from west to east, usually with one depression following another. They are caused by horizontal waves in the jet stream, called Rossby waves, at the top of the front between tropical and polar air. Rossby waves produce similar waves at surface level. A crest in a Rossby wave, projecting into cold air on the polar side of the jet stream, produces a ridge of high pressure. A trough in the jet stream, projecting into warm air on the equatorial side, produces low pressure. These ridges and troughs are where the winds are fastest. At the surface, anticyclones usually occur downwind of the ridges and cyclones downwind of the troughs. A mid-latitude depression tends to develop a characteristic arrangement of warm and cold fronts. The system begins as a simple boundary between warm and cold air, with the air on either side flowing in opposite directions. As it travels eastwards, separate fronts develop. Because cold fronts move more quickly than warm ones, they eventually merge, form an occlusion, and then dissipate.

DEPRESSION DIARY
These illustrations show the development, movement, and decay of a mid-latitude depression. A front develops, and then deepens. Over time, the cyclonic circulation results in occlusion of the front, which then dissipates.

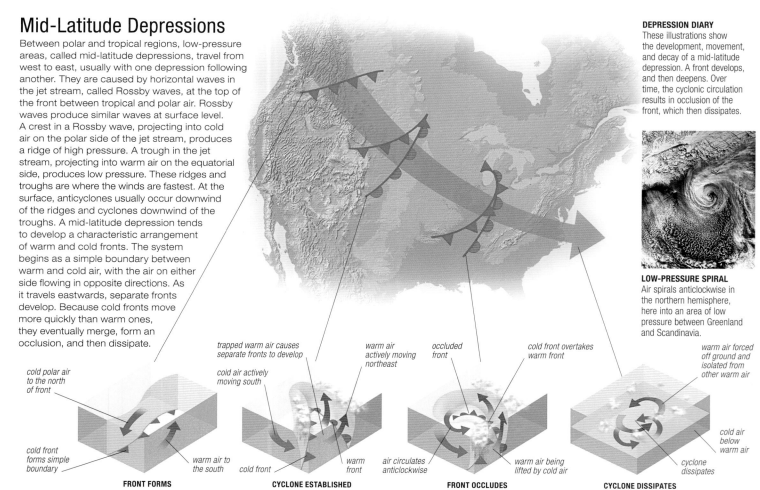

LOW-PRESSURE SPIRAL
Air spirals anticlockwise in the northern hemisphere, here into an area of low pressure between Greenland and Scandinavia.

cold polar air to the north of front

trapped warm air causes separate fronts to develop

cold air actively moving south

warm air actively moving northeast

occluded front

cold front overtakes warm front

warm air forced off ground and isolated from other warm air

cold front forms simple boundary

warm air to the south

cold front

warm front

air circulates anticlockwise

warm air being lifted by cold air

cold air below warm air

cyclone dissipates

FRONT FORMS

CYCLONE ESTABLISHED

FRONT OCCLUDES

CYCLONE DISSIPATES

Tropical Cyclones

A tropical cyclone, known also as a hurricane or typhoon, is a low-pressure system that develops only where a large area of the sea's surface has a temperature no lower than 27°C (81°F) and the wind direction or speed changes with height, to remove rising air. The Coriolis effect makes the winds spiral. These conditions occur mainly in late summer, when the sea is warm, between latitudes 5° and 20°. A disturbance in the wind pattern starts the air turning, producing low surface pressure with air accelerating into it. The high temperature causes high evaporation, so the rising air is very moist. As it cools, condensation releases latent heat, increasing instability and generating towering storm clouds.

HURRICANE PAULINE
This hurricane struck Mexico in October 1997. Wind speeds in the cloud wall surrounding the eye reached over 200kph (125mph).

TYPHOON ODESSA
The eye of typhoon Odessa, over the western Pacific, is evident from the "hole" where warm, dry air is sinking, causing cloudless skies.

ANATOMY OF A HURRICANE
Warm, sinking air forms the calm area known as the eye, around which there is a cloud wall where the strongest winds and the heaviest rain occur. Rising air produces high pressure aloft, while subsiding air produces spiral bands of cloud with clear skies in between. Strong convection and condensation inside the clouds power the hurricane.

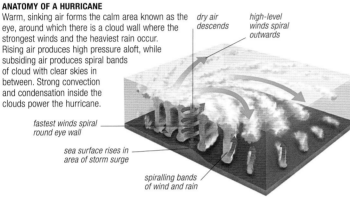

dry air descends

high-level winds spiral outwards

fastest winds spiral round eye wall

sea surface rises in area of storm surge

spiralling bands of wind and rain

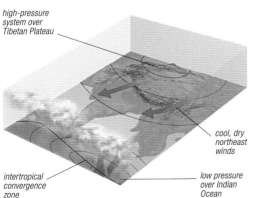

end of hurricane track

start of hurricane track

HURRICANE TRACK
Hurricanes travel westwards, and swing to the right as they approach land. How soon this happens will determine whether or not they cross land. This is the track of Mitch, which struck Central America in October 1998. Its winds blew continuously at 290kph (180mph) for 15 hours.

Monsoon Systems

Monsoon weather occurs in some parts of the tropics due to a seasonal change in the direction of the prevailing wind. During the winter, very dry air flows outwards from high pressure in the continental interior. In summer, as the land warms, the high pressure gives way to low pressure. Over the ocean, however, pressure remains relatively high because water warms much more slowly. The winds reverse direction, bringing moist air over the land. As it crosses onto the land, the air rises, producing huge clouds and the torrential rains of the monsoon.

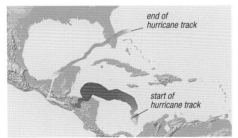

high-pressure system over Tibetan Plateau

cool, dry northeast winds

intertropical convergence zone

low pressure over Indian Ocean

WINTER
In winter, a high-pressure system develops over the Tibetan Plateau, giving rise to very dry winds that flow towards a low-pressure system over the Indian Ocean.

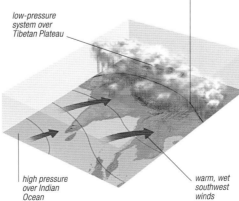

low-pressure system over Tibetan Plateau

intertropical convergence zone

high pressure over Indian Ocean

warm, wet southwest winds

SUMMER
In summer, low pressure over the Tibetan Plateau and northward movement of the intertropical convergence zone draw in moist air from the Indian Ocean over Asia.

THE ONSET OF THE MONSOON

People in India and elsewhere in Asia are dependent on the monsoon for growing their crops. If the monsoon is early, the fields may not be ready for planting, but if it is late, planting is delayed and the crop yield is poor. This can be devastating for subsistence farmers. In drought years, people go hungry and their stock starves. Today, the increasing accuracy with which the onset of the monsoon is predicted allows some farmers to plan which crops to plant. If they know when the rains will arrive, they also know the length of the growing season. The monsoon starts with a spectacular thunderstorm, and torrential rain that lasts for a week or more. Heavy rain can cause flooding, but people still celebrate the first rains of the monsoon.

ATMOSPHERE

LIVING WITH HURRICANES

Although tropical cyclones – also known as hurricanes and typhoons – are generated at sea, they can inflict widespread devastation and loss of life if they reach coasts and travel inland. These storms also trigger storm surges, flooding, and catastrophic landslides. Attempts to defuse hurricanes by cloud-seeding – a method of inducing rain – have met with limited success. In recent years, however, improvements in hurricane and typhoon warning systems have saved thousands of lives.

Wind and Water

Hurricane strength is measured using the Saffir-Simpson scale, which has a maximum rating of five. In a category-five hurricane, wind speeds can reach over 240kph (155mph), and, because the wind's impact increases exponentially with its speed, it can tear off roofs and smash wooden buildings apart. This destructive power is made worse by the fact that the wind blows in the opposite directions on either side of the hurricane's centre, or eye. As the eye passes overhead, the wind stops and then reverses, ripping out anything that has been loosened in the initial blast.

Violent though it is, wind takes second place to water as a cause of cyclone-related deaths. Where land is low-lying, seawater is forced inland in storm surges, which can be more than 8m (26ft) high. Such surges are a hazard in the Gulf of Mexico, where they played a deadly role in the hurricane that hit the Texan port of Galveston in 1900, but they can also strike the USA's Atlantic coast, from Florida to the Carolinas. They are also a particular risk in the Ganges delta, in the northern reaches of the Bay of Bengal. This region is densely populated, and the land is only a few metres above sea-level during a normal high tide. To make matters worse, the gentle slope of the sea bed increases the chance of large storm surges forming when winds blow towards the coast. Since 1980, storm surges in the Bay of Bengal have claimed over half a million lives.

When Hurricane Katrina struck New Orleans in August 2005, it was also the storm surge that did the real damage. However, investigations show that flood control measures can make the problem worse if not well-maintained.

New Orleans is cut through by a channel to take shipping from the river to the sea, flanked by levees. Some of the levees were all too quickly breached to swamp parts of the city as the channel became a superhighway for the surge.

Advance Warning

Geostationary and polar-orbiting satellites are used to track developing storms in cyclone-generating regions. Once a potential threat has been identified, regional forecasting systems estimate a storm's strength, path, and eventual arrival on land. Hurricanes and typhoons make slow progress across the sea, so storm alerts can give several days' advance warning – vital preparation time for people likely to be affected.

It is relatively easy to track storms, but much more difficult to forecast their future path. Hurricanes and typhoons follow curving tracks, moving outwards from the tropics, but local conditions such as surface-temperature variations often make them deviate from their original course. Autumn 2012's Hurricane Sandy, which blasted the northeastern USA, looked like a weak tropical storm when it was in the Caribbean, but turned suddenly into a giant as it merged with a mid-latitude depression.

CLOSING IN
Seen by satellite, Hurricane Fran spirals near the USA's Atlantic coast in 1996. The hurricane inflicted extensive damage in North Carolina.

Warming of The Seas

Hurricanes develop best in hot conditions, and form in regions where the sea's surface temperature exceeds about 27°C (81°F). In the North Atlantic, heat stored during the summer produces a peak hurricane season from August to October, while off northern Australia the cyclone season reaches its height between January and March. Once the sea starts to cool, violent cyclonic storms are rare, but with an increase in global warming (see pp.452–53), zones of warm surface water will become larger and more persistent. As a result, the threat from hurricanes and typhoons is almost certain to grow.

With more heat available to generate rising air, cyclonic storms are likely to become more frequent, and the strength of each storm may increase. The amount of moisture that it carries will grow. Hurricanes may also wander further from the tropics than they do now, threatening areas as far north as New England relatively often. Hurricane Sandy gave an uncomfortable warning that it may well be happening already.

STORM SURGE
This town in the Dominican Republic was flooded in 1998 by Hurricane George, which caused millions of dollars' worth of damage.

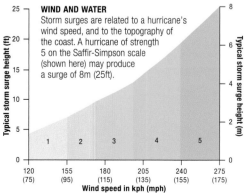

WIND AND WATER
Storm surges are related to a hurricane's wind speed, and to the topography of the coast. A hurricane of strength 5 on the Saffir-Simpson scale (shown here) may produce a surge of 8m (25ft).

SANDY'S AFTERMATH
In 2012, a tropical cyclone, Hurricane Sandy, merged with fast-moving cold air in the north Atlantic. When the resulting superstorm came ashore, it caused great damage, as seen here at Breezy Point, New York, USA.

Precipitation and Clouds

106 The global water cycle

108 Water in the atmosphere

445 Airborne water

452 Moisture and the oceans

466–69 Air masses and weather systems

Clouds form when water vapour in rising air condenses into droplets or freezes directly into ice crystals. The height at which this occurs depends on the stability of the air and the amount of moisture present. A typical cloud droplet, or ice crystal, is about 0.01mm (0.0004in) across. Cold clouds formed at high altitude contain only ice crystals, lower-altitude warm clouds contain only water droplets, and mixed clouds contain both. Snowflakes form if ice crystals and water droplets cool to below freezing. Water evaporates from the droplets and is deposited on the ice crystals, which collide and form snowflakes. Fog and dew form when condensation occurs at ground level, but if temperatures fall below freezing, dew is replaced by frost and rime.

STORM CLOUD
Clouds like this form in unstable air. Water vapour condenses, releasing latent heat so the air continues to rise. Violent vertical currents in the cloud may produce a thunderstorm.

Stable and Unstable Air

Air is warmed when it is in contact with the ground. This causes it to become less dense than the cooler air above and so it starts to rise. As air moves upwards, its temperature falls at a set rate, the lapse rate, making it denser and heavier as its height increases. Eventually the rising air reaches a point at which it is more dense than the surrounding air and it starts to sink again. This air is said to be stable. As air rises and cools, the amount of water vapour it can hold decreases until it becomes saturated and water droplets form. The temperature at which this happens is referred to as the dew point. The height of the dew point is not fixed and varies according to air temperature.

When water vapour condenses into water droplets, it releases latent heat (see below). If the saturated air is cooling at a slower rate than the surrounding air, it will also be warmer and so it will continue to rise. This air that rises without being forced is referred to as unstable. Clouds form at the dew point and may attain great depth under these conditions.

UNSTABLE CONDITIONS
In these conditions, a rising pocket of air is always warmer than that surrounding it. At the dew point, the air becomes saturated with water vapour and clouds start to form.

	°C	°F		°C	°F	
warm air continues to rise						5,000m (16,500ft)
	-8	18		8	46	4,000m (13,0001ft)
clouds form at dew point	4	39		14	57	3,000m (10,000ft)
air pocket less dense than environment	16	61		20	68	2,000m (6,500ft)
air cools as it rises	28	82		30	86	1,000m (3,300ft)
pocket of air at ground temperature	40	104		40	104	0

Condensation

Condensation is the change that takes place when a gas, such as water vapour, changes into a liquid, such as water. A water-vapour molecule consists of elements, two hydrogen and one oxygen, that move freely among the molecules of the other atmospheric gases. When two water molecules collide, they bounce off each other. However, as the temperature falls, the molecules have less energy and move more slowly. When they reach the dew-point temperature, at which water condenses, water-vapour molecules join together when they meet. A hydrogen bond forms between one of the hydrogen atoms of one molecule and the oxygen atom of another, forming them into short strings. The water vapour has then condensed into a liquid. As they join, the molecules fall into a state that requires less energy to sustain it, and so they give up some of their energy as heat. This is called latent heat. Its release warms the surrounding air and may be absorbed by liquid molecules, allowing them to vaporize. Thus, groups of water molecules are constantly breaking and re-forming. If the water droplets grow large enough, they eventually fall as rain.

RAINBOW IN HAWAII
Light is refracted (bent) as it enters a raindrop and again as it leaves it. This separates the light into its constituent colours and forms the rainbow.

STORM CLOUD
Cumulonimbus clouds are among the largest of all clouds and have great vertical height, but they are short-lived, rarely lasting for more than an hour.

Lifting and Rainfall

Clouds form when moist air cools to its dew-point temperature. Air becomes cooler if it is made to rise, and so cloud often forms in rising air. The height at which water vapour condenses in rising air marks the cloud base. Air can be made to rise in three ways. If it is heated from below, due to contact with a warm surface, air rises by convection, forming small fair-weather cumulus or cumulonimbus clouds. If moving air has to pass over a mountain, it must rise. This is called orographic lifting and results in orographic cloud formation. Warm air also rises when cooler air pushes beneath it at a weather front. This is frontal lifting, producing frontal clouds. If clouds are thick enough, water droplets and ice crystals link together until they are heavy enough to fall as precipitation. Convection produces showers, and brief storms. Orographic and frontal clouds produce more persistent rain and drizzle.

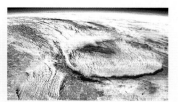

OROGRAPHIC CLOUD
Clouds form in the rising air passing over this mountain in Greenland. Orographic clouds often form a cap shrouding the mountain summit and extending a short distance downwind.

LAKE EFFECT CLOUDS
In winter, large lakes, such as Lake Superior, USA, are slow to freeze. Air crossing them gathers moisture that condenses into clouds as it passes over cold land.

RAIN-CLOUD FORMATION
Moist air that is made to rise above the dew point forms clouds. They form where warm air rises over cool air at a front (frontal rain); where air rises by convection over warm ground, producing a local area of low pressure into which air converges and rises (convergence rain); and where air rises over mountains (orographic rain).

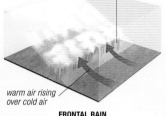
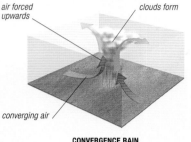
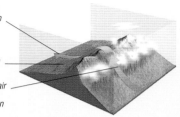

rain falls along front

warm air rising over cold air

FRONTAL RAIN

air forced upwards

clouds form

converging air

CONVERGENCE RAIN

airflow continues in lee of mountain

leeward slope receives little rain

mountains force air to rise and cool, resulting in rain on windward slope

OROGRAPHIC RAIN

No-Freeze Rainfall

Rain feels warm if it falls from a cloud that consists entirely of water droplets. Such clouds, which form only in air that is above freezing point, are known as warm clouds, and the rain they produce is called no-freeze rainfall. The lower parts of clouds are often warm, but outside the tropics the freezing level is rarely high enough for deep warm clouds to form and produce rain. Raindrops form in warm clouds by the merging of cloud droplets, which vary greatly in size. Bigger ones, being heavier, fall faster than small ones and overtake them. When droplets of similar size collide, they merge and break into two again, but when a big droplet meets a small one, they coalesce. The droplet grows bigger, collects more small droplets, and finally falls as rain.

HEAVY RAIN
Warm clouds produce big, heavy raindrops. Regular, heavy rainfall is a feature of regions with tropical climates.

TROPICAL RAINSTORM
Water evaporates quickly in the tropics. It rises fast, forming warm clouds that produce frequent rainstorms.

SATELLITE DATA

This is a false-colour image produced as part of a project called the Tropical Rainfall Measuring Mission. The aim of the project is to measure the energy released by tropical rainfall, and to calculate the effect of this energy on the global climate. Data is gathered using instruments carried on satellites, including radar to measure raindrop formation in clouds, and microwave receivers to measure air temperature. When the images are received, computers colour parts of them differently to enhance any variation. In this image, pale blue indicates light rainfall. Areas of heavy rainfall are coloured orange and red.

ATMOSPHERE

Rain

Water that reaches the ground as droplets is known as rain or drizzle. Drizzle consists of drops that are less than 0.5mm (1/50in) in diameter and very close together. These droplets form near the base of stratus cloud, in which there are no upcurrents that would allow them to coalesce with other droplets and grow larger. They sink slowly, but because the cloud base is so low they have no time to evaporate before reaching the ground. Drops larger than those of drizzle fall as rain. Raindrops vary in size, but within limits. The smaller the drop, the more slowly it falls, and therefore the longer it has to evaporate in the air between the cloud base and the ground. This means that a raindrop must be at least 0.5mm (1/50in) in diameter if it is to reach the ground. Most raindrops are 2–5mm (1/25–1/5in) in diameter. This is between 1 million and 2.5 million times bigger than an average cloud droplet, so many cloud droplets must merge together before they are big enough to fall as rain. Raindrops fall at 23–33kph (14–20mph). Friction with the air breaks up drops over 5mm (1/5in).

RAINY SEASON
Seasonal rainfall, such as this in the Serengeti in Tanzania, triggers the regrowth of plants, which in turn supports a wide variety of wildlife.

DISTANT RAINFALL
Rain seen falling in the distance looks like a grey curtain under cloud. Sometimes the curtain does not reach the ground, because raindrops evaporate in the dry air beneath the cloud.

SAHEL DROUGHT
The semi-arid Sahel region lies along the southern edge of the Sahara Desert. Usually there is a short rainy season in summer, when the intertropical convergence zone moves north, which revives the pasture for livestock and allows some crop cultivation. However, the rains are unreliable, and the region has endured many droughts. One of the worst droughts of modern times began with light rains in the 1960s and complete failure of the rains in 1972 and 1973. Rainfall did not return to normal until the 1980s. Up to 200,000 people and 4 million cattle died.

Hail and Snow

Most hailstones are hard, roughly spherical pellets of ice, 5–50mm (1/5–2in) across. Soft hail forms when rime (see opposite page) collects on snow crystals, producing pellets up to 5mm (1/5in) in diameter that flatten when they hit the ground. A hailstone begins as a raindrop near the bottom of a large cumulonimbus cloud. Carried aloft by air currents, it freezes when it is above the freezing level, then falls in downcurrents near the top of the cloud. The hailstone grows by encountering supercooled water droplets that freeze onto it. Its final size depends on the number of circuits it has made between the top and bottom of the cloud. Snowflakes form when ice crystals adhere to one another, and they are 1–20mm (1/25–4/5in) across. The biggest snowflakes develop between 0° and -5°C (32–23°F). At this temperature, water coats the crystals and freezes as soon as another crystal touches it. All snowflakes are six-sided, but the random way they form ensures no two are identical. Ice pellets, or sleet, are melted snowflakes or raindrops, 2–5mm (2/25–1/5in) across, that freeze between the cloud base and the ground in air that is below freezing temperature. Ice particles smaller than 1mm (1/25in) are called snow grains.

INSIDE A HAILSTONE
In this section through a hailstone, viewed in polarized light, the alternating layers of clear ice and opaque rime that form the hailstone are clearly evident.

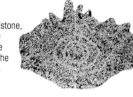

SHAPED BY THE WIND
An ice storm in New Hampshire, USA, has decorated this sign. Freezing raindrops, driven by a strong wind, struck the sign, spread out, and then froze. Later drops spread over earlier ones, producing this shape.

SNOWFLAKE
Snowflakes, such as the one shown here, form when ice crystals collide with one another. Each has a unique shape, but all are six-sided.

HAILSTONES
Hailstones form only in large cumulonimbus clouds, where they are carried up and down repeatedly, growing all the time. The bigger the cloud, the larger the hailstones become.

BLIZZARD CONDITIONS
Heavy snow, low temperatures, and strong winds are characteristic of blizzards. Here, Emperor Penguins sit out a blizzard in Antarctica.

Fog and Dew

Fog and mist occur when a layer of moist air, near ground level, cools to below its dew-point temperature. On high ground, fog and mist are often stratus cloud with a base lower than the hilltops. Dew forms late at night, when objects on the ground lose the warmth they absorbed by day. If the air is moist, contact with cold surfaces can cause water vapour to condense, coating them with liquid droplets. Cooling the surface also produces radiation fog because the ground cools by radiating away its warmth. Valley fog is a type of radiation fog. Advection fog forms if a wind carries warm, moist air over a cold surface.

RADIATION FOG
Early-morning fog on the Vermilion Lakes, Canada, formed by condensation when the Earth's surface cooled and reached the dew point.

ICE DEPOSITS
Freezing raindrops fell onto these pine needles, which were below freezing temperature, and froze almost instantly into clear ice.

RIME ON FORESTED HILLSIDE
These trees near Bamboo, North Carolina, USA, are coated with sparkling, white rime frost.

Frost and Rime

Frost is a layer of ice crystals that covers objects such as plants, parked cars, and windows. It also forms on the surface of fallen snow and in the air spaces inside it. If the dew-point temperature is below freezing, water vapour will be deposited on surfaces as ice crystals. This is how hoar frost, the commonest type of frost, forms. Dew may also freeze if the temperature falls. It may form clear ice, with rounded surfaces, or fern frost, which covers windowpanes with patterns resembling fern fronds. Fern frost develops when the dew on the window cools below freezing, but remains liquid. Ice crystals form between the droplets, then the droplets freeze onto the crystals, rapidly covering the window with frost. Rime frost is white and rough to the touch, forms thicker layers than hoar frost, and may grow into feathery shapes. It forms in cold, foggy conditions when supercooled water droplets in freezing fog or drizzle freeze on contact with a very cold surface, or when water vapour is deposited directly as ice crystals. Once the surface is coated with ice crystals, more water freezes onto the existing crystals. If a wind is blowing, rime frost forms only on the side of structures, such as radio and ships' masts, that are facing into the wind.

LUKE HOWARD

Luke Howard (1772–1864) was an English pharmacist with a keen interest in meteorology. He began keeping weather records when he was 11 years old. Howard wrote the first study of urban climate and the first book on meteorology. His cloud classification was strongly influenced by the system of two-part names used for biological classification.

GLOBAL CLOUD COVER
This satellite image shows that clouds occur worldwide, but are not uniform. Different clouds form under different climatic conditions.

Cloud Types

Clouds differ in appearance and in the height at which they develop. These differences provide the basis for an international system for classifying them. The classification is derived from the one proposed in 1803 by Luke Howard (see panel, left), and it uses the names he coined: combinations of cirrus (curl), stratus (layer), nimbus (rain), and cumulus (heap). It groups clouds as high, middle, or low level, according to the height at which the cloud base most often occurs. In middle- and high-level clouds, this varies with latitude, the height and range being lower at the poles than in temperate latitudes and highest at the equator. Clouds are grouped into ten basic types: cirrus, cirrostratus, and cirrocumulus (high clouds); altocumulus and altostratus (medium clouds); and nimbostratus, cumulonimbus, cumulus, stratus, and stratocumulus (low clouds). Some cloud types are further subdivided using 14 specific terms, such as humilis and congestus (with cumulus), which refers to the character of the cloud, or lenticularis and undulatus (with altocumulus), which further defines the shape.

cirrus
cirrocumulus
cirrostratus
altocumulus
altostratus
stratocumulus
stratus
cumulus
nimbostratus
cumulonimbus

Tropopause
high level, above 6,000m (20,000ft)

middle level, 2,000– 6,000m (6,500–20,000ft)

low level, 0–2,000m (0–6,500ft)

Height of cloud base

CLOUD TYPES
The ten basic types of cloud, which are classified by their appearance, are shown here. There is a clear division between high-, middle-, and low-level clouds. Only the cumulonimbus cloud has great vertical height.

CLOUD PROFILES

The pages that follow contain profiles of the ten basic cloud types. Each profile begins with the following summary information:

LEVEL Cloud base: low level, 0–2,000m (6,500ft); middle level, 2,000–6,000m (6,500–20,000ft); high level, above 6,000m (20,000ft)

DISTRIBUTION Global distribution

OCCURRENCE Season and conditions in which it forms

SUPERCELL
This impressive storm cloud, called a supercell, developed in Montana, USA. It produced a rotating updraft that sucked up dust and loose material from the landscape while also delivering torrential rain in its centre.

ATMOSPHERE

Stratus

| DISTRIBUTION Worldwide, more common near mountains and coasts |
| OCCURRENCE All year, more common in winter |

Low cloud that forms a featureless, grey layer covering most or all of the sky is called stratus. It consists entirely of liquid droplets, typically containing 0.05–0.5 grams of water per cubic metre (0.00005–0.0005 ounces per cubic foot). The cloud base is usually

ACID RAIN

Rain is slightly acid because carbon dioxide and other atmospheric gases dissolve in cloud droplets. Industrial emissions, especially of sulphur dioxide, increase its acidity. The acid reaches the surface as acid rain, snow, or mist, or as dry particles. It can damage plants directly as shown here, or  indirectly by altering chemical reactions in the soil. Mist and dry deposition cause more harm to plants than rain, as rain tends to run off leaves.

lower than 2,000m (6,500ft), and if it is at ground level then stratus occurs as fog. Hill fog has a base lower than the hilltop it covers. Fog clears by the evaporation of its lowest layer, and any remaining fog becomes stratus cloud. As there is no vertical air movement within stratus, it cannot produce rain, snow, or hail, but drizzle or snow grains sometimes fall. If the cloud is thin enough for the Sun or Moon to be visible, they appear as sharply defined discs. Stratus often forms overnight in fine weather, especially over water. It dissipates quickly during the morning as the temperature rises and the cloud droplets evaporate. Stratus also forms in valleys. Frontal stratus forms along warm fronts, where warm, stable air is being slowly raised up by cooler air. Cirrostratus and altostratus appear first, followed by stratus, which is the most likely to produce precipitation.

STRATUS OPACUS
This type of stratus forms an opaque layer of cloud that is thick enough to obscure the Sun and Moon completely.

STRATUS TRANSLUCIDUS
This sheet of stratus cloud forms a thin, translucent layer through which the Sun or Moon is visible as a sharply defined disc.

FOG OVER THE GOLDEN GATE BRIDGE
Moist air approaching San Francisco, USA, from the ocean, crosses the cold California current. This chills the air, and its moisture condenses to produce advection fog, which is identical in composition to stratus cloud.

Stratocumulus

| DISTRIBUTION Worldwide |
| OCCURRENCE All year |

Stratocumulus is low cloud, usually with a base below 2,000m (6,500ft), composed entirely of liquid droplets. It forms patches, sheets, or extensive layers of white or grey cloud. Rolls, or rounded masses of darker cloud, give stratocumulus a textured appearance. Stratocumulus forms when rising air meets a layer of warmer air and is flattened against its underside. In winter, this produces an overcast sky that lasts for several days, but in summer the sky usually clears quickly. Gaps in the clouds allow the Sun to shine through at intervals. At dusk and dawn, the sunlight shining through these gaps may illuminate dust particles, producing rays that converge onto the cloud from below.

BANDS OF STRATOCUMULUS CLOUDS

Cumulus

| DISTRIBUTION Worldwide, common in humid regions |
| OCCURRENCE All year, but more common in summer |

This cloud forms by convection, with a base that is usually below 2,000m (6,500ft). Cumulus varies greatly in size and can be up to 10km (6 miles) across and 20km (12 miles) high. The cloud is fleecy and very bright when seen from above, because of the amount of sunlight it reflects. The clouds are usually separate from each other, with blue sky in between. This allows the Sun to shine directly on the clouds, so

they appear very white with clearly defined edges. Cumulus can be embedded in clouds of other types, making them more difficult to identify. Cumulus clouds form in columns of rising air, or thermals. Their base marks the level at which water vapour condenses, which is why all cumulus clouds in one area are at the same height. As it rushes upward, a thermal draws in surrounding air at its sides and at its base. This mixes the cloud with drier and cooler air. Cloud droplets evaporate in the drier air, and fragments of cloud sink in the cooler air. This prevents the cloud from growing wider, and also prevents other cloud from forming close to it.

CUMULUS CONGESTUS
This large cumulus cloud is growing rapidly, mainly by expanding upwards, giving the top a cauliflower-like appearance.

SCATTERED CUMULUS CLOUDS
These detached, dense cumulus clouds in the Great Karoo, South Africa, have a typical horizontal cloud base and extend upwards due to convection.

Cumulonimbus

DISTRIBUTION Worldwide except Antarctica; some types are common in tropical regions

OCCURRENCE All year, associated with tropical cyclones and thunderstorms

Low-level cloud that extends vertically to a great height, sometimes as far as the tropopause or beyond, is called cumulonimbus. It is composed of liquid cloud droplets in the lower regions and ice crystals near the top. These clouds produce thunderstorms, hailstorms, and tornadoes, as well as torrential rain or snow. Because of its great depth, cumulonimbus is very dark when viewed from below. Light encounters many ice crystals and cloud droplets in passing through the cloud. These reflect and scatter it, reducing the amount of light that reaches the ground. Cumulonimbus cloud is created by convection currents in very unstable air. Warm, moist air rises rapidly. As it cools, its moisture condenses, releasing latent heat (see p.472) that warms the air, so it

LIT UP BY LIGHTNING
This cumulonimbus cloud in Wyoming, USA, is illuminated by lightning sparking between two regions within it.

MAMMATUS CLOUDS
Distinctively shaped mammatus clouds, such as these in Utah, USA, form on the underside of the "anvil" (see right) in cumulonimbus clouds.

continues to rise. Air is drawn into the cloud to feed the upcurrents, which can reach a vertical speed of 160kph (100mph). Ice crystals and hailstones falling through the cloud produce downcurrents that exit the base of the cloud as strong winds, cooling the air in the adjacent upcurrents. Eventually this suppresses the upcurrents, convection ceases, and the cloud

THREAT FROM THE AIR
Cumulonimbus cloud is typically deep and large, with a dark, menacing appearance. Here, there are scattered cumulus clouds in the background.

dissipates. Cumulonimbus cloud rarely survives more than an hour. As it dissipates, it may release all of its moisture at once – up to 275,000 tonnes in a really big storm cloud. In some cumulonimbus, the upcurrents and downcurrents separate. This produces a "supercell" in which the downcurrents no longer interfere with upcurrents, allowing the cloud to exist for several hours. If the wind speed or direction changes with height, the centre of the cloud may start rotating, and extend downwards to become a tornado.

CLOUD SEEDING

Rain can be induced from clouds by "seeding" from an aircraft, such as the one shown below. Silver iodide is burned above a cloud. As the smoke cools, the silver iodide crystals re-form and fall into the cloud. The water vapour freezes onto the crystals, which may induce the clouds to produce rain when otherwise they would not.

TOWERING STORM CLOUD
The height of this cumulonimbus cloud over the European Alps is readily apparent. The top has spread out into an anvil shape, and it is being lit from within by lightning.

ATMOSPHERE

ATMOSPHERE

LOW LEVEL

Nimbostratus

DISTRIBUTION Worldwide

OCCURRENCE All year except in Antarctica

Nimbostratus is dark and shapeless, but may appear to be illuminated from inside because gaps in the cloud reveal the paler stratus cloud above. It has a low base, but usually extends upwards to more than 2,000m (6,500ft), and so is thick enough to obscure the Sun and Moon completely, resulting in dull days and dark nights. It may produce steady, persistent rain, or snow. If the cloud temperature is above -10°C (14°F) throughout, the cloud will consist mainly of water; from -10° to -20°C (14°F to -4°F), it will be a mixture of water at the bottom of the cloud and ice at the top; and below -20°C (-4°F) it will consist mainly of ice. The mixture of water and ice favours the formation of larger ice crystals that fall as snow, or melt and fall as rain.

NIMBOSTRATUS CLOUDS
This picture taken over Utah, USA, reveals the vertical extent of the cloud and the darkness of its base.

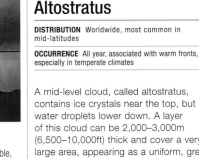

NIMBOSTRATUS PRAECIPITATIO
This nimbostratus cloud has precipitation falling from it. The precipitation is clearly visible, resembling a pale curtain beneath the dark cloud.

MIDDLE LEVEL

Altostratus

DISTRIBUTION Worldwide, most common in mid-latitudes

OCCURRENCE All year, associated with warm fronts, especially in temperate climates

A mid-level cloud, called altostratus, contains ice crystals near the top, but water droplets lower down. A layer of this cloud can be 2,000–3,000m (6,500–10,000ft) thick and cover a very large area, appearing as a uniform, grey or slightly blue sheet of cloud, or as a thin veil of fibrous cloud. The Sun and Moon may be faintly visible through it, but more often the cloud is thick enough to obscure them completely. Altostratus forms at warm fronts, where warm, moist air is being lifted above cooler air. Cirrostratus (see opposite page) usually appears ahead of it and merges with the altostratus as the cloud base becomes lower. Its appearance usually heralds rain or snow, and the altostratus itself may produce light snow or drizzle.

ALTOSTRATUS CLOUDS

MIDDLE LEVEL

Altocumulus

DISTRIBUTION Worldwide

OCCURRENCE All year, associated with cold fronts especially in temperate climates

Altocumulus is a mid-level cloud, with a base ranging from 2,000–4,000m (6,500–13,000ft) over polar regions to 6,000m (20,000ft) over the tropics. It is composed mainly of water droplets. Its appearance is very variable, but it often tends to form rolls, arranged in lines or waves or distinctive rounded masses. The clouds are sometimes shaded around the edges, which defines them clearly, but they may also form so close together that they merge into a continuous sheet. Altocumulus cloud is white, grey, or a mixture of both. Its structure results from small, vertical air movements. It absorbs warmth at its base and also at its top. This causes air to rise and water vapour to condense, but when the cloud droplets

LAMMERGEIER
Many of the larger carnivorous birds use gliding flight, which relies on the same thermal currents that produce altocumulus clouds to give them lift.

reach the top of the cloud, where they are exposed to direct sunshine, they evaporate again. Altocumulus often forms at night, when the radiation balance is different to its daytime state. The cloud still absorbs heat radiated from the surface below, but it also radiates away its warmth at the cloud top. Air rises, water vapour condenses into droplets, and at the top these are chilled and sink back into the cloud. This type of altocumulus dissipates in the morning, when the Sun warms the top of the cloud and the droplets evaporate.

TOPOGRAPHIC BARRIER
Altocumulus lenticularis cloud forms when air rises rapidly to pass over a topographic barrier, such as Mount Rainier, Washington State, USA.

MACKEREL SKY
Waves in the air sometimes produce cloud near their crests, with clear sky in the troughs. This produces a pattern in altostratus and cirrocumulus clouds reminiscent of the markings on a mackerel.

ALTOCUMULUS CLOUD COVER
These clouds often indicate fair weather in temperate latitudes. If enough of these clouds form, they may merge so that little blue sky is visible.

HIGH LEVEL

Cirrus

DISTRIBUTION Worldwide

OCCURRENCE All year

High-level cloud that consists entirely of ice crystals is called cirrus. It is thin, and wispy, and has a fibrous appearance. Once it starts forming, its ice crystals grow until they reach a maximum size of about 0.05mm ($^1/_{500}$in) across. Once the crystals are heavy enough to start falling, they sometimes drift downwards into a wind flowing from a different direction. This stretches the cloud into long tails that often curve – cirrus means "curl". They are then known as mares' tails or cirrus uncinus. Their appearance usually means that the ground-level wind will soon strengthen. When the top of a towering cumulonimbus cloud reaches the tropopause and spreads out beneath the boundary of air that is less dense than itself, it forms an incus, or anvil shape, composed of ice crystals. This is a type of cirrus, and when the storm cloud dissipates its incus often remains. Cirrus is most often produced by the lifting of

CIRRUS FIBRATUS
The cirrus fibratus clouds shown in this photograph are among the highest of all clouds. They rarely cover the sky, allowing sunlight to shine through.

THIN AND WISPY
Cirrus clouds are the most common type of high-level cloud. Winds can draw out these clouds into wisps, known as mares' tails.

stable air. This happens along a weather front. As the air rises, different cloud types form in sequence as more and more moisture condenses. Cirrus is the last to form and represents almost the last of the water vapour. The air above the cirrus is extremely dry. Although the last to form, cirrus is the first cloud to be seen as a front approaches. Lifting also occurs close to the core of a jet stream. The location of a jet stream is sometimes revealed by the cirrus along its track. This type of cloud also forms as rapidly moving air masses cross high mountains.

CONTRAILS

Condensation trails, or contrails, form behind aircraft (particularly jet aircraft) flying above a certain height. Water vapour is produced whenever fossil fuel burns and it is one of the exhaust gases discharged by engines. The engine exhaust cools in the air behind an aircraft. If the air is close to saturation, the water vapour condenses as a trail that commences some distance behind the aircraft. The trail dissipates, but on major routes there are often many contrails, and these may spread out to form a layer of cirrus.

HIGH LEVEL

Cirrocumulus

DISTRIBUTION Worldwide

OCCURRENCE All year, associated with cold fronts especially in temperate climates

Cirrocumulus is a high cloud, consisting entirely of ice crystals, that appears as white patches, or approximately spherical masses, called elements. These are arranged in regular patterns, often resembling ripples on a sandy beach or, less frequently, straight lines. Small, vertical air movements can also produce a mackerel sky (see opposite

CIRROCUMULUS CLOUDS
Occurring at high altitude, cirrocumulus appears as thin sheets or layers with varying degrees of regularity.

CIRROCUMULUS UNDULATUS
The ripples seen here are due to cloud forming at the crests of air waves. The clouds below cirrocumulus are small patches of altocumulus.

page) made of cirrocumulus. Like mares' tails, a cirrocumulus mackerel sky is an indication of approaching strong surface-level winds. Like all cirrus-type cloud, cirrocumulus has a fibrous appearance. This betrays its origin, because cirrocumulus is usually cirrus or cirrostratus that has been reshaped by air movements.

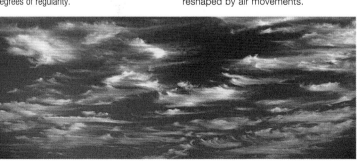

HIGH LEVEL

Cirrostratus

DISTRIBUTION Worldwide

OCCURRENCE All year, associated with approaching warm fronts

When the Sun shines on high-level cirrostratus at dawn and dusk, it sometimes makes the cloud appear pink, producing spectacular sunrises and sunsets. At other times of day, thin cirrostratus appears as a white veil that makes the sky look milky. If the cloud is thicker, it looks like many tangled filaments. Cirrostratus consists entirely of ice crystals. It is not thick enough to obscure the Sun or Moon, but it can produce a halo around them, which is caused by the refraction (bending) of light as it passes through the ice crystals. Haloes are quite large: most have a diameter of 22° (measured as an angle at the eye of the observer, formed by two lines drawn from opposite sides of the halo), but less

commonly 46°, depending on the route taken by the light through the ice crystals. A halo is mainly white, but sometimes it is faintly coloured, changing from yellow at the centre to green, white, and blue at the outer edge. A halo often means rain is approaching, because the cirrostratus occurs at the top of a warm front. This is especially likely if cirrus appears first, and cirrostratus gradually replaces it. If a halo is seen around the Moon, it often means rain will arrive by morning. The cloud forms when an entire layer of air is lifted, as colder air moves beneath it at a front. The air rises slowly, at no more than 20cm (8in) per second. When cirrostratus is associated with either nimbostratus or altostratus, prolonged rainfall is likely.

CIRROSTRATUS CLOUDS
This uniform and featureless sheet of cloud is characteristic of high-level cirrostratus. Its appearance is usually a sign of approaching rain.

ATMOSPHERE

Wind

394–95 Circulation and currents
450–53 Atmospheric circulation
466–69 Air masses and weather systems
470–71 Living with hurricanes
472–81 Precipitation and clouds

Wind is the movement of air away from regions of high pressure towards regions of low pressure. The air continues to move until the pressure difference is eliminated and an equilibrium is reached. Pressure differences are associated with large-scale weather systems, but they can also occur and produce winds on a smaller scale. A towering storm cloud, for example, draws in air to feed its upcurrents and expels air from its downcurrents, producing winds with gusts that can reach gale, or even hurricane, force.

Pressure and Wind Speeds

Wind speed is proportional to the pressure gradient, which is the rate of pressure change over a horizontal distance. However, it flows parallel to the gradient due to deflection by the Coriolis effect (see p.450). Air spirals into a low-pressure centre. The closer it gets, the smaller the radius of the spiral. The conservation of its angular momentum accelerates the wind in the same way a pirouetting dancer speeds up by drawing her arms into her sides. This means that deep depressions, drawing air from a large area towards a small centre, generate stronger winds than mild ones. It is also why cyclones and tornadoes, with small cores, generate such strong winds.

HIGH WINDS IN THE TROPICS
In the tropics, strong surface heating produces intense low pressure and violent weather, as seen here in Kerala state, India.

THE BEAUFORT WIND FORCE SCALE

This wind force scale was devised by Francis Beaufort, an English naval officer, in 1805. Initially devised for sailors, who used visual features of the sea to judge wind strength, it was later modified for use on land by substituting features such as trees and cars.

Beaufort number	Wind KPH	Speed MPH	Description
0	0–2	0–1	Calm – smoke rises vertically, air feels still
1	2–6	1–3	Light air – smoke drifts
2	7–11	4–7	Slight breeze – wind detectable on face, some leaf movement
3	12–19	8–12	Gentle breeze – leaves and twigs move gently
4	20–29	13–18	Moderate breeze – loose paper blows about
5	30–39	19–24	Fresh breeze – small trees sway
6	40–50	25–31	Strong breeze – difficult to use an umbrella
7	51–61	32–38	High wind – whole trees bend
8	62–74	39–46	Gale – twigs break off trees, walking into wind is difficult
9	75–87	47–54	Strong gale – roof tiles blow away
10	88–101	55–63	Whole gale – trees break and are uprooted
11	102–119	64–74	Storm – damage is extensive, cars overturn
12	120+	75+	Hurricane – widespread devastation

HIGH-ALTITUDE WINDS
At altitude, wind speed and strength can be especially high. This storm cloud over Mount Chephren, Canada, is being sculpted by strong winds.

Oceanic Winds

The wind is always stronger over the sea than it is over land. This is because there is less friction over water. Hills, trees, buildings, and other obstructions slow the wind on land. The distance the wind blows without interruption is called the fetch, and the long fetch over an ocean alters the characteristics of air masses crossing it from a continent. Continental air is dry, and water will evaporate into it as it crosses the ocean. If the air was still, evaporation would be slow, as the lowest layer would soon become saturated. Because the wind constantly replaces the surface layer with dry air drawn down from above, it greatly increases the rate at which the air gathers moisture, and changes it from a continental to a maritime air mass. The longest fetch on Earth is in the Southern Ocean, where the wind blows, uninterrupted by land, all around the world. That is where the wind, and the waves raised by the wind, are strongest. Sailors refer to the latitudes to the south of South America as the roaring forties, furious fifties, and screaming sixties.

TREE SCULPTURE
Salt-laden winds have dried out and destroyed the buds on the exposed side of this tree on the British coast.

POLAR WINDS
The Greenland and Antarctic ice-sheets are higher at the centre than the edges. Sinking cold air flows down to the coast, producing almost constant wind, such as here at Marsh Base, Antarctica.

HARBINGER OF CHANGE
Altostratus clouds (see p.480) form where warm, moist air is lifted above cooler air. They mark the arrival of wet weather.

Seasonal Winds

Some winds are associated with particular seasons. A few such winds are so constant that they are given individual names. Most occur because the high- and low-pressure systems that create wind change position during the year. The sirocco, for example, most often occurs in spring, and is drawn northwards by the low-pressure systems that develop in the Mediterranean. It blows ahead of depressions, gathering dust in the Sahara and moisture from the sea, and arriving over southern Europe as a warm, moist wind that sometimes delivers dust-laden rain. In contrast, the mistral may occur at any time, but is most common in winter when there is a high-pressure system in the North Atlantic and low pressure over Europe. It is a cold, northerly wind that blows over southern Europe, increasing in speed as it is funnelled down the lower Rhône Valley. Other cold, winter winds include the buran, a fierce blizzard that blows across the Russian plains.

SEASONAL WINDS IN EUROPE
Depending on the season, different winds occur in the Mediterranean. The cold mistral, and bora (blue arrows), and warm föhn (orange) flow in winter; the hot leveche and sirocco (red) flow in summer.

Daily Winds

Some local winds occur at particular times of day. These daily winds, as they are known, include land–sea breezes (see right), and katabatic (see below) and anabatic winds, which flow up and down the sides of valleys due to changes in temperature. The haboob is a dust-storm that occurs in northern Sudan, most often late on a summer day, when a squall, producing strong downdrafts, moves over ground covered with loose sand and dust. It can raise pillars of dust to a great height, and when these merge, they form a wall of dust up to 24km (15 miles) long, advancing at about 56kph (35mph). The harmattan blows from the north or northeast, over west Africa to the south of the Sahara at any time of year but only by day. At night, the wind dies down. The northeasterly trade wind becomes very hot, dry, and dusty as it blows across the desert. Ordinarily hot air rises by convection, drawing in cooler air, but if the wind blows under a temperature inversion, where warmer air lies below cool air, the hot air is trapped, and that is when it becomes a harmattan.

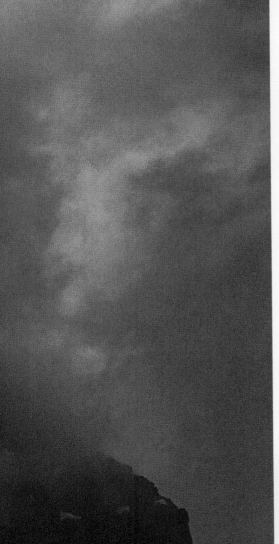

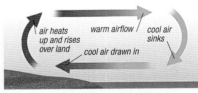

DAYTIME SEA BREEZE

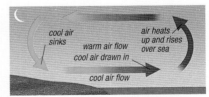

NIGHT-TIME LAND BREEZE

LAND–SEA BREEZES
During the day, land heats faster than water. Warm air rises over land, and cool air flows in to replace it, creating a sea breeze. At night, land cools more quickly than water, and the airflow is reversed. Similar winds develop over large lakes, such as the Great Lakes between Canada and the USA.

KATABATIC WIND FOG
At night, the ground cools, the temperature drops, and moisture condenses. The cold fog creates a katabatic wind by sinking into the valley, displacing warmer air, as here in the Grand Canyon, USA.

ATMOSPHERE

Thunderstorms

Thunderstorms are generated inside large cumulonimbus clouds (see p.479). Hailstones and ice crystals, colliding as they move up and down through the cloud, cause particles with positive electric charge to accumulate near the top of the cloud. Those with negative charge accumulate near the base and then induce a positive charge on the ground below. When the charge builds beyond a certain point, it discharges as a spark of lightning. Lightning flashing between different regions inside a cloud or between two clouds is called sheet lightning. Forked lightning flashes between a cloud and the ground. Lightning heats the adjacent air by up to 30,000°C (54,000°F) in less than a second, causing it to explode violently. Thunder is the sound of that explosion.

STORM CLOUDS
Frontal rain is a feature of temperate climates in spring. This front, passing over Colorado, USA, is accompanied by a line of thunderstorms.

THUNDERSTORM OVER BRAZIL
The high humidity and warm temperatures found near the equator provide ideal conditions for the formation of cumulonimbus cloud that results in thunderstorms.

Squall Lines

A squall line develops when active cumulonimbus clouds (see p.479) merge to form a continuous belt of storms, up to 1,000km (600 miles) long. The storms advance at right angles to the squall line. They begin along a cold front that is moving rapidly beneath warm, moist air. The warm air is lifted, and becomes unstable, producing towering clouds. Wind speed increases with height, pushing the tops of the clouds forwards so they overhang their bases as anvils. Air is drawn in ahead of the line, and cold air leaves as downdraughts in the rain behind the storms. Some of the cold air flows beneath the cloud, undercutting the updraught and producing what is known as a gust front. This front is so vigorous that it makes the squall line move faster than the cold front that gave rise to it. The two then become separated, with the squall line travelling ahead of the cold gust front.

FULGURITE
This curiously shaped mineral is a type of glass made from sand grains fused together by the heat of a lightning strike.

STORMY WEATHER
A squall line in the Atlantic Ocean, bringing heavy precipitation, gale-force winds, and thunderstorms. It may travel a huge distance before finally expending all of its energy.

Dust-Devils and Sandstorms

A dust-devil is a column of rapidly spinning air that rises without warning and dies down just as suddenly. Most are about 30m (100ft) tall, but they can rise to 100m (300ft) and occasionally even to 1,800m (6,000ft). They last for only a few minutes, snaking and moving erratically over the ground, but as one dies another often rises nearby. Dust-devils are caused by the uneven heating of the ground. Rocks heat at different rates, so by early afternoon in a desert some patches of ground can be up to 17°C (30°F) hotter than others. Air rises over the hot patches, and cooler air spirals in to replace it. The wind lifts dust and sand, making the dust-devil visible. As the cool air reduces the surface temperature, it is not long before the temperature of the hot patch drops to match that of its surroundings. Deprived of energy, the dust-devil disappears. A sandstorm occurs when a wind of 55kph (35mph) or more coincides with very unstable air, usually at a front. The wind raises dust and sand, but not high enough for it to travel very far. However, if the air is unstable, vertical air currents lift the dust and sand to a great height, over a vast area. In March 1998, a sandstorm passed over Egypt and parts of Lebanon and Jordan, reducing the visibility to about 180m (600ft). It meant that both Cairo Airport and the Suez Canal had to be closed.

DUST-DEVIL
Pockets of circulating air become visible as they collect fine particles of dust. This dust-devil has developed on a hot patch of ground in the savanna region of Kenya.

DUST-STORM OVER THE SAHARA
This dust-storm, seen from space, covers a vast area of Libya and Algeria. Visibility on the ground will be reduced to almost zero.

WALKING INTO THE WIND
This male lion in South Africa walks into the wind and dust, possibly to maximize his chance of scenting prey and stalking it successfully.

Tornadoes

A tornado is the fiercest wind on Earth, reaching speeds of up to 480kph (300mph). Tornadoes can happen anywhere, but most occur on the Great Plains of the USA. When the wind changes speed or direction near the top of a cumulonimbus cloud (see p.479), it tilts the upcurrents. Cold downcurrents then fall to the side of the rising air without cooling it. This allows the cloud to grow bigger and last longer than an ordinary storm cloud. The rising air starts rotating slowly, starting near the top of the cloud and extending downwards, getting narrower as it does so. As the radius of the spiral decreases, the rate of rotation increases. The rotating air continues downwards, forming a funnel below the cloud. Air drawn into the funnel enters an area of much lower pressure, so it expands, cools, and gives up its moisture. A funnel becomes a tornado on touching the ground.

TORNADO
The funnel of this developing tornado is snaking down from a cumulonimbus cloud over Colorado, USA. If the funnel touches the ground, it will become a tornado, changing colour as dust and debris are drawn into the vortex.

TORNADO FORMATION
A tornado begins to form as a funnel, emerging beneath a cloud (1). It extends downwards (2) and narrows at the base (3). On touching the ground (4), it raises a cloud of dust and debris.

TORNADO APPROACHING
This tornado is in an area called Tornado Alley, a region of the Great Plains, USA, that has more tornadoes than anywhere else in the world.

ATMOSPHERE

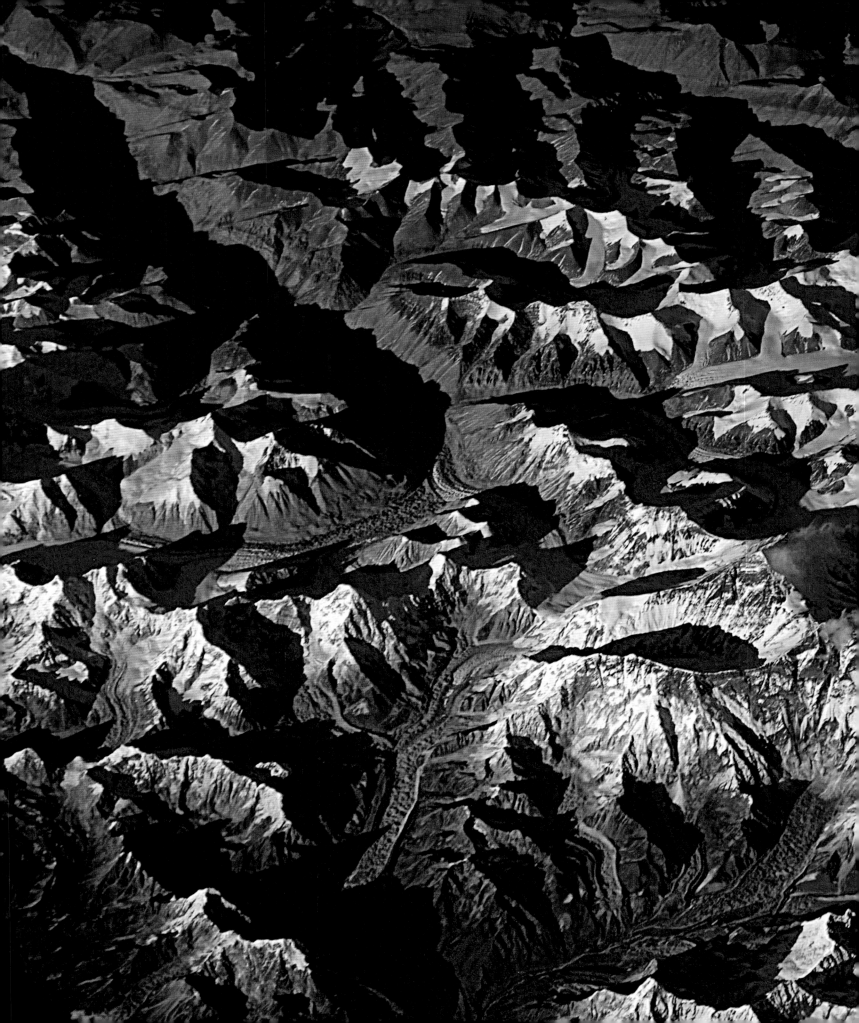

TECTONIC EARTH

The Earth's Plates

40–41 The crust

86–87 Tectonic plates

88–89 Plate boundaries

122–35 Fault systems

136–53 Mountains

154–77 Volcanoes

386–89 Ocean tectonics

400–01 Oceans of the world

The shape and geology of the Earth's surface is mainly a result of billions of years of slow shifting, making, and breaking of the Earth's tectonic plates. There are seven large plates (the Pacific, African, North American, South American, Eurasian, Australian, and Antarctic) and a dozen or so minor ones. Six of the major plates carry a continent, meeting adjacent plates under the ocean. Oceanic margins have spreading ridges, where plates grow, and subduction zones, where one plate is consumed. Of the large plates, only the Pacific lies entirely underwater, as do most of the minor plates.

Plate Movement

Tectonic plates have been moving over the Earth's surface since they first formed billions of years ago. Today, they are moving at about 2–20cm (³⁄₄–7³⁄₄in) a year, the rate at which fingernails grow, and future geological time will see continuing changes to the positions of the continents. The most striking movements are the widening of the Atlantic, which is pushing the Americas west; the splitting of Arabia from Africa down the Red Sea; and the pushing of India up against Asia, which has thrust up the Himalayas. Most dramatic of all is the rapid subduction of the western edge of the Pacific plate, which makes eastern Asia uniquely prone to violent volcanic eruptions and earthquakes.

VOLCANIC ACTIVITY
Volcanoes, such as these in the Galapagos Islands, are fed by a hotspot, the lava rising from within the deep mantle.

FOLDING
Rock strata buckle and fold when plates collide, as here in Utah, USA. Such collisions thrust up long mountain chains at continental margins.

FAULT LINE
This ancient fault line in Arizona, USA, is not close to the edge of a tectonic plate. Rather it is in a broad zone of distributed tectonic activity.

EARTH GEOLOGY AND TECTONICS
Although tectonic activity constantly alters the Earth's surface, each continent has a stable core of ancient, crystalline rock, called a craton, forming a base on which the younger rocks are welded. Where exposed at the surface, cratons often form vast, flat plains in continental interiors. They are usually covered by a thin layer of younger sediments. Around the margins of continents, volcanoes add new igneous rocks, and mountain ranges are uplifted as rock strata buckle under pressure caused by the collision of tectonic plates.

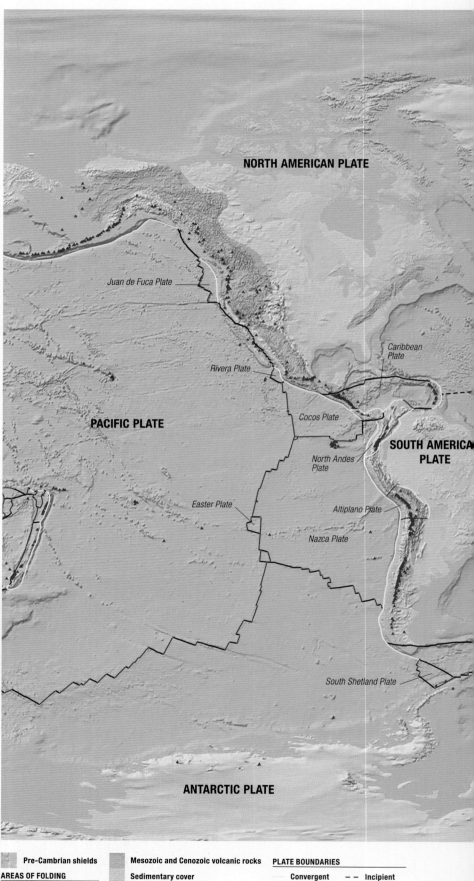

NORTH AMERICAN PLATE

Juan de Fuca Plate

Rivera Plate

Caribbean Plate

Cocos Plate

PACIFIC PLATE

North Andes Plate

SOUTH AMERICA PLATE

Easter Plate

Altiplano Plate

Nazca Plate

ANTARCTIC PLATE

South Shetland Plate

Pre-Cambrian shields

AREAS OF FOLDING

Paleozoic

Mesozoic

Cenozoic

Mesozoic and Cenozoic volcanic rocks

Sedimentary cover

▲ Active volcanoes

PLATE BOUNDARIES

Convergent   – – Incipient

— Divergent

— Transform

– – Uncertain

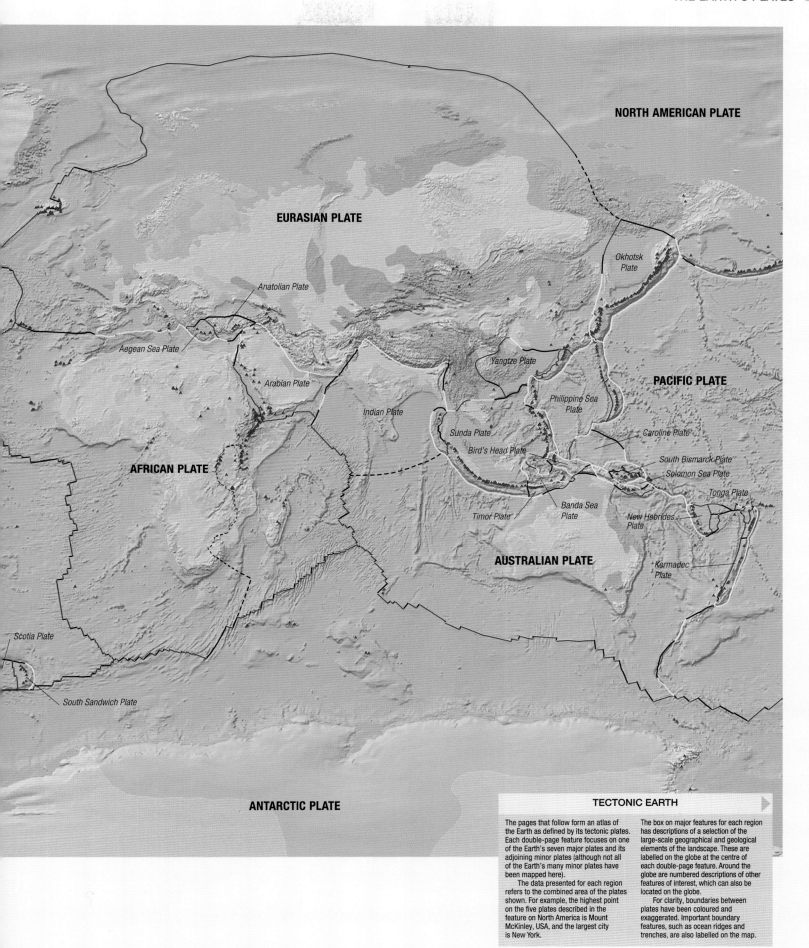

NORTH AMERICAN PLATE

EURASIAN PLATE

Okhotsk Plate

Anatolian Plate

Aegean Sea Plate

Yangtze Plate

PACIFIC PLATE

Arabian Plate

Indian Plate

Philippine Sea Plate

Caroline Plate

Sunda Plate

Bird's Head Plate

South Bismarck Plate

Solomon Sea Plate

AFRICAN PLATE

Tonga Plate

Banda Sea Plate

Timor Plate

New Hebrides Plate

AUSTRALIAN PLATE

Kermadec Plate

Scotia Plate

South Sandwich Plate

ANTARCTIC PLATE

TECTONIC EARTH

The pages that follow form an atlas of the Earth as defined by its tectonic plates. Each double-page feature focuses on one of the Earth's seven major plates and its adjoining minor plates (although not all of the Earth's many minor plates have been mapped here).

The data presented for each region refers to the combined area of the plates shown. For example, the highest point on the five plates described in the feature on North America is Mount McKinley, USA, and the largest city is New York.

The box on major features for each region has descriptions of a selection of the large-scale geographical and geological elements of the landscape. These are labelled on the globe at the centre of each double-page feature. Around the globe are numbered descriptions of other features of interest, which can also be located on the globe.

For clarity, boundaries between plates have been coloured and exaggerated. Important boundary features, such as ocean ridges and trenches, are also labelled on the map.

TECTONIC EARTH

North American

The sixth largest plate, the North American Plate was joined to the Eurasian, African, and South American Plates, until the former supercontinent of Pangea started to break up and the Atlantic Ocean began to open. The relatively young mountains of western North America reflect on-going subduction and collision with the Pacific Plate. The more ancient Appalachian mountains of eastern North America reflect older Paleozoic collisions in a pre-Pangean Atlantic Ocean. The Cocos and Rivera plates are currently subducting beneath the southern edge of North America to produce the active Mexican Volcanic Belt.

PLATES ❶ North American Plate
❷ Caribbean Plate ❸ Cocos Plate
❹ Rivera Plate ❺ Juan de Fuca Plate

COMBINED AREA 62 million square km
(24 million square miles)

HIGHEST POINT Mount McKinley, USA,
6,194m (20,320ft)

LOWEST POINT Puerto Rico Trench
8,648m (28,374ft) below sea-level

LARGEST CITY New York (8.5 million)

① **MOUNT MCKINLEY**
The highest peak in North America is Mount McKinley in the Alaska Range, USA. Its granite outline has been shaped by glaciers that have blanketed the region for the last 2 million years.

② **BADLANDS**
This part of South Dakota, USA, is made up of ridges, separated by deep ravines and gullies, that formed as violent cloudbursts eroded away the soft clay rock. The area covers 2,600 square km (1,000 square miles).

③ **BRYCE CANYON**
At Bryce Canyon in Utah, USA, rivers cut deep into the Paunsaguant Plateau, and carved out tall, thin ridges, known as fins. Over time they were eroded into spectacular sandstone pinnacles and spires called hoodoos.

④ **YOSEMITE**
The granite mountains of the Sierra Nevada, USA, were shaped by glaciers in the last ice age, creating spectacular rock formations, and cliffs up to 1,000m (3,300ft) high.

⑤ **DEATH VALLEY**
This is the lowest, hottest, and driest place in North America, descending to 86m (282ft) below sea-level, with shade temperatures up to 57°C (134°F) and years with no rain at all.

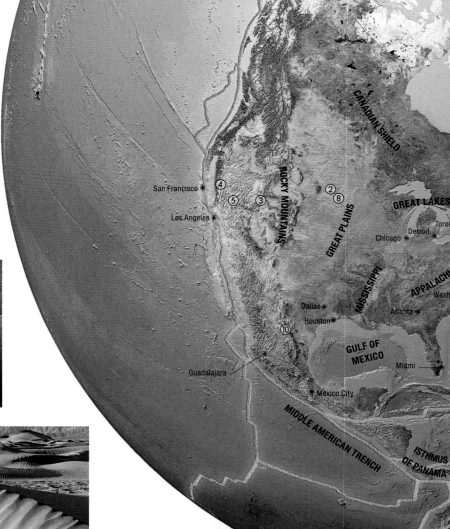

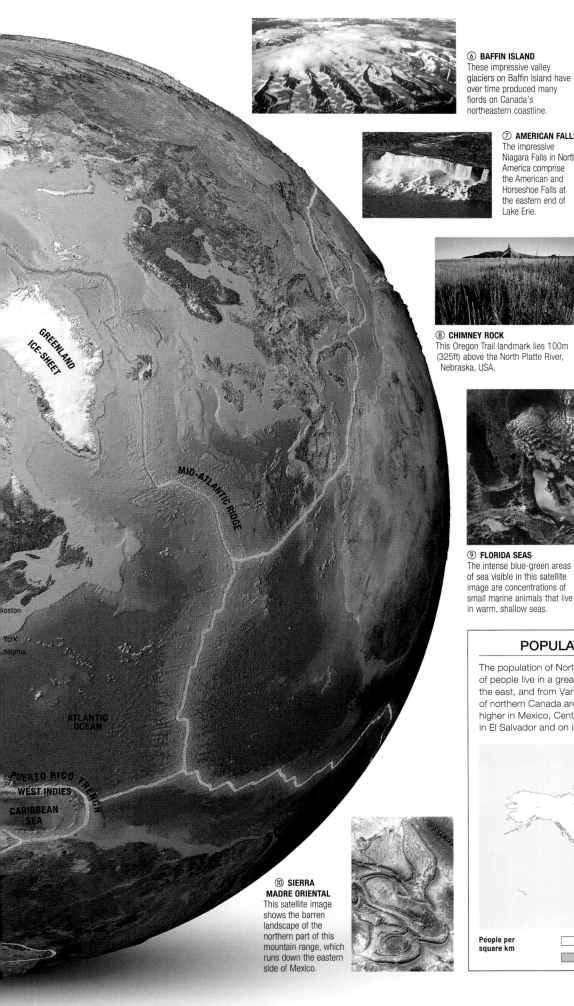

GREENLAND
ICE-SHEET

MID-ATLANTIC RIDGE

ATLANTIC
OCEAN

PUERTO RICO TRENCH

WEST INDIES

CARIBBEAN
SEA

oston

York
delphia

⑥ **BAFFIN ISLAND**
These impressive valley glaciers on Baffin Island have over time produced many fiords on Canada's northeastern coastline.

⑦ **AMERICAN FALLS**
The impressive Niagara Falls in North America comprise the American and Horseshoe Falls at the eastern end of Lake Erie.

⑧ **CHIMNEY ROCK**
This Oregon Trail landmark lies 100m (325ft) above the North Platte River, Nebraska, USA.

⑨ **FLORIDA SEAS**
The intense blue-green areas of sea visible in this satellite image are concentrations of small marine animals that live in warm, shallow seas.

⑩ **SIERRA MADRE ORIENTAL**
This satellite image shows the barren landscape of the northern part of this mountain range, which runs down the eastern side of Mexico.

MAJOR FEATURES

GREENLAND ICE-SHEET
Containing a tenth of the world's ice, the Greenland Ice-sheet is up to 3,000m (9,842ft) thick in places.

CANADIAN SHIELD
Formed some three billion years ago, the ancient crystallized rocks of the Canadian Shield cover a vast area of North America.

GREAT LAKES
Water from glaciers created the Great Lakes at the end of the last ice age, 15,000 years ago.

GREAT PLAINS
Up to 1,000km (625 miles) wide, the Great Plains are a vast area that stretches down through the middle of the continent.

ROCKY MOUNTAINS
These mountains run from California to Alaska, a distance of about 6,000km (3,700 miles).

APPALACHIANS
These are among the oldest of North America's mountain ranges, formed in Paleozoic times, over 250 million years ago.

MISSISSIPPI
The second longest river in the USA after its tributary the Missouri, the Mississippi flows 3,780km (2,350 miles) from Minnesota to the Gulf of Mexico.

ISTHMUS OF PANAMA
A strip of land just 58km (36 miles) wide that appeared about 3 million years ago, linking North and South America.

POPULATION AND BOUNDARIES

The population of North America is unevenly spread. Huge numbers of people live in a great arc of cities from Chicago to Washington in the east, and from Vancouver to Los Angeles in the west. Vast areas of northern Canada are virtually uninhabited. Average densities are higher in Mexico, Central America, and the Caribbean, being highest in El Salvador and on islands such as Barbados and Haiti.

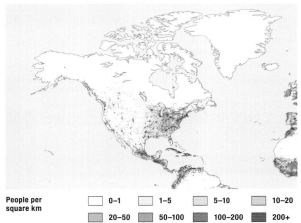

People per square km				
0–1	1–5	5–10	10–20
20–50	50–100	100–200	200+

TECTONIC EARTH

South American

The smallest of the seven large tectonic plates, half of the South American Plate is under the Atlantic. Its northern boundary is washed by the waters of the Caribbean, but its southern tip at Cape Horn is less than 1,000km (620 miles) from the Antarctic Circle. Down the west coast are the Andes, thrust up as the South American Plate moves west over the subducting Nazca Plate that lies beneath the Pacific. The Nazca Plate is the fastest moving of all plates, and as it dives steeply under South America it generates earthquakes and volcanoes along the length of the Andes.

PLATES ❶ South American Plate
❷ South Sandwich Plate
❸ Scotia Plate ❹ Nazca Plate
❺ Juan Fernandez Plate ❻ Easter Plate
❼ Panama Plate ❽ North Andes Plate
❾ Altiplano Plate

COMBINED AREA 60 million square km (23 million square miles)

HIGHEST POINT Mount Aconcagua, Argentina, 6,960m (22,834ft)

LOWEST POINT South Sandwich Trench, 9,334m (30,615ft) below sea-level

LARGEST CITY São Paulo (23.08 million)

MAJOR FEATURES

AMAZON BASIN
The rainforest in the Amazon Basin covers an area larger than western Europe.

ANDES
Winding all the way down the west side of South America, the Andes are the world's longest mountain range.

ATACAMA DESERT
This windswept plain, 1,000m (3,300ft) above sea-level, is the world's driest desert. Some areas have no record of any rainfall.

PAMPAS
The extensive plains in Argentina and Uruguay are used for growing corn and for grazing.

PATAGONIA
Bitterly cold and windswept, Patagonia is virtually a desert, where almost nothing grows except in the river valleys.

TIERRA DEL FUEGO
The southern tip of this island, called Cape Horn, is South America's most southerly point.

SCOTIA SEA
Along the northern edge of the Scotia Sea, the Scotia Plate is moving beneath South America, creating a trench in the ocean floor and an arc of volcanic islands.

PERU–CHILE TRENCH
Where the Nazca Plate meets the South American Plate, the Peru–Chile trench has developed. It is 6,000km (3,700 miles) long and up to 8,065m (26,460ft) deep.

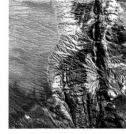

① ALTIPLANO
This high plateau (seen from directly overhead in this satellite image) extends from Peru through Bolivia. Its rolling plains lie between the two mountain ranges of the Andes at about 3,660m (12,000ft) and is covered with pyroclastic deposits.

② ACONCAGUA
The world's highest peak outside the Himalayas, the Andean mountain of Aconcagua in Argentina was uplifted by the subduction of the Nazca Plate below the South American Plate.

③ PATAGONIA GLACIER
This thermal satellite image shows part of southern Patagonia. A vast glacier, coloured white, has carved deep valleys in the landscape.

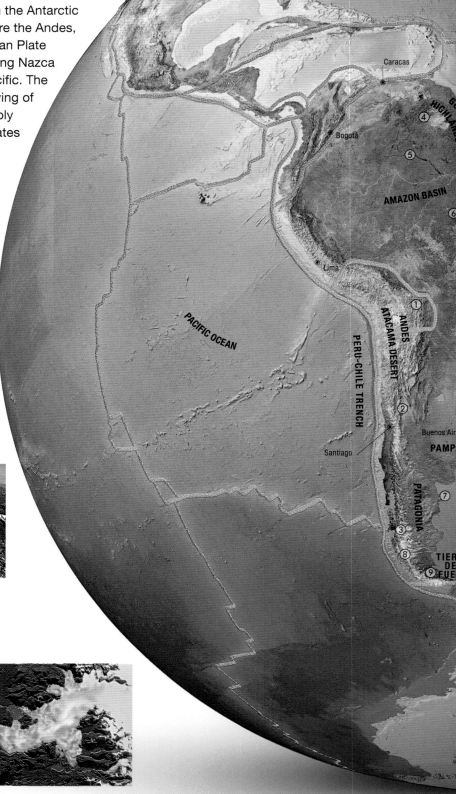

Caracas

Bogotá

HIGHLAND

AMAZON BASIN

Lima

PACIFIC OCEAN

ANDES

ATACAMA DESERT

PERU–CHILE TRENCH

Santiago

Buenos Air

PAMP

PATAGONIA

TIER DE FUE

④ TEPUIS

In Venezuela, ancient rocks form gigantic tablelands, known as tepuis. Spectacular waterfalls plunge over steep cliffs, which typically have pink, sandstone rock faces exposed by erosion.

⑤ RIO NEGRO

Stained black by rotting vegetation from Colombian swamps, the Rio Negro is the main tributary of the Amazon.

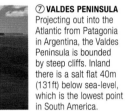

⑥ DEFORESTATION

Although the Amazon basin is the world's greatest region of tropical rainforest, it is being destroyed for its timber, by mining activities, and to make way for farms. In this picture, one side of the river is dominated by farmland (upper half) whereas the other side is still covered in rainforest (lower half).

⑦ VALDES PENINSULA

Projecting out into the Atlantic from Patagonia in Argentina, the Valdes Peninsula is bounded by steep cliffs. Inland there is a salt flat 40m (131ft) below sea-level, which is the lowest point in South America.

⑧ TORRES DEL PAINE

One of the world's last remaining wildernesses, the Torres del Paine in southern Chile has a unique range of wildlife, including Guanacos, Nandus, and Condors. The spectacular Torres del Paine are twin peaks of towering granite.

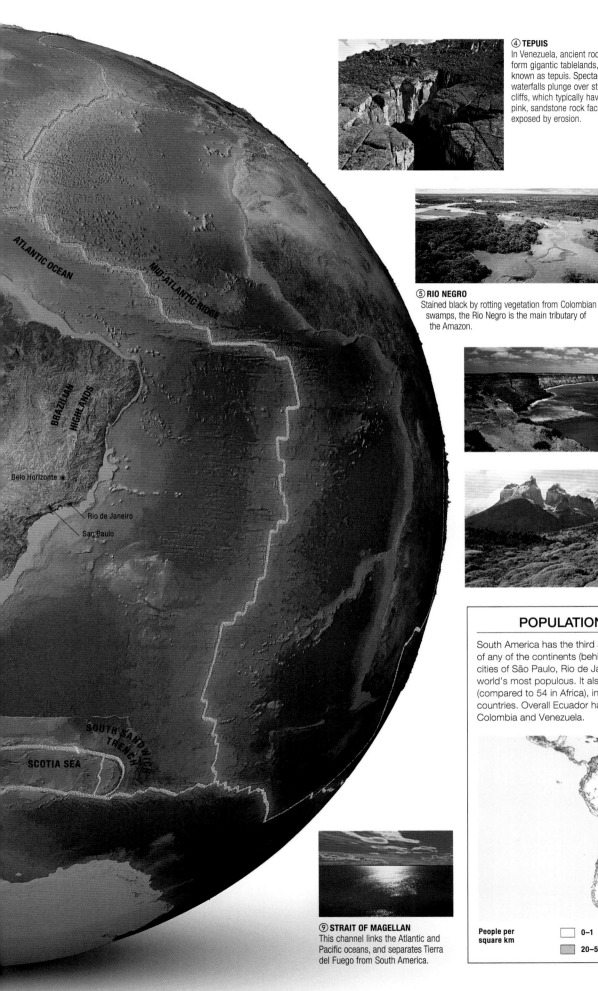

ATLANTIC OCEAN

MID-ATLANTIC RIDGE

BRAZILIAN HIGHLANDS

Belo Horizonte ◉

Rio de Janeiro

Sao Paulo

SOUTH SANDWICH TRENCH

SCOTIA SEA

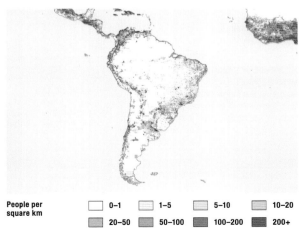

⑨ STRAIT OF MAGELLAN

This channel links the Atlantic and Pacific oceans, and separates Tierra del Fuego from South America.

POPULATION AND BOUNDARIES

South America has the third smallest population (nearly 400 million) of any of the continents (behind Australia and Antarctica), but the cities of São Paulo, Rio de Janeiro, and Buenos Aires are among the world's most populous. It also has the fewest countries – just 12 (compared to 54 in Africa), including Brazil, one of the world's largest countries. Overall Ecuador has the densest population, followed by Colombia and Venezuela.

People per square km	0–1	1–5	5–10	10–20
	20–50	50–100	100–200	200+

Eurasian

The Eurasian plate is the largest tectonic plate and has the most complex history of formation. To the west, the North American Plate is pulling away to open up the Atlantic Ocean. To the east, the Pacific and Philippine Sea plates are subducting beneath the Eurasian Plate, creating an arc of volcanic islands that includes southern Japan and the Philippines. To the south, the Indian and Australian plates are moving north. The collision of the Indian Plate with the Eurasian Plate has pushed up the world's highest mountain range, the Himalayas, and is continuing to do so.

PLATES ❶ Eurasian Plate
❷ Okhotsk Plate **❸** Philippine Sea Plate
❹ Mariana Plate **❺** Okinawa Plate
❻ Yangtze Plate **❼** Bird's Head Plate
❽ Banda Sea Plate **❾** Timor Plate
❿ Sunda Plate **⓫** Burma Plate
⓬ Indian Plate **⓭** Anatolian Plate
⓮ Aegean Sea Plate

COMBINED AREA 90 million square km
(35 million square miles)

HIGHEST POINT Mount Everest, Nepal,
8,850m (29,035ft)

LOWEST POINT Galathea Deep,
10,540m (34,580ft) below sea-level

LARGEST CITY Tokyo (37.7 million)

MAJOR FEATURES

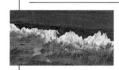

HIMALAYAS
These mountains started to form about 60 million years ago.

ALPS
The Alps are young in geological terms. They formed 10–25 million years ago as the African Plate collided with Europe.

MEDITERRANEAN SEA
About 5 million years ago, the Straits of Gibraltar appeared; water flowed in from the Atlantic to form the Mediterranean.

YANGTZE
Asia's longest river flows 6,300km (3,900 miles) from the Tibetan Highlands to the China Sea, near Shanghai.

DECCAN PLATEAU

The basaltic rocks covering the Deccan Traps erupted about 65 million years ago.

GOBI DESERT
This is named after the small stones, called gobi, that remain on the surface after the wind has removed any finer material.

ISLAND ARC OF SUMATRA, JAVA, AND THE PHILIPPINES
Along the southeastern edge of the Eurasian Plate is an island arc with some of the most violent subduction-related volcanoes in the world, such as Krakatau.

BOREAL FOREST
Also known as taiga, this is the climax vegetation of northern Asia.

① MOSELLE
The valley of the Moselle River follows a narrow, winding course through western Germany to join the Rhine at Koblenz.

② METEORA
The rock pinnacles of Metéora in Greece are all that remains of a high sandstone plateau that was traversed by innumerable faults and eroded over millions of years.

③ MOUNT EVEREST
Satellite measurements of the world's highest peak show that it is still being uplifted due to tectonic plate movement.

④ TIBETAN PLATEAU
The crust beneath this vast plateau is over 80km (50 miles) thick, making it the thickest part of the Earth's crust.

London
Essen
Paris
Berlin
Madrid
St Petersburg
URALS
ALPS
Moscow
MEDITERRANEAN SEA
Istanbul
BLACK SEA
Tehran
Lahore
Karachi
Ahmedabad
Mumbai
INDIAN OCEAN

⑤ FIORDS IN NORWAY

All along the coast of Norway there are deep sea inlets, called fiords, gouged out by glaciers during the last ice age.

⑥ BIALOWIESKA FOREST

The Bialowieska Forest on Poland's border with Belarus is one of the last remnants of the vast primeval forests that once covered much of Europe. It provides a refuge for wolves, bears, deer, wild boar, and European Bison.

⑦ LENA DELTA

Frozen solid during the bitter Siberian winter, the River Lena is transformed in June into a surging flood as the ice melts, and water flows through its green delta into the Laptev Sea.

⑧ TURPAN DEPRESSION

Created as a large block of the Earth's crust gradually slipped down between parallel faults, this rift valley is northern Asia's lowest point, descending to 154m (505ft) below sea-level.

⑨ GUILIN KARST

The bewitching pinnacles of the Guilin Hills in China are the world's best example of karst landscape. They were formed as the limestone rock was gradually dissolved by naturally acidic rainwater.

POPULATION AND BOUNDARIES

About 70 per cent of the world's population lives on the Eurasian Plate, but its distribution is very uneven. The vast majority live in Europe, the Indian subcontinent, southeast Asia, or China, and these are the most densely populated parts of the world. Up until the 1990s, much of northern Asia fell within the boundaries of the Soviet Union, but with its collapse, an array of independent states has emerged.

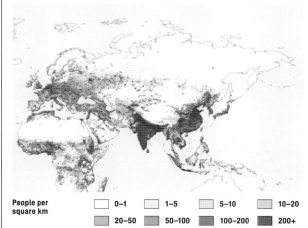

⑩ GANGES FLOODPLAIN

Flooding of the Ganges has created this fertile floodplain, which is intensively cultivated.

People per square km				
0–1	1–5	5–10	10–20	
20–50	50–100	100–200	200+	

Map labels: SIBERIA, GOBI DESERT, Harbin, Shenyang, Beijing, Seoul, Tianjin, YANGTZE, Shanghai, Wuhan, Chongqing, Guangzhou, Hong Kong, Taipei (Taibei), GALATHEA DEEP, Manila, PHILIPPINES, Nagoya, Tokyo, Osaka, HIMALAYAS, Dhaka, ECCAN ATEAU, Kolkata, Yangon, Bangkok, Ho Chi Minh City, Hyderabad, aloru, Chennai, Jakarta, SUMATRA, JAVA

African

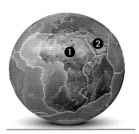

As well as the African continent, the African Plate includes parts of the Atlantic, Indian, and Southern oceans. Unlike other major plates, it has no great range of fold mountains, only the ancient Atlas range in the northwest. Much of the continent has been warped into saucer-shaped basins, and highlands. The south and east is mostly a vast high plateau, broken by a few mountain regions and rift valleys, notably the Great Rift Valley. The north and west is much lower and flatter. To the northeast lies the Arabian Plate, recently split from the African Plate by the opening of the Red Sea, which is growing wider all the time.

PLATES ❶ African Plate ❷ Arabian Plate

COMBINED AREA 83 million square km (32 million square miles)

HIGHEST POINT Mt. Kilimanjaro, Tanzania, 5,895m (19,340ft)

LOWEST POINT Somali Basin, 5,826m (19,335ft) below sea level

LARGEST CITY Cairo (20.2 million)

MAJOR FEATURES

ATLAS MOUNTAINS
These mountains run parallel to the northwestern edge of Africa, passing through Morocco, Algeria, and Tunisia.

NILE
The second longest river in the world, the Nile flows north from several main sources, including the Luvironza River in Burundi, to the Mediterranean Sea.

SAHARA DESERT
The largest desert in the world, the Sahara covers almost a third of the African continent.

ARABIAN PENINSULA
Almost the whole of the Arabian Plate is a vast, sandy desert. The Red Sea borders its western edge.

GREAT RIFT VALLEY
This valley in east Africa has opened up over the last 30 million years as the continent has slowly started to split apart.

CONGO BASIN
These forest elephants live in the tropical rainforests of the Congo Basin in central Africa.

MADAGASCAR
Separated from Africa 95 million years ago, Madagascar has a unique fauna, including its lemurs.

MID-ATLANTIC RIDGE
This long seafloor-spreading ridge marks the western boundary of the African Plate.

① NEFTA OASIS
A green island in the desert, the Nefta oasis in Tunisia has 152 natural springs that are artesian wells.

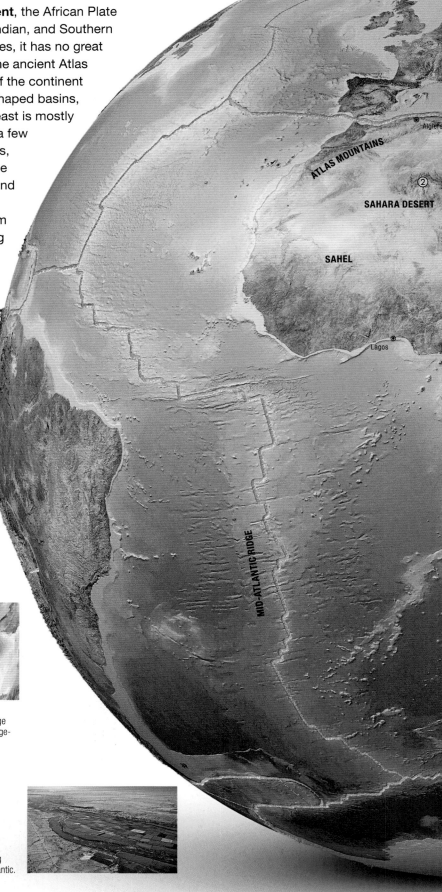

② AHAGGAR MOUNTAINS
This vast lunar landscape in Algeria is formed from rocks about 2 billion years old — some of the oldest in Africa.

③ ACACUS–AMSAK REGION
This satellite image shows three large rock massifs exposed between orange-coloured sand dunes: Tassili (left), Tadrart–Acacus (centre), and the crescent-shaped Amsak (right).

④ ORANGE RIVER
This is the longest river in southern Africa, flowing from the Drakensberg Plateau in Lesotho west into the Atlantic.

ATLAS MOUNTAINS

Algiers

②

SAHARA DESERT

SAHEL

Lagos

MID-ATLANTIC RIDGE

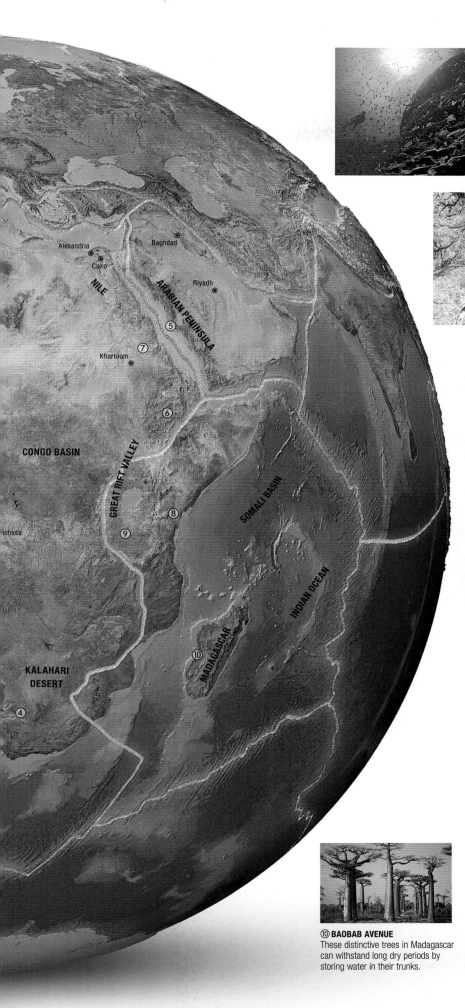

Alexandria
Baghdad
Cairo
NILE
ARABIAN PENINSULA
Riyadh
⑤
Khartoum ⑦
CONGO BASIN
GREAT RIFT VALLEY
⑥
SOMALI BASIN
⑧
⑨
shasa
INDIAN OCEAN
MADAGASCAR
⑩
KALAHARI
DESERT
④

⑤ RED SEA
This is one of the world's warmest seas, growing as the Arabian and African plates split apart. Its clear waters provide an ideal environment for corals.

⑥ ETHIOPIAN HIGHLANDS
Although split by the Great Rift Valley, the Ethiopian highlands form the largest mountainous region in Africa. Many areas are higher than 4,000m (13,000 ft).

⑦ KERAF SUTURE
This radar image reveals an approximately north–south line, known as the Keraf Suture, in northern Sudan. It shows where two ancient continents collided about 650 million years ago.

⑧ MOUNT KILIMANJARO
The highest peak in Africa, Mount Kilimanjaro formed about 1.8 million years ago, growing to a height of 5,895m (19,340 ft). Today there are three volcanoes, Shira, Mawenzi, and Kibo, nested in its summit.

⑨ EAST AFRICAN SAVANNA
The tropical grasslands that cover much of east Africa are famed for their wildlife, which includes lions, elephants, and vast herds of antelope and zebra.

POPULATION AND BOUNDARIES

In terms of area and population, Africa is the second largest continent, but vast regions are almost completely empty. Most of the main population centres are in countries near the sea, for example, along the north coast and down the Nile Valley, including Cairo. The one major inland population centre is around lakes Victoria and Tanganyika. Africa has 54 countries, more than any other continent.

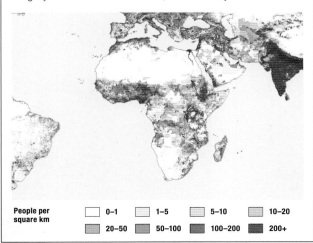

⑩ BAOBAB AVENUE
These distinctive trees in Madagascar can withstand long dry periods by storing water in their trunks.

People per square km	0–1	1–5	5–10	10–20
	20–50	50–100	100–200	200+

Australian

Australia, Antarctica, and New Guinea split away from the other continents about 200 million years ago. Then Antarctica broke away from Australia and New Guinea about 50 million years ago, leaving them isolated. The gap between them is still widening, along the Indian Ocean Ridge. Due to its isolation, Australia is one of the most stable of all the continents, made mostly of ancient rocks. The western plateau is a Precambrian shield, 570–3,500 million years old. Even the Great Dividing Range of mountains, down the east side, formed hundreds of millions of years ago and is now extremely eroded.

PLATES ❶ Australian Plate
❷ Maoke Plate ❸ Woodlark Plate
❹ Futuna Plate ❺ Niuafo'ou Plate
❻ Tonga Plate ❼ Kermadec Plate

AREA 46 million square km
(18 million square miles)

HIGHEST POINT Mount Wilhelm, Papua New Guinea, 4,509m (14,794ft)

LOWEST POINT Java Trench,
7,450m (24,440ft) below sea-level

LARGEST CITY Sydney (4.7 million)

① **LAKE CARNEGIE**
Australia's interior is so flat that rivers drain into inland lakes, rather than to the sea. With barely enough rainfall to keep rivers flowing, many of these lakes are ephemeral. This satellite image of Lake Carnegie, in Western Australia shows one of the rare occasions when the lake contains water.

② **PINNACLES**
The dry conditions of Western Australia have produced many extraordinary rock formations; none more so than the Pinnacles, 150,000 limestone stacks up to 4m (13ft) tall near Cervantes.

POPULATION AND BOUNDARIES

Australia is the most sparsely populated of all the continents except Antarctica, with a total of more than 20 million people and an average of less than 4 people per square kilometre (0.4 square miles). The interior of Australia is almost empty; most people live near the southeast coast or in a small area around Perth in the southwest. The western part of the island of New Guinea is more densely populated.

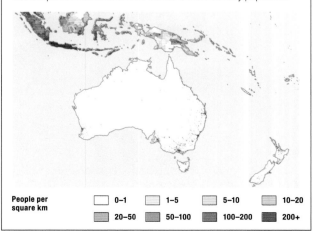

People per square km	0–1	1–5	5–10	10–20
	20–50	50–100	100–200	200+

JAVA TRENCH

INDIAN OCEAN

GREAT SA
DESERT

① GIBSON
DESERT

②

INDIAN OCEAN RIDGE

③ **SAND DUNES**
As well as being famous for its caves and karst scenery, the Nullarbor Plain has some spectacular sand dunes along its southern edge, such as these on the coast of South Australia at Eucla.

④ BUNGLE BUNGLES
In Western Australia, there is an extraordinary landscape of beehive-shaped, orange and black sandstone pinnacles up to 200m (660ft) tall, called the Bungle Bungles.

MAJOR FEATURES

BARKLY TABLELAND
This flat, grassland region in the Northern Territory has become Australia's foremost area for rearing beef cattle.

GREAT DIVIDING RANGE
Running down the east side of Australia, the Dividing Range is so called because it separates rivers that flow east to the coast from those that run west to the interior.

GREAT SANDY DESERT
This is one of the many deserts of Australia's interior. Rainfall is patchy, falling mainly in an isolated downpour, and there are many drought years.

GREAT BARRIER REEF
The world's largest living structure, this reef extends 2,570km (1,600 miles) along the coast of Queensland.

MURRAY–DARLING
Rising in the Snowy Mountains, the Murray is Australia's longest river. The Darling flows only sporadically.

NULLARBOR PLAIN
This vast, flat, limestone plain in South Australia is an ancient sea-floor that was uplifted 25 million years ago.

TASMAN SEA
This is one of the world's stormiest seas, due to its proximity to a wind belt called the Roaring Forties.

⑤ COPPER AND GOLD
Papua New Guinea has some of the world's richest copper and gold deposits. This giant quarry is in the Star Mountains.

⑥ CAPE YORK
The northernmost point of the Australian mainland, Cape York is at the head of a peninsula, covered in dense tropical rainforest on the east and grassland to the west.

⑦ ULURU
The world's largest freestanding rock, Uluru formed when beds of sandstone, about 450 million years old were tilted 90 degrees by crustal movement, then eroded by wind and water.

⑧ BUSH FIRES
Bush fires, here fanned by prevailing winds, spread rapidly through the dry eucalypt vegetation of New South Wales. People living on forest fringes are particularly at risk.

⑨ WHITE ISLAND
White Island in the Bay of Plenty, off the coast of New Zealand's North Island, is the tip of a huge volcano that rises from the sea-floor.

⑩ MOUNT COOK
New Zealand's highest peak, Mount Cook is in the Southern Alps on South Island. Standing 4,744m (12,284ft) high, it was forced up by movements of the Australian and Pacific plates.

TECTONIC EARTH

Pacific

About 85 million years ago, the Pacific Plate was one of several in the Pacific Ocean, but as it widened and spread northwestwards, the other plates were mostly subducted under the Americas. The Pacific Plate moves northwest at about 100mm (4in) a year. Its leading edge turns down sharply under the Eurasian, Philippine Sea, Australian, Caroline, and Bismarck plates, creating major subduction zones and the world's deepest ocean trenches. Much of this plate is featureless, but volcanic islands and undersea volcanoes occur where plumes of magma have burst through the plate.

PLATES ❶ Pacific Plate
❷ Balmoral Reef Plate
❸ Conway Reef Plate
❹ New Hebrides Plate
❺ Solomon Sea Plate
❻ South Bismarck Plate
❼ North Bismarck Plate
❽ Caroline Plate

AREA 108 million square km
(42 million square miles)

HIGHEST POINT Mauna Kea, Hawaii,
4,205m (13,796ft)

LOWEST POINT Challenger Deep,
11,033m (36,198ft) below sea-level

LARGEST CITY Honolulu (337,000)

① MOUNT JUMULLONG
This mountain is inland from Cetti Bay in Guam, one of the Mariana Islands. These volcanic islands have formed along the edge of the Philippine Plate as the Pacific Plate is thrust down beneath it.

② MARSHALL ISLANDS
Many of the 34 Marshall Islands are ring-shaped reefs, called atolls, that form as corals grow in the waters around a volcanic island, now submerged beneath the ocean surface. The islands include some of the world's largest atolls.

③ TONGA
This is the Vivau Island Group to the east of Tonga, which are coral islands. To the west, volcanic islands formed along the edge of the Australian Plate as the Pacific Plate subducted beneath it.

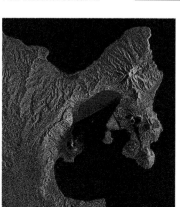

④ RABAUL VOLCANO
This satellite image shows Blanche Bay, New Britain, a semi-circular body of water formed from the submerged caldera of Rabaul volcano. New cones have developed and are visible around the margin of the old caldera.

⑤ NEW ZEALAND FIORD
Milford Sound, on the southwest coast of the South Island, is the most spectacular of New Zealand's fiords. It is a drowned valley that was carved out by a glacier about 15,000 years ago.

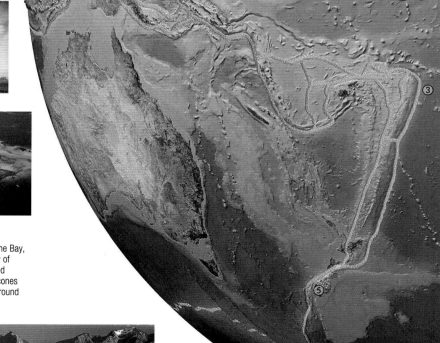

RING OF FIRE

RING OF FIRE

RING OF FIRE

HAWAIIAN-EMPEROR CHAIN

MARIANA TRENCH

CHALLENGER DEEP

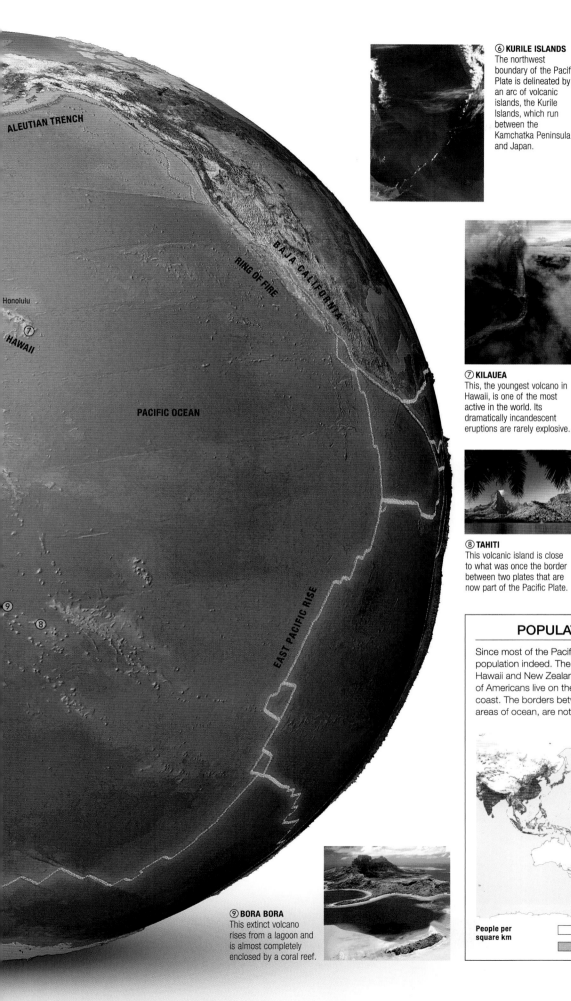

ALEUTIAN TRENCH

BAJA CALIFORNIA

RING OF FIRE

Honolulu

⑦

HAWAII

PACIFIC OCEAN

EAST PACIFIC RISE

⑨

⑧

⑥ KURILE ISLANDS

The northwest boundary of the Pacific Plate is delineated by an arc of volcanic islands, the Kurile Islands, which run between the Kamchatka Peninsula and Japan.

⑦ KILAUEA

This, the youngest volcano in Hawaii, is one of the most active in the world. Its dramatically incandescent eruptions are rarely explosive.

⑧ TAHITI

This volcanic island is close to what was once the border between two plates that are now part of the Pacific Plate.

⑨ BORA BORA

This extinct volcano rises from a lagoon and is almost completely enclosed by a coral reef.

MAJOR FEATURES

RING OF FIRE

The Pacific Plate is nearly encircled by a string of active, often violently explosive volcanoes, called the Ring of Fire.

BAJA CALIFORNIA

Extending 1,300km (800 miles) south into the Pacific, this rugged peninsula is mainly desert.

HAWAIIAN–EMPEROR CHAIN

About 100 volcanic islands and seamounts developed over a fixed mantle hotspot, as the Pacific Plate moved northwest over it.

CORAL REEFS

The warm, shallow waters of the South Pacific are ideal for coral growth. There are both long, fringing reefs and atolls.

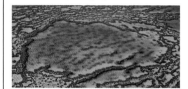

SOUTH ISLAND, NEW ZEALAND

The Pacific Plate is riding over the Australian Plate to create this island.

EAST PACIFIC RISE

Situated between the Pacific and Nazca plates, this rise contains one of the world's fastest spreading ridge systems.

MARIANA TRENCH

The lowest point on Earth is in the Mariana Trench, where the Pacific Plate starts to pass under the Philippine Plate.

POPULATION AND BOUNDARIES

Since most of the Pacific Plate is covered by ocean, it has a very small population indeed. The only islands with any sizeable population are Hawaii and New Zealand's South Island. However, a huge number of Americans live on the Pacific Plate's margins along the California coast. The borders between island groups, because they include areas of ocean, are not well defined other than by a reference map.

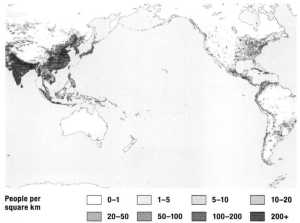

People per square km	0–1	1–5	5–10	10–20
	20–50	50–100	100–200	200+

TECTONIC EARTH

Antarctic

PLATES ❶ Antarctic Plate
❷ South Shetland Plate

COMBINED AREA 58 million square km
(22 million square miles)

HIGHEST POINT Vinson Massif,
4,897m (16,067ft)

LOWEST POINT Bentley Subglacial
Trench, 2,538m (8,327ft) below sea-level

LARGEST CITY None

Larger than Australia, the continent of Antarctica lies in the centre of the Antarctic Plate. High mountain peaks are visible, but the continent is almost entirely covered by ice, formed over 15 million years and now extending beyond the land to form floating ice shelves. Beyond the the ice, the Antarctic plate extends into the Southern Ocean for 1,500km (1,000 miles) or more in every direction. In fact, Antarctica is entirely surrounded by oceanic spreading ridges. The Antarctic continent was once two plates, West Antarctica and East Antarctica, which have now bonded along a line joining the Ross and Weddell Seas.

MAJOR FEATURES

ANTARCTIC PENINSULA
This peninsula stands out as a thin tail of land, projecting from an otherwise almost circular mass of land and ice.

TRANSANTARCTIC MOUNTAINS
These mountains include two (Erebus and Buckle Island) of Antarctica's three volcanoes that have erupted since 1900.

ROSS ICE-SHELF
This is the largest of the ice-shelves floating around the coast of Antarctica. It is about the same size as France.

GEOMAGNETIC SOUTH POLE
Situated near Russia's Vostok Station, the geomagnetic south pole is where magnetic currents in the upper atmosphere are vertical. The South Pole is at 90° south.

LAMBERT GLACIER
Not only is the Lambert Glacier the world's largest glacier, it is also the fastest moving, flowing 1,200m (4,000ft) or more a year.

ELLSWORTH MOUNTAINS
The highest peak in Antarctica, Vinson Massif, is found in these mountains at the head of the Ronne Ice-Shelf.

OZONE HOLE
The ozone hole forms each spring and is increasing in size. Gases released by human activity may be the cause.

① PALMER LAND
The narrow Antarctic Peninsula is formed of low mountains, such as those of Palmer Land. They were created by the same tectonic events that threw up the Andes mountains. These low mountains are permanently covered by a thin sheet of ice.

② MOUNT EREBUS
The world's most southerly active volcano, Erebus is the largest of three volcanoes on Ross Island. Daily, it ejects lava bombs and jets of steam from its icebound summit.

③ MCMURDO BASE
The largest settlement in Antarctica, McMurdo station is like a small town, with shops and banks. Even in winter about 200 people stay here, and in summer, visitors push the population up to 1,000.

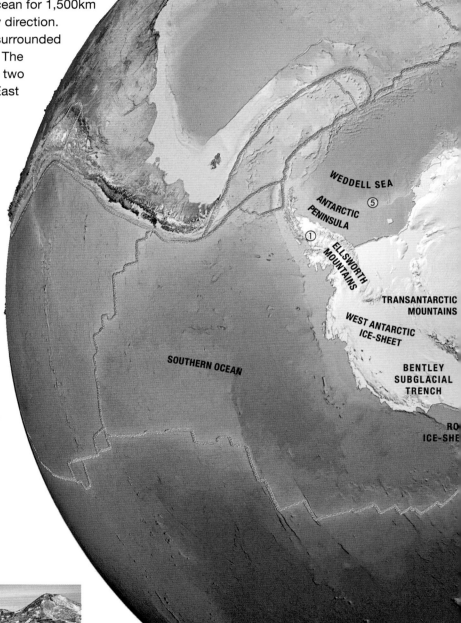

WEDDELL SEA

ANTARCTIC PENINSULA

⑤

①

ELLSWORTH MOUNTAINS

TRANSANTARCTIC MOUNTAINS

WEST ANTARCTIC ICE-SHEET

SOUTHERN OCEAN

BENTLEY SUBGLACIAL TRENCH

RO ICE-SHE

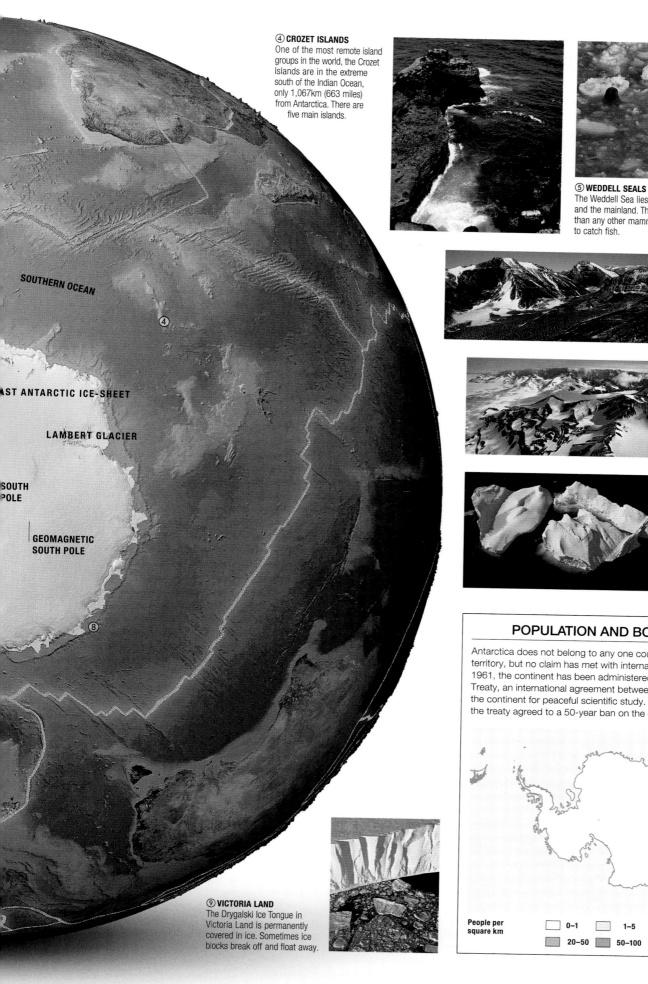

SOUTHERN OCEAN

④

ST ANTARCTIC ICE-SHEET

LAMBERT GLACIER

SOUTH POLE

GEOMAGNETIC SOUTH POLE

⑧

④ CROZET ISLANDS
One of the most remote island groups in the world, the Crozet Islands are in the extreme south of the Indian Ocean, only 1,067km (663 miles) from Antarctica. There are five main islands.

⑤ WEDDELL SEALS
The Weddell Sea lies between the Antarctic Peninsula and the mainland. The Weddell Seal lives further south than any other mammal, diving through holes in the ice to catch fish.

⑥ WRIGHT VALLEY
One of very few areas of land in Antarctica that is sometimes completely free of ice, the Wright Valley appears to be devoid of life but algae, bacteria, and fungi are all found there.

⑦ TERRA NOVA
Located at the foot of the Transantarctic Mountains in Victoria Land, Terra Nova overlooks a bay with the same name in the Ross Sea. The area is renowned for its spectacular ice caves.

⑧ ICEBERGS
These icebergs floating off the Adelie coast are huge. Usually bigger than Arctic icebergs, they last much longer, surviving up to ten years. The largest to date was 700km (435 miles) long and broke away from the Ross Ice-Shelf in 2002.

POPULATION AND BOUNDARIES

Antarctica does not belong to any one country. Seven nations claim territory, but no claim has met with international agreement. Since 1961, the continent has been administered under the Antarctic Treaty, an international agreement between 43 countries to preserve the continent for peaceful scientific study. In 1991, the members of the treaty agreed to a 50-year ban on the exploitation of minerals.

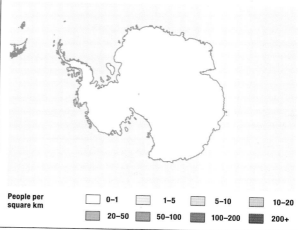

People per square km

| 0–1 | 1–5 | 5–10 | 10–20 |
| 20–50 | 50–100 | 100–200 | 200+ |

⑨ VICTORIA LAND
The Drygalski Ice Tongue in Victoria Land is permanently covered in ice. Sometimes ice blocks break off and float away.

GLOSSARY

In this glossary, terms defined within a larger entry are highlighted in **bold** type, while references to terms defined in other entries are identified with *italic* type.

A

ABLATION The loss of ice from a glacier due to melting, evaporation, or calving. The region of a glacier where there is a net loss of ice is known as the **ablation area**. See also *accumulation area*.

ABYSSAL Relating to the deep ocean floor and its environment, beyond the continental shelves and slopes. Abyssal depths are deeper than bathyal depths but not as deep as deep-sea trenches. The **abyssal plain** is the flat, sediment-covered plain at about 4,000–6,000m (13,000–20,000ft) depth that forms the bed of most of the world's oceans. See also *bathyal, deep-sea trench*.

ACCESSORY MINERAL A mineral present in only small quantities in a particular rock.

ACCRETION The gravitational coming-together of small particles, for example in the early Solar System, resulting in the formation of the planets.

ACCUMULATION AREA The part of a glacier where snowfall exceeds losses by melting, evaporation, and calving. See also *ablation*.

ACID RAIN Any precipitation (snow as well as rain) that contains dissolved acids. Most rain is slightly acidic due to dissolved carbon dioxide. More strongly acidic rain also occurs due to atmospheric pollution, or sometimes from gases released by volcanic eruptions.

AEROSOL Tiny particles (about a millionth of a millimetre across) of dust or liquid suspended in the air.

AGNATHAN A primitive jawless fish. Present-day agnathans include lampreys and hagfishes.

AIR MASS A mass of air of uniform characteristics that may extend over thousands of miles.

ALBEDO The extent to which a part of the Earth's surface reflects incoming radiation from the Sun, usually expressed as a decimal fraction.

ALLUVIUM Sedimentary material deposited by rivers. Any particular accumulation of alluvium is called an **alluvial deposit**. An **alluvial fan** is a deposit laid down where a stream leaves a highland area and spreads out over a plain. **Alluvial soil** is soil derived from alluvium.

ANGIOSPERM A flowering plant (as distinct from a conifer, fern, moss, or other plant). Angiosperms also include many tree species. See also *coniferous*.

ANNUAL A plant that lives and dies within one year.

ANOXIA Lack of molecular oxygen (O_2), which many (but not all) organisms require for respiration.

ANTICLINE An arch-like, upwards fold of originally flat strata due to horizontal compression. See also *syncline, fold*.

ANTICYCLONE A weather system in which winds circle around an area of high pressure. See also *cyclone*.

AQUIFER A layer of porous rock that can hold and transmit water.

ARCHIPELAGO A group of islands forming a chain or cluster.

ARÊTE A narrow mountain ridge separating two adjacent cirques. See also *cirque*.

ASTEROID One of thousands of bodies of rocky material orbiting the Sun that are smaller in size than the planets.

ASTHENOSPHERE The layer of the mantle immediately below the rigid lithosphere. It is sufficiently non-rigid to flow slowly in a solid state and plays a key part in the movement of tectonic plates. See also *lithosphere, tectonic plate, mantle*.

ATOLL see *coral reef*.

AUREOLE The area around an igneous intrusion where thermal alteration (metamorphism) of country rock has taken place. See also *igneous rock, metamorphism, country rock*.

AURORA BOREALIS Flickering lights in the sky that are sometimes visible in northern polar regions. They are caused by the interaction between high-energy particles from the Sun and the Earth's magnetic field. Their southern hemisphere equivalent is called the **aurora australis**.

B

BACKWASH The movement of water down a beach after waves have broken. See also *swash*.

BAR A long, sandy deposit built up by sea or river action. An **offshore bar** is a coastal bar not attached to the land. A **baymouth bar** is a spit that has grown so it extends right across the mouth of a bay. A **longshore bar** lies between the high- and low-water marks, parallel to the shoreline. A **barrier bar** is a long bar parallel to the coastline. A **point bar** is a bar deposited by a river on the inside of a meander. See also *spit, meander*.

BARCHAN see *dune*.

BARRIER BAR see *bar*.

BARRIER ISLAND A long island of sediment parallel to a coastline. It is normally higher than a barrier bar. See also *bar*.

BASALT The Earth's most common volcanic rock, which usually originates as solidified lava. Basaltic lava can erupt on continents, but basalt also forms the oceanic crust. Basalt is glassy to fine-grained (composed of very small crystals). See also *igneous rock, flood basalt, crust*.

BATHOLITH A very large igneous intrusion, 100km (60 miles) or more across, which originates deep underground. The Earth's great batholiths are known to science only if exposed on the surface by weathering. See also *igneous rock*.

BATHYAL A term referring to ocean depths of about 200–2,000m (660–6,500ft). See also *abyssal*.

BAYMOUTH BAR see *bar*.

BEACH An accumulation of sand, shell debris, or larger particles along a shoreline. A **beach face** is the steeply sloping part of a beach below a berm. See also *berm*.

BEDDING The manner in which a sedimentary rock was originally laid down in layers. A **bedding plane** is any plane in the rock that is parallel to these layers. See also *sedimentary rock*.

BERGSCHRUND A deep crack formed at the back of a cirque glacier caused by the glacier moving away from the cirque wall. See also *cirque*.

BERM A ridge of shingle on a beach, usually marking the point of the highest high tides. See also *beach*.

BIODIVERSITY The total variety of living things, either on the Earth as a whole or in the region described.

BIOLUMINESCENCE The production of light by living organisms.

BIOME A biological community existing on a large scale and defined mainly by vegetation type (for example, rainforest). The extent of any particular biome is determined by climatic conditions.

BLACK SMOKER see *hydrothermal system*.

BLOCKFIELD An area of broken rocks, usually in a mountainous region.

BOG A peat-accumulating wetland that receives water mainly from rainfall. Bogs are acidic, extremely poor in nutrients, and most support an abundance of *Sphagnum* mosses. The soils are composed almost entirely of dead plant matter, and are waterlogged rather than completely covered by water. See also *fen, marsh*.

BOREAL Relating to or coming from the colder parts of the northern hemisphere, between the Arctic and temperate zones.

BOSS An igneous intrusion that is roughly circular in horizontal cross-section. See also *igneous rock*.

BRAIDED STREAM A stream or river where water flows in many shallow channels that constantly separate and rejoin.

BRECCIA A sedimentary rock made up of angular fragments of minerals and other rocks.

BROADLEAF Of trees: having broad leaves (such as oak and chestnut), as distinct from needle-like leaves. Broadleaved trees are mainly angiosperms rather than conifers, and often deciduous rather than evergreen. See also *angiosperm, coniferous, deciduous, evergreen*.

C

CAATINGA A thorny, dry, tropical woodland found in northeast Brazil.

CALCITE A common mineral form of calcium carbonate.

CALDERA A bowl-shaped volcanic depression larger than a crater, typically greater than 1km (0.6 miles) in diameter caused by the collapse of a volcano into its emptied magma chamber after it has been evacuated by a volcanic explosion.

CALVE Of a glacier: to create icebergs by shedding ice blocks into a sea or lake.

CANYON A deep valley, usually not as narrow as a gorge; also a deep, irregular passage in a cave system. See also *gorge, tube*.

CARAT A term used to describe the proportion of gold in a gold alloy (for example, 24-carat gold is pure gold). It is also a unit of weight used for diamonds and other precious stones.

CATCHMENT AREA The area within which all precipitation is drained by a single river system.

CERRADO A type of South American savanna dotted with small, gnarled trees. See also *savanna*.

CHEMOSYNTHESIS The biological process of extracting food energy from inorganic (non-living) chemicals. Many micro-organisms rely on chemosynthesis. See also *photosynthesis*.

CHLOROPHYLL The green pigment that plants require for photosynthesis. See also *photosynthesis*.

CHLOROPLAST A structure, found within the cells of green plants, in which chlorophyll is found and photosynthesis takes place. See also *photosynthesis*.

CINDER CONE see *volcano*.

CIRQUE A steep-sided, rounded hollow carved out by a glacier. Many glaciers have their origin in a mountain cirque, from which they flow to lower ground.

CLAST A fragment of rock, especially when incorporated into another newer rock. See also *grain*.

CLAY Mineral particles smaller than about 0.002mm (0.00008in) in diameter that are found in soils and other earthy deposits. A clay soil contains an abundance of clay particles and transmits water slowly.

CLEAVAGE The characteristic plane (or planes) along which a particular rock or mineral splits. See also *foliation*.

CLIMAX VEGETATION The vegetation that in time would naturally dominate any given site or region.

CLOUD CONDENSATION NUCLEUS A tiny particle in the air around which a raindrop or snow crystal may begin wto form.

CLOUDFOREST A damp forest almost constantly under clouds, especially in highland regions.

CLOUD-SEEDING A method of creating rain by scattering tiny particles onto clouds.

COALESCENCE In meteorology, the fusion of tiny droplets within clouds until they are big enough to fall as rain.

COAST The boundary between land and sea. A **drowned coast** occurs where, over the centuries, the sea-level has risen relative to the land; its opposite is an **emergent coast**.

COLD DESERT A desert in high-altitude, high-latitude, or temperate regions that may become very cold during the winter months.

COMET A body composed of rock and ice that orbits the Sun. Vaporization of the ice produces the spectacular tail seen during close passage to the Sun.

CONDENSATION The conversion of a substance from the gaseous to the liquid state – for example, water vapour condensing to liquid water. See also *cloud condensation nucleus*.

CONFLUENCE A place where two rivers, streams, or glaciers meet.

CONIFEROUS Cone-bearing. Typical coniferous trees, such as pines and firs, have needle-like leaves and are usually evergreen. See also *evergreen*.

CONSERVATIVE BOUNDARY see *plate boundary*.

CONSTRUCTIVE BOUNDARY see *plate boundary*.

CONTINENTAL MARGIN The continental shelf and continental slope taken together.

CONTINENTAL RISE The slightly sloping area around the edge of the deep sea floor where it abuts the continental slope.

CONTINENTAL SHELF The gently sloping, submerged portion of continental crust seaward of most continental coasts. See also *crust*.

CONTINENTAL SLOPE The sloping ocean bottom between the edge of the continental shelf and the continental rise.

CONVECTION The movement and circulation of fluids (gases, liquids, or hot and ductile rocks) in response to differences in temperature, which in turn cause density variations between different parts of the fluid.

CONVERGENCE In meteorology, the situation where two air masses are moving towards one another. (For another sense, see *plate boundary*.)

CONVERGENT BOUNDARY see *plate boundary*.

COOLING JOINT see *joint*.

CORAL Any one of various simple animals related to sea-anemones. They are often colonial and can secrete skeletons to support themselves. **Hard corals** (sometimes called true corals) eventually create reefs out of the calcium-carbonate skeletons they lay down.

CORAL REEF A structure in shallow tropical seas built up by the activity of coral animals and other organisms over many years. A **fringing reef** is attached to the shore with little or no water between the reef and shore. A **barrier reef** is parallel to the shore but separated from it by a lagoon. A reef that has grown on top of a sunken extinct volcano forms a circular island called an **atoll**.

CORE The innermost part of the Earth. It consists of a liquid outer core and a solid inner core, both made of nickel-iron. See also *mantle*, *crust*.

CORIOLIS EFFECT The tendency for winds and currents moving in a northerly or southerly direction to be deflected and move at an angle because of the effect of the Earth's rotation. The deflection is to the right in the northern hemisphere and to the left in the southern hemisphere.

CORRASION Erosion of rocks by scraping – for example by rock-laden glacial ice.

COUNTRY ROCK Existing rock into which a new body, such as an igneous intrusion, is emplaced. See also *igneous rock*.

CRATER A bowl-shaped depression through which an erupting volcano discharges gases, pyroclasts, and lava. The crater walls form by accumulation of ejected material. This term also refers to a circular depression in the landscape caused by a large meteorite impact.

CRATON A stable area of the Earth's continental crust, made of old rocks largely unaffected by mountain-building activity since the Precambrian Eon. Also called a shield.

CRUST The rocky outermost layer of the solid Earth. The continents and their margins are made of thicker but less dense **continental crust**, while thinner, denser **oceanic crust** underlies the deep ocean floors. See also *mantle*, *tectonic plate*.

CRYOTURBATION The churning up of the land surface by the action of ice below ground level.

CRYPTOCRYSTALLINE Made of crystals too small to be seen with the naked eye.

CRYSTAL Any solid in which the individual atoms or molecules are arranged in a regular geometrical pattern. Most pure substances can form crystals. There are seven basic patterns of crystal growth, called **crystal systems**.

CURRENT In oceanography, a flow of ocean water. Both surface and deep-sea currents exist. Some are driven by the wind; others (**thermohaline currents**) by differences in temperature and salinity that create differences in density between water masses.

CYCLONE A pressure system in which air circulates around an area of low pressure. A **tropical cyclone** is another name for a hurricane or typhoon. See also *pressure system*, *hurricane*, *anticyclone*.

D

DECIDUOUS Of trees and shrubs: having leaves that fall at a particular time of year, such as winter or a dry season. See also *evergreen*.

DEEP-SEA FAN A fan-shaped deposit on the ocean floor built up by turbidity currents. See also *turbidity current*, *alluvial fan*.

DEEP-SEA TRENCH A canyon-like, linear depression in the ocean floor, the site of subduction of one tectonic plate below another. Trenches are the deepest regions of the oceans. See also *subduction*, *plate boundary*.

DEFORMATION The ductile flow or brittle fracture of existing rocks due to geological movements.

DELTA The area of gently sloping sediment built up by many rivers when they enter the sea or a lake. Its overall shape depends on factors such as the amount of sediment brought down by the river, and the currents, wave action, and tides that it meets. See also *estuary*.

DENDRITE A mineral occurring naturally as a branching, tree-like form within rocks. A dendritic crystal is a crystal that has grown in this way.

DEPOSITION The laying down of material such as sand and gravel in new locations, usually by wind or moving water or ice.

DESERT PAVEMENT A rocky or stony surface layer found in many deserts. A **hammada** is a solid-rock surface with weathered rock fragments on top.

DESERT VARNISH A black or brown glossy coating sometimes found on rock surfaces in a desert that have been exposed to the atmosphere for long periods of time.

DESERTIFICATION The transformation of a formerly more fertile region into desert.

DESTRUCTIVE BOUNDARY see *plate boundary*.

DEW POINT The temperature at which, under given conditions, liquid water will start to condense out of the air.

DINOSAUR A member of a dominant group of reptiles that died out between 145–65 million years ago. Their closest living relatives are birds and crocodiles.

DIPTEROCARP A member of a family of tall, mainly evergreen trees native to Africa and south Asia.

DISTRIBUTARY A branch of a river that flows away from the main stream and does not reconnect with it later.

DIVERGENT BOUNDARY see *plate boundary*.

DOLINE A depression in the surface of a karst landscape. It often leads down to an underground drainage system. Also called a sinkhole. See also *karst*.

DOME VOLCANO see *volcano*.

DRAINAGE BASIN see *catchment area*.

DROWNED COAST see *coast*.

DRUMLIN A hill-sized, usually streamlined mound of debris left behind by a retreating glacier. See also *moraine*.

DRY FOREST A forest growing in an area with a predominantly dry climate.

DUNE A mound or hill of sand, either in a desert, a river bed, or by the shore of a sea or lake. There are several varieties of desert sand dunes. A **barchan** is a crescent-shaped dune. A **linear** or **seif dune** can run for long distances roughly parallel to the prevailing wind. **Crescentic dunes** form ridges with slightly wavy edges. **Parabolic dunes** have long, trailing arms, while **star dunes** have arms stretching in several directions.

DYKE A sheet of intrusive igneous rock at a steep angle to the surface. A large number of dykes in a given area is termed a **dyke swarm**. See also *country rock*, *sill*.

E

ECOSYSTEM The entirety of living and non-living features of a region and the interactions between them. Organisms include plants, animals, and micro-organisms; non-living features include light, water, nutrients, and the environment. An ecosystem can be exceedingly small or as large as the Earth itself.

EKMAN SPIRAL The tendency of deep-ocean currents to change direction as they rise to the surface. The term also refers to the tendency of winds lower in the atmosphere to blow at an angle compared with winds higher up. The causes of the Ekman spiral include frictional effects and the Coriolis effect.

EL NIÑO The phenomenon in which the normally strong currents of the equatorial Pacific weaken and warm water that has piled up in the western Pacific floats eastwards, making water in the eastern Pacific warmer than usual. It is part of a larger phenomenon that can trigger changes to weather patterns worldwide. See also *La Niña*.

EMERGENT Of a forest tree: growing taller than most other species of tree in the forest. Of a wetland plant: growing taller than the water surface.

EMERGENT COAST see *coast*.

ENDEMIC Of animals and plants: native to a particular region, and found nowhere else.

EPEIROGENESIS The raising or lowering of parts of the Earth's crust due to vertical movements only, rather than the sideways movements involved in orogenesis. See also *orogenesis*.

EPHEMERAL Appearing for only a short time. In deserts, both plant life and rivers are frequently ephemeral.

EPIPHYTE A plant, especially a non-parasitic one, that grows on another plant.

ERG A large expanse of sand dunes within a desert.

EROSION The processes by which rocks or soil are loosened and worn or scraped away from a land surface. The main agents of erosion are wind, water, and moving ice, and the sand grains and other rock particles that they carry. See also *transport*, *weathering*.

ERRATIC A rock that has been transported from its original location, usually carried by ice.

ERUPTION The discharge of lava, pyroclasts, gases, and other material from a volcano. **Hawaiian eruptions** involve large flows of very liquid lava but little explosive activity. **Strombolian eruptions** involve frequent ejections of incandescent clasts and gases, but without major explosions. **Surtseyan eruptions** take place in shallow water and are partly powered by the conversion of water to steam. **Vulcanian eruptions** involve a series of explosive events that feed significant eruption columns. **Plinian eruptions** involve massive explosions to form towering eruption columns, in which much of the volcanic cone may be destroyed. **Fissure eruptions** take place through long cracks in the ground rather than via a volcanic crater. See also *volcano*.

ESCARPMENT The steep slope at the edge of a plateau, or at the edge of an area of exposed, gently sloping strata. An escarpment is also called a scarp.

ESKER A long line of raised debris left behind by a glacier. Eskers are thought to mark the position of former meltwater channels. See also *meltwater*.

ESTUARY The broad, funnel-shaped stretch of water found where many large rivers meet the sea. Most present-day estuaries owe their form to the drowning of river valleys when sea-level rose at the end of the last ice age. The term is also used more broadly to include all bays and inlets where seawater and fresh water mix and muddy sediments are deposited.

EUKARYOTE Any living organism built out of a complex cell or cells containing nuclei (which are enclosed in a membrane). All complex animals and plants are eukaryotes. See also *prokaryote*.

EUTROPHICATION The enrichment of natural waters with extra nutrients, especially as a result of human action. It often has drastic deleterious effects on aquatic ecosystems.

EVAPORITE A chemical sedimentary deposit created by the progressive evaporation of salty water to form various minerals such as halite and gypsum.

EVERGREEN Having leaves all year round. See also *coniferous*, *deciduous*.

EXFOLIATION A weathering process that involves the splitting off of outer layers of rocks, like the layers of an onion.

F

FAULT A fracture where the rocks on either side have moved relative to one another. If the fault is at an angle to the vertical, and the overhanging rock has slid downwards, it is called a **normal fault**. If the overhanging rock has slid upwards (in relative terms) it is called a **reverse fault**. A **strike-slip fault** is one where movement is horizontal. A **transform fault** is a large-scale strike-slip fault associated with oceanic spreading.

FELSIC ROCK An igneous rock rich in silica. The name reflects the abundance of the mineral feldspar and the high silica content. See also *silica*, *igneous rock*.

FEN A peat-accumulating wetland that receives water mainly from groundwater seepage. Fens are typically alkaline and nutrient poor (although not extremely so). They support a variety of vegetation types including *Sphagnum* mosses, sedges, and conifer trees. The soils are composed almost entirely of dead plant matter, and they are sometimes completely covered by water.

FERREL CELL A large-scale circulation within the atmosphere, in which air rises at around 60° North or South and flows towards the Equator before descending at around 30°. See also *Hadley cell*.

FETCH The extent of open water across which a wind or water wave has travelled.

FIORD A former glacial valley on the coast that has become an inlet of the sea.

FISSURE ERUPTION see *eruption*.

FLASH POINT The point at which super-heated water changes to steam. See also *geyser*.

FLASH-FLOOD A sudden flood occurring after heavy rain.

FLOOD BASALT An extensive area of basalt resulting from massive volumes of lava erupting and flowing widely over the landscape in relatively short intervals of geological time. It is resistant to erosion and commonly forms extensive plateaus after its formation.

FLOODPLAIN A flat plain next to a river that is liable to be covered with water when the river floods. See also *alluvium*.

FLOWSTONE A smooth coating of younger limestone precipitated over surfaces in caves. See also *calcite*.

FOG Condensed droplets of water forming in the air at ground level. **Radiation fog** is fog caused by the ground surface radiating heat at night, becoming cooler and also cooling the air above it.

FOLD A geological structure in which originally flat-lying rocks appear to have been flexed and bent. In reality, the rocks behaved in a ductile manner and flowed into these shapes, like a warm caramel bar. They may bend upwards in the middle to form a ridge (anticline) or downwards to form a trough (syncline). A **symmetrical fold** has slopes that are the same steepness on either side of the vertical; an **asymmetrical fold** does not. A **recumbent fold** is a fold that is lying on its side. In an **overturned fold**, a syncline is partly tucked underneath a neighbouring anticline. See also *anticline*, *syncline*.

FOLIATION The arrangement of platy minerals in parallel bands in some deformed metamorphic rocks. See also *cleavage*, *lamination*.

FOSSIL The remains or traces of an organism that have been buried and often petrified (with its original tissues replaced by minerals). Some fossils, such as preserved footprints and coprolites (fossilized droppings), also record activities.

FOSSIL FUEL Any fuel such as coal or oil that is derived from the remains of once-living organisms that were deeply buried beneath the ground.

FRONT In meteorology, the forward-moving edge of an air mass. See also *air mass*.

FUMAROLE In volcanic regions, a small opening in the ground through which hot gases can escape.

FUNNEL A tube of whirling air descending from a cloud. It becomes a tornado when its lower end touches the ground. See also *tornado*.

G

GALAXY An aggregation of millions of stars, often formed into an immense spiral shape. The Solar System lies within the Milky Way Galaxy.

GEODE A hollow cavity found within a rock and lined with crystals.

GEOTHERMAL ENERGY Energy obtained by tapping the heat generated in the Earth's interior.

GEYSER A jet of boiling water and steam that rises at intervals from the ground. It is powered by hot rocks heating groundwater. See also *hot spring*.

GLACIER A mass of semi-permanent ice capable of flowing downhill. Glaciers come in many varieties. The largest are **ice-sheets**, such as the Antarctic ice-sheet. Similar, but slightly smaller, are **ice-caps**. A **valley glacier** is smaller and flows down a valley, eroding rock as it does so. An **outlet glacier** is one that flows down from an ice-cap or ice-sheet. A **piedmont glacier** spreads out from higher ground onto a broad plain. A **surge-type glacier** is one that periodically increases its speed much above normal. A **cirque glacier** does not spread behind the mountain hollow (cirque) that it has carved out. In a **polythermal glacier**, a cold surface layer of ice overlies much warmer ice.

GORGE A narrow, deep valley, usually with vertical cliffs on either side. See also *canyon*.

GRABEN see *rift valley*.

GRAIN A term referring to the texture of a rock: rocks can be either fine- or coarse-grained. The term also refers to a single small particle (for example, of sand).

GREASE ICE see *sea-ice*.

GREENHOUSE EFFECT The tendency of the atmosphere to contribute to warming of the Earth, by letting through the Sun's radiation but absorbing some of it that is re-radiated by the Earth. A **greenhouse gas** (such as water vapour, carbon dioxide, or methane) is any gas that promotes this process, whether naturally or via human action.

GROUNDMASS The fine-grained material surrounding larger crystals or clasts. The term is most commonly applied to volcanic rocks and some heterogeneous sedimentary rocks. See also *matrix*.

GROUNDWATER Water occurring below the surface of the land and held between grains or in the interstices of rocks. The upper limit of such a groundwater zone is called the water-table. Water held in the soil is not usually counted as groundwater. See also *aquifer*.

GUST FRONT An area of strong winds that travels ahead of a line of storm clouds. See also *front*.

GUYOT A flat-topped seamount. See also *seamount*.

GYRE A large-scale circular movement of ocean currents.

H

HABITAT Any area that can support a particular group of living things.

HADLEY CELL A major element in the Earth's atmospheric circulation, in which warm air rises at the Equator and flows north and south at high level before sinking and flowing at the surface, mainly back to the Equator.

HALO A hazy ring sometimes seen around the Sun or Moon, caused by refraction of light by high-level clouds.

HAMMOCK A small area of vegetation growing on raised ground. Hammocks are common in the Florida Everglades.

HANGING VALLEY A valley, usually carved by a glacier, that enters high up the side of a larger valley.

HEADWATER The upper portion of any river or stream, close to its source.

HORN, GLACIAL A steep-sided mountain left behind after erosion by glaciers occupying cirques. See also *cirque*.

HOTSPOT A long-lived zone of volcanic activity thought to originate deep in the Earth's mantle. Tectonic plates passing over hotspots are marked by linear chains of volcanoes that become progressively older with increased distance from the hotspot. See also *mantle plume*.

HOT SPRING A bubbling up of hot water and steam from the ground, caused by hot rocks heating the groundwater beneath. See also *geyser*.

HUMUS A dark-coloured substance found in soils, derived from dead plants, micro-organisms, and animals.

HURRICANE A large-scale low-pressure system of tropical regions, involving powerful circulating winds and torrential rain. It is also termed a **tropical cyclone** and (especially in east Asia) a **typhoon**. Its energy comes from the latent heat of water that has evaporated from warm oceans and later condenses. See also *latent heat*.

HYDROCARBON A chemical compound of carbon and hydrogen. Petroleum is a complex mixture of hydrocarbons.

HYDROTHERMAL SYSTEM Any natural system involving the heating and circulation of underground water and steam, powered by nearby hot or molten rocks. A **hydrothermal vein** is a mineral vein laid down by past hydrothermal action. A **hydrothermal vent** is an outlet for heated water from a hydrothermal system, especially one at the bottom of an ocean. When the heated water is coloured with dark material, sulphides, and other minerals precipitated when the hot fluid comes into contact with cold seawater, the vent is known as a **black smoker**. See also *vein*, *geyser*, *hot spring*.

I

ICE LEAD A open channel formed in sea-ice.

ICEBERG A floating mass of ice derived from a glacier.

ICE-CAP see *glacier*.

ICE-FIELD A large expanse of ice at high altitude, partly hemmed in by mountains.

ICE-HEAVE The growth of ice within the soil in periglacial regions, leading to disruption of the land surface and the creation of ice wedges and pingos. See also *periglaciation*, *ice wedge*, *pingo*.

ICE-SHEET see *glacier*.

ICE-SHELF A large area of floating ice attached to land and originally derived from a glacier, especially from an ice-sheet.

ICE-STREAM A region of an ice-field or ice-sheet within which faster flow is taking place.

ICE WEDGE An irregular mass of ice growing within a soil layer in a periglacial region. See also *ice-heave*.

IGNEOUS ROCK Rock that originates from the solidification of molten magma. **Extrusive** (volcanic) igneous rocks have solidified on the Earth's surface after volcanic activity and commonly have small, hardly visible crystals. **Intrusive** (plutonic) igneous rocks have solidified below the surface, cooling slowly enough to allow larger crystals to form. A body of intrusive igneous rock is called an **igneous intrusion**.

INSELBERG An isolated, steep-sided hill rising up out of a flat plain.

INTERTROPICAL CONVERGENCE ZONE The region where the main air masses north and south of the Equator come into contact. See also *Hadley cell*.

INTRUSION see *igneous rock*.

INVERTEBRATE An animal without a backbone, such as an insect, snail, or worm.

IRRIGATION The artificial supply of water to agricultural crops.

ISLAND ARC, VOLCANIC A line of volcanic islands associated with an ocean–ocean subduction zone. See also *subduction*.

ISOBAR A line on a weather map that joins points with equal atmospheric pressure.

ISOSTASY The concept that all columns of rock on the Earth have the same weight at some depth in the mantle. Continents rise high above the sea floor because continental crust is less dense than oceanic crust.

JK

JET STREAM Any of several winds that blow for long distances high in the troposphere. See also *troposphere*.

JOINT A crack running through a rock. Unlike a fault, the rocks on either side of a joint remain in the same position with respect to each other. A **cooling joint** results from an igneous rock shrinking as it cools. See also *fault*.

KARST A characteristic landscape formed in regions built of water-soluble rocks, especially limestone. A **karst coast** occurs where limestone has been eroded and later inundated by the sea.

KATABATIC WIND A wind that blows downwards from a glacier or cold valley at night.

KETTLE LAKE A lake occupying a **kettle hole**, a depression left after an ice block from a former glacier melted.

KT BOUNDARY The boundary between the Cretaceous and Tertiary Periods. It dates to about 65 million years ago and coincides with the extinction of some dinosaurs and many other life-forms.

L

LA NIÑA A situation in which the waters of the eastern Pacific become unusually cold; the opposite of the El Niño phenomenon. See also *El Niño*.

LAGOON An area of sheltered seawater, almost cut off from the open ocean. The term includes both shallow coastal lagoons and the lagoons of atolls.

LAHAR A muddy flow of water mixed with volcanic ash and other volcanic debris. See also *mass movement*.

LAMINATION The occurrence of thin, distinct, and generally parallel layers within rocks.

LAPSE RATE A rate at which air temperature decreases with height. See also *latent heat*.

LATENT HEAT The heat that is input when a liquid such as water turns into vapour. It is released again when the vapour condenses.

LAVA Molten rock that has reached the Earth's surface. Basaltic **a'a lava** (45–52 per cent by weight of silica) forms a rough-textured rock when it cools. **Pahoehoe lava**, of similar basaltic composition, flows easily and creates a smooth or ropy-textured surface. Basaltic **pillow lava** is lava ejected underwater, forming pillow-shaped mounds of rock. More silica-rich and andesitic-dacitic lavas (more than 57 per cent by weight of silica) typically form **block lava**, jostling masses of angular blocks, some of them many metres across.

LEACHING The removal of minerals from soil or rock by downward-percolating water.

LEVEE A raised bank of a river that is higher than the surrounding active floodplain. Levees commonly mark former floodplains prior to major downcutting events (when the river rapidly erodes and lowers its bed).

LIGHTNING The visible flash that accompanies electrical discharges from storm clouds. Discharges within or between clouds are **sheet lightning**; those between clouds and the ground are **forked lightning**.

LITHOSPHERE The Earth's crust together with the rigid uppermost layer of the underlying mantle. Each of the Earth's tectonic plates is made of a section of lithosphere. See also *crust*, *mantle*, *asthenosphere*, *tectonic plate*.

LITTORAL Relating to the shoreline, especially between the high- and low-water marks. See also *tide*.

LONGSHORE BAR see *bar*.

LONGSHORE DRIFT A current that flows parallel to a shoreline, usually picking up and transporting sediment as it does so.

M

MAAR A wide volcano crater formed by explosive eruptions that have excavated into underlying bedrock, which is exposed in the crater walls.

MAGMA Molten rock rising from the interior of the Earth.

MAGMA CHAMBER A region just below the Earth's surface where magma (molten rock) has collected.

MAGMATIC DIFFERENTIATION The various processes by which magma changes composition, typically towards higher silica contents. The processes include gravitational separation of early formed low-silica minerals and assimilation of high-silica crustal rocks.

MANTLE The rocks lying between the Earth's crust and its core. It makes up 84 per cent of the Earth's volume. See also *crust*, *core*.

MANTLE PLUME A column of hot rocks rising through the mantle and crust, giving rise to a hotspot at the Earth's surface. See also *hotspot*.

MARSH A freshwater or saltwater wetland that is dominated by soft-stemmed plants such as grasses or reeds, but not by woody plants such as trees or shrubs. Marshes are characterized by frequent or continual flooding. See also *swamp*, *salt marsh*.

MASS MOVEMENT The movement of rocks, soil, or mud down a slope in response to gravity. See also *transport*.

MASSIF A well-defined mountainous region whose rocks and landforms tend to be similar across the region.

MASSIVE Of a rock: having a structure that does not show layering or other divisions into smaller segments.

MATRIX The mass of relatively fine-grained material that binds together larger particles in some heterogeneous sedimentary rocks and larger crystals in volcanic rocks. See also *groundmass*.

MEANDER A loop or bend in a river. Meandering rivers gradually change their course, as erosion occurs on the outside of the bend and deposition on the inside.

MELTWATER Water flowing within or from a glacier. A **meltwater channel** is a channel carved by meltwater running under or near a glacier.

MESA A tall, usually flat-topped outcrop of rock, especially in desert regions, left behind after the rest of the former land surface has been eroded. It is usually capped by a layer of resistant rock.

MESOSPHERE The layer of the Earth's atmosphere between the stratosphere and thermosphere, at an altitude of about 50–80km (30–50 miles).

METAL Any of numerous chemical elements that are typically hard, shiny, malleable in the solid state, and good conductors of heat and electricity; also any alloy (mixture) of such elements.

METAMORPHIC ROCK A rock that has been transformed underground by heat or pressure to a new texture or new set of minerals. For example, marble is metamorphosed limestone.

METAMORPHISM The processes by which rocks are transformed by heat, pressure, or chemical reactions underground. In **thermal metamorphism**, the agent is heat. **Dynamic metamorphism** involves differential stresses that can impart new textures and structures to the rock. **Regional metamorphism** affects broad areas the size of mountain belts. **Contact metamorphism** results from the effects of hot magma on surrounding rocks.

METASOMATIC ROCK A rock that has been transformed by hot fluids penetrating it and changing its bulk composition and mineral assemblage.

METEOR A small mass of rock from space that vaporizes completely as it falls through the Earth's atmosphere, glowing as it does so.

METEORITE A rock from space that has fallen to the Earth's surface without completely burning up.

METEOROID A potential meteor or meteorite before it has entered the Earth's atmosphere.

MICROCLIMATE The distinctive climate of a particular place, such as a narrow valley or hillside.

MID-OCEAN RIDGE Any of the submerged mountain ranges running across the floors of the major oceans (although not all occur exactly in mid-ocean). They are sites where two tectonic plates are spreading apart and material is rising up from the mantle to form new oceanic crust. Mid-ocean ridges vary widely in their spreading rates, from fast- to slow-spreading ridges. See also *plate boundary*, *tectonic plate*.

MINERAL Any solid, naturally occurring inorganic material in the Earth that has a characteristic crystal structure and well-defined chemical composition. Most rocks are mixtures of more than one mineral. See also *rock*.

MIRE A peat-accumulating wetland such as a bog or fen.

MONSOON A pattern of winds, especially in southern Asia, that blow from one direction for about half the year, and from the opposite direction for the other half. The term is also used to refer to the heavy rains carried by these winds at certain times of the year.

MONTANE Of a forest: a type of woodland found in mountainous areas that includes conifers and other species such as beech.

MORAINE An accumulation of rock debris resulting from glacial action. Active valley glaciers create **lateral moraines** at their edges, **medial moraines** where two glaciers merge, and **terminal moraines** at their tips. These features often remain after the glacier disappears. See also *till*.

N

NATIVE ELEMENT A chemical element that is found in its pure (uncombined) state in nature. Gold is always found native, other elements, such as sulphur and copper, sometimes occur in native form.

NEEDLE-LEAF A needle-shaped leaf, typical of pine trees and other conifers. See also *coniferous*.

NOCTILUCENT CLOUD A very high-level cloud of ice crystals or dust particles that has a shiny appearance at night.

NUNATAK A mountain-top that rises above an ice-sheet that is otherwise covering the land.

O

OASIS A fertile area within a desert, normally fed by a spring arising from an aquifer. See also *aquifer*.

OCCLUSION In meteorology, the situation where a moving mass of colder air catches up with a warmer air mass, pushing the latter away from the surface of the Earth. See also *front*.

OFFSHORE BAR see *bar*.

OOZE The fine sediment that covers much of the deep sea floor. The names of some kinds of ooze refer to the particular tiny organisms whose skeletons or remains dominate: for **globigerina ooze**, single-celled organisms (foraminifera) of the genus *Globigerina*; for **diatom ooze**, certain algae (diatoms); for **radiolarian ooze**, other single-celled organisms called radiolarians; and for **pteropod ooze**, the shells of certain free-swimming snail-relatives.

ORE A rock from which a metal can be profitably mined.

OROGENESIS The process of mountain building, resulting from pressures generated by the horizontal movement of tectonic plates. An **orogeny** is a particular episode of mountain building. See also *epeirogenesis*.

OROGRAPHIC LIFTING The rising of an air mass as it moves over mountains, often resulting in clouds being formed.

OUTLET GLACIER see *glacier*.

OXIDE A compound of oxygen with one or more chemical elements. Examples of oxides include ferrous oxide (FeO) and silicon dioxide (SiO_2).

PQ

PACK ICE see *sea-ice*.

PANCAKE ICE see *sea-ice*.

PEAT Dead plant material that accumulates in bogs, fens, and other wetland habitats, remaining largely undecomposed. Peat soils are composed almost entirely of plant matter, particularly *Sphagnum* mosses, with little mineral content. See also *bog*.

PEGMATITE A plutonic igneous rock made up of unusually large crystals. **Pegmatitic veins** are narrow sheets or lenses of such rock. They are sometimes the source of valuable ores. See also *vein*.

PENINSULA An area of land jutting out into the sea or a lake.

PERENNIAL A plant that lives for three or more years. See also *annual*.

PERIGLACIATION Any of various features peculiar to landscapes that are subject to very cold conditions, while not having a permanent ice cover. See also *permafrost*, *pingo*, *ice wedge*, *tundra*.

PERMAFROST Technically, land that remains frozen for at least two years. It is characteristic of glacial and many periglacial regions. In tundra, the top layer of the soil generally thaws in spring and summer, while permafrost may remain deeper down. See also *periglaciation*, *tundra*.

PHENOCRYST A large crystal surrounded by smaller ones in an igneous rock. See also *porphyry*.

PHOTOSYNTHESIS The process by which green plants and some micro-organisms use the energy of sunlight in the presence of chlorphyll to convert water and carbon dioxide into food.

PHREATIC ERUPTION A type of volcanic eruption or geyser activity where groundwater is turned to steam by contact with hot rocks near the surface.

PHYTOPLANKTON The tiny, mostly single-celled plants (algae) that float near the surface of oceans and lakes and are the foundation of most aquatic food chains.

PIEDMONT GLACIER see *glacier*.

PILLOW LAVA see *lava*.

PINGO In periglacial areas, a hillock with a core of ice that has pushed up above the surrounding landscape.

PLACER DEPOSIT An accumulation of sediments deposited by running water that contains minerals that can be profitably mined.

PLANET A large astronomical body that orbits a star and, unlike a star, is not heated by thermonuclear reactions taking place within it.

PLANETESIMAL One of millions of rocky objects of variable size believed to have been present in the early history of the Solar System and which later came together to create the planets.

PLATE see *tectonic plate*.

PLATE BOUNDARY A boundary between two tectonic plates. The plates concerned can be diverging (**constructive boundary**), converging (**destructive boundary**), or sliding past one another (**conservative boundary**).

PLATEAU A large area of relatively flat land that stands topographically above its surroundings.

PLATY A term used to describe thin, flat crystals.

PLAYA A level plain in a desert basin, sometimes occupied by a temporary lake that fills and empties seasonally.

PLINIAN ERUPTION see *eruption*.

PLUTON A large body of igneous rock that solidifies slowly underground to form a coarse-grained (plutonic) rock such as granite. A mass of many hundreds of adjacent plutons is called a batholith. See also *batholith*.

PLUTONIC Relating to igneous processes or materials formed deep below the Earth's surface. The term is derived from Pluto, the Roman god of the underworld.

PODOCARP A member of a family of coniferous trees found mainly in the southern hemisphere.

POINT BAR see *bar*.

POLDER An area of low-lying land reclaimed from the sea.

POLYMORPH A substance that can exist in different mineral or crystalline forms, all sharing the same chemical composition. For example, pure carbon can exist as both diamond or graphite.

POLYNA A stretch of open water in an ice-covered sea.

POLYTHERMAL GLACIER see *glacier*.

PORPHYROBLAST A large crystal surrounded by smaller crystals found in some metamorphic rocks.

PORPHYRY Igneous rock that contains large phenocrysts, embedded in finer-grained material. See also *phenocryst*.

PRECIPITATE Any material that has come out of solution to form a solid deposit.

PRECIPITATION Water that reaches the Earth's surface from the atmosphere, including rain, snow, hail, and dew.

PRESSURE SYSTEM Any pattern of weather in which air circulates around an area of high or low pressure. The **pressure gradient** is the pressure difference between two given points. See also *cyclone*, *anticyclone*, *isobar*.

PREVAILING WIND The source direction of wind that tends to blow most commonly at a particular location.

PRIMARY MINERAL A mineral that derives from the cooling of molten igneous rock, without being altered subsequently. See also *secondary mineral*.

PROKARYOTE Any single-celled organism, such as a bacterium, whose cell has a simple structure that lacks a well-defined nucleus. See also *eukaryote*.

PUMICE A light-coloured, low-density volcanic rock containing innumerable bubbles formed by expanding gases, ejected during some volcanic eruptions.

P-WAVE see *seismic wave*.

PYROCLASTIC Consisting of, or containing, volcanic rock fragments. **Pyroclastic flows** are fast-moving, sometimes deadly clouds of hot gases and debris. See also *pumice*, *scoria*, *tuff*.

R

RADIATION A flow of high-energy particles or waves. **Electromagnetic radiation** consists of electromagnetic waves: listed from long-wavelength to short-wavelength forms, these include radio waves, microwaves, infra-red rays, visible light, ultraviolet light, X rays, and gamma rays. Short-wavelength electromagnetic radiation has the highest energy.

RADIATION FOG see *fog*.

RAINBOW A large-scale, prism-like light effect caused by the Sun's rays being refracted (bent) as they pass through raindrops. See also *refraction*.

RAIN SHADOW The occurrence of low rainfall downwind of a mountain range, caused by the air shedding its moisture as it passes over the mountains.

RAINFOREST A forest, usually tropical, with high rainfall, high humidity, and high temperatures all year round.

REEF see *coral reef*.

REFRACTION The tendency for waves of any kind to be bent when they pass from one medium to another of different properties. For example, light waves can be refracted by passage from air to water. The amount of refraction depends on wavelength.

REGOLITH A general term for all the materials that are not solid rock covering the surface of the Earth, Moon, and other planetary bodies. On Earth, it includes soil and debris from glaciers. The lunar regolith refers to the thick blanket of rock dust produced by meteorite impacts.

RELATIVE HUMIDITY The amount of water vapour in the atmosphere, relative to the total amount that the atmosphere can hold at that particular temperature. Above 100 per cent relative humidity, water vapour will tend to condense as liquid water. Warm air can hold more water vapour than cold air. See also *dew point*.

REPLACEMENT DEPOSIT A mineral deposit in which the original material has been replaced by other minerals.

RESURGENCE The emergence of water from an underground aquifer at the surface as a spring.

RIA COAST A type of drowned coast indented by former river valleys, known as **rias**, that are now inlets of the sea. See also *coast*.

RIDGE OF HIGH PRESSURE A long, narrow area of high pressure extending out from an anticyclone. See also *anticyclone*.

RIFT VALLEY A large block of land that has dropped vertically downwards compared with the surrounding regions, as a result of horizontal extension and normal faulting. A rift valley is also called a graben. See also *fault*.

RIMSTONE A deposit of calcite around an underground pool in a cave. See also *calcite*, *flowstone*.

ROCK Any solid material made up of one or more minerals and occurring naturally on the Earth or other planetary bodies. See also *mineral*.

ROSSBY WAVE A wave that forms in a jet stream at temperate latitudes, giving rise to areas of high and low pressure. See also *jet stream*, *pressure system*.

S

SALINITY The concentration of dissolved salts in, for example, water or soil.

SALT MARSH A wetland that supports salt-tolerant plants that grow on flat coastal areas regularly inundated by the tides.

SARGASSUM A floating seaweed common in the Sargasso Sea in the North Atlantic.

SASTRUGI The sculpted appearance of some ice surfaces, caused by erosion from wind-blown ice crystals.

SAVANNA A general term for all tropical natural grasslands. Most savannas also have scattered trees.

SCARP see *escarpment*.

SCORIA A dark-coloured volcanic rock containing innumerable bubbles formed by expanding gases ejected during some basaltic volcanic eruptions. Scoria is also known as cinder. See also *pyroclastic*.

SEA ARCH A natural arch formed in a sea cliff by erosion.

SEA CAVE A cave eroded into a sea cliff by the action of the sea.

SEA-ICE Ice that arises from seawater freezing. (It does not include icebergs, which originally come from land.) The sea freezes in several stages. At first, separate crystals form. These begin to join together, forming a thick, oily texture called **grease ice**. Thin slabs of ice called **pancake ice** may form. When the sea-ice forms a continuous sheet it is called **fast ice**. Winds and storms may later break this up into fragments called **pack ice**.

SEAMOUNT An undersea mountain, usually volcanic in origin. See also *guyot*.

SEA STACK A tall column of rock rising out of the sea near a coastline. It is a resistant remnant of a formerly higher land surface that has otherwise disappeared due to erosion.

SECONDARY MINERAL A mineral that has replaced a primary igneous mineral, one originally solidified from a molten state. See also *primary mineral*.

SEDIMENT Solid particles that have been transported by water, wind, volcanic processes, or mass movement, and later deposited.

SEDIMENTARY ROCK Rock formed when small particles are deposited – by wind, water, volcanic processes, or mass movement – and later harden.

SEISMIC WAVE A shock wave generated by an earthquake. **P-waves** can travel through both solid and liquid portions of the Earth's interior, **S-waves** only through the solid parts.

SEMI-DESERT A dry region that has enough precipitation to support some plant life.

SERAC A pinnacle of ice in a glacier.

SHIELD see *craton*.

SHIELD VOLCANO see *volcano*.

SILICA Silicon dioxide (SiO_2), the most common mineral form of which is quartz, a hard mineral that is the dominant component of sand.

SILICATE Any rock or mineral composed of groups of silicon and oxygen atoms in chemical combination with atoms of various metals. Silicate rocks make up most of the Earth's crust and mantle.

SILL A roughly horizontal, sheet-like igneous intrusion that usually forms when igneous rock forces its way between layers of existing sedimentary rocks. See also *dyke*.

SINKHOLE see *doline*.

SKIN FLOW Creeping movements of soil in periglacial regions following rapid thawing of the surface layers. See also *periglaciation*.

SOFAR CHANNEL Short for Sound Fixing and Ranging channel. Sound achieves a much longer range at particular depths in the ocean, typically at 1,000m (3,300ft). This layer allows scientists with underwater microphones to detect sounds thousands of kilometres away.

SOIL HORIZON A particular layer of a soil with distinctive appearance and physical or chemical properties.

SOIL PROFILE A vertical section through the soil showing all of its horizons.

SOLAR BUDGET The overall flow of radiant energy from the Sun to the Earth, and its ultimate re-radiation into space. See also *radiation, greenhouse effect*.

SOLAR CONSTANT The amount of energy from the Sun that reaches a unit area at the top of the Earth's atmosphere. It is not truly constant, but nearly so.

SPECIFIC GRAVITY A measure of the density of a substance, in terms of its weight for a given volume compared to the same volume of a reference material, usually pure water.

SPIT A peninsula of sand or shingle projecting from a shore, usually where the coastline changes direction. It is created by longshore drift. See also *bar, tombolo, longshore drift*.

SPUR A mountain ridge cut off as a result of past glacial action or tectonic faulting.

STALACTITE A deposit of calcite hanging down from the roof of a cave or underground passage. See also *calcite*.

STALAGMITE A deposit of calcite rising up from the floor of a cave or underground passage. See also *calcite*.

STAR A self-luminous astronomical body, such as the Sun, that generates energy by nuclear reactions, mainly involving the conversion of hydrogen to helium.

STEPPE Temperate grassland habitats, especially in regions with hot, dry summers and cold winters.

STRATOSPHERE A layer of the Earth's atmosphere extending from the top of the troposphere, at 8–16km (5–10 miles), up to about 50km (30 miles). See also *troposphere, mesosphere*.

STRATOVOLCANO see *volcano*.

STRATUM (pl. STRATA) A layer of sedimentary rock.

SUBDUCTION The descent of an oceanic tectonic plate under another plate when two plates converge. Subduction zones can be classified as either ocean–ocean or ocean–continent depending upon the nature of the two converging plates. See also *deep-sea trench, plate boundary, tectonic plate, crust*.

SUBLITTORAL Below the low-tide mark.

SUBMONTANE Of a forest: a type of woodland found in lowland and hilly sites that consists mainly of deciduous, broadleaved trees.

SUCCESSION A gradual change with time from one community of plants to another in a particular location – for example, from grassland to forest.

SUCCULENT A plant having thick, juicy leaves or stems for storing water.

SURGE-TYPE GLACIER see *glacier*.

SWAMP A freshwater or saltwater wetland that is dominated by trees. See also *marsh*.

SWASH The movement of turbulent water up a beach when a wave breaks. See also *backwash*.

S-WAVE see *seismic wave*.

SYNCLINE A downwards fold of originally flat strata as a result of horizontal compression. See also *anticline, fold*.

T

TAIGA The coniferous forest (also called boreal forest) that covers much of Europe, Asia, and northern North America.

TECTONIC PLATE Any of the large rigid sections into which the Earth's lithosphere is divided. The relative motions of different plates leads to earthquakes, volcanic activity, continental drift, and mountain-building. See also *plate boundary, subduction, deep-sea trench, mid-ocean ridge, lithosphere*.

TEKTITE A glassy particle sometimes formed when a large meteorite strikes the Earth. Tektites originate when melted target rock is thrown great distances from the site of impact.

TEMPERATE Relating to the regions of the Earth between the tropics and the polar regions.

TERRACE A flat region in a river valley that is higher than the present floodplain. It represents a former floodplain created when the river ran at a higher level. See also *floodplain*.

THERMOHALINE see *current*.

THERMOPAUSE The upper edge of the thermosphere, about 640km (400 miles) above the surface.

THERMOSPHERE The layer of the Earth's atmosphere above the mesosphere. It extends over altitudes of about 80–640km (50–400 miles).

TIDAL BORE A single large wave sometimes created when a rising tide enters a narrowing channel such as an estuary.

TIDE The rise and fall (usually twice each day) of water on the shore caused by the gravitational pull of the Moon and Sun. The difference in level between high and low tide is called the **tidal range**, and tends to vary on a monthly basis: **spring tides** (when the effects of the Moon and Sun reinforce each other) produce the highest high tide and the lowest low tide; **neap tides** (when the effects of the Moon and Sun oppose each other) have the smallest tidal range.

TILL Solid material laid down or left behind by a glacier. It typically consists of rock fragments of many sizes. Till left behind by glaciers from the last ice age covers many regions today. See also *moraine*.

TOMBOLO A type of spit connecting a small island to the mainland. See also *spit*.

TORNADO A narrow, rapidly whirling tube-shaped column of air, which is often highly destructive.

TRANSPIRATION The process by which water is drawn through plants. The evaporation of water vapour from leaves causes more water to be drawn in through the roots to compensate.

TRANSPORT The conveying of weathered and eroded material to another location, for example by wind, water, or ice action. See also *weathering, erosion*.

TRAVERTINE A rock, mainly consisting of calcium carbonate, that is deposited around the edges of hot springs.

TREE-LINE The latitude or altitude above which conditions are too harsh for trees to grow in a particular location.

TRIBUTARY Any river or stream that flows into a larger river.

TROPICAL Relating to the warm regions of the Earth that lie between the Equator and the tropics of Cancer and Capricorn, at latitudes of 23.5° North and South, respectively.

TROPOPAUSE The boundary between the troposphere and the stratosphere. Air temperature starts to increase with height above the tropopause, which varies from about 16km (10 miles) at the Equator to 8km (5 miles) over the poles.

TROPOSPHERE The lowest, densest layer of the atmosphere, where most weather phenomena occur. The height of the upper limit of the troposphere (known as the tropopause) varies from the Equator to the poles. See also *tropopause*.

TROUGH In meteorology, a long, relatively narrow area of low pressure. See also *front*.

TSUNAMI A fast-moving, often destructive sea wave generated by earthquake activity; popularly but incorrectly known as a tidal wave. It rises in height rapidly as it reaches shallow water.

TUBE A tunnel-shaped passage within a cave system through which water flows or has flowed. See also *canyon*.

TUFF A rock made of fine-grained pyroclastic material produced by an explosive volcanic eruption. See also *pyroclastic*.

TUNDRA A treeless habitat of low-growing, cold-tolerant plants widespread in the far north of North America and Siberia.

TURBIDITY CURRENT A flow of sediment-containing water from the continental margin down onto the ocean floor. Sediment deposited by such a current is known as **turbidite**.

TWINNING A form of crystal growth in which the same crystal grows in more than one orientation, giving the appearance of two interpenetrating crystals separated by a mirror plane of symmetry.

TYPHOON see *hurricane*.

UV

ULTRAVIOLET see *radiation*.

UNDERSTOREY The layer of smaller trees and shrubs growing beneath the main trees of a forest.

VALLEY FOG A type of radiation fog occurring in valleys. See also *fog*.

VALLEY GLACIER see *glacier*.

VEIN A thin, sheet-like body of rock penetrating through other rocks. See also *hydrothermal system*.

VENTIFACT A rock or pebble polished by wind-blown sand in a desert.

VERTICAL TRANSPORT The upwelling or downwelling of nutrient-rich water in the oceans.

VISCOSITY Resistance to flow in fluids. The higher the viscosity of a fluid, the more sluggishly it flows.

VOLATILE In volcanology: a term used to describe water, carbon dioxide, and other potentially gaseous compounds that are dissolved in molten rocks.

VOLCANO An opening in the Earth's crust where magma reaches the surface; also, a mountain that contains such an opening. A **shield volcano** has shallow slopes and is built from lava that flowed easily. A **stratovolcano** is steeper and built from alternate layers of ash and lava. A **dome volcano** is rounded, steep, and built from viscous lava. A **cinder cone** is built from scoria that has fallen from the explosion clouds of an eruption.

VULCANIAN ERUPTION see *eruption*.

WXYZ

WADI A usually waterless river valley in a desert.

WATER-TABLE The upper surface of the groundwater zone in places where it is not confined by impermeable rocks. Water soaking into the ground will tend to sink down until it reaches the water-table. See also *groundwater*.

WATERSHED The divide between two catchment areas.

WAVE A regular motion or disturbance that transfers energy. For a wave crossing the open ocean, the water itself does not move significantly except up and down as the wave passes. The high point of a wave is its **crest** and the low point its **trough**. For waves breaking on shores (**breakers**), water motion becomes more complex and chaotic (**turbulent**).

WAVE REFRACTION see *refraction*.

WAVELENGTH The distance between successive crests (or troughs) of a wave.

WEATHERING The alteration of rocks caused by their being exposed at or close to the Earth's surface. Usually the original rock eventually crumbles or becomes weakened. See also *erosion*.

XENOLITH A fragment of older rock incorporated within a younger igneous rock. See also *igneous rock*.

YARDANG A wind-sculpted ridge of rock in a desert.

INDEX

Page numbers in **bold** indicate feature profiles or extended treatments of a topic. Page numbers in *italic* indicate pages on which the topic is illustrated.

A

a'a (basaltic lava) 70, 155
Abereiddy Bay *76*
Aboriginal peoples, Australian 105
Abu Simbel 217
abyssal plains 392, 393
acacia trees *335, 461*
Acacus-Amsak Region *496*
Acadia National Park *407*
Açai Palm 307
acapulcoites 83
achondrites **83**
Aconcagua 140, *492*
Acorn Banksia 114
actinolite 59, 75
Acuri Palms 324
adiabatic cooling and warming 450
Adriatic Sea 146, 414, 415
Aegean islands *414*
Aegean Sea 414
aegirine 58
Aeolian Islands 192
aerosol particles and deposits 95, *445*
Afar Triangle 128
Africa 21
  agriculture *344*, 349
  caves 247
  cities 366, 367
  desertification 290
  deserts *276, 278*, 279, 280, 281, 288–92
  early hominids 25, 28, 31, 128, 289
  fault systems 128–29
  glaciers 268, 455
  grasslands 330, 331, 334–35
  horticulture 349
  hot springs and geysers 192
  igneous intrusions 184
  lakes 217, 218, 234–36
  meteorite impacts 104
  mountains 147, 148, *149*
  plateaus 218, 234, 236
  population 497
  rainforest 313
  rivers 217–19
  volcanoes 171–73
  wetlands 326, 327
African Fish-Eagle 326
African Mahogany 313
African Plate 27, 86, *129*, 143, 144, 146, 147, 488, 494, **496–97**
African Shield 12
African Whitebacked Vulture *292*
African–Nubian Plate 128
African–Somalian Plate 128
Agadir 147
Agassiz Glacier 261
agate 68
agriculture
  agricultural areas **341–51**
  commercial 342, 344, 345, 348–51
  commercial livestock 344, 349, 351
  on the Great Plains 332, 343, 485, 491

agriculture *cont.*
  and global warming 459
  landscape and 343
  mechanized *342*
  Mediterranean-type 351, 462
  mixed farming 343, 351
  monoculture 345
  pastoralism 344
  ploughing 445
  scale and intensity of 342
  shifting/nomadic 342, 343, 344, 348
  slash and burn 344
  soil and 343
  subsistence 342, 344, 345, 348, 350, 351
  wildlife and 343
agrochemicals 343, 345
Ahaggar Mountains *288, 289, 496*
AIDS virus *112*
Ailuravus 26
air; *see also* atmosphere
  atmospheric circulation 450–53
  atmospheric particles 445
  pollution 91, 107, 361, 445, 447
  sampling 445
  stable and unstable 472
Aïr Massif 289
Aïr Mountains **184**
air traffic 377
Aitken Basin 17
Akiyoshi-dai Plateau 249
Akiyoshi-do **249**
Alabama Hills (California, USA) *66*
alabaster 55
Aland Islands *413*
Alaska 158, *250*, 256, 257, 260, 261, *339*, 398, 422
  fiords 422
  hot springs **188**
  ice caves *252*
  natural gas fields 139
  oil fields 139, 375
Alaska Current 422
Alaska Stream 422
Albatross, Light Mantled Soot 428
albedo 253, 263, 270, 447, 456
albite 60
Alcántara dam 212
alder trees 311, 312
Aldeyarfoss *64*
Aleppo Pine 462
Aletsch Glacier 144, *254*, **267**
Aleutian Trench 422
Aleutians 86, 137, 422
Alexander Island 153
Alexandra's Parrot *297*
Alexandria *415*
algae 18, 21, *107*
All-American Canal 206
Allalin Glacier **268**
Alligator Snapping Turtle 322
alligators 321, 322, 323; *see also* caimans, crocodiles
allosaurs 23
Alpaca *229*
alpine meadows *144*, 149
Alpine Parrot *152*
alpine sports 264, 267, *274*
Alpine Himalayan belt 27
Alps 26, *86*, **144**, 267, 268, 494
Altai Mountains **148**, 294, *295*
alternative energy systems
  biomass 375

alternative energy systems *cont.*
  geothermal energy and power 185, 189, 192, 193, *194*, **195**, 375
  hydroelectric power 266, 267, 375
  solar power 375; *see also* solar panels *458*
  tidal power 375
  wave power 375
  wind power and wind farms 375, 413
Altiplano plateau *140*, 141, 229, *492*
altocumulus clouds *108*, 475, **480**
altostratus clouds 478, **480**, 483
aluminium 82, 376
alumino-silicates 55, 56, 57, 60, 75
Alvarez, Luis and Walter 25
Amazon **210**, *211*, 307, 392
Amazon Basin, tropical rainforest 210, *211*, **307**, *447, 492, 493*
Amazon Rainforest **307**
amber *26*, 79
Amerasia Basin 402
American Indigenous peoples *181, 182, 330, 332, 360, 408
American Plate 122
ammonites 24, 113
amoeba
  giant *112*
  single-celled *112*
amphibians 22; *see also* particular amphibians by name
amphiboles 45, **59**, 60, 66, 68, 69, 71, 74, 75, 91
amphibolite 57, 59, **75**
  Switzerland *75*
amphipods *387*
Amu Darya 293
amulets 57
Amundsen, Roald 273, 402
Amundsen Sea 270
An Teallach *126*
anabatic winds 483
Anaconda 321, 324
Anak Krakatau *47, 176*
analcime 60
Anatolian Plate 127
anchovies 424
andalusite 56, **57**
Andaman Islands 419
Andaman Sea 419
Andes 26–27, 89, *136*, 137, **140–41**, 162, 229, 264, *284*, 310, 492
andesite 59, **69**, 70, 75
Ang Thong National Park *424*
Angel, Jimmie 209
Angel Falls 209
angiosperms 23
Angkor Wat *91*
anglerfish 392
anglesite 55
angrites 83
Anhinga 323
anhydrite 44, **54**, 55
animal life
  co-existing species *113*
  in deserts *113*, 280, 282, 283, 285, 292, 294
  extinctions 23, 31, *101*, 102, 103, 104, 116

animal life *cont.*
  in forests 303, 304, 305, 306, 307, *310*, 311, 312, 313, 315, 316, 317
  fossil evidence 18–24, 26–27
  glaciers 261, *273*
  in grasslands *328*, 331, 332, 333, 334–35, 336, 337
  habitats 114,
    loss of 402
  in ice-age adaptations 30–31
  nutrient cycles 115
  around volcanoes 159
  in wetlands 321–27
Altiplano plateau 140
Annapurna 150
Antarctic Circumpolar Current 428, 452
Antarctic Ice-sheet 29, *65*, 107, *153*, **272–73**, 463
Antarctic Peninsula 429, 502
Antarctica
  climate 452, 463
  glaciers **272–73**
  ice-shelves *153*, 273, 398, 429
  igneous intrusions 184
  mountains 153
  ozone hole 449, 502
  population 503
  reflected sunlight *447*
  volcanoes 177
Antarctica Plate 86, 429, 488, 498, **502–503**
Anteater, Giant 333
anthracite 79
anticlines *136*, 148
anticyclones 466, 467
Antrim Plateau **163**
ants 303
  Weaver Ant *303*
apatite 44, **55**, 68
apes 27–28
  Chimpanzee 28
Apollo Butterfly 146
Appalachians 138, **140**, 142, 306, 375, 490, 491
Appenines 146
aquaculture 427
aquamarine 58
aquifers 109, 110, 280
Ar Rub'al Khali (Empty Quarter) 285
Arabian Peninsula 281, **285**, 496
  fault systems 71
  fishing 419
  oil fields 285, 375
  wadis *280*
Arabian Plate 127, 128, *129*, 239, 496
Arabian Sea 368, 416, **419**
Arafura Sea 425
aragonite crystals *241*
Aral Basin 40
Aral Sea **237**
Aravali Hills 293
archaea 112
archaeological remains;
    see also fossils
  deserts 284, 289
  early hominid discoveries 128, 289
    Hadrian's Wall 184
    ice mummy 144
    Neanderthal 143
    Roman lead 51
  Rome **366**
  sea-floor treasure 415, 424
archaeopteryx *24*

architecture
  Berlin Reichstag Building *364*
  Cairo mosque and citadel *366*
  Chrysler Building *357*
  Delhi **368**
  earthquake-proof 131
  Kuala Lumpur, Petronas Twin Towers *371*
  London, Lloyd's Building *365*
  London, Parliament Buildings *365*
  Manila Cathedral *370*
  Moscow, St Basil's Cathedral *364*
  Mumbai 368
  Paris, La Grande Arche *365*
  San Francisco, Transamerica Building 131
  skyscrapers 352, 357, 360, *365, 366, 368, 369, 370*
  Toronto, CN Tower 360
Arctic
  glaciers *262, 263*
  ice-cap 30, 402
  ice-shelves 398
  igneous intrusions 180
  permafrost zones 338
  tourism *379*
Arctic Basin 399, 402, 405, 455
Arctic currents and water circulation 399
Arctic Fox 30, *339*
Arctic Ocean 107, 398, 400, *401*, **402–405**, 452, 463
Arctic poppies *339*
Arctic Willow *339*
Ardennes *311*
Ardipithecus 28
Ardnamurchan **183**
Ardoukoba Volcanic Field *128*
Arenal volcano *161*
arêtes 93, *254*, 255
Arima Takatsuki Line 132
Aristotle 460
Arizona desert 103, *279*
arkose **76**
Arkwright coal mine *81*
armadillos *331*
aroids 306, 317
  Arrerrent Aboriginal people 105
art, prehistoric 240
artesian wells 110, *496*
arthropods 20, 21–22
Aruba 412
asbestos 59, 60
Ascension Island 407
ash trees 311, 316
Asia
  agriculture 344, 345, 349, 351
  bamboo forest 313
  boreal forest 312, 494
  caves 247–49
  cities 367–71
  deciduous forest 301
  desertification 290
  deserts 278, 279, 280, 281, 285, 293–95
  fault systems 132
  forests 301, 303, 313, 316
  glaciers 268, 269
  grasslands 330, 331, 336, 338
  horticulture 349
  hot springs and geysers 192
  lakes *224*, 236–39

Asia cont.
  meteorite impacts 104, 105
  mountains 148–51
  plateaus 238
  population 495
  rainforest 300, *314*, 315, 316
  rice cultivation 345
  rivers 219–23
  tundra 339
  volcanoes 173–77
Asian Water Buffalo 321,
  345, 349
Asiatic wild ass (Khulan) 294
asphalt 79
asteroids 83
asthenosphere 38, *40*, 41
Aswan Dams 214, 217, 366
Atacama Desert 140, 280, **284**,
  463, 492
Ataturk Dam 219
Athabasca Glacier **264**
Atlantic Convergence 428
Atlantic Mudskipper *321*
Atlantic Ocean 142, 143, *382*,
  398, 400, 402, 406, **407–**
  15, 490, 494, 496
  hot springs and geysers
    188, 189
  rifting 25
  temperature and depth 385
  volcanoes 163, 171
Atlantic White Cedar 322
Atlas Mountains **147**, 496
atmosphere 17, *40, 41*; *see also*
  air, greenhouse effect
  carbon dioxide 447, *454*, 459
  circulation 450–53
  circulation cells *451*
  composition 445
  energy 446–47
  layers 444
  oxidation 18
  structure 444–45
  water vapour 108, 445, 452
atolls 397, *500*
atoms 14
aubrites 83
augite *58*
aurorae *37, 444*
  solar activity and 35
Ausable *92*
Australasia
  agriculture 333, 348, 349,
    350
  caves 249
  cities 371
  deserts 279, 281, 296, 297,
    462
  endemic species 24, 26, *114*
  eucalypt forest 308, 316, *499*
  fault systems 133
  glaciers 274–75
  grasslands *330*, 331,
    337, *499*
  hot springs and geysers 193
  lakes 239, *498*
  meteorite impacts 105
  mountains 152
  New Zealand mixed forest
    303, 316
  Northeast Australian
    rainforest 317
  rivers 223
  sheep farming 348
  volcanoes 177
  wetlands 327
Australasian Bittern 327
Australia, population 498
Australian Mountain Ash 316
Australian Plate 86, 133, 425,
  488, 494, **498–99**, 500
Australopithecine 29
  Paranthropus 28, 29
Australopithecus 28

Australopithecus afarensis 28
avalanches and landslides 96,
  97, 98, *99, 453*
Avalonia 126, 142
avocets 325
Aye-aye 313
Azores 387, 407
azurite 43, 54

B

bacteria 112
bacterium 112
Back-swimmer, Common *106*
Bactrian Camel 294
badgers 303
Badlands National Park
  332, 490
Baffin Bay **405**
Baffin Island *491*
Bagley Ice field *252*, 257
Bahia De Los Angeles *422*
Bahr el Ghazal 234
Baikal Seal 236
Baitoushan volcano **174**
Baja California 501
Bald Cypress 322
Bald Eagle *139*
Balearic Islands 146
Balsam Fir 304
Baltic Sea **413**, 433
Baltoro Glacier **269**
bamboo 302
bananas 316, 345
Banda Sea **425**
Banded Sea Snake *416*
Bandicoot, Rabbit-eared *296*
Bangiomorpha 18
Bangkok 424
Banks, Joseph 316
Banksia trees *114*
Baobab trees *337*, *497*
Bar-headed Goose 150
Barbados, wind-borne
  deposition 95
Barbary Ape *147*
Barents Sea **404**
Barisan mountains 239
Barkly Tableland *337*, 499
barley *350*
barophilic bacteria 387
Barred Owl 322
barrier islands 437
barrier reefs *396*, 397, 425, 499
Barringer Crater *see* Meteor
  Crater
Barringer, Daniel Moreau 103
basalt 40, 56, 57, 60, 64, **68**,
  74, 75, 83
basaltic lavas 25, 41, *56*, 60, *63*,
  64, 68, 70, 148, 155, 158,
  160, 161, 162, 163, 166,
  171, 172, 173, 174, *176*,
  178, 180, 183, 184
Basin and Range fault
  system **125**
Basking Shark *407*
Basle, earthquake 127
Bass Strait *394*, 425
Basswood *306*
Bateke Plateau 218
batholiths 66, *178, 179*,
  182, 183, 184
bats 242, *248*, 283
bauxite **82**, 376
Bavarian Forest 311
Bay of Bengal 416, **419**, 470
Bay of Biscay **413**
Bay of Fundy *92*, **408**, *434*
Bay of Islands 71
baymouth bars 437
Bear Creek 228
Bear River Valley 188

Beardmore Glacier 272
bears 139, 303
  Black Bear 305, 322
  Brown Bear 143, *225*, 312
  Florida Black Bear 322
  Mazaalai (Gobi Bear) 294
  polar bears 30
  White-clawed Bear 149
Beaufort, Francis 482
Beaufort Sea *106*, **405**
Beaufort Wind Force Scale 482
beavers 321, 325
Becquerel, Henri 52
Bedouin people *344*
bee-eaters *462*
beech trees 114, *306*, 311
  Japanese Beech 313
beetles 335
  Darkling Beetle 280
  Domino Beetle 289
  dung 335
Beijing **369**
  Olympics (2008) 369
Belle Ile (France), coastline *413*
Belolakaya *146*
Beluga (White Whale) *408*
bennettitaleans 23
Beppu, hot springs **192**
Bergen Geophysical
  Institute 466
Bering Glacier **257**
Bering land bridge 31
Bering Sea **422**
Bering Strait 402, 405
Berlin **364**
Berlin Reichstag Building *364*
Bermuda 407
beryl **58**, 67
beryllium 58, 67
Betic Cordillera 146
Bialowieska Forest 311, *495*
Biebrza Marshes **325**
Big Bang 14
Big Bend canyon system 208
Big Bluestem grass 332
Big Obsidian Flow 69
Big-Leaf Mahogany 307
Bilberries 312, 339
Bilby, Greater *296*
Bingham, Hiram 141
Bingham Quarry 374, *376*
biodiversity, threats to **116**, *117*
biogeography 114
biomass, power generation and
  375
biomes 114, *115*, 304,
  320, 457
biosphere 34, 40
biotechnology, food
  provision 362
biotite 59, 69, 73, 74, 75
birch trees 114, 304, 311, 312,
  313, 339
  dwarf birch 31
  Erman's Birch 313
birds; *see also* particular birds by
  name
  Cenozoic Period 24
  coastal oil spillage 439
  desert fowl *280*
  maniraptoran dinosaurs 24
  migration of 285
birds' nests, Gunung Mulu caves
  248
Bismark Plate 500
bison; *see also* buffaloes
  American *186*, *330*, 332, *348*
  European 311
Bittern, Australasian 327
Bjerknes, Vilhelm 466
Black Bear 305, 322
Black Fir 316
Black Forest 127
Black Gum (Tupelo) tree 322

Black Hills (South Dakota, USA)
  303
Black Rapids Glacier **256**
Black Rhinoceros *116*
Black Rock Desert 188
Black Sea **413**, 415
black smokers 386, *387*
Black Spruce 304
Black-footed Ferret *331*
Blackbrush 282
Blacktailed Gazelle 294
bladderworts 322, 326
blizzards *474*
block lava *155*, *160*
Blue Heron *323*
Blue Gum tree 316
Blue Mesa (Arizona, USA)
  petrified forests 61
Blue Mountains *152, 316*
Blue Nile 217
Blue Whale 383
bluebell woods *302*
Blyde River *148*
Boar, Wild 143, 311
Bobcat 322
bog iron ore 53
Bogbean *324*
Bogda Feng 149
Bogda Shan 149
Bogdanovich Glacier **268**
bogs 320
  blanket bogs 324
  cushion bogs 327
  quaking bogs 322
  raised bogs 325
Bohol Sea *108*
Bombay *see* Mumbai
Bonaire 412
Bongo *461*
Bonobo 313
Bonplan, Aimé 141
Boojum Tree *116*
Bora Bora *396, 501*
bora wind *483*
borates 44, 54
borax 44, **54**
boreal forests 303, 304, 312,
  494
Borlaug, Norman 343
Borneo, rainforest *314*
bornite **50**
boron 67
Bosporus 413
Bosumtwi Crater **104**
Boteti River 326
Bottlenose Dolphin *400*
boundary currents 394
Bowhead Whale *402*
Boyoma Falls 218
Brahmaputra 151, 222
braided rivers 201, *210*, 213,
  255, *274*, 320, 325
Braldu River 269
Brasilia 354, 361
Brazil Nut tree 307
Brazza, Pierre 218
Bread Knife (Warrumbungle
  Range, Australia) *152*
breccia **76**,182
bridges
  Brooklyn Bridge *356*
  George Washington Bridge
   180
  Golden Gate Bridge *478*
  Howrah Bridge *367*
  Sydney Harbour Bridge *371*
Brine Shrimp 284, 325
Bristlecone Pine 282, 305
British Antarctic Survey 449
British East India Company 371
brittlestar fish *392*
bromeliads 303, 306, 307
Brooklyn Bridge *356*
Brown Bear 143, *225*, 312

Brunt Ice-shelf *398*
Bryce Canyon *65*, *490*
bryozoans 21
Buachaille Etive Mor *142*
Buch, Christian Leopold
  von 67
Buckle Island 502
Buenos Aires **364**, 493
Buffalo Grass *330*
buffaloes; *see also* bison
  Asian Water Buffalo 321,
   345, 349
building materials 377
  clays 377
  granite 66, *179*
  marble 71
  mica 59
  slate 72
Bukhan Range 368
bulls, Camargue 325
Bungle Bungles *499*
buran wind 483
Burchell's Zebra *113*, *321*
Burgess Shale 13, 20
Burkino Faso, drinking
  water *230*
Burrowing Owl *331*
Bushveld Complex 67
butterflies 324
  Apollo Butterfly *146*
  Common Brown Butterfly *331*
  Rajah Brooke's Birdwing
   Butterfly *316*
buttes *97*, 279
Byrd Glacier 272

C

caatinga region (Brazil) 307
Cabot Strait 408
cacti *206*, 281, 293
Cactus Finch *114*
cadmium 50
caimans 324; *see also* alligators,
  crocodiles
Cairo **366**
  New Administrative Capital
   (NAC) 366
calcite 68, 71, 78, 82
  cave pearls *241*
  crystals 43, 44, 53, **54**, *249*
  deposits 78, 240, 241, 242,
   246, 247, 248, 249
  double refraction 54
  dripstone decorations
   242, *247*
  flowstone 54, 78, 91, 241,
   243, 247
calcium 71
calcium carbonate 77, 91
calderas 158, 159, 160, 161,
  166, 170, 171, 173, 174,
  175, 177
  formation 155
  hot springs 189
  lakes 158, *161*, 172, 174,
   175, 224, 239
  Sonoran Desert *282*
Caledonian Mountains 87, 138,
  140, **142**
Caledonian Orogeny 126
California
  earthquakes 124, 125, 131
  El Niño effects *453*
  Silicon Valley 378
  wind turbines *375*
California coniferous forest **305**
call centres *379*
Camargue 213, **325**
camels
  Bactrian Camel 294
  Dromedary Camel *276*, 280,
   289, 296, 343

Cameroon Mountain 172
Cameroon Volcanic Line 172
Canadian Shield 102, 180, 225, 304, 408, 491
canals
　All-American Canal 206
　in Berlin 364
　Erie Canal 204
　Grand Canal (China) 220, 369
　Jonglei Canal 327
　Karakum Canal 293
　Panama Canal 412
　in Shanghai 370
　Suez Canal 368, 414, 415, 418
Canaries Current 409
Canary Islands, volcanoes 97, 168, 463
Cango Caves 247
Canton see Guangzhou
Canton Observation Tower 370
canyons and gorges
　Big Bend canyon system 208
　Blyde River 148
　Colorado River 138, 206, 207
　Congo River 218
　Danube River 216
　Dinaric Alps 146
　erosion of 92
　Grand Canyon 12, 103, 108, 206, 207, 483
　Murray River 223
　Nile 217
　Paraná River 208
　Rhône River 213
　Rio Grande 208
　Santo Domingo 243
　underground 240, 243, 246, 247, 248
　undersea 392, 416, 419
　White Rock Canyon 208
　Yangtze River 214, 220
Cape Barren Geese 327
Cape Bon 414
Cape Horn 429
Cape Spencer 422
Cape York (Australia) 499
Cape York (Greenland) 262
Capercaillie 143, 303, 312
Capuchin Monkey 324
Capybara 321, 324
Caranda Palms 324
carbon cycle 115
carbon dating 13
carbon dioxide 447, 454, 459
carbonates 44, 45, 45, 53, 54, 63, 68, 71, 77, 78
Caribbean Sea 400, 412, 491
Caribou 31, 139, 204, 339
Carlsbad Cavern 242
carnotite 55
Caroline Plate 500
Carpathians 146
Carpenter Ant 333
Carrara marble 71
Cascade Range 86, 98, 120, 137, 158, 159, 305
Cascadia subduction zone 34
Cashew trees 306
Casiquiare Channel 209
Caspian Sea 238, 375
cassiterite 52, 55
Castle Rock (Edinburgh, Scotland) 183
catfish 322
cats; see also lions, lynx
　big cats 26, 27
　Bobcat 322
　Geoffroy's Cat 333
cattle ranching 348; see also dairy farming
　Llanos 333
　Pampas 333
Caucasus 146

cave art 29
　Cango Caves 247
　Lascaux 246
caves, rivers and 197 individual caves are listed under underground rivers and caves
Cave of Crystals 245
Cecropia trees 306
Cedar Pine 316
　cedar trees 147, 313
　Atlantic White Cedar 322
　Incense Cedar 305
　Red Cedar 317
cement production 377
Cenozoic Era 25–31
Central America
　caves 243
　isthmus 491
　lakes 232
　population 491
　rainforest 232, 306
　volcanoes 161
cephalopods 21
cereal cultivation 343, 350
Cerrado landscapes 307, 330, 333
Cerro Azul–Quizapu 162
Cerro Negro (Nicaragua) 155, 455
Cerro Pirámide 264
cerussite 55
Cetti's warbler 457
chaco regions (South America) 307
Chaîne des Puys 163, 189
chalcedony 61
chalcopyrite 50
Challenger Deep 387, 420
Chambers Pillar 296
Chameleon, Parson's 313
Chamois 143, 213
chamosite 60
Champagne Pool, hot springs 193
Chang Jiang River 370
Chao, Edward 100, 103
Chaobai river 369
chaparral 462
Chari River 234
Chaudes-Aigues, hot springs 189
Cheju Island 424
chemical industry 377
　agrochemicals 343, 345
　petrochemicals 377
cherry trees 313
chert 78, 82
Chesapeake Bay Crater 102, 105
Chestnut Mandibled Toucan 232
chestnut trees, American Chestnut 306
chiastolite 57
chiastolite hornfels 72
Chiba 369
Chicago 352, 360
　water usage 226
chicken farms, intensive 351
Chicle tree 306
Chicxulub Crater 24, 25, 103
Chignecto Bay 408
Chihuahuan Desert 283
　Carlsbad Cavern 242
Chile
　coastal desert 280
　earthquake (1960) 122
　fiords 435
Chimney Rock 491
Chimpanzee 28
China
　industrialization and congestion 363

China cont.
　industry 373, 374, 376
　Qianjiang Tidal Bore 432
　rice cultivation 340
　silk production 378
　textile and clothing industry 378
　tourism 379
Chincoteague Inlet 320
chinook wind 483
Chinstrap Penguin 401
chlorite 60, 66, 73
chlorofluorocarbons (CFCs) 449
Chocó rainforest 306
chondrites 83
Christchurch earthquake 130
chromite 44, 52, 184
chromium 82, 184
Chrysler Building (New York City, USA) 357
chrysotile asbestos 60, 71
Chuckwalla 282
Chugach Mountains 257
Chukchi Sea 405
Chuquicamata 182
Churchill Falls Dam 214
cichlid fishes 235, 236
cinnabar 50
Circumpolar Deep Water 428
Cirio Tree 116
Cirque du Gavarnie waterfall 143
cirques 93, 143, 147, 252, 254, 255
cirrocumulus clouds 475, 481
cirrostratus clouds 475, 478, 481
cirrus clouds 108, 475, 481
citrine crystal 61
civilizations, early 223
cladistic system 112
clams 21, 387
clay 76, 377
claystone 77
cliff slides 96, 98
climate 442–463; see also weather
　carbon dioxide variation 447, 454
　classification 460
　climate change 454–57
　cold climates 463
　cool-temperate continental 463
　cool-temperate maritime 462
　desert climates 442, 460, 462, 463
　desertification and 290, 454
　El Niño effect 98, 450, 453, 455
　greenhouse effect 214, 345, 447, 454, 459
　Greenland Climate Network project 263
　hot dry 462
　hot with seasonal rain 461
　hot with year-round rain 461
　maritime 462, 463
　Mediterranean climate 462
　Mesozoic–/Cenozoic climate change 24
　mid-latitudes 462–63
　monsoon climates 111, 150, 300, 337, 416–17, 419, 424, 461, 469
　mountain climates 463
　oceans and 452
　oscillations 452, 453
　Paleozoic climate change 22
　past climates 153, 454
　polar regions 463
　records and trends 455
　regions 460–63

climate cont.
　Sun and 455
　temperate 460, 462
　tropical (equatorial) 460, 461
　urban areas and 354, 355
　volcanic particles and 95
　warm temperate 462
clinochlore 60
Closed Anemone 425
Cloudforest 298, 306
Cloudina 19
clouds
　altocumulus 108, 475, 480
　altostratus 478, 480, 483
　cirrocumulus 475, 481
　cirrostratus 475, 478, 481
　cirrus 108, 475, 481
　cloud condensation nuclei 445
　cloud types 475–81
　condensation 472
　contrails 481
　cumulonimbus 472, 475, 479, 484
　cumulus 466, 475, 478
　global cloud cover 475
　lifting and rainfall 473
　mammatus 479
　nimbostratus 480
　noctilucent 445
　orographic 473
　precipitation and 472–81
　reflection of sunlight 447, 452
　seeding 479
　solar activity and 455
　stable and unstable air 472
　storm clouds 447, 450, 452, 464, 472, 476–477, 479, 484
　stratocumulus 475, 478
　stratus 467, 474, 475, 478
　topographic barrier 480
clubmosses 23
CN Tower (Toronto, Canada) 360
coal 63, 79, 115, 147, 148, 374, 375
　iron and 376
　open-cast mining 81, 126, 374, 375
Coast Redwood trees 301, 305
coasts 431–39; see also oceans and seas, tides, waves
　beaches 94, 412, 433
　caves 436
　coastal climates 463
　coastal coals 79
　coastal dunes 95, 279, 292, 433
　delta coasts 434
　depositional 431, 434, 436, 437
　drowned coasts 434, 435
　emergent 434, 436
　erosion 92, 249, 412, 416, 430, 431, 433, 436
　glaciers 250, 256, 431
　karst coasts 434
　longshore drift 436
　marine terraces 436
　pollution 438, 439
　raised beaches 434
　rocky 407, 413, 423, 430
　sea defences 431, 432
　sea-level and 41, 434–35
　tides 432
　types of 434
　volcanic 434

Coelacanth 419
coesite 100, 103
coffee 345
Colima volcano 155, 160
Colorado Desert 278
Colorado Plateau 138, 206
Colorado River 110, 138, 206, 297, 214, 215, 422
Columbia Bay 260
Columbia Glacier 260, 264
Columbia River 57, 68, 158
　lava 25
Columbia River Plateau 84, 158
Columbia terminus 260
Comb Sedge 327
comets 100, 103, 105
Common Back-swimmer 106
Common Brown Butterfly 331
computer imaging, meteorite impact craters 105
Concepción volcano 232
conches 412
condensation 472
　latent heat and 447, 452
Condors 141, 493
conglomerate 73, 76
Congo Basin 496
Congo River 218, 236, 313
coniferous forests 142, 300, 301, 302, 304, 305, 310, 312; see also evergreen forests
conifers 23
conodonts 20, 21
continental climates 460, 463
continental crust 34, 35, 40, 40, 41, 41, 86
Continental Divide (North America) 198, 206
continental ice-sheets 29–30, 65, 153, 254, 263, 272–273, 463
continental lithosphere 41
continents
　collisions 137, 142
　formation 17
Cook, Captain James 316
Cooksonia 21
Coon Butte see Meteor Crater
Coorong 327
Copenhagen 354
copepods 402
copper 46, 71, 82, 141, 142, 284, 376
　open-cast mines and quarries 374, 376, 499
　ores 50, 54, 182
coral islands 421, 425, 500, 501
Coral Sea 425
Coral Sea Basin 425
Coral Sea Plateau 425
coral terraces 29
corals and coral reefs 21, 23, 128, 379, 396–97, 410, 412, 417, 424, 425, 501
　cold-water 396
　threats to 411
　true colours 384
　warm-water 396, 411
Corcovado National Park 306
Cordgrass 332
Cordillera Real 140
core, of the Earth 36–37
Coriolis, Gaspard-Gustave de 450
Coriolis effect 394, 409, 450, 452, 466, 469, 482
cork, harvesting 310
Cork Oak 310
corn (maize) 332, 350, 463
Corsica 414
corundum 52
Cosmic Microwave Background Radiation (CMBR) 14

Costa del Sol 379
Costa Rica, deforestation 308
Cotopaxi **162**, *445*
cotton 345, 368, 378
cotton-grass 114
Couch's Spadefoot Toad *282*
Cougar *see* Mountain Lion
Cousteau, Jacques 414
Crabeater Seal *399*
crabs
　mud crabs 327
　Vent Crab *386*
Cranberries 312
cranes *325*
Crater Lake **158**
crater lakes 102, 104, 105, 158,
　161, 172, 174, 175, 177
craters *see* meteorite craters
cratons *488*
Creosote Bush 283
Crete 414
crickets *117*
Crimson Anemone *401*
crinoids 21
crocidolite 59, 60
crocodiles 23, 26, 321, 323; *see
　also* alligators, caimans
　Nile Crocodile *217*
crops
　production ranking *342*
　slash-and-burn agriculture
　　344
crossbills 303
Crowberries 312, 339
Crown of Thorns Starfish 397
Crozet Islands *503*
crust **40–41**; *see also* Earth,
　rocks
　composition 35
　continental 40, 86
　epeirogenic movements 138
　formation 17
　movement 33, 137, 138, 198
　oceanic 41, 86
　stratigraphic record 65
Cuba 412
Cubango River 326
Cuillin Hills, igneous intrusion *64*
cumulonimbus *472*, 475,
　479, 484
cumulus clouds *466*, 475, **478**
Curaçao 412
Curie, Marie 52
Cushion Plant 327
Cutlass Fish 424
cycads 23
cyclones *466*, 457, *468*, 482
　tropical 469, *470*, *471*
Cypress National Reserve, Big
　323
cypress trees
　Bald Cypress 322
　cypress marshes 323
　Patagonian Cypress 310
Cyprus 414

D

dacite **69**, 70
Daintree Rainforest *317*
dairy farming **349**; *see also*
　cattle ranching
Dallol volcano *190*
dalmatian coasts 435
Damodar Valley, coal 375
Damour, Augustine 75
dams **214–15**
　Colorado River 206, 214, *215*
　Euphrates River 219
　hydroelectric power 214, *215*
　Nile River 214, 217
　Tagus River 212
　Vaiont Dam disaster 98

dams *cont.*
　Volga River 216
　Yangtze River 214, 220
　Zambezi River 219
Dana, James Dwight 396
Danakil Desert 128
Danube **216**
Dardanelles 413
Darjeeling 444
Darjiling Mountaineering Institute
　150
Darkling Beetle 280
Darling River 152, 499
Darwin, Charles 28, 396, 397,
　420
Darwin's Finch *114*
Darya-ye-Sefid *238*
Davis, William Morris 92
Davis Strait *405*
Dead Sea 106, 128, **239**, *434*
Death Valley 78, *125*, 282, *442*,
　490
Deccan Plateau 24, 337, *494*
Deccan Traps 24, *494*
Deception Island 153
deciduous forests 114, 300,
　301, 302, 303, 306, 308,
　310, 311, 312
deer 303
　Fallow Deer 311
　Musk Deer 316
　Red Deer 311
　White-tailed Deer 322
Deere, John 350
deforestation *146*, 210, 290,
　294, 308–309, 310, 348,
　447, *493*
Delaware Basin 242
Delhi **368**
deltas 94, 201
　Chang Jiang River 370
　Colorado 206
　Danube 216
　delta coasts 434
　Ganges 94, 222, 470
　Indus 223, 367
　Irawaddy 221
　Lena *495*
　Mackenzie *204*
　Mekong 221
　Mississippi *201*, 205
　Niger 219
　Nile 94, *201*, 217
　Ob' 221
　Okavango Delta *321*, 326
　Orinoco *209*
　Paraná *208*
　Pearl River 355
　Rhône 213, 325
　Rio Grande 208
　Volga 216
　Yangtze 220
　Yellow River (Huang-He) 221
　Yukon River *203*
Denali National Park *139*, *339*
Dengue fever 459
Denisova cave, Russia 31
Denisovans 31
Déodat de Dolomieu 53
deposition **94–95**
　aerosol deposits 95
　coastal 431, 434, 436, **437**
　glaciers and 65, 95, 147,
　　255
　in lakes 63, 224, 236
　ores 45, 50, 51, 53, 54, 55,
　　71, 78, 182, 184, 376
　organic **79**
　of sediment (river systems)
　　63, 65, 94, 199, 201, 205,
　　206, 208, 221, 222, 416,
　　419, 422, *423*, 431
　sediment sampling 94

deposition *cont.*
　volcanic 63, 65, **70**, 95,
　　157, *158*
　wind-borne 95
depressions, weather 467, 482
Desert Tortoise 282
Desert Horned Viper 289
Desert Rose (gypsum) 55, *289*
desertification *146*, 210, 283,
　290, *291*, *454*
deserts **276–97**
　animal life 280, 282, 283,
　　285, 292, *292*, 294
　Arabian Peninsula *280*, 281,
　　285
　archaeological remains
　　284, 289
　Arizona Desert 103, *279*
　Atacama Desert 140, 280,
　　284, *463*, *492*
　Black Rock Desert 188
　caravans *276*, 289
　Chihuahuan Desert 242, **283**
　climates *442*, 460, 462, 463
　Colorado Desert *278*
　Danakil 128
　environmental issues 284,
　　289, 296
　ergs 95, 279, 289
　Gibson Desert *297*
　Gobi Desert 95, 281, **294**,
　　295, *494*
　Great Basin 278, **282–83**
　Great Sandy Desert **296**, 499
　Great Victoria Desert *297*
　Kalahari Desert 279, **292**, 326
　Karakum Desert 238, **293**
　Mojave Desert *282*
　Namib Desert *278*, *279*, 280,
　　292, *451*, *463*, *468*
　nomadic dwellers 289, 292,
　　294, *344*
　nuclear storage 283
　Patagonia Desert **283**
　pavement 279, 294
　Pinnacles Desert *462*
　plant life 281, 282, 283, 284,
　　285, 289, 292, 293, 294,
　　296, 297
　Sahara Desert *93*, 95, *147*,
　　276, 279, 280, 281, 285,
　　289, *445*, 460, 462, 474,
　　483, *484*, 496
　saltwater wetlands 320
　Simpson Desert 279, **296**
　Sind 367
　Sonoran Desert *281*, **282**
　Taklamakan Desert 281, **293**
　Tanami Desert **296**
　Thar Desert 279, 280, 281,
　　293
　types of 278
　varnish 279
　world distribution 281
dew 452, 472, 475
　erosion and 278
　water source 280
Dhaka **367**
　Lalbagh Fort *367*
　Narayanganj Boat Market *367*
diamonds 43, **47**, 68, 377
　mining of 47, 361
　　Koidu, Sierra Leone *47*
　Namibia *12*
diatoms *393*
dickite 60
Dieback disease 316
Dinaric Alps **146**

Dingo 296
dinosaurs 23–24
　extinction of 23, *101*, 102,
　　103, 104, 113, 116, 173
　fossils 13, 24, 77
diogenites 83
diorite 52, **67**, 75
Dipterocarpaceae 315, 316
Disenchantment Bay *250*, 256
Disko Bay *403*
Disneyland (Florida, USA) 379
Doda river 269
Dogger Bank 413
dogs 26
Doi Inthanon *316*
doldrums 451
dolerite *58*, **67**, 179, 180, 183,
　184
dolines 240, 241, 243, 247, 249;
　see also sinkholes
dolomite **53**, 71, **77**
Dolomite Mountains *53*
dolphins 26, *220*
　Bottlenose Dolphin *400*
　military sonar and 385
　Yangtze River Dolphin 220
Dombay-Ul'gen *146*
Domino Beetle *289*
Douglas Fir 301, 303, 305
Drake Passage 429
Drakensberg Plateau
　(Ukhahlamba) *148*, *496*
Dropseed Grass 330
Dromedary Camel *276*, 280,
　289, 296, *343*
drought 289, 290, 456, 474
drumlins 255
Dry Valley (Antarctica) *153*
d'Sousa, Manuel 57
dubh lochans *324*
Ducie Island, marine litter 439
ducks 321, 339
　Wood Duck 322
Dumbo Octopus *384*
dunes
　Acacus-Amsak Region *496*
　barchan 95, 279
　Chihuahuan Desert *283*
　coastal 95, 279, *292*, 433
　Coorong 327
　crescentic 279, 289, 293
　domes 95
　formation of 93, 95
　Great Sandy Desert *296*
　Great Victoria Desert 297
　gypsum *55*
　lunette 297
　Namib Desert *278*, *292*, *451*
　Nullarbor Plain *498*
　parabolic 279, 293
　Sahara Desert 289
　seif (linear) 279, 289, 292,
　　293
　Simpson Desert 296
　slope stability and *96*
　Sossuvlei *278*
　stabilization 281
　star 95, 279
　Thar Desert 293
　Tularosa Basin *283*
dung beetles 335
dunite 56, 184
durian 316
Durung Drung Glacier **269**
dust storm 286–87
dustbowls 343
Dwarf Hamster 294
Dwarf Willow 339
Dwyka tillite 148
Dygalski Ice Tongue *503*
dykes, igneous rocks 64, 66, 67,
　68, 178, 179, 180, 182,
　184

E

eagles 143
　African Fish-Eagle 326
　Bald Eagle *139*
　Harpy Eagle 306
　Spanish Imperial Eagle *310*
　Stellar's Sea Eagle *423*
early civilizations 223
Earth
　anatomy and structure **32–41**
　Cenozoic Era **25–31**
　core **36–37**
　geology and tectonics *488*
　global water cycle 109
　history **11–31**
　layers *16*, 33, 35
　magnetic field 35, 37
　Mesozoic Era **23–24**
　and the Moon **30–31**
　orbital cycles *454*
　origins **14–15**
　Paleozoic Era **20–22**
　Precambrian Period **16–19**
　shape and form 34
　snowball earth 19
　in space *34*, *109*
　and the Sun *454*
　surface change and
　　movement 85
earthquakes 27, 98, 121, 122,
　123, 124, 125, 127, *130*,
　132, 140, 147, 149, 189,
　369, 488, 492
　offshore 390
　prediction 122, 125, 127, 131
Ease Gill Caverns 246
East African Plateau 236
East African Ridge System 418
East Antarctic Ice-Sheet
　(EAIS) 270
East Australia Current 425
East China Sea 423, **424**
East Pacific Rise 386, 420, 501
East Rongbuk Glacier **269**
East Siberian Sea *405*
ebony 313
echinoderms 21
eclogite 56, **74**, 182
ecosystems 114–15
Edelweiss *149*
Ediacaran fossils 19
eels *409*
egrets 321
Eiffel Tower *365*
Eiger *144*
ejecta 95, 101, 105, *502*
Ekman transport 394
El Capitan (California, USA) **182**
El Chichón *161*
El Gezira, Sudan
　cotton fields *346–47*
El Niño 98, 450, **453**, 455, 461
El Niño-Southern Oscillation
　(ENSO) 453
El Tatio geysers **188**
Elburz Mountains 238, 367
electrical power *see* power
　generation
Elephant Grass 114, *337*
Elephant Island *429*
elephants 30, 292, 313, 326,
　327, 496
Elk 159, 325
Ellesmere Island *398*, *405*
Ellsworth Mountains 502
Elm trees 311
emeralds 58
emery, industrial abrasive 52
Emperor Penguin *273*, *398*
Emperor Seamounts 500, 501
Empty Quarter *see*
　Ar Rub'al Khali

INDEX

Endeavour space shuttle *173*
Engabreen Glacier 266
Engelmann Spruce 305
Enhanced Geothermal Systems (EGS) 195
environmental issues; *see also* dams, earthquakes, erosion, irrigation
    atmospheric pollution 91, 107, 361, 445, 447
    biodiversity threats **116**, *117*
    cattle ranching 310, 348
    cetaceans at risk 385
    coastal pollution *438*, **439**
    coral reefs 397, **411**, 412
    dams 214
    deforestation *146*, 210, 290, **308–09**, 310, 348, 447, *493*
    desertification *146*, 210, **290**, *291*, *454*
    deserts 284, 289, 296
    dustbowls 343
    El Niño effects 453
    fertilizer runoff 409
    flotsam rubbish 428
    fossil fuels *80*, *81*
    global warming 171, 195, 253, 339, 345, 362, 411, *447*, 455, 459
    greenhouse gases *214*, 345, 447, 454, 459
    ground-level ozone 449
    industrial landscapes 374
    industrial pollution 213, 231, 325
    introduced species 296
    lakes 225, 226, 228, 237, 238, 239
    landfill sites *374*
    landslides 98, *99*
    Lascaux 246
    living with volcanoes 169
    Manila rubbish dump *370*
    methane and rice 345
    mining 284, 296
    Mount Everest 151
    nuclear storage 283
    nuclear waste dumping 375, 404
    oil pollution 81, 439
    open-cast coal mines 374
    overcultivation 92
    overfishing 427
    overgrazing *146*, 210, 289, 290, 310, 343, 348
    ozone layer depletion *448*, **449**, 502
    population growth **362**, *363*
    quarries 374
    recycling 378
    river systems 198, 201, 205, 213, 220, 223
    salinization 349
    savanna 335
    Southern Ocean 428
    tourism 379, 411
    urban areas 354, 355
    water contamination by pesticides 324
    waste disposal 231, 357, *374*, 378
    water pollution 231, 376
    wetlands 319, 323
Environmental Agency (UK) 200
epeirogenic movements 138
epidote group **57**
epiphytes *298*, 303, 306, 313, 317
Eqip Sermia Glacier *263*
equatorial climates 460, **461**
ergs 95, 279, 289
Erie Canal 204

Erman's Birch 313
erosion 34, 85, **92–93**; *see also* environmental issues, weathering and erosion
    rivers *84*, 92, 138, 197, 198, 199, *200*
    coastal 92, 249, *412*, *416*, *430*, 431, 433, **436**
    deforestation and *309*
    deserts 278, 279, *288*, 289
    erosional landforms 436
    farming and *343*
    forests and 299
    glacial 93, *140*, *142*, 144, *145*, 224, 225, 226, 233, 255
    marine 249, *416*, *430*
    overgrazing 343, 348
    sculpted rocks *416*
    sedimentary rocks 63, 65, 92, 181, 184
    stone arches *93*
    underground 240
    wind erosion 93
erratics 255
Erta Ale 128, **171**
eruptions, volcanic 33; *see also* rocks, igneous
    Andes 140
    atmospheric particles 445
    climate and *455*
    fissure *156*, 160, 163, 166, 171
    Hawaiian *39*, 156
    landslides and 98
    Mount St. Helens 69, *120*, 169
    phreatic *156*, 163, 172, 177
    Plinian 156, *158*, *159*, 160, 161, 170, 175
    Strombolian *156*, 163, *167*, *176*, 177
    subglacial 265
    sulphur ores and 47
    Surtseyan 156
    Vulcanian 156, 161, 170, *176*
Erzincan, Turkey, earthquake 127
eskers 95, 255
estuaries
    Amazon 210
    Everglades 323
    Loire 212
    Rhine 213
    River Plate 208
    Severn 212
    Sumida 369
    Tagus 212
    Thames 212
Ethiopian flood basaltic lava 25
Ethiopian Highlands *497*
Etosha National Park *112–13*
eucalyptus trees 308, 316, *330*, *337*, *499*
eucrites 83
eukaryotes 17, 18, *19*, 112
Euphrates **219**, 419
Eurasia, tundra 339, 494
Eurasia Basin 402
Eurasian Plate 86, *88*, *122*, 127, 132, 143, 144, 416, 488, **494–95**, 500
Europe
    agriculture 344, 348, 351
    boreal forest 312, 494
    caves 246, 247
    cities 364–66
    coal reserves 375
    coniferous forests 308, 312
    deciduous forests 308, 312
    fault systems 126, 127
    glaciers 265–68
    horticulture 349

Europe *cont.*
    hot springs and geysers 173, 191, 192
    igneous intrusions 183, 184
    industry 373
    lakes 232, 233
    meteorite impacts 104
    mixed forest 301, 302, 303, 311
    mountains 142–46
    nomadic herding 344
    population 495
    rivers 212–16
    tundra 339
    volcanoes 163–70
    wetlands 213, 324, 325
European Plate 143
evaporites 54, **78**
Everglades 228, **323**
evergreen forests 300; *see also* coniferous forest
    Mediterranean **310**, *462*
evolution **113**; *see also* life
    Cambrian explosion 20–22
        yunnanzoans 20
    karyotic 18
    humans **27–29**, *29*
    mammals and birds 24, 26
    marine life 26
    plant life 21–22, 23
    reptiles 22
    speciation 113
    vertebrates 26
extinctions of species **113**
    dinosaurs 23, *101*, 102, 103, 104, 113, 116, 173
    Great Dying, the 113
    human influence on 31
    K-PG extinction event 24, 103, 104
    Permo-Triassic extinction event 22–23, 173
    recent extinctions 31
Exxon Valdez oil spill 422

F

Fair Head Sill **183**
Fallow Deer 311
farming *see* agriculture
fault systems 121, **122–33**
    Arabian Peninsula 71
    Atacama Fault *284*
    Atlas Mountains 147
    Basin and Range **125**
    earthquakes and 122, 123, 125, 131
    fault lakes 224, 228, 232, 233, 235, 236
    fault planes 122
    faults and joints 122
    Great Alpine Fault (New Zealand) **133**
    Great Rift Valley 25, 28, *34*, **128–29**, 138, 171, 172, 217, 234, 235, 238, 418, 488, 496, *497*
    Japan 192
    Keraf Suture *497*
    Midland Valley **126**
    Moine Thrust **126**
    Nojima Fault **132**
    North Anatolian Fault **127**
    North Sea Basin **126**
    Peel Fault 127
    reverse faults 126
    rifting 25, *122*, **127**, **128–29**, 163
    San Andreas Fault *88*, 123, **124–25**, 127
    Southeast Korea Fault Zone **132**
    strike-slip faults 123, 132

fault systems *cont.*
    tectonic plates and *488*
    Tien Shan Mountains 149
    transform fault *88*
    types of fault 123
    Yangsan Fault 132
    Zagros Mountains 148
favelas 361
feather stars *404*
feldspar 45, **60**, 66, 67, 68, 69, 73, 74, 75, 76, 77, 91, 179, 376
Fennec Fox 289
fens 320
ferns 23, 313, 315
    Log Fern 322
    Tangle Fern 327
    tree ferns 306
    Water Fern 321, 324
Ferret, Black-footed *331*
ferro-manganese nodules 82, *393*
Fertile Crescent 290
fig trees 324
financial services 379
finch
    Cactus Finch *114*
    Darwin's Finch *114*
    Ground Finch *114*
Fingal's Cave, basalt 68
Finger Lakes 228
fiords 435
    Alaska 422
    Chile 435
    New Zealand 152, 435, *500*
    Norway *142*, 435, *495*
fir trees 311, 313
    Balsam Fir 304
    Black Fir 316
    Douglas Fir 301, 303, 305
    Nikko Fir 313
    Siberian Fir 312
    Silver Fir *302*
    White Fir 305
fire
    air pollution and 445
    bush fires, Australia *499*
    climate and *455*
    El Niño and 453
    forests 303, 305, *455*
    grasslands 330, 334
    rainforests 315
    wildfires 322
fish; *see also* particular fish by name
    anglerfish *392*
    Devonian jawless fish 21
Fish-Eagle, African 326
fishing **427**
    Arabian Sea 419
    Aral Sea 237
    Bay of Bengal *419*
    Bay of Biscay 413
    East China Sea 424
    fish ponds 351
    fish-farming 427
    Great Slave Lake 225
    Gulf of Mexico 409
    Gulf of St. Lawrence 408
    Indian Ocean *417*
    industrial fishing *426*, 427
    Lake Baikal 236
    Lake Nakuru 235
    Lake Nyasa 235
    Lake Titicaca 229
    Lake Turkana 234
    Lake Victoria 235
    mangrove removal and 308
    Mediterranean Sea 415
    North Atlantic *407*
    North Sea 413, *426*
    Norwegian Sea *382*
    overharvesting 427

fishing *cont.*
    Persian Gulf 419
    Sea of Okhotsk 423
    Southern Ocean 429
    stilt fishing *417*
    Sulu Sea 425
    Yangtze River *220*
Fitzroy Range *140*
flamingoes 235, *235*, 292, 325, 326
flint and flint tools 82
floating islands, Lake Titicaca 229
Flohn, Hermann 460
flooding 200, 201, 205, 459
    Africa 217, 218, 236
    Asia 214, 220, 221, 222, *459*, *495*
    Australasia *318*
    in deserts 280, 292
    El Niño and *453*
    Europe 212, 213, 216
    Ganges floodplain *495*
    glaciers and 265
    Jakarta 371
    London 365
    North America 205, 208, *453*
    paved landscapes and 109
    Rome 366
    seasonal *318*, 321, 324, 326
floodplains 94, 200, *201*, 214, *318*, 320, 324
Florida Black Bear 322
Florida Cottonmouth Snake 322
Florida Keys *409*
Florida seas *491*
Flow Country **324**
Flowerpot Rocks *408*
fluorides 44
fluorite 44, **53**
Fly Agaric *311*
Fly Geyser **188**
fog 452, 472, 475, 478
    radiation fog *444*
    source of moisture in deserts 280, 292
fohn wind *483*
food processing industry **378**
foraminiferans 30
    unicellular *454*
forests 298, **299–317**; *see also* deforestation
    alpine and submontane 302, 311
    animal life 303, 304, 305, 306, 307, *310*, 311, 312, 313, 315, 316, 317
    Ardennes *311*
    Asian mixed 313, 316
    Australian eucalypt 308, 316, *499*
    Bavarian Forest 311
    Bialowieska Forest 311, *495*
    Black Forest 127
    Black Hills (South Dakota, USA) 303
    boreal forest 303, 304, 312
    canopies 300, 303, 313, *314*, 317
    coniferous *142*, 300, 301, *302*, 304, 305, 308, 310, 312
    coppicing 311
    deciduous 300, 301, 303, 306, 310, 311, 312
    Devonian Period 21
    distribution of 303
    dry forest 300, 310
    European mixed 311
    evergreen 300, 310
    fires 303, 305, 315, *455*

forests cont.
glaciers and 274
ground flora 302, 303
inhabitants 299
New Zealand mixed 303, 316
North America **304–306**, 308
Pacific Northwest
rainforest 305
petrified forests 13, 61
plant life 300–302, 303, 306,
307, 310, 311, 312, 313,
315, 316, 317
regeneration 303
San Juan Forest (Colorado,
USA) *460*
sizes of 303
swamp forest *307*
taiga 299, 301, 312, 315,
339, 494
temperate 114, 301, 303,
308
temperate rainforests 301,
305, 310
tropical *see* tropical rainforest
zones 302
forsterite crystals 56
fossil fuels 80, **81**, *115*, 375
greenhouse gases and 447,
454, 459
fossils; *see also* archaeological
remains
Antarctic trees *22*
Antarctica *22*, 153
chemical 17, 18
coniferous trees 153
dinosaur footprints 24, 77
Ediacaran 19, 21
geological time and 12, 13
Himalayas 150
hominids 25, 28, *31*, 128, 369
in iron deposits 78
limestone and 54
in marble 71
in metallic nodules 82
in mudstone and shale 77
in peat 320
petrified forests 13, 61
reptile ancestors 22
Sahelanthropus tchadensis
28, 289
in sandstone 77
savanna animals 285
in siltstone 76
in slate 72
soft-bodied 19, 20
tectonic plate movement
and 87
Foster, Norman 364
Fourwing Saltbush *281*
Foxe Basin 408
foxes 303
Arctic Fox 30, *339*
Fennec Fox 289
Foxtail Grass *332*
fracking 81
fracture, of minerals *43*
Fram Strait 404, 405
Franz Josef Glacier 259, **274**
Franz Josef Land 339
Fresh Kills landfill site 357
Fringed Pinks *143*
fringing reefs 396
frogs
Red-eyed Treefrog *306*
frost 90, 96, 180, 452, 475
Fujita, Theodore 482
Fujiyoshida Fire Festival 174
fulgurite *484*
fumaroles 185, 186, 188, *189*,
192
fungi 112, *311*

G

gabbro 40, 52, 56, 58, 60, 61,
64, **67**, 68, 75, 179, 180,
184
Galah *223*
Galápagos Islands *82*, 420, *488*
galaxies 15
galena 43, **51**, 55
Galeras volcano **162**
Galilee, Sea of **238**
Ganges 94, **222**, 367, 392, 416,
419, 470
Ganges Plain 150, *495*
Gansu, China 98
Garlic, Wild *311*
garnet **56**, 66, 68, 71, 73,
74, 75
garnet augen gneiss 74
garrigue *462*
gas *see* methane, natural gas;
see also atmosphere,
composition
Gazelle, Blacktailed 294
Gecko, Gobi 294
geese 327, 339
Bar-headed Goose 150
Gemsbok *113*, 292
gemstones 45, 47, 50, 51, 52,
54, 55, 56, 57, 58, 60, 61,
377; *see also* jewellery
General River *306*
gentians 327
geodes *43*, *61*
Geoffroy's Cat 333
geoid 34
geological time 12, 13
Geomagnetic South Pole 502
George Washington Bridge *180*
geothermal energy and power
185, 189, 192, 193, **194**,
195, 375
geothermal systems 185
gerbils 336
geysers 107, 185; *see also* hot
springs, volcanoes
El Tatio geysers **188**
Fly Geyser **188**
Geysir *88*, **188**
Lake Bogoria **192**
Soda Springs (Idaho,
USA) **188**
Prince of Wales Feathers
(Rotorua, New Zealand)
193
Rotorua **193**, 379
Strokkur Geyser **189**
Tatsumaki jigoku geyser 192
Waimangu **193**
Yellowstone 185, **186**, *187*
Geysers (California, USA),
geothermal power
plant 195
Geysir *88*, **188**
Ghaf tree 285
Giant Anemone *418*, 425
Giant Anteater 333
Giant Kelp forests *420*
Giant Tortoise *116*
Giant's Castle 148
Giant's Causeway 25, 68, 163
Gibson Desert **297**
Gilbert, G. K. 105
Gill Polypore *303*
ginger 317
ginseng 316
Gippsland Basin 425
giraffes 327
glaciation; *see also* ice ages
deposition 95, 147, 255
erosion 93, *140, 142, 145*,
224, 225, 226, 233, 255
glacial rebound 434

glaciation cont.
glacial striae *274*
glaciated landscapes 143,
144, *145*, 147, 148, 152,
183, *254*, 255, 264
periglacial landforms 338,
339
post-glacial lakes *141*, 204,
224, 225, 226, 228,
232, 233
post-glacial uplift 144,
204, 226
Precambrian 16, 19
sedimentation 76
volcanoes and 158, 163, 171
Glacier Bay 98, 260, 422
Glacier d'Argentière *267*
glaciers 107, **250–75**; *see also*
ice-sheets, continental
aerial mapping 257
Agassiz Glacier 261
Aletsch Glacier 144, *254*, **267**
Allalin Glacier **268**
animal life 261, *273*
Antarctic Ice-sheet 29, 65,
107, 153, **272–73**, 463
Athabasca Glacier **264**
Bagley Ice field *252*, 257
Baltoro Glacier **269**
Beardmore Glacier 272
Bering Glacier 257
Black Rapids Glacier **256**
Blomstrandbreen
Glacier *253*
Bogdanovich Glacier **268**
Byrd Glacier 272
cirque glaciers 93, 254
coastal and tidewater glaciers
250, 256, 431
Columbia Glacier **260**, 264
Durung Drung Glacier **269**
East Rongbuk Glacier **269**
Engabreen Glacier 266
equilibrium zone *253*
erosion by 93, *140, 142*,
144, 224, 225, 226,
233, 255
formation 253
Franz Josef Glacier 259, **274**
glacial (blue) ice 253
Glacier d'Argentière **267**
global warming and 253,
263, 270, 455
Greenland Ice-sheet *254*,
263, 270, 398, 402,
463, 491
Grinnel Glacier *270*
Hayden Glacier 261
Helheim Glacier 270
Hubbard Glacier *250*,
256, 422
ice collapse *257*
ice mummy 144
Jakobshavn Glacier 263
Jostedalsbreen Ice-field **266**
Juneau Ice-field 260
Kangerdlugssuaq Glacier 270
Kaskawulsh Glacier **264**
Kennicott Glacier **256**
Khumbu Glacier **269**
Kilimanjaro Ice-cap **268**
Kjer Glacier 270
Kongsvegen Glacier **266**
Kronebreen Glacier *266*
Lambert Glacier 272, 502
Malaspina Glacier *254*, **261**
Margerie Glacier **260**
Marvine Glacier 261
melting 255
Mendenhall Glacier **260**
Mer de Glace **267**
movement 253
Muir Glacier *252*

glaciers cont.
outlet glaciers 254, 260, *262*,
263, 264, 265, 266, 272
Palisade Glacier **264**
Pastoruri Glacier **264**
Perito Moreno Glacier **265**
piedmont lobes 254, 257,
261
Pine Island Glacier 270
plant life 257, *261*
polythermal glaciers 266
rainforest and 274
Rhine Glacier 233
Rhône Glacier *253*, **268**
sea-level and 434
Seward Glacier 261
South Engilchek **149**, **268**
Southern Patagonian Ice-field
264, 265, *492*
Spitsbergen Ice-cap *254*
structure 252
surge-type glaciers 253, 256,
266, 269
Taku Glacier 253
Tasman Glacier 152, **274**,
275
Thwaites Glacier 270
terminus and terminal walls
250, 252, 255, 260, 261,
265, 266, *268*
Trapridge Glacier **256**
Tschierva Glacier **267**
types 254
valley glaciers 254, 260, 267,
268, 269
Vatnajökull Ice-cap 163, **265**
Waiho 274
West Svartisen Ice-cap **266**
world distribution 255
Worthington Glacier **260**
Glen Canyon Dam 206, *215*
Gletscherrandsee 268
global warming 171, 195, 273,
339, 362, *447,* 456–457,
458–59
coral reefs and 411
glaciers and 253, 263, 455,
456
rice and 345
globigerina *393*
Glory Bush 306
GM crops 350
gneiss 52, 56, 57, 58, 61, 63,
74, 183
Goanna 297
goats
milk production 349
mountain goats 139
Gobi Desert 95, 281, **294**, *295*,
494
Gobi Gecko 294
Gobi Gurvansaikhan
(Three Beauties)
National Park *295*
Goethe, Johann Wolfgang
von 53
goethite **53**, 78
Golan Heights 238
Golconda Cavern (UK) *241*
gold 43, 44, **46**, 147, 376
Gold Rush, the 46
mining 46, 104, 141, 296,
324, 361, 367, *372*,
376, *499*
placer deposits *221*
Golden Barrel Cactus *281*
Golden Gate Bridge *478*
Golden Trumpet *322*
Gondwanaland 19, 20, 21, 138,
148, 416
Goodall, Jane **28**, 334
Goosenecks State Park *138*
Gorda ridge 34

gorges *see* canyons and gorges
gorillas 28
Mountain Gorillas *313*
Gosses Bluff Crater
(Australia) **105**
Gotland, wind farm 413
Gotzen, Adolf von 172
Gouffre Mirolda caves 247
Gournier cave lake *246*
Grampians 142, 232
Grand Canal (China) 220, 369
Grand Canyon *12*, *103*, *108*,
206, *207*, *483*
Grand Tetons 139
granite 41, 52, 59, 60, 61, 62,
63, 64, **66**, 178, 179, 182,
183, 184, *493*
granodiorite **67**, 182
Grant's Golden Mole 292
granulite **74**
grapevines *168, 233,* 351, *462*
graphite **47**
graptolites 20, 21
Grasmere 232
Grass Peaks (Lesotho) *148*
grass snakes 321
grasslands 114, **329–33**; *see*
also savanna
animal life *328,* 331, 332,
333, 334–35, 336, 337
distribution of 331
fires 330, 334
grass burning 331
Great Plains (prairies) 225,
330, 331, **332**, 343, 485,
491
Indian savanna *330,* 331, **337**
Pampas 114, *330,* **333**, 350
savanna 330, 331, 333, 334,
337, *497*
Serengeti Plains **334–35**,
461, 474
steppes *148*, 331, **336**
temperate 330, 331
tourism 335
trees 330, *333*, 336, 337
Great Alpine Fault (New Zealand)
133
Great Artesian Basin
(Australia) 110
Great Barrier Reef 397, 425, 499
Great Basin 278, **282–83**
Great Basin Rattlesnake 282
Great Bear Lake 204, **225**
Great Bear River 225
Great Blue Hole *412*
Great Dismal Swamp *322*
Great Divide basins 139
Great Dividing Range **152**, 337,
498, 499
Great Dying, the 113
Great Dyke (Zimbabwe) **184**
Great Glen 126, 232
Great Grey Owl 304
Great Lakes 204, **226–27**, 304,
360, *483*, 491
Great Plains (North America)
138, 225, 330, 331, **332**,
343, 485, 491
Great Pyramids 366
Great Rift Valley 25, 28, *34*,
128–29, 138, 171, 172,
217, 234, 235, 238, 418,
488, 496, 497
Great River Otter 210
Great Salt Lake **228**
Great Sandy Desert **296**, 499
Great Slave Lake 204, **225**
Great Snipe *325*
Great Spotted Woodpecker *303*
Great Victoria Desert **297**
Great Wall of China 369
Great Water Lilies *210*
Greater Antilles 137, 412

Greater Bilby *296*
Green Revolution 343
Green River 243
Green Turtle *423*
greenhouse effect and
greenhouse gases *214*,
345, 447, 454, 459; *see
also* atmosphere
Greenland
gabbro *67*
Greenland Climate Network
project 263
Greenland Current, East and
West 405
Greenland Halibut 408
Greenland Ice-sheet *254*, **263**,
270, 398, 402, 463, 491
Greenland Sea *401*, **405**
Grey Teal 327
Grey Wolf 311
Grímsvötn volcano **163**, 265
Grossglockner *144*
Grotte Casteret **246**
Ground Finch *114*
groundwater 91, 107, 109, 110,
198, 280, 452
geothermal systems 185, 195
groynes 431, 433
Gruta do Janelão **243**
Gryllefjord *255*
Grypania 18, *19*
Guaira Falls 208
Guanacos *493*
Guangzhou (formerly Canton)
355, 369, **370**
guano deposits *82*
Guanyuan Cave System 247
Guaymas Basin 422
Guilin karst *495*
Gulf of Aden 418
Gulf of Alaska *250*, 257, **422**
Gulf of Aqaba *129*, 418
Gulf of Bo Hai 423
Gulf of Bothnia 413
Gulf of California **422**
Gulf Coastal Plain 208
Gulf of Gabes 414, 415
Gulf of Mexico 208, 323,
375, **409**
hurricanes 470
Gulf of Oman *419*
Gulf of St. Lawrence **408**
Gulf of Suez *129*, 418
Gulf of Thailand 424
Gulf of Tongking 424
Gulf Stream 409, *452*
Gulf War (1991), oil well fires *80*
gum trees; *see also* eucalyptus
trees
Australian Mountain Ash 316
Black Gum (Tupelo) 322
Blue Gum 316
Red Gum *316*
Gunung Mulu **248**
Gunung Mulu National Park 248
pinnacle karst *241*
Gutenberg, Beno 122
gypsum 54, **55**, 77, 78
Carlsbad Cavern 242
Desert Rose 55, *289*
gypsum flowers 243
Optimisticheskaya 247
gyres 394, 407, 420, 452

H

Ha Long Bay **249**, *434*
haboob wind 483
Hadley cells 451
Hadrian's Wall 184
hailstones 108, 452, 474,
479, 484
Hajar Mountains 285

Half Dome (California, USA) **182**
Halibut, Greenland 408
halides 44, 53
halite 43, **53**, 78
Hamersley Range *45*, 78
Hamster, Dwarf or Desert 294
Han River 368
hanging valleys 144, 255
Hardy Reef *425*
hares 30, 304
harmattan 289
harmattan wind 483
Harp Seal *404*
Harpy Eagle 306
Haughton Crater (Canada) **102**
Haüy, René 57
Hawaiian honeycreeper *113*
Hawaiian Islands 41, 89,
160, 500
volcanic hotspot *32*, 155,
387, 500
Hawaiian Koki'o *113*
Hawaiian-Emperor chain
500, 501
Hayden Glacier 261
heat transfer, core and mantle 39
heat wave 456
Hedgehog, Long-eared
Desert 294
Helheim Glacier 270
hematite 43, 45, **51**, 78
hemlock 303, 305, 313
Herculaneum 70, 170
herons 321
herring, collapse of stocks *426*,
427
Herzog, Maurice *150*
hickory trees 306
high-technology industries **378**
Highland Boundary Fault 126
Hillary, Edmund 150
Himalayan poppies *151*
Himalayas 25, 26, 27, 86, 89,
137, **150–51**, 269, 337,
445, 488, 494
hippopotamuses 26, 30, *219*,
321, 327
Hispaniola 412
Hô Chi Minh City 424
Hoba West meteorite *100*
Hochstetter Ice-fall 274
Hoggar Massif *see* Ahaggar
Mountains
Hokkaido 423
Hokou Falls 221
Hollywood (Los Angeles,
USA) *360*
Holm Oak 310
hominids 25, 28; *see also*
humans
early discoveries 128, 289,
369
hominoids 26, 27
Homo antecessor 29
Homo erectus 28
Homo floresiensis 29
Homo habilis 28
Homo heidelbergensis 28
Homo neanderthalensis 29
Homo sapiens 28, 29
Honey Possum 316
Hong Kong *354*, **370**, 424
container traffic *377*
land reclamation 94
landfill site *374*
Hood River *96*
hoodoos *488*
Hooghly 222, 367
Hoover Dam 206, *214*
Hopewell Cape *408*
Hornbeam *302*, 311
hornblende 59, 66, 69, 74, 75
hornfels *41*, 57, **72**
hornitos 172

horns, glacial 93, *254*, 255
horses 26
Camargue 325
cattle ranching and *348*
Przewalski's Horse 336
horsetails 23
horticulture **349**
cut-flower cultivation *349*
Hveragerdhi, Iceland *195*
Hortobágy **325**
Hortobágy National Park **325**
hot rocks 195
hot springs **185–93**; *see also*
geysers, volcanoes
algae *107*
Beppu *192*
Bogoria **192**
Champagne Pool *193*
Chaudes-Aigues **189**
geothermal energy 195
Lardarello **189**, 195
Mammoth Hot Springs 186
Nagano **192**
Pamukkale *107*
Prismatic Pool *187*
Punchbowl Springs *185*
Rotorua **193**, 379
Soda Springs **188**
Solfatara *192*
Valley of Ten Thousand
Smokes **188**
Waimangu **193**
Yellowstone (USA) *45*, **186–87**
housing developments 323
Howard, Luke 475
Howard-Bury, Charles 149
Howler Monkey 324
Howrah Bridge *367*
Huang-He River *see* Yellow River
Huangpu River 370
Huascarán *140*
Hubbard Glacier *250*, **256**, 422
Hudson, Henry 204, 408
Hudson Bay 225, **408**
Hudson River 180, **204**, 356
Hudson Strait 408
humans *106*; *see also* hominids
fertility and population growth
362, *363*
migration 31
origins **27–29**, *29*
Humboldt, Alexander von 83, 141
humidity 108
Hunter Valley (New South Wales,
Australia) *462*
hunting and gathering 344
hurricanes 407, *412*, *424*, 452,
469
impact of 98, *451*, 470, *471*
Hurricane Fran 470
Hurricane George 470
Hurricane Katrina 470
Hurricane Sandy 470, *471*
Hurst Castle, spit *437*
Hutton, James 70
Hyacinth Macaw 324
hydrocarbons 79; *see also* coal,
fossil fuels, natural gas, oil
hydroelectric power 214, *215*,
266, 267, 375
hydrosphere 40
hydrothermal reservoirs 195
hydrothermal veins *45*, 50, 51,
52, 53, 55, 61
hydrothermal vents 386, 418
hydroxides 44, 53

I

Iapetus Ocean 142
Iberian Peninsula 143
Iberian Plate 143

ice; *see also* glaciers, water
blue ice 253, 260
calving ice 253, *253*, 257,
260, 261, *263*, 274, 284
fresh water and 107, 447
frost and rime 452, 475
grease ice 399
hailstones 108, 474, 479, 484
ice leads and polynyas *399*
ice-core analysis 273
pack ice *106*, *429*
pancake ice *398*, 399
sea-ice 399, 402, 405, *422*,
429, 455, 456
sea-ice shrinkage *456*
seracs 259, 269
ice ages 21, **29–31**, 267, 268,
454, 455; *see also*
glaciation
ice-breakers 402, 404
ice-caps 30, 107, 254, 263,
265, 266, 402
ice caves; *see also* underground
rivers and caves
Grotte Casteret 246
Mount Erebus *177*
Muir Glacier *252*
ice factories, Arctic *403*, 404
ice-falls 269, 274
ice islands 402
ice-sheets, continental 29, *65*,
107, *153*, 254, 263, 272–
73, 398, 413, 434, 463,
491; *see also* glaciers
ice-shelves 272, 273, 398, 402,
429, 502; *see also* shelves
and plains, sea floor regions
ice wedges 338
ice worms 261
icebergs 107, *253*, 257, 260,
261, *262*, 263, 265, 272,
274, *275*, 398, 402, *403*,
405, 422, *428*, *429*, 452,
503
Iceland 88, 407, 434
geothermal energy *194*, 195
geysers 188, 189
Laki fissure 163
Vatnajökull Ice-cap 163, 265
Icelandic Ridge *41*
ICEsat 263
ichthyosaurs 23
Idiot Fruit 317
igneous extrusion *63*, 64
igneous intrusions *63*, 64, **178–
84**
Aïr Mountains **184**
Ardnamurchan **183**
Castle Rock (Edinburgh,
Scotland) **183**
Chuquicamata **182**
Devil's Tower (Wyoming,
USA) **181**
El Capitan (California,
USA) **182**
Fair Head Sill **183**
Great Dyke (Zimbabwe) **184**
Half Dome (California,
USA) **182**
Mackenzie Dykes **180**
Palisades Sill **180**, 204
Salvesen Mountains **184**
Ship Rock (New Mexico,
USA) **182**
Skaergaard **180**
Sugar Loaf (Rio de Janeiro,
Brazil) **183**, *361*
Whin Sill **184**
igneous rocks 40, 41, 45, 51,
52, 56, 57, 59, 60, 61,
62, 63, 64, **66–70**, 83
crystal structure 179
joints 122, 181, 183, *184*

ignimbrite *62*
Iguaçu Falls 208
iguanas 26
Marine Iguana *420*
illite 60, 377
ilmenite **52**, 67, 68
Imperial Dam 206
Incas 141, 229
Incense Cedar 305
India
coal reserves 375
transport networks 377
Indian Cobra 337
Indian Grass 332
Indian Ocean 400, **416–19**, 417,
496, 498, *503*
Indian Ocean Ridge 498
Indian Plate 27, 86, 150, 416,
425, 494
Indian Rhinoceros *337*
Indian savanna grasslands *330*,
331, **337**
Indonesia
oil fields 375
rainforest *314*, 315
Indricotherium 26
Indus 151, **223**, 269, 367, 416
industrial areas **373–79**
recent developments 374
Industrial Revolution 373, 376
industrialization *363*, 373–79
industrial pollution 213, 231, 325
industry; *see also* agriculture,
coal, nuclear, oil
power generation 375
primary industry 374, 375–77
secondary industry 374,
377, 378
tertiary industry 273, 377
Inge Lehmann 36
Ingleton, slaty cleavage 64
insects, evidence for past
climates 454
inselbergs, Sahara Desert 289
International Union for the
Conservation of Nature
and Natural Resources
(IUCN) 116
International Year of Freshwater
(2003) 110
Intertropical Convergence Zone
(ITCZ) *451*
Inuit people 408
invertebrates 112; *see also*
particular invertebrates by
name
underwater *404*
Irawaddy 221
Iroko wood 313
iron 46, 71, 142, 147, 374
layered *18*
magnetism 35
meteorite **83**
ores 45, 51, 78, 376
production 376
sedimentary formations **78**
Iron Gate gorge 216
Irrawaddy **221**
irrigation
Africa 234
Asia 219, 221, 237, 293, 453
Australasia 327
Europe 212
Mesopotamia 290, 343
North America 206, 208
salinization and 349
South America 453
water resources and 110,
290
Irtysh River 148, 221
islands
Aeolian Islands 192
Aland Islands 413
Aleutians 86, 137, 422

islands *cont.*
Alexander Island 153
Andaman Islands 419
artificial 420, 431
Ascension Island 407
Atlantic Ocean 407
atolls 397, *500*
Azores 387, 407
Baffin Island *491*
Balearic Island 146
barrier islands 437
Bora Bora *396*, *501*
Buckle Island 502
Canary Islands 97, *168*, 463
Cheju Island 424
coral islands *421*, *500*, *501*
Coral Sea 425
Crozet Islands *503*
Deception Island 153
Devon Island *101*
Ducie Island 439
Elephant Island *429*
Ellesmere Island 398, 405
floating islands, Lake Titicaca 229
Galápagos Islands 420
Hawaiian Islands 41, 89, 155, 160, 387, 500
ice islands 402
Iceland *64*, *69*, *84*, *122*, 163, 188, 189, *194*, 195, 265, 407, 434
Indian Ocean 417, *503*
island arcs 387, 388, *388*
island chains 387
Jan Mayen *401*
Japanese 137, 169
Kerguelen *416*
Kodiak Island 422
Kurile 423, *501*
Livingston Island 439
Macquarie Islands *428*
Maldives *397*, *400*, 419
Maluku Islands (Moluccas) 401, 425
Managaha Island *387*
Manhattan Island *180*, 204, 356, 357
Marianas 137, *387*
Marshall Islands 500
Mediterranean Sea 414–15
Micronesia, coral islands *421*
Mount Desert Island *407*
Neijalakka Island *425*
Nicobar Islands 419
Pacific Ocean *401*, *421*
Palau *396*, *421*
Palawan 425
Ryukyu 424
Samosir Island (Lake Toba) 239
Seychelles *400*, *416*, 419
Sicily 414
Solomons 137, *401*, 463
South China Sea 424
South Georgia 184, *428*, 429
South Sandwich Islands 153, 429
South Shetland Islands 429
Staten Island (USA) 357
Surtsey *156*, 386
Tahiti *501*
Taiwan 424
Tristan da Cunha 162, *407*
Vanuatu 137, 425
volcanic 41, 86, 89, 137, 153, 158, 232, 234, 239, *387*, *401*, *407*, 425, 429, 494, *501*, 502
West Indies 412

South Shetland Islands *cont.*
White Island (New Zealand) *499*
Wizard Island (Crater Lake) *158*
isostasy 41, *137*
Istanbul **366**
Itacarambi Reserve 310
Itaipu Dam 214
Italy
fertility rate 362
hot springs 189, 192, 195
marble *71*
Izmit, earthquake 127, 131

J

Jack Fish *422*
Jackrabbits 282
jade (nephrite) 59
artefacts 75
jadeitite 58, **75**
jaguars *321*, 324
Jakarta **371**
Jakobshavn Glacier 263
Jamaica 412
Jan Mayen island *401*
Japan 249, 420, *501*
earthquakes *130*, 131, 132, *134*, 369
geothermal energy 195
hot springs **192**
industry 376
volcanoes *169*
Japanese Beech 313
Japanese islands 137, 169
Japanese Macaque 192
Java, volcanoes 169
Java-Sunda Trench 38, 387
Jays, Siberian 312
jellyfish 106
Colobonema *384*
Jenolan Caves **249**
Jenolan River 249
Jerboa *285*
jet streams 450, 451, 453, 467, 468
jewellery
obsidian 69; *see also* gemstones
Johannesburg 354, **367**
Johnson, Don, hominid remains 128
Jonglei Canal 327
Jordan River 238, 239
Joshua Tree *282*
Jostedalsbreen Ice-field **266**
Jotenheimen National Park *198*
Jotunheimen Range 142
Juan de Fuca Plate 86, 500
Juneau Ice-field 260
Jungfrau 267
Junggar Pendi depression 148, 149
juniper 31
Jurassic limestone 72

K

K2 *150*, 269
Kahitakatea trees 316
Kaikoura Oregeny 133
Kaiserstuhl, ancient volcanoes *127*
Kakamega Forest *461*
Kakapo *152*
Kalahari Desert 279, **292**, 326
Kamahi trees 316
Kamchatka Peninsula 173, 268, 423, *501*
Kangaroo Grass *330*

kangaroos and wallabies
Red Kangaroo 152, 297, *337*
Rufous Hare Wallaby 296
tree kangaroos 317
Kangerdlugssuaq Glacier 270
Kansai Airport 420
kaolin (China clay) 66, 374
kaolinite 43, *45*, **60**, 91, 377
Karakum Canal 293
Karakum Desert 238, **293**
Kara Sea 104, **404**
Kara and Ust Kara Craters **104**
Karachi **367**, *367*
Karakorum Range 269
Kariba Dam 219
Karri trees 316
karst 78, 91, 146, 241, 242, 243, 246, 247, 249, 434, *495*
Kaskawulsh Glacier **264**
katabatic winds 483
Kati Thanda-Lake Eyre **238**
Katmai 158
katydids *117*
kauri pine 317
Kawasaki 369
Kea *152*
Keeling curve 459
kelp, Giant Kelp forests 420
Kennicott Glacier **256**
Kennicott River **256**
Kenyan Giant Fig *461*
Keraf Suture *497*
Kerguelen *416*
kettle holes 255, 261
kettle lakes 224, 255
Khan-Tengri *149*
Khumbu Glacier **269**
Kilauea volcano 70, *155*, *156*, **160**, *501*
Kilimanjaro Ice-cap **268**
Kimberley region (Australia) 68
Kimberley (South Africa) 47, 68
kimberlites 47, **68**, 74, 377
Kingfishers 326
kittiwakes, Black-legged *405*
Kiwis *112*
Kjer Glacier 270
Kliuchevskoi *173*, 268
Kluane Lake 264
Ko Panyee *419*
Koalas 316
Kobe, earthquake 132
Kodiak Island 422
Kolyma River 405
Konkan coast 368
Konkordiaplatz Ice Plateau 267
Koonalda Caves **249**
Köppen, Wladimir 460
Korea
high-technology industries 378
industrialization 374
Korea Bay 423
Korean Peninsula 423
Southeast Korea Fault Zone **132**
Krafft, Maurice and Katia 174
Krakatau **176**
Kras region, Slovenia 247
krill 428
Kronebreen Glacier *266*
Krubera **247**
K-PG extinction
Event/Boundary 24, 103, 104
Kuala Lumpur **371**
Kuna Yala reserve 306
Kurile Islands 423, *501*
Kuroshio Current 424
kyanite 56, **57**, 66, 73, 74

L

La Digue *416*
La Grande Arche (Paris) *365*
La Niña 453
La Paz 140
La Pérouse Strait 423
La Planada, cloudforest *306*
La Sal Mountains *463*
La Tigra National Park 306
labradorite 60, 67
Lac Pavin Maar 163
laccoliths 179
lacustrine deltas 201, 237
Laetoli, hominid footprints 28
Lago Argentino 265
Lago Martin 264
Lago Viedma 264
lagoons, salt marshes and 320, 325, 327
Laguna Verde *140*
Lagos (Nigeria) 355, **366**
lahars 97, 98, *156*, 158, 161, 162, 175, 176, 177
Lake Albert 217
Lake Alexandrina 327
Lake Athabasca 225
Lake Baikal 224, **236**
Lake Balkhash 238
Lake Bogoria, hot springs **192**
Lake Bonneville 228
Lake Carnegie *498*
Lake Chad **234**
Lake Como 144, **233**
Lake Constance 233
Lake Disappointment 297
Lake District (England) *232*
Lake Drummond 322
Lake Erie 226
Lake Eyre, Kati Thanda **238**
Lake Garda 144
Lake Geneva **233**
Lake Huron 226
Lake Istokpoga 228
Lake Ladoga 232
Lake Malawi *see* Lake Nyasa
Lake Manyara 128, **235**
Lake Messel 26
Lake Michigan 226, 360
Lake Nakuru **235**
Lake Natron 128, *172*
Lake Nicaragua **232**
Lake Nyasa 128, **235**
Lake Nyos **172**
Lake Okeechobee **228**, 323
Lake Onega 404
Lake Ontario 226, 360
Lake Poopo 229
Lake Rudolph
see Lake Turkana
Lake St Clair 226
Lake Seneca **228**
Lake Shastina *447*
Lake Superior 78, 226, *226*, *227*
Lake Tahoe **228**
Lake Tana 217
Lake Tanganyika 128, 218, 224, **236**
Lake Texcoco 361
Lake Tianchi ("Sky Lake") 174
Lake Titicaca 140, **229**
Lake Toba **239**
Lake Turkana **234**
Lake Victoria 217, **235**
Lake Vostok *273*
Lake Windermere **232**
Lake Winnipeg **225**
lakes 110, 197, **224–39**; *see also* individual lakes by name
age 224
crater lakes 104, 105, 158, *161*, 172, 174, 175, 177

lakes *cont.*
environmental issues 225, 226, 228, 237, 238, 239
formation 201, 224, 225, 232
glacial lakes *141*, *252*, 261, 274, *275*
inflow and outflow 224
kettle lakes 224, 255
lacustrine deltas 201, 237
lake effect clouds *473*
moraine lakes 144, 224, 225, 226, 228
playas 282, 285
post-glacial 204, 224, 225, 226, 228, 232, 233
saline 106, 224, 228, 235, 237, 238, 239, 297
subglacial 265, *273*
types of 158, 201, 224
underground 242, *246*, 249
Laki fissure 163
Lambert Glacier 272, 502
Lammergeier 143, *480*
lamproite **68**
lamprophyre **68**, 182
land reclamation 323
Hong Kong 94
Japan 369
Netherlands *413*
Landsat 7 *261*
landslides and avalanches 96, 97, 98, 99, *453*
Lanzarote, grapevines *168*
Laptev Sea **404**, *495*
larch trees *301*, *302*, 312
Lardarello, hot springs **189**, 195
Lascaux caves **246**
latent heat, condensation and 447, 452
laterite 82
laurels 26
Laurentia 78, 126, 142
Laurentian Plate 22
Lautaro volcano 265
lavas 63, 75, 154, 157
Aegean islands *414*
alkaline 177
andesite 59, 69, 70
basaltic 25, 41, *56*, 60, *63*, *64*, 68, 70, 148, 155, 158, 160, 161, 162, 163, 166, 171, 172, 173, 174, *176*, 178, 180, 183, 184
carbonatite 172
dacite 69, 70
diversion of 166
domes 25, *155*, 158, 160, 161, 162, 163, 174, 175, 176
lava bombs 70, *157*, 161, 163, *167*, 177, 502
lava flows *32*, 156, 157, 158, 159, 161, 162, 163, 239, 431
lava-lakes 160, 177
pyroclastic flows and material *32*, 70, 156, 157, 158, 159, 161, 162, 163, 170, *173*, 174, 175, 176, 177, 183, 184
rhyolite 60, 61, **69**, 70, 181
sampling 160
types 155
Lavoisier, Antoine 83
Lawick, Hugo van 334
lead 51, 55, 71, 376
Leakey, Louis and Mary, hominid discoveries 128
Leakey, Richard 128
leaves, types of 300, 301
Lechwe 321, *327*
Leck Fell *246*
Lemming, Steppe 336
Lemon Cichlid **236**

INDEX

lemurs 27, *87*, *303*, 313
Lena River *495*
L'Enfant, Pierre 368
Lenga trees 310
lens
  granite 66
  igneous intrusions 179, 184
leopards *128*, *148*, 326
lepidolite 59
Lesbos *414*
Lesser Antilles 137, 412
leucite 68
leveche wind *483*
levees 201, 205, 220, 221
Leviathan Cave *241*
Levy, David 100
lichens 91, 304, *310*, 339
  life **112–17**; *see also* animal
    life, evolution, plant life
    adaptation to changing
      climates 30–31
    diversity of species 112
    earliest evidence for 17–22
    kingdoms of 112
    maintenance of 40
    marine life 20–21, 23, 26
    molecular clocks 20
    multicellular organisms 18
    origins *10*, 11
    properties 106
    reproduction 112
    water and 106, *113*
light
  penetration in ocean
    water 384
  sunlight 384, 446
  urban lighting *354*, *355*, *369*
Light Mantled Soot
  Albatross *428*
light studies
  mineral properties 43
  refractive index 54
  urban lighting *354*, *355*
Lighthouse Reef *412*
lighthouses *401*, 407
lightning 484
lignite 79
lime trees (linden) 311, 316
limestone *41*, 45, 50, 65, *78*, 79,
    94
  acidic rain and 91, 107
  bauxite and 82
  caves 143
  cliff *425*
  coral reefs and 396
  joints *122*
  Jurassic 72
  karst 78, 91, 146, 241, 242,
    243, 246, 247, 249, *495*
  karst coasts 434
  marble and 54, 57, 71, *136*
  metamorphosed 54, 55, 56,
    57, 58
  pavements *122*, 241
  phosphates and 82
  pinnacles *498*
  reef limestones 53
  rock outcrops *419*
  sedimentary 54, *122*
  Sierra Madre *140*
  underground rivers and
    caves 240–41, 249
  undersea sinkhole *412*
  valleys *200*
Limmen River *318*
Limoncocha tropical
  rainforest *461*
limpets 387
linear valleys *125*
Linnaeus, Carolus 27
lions 326, *334*, *484*
  Mountain Lion 139, 305
Lipari 70, **192**
lithic sandstone *76*

lithium 67
lithosphere, earth's 38, 40
Little Ice Age 267, 268,
    455, 461
Livingston Island, marine litter 439
Livingstone, David 236
Lizard, Cornwall *71*
lizards 297
  Chuckwalla 282
  desert lizards 280, 282, 294
  Goanna 297
  Gobi Gecko 294
  Thorny Devil 297
Llanos wetlands 209, **324**, 333
Lloyd's Building (London) *365*
Lobster Cove (Canada) *68*
lobsters *392*
  spiny lobsters 412
Loch Ard Gorge, wave eroded
  arch *436*
Loch Lomond 126
Loch Ness 126, *224*, **232**
Lodgepole Pine *186*, 305
loess deposits 65, **77**, *92*,
    95, 221
Lofoten Maelstrom 413
logging 307, 310, 315, 321
Log Fern 322
Loire **212**
Lomonosov Ridge 402
Lonar Crater **105**
London 212, 353, *365*, 379
longshore drift 436, 437
Longtan Dam 214
lopoliths 179, 184
Lord Howe Island *425*
Lord Howe Rise *425*
Los Angeles 355, **360**, 462
Lough Derg **232**
lusakite 57
Lutyens, Sir Edwin 368
Lyme Regis, sedimentary rocks
  65
lynx 143, 304

M

maars 163, 172, 189
MacAllister Range 317
Macaroni Penguin *416*
magnesium silicate perovskite
    39, *39*
McGee Creek *254*, 255
Machu Picchu 141
Mackenzie Dykes **180**
Mackenzie River **204**, 225, 405
McKeown, James 249
mackerel 424
mackerel sky *480*
McMurdo Base (Antarctica) *502*
McMurdo Sound 177, *429*
Macquarie Islands *428*
Madagascar 419, 496, *497*
  intensive farming and
    erosion *343*
  rainforest *300*, *303*, 313
mafic rocks 68
Magellan, Ferdinand 401
magma 38, 39, *39*, 60, *62*, 63,
    64, 66, 67, 69, 70, 133,
    137, 154, 155, 156, 158,
    159, 161, 169, 386, 387,
    500
  hot springs and geysers 186,
    189, 192
  igneous intrusions *64*, 178–
    79, 180, 181, 182, 183,
    184
magnesium 53, 71
magnetic field
  of the Earth 21, 35, 88
  Ordovician 21
  of volcanoes 169

magnetic reversals 88
magnetism 35
magnetite 35, 43, **51**, 52, 53,
    68, 69, 73, 78
magnolias 26
mahogany
  African 313
  Big-Leaf Mahogany 307
maize 332, *350*, *463*
Majorca 414
Makgadikgadi Pans 326
malachite **54**
Malaspina, Alessandro 261
Malaspina Glacier *254*, **261**
Malaysia
  industrialization 374
  rainforest *303*
Maldives *397*, *400*, 419
Malebo Pool 218
Maluku Islands (Moluccas) 401,
    425
mammals; *see also* particular
    mammals by name
  Cenozoic Period 24, 26
mammatus clouds *479*
Mammoth Cave **243**
Mammoth Hot Springs 186
mammoths 31
Managaha Island *387*
Manatees, conservation of 412
Maned Wolf *333*
manganese nodules 82, *393*
mangoes 316
mangrove swamps 306, *315*,
    320, 323, 371, 457, *457*
mangrove trees and forests 308,
    315, 320, 323
Manhattan Island *180*, 204, 356,
    357
Manicouagan Crater **102**
Manila **370**, 424
Manila Cathedral *370*
Manson Crater (Iowa, USA) **102**
Manta Ray *424*
mantle, of the Earth 38–39
mantle 88
  hotspots 154, 155, 157, 162,
    166, 173, 387
mantle plume 39
Manuks 327
Maori people 193
maple trees 114, 313, 316
  Queensland Maple 317
  Sugar Maple 304, 306
maquis 462
Mara River, meanders *196*
marble 54, 56, 57, **71**, *136*
marcasite 78
Margerie Glacier **260**
Mariana Trench 387, 388,
    420, 501
Marianas 137, *387*
Marine Iguana *420*
marine life 20–21, 23, 26
marine terraces 436
Mars
  expeditions to 102
  meteorites from 83
Marsh Marigold *321*
Marshall Islands *500*
marshes 320
  cypress marshes 323
  salt marsh 320, 323, 437
marsupials 24, 26, *114*
Marvine Glacier 261
Masai people 331, 335
Masaya volcano **161**
Masoala National Park *313*
mass damper *131*
mass movement, of rock
  materials 76, **96–99**
mass-transit systems *355*
Matterhorn *93*, *145*
Mau Escarpment 128

Mauna Kea *155*
Mauna Kea Observatory,
    measuring plate
    movement 87
Mauna Loa **160**
Maunder Minimum 455
Mayombe Forest Reserve *313*
Mayon, Plinian eruption *156*
Mazaalai (Gobi Bear) 294
Meadowlark *332*
Median Tectonic Line (MTL)
    132
medicinal plants 306, 307,
    313
Mediterranean climate **462**
  agriculture 351, 462
Mediterranean region
  evergreen forests **310**
  seasonal winds 483
Mediterranean Sea 147, *400*,
    414–15, 418, 494
Meerkat 292
Megazostrodon 24
Mekong 221
Melinau River 248
meltwater *107*, 110, 147, 163,
    171, 200, 232, 260
  fluvioglacial deposits 95
  glacial lakes *252*, 261,
    274, *275*
  streams *252*, *254*, 260
  in tundra *339*
Mendenhall Glacier **260**
Mer de Glace *267*
Merapi volcano **176**
mercury 50
mesas 279
mesosphere 444
metals; *see also* individual
    metals by name
  related industries **376**
Meteor Crater 100, *101*, **103**
Metéora *494*
meteorite craters 101–105
  Bosumtwi **104**
  Chesapeake Bay **102**, 105
  Chicxulub 24, *25*, **103**
  Gosses Bluff (Australia) **105**
  Haughton (Canada) **102**
  Kara and Ust Kara **104**
  Lonar **105**
  Manicouagan (Canada) **102**
  Manson (Iowa, USA) **102**
  Meteor Crater 100, *101*, **103**
  Ngorongoro **334**, 335
  Popigay **105**
  Ries **104**
  Steinheim **104**
  Sudbury (Canada) **102**, 104
  Tunguska **105**
  Vredefort (South Africa) **104**
  Woodleigh (Australia) **105**
  Wolfe Creek *100*
meteorite impacts 17, **100–105**
  cataclismic 101, 103, 105
  causes of 100
  Chicxulub 24, *25*, **103**
  craters 101–05
  indentifying sites 100
  K-PG extinction event 24,
    103, 104
  post-impact expectations 101
  shatter cones 101
meteorites 17, 65, **83**
meteorology *see* weather
methane 17, 214, *456*; *see also*
    natural gas
  and global warming 339, 345
methane hydrate *392*
Meuse 213
Mexican Volcanic Belt 160
Mexico, high-technology
    industries *378*
Mexico City 355, **361**

Mesozoic Era **23–24**
mica 57, **59**, 66, 67, 68, 69, 71,
    72, 73, 74, 91, 133, 182
Michoacán-Guanajuato volcanic
    field 160
Micronesia, coral islands *421*
Mid-Atlantic Ridge 88, 140, 386,
    387, 407, 416, 496
Mid-Indian Ridge 416
mid-latitude depressions 468
Midland Valley fault system **126**
Midnight Sun *429*
migmatite **72**
migration
  birds 285
  Caribou *31*
  human 31
Milankovitch, Milutin 454
milk production 349
Milton, Daniel 103
Miltonia Orchid *303*
Minas Gerais 361
mineral terraces *107*
minerals **42–61**; *see also* metals,
    individual metals and
    minerals by name
  classifying 44
  cleavage *43*
  coal 63, 79, 81, 126, 147,
    148, 374, 375, 376
  colour of *43*
  crystalline structure and
    shape 42
  formation 42, *45*, *47*
  fracture *43*
  gemstones 45, 47, 50, 51,
    52, 54, 55, 56, 57, 58, 60,
    61
  hardness *43*
  identification tests 43
  lustre *43*
  non-metallic industries **377**
  ore-deposits 45, 50, 51, 53,
    54, 55, 71, 78, 182,
    184, 376
  rock-forming 45
  streak *43*
  transparency *43*
mining
  beryl 58
  Chuquicamata copper mine
    182, 284
  coal 81, 126, 374
  diamonds 47, 68, 361
    Alrosa 47
    Koidu, Sierra Leone *47*
  gold 104, 141, 296, 361,
    367, *372*, 376, *499*
  kimberlite 68
  meteorite craters 104
  nickel 60, 102
  salt 53
  uranium 184
Minnesota marshes *205*
Minoan civilization, collapse
    of 170
mires 320, 322
Mississippi 94, *201*, 205,
    205, 491
  navigation channel *201*
  water from Chicago 226
Missouri 205
mistral wind *483*
Mitre Peak 152
Moa River 45, *47*
Moab (Utah, USA), stone
    arch *93*
Mohorovicic seismic
    discontinuity 40, 41
Mohs, Friederich 53
Moine Thrust fault system **126**
Mojave Desert **282**
  coral bleaching *411*

moldavite 104
Mole, Grant's Golden *292*
molecular clocks 20
Molina, Mario 449
molluscs 21, 422
Molucca Sea 425
Moluccas *see* Maluku Islands
molybdates 44, 55
molybdenum 55
Mönch Peak 267
Monfragüe National Park *310*
Mongol peoples 294
Mongolian Gazelle 294
Monkey Puzzle trees 310
monkeys 27
    baboons *114*
    Barbary Ape *147*
    Capuchin Monkey 324
    Howler Monkey 324
    Japanese Macaque 192
    Old World and New World
        species *114*
    Woolly Monkey *114*
monotremes 24, 26
monsoon climates 111, 150,
    214, 300, 337, 416–17,
    419, 424, **461**, 469
plate movements and 25, 27
Mont Blanc 267
Mont-aux-Sources 148
Monte Fitzroy 264
Monte Somma 170
montmorillonite 377
Montreal Protocol (1987) 449
Montserrat, volcanic
    ash *157*, 161
Monument Valley *97, 279*
Moon,
    craters 17, 104
    Earth and **30–31**
    formation of 16
    tides and 432
moose 139
moraines 76, 95, 252, 254, 255,
    256, 261, 264, 267, 274,
    413, 434
moraine lakes 144, 224, 225,
    226, 228
Moscow **364**
Moselle *494*
Moskva (Moscow) River 364
mosses 339
mosquito
    Aedes 459
moths 327
    Peppered Moth *113*
Mount Adams *86*
Mount Assiniboine *139*
Mount Belukha 148
Mount Chephren *482*
Mount Cook 152, 274, *275,*
    499
Mount Damavand 367
Mount Desert Island *407*
Mount Elbert 139
Mount Elie de Beaumont 274
Mount Erebus **177**, 502
Mount Etna **166**, *167*, 169, *445*
Mount Everest *137, 150,* 269,
    494
Mount Fuji *155,* **174**
Mount Jumullong *500*
Mount Kenya 270
Mount Kilimanjaro **171**, 268,
    270, *497*
Mount Kirkpatrick 153
Mount Kosciusko 152
Mount McKinley *139, 339, 490*
Mount Mazama 158
Mount Pelée 70
Mount Pinatubo 69, 95, *156,*
    169, **175**
Mount Rainier **158**, *480*
Mount Robson 304

Mount Robson Provincial
    Park *304*
Mount St. Helens 69, 96, *120,*
    155, **159**, 169
Mount Tarawera 193
Mount Tasman 274
Mount Unzen **174**
Mount Whitney *139*
Mount William Range *153*
mountain goats 139
Mountain Gorillas *313*
Mountain Lion (Cougar)
    139, 305
mountain sheep 139
mountains 62, **121–53**; *see also*
    volcanoes
    Aconcagua *140, 492*
    Alps 26, *86,* **144–45**, 267,
        268, *494*
    Altai Mountains **148**, 294,
        295
    An Teallach *126*
    Andes 26–27, 89, *136,* 137,
        140–41, *162,* 229, 264,
        284, 310, 492
    Annapurna *150*
    Appalachians 138, **140**, 142,
        306, 375, 490, 491
    Appenines 146
    Assiniboine *139*
    Atlas Mountains **147**, 496
    Barisan 239
    Belolakaya *146*
    Belukha 148
    Betic Cordillera 146
    Blue Mountains (Australia)
        152, 316
    Bogda Feng 149
    Bogda Shan 149
    Bread Knife (Warrumbungle
        Range, Australia) *152*
    Buachaille Etive Mor *142*
    Bukhan Range 368
    Caledonian Range 87, 138,
        140, **142**
    Carpathians **146**
    Cascade Range *86, 98, 120,*
        137, 158, 159, 305
    Caucasus **146**
    Cerro Negro *155, 455*
    Cerro Pirámide 264
    Chugach Mountains 257
    climates **463**
    Cordillera Real *140*
    Damavand 367
    Dinaric Alps *146*
    Doi Inthanon *316*
    Dolomites *53*
    Dombay-Ul'gen *146*
    Drakensberg Plateau
        (Ukhahlamba) **148**, *496*
    Eiger *144*
    Elburz 238, 367
    Ellsworth Mountains 502
    Ethiopian Highlands *497*
    Everest *137, 150,* 269, *494*
    Fitzroy Range *140*
    folding 89, *136–37,* 143, 144,
        146, 147, 148, 488
    formation 26–27
    Giant's Castle 148
    glacier cover *149,* 151
    Grampians 142, 232
    Grand Tetons 139
    Great Dividing Range **152**,
        337, 498, 499
    Grossglockner *144*
    Hajar 285
    Hamersley Range *45, 78*
    Himalayas 25, 26, 27, 86, 89,
        137, **150–51**, 269, 337,
        445, 488, 494
    Huascarán *140*
    Jotunheimen Range 142

mountains *cont.*
    Jungfrau 267
    K2 *150*, 269
    Karakorum Range 269
    Khan-Tengri *149*
    La Sal Mountains *463*
    MacAllister Range 317
    Matterhorn 93, *145*
    Mitre Peak *152*
    Mönch peak 267
    Mont Blanc 267
    Mont-aux-Sources 148
    Monte Fitzroy 264
    Monte Somma 170
    Mount Adams *86*
    Mount Chephren *482*
    Mount Cook 152, 274,
        275, 499
    Mount Elie de Beaumont 274
    Mount Kirkpatrick 153
    Mount Kosciusko 152
    Mount McKinley *139,*
        339, 490
    Mount Robson *304*
    Mount Tasman 274
    Mount William Range *153*
    Ntlenyana *148*
    orogenic belts *126,* 138, 147
    Pamirs 237
    Pontic *146*
    Pyrenees **143**
    Qilian Shan 294
    Qin Ling Mountains 313
    rifting 138
    Rocky Mountains 26, 138,
        139, 198, 264, 282,
        332, 491
    Royal Society Range
        (Antarctica) *136,* 153
    St. Elias Mountains 260, 261
    Salvesen Mountains **184**
    seamounts 387
    Sierra Madre 139, **140**, *491*
    Sierra Mazateca 243
    Sierra Nevada Range (North
        America) *179,* 182, *254,*
        264, 282, *442, 490*
    Sierra Nevada Range
        (Spain) *146*
    Sikhote-Alin Range 316
    Sneffels Mountains (Colorado,
        USA) *460*
    snow line 445
    Snowy Mountains 499
    Southern Alps
        (New Zealand) **152**
    Star Mountains *499*
    submarine 138, 386, 387
    Sugar Loaf (Rio de Janeiro,
        Brazil) 361
    Tatra Mountains 146
    tectonic plates and 488, 494
    Tien Shan **149**, 268
    Torres del Paine *141, 493*
    Toubkal 147
    Transantarctic **153**, 272,
        502, *503*
    Urals 56, **142**
    Vinson Massif 502
    Vosges 127
    Warrumbungle Range
        (Australia) *152*
    Western Cordillera (North
        America) 137, 139
    Whitney *139*
    Wrangell Mountains 256
    Yen Mountains 369
    Yucca Mountain 283
    Zagros **148**
Mozambique Channel **419**
Mozambique Current 419
mud crabs 327
mudflats, Alaska *437*
mudflows 97, 98

Mudskipper, Atlantic *321*
mudstone 72, **77**
Muir Glacier *252*
mulga trees 297
Mumbai (formerly Bombay) **368**
Murchison Falls 217
Murchison, Roderick 142
Murray River 152, **223**, 327,
    499
muscovite 43, 59
Musk Deer 316
Muskrat 321, 325
mylonite **75**

N

Nabuyatom volcanic cone *234*
Nagano, hot springs **192**
Namaqua Sandgrouse *280*
Nambung National Park *462*
Namib Desert *278, 279,* 280,
    292, *451, 463, 468*
Namibia, Hoba West
    meteorite *100*
Nandus *493*
Nankai Trough 132
Nanxu Arch **247**
nappes 144
Naruto Whirlpool *395*
native elements 44, 46, 47
natrolite 60
natural gas 79; *see also*
        methane
    extraction 81, 375
    power generation 375
natural gas fields
    Alaska 79
    Caspian Sea 238
    continental-shelf areas 392
    Gulf of Mexico 409
    North Sea 413
    Ob'-Yenisey region 404
    Persian Gulf 419
    Sea of Okhotsk 423
natural selection 113
Nazca Plate *89,* 141, 420, 492
Neanderthals 28–29, 31, 143
Nefta Oasis *496*
Neijalakka Island *425*
nematodes (worms) 21
Nenet people 312
nephrite (jade) *see* jade
Neretva River 146
Netherlands *413*
Nevado del Ruiz 98, **162**
New Administrative Capital
    (NAC) 366; *see also* Cairo
New Caledonia Depression 425
New Cornelia Tailings 374
New Forest (England) *311*
New Hebrides *see* Vanuatu
New York City 353, 354, *355,*
    356–57, 379, *455*
New Zealand
    agriculture 349
    earthquake, Canterbury 274
    fiords 152, 435, *500*
    forests 303, 316
    Great Alpine Fault **133**
    hot springs and geysers
        193, 379
    tectonic plates 501
New Zealand Geological Society
    133
Newberry (Oregon, USA), Big
    Obsidian Flow 69
Newfoundland 408
Ngorongoro Crater *334,* 335
Niagara River and
    Falls 226, *491*
nickel 60, 102, 376
Nicobar Islands 419
Niger River **219**

Nigeria 355, **366**
    oil fields 375
Nikko Fir 313
Nile 94, *201,* 214, *217,* **217**,
    235, 366, 496
Nile Crocodile *217*
Nile Perch 235
nimbostratus clouds **480**
nitrogen fixation 115
nodules **82**, *393*
Nojima Fault **132**
Nördlingen *104*
Norfolk Bank 413
Norgay, Tenzing 150
North America
    agriculture 333, 348, 350,
        351
    air masses *466*
    boreal forest 304
    cattle ranching 333, 348
    caves 242, 243
    cities 356–60
    coal 375
    coniferous forest 304,
        305, 308
    Continental Divide 139,
        198, 206
    deciduous forest 301,
        306, 308
    desertification 290
    deserts 278, 279, 281, 282–
        83
    fault systems 124–25
    forests 303, 304–306
    geysers and hot springs 186–
        88
    glaciers *250,* 256–61, 264
    grasslands 330, 331, 332
    horticulture 349
    hot springs and geysers
        186–88
    igneous intrusions 180–82
    lakes 204, 224, 225–28
    meteorite impacts 102–103
    mountains 139, 140
    plateaus 206
    population 491
    rivers 204–208
    temperate rainforest 301, 305
    tundra 304, *338,* 339
    volcanoes 158–61
    wetlands 228, 322, 323
North American Plate *86,* 88,
    124, 125, 422, 488, **490–**
    91, *494*
North Anatolian Fault **127**
North Atlantic Current 413
North Atlantic Oscillation (NAO)
    453, 455
North Equatorial Current
    409, 412
North German Plain 144
North Pole 402, 452
North Saharan Basin 147
North Sea 126, 375, 413, **413**,
    426
North Sea Basin fault
    system **126**
North Water polynya 405
Northern Flying Squirrel 304
Northern Lights *see* aurorae
Northern Spotted Owl 305
Norway, fiords *142,* 435, *495*
Norway Current 413
Norway Spruce 312
Norwegian Sea *382,* **413**
Nova Scotia 408
Novarupta **158**
Novaya Zemlya 339
Ntlenyana *148*
Nubian Plate 128
nuclear power 375
nuclear storage 283
nuclear tests 103

nuclear waste dumping 375, 404
Nullarbor Plain 249, *498*, 499
nunataks *254*
Nyiragongo volcano **172**

O

oak trees 114, 147, 151, *302*, 306, 310, 311, 313
oases 280, 289, *496*
  man-made *110*
Ob' River 148, **221**, 404
Ob'-Yenisey region 404
observatories, Pic du Midi Observatory 143
obsidian **69**
oceanic crust 34, *35*, 41, 86
  collisions with continental plates 137
  collisions with oceanic plates 133
oceanic lithosphere 40, 386
oceanography
  submersibles 386
  underwater pressure 384
oceans and seas 41, **382–429**; *see also* coasts, tides, waves
  Adriatic 146, 414, 415
  Aegean 414
  Andaman Sea 419
  Arabian Sea 368, 416, **419**
  Arafura Sea 425
  Arctic Ocean 107, 398, 400, *401*, **402–405**, 452, 456, 463
  Atlantic Ocean 142, 143, 163, 171, 188, 189, *382*, 384, 398, 400, 402, *406*, **407–15**, 490, 494, 496
  Baffin Bay **405**
  Baltic Sea **413**, 433
  Banda Sea **425**
  Barents Sea **404**
  Bay of Bengal 416, **419**, 470
  Bay of Biscay **413**
  Bay of Fundy *92*, **408**, *434*
  Beaufort Sea *106*, **405**
  Bering Sea *203*, **422**
  Black Sea **413**, 415
  Bohol Sea *108*
  Caribbean Sea 400, **412**, 491
  Chukchi Sea **405**
  circulation and currents **394–95**, 399, 436, 452
  climate and 452
  coastal pollution *438*, **439**
  Coral Sea **425**
  deep-water currents 395
  East China Sea 423, **424**
  East Siberian Sea **405**
  eddy currents *452*
  environmental issues 404, 409, 422, 428
  Florida seas *491*
  Greenland Sea *401*, **405**
  Gulf of Alaska 250, 257, **422**
  Gulf of California **422**
  Gulf of Mexico **409**
  Gulf of St. Lawrence **408**
  gyres 394, 407, 420, 452
  Hudson Bay **408**
  Iapetus Ocean 142
  Indian Ocean 400, **416–19**, 417, 496, 498, *503*
  Kara Sea 104, **404**
  Laptev Sea **404**, *495*
  marine litter 439
  marine pollution 411, 412, 415, 422
  Mediterranean Sea 147, 400, **414–15**, 418, 494

oceans and seas *cont.*
  mid-ocean ridges 41, 386–87
  Molucca Sea 425
  Mozambique Channel **419**
  North Sea **413**
  Norwegian Sea *382*, **413**
  ocean-floor trenches 141, 386, 387, 388–89, 420, 422, 425, 492, 500
  oceanic winds 483
  oil shale 81, 375
  oil spillage 439
  Pacific Ocean 97, 160, 398, 400, *401*, 402, **420–25**
  Persian Gulf **419**
  polar oceans **398–99**, 400, *401*
  Red Sea 128, *129*, 414, **418**, 488, 496, *497*
  Ross Sea 273, **429**, 502, *503*
  salinity 106, 384, 395
  Sargasso Sea 400, **409**
  Scotia Sea **429**
  Sea of Japan **423**
  Sea of Okhotsk **423**
  sea-level **435–35**
  seamounts 387
  sewage and industrial waste 411, 439
  shelves and plains **392–93**
  sound underwater 385
  South China Sea **424**
  Southern Ocean 398, 399, 400, *401*, **428–29**, 452, 496, 502
  storms *406*, 407
  submarine avalanches 97, 160
  Sulu Sea **425**
  Tasman Sea **425**, 499
  temperature and depth 385
  Tethys Ocean 27, 71, 144, 146, 147, 151
  tropical cyclones and 470
  undersea canyons 392, 416, 419
  vertical transport currents 394
  water 384–85
  waterspouts *409*, *447*, *452*
  Weddell Sea *401*, **429**, 502, *503*
  White Sea **404**
  wrecks *392*
  Yellow Sea **423**
octopus *414*
  Dumbo Octopus *384*
Oder River **213**
Ogallala Aquifer 110
O'Gorman, Juan, Central Library, Mexico City *361*
Ohio River 205
oil 79, *115*
  extraction *80*, 81, 375
  fracking 81
  pipelines 139, *374*
  power generation 375
  refineries *374*
  spillage 439
oil fields
  Alaska 139, 375
  Arabian Peninsula 285, 375
  Caspian Sea 238, 375
  Chicxulub Crater 103
  continental shelf areas 392
  Great Sandy Desert 296
  Gulf of Mexico 375, 409
  Indonesia 375
  Nigeria 375
  North Sea 126, 375, 413
  Ob'-Yenisey region 404
  Persian Gulf 419
  Sea of Okhotsk 423
  Tasman Sea 425

oil fields *cont.*
  Texas 375
  Venezuela 375
  Zagros Mountains 148
Okavango Delta *321*, **326**
Okefenokee Swamp **322**
Okhotsk Plate 132
Okinawa Trough 424
Oku Volcanic Field 172
Old Delhi traffic congestion *368*
Ol Doinyo Lengai volcano 128, **172**
Old Faithful 186, *187*
olives 351, 462
olivine 38, 44, 45, **56**, 57, 59, 60, 66, 67, 68, 71, 83, 91, *178*
Oman, serpentinites 71
oolitic limestone 78
opal 43
open-cast mining
  coal 81, 126, 374, 375
  gold *48*
  copper 374, *376*, *499*
    phosphates *377*
opossums 26
Oppenheimer Diamond *47*
Optimisticheskaya **247**
orang-utans *315*
Orange River 148, *496*
orchids 303, 306, 313, 315
Ordesa National Park *143*
ore-deposits 45, 376
organic deposits *79*
Orinoco **209**, 324, 333
orogenic belts 126, 133, 138, 147
orographic lifting, of air *473*
Orrorin 28
ostriches *113*
Oswego River 228
otters 26, 322
  Great River Otter *210*
  Sea Otter *114*, 422
owls 305, 311
  Barred Owls 322
  Burrowing Owl *331*
  Great Grey Owl *304*
  Pel's Fish-owl *326*
ox-bow lakes 201
oxides 44, 51–52
oyster culture 413
Oystercatcher, Variable *327*
Ozark Plateau *109*
Ozero Issyk-Kul' lake 149
ozone layer 16, *448*, **449**, 502

P

Pacific Basin 131
Pacific Northwest rainforest 301, **305**
Pacific Ocean 398, 400, *401*, 402, **420–25**
  gyre formation *394*
  islands *401*, *421*; *see also* islands and groups by name
  Ring of Fire 137, 153, 157, 420, 501
  subduction zone 27, *39*, 41, 86, 89, 123, 132, 133, 494, 500
  volcanoes 160, *401*
Pacific Plate 27, 86, 124, 132, 133, 160, 420, 422, 488, 490, 494, **500–501**
Pacific Region, population 501
Pacific-Antarctic Basin 429
Pacific-Antarctic Ridge 416
pack ice *106*, *429*
Pagoda Pillars *78*
pahoehoe (basaltic lava) 70, 155

Palau *396*, *421*
Palawan island 425
Paleoclimatic data 455
Paleozoic Era **20–22**
Palisade Glacier **264**
Palisades Sill **180**, 204
Pallas, Peter 83
pallasites **83**
palm oil 345
palms 313, 315
  Açaí Palm 307
  Acuri Palm 324
  Caranda Palm 324
  rattans 316
  Walking Stick Palm 317
Pamukkale hot springs *107*
Pamirs 237
Pampas *330*, 333, 350, 492
Pampas Deer 333
Pampas Grass 114, *330*, 333
Panama Canal 412
Pangea 20, 86, 87, 126, 142, 143, 490
Pantanal wetlands 208, **324**, 333
paper-making 377
paperbark trees *330*
papyrus 326, 327
Papua-New Guinea 29, *410*, *499*
Paraguay River 324
Paraná River *208*
Paraná Plateau **162**
Paricutín volcano **160**
Paris **365**
Paris Basin 212, 365
Parliament Buildings (London) *365*
parrots
  Alexandra's Parrot *297*
  Alpine Parrot *152*
Pars Snow Fields *148*
particle tracks *14*
Passamaquoddy Bay 408
pastoralists, nomadic 344, 348
Pastoruri Glacier **264**
Patagonia 492
  cypress trees 310
  desert **283**
  pampas *333*
  Southern Patagonian Ice-field **264**, 265, *492*
  Torres del Paine *141*, *493*
Patterson, Clair 17
Pau d'Arco 307
paved landscape 109
Peace River 225
peacock ore 50
pearl fishing 425
Pearl River 355
peat 79, 320, 322, 324, 327
  County Kerry, Ireland *79*
Peel Fault 127
pegmatites 52, 55, 56, 57, 58, 59, 61, 64, **67**, *178*
Peking Man 369
pelicans, Australian 327
Pel's Fish-owl 326
penguins *401*, 416
  Chinstrap Penguin *401*
  Emperor Penguin *273*, *398*
  Macaroni Penguin *416*
Pennsylvanian Epoch 22
Pennyroyal Plateau 243
peridotite 38, *40*, *41*, 52, 56, 67, **68**, 71, 182, 184
periglacial landforms 107, 338, 339
Perito Moreno Glacier **265**
permafrost 30, 81, 319, 338
Permo-Triassic extinction event 22–23, 173

Persian Gulf **419**
Peru–Chile Trench 141, 492
Peruaçu nature reserve 243
Peruvian Anchoveta fishery 427
Petitcodiac River 408
petrified forests 13, 61
petrochemicals 377
petroleum *see* oil
Petronas Twin Towers (Kuala Lumpur, Malaysia) *371*
Peyto Lake *224*
Phanerozoic Eon 12
Phang Nga, limestone cliffs *419*
Philippine Plate 132, 494, 500
phlogopite 68, 71
phonolite 181
phosphates 44, 55, **82**, 147, *377*
phosphorites 82
photosynthesis 115, 300
phyllite 57, 59, **73**
Pic du Midi Observatory 143
Pico de Teide **171**
piedmont lakes 225
piedmontite 57
piezo-electric crystals 58
pig farming, intensive 351
pillow lavas 155
Pine Island Glacier 270
pine trees 147, 151, 313, 322
  Aleppo Pine 462
  Bristlecone Pine 305
  Cedar Pine 316
  kauri pine 317
  Lodgepole Pine *186*, 305
  Ponderosa Pine 305
  Rimu (Red Pine) 316
  Scots Pine 312
  Siberian Stone Pine 312
  Sugar Pine 305
pingos 204, 338, 339
  Mackenzie River delta *204*
pinnacles *241*, *462*, *494*, *498*, *499*
Pinnacles Desert (Western Australia) *462*
piranhas 324
Pirogue boats *419*
pitchblende 52
pitcher plants 322
Piton de la Fournaise volcano **173**
Pittsburgh 354, 374
placer deposits 45
  beryl 58
  cassiterite 52
  garnets 56
  gold *221*, *372*
  ilmenite 52
  magnetite 51
  rutile 52
  topaz 57
  uraninite 52
  zircon in 56
placerias 23
plagioclase 67, 68, 69, 83
plankton 394, 404, 409, 413, *420*, 422, 428
  phytoplankton 115, 384
plankton mass 456
plant life *40*, 112; *see also* life
  adaptation to climate change 31
  Banksia species *114*
  biological weathering 91
  deserts 281, 282, 283, 284, 285, 289, 292, 293, 294, 296, 297
  evolution 21–22, 23
  forests 300–302, 303, 306, 307, 310, 311, 312, 313, 315, 316, 317
  glaciers 257, *261*
  habitats 114

plant life *cont.*
  nutrient cycles 115
  volcanoes 159, *171*
  wetlands 321–27
plantation agriculture 345
plastics 377
Plateau of Tibet 25, 27, *40*, 151, *494*
plateaus 138
  Akiyoshi-dai Plateau 249
  Altiplano *140*, 141, 229, *492*
  Antrim Plateau 163
  Barkly Tableland *337*, 499
  Bateke Plateau 218
  Colorado Plateau 138, 206
  Columbia River Plateau 68, 158
  Coral Sea Plateau 425
  Deccan Plateau 24, 236, 494
  Drakensberg (Ukhahlamba) 148, *496*
  East African Plateau 236
  Konkordiaplatz Ice Plateau 267
  Metéora *494*
  Ozark Plateau *109*
  Paraná Plateau 162
  Pennyroyal Plateau 243
  Plateau of Tibet 25, 27, *40*, 151, *494*
  tepuis *493*
  Ubangi Plateau 234
  Ustyurt Plateau 238
plateau basalts 64
platinum 45, 141, 142, 376
platypuses 152
plesiadapiforms 27
plesiosaurs 23
Plinian eruptions 156, *158*, *159*, 160, 161, 170, 175
Pliny the Elder 51, 57
plutons, igneous rocks 64, 67, 68
poaching, wildlife 324
podocarp trees 274, 306, 310, 316
Podol'sky flatlands 247
Poland
  Ustronie Morskie *433*
polar climates 460, **463**
polar oceans **398–99**, 400, *401*; *see also* Arctic Ocean, Southern Ocean
polar winds *483*
polders 413
poljes 146
pollen grains
  atmospheric particles 445
  climate and 454
  fossils 13, 454
polygons, tundra soils *338*, 339
polynyas *399*
polythermal glaciers 266
Pompeii, destruction of 169, 170
Ponderosa Pine 305
Pontic Mountains *146*
Popigay Crater **105**
Popocatépetl volcano **161**, 361
poppies *151*, *339*
population
  Africa 497
  Antarctica 503
  Asia 495
  Australia 498
  Europe 495
  growth in urban areas 353, 355, **362**, *363*, 366, 368, 369
  North America 491
  Pacific Region 501
  South America 493
Port (Co. Donegal, Ireland) *430*
Porto do Moniz (Madeira) *406*
Possum, Honey 316

potassium 195
potholes *199*
pottery industry 377
poultry farming, intensive 351
Prairie Coneflower *330*
prairie dogs *331*, 332
prairies *see* Great Plains (North America)
Precambrian Period **16–19**
prehistoric art 240
priapulids 21
primates 27
Prince of Wales Feathers (Rotorua, New Zealand) *193*
Prince William Sound 260, 422
Prismatic Pool hot springs 187
Proconsul 27
prokaryotes 17, *19*, 112
Pronghorn 332
Proterozoic Era 19, 21
protists 112
Przewalski's Horse 336
Ptarmigan 143
pteropods *393*
pterosaurs 23
Puerto Rico 412
pumice 69, 70
Punchbowl Spring *185*
Purgatorius 27
Purple Gallinules 323
Puyehue volcano *165*
Pyrenean Ibex *143*
Pyrenees **143**
pyrite **50**, 53, 73, 77, 78, 82
pyroxene 38, 45, **58**, 66, 67, 68, 69, 71, 74, 83
pyroxene hornfels 72
pyroxenite 61
pyrrhotite *35*

Q
Qattara Depression 289
Qilian Shan mountains 294
Qin Ling Mountains 313
Qinghai Hu **239**
Quaking Aspens 304
quarries
  Bingham Quarry 374, *376*
  copper and gold deposits *499*
  granite *179*
  talc *59*
quartz 43, 45, 52, **61**, 66, 67, 69, 73, 74, 76, 82, 179
  artefacts 61
  milky quartz 43
  shocked and shattered quartz *100*, 102, 105
  smoky quartz 43
quartz groundmass *46*
quartz sandstone *65*
quartzite **64**, **73**
quartzite prisms *73*
Quaternary ice ages 29, 30
Queen Maud Land *153*
Queensland Maple 317
quicksilver 50
Quizapu volcano 162

R
Rabaul volcano *154*, **177**, *500*
rabbits 296, 333
  Jackrabbits 282
Rabbit-eared Bandicoot *296*
raccoons 322
radiation
  solar 446–47
  ultraviolet 444
radiolarian *393*

radiometric dating 12, 13
radium 52
Raffles, Sir Stamford 371
Rafflesia Arnoldii 315
Rajah Brooke's Birdwing Butterfly *316*
rainbows *445*, *472*
rainfall 108, 452, 473, 474
  acidic 91, 107, 478
  convectional thunderstorms 210
  in deserts 280, 282, 289, 296, *297*
  drought 290, 307, 474
  frontal systems and 467, 473
  global water cycle 109
  monsoon 150, 300, 337, 461, 469
  no-freeze rainfall 473
  rainstorms *108*, *447*, *473*
  Tropical Rainfall Measuring Mission 473
  tropical rainstorms *473*
Ramon (Israel), shattered quartzite *73*
rangeland, Australia *330*, 331
rattan 316
Rattlesnake, Western Diamondback *283*
Ray, Manta *424*
realgar **50**
Red Cedar 317
Red Deer 311
Red Gum *316*
Red Kangaroo 152, 297, *337*
red ochre 51
Red River 225
Red Sea 128, *129*, 285, 414, **418**, 488, 496, *497*
Red Snapper 409
Red-cockaded Woodpecker 322
Red-eyed Treefrog *306*
Redfin Pickerel 322
reeds 321, *325*, 327
  papyrus 326, 327
  phragmites 326
reindeer *31*, 312, 339
  herding *342*, 344
Reka River 247
renewable energy sources *see* alternative energy systems
reptiles 22; *see also* particular reptiles by name
rheas 333
Rhine River **213**, 233
Rhine Falls 213
Rhine Glacier 233
Rhine Rift **127**, 163
rhinoceros 26
  Black Rhinoceros *116*
  Indian Rhinoceros *337*
rhodochrosite **53**
rhododendrons 151
Rhône **213**, 233, 268, 325
Rhône Glacier *253*, **268**
rhyolite 60, 61, **69**, 70, 181
Rhyolite Hills *69*
ria coasts 435
rice cultivation **345**, *350*
  Camargue 325
  China *340*
  global warming 345, 459
  terracing *340*
Richter, Charles 122
Richter Scale 122
riebeckite 59
Ries Crater **104**
Rift Shrimp *386*
Rimu (Red Pine) 316
Rimu trees 316
rinderpest 348
Ring of Fire 137, 153, 157, 420, 501

Rio Cueiras 211
Rio de Janeiro 183, *361*, 493
Rio Grande **208**
Rio Negro 209, *211*, *493*
River Plate **208**, 333, 364
River Shannon 232
river systems
  dendritic 198
  environmental issues 198, 201, 205, 210, 213, 220, 223
  sediment deposition 63, 65, 94, 199, 201, 205, 206, 208, 221, 222, 416, 419, 422, *423*, 431
river-beds, dredging *201*, 208
Rivera Plate 490
rivers 110, **197–223**, 452; *see also* deltas, estuaries, flooding, underground rivers and caves
  age 197
  Amazon **210**, *211*, 307, 392
  Amu Darya 293
  Ausable *92*
  Bear Creek 228
  Blue Nile 217
  Boteti 326
  Brahmaputra 151, 222
  braiding 201, *210*, 213, 255, *274*, 320, 325
  Braldu 269
  catchment areas 198
  Chang Jiang 370
  Chaobai 369
  Chari 234
  Colorado 110, 138, **206**, *207*, 214, *215*, 422
  Columbia 57, 68, 158
  Congo **218**, 236, 313
  Cubango 326
  Danube **216**
  Darling 152, 499
  Darya-ye-Sefid *238*
  Doda 269
  erosion by *84*, 138, 197, 198, 199, *200*
  Euphrates **219**, 419
  Ganges 94, **222**, 367, 392, 416, 419, 470
  General *306*
  Great Bear 225
  Green River 243
  Han 368
  headwaters *198*
  Hood *96*
  Hooghly 222, 367
  Huangpu 370
  Hudson 180, **204**, 356
  Indus 151, **223**, 269, 367, 416
  Irrawaddy **221**
  Irtysh 148, 221
  Jenolan 249
  Jordan 238, 239
  Kennicott 256
  Kolyma 405
  Lena *495*
  Limmen *318*
  Loire **212**
  lower courses 201
  Mackenzie **204**, 225, 405
  Manicouagan Crater and 102
  Mara *196*
  meanders *196*, 201
  Mekong **221**
  Melinau 248
  Meuse 213
  Mississippi 94, *201*, 205, **205**, 226, 491
  Missouri 205
  Moa *45*
  Moselle *494*
  Moskva (Moscow) 364

rivers *cont.*
  mud *45*
  Murray 152, **223**, 327, 499
  Neretva 146
  Niagara 226
  Niger **219**
  Nile 94, *201*, 214, *217*, **217**, 235, 366, 496
  Ob' 148, **221**, 404
  Oder 213
  Ohio 205
  Orange 148, *496*
  Orinoco **209**, 324, 333
  Oswego 228
  Paraguay 324
  Paraná 208
  Peace 225
  Pearl 355
  Plate **208**, 333, 364
  point bars 199
  Red 225
  Reka 247
  Rhine **213**, 233
  Rhône **213**, 233, 268, 325
  Rio Grande **208**
  Rio Negro 209, *211*, *493*
  Saaser Vispa 268
  St. Lawrence **204**, 226, 408
  San Juan 138
  Saskatchewan 225
  Seine **212**, 365
  Severn **212**
  Shannon 232
  Slave 225
  Slims 264
  sloughs (backwaters) *205*
  Snowy 223
  Sutlej 293
  Suwannee 322
  Syr Darya 237
  Tagus (Tajo) **212**
  Thames **212**
  Thjorsa *84*
  Tiber 366
  Tigris 219, 419
  Tisza 325
  tributaries 198
  Tugela 148
  upper courses 200
  Ural 238
  Uruguay 208
  Vaal 104
  Volga **216**, 238, 364
  Weber 228
  White Nile 217, 234, 327
  Yamuna 368
  Yangtze 214, **220**, 423, 424, 494
  Yellow 92, **221**, 423
  Yenisey 404
  Yongding 369
  Zambezi 128, **219**, 235, 416, 419
  Zanskar 269
Rjukandi waterfall *98*
roadstone 68
Robert-Bourassa Dam 214
rock art, Southern Africa 149
rock climbing 181
rock crystal 43
Rock Hyrax *148*
rock salt 53, 78
rocks 33, 45, **62–83**
  chemical weathering 91
  country rocks 178
  erratics 255
  frost shattering 90, 96, 180
  igneous 45, 51, 52, 55, 56, 57, 59, 60, 61, *62*, 63, 64, **66–70**, 83, 178
  mafic 68
  metamorphic 17, 45, 51, 52, 56, 57, 58, 59, 60, 61, 63, 64, 66, **71–75**, 183

rocks *cont.*
  meteorites 65, **83**
  minerals and 62
  physical weathering 90
  saltation 93
  sedimentary 33, *40*, 45, 55,
    58, 59, 60, 61, 62, 63, 65,
    76–82
  tectonics and 33, *488*
  ultra mafic 68
  ventifacts 93
Rocky Mountains 26, 138, **139**,
  198, 264, 282, 332, 491
rodents 26
Rodinia 19, 87
Rogers, Richard, Lloyd's
  Building, London *365*
Rome **366**
Ronne Ice-shelf 273, 502
Ronne-Filchner Ice-shelf
  398, 429
Roraima State (Brazil), slash and
  burn agriculture *344*
Rose, Gustav 83
Roseate Spoonbill *409*
Ross, James Clark 177
Ross Ice-shelf *153*, 273, 398,
  429, 502
Ross Sea 273, **429**, 502, *503*
Rossby waves 467
Rosy Periwinkle 313
Rotorua, geysers and hot
  springs **193**, 379
rowan trees 312
Rowland, Sherwood 449
Royal Society Range (Antarctica)
  136, *153*
Royal Spoonbill 327
Ruapehu volcano **177**
Rub'al Khali 285
rubber trees 307, *345*
rubies 52
Ruffed Lemur *303*, 313
Ruhr Valley 354, 374, 376
rushes 327, 339
Russell Fiord (Alaska) 256
Russia
  dams 214
  fertility 362
Rutford Ice-stream 273
rutile **52**
rye 350
Ryukyu Islands 424

S

Saaser Vispa river 268
Sable 312
Saffir-Simpson scale 470
Sagebrush 282
Saguaro cactus 293
Sahara Desert *93*, *95*, 147,
  276, 279, 280, 281,
  285, **289**, *291*, 445,
  460, 462, 474, 483,
  484, 496
Sahel 219, 234, 289, 290, 344,
  454, 474
Sahelanthropus tchadensis
  28, 289
Saiga 336
Saiga Antelope 336
St. Basil's Cathedral (Moscow)
  364
St. Elias Mountains 260, 261
St. Lawrence River **204**,
  226, 408
St. Lawrence Seaway 204, 356,
  360
Salamander, Marbled *140*
salmon 408
  processed food *378*
  Sockeye Salmon *98*

salt
  crystals 42, 78
  deposits 65
  production of 320, 377
  weathering *91*
salt flats
  Arabian Peninsula 285
  Atacama Desert 284
  Atlas Mountains 147
  Death Valley *78*, *282*
  Kalahari Desert 292
  Valdes Peninsula *493*
salt marsh 320, 323, 325, 327,
  437
  North America 228, *408*
salt pans, Okavango Delta 326
saltation 93
Salton Sink 206
Saltstraumen whirlpool 413
Salvesen Mountains **184**
Samara Bend (Volga River) 216
Samosir Island (Lake Toba) 239
San Agustin, Mexico,
  dolines 243
San Andreas Fault *88*, 123,
  124–25, 127
San Francisco, earthquake 124,
  125, 131
San Juan Forest (Colorado,
  USA) *460*
San Juan River *138*
San people 292, *344*
sand banks, in rivers *199*
sand bars, offshore 437
Sand Crocus 325
Sandalwood trees 337
sandpipers 339
sandstone *41*, 62, *63*, *64*, 65,
  76, **77**, 79, *94*, 183, 184
  arkose 76
  layered *95*
  Permian 22
  pinnacles *494*, *499*
  quartzite and 72
  tepuis *493*
Santo Domingo river 243
Santorini volcano *170*, *400*
Sao Paulo **364**, *493*
Sapele wood 313
sapphires 52
Sardinia 414
sardonyx *61*
Sargasso Sea *400*, **409**
Saskatchewan River 225
sastrugi, ridged pack ice *429*
satellite imaging; *see also* space
  shuttle
  dalmatian coasts *435*
  fault systems *124*, *127*
  forests 303
  Gulf Stream *452*
  hurricane warnings 470
  ICEsat 263
  igneous intrusions *184*
  Lake Carnegie *498*
  Landsat 7 *261*
  meteorite impacts *102*, *105*
  rainfall measurement 473
  rock massifs *496*
  Sahara dust storm *484*
  seas *404*, *423*
  tsunami warnings 131
  urban lighting *354*
  volcano early warnings 169
  weather satellites *468*, 470
Satin Bowerbird *317*
satin spar 55
sauropods 23
Saussure, H. B. de 53
savanna; *see also* grasslands
  Australia *330*, 331, 337
  East Africa *497*
  Indian *330*, 331, 337
  South American 331, 333

sawgrass 323
saxaul, desert shrub 293, 294
saxifrages 339
  purple saxifrage *30*
Scarlet Ibis *308*, 324
schist 52, 56, 57, 58, 61, 63, **73**
  Glen Nevis, Scotland *73*
Schistomiasis 459
scolecite 60
Scorpions, Desert *280*, 282
Scotia Plate 492
Scotia Ridge 429
Scotia Sea **429**, 492
Scots Pine 312
Scott, Robert Falcon 273
scree 96, *97*
  deserts *278*
sea anemones *404*
  Closed Anemone *425*
  Crimson Anemone *401*
  Giant Anemone *418*, *425*
  Wood Anemone *302*
Sea of Azov 413
Sea of Cortez 422
sea cucumbers 387, *404*
Sea Daffodil 325
sea mouse *392*
Sea of Japan **423**
Sea of Marmara 413
Sea of Okhotsk **423**
Sea Otter *114*, 422
Sea Rose *418*
sea snails 21
sea stacks 92, *437*
sea-level **434–35**
  sea-level change *41*, 434
sea-walls 431
Seahorse, Yellow *401*
seals 26, 405
  Baikal Seal *236*
  Crabeater Seal *399*
  Harp Seal *404*
  Weddell Seal *429*, *503*
seas *see* oceans and seas
seaweed
  Giant Kelp forests *420*
  seaweed mats 409
Secretary Bird *335*
Sederholm, Jakob Johannes 72
sedges 327, 339
sediment
  ocean sediments 392, 393
  river systems 63, 65, *94*,
    199, 201, 205, 206, 208,
    221, 222, 416, 419, 422,
    423, 431
sedimentary rocks 33, *41*, 45,
  55, 58, 59, 60, 61, 62, 63,
  65, **76–82**
  erosion 63, 65, 92, 181, 184
  folding *64*, 184
  igneous intrusions 178, 179
  joints *122*
Seine **212**, 365
seismic activity prediction 122,
  125, 127, 131, 157
seismic tomography 38, *38*
selenite 55, 245
Semonyov, Peter 149
Seoul **368**
septarian concretion **82**
Sequoia, Giant 301, 305
Sequoia National Park *305*
Serengeti Plains 331, **334–35**,
  461, *474*
serpentine 60, 68
serpentinite 52, 56, 60, **71**, 74,
  184
service industries **379**
Sessile Oak 311
Severn **212**
sewage and industrial waste
  357, 370, 411, 439
Seward Glacier 261

Seychelles *400*, *416*, 419
Shackleton, Ernest Henry
  184, 273
shale 63, 65, 72, **77**, 79, 182
  Burgess Shale 13, 20
Shanghai 369, **370**
shanty towns
  Johannesburg *367*
  Mexico City *355*
  Rio de Janeiro 361
Shard, the *365*
Shark Bay (Western Australia) *10*
sharks 21, 23
  Basking Shark *407*
  Whale Shark *424*
Shatt al 'Arab 219, 419
sheep
  milk production 349
  mountain sheep 139
  sheep farming 348
  wool production 378
shelves and plains, sea floor
  regions **392–93**
  abyssal plains 392, 393
  continental margins 392
  ocean sediments 392, 393
  submarine canyons 392
shifting sand desert 293
Shimla *461*
Shining Rock Wilderness, quartz
  outcrop *61*
Ship Rock (New Mexico,
  USA) **182**
shipwrecks 415, 424
shock wave pattern 36
Shoebill stork *327*
Shoemaker, Eugene and Carolyn
  100, 103
Shoemaker-Levy 9 comet 100
shopping malls *379*
shrews 26
shrimps 424
  amphipods *388*
  Brine Shrimp 284, 325
  Rift Shrimp *386*
Siberia
  coal reserves 375
  Eurasian Plate *494*
  Permo-Triassic extinction
    event 22–23, 173
Siberian Fir 312
Siberian Spruce 312
Siberian Stone Pine 312
Siberian Tiger 316
Siberian Traps 22, **173**
Siccar Point (Scotland) *12*
Sicily 414
Side-Winding Adder *280*
siderite 78
Sierra Madre 139
Sierra Madre Oriental **140**, *491*
Sierra Mazateca 243
Sierra Nevada (USA) 64, *125*,
  179, 182, *254*, 264, 282,
  442, *490*
Sierra Nevada (Spain) **146**
Sierra Pelada, gold mine *372*
Sikhote-Alin Range 316
silicates 44, 45, 56–**61**
Silicon Valley 378
silk production, China *378*
Silver Fir *302*
Silliman, Benjamin 56
sillimanite **56**, 73
sills
  igneous rocks 64, 67, 68, 178,
    179, 180, 183, 184, 204
  underwater 414
siltstone **76**
silver **46**, 141, 147
Simpson Desert 279, **296**
Sinai Peninsula 418
Sinai Subplate 239
Sind Desert 367

Singapore **371**, 424
  high-technology industries
    378
sinkholes 240, 241, 243, 247,
  249; *see also* dolines
  in glaciers 252
  Great Blue Hole *412*
Sipiso-Piso waterfall 239
sirocco wind 483
sisal 345
Sistema Huautla **243**
Sitatunga 326
Sitka Spruce 305
Siwalik Hills 150–51
Skaergaard 64, **180**
Skagerrak 413
skarn **71**
Skeleton Coast *292*
Skocjanske Jame **247**
Sky Lake *see* Lake Tianchi ("Sky
  Lake")
skyscrapers
  Cairo 366
  Chicago *352*, *360*
  Johannesburg *367*
  Kuala Lumpur *371*
  London *365*
  New York City 357
  Seoul *368*
  Tokyo *369*
slate 63, 64, **72**
  Stonesfield slate 72
Slave River 225
slickensides *122*
Slims River 264
slopes
  stability of 96
  types of *97*
Sloth, Southern Two-Toed *307*
sloughs (backwaters), on the
  Mississippi 205
slushball event *see* snowball
  event
Smith, William 79
Smithsonian Institution 7, *47*
Smithsonian Tropical Research
  Institute 306
smog 355, 360
Snaefellsnes Peninsula (Iceland)
  434
Snail Kites 323
snakes 26, 280, 293, 294
  Anaconda 321, 324
  Banded Sea Snake *416*
  Desert Horned Viper 289
  Florida Cottonmouth 322
  grass snakes 321
  Great Basin Rattlesnake 282
  Indian Cobra 337
  Side-Winding Adder *280*
  Tatar Sand Boa 294
  Western Diamondback
    Rattlesnake *283*
Snapper, Red 409
Sneffels Mountains (Colorado,
  USA) *460*
Snezhnaya cave 146
Snipe, Great *325*
snow 108, 452, 474
  blizzards and snowstorms
    108, 474
  snowflake crystals *472*, *474*
  watermelon snow *264*
snowball event 19
Snow Leopard *148*
Snowshoe Hare 304
Snowy Mountains 499
Snowy River 223
soapstone (steatite) 59
Sockeye Salmon *98*
Soda Springs (Idaho, USA) **188**
soils; *see also* peat
  mass movement of 96, 97,
    98, *99*

soils *cont.*
of temperate forests 303, 311
tundra polygons *338*, 339
types 114
volcanic 161, *168*, 169, 333
solar activity *446*
solar budget 446
solar energy 446–47
solar power 375, 446, *458*
solar radiation 270, 446–47; *see also* greenhouse effect
absorption variables 447
Solar System birth of 15
meteorites 65, 83
Solfatara, hot springs **189**
solifluction 97
Solomon Islands 137, *401*, 463
Sonoran Desert *281*, **282**
Sossuvlei dunes *278*
Southern Two-Toed Sloth *307*
Soufrière Hills, volcanoes 70, **161**
South Africa 67, *148*, 331, 376
diamonds 47, 68, 377
Vredefort Crater **104**
South America 21
agriculture 333, 344, 348, 351
caves 243
cities 361–64
coniferous trees 310
deciduous trees 310
deserts *278*, 280, 281, 283, 284
dry forest 300, 310
glaciers 264, 265
grasslands *330*, 331, **333**
horticulture 349
hot springs and geysers 188
igneous intrusions 182, 183
lakes *224*, 229
mountains 140–41
plateaus 229, *278*, 283
population *493*
rivers 208–11
slash-and-burn agriculture 344
temperate rainforest 303, 310
tropical rainforest 210, *211*, **306, 307**
volcanoes 162
wetlands 208, 209, 324
South American Plate 27, 86, *89*, 141, 429, 488, **492–93**
South China Basin 424
South China Sea **424**
South Engilchek Glacier *149*, **268**
South Equatorial Current 425, 453
South Georgia 184, *428*, 429
South Island Fernbird 327
South Pole 21, 452, 502
South Sandwich Islands 153, 429
South Shetland Islands 429
Southeast Asian monsoon 25, 27, 214
Southeast Korea Fault Zone **132**
Southern Alps (New Zealand) **152**
Southern Lights *see* aurorae
Southern Ocean 398, 399, 400, *401*, **428–29**, 452, 496, 502
Southern Patagonian Ice-field **264**, 265, *492*
Southern Upland Fault 126
space shuttle; *see also* satellite imaging
image of Sierra Madre *140*
image of Tibetan Plateau *151*
Spain 362, 379

Spanish Imperial Eagle *310*
Spanish Moss *298*, *322*
speciation 113
specific gravity, of minerals *43*
Sphagnum Moss *321*, 324
sphalerite **50**, 55
sphene 52
spiders, tarantulas 283
spinel **51**, 52, 68
spinifex 296, 297, 337
spits, formation of 437
Spitsbergen 266
Ice-cap *254*
sponges 21, 415, 422
spoonbills 327, *409*
Springbok 113, 292
Spruce Grouse 303
spruce trees *302*, 311, 313, 316
Black Spruce 304
Engelmann Spruce 305
Norway Spruce 312
Siberian Spruce 312
Sitka Spruce 305
White Spruce 304
squall lines 484
squid 423
squirrels
Northern Flying Squirrel 304
Sri Lanka, stilt fishing *417*
Staghorn corals *411*
stalactites 54, 91, 240, 242, 243, *249*
stalagmites 54, 78, 240, 241, 242, 247, *249*
lava *70*
Stanley, Henry Morton 218, 236
star cabochon *52*
Star Mountains *499*
starfish *392*, *404*, 422, *429*
Crown of Thorns Starfish 397
stars 15
Staten Island (USA), Fresh Kills landfill site 357
Statue of Liberty 356, *357*
staurolite **57**, 73
steatite (soapstone) 59
steel 376
Steinheim Crater 104
Stellar's Sea Eagle *423*
Steno, Nicolaus 65
steppes, Central Asian *148*, **336**
Sterlet *238*
stibnite 44, **50**
Stilts, Black-winged 325
stishovite 103
stocks, igneous rocks 64, 179
Stonesfield slate 72
storks
White Stork 325
Black Stork 325
storms; *see also* wind
Atlantic *400*, *406*, 407
Baltic Sea 413
cold fronts and *468*
convectional thunderstorms 210
cyclones *406*, 457, 466, *468*, 469, 470, *471*
damage from *451*
Indian Ocean 416–17
rainstorms *108*, *447*
sandstorms 95, 484
squall lines 484
storm surges 470
thunderstorms 210, 446, *464*, *472*, 479, 484
tropical rainstorms *473*
Strahler, Arthur 460
Strait of Belle Isle *408*
Strait of Hormuz *419*
Straits of Florida *409*
Straits of Gibraltar 414, *415*
Straits of Mackinac 226

Straits of Magellan 401, *493*
strangler figs 317
stratocumulus clouds 475, **478**
stratosphere 444, 467
stratus clouds *467*, 474, 475, **478**
Strokkur Geyser **189**
stromatolites 10, 17, *18*
Stromboli volcano **163**
eruptions 156, 163, *167*, *176*, 177
Strzelecki, Paul 152
Sturt's Desert Pea *297*
subduction zones 27, 41, 86, 89, 123, 132, 133, 494, 500
subglacial lakes 265, *273*
submersibles 386
submarine volcano 386, *386*
Sudbury Crater (Canada) **102**, 104
Sudd, the 217, **327**
Suess, Eduard 138
Suez Canal 368, 414, 415, 418
sugar 345, 361
Sugar Loaf (Rio de Janeiro, Brazil) **183**, 361
Sugar Maple 304, 306
Sugar Pine 305
sulphates 44, 54, 55
sulphides 44, 50, 51, 71, 142
sulphur **47**
crystals *47*, *185*
Vancouver Harbour Sulphur Mine *377*
Sulu Sea **425**
Sumida estuary 369
Sun
climate and weather and 109, 455, 465
life on Earth and 17
solar energy 446–47, *458*
tides and 432
Sunda Shelf 424
Sundarbans 222
sundews *322*, 327
sunflowers *332*, *447*
sunspots 455
supernovas *15*
surface tension 98
Surtsey
islands 386
phreatic eruption *156*
susliks 336
Supercell *476–477*
Sutlej 293
Suwannee River 322
Svalbard, solifluction *97*
Svartsengi geothermal power plant *194*
Swampfish 322
swamps 320, 322
swamp forest *307*, 315
Swan, Black 327
Sydney **371**
Sydney Harbour Bridge *371*
synclines 136
Syr Darya 237

T

Taal volcano **175**
Tagus (Tajo) **212**
Tahiti *501*
taiga *142*, 299, 301, 312, 339, 494
Taipei 101 *131*
Taiwan 424
high-technology industries 378
industrialization 374
Taiwan Strait 424
Taj Mahal, use of marble *71*

Tajo *see* Tagus
Taklamakan Desert 281, **293**
Taku Glacier *253*
talc 43, **59**
Tamarisk *293*, 325
Tambopata-Candamo Reserve *307*
Tambora volcano **175**
Tanami Desert **296**
Tangle Fern 327
tanzanite 57
tapirs 324
tar pits, Trinidad *79*
tar sands 81, 375
Tarbush 283
Tarim river system 293
tarmac 109
tarsiers *26*, 27
Tashkent, fault line *123*
Tasman Basin 425
Tasman Glacier 152, **274**, *275*
Tasman Sea *425*, 499
Tatar Sand Boa 294
Tatra Mountains 146
Tatsumaki jigoku geyser 192
Taupo Volcanic Zone, New Zealand 193
Tawa trees 316
tea plantations 345
Teak 316
Teal, Grey 327
tectonic plates 16, 17, 20, 21, 34, 38, **86–89**, 239, 369, 416, 418, 420, 422, 425, 429
atlas of **488–503**
boundaries **88–89**
emergent coasts and 434, 436
fault systems **122–33**, 488
geothermal systems and 185, 192
hydrothermal vents 142, 389, 418
Median Tectonic Line (MTL) 132
mountains and 71, 121, 126, 136–38, 141, 143, 144, 146, 147, 149, 494
movement of 33, 85, 87, 488
ocean floors **386–87**
population and plate boundaries 491, 493, 495, 497, 498, 501, 503
serpentinite and 71
subduction 27, 41, 69, 74, 86, 89, 123, 132, 133, 137, 141, 142, 184, 490, 494, 500
volcanoes and 25, 27, 155, 157, 160, 265, 488
Tehran-Karaj **367**
Tek Sing porcelain 424
tektites 100, 102, 104
telecommunications industry *378*
Telluride (Colorado, USA), water pollution 376
temperate climates 460, 462
Tenerife, volcanic complex 171
Tengger volcanic complex *154*
Tenzing Norgay 150
Tephra 157, 172, 174
Tepuis *493*
Terai-Duar grasslands 337
termite mounds 330
Terra Nova (Antarctica) *503*
terracing 343
rice cultivation *340*
Tethys Ocean 27, 71, 144, 146, 147, 151
Texas 375
textiles industry **378**
Thailand 379, *417*

Tham Hinboun **249**
Thames **212**
Thames Barrier *212*, 365
Thar Desert 279, 280, 281, **293**
therapsids 23
thermohaline conveyor belt 395
thermopause 444
thermosphere 444
Thesiger, Wilfred 285
Thingvellir Rift *122*
Thingvellir Rift Valley *386*
Thjorsa River *84*
Thomson's Gazelles 335
thorium 195
Thornthwaite, Charles 460
Thorny Devil 297
Three Gorges Dam 214, 220
Three Gorges Project (Yangtze River) 214
Thunder Bay (Lake Superior) *227*
Tiber 366
Tibesti Massif 289
tides **432**; *see also* coasts, oceans and seas, waves
bores *432*
Qianjiang Tidal Bore *432*
intertidal organisms *432*
local currents and 394
Mediterranean Sea 415
Moon and 432
range 432
red tides *420*
sea-level change and 434
Sun and 432
tidal power 375
Tien Shan mountains *149*, 268
Tierra del Fuego 141, *492*
Tiger, Siberian 316
Tigris 219, 419
Tiksi harbour *404*
till 95, 255, 256
tillite 65
tin 52
Tisza River 325
titanium 52, 82
TJ event, disappearance of species 102, 103
Toad, Couch's Spadefoot *282*
Tokyo 132, 353, 354, 355, **369**
earthquake 132, 369
Toledo 212
tombolos 437
Tombouctou *343*
Tonga *500*
Tonga trench 34
toothfish 428
topaz **57**
tornadoes *409*, 452, 479, 482, **485**
Toronto **360**
Torres del Paine 141, *493*
tortoise
Argentine Tortoise 333
Desert Tortoise 282
Giant Tortoise 116
Totara trees 316
Toubkal 147
Toucan, Chestnut Mandibled *232*
tourism **379**
air traffic 377
caves 242, *248*, 249
ecotourism 326, 335
effect on coral reefs 411, 412
grasslands 335
marine pollution 411, 412, 415, 422
Okavango Delta 326
Rio de Janeiro 183
Serengeti and Ngorongoro 335
volcanoes 173
tourmaline 52, **58**, 67
Trachypogon grass 333
trade winds 452

traffic congestion
China 363
Old Delhi 368
Seoul 368
Transamerica Building (San Francisco) 131
Transantarctic Mountains 153, 272, 502, 503
Transpolar Drift 399, 404
transport, rock and sediment (river systems) 63, 65, 85, 94, 199, 205, 206, 208, 221, 222, 419, 422, 423
transport networks 377
Trapridge Glacier 256
tree ferns 306
trees; see also forests, tropical rainforest, individual tree genera by name
breakdown of rocks 91
buttresses 307
Cenozoic flowering 26
coppicing 311
drought resistant 290
effect of salt-laden winds 483
medicinal and food use 306
rubber trees 307, 345
in savanna 333, 337
in temperate grassland 330, 336
tree ring evidence for climate change 454
in tundra 339
tremolite 59, 71, 75
trilobites 20, 21, 113
Tristan da Cunha 162, 407
tropical grasslands, South America 333
Tropical Rainfall Measuring Mission 473
tropical rainforest 300, 303, 461
see also deforestation
Africa 218, 303, 313, 461
Amazon Basin 210, 211, 307, 447, 492, 493
Asia 300, 303, 314, 315, 316
Borneo 314
canopy 300, 313, 314, 317, 447
Central America 232, 298, 303, 306
Congo Basin 218, 313
Costa Rica 117
Daintree Rainforest 317
fires 315
ground vegetation 315
igapó forests 307
Madagascar 300, 303, 313
Malaria 459
Malaysia 303
montane 316
Northeast Australia 303, 317
Pennsylvanian Epoch 22
South America 210, 211, 303, 306, 307, 461
swamp forest 307, 315;
tropopause 444, 467
troposphere 444, 450, 451, 467
Tschierva Glacier 267
Tsugaru Strait 423
tsunamis see waves
Tsushima Current 423
Tuareg people 289
Tugela 148
Tularosa Basin 283
tuna 424
tundra 329, 338–39
Eurasia 339, 494
North America 304, 339
periglacial landforms 107, 338, 339
permafrost zones 338

tungstates 44, 55
tungsten ores 55, 71
Tunguska Crater 105
turbidity currents 94, 392
Turkey, earthquakes 127, 131
Turner, J.M.W. 175
Turpan Pendi depression 149, 495
turquoise 55
turtles
Alligator Snapping Turtle 322
freshwater 321
Green Turtle 423
marine, Cretaceous 23, 26
Twelve Apostles, sea stacks 437
Twin Sisters (Columbia River Plateau, USA) 158
typhoons 175, 354, 424, 452, 469, 470

U

Ubangi Plateau 234
Ugab Valley 64
ultra mafic rocks 68
ultraviolet radiation 444
Ulu Burun, undersea archaeology 415
Uluru (formerly Ayers Rock) 279, 296, 499
underground rivers and caves 240–49; see also coasts, caves, ice caves, rivers
Akiyoshi-do 249
Cango Caves 247
Carlsbad Cavern 242
cave paintings 143
Dinaric Alps 146
Ease Gill Caverns 246
erosion 92, 240
formation 240–41
Gouffre Mirolda 247
Grotte Casteret 246
Gruta do Janelão 243
Guanyuan Cave System 247
Gunung Mulu 248
Ha Long Bay 249
Jenolan Caves 249
karst 241, 242, 243, 246, 247, 249
Koonalda Caves 249
Krubera 247
Lascaux 246
Leviathan Cave 241
Mammoth Cave 243
Nanxu Arch 247
Optimisticheskaya 247
Pyrenees 143
Sistema Huautla 243
Skocjanske Jame 247
Snezhnaya 146
Tham Hinboun 249
tourism 242, 248, 249
tubular caves 243, 248
Vercors 246
Windsor Caves 243
unicellular foraminiferan 454
United Nations, International Decade for Natural Disaster Reduction 176
United States of America
high-technology industries 378
tourism 379
transport networks 377
United States Geological Survey 20, 125
galaxies 15
origins 14–15
stars 15
Ural River 238
Uralian Ocean 142

Urals 56, **142**
uraninite **52**
uranium 55, 184, 195, 241
uranium oxide 52, 55
urban areas **353–71**
Bangkok 424
Beijing **369**
Berlin **364**
Brasilia 354, 361
Buenos Aires **364**, 493
Cairo **366**
Chiba 369
Chicago 352, **360**
climate and 354, 355
Copenhagen **354**
Delhi **368**
Dhaka **367**
green belts 365
Guangzhou (formerly Canton) 355, 369, **370**
Hô Chi Minh City 424
Hong Kong 94, 354, **370**, 374, 424
internal zoning 354
Istanbul **366**
Jakarta **371**
Johannesburg 354, **367**
Karachi 367, **367**
Kawasaki 369
Kuala Lumpur **371**
Lagos 355, **366**
lighting 354, 355, 369
location of 354
London 212, 353, **365**, 379
Los Angeles 355, **360**, 462
Manila **370**, 424
Mexico City 355, **361**
Moscow **364**
Mumbai **368**
New York City 353, 354, 355, **356–57**, 379
Paris **365**
Pittsburgh 354, 374
planning 354, 365
population growth 353, 355, 366, 368
Rio de Janeiro 183, **361**, 493
Rome **366**
Ruhr Valley 354, 374, 376
Sao Paulo **364**, 493
Seoul **368**
Shanghai 369, **370**
shanty towns 355, 361, 367
Singapore **371**, 424
skyscrapers 352, 357, 360, 365, 366, 368, 369, 370
Sydney **371**
Tehran-Karaj **367**
Tokyo 132, 353, 354, 355, **369**
Toronto **360**
water rights 360
Yokohama 369
ureilites 83
Uruguay River 208
Ustyurt Plateau 237, 238

V

Vaal 104
Vaiont Dam, landslide disaster 98
Valdes Peninsula 493
Valle de Luna 89, 284
Valley of the Moon 89, 284
Valley of Ten Thousand Smokes 158, **188**
vanadates 44, 55
vanadium 55
Vancouver, Captain George 260
Vancouver Harbour Sulphur Mine 377
Vantanea trees 306

Vanuatu 137, 425
Vanuatu Trench 425
Variable Oystercatcher 23
Variscan origeny 147
Vatnajökull Ice-cap 163, **265**
vegetation see plant life
veld 331
Venezuela 375
Vent Crabs 386
Venus Flytrap 321
Vercors caves 246
vermiculite **60**, 377
Vermilion Lakes 475
vertebrates 26, 112
Vesuvius 70, 156, 169, **170**
Victoria Falls 219
Victoria Land (Antarctica) 503
Victoria Water Lily 324
Vicuña 229
Vindeby, wind farm 413
Vinson Massif 502
viruses 112
viticulture see grapevines
Vitus Lake 257
Vladivostok, climate 463
volcanic activity
Andes 89, 140–41, 492
Atacama Desert 284
Atlantic Basins 407
in the Cenozoic Era 25
climate and 95
deposition 63, 70, 72, 95, 157
erupting magma 69, 70, 137, 154, 155, 156, 387
eruption styles see eruptions, volcanic
hot springs and geysers **185–93**
hotspots 148, 154, 155, 157, 158, 162, 166, 173, 387, 420
island formation 170, 386, 422
islands 41, 86, 89, 137, 153, 158, 232, 234, 239, 387, 401, 407, 425, 429, 494, 500, 501, 502
lake formation 224, 232, 239
landslides and 98
lavas see lavas
Pacific region 137, 153, 157, 420, 422, 494, 500
prediction 157, 169
and respiratory diseases 175
Rocky Mountains 139
Sahara Desert 288
Sonoran Desert 282
submarine avalanches 97, 160
Sumatra and Java arc 416, 494
Taupo Volcanic Zone (New Zealand) 193
tectonic movement and 488
undersea 386, 500
volcanism 154
Zagros Mountains 148
volcanoes 27, 62, 120, 121, **154–77**; see also geysers, hot springs, mountains
Anak Krakatau 47, 176
Antrim Plateau 163
Ardoukoba Volcanic Field 128
Arenal 161
Baitoushan 174
Cameroon Mountain 172
Canary Islands 97, 168, 463
Cerro Azul–Quizapu 162
Cerro Negro 155, 455
Chaîne des Puys 163
cinder cone 68, 155, 158, 160, 162, 163, 189

volcanoes cont.
Colima 155, **160**
collapse of 96
Columbia River Plateau **158**
Concepción volcano 232
Cotopaxi **162**, 445
Crater Lake 158
cultivation and 169
Dallol 191
dome volcanoes 25, 155, 158, 160, 161, 162, 163, 174, 175, 176
El Chichón 161
Erta Ale 128, **171**
Galeras 162
Grímsvötn 163, 265
Japan 169
Java 169
Kaiserstuhl 127
Katmai 158
Kilauea 70, 155, 156, **160**, 501
Kliuchevskoi 173, 268
Krakatau 176
lake islands 158, 232, 234, 239
Lake Nyos 172
Lipari 70, **192**
living with 169
Masaya 161
Mauna Kea 155
Mauna Loa 160
Merapi 176
Mount Erebus 177, 502
Mount Etna 166, 167, 169, 445
Mount Fuji 174
Mount Jumullong 500
Mount Kilimanjaro 171, 268, 270, 497
Mount Pinatubo 69, 95, 156, 169, 175
Mount Rainier 158, 480
Mount St. Helens 69, 96, 120, 155, 159, 169
Mount Tarawera 193
Mount Unzen 174
Nevado del Ruiz 98, 162
Novarupta 158
Nyiragongo 172
Ol Doinyo Lengai 128, 172
Paraná Plateau 162
Paricutín 160
Pico de Teide 171
Piton de la Fournaise 173
plant and animal life 159, 168, 171
plate boundaries and 88
Popocatépetl 161, 361
predicting seismic activity 157, 169
Puyehue 165
Quizapu 162
Rabaul 154, 177, 500
Ruapehu 177
Santorini 170, 400
shield volcanoes 155, 156, 160, 173
Siberian Traps 22, 173
soils 161, 168, 169, 333
Soufrière Hills 70, 161
stratovolcanoes 155, 163, 172, 173, 174, 175
Stromboli 163
Taal 175
Tambora 175
tourism 173
vent-gas analysis 169
Vesuvius 70, 156, 169, 170
volcanic complex 154, 158, 160, 161, 170, 171, 174, 175
volcano, submarine 386

volcanoes cont.
Vulcano 70, 169, **170**, *185*,
192
West Eifel Field **163**
world distribution 157
volcanology 157, 160, *170*, 174,
176
voles 303
Volga **216**, 238, 364
Volga-Baltic Waterway 216
Vosges 127
Vostok Station (Antarctica) 502
Vredefort Crater **104**
Vulcano 169, **170**, *185*
eruptions 156, 161, 170, *176*
Vulture, African
Whitebacked *292*

W

Wadi al-Mujib delta *434*
wadis 278, *280*, 289
Waiho 274
Waimangu, hot springs **193**
Wairakei, geothermal field 193
Waituna Lagoon **327**
Awarua Wetland 327
Wakefield, fossil fuel power
station *375*
Walcott, Charles 20
Walking Stick Palm 317
wallabies; *see also* kangaroos
and wallabies
Rufous Hare Wallaby 296
Wallace, Alfred 28
walnut trees 316
walruses *401*, 405
Walvis hotspot 162
Warrumbungle Range (Australia)
152
Wasatch Range 228
water **106–11**; *see also* ice,
meltwater, rainfall, water
resources
acidic 91, 107, 478
as agent of change *84*,
85, 91
in the atmosphere 108, 109,
445, 447, 452, 473
changes in state 447
fresh water *40*, 107, *109*,
230, **231**
freshwater springs *109*
global water cycle 109
groundwater 107, 109, 110,
185, 195, 198, 280
and life 106, *113*
local water cycle 109
oceans and seas 384–85
paving and run-off 109
salinity 106, 384, 395
salt water 106
vanadium- and uranium-
enriched 55
water birds 321
Water Fern 321, 324
Water Hyacinth 321, 324, 327
water meadows *325*
water resources **110**; *see
also* lakes, meltwater,
rivers, water
dams and 214
deposits 63, 65, 185
desalination 110, 419
desert wells *343*
in deserts 280
eutrophication 231
fresh water *40*, 107, *230*, **231**
pollution 231, 376
recreational use *110*
shortages *111*, 147, 371
surface water 109, 110
wars over 110

water resources cont.
water rights 360
waterborne disease *230*, 231
waterfalls and rapids
California *200*
Cirque du Gavarnie 143
Congo 218
frozen *204*
Iguacu River 208
Niagara 226, *491*
Nile 217
Orinoco Churun 209
Paraná 208
Rhine 213
Rjukandi waterfall *98*
Siposo-Piso 239
St. Lawrence 204
underground 240, 243, 246,
247, *249*
Yellow River 221
Yellowstone River *199*
Zambezi 219
watersheds 198
waterspouts *409*, *447*, *452*
Wattled Jacana *209*
wavelengths, solar radiation *446*,
447
waves **432–33**; *see also* coasts,
oceans and seas, tides
coastal deposition 437
coastal erosion 436
refraction 436
sea-level change and 434
storm surges 470
tsunamis (tidal waves) 103,
123, 131, 174, 176, *263*,
388–89, *390–91*, **433**
wave power 375
weather **464–85**; *see also*
climate, clouds, storms,
wind
air masses 466
changes in *464*, 465
depressions 467, 482
effect of the sun 465
fronts *466*, 467, 468, 473
Greenland Climate Network
project 263
jet streams 450, 451, 453,
467, 468
pressure systems 461,
466–469, 482
records and trends 455
weathering and erosion 85,
90–91
biological 91
chemical 91
coastlines 92, *249*, *412*, *416*,
430, 431, 433, **436**
deserts 278, 279, *288*, 289
fault systems 125
physical 90
rock mass movement and 96
sedimentary rocks 63, 65,
92, 181, 184
spheroidal *66*
stone arches *93*, 436
weather station, automatic *263*
Weaver Ants *303*
Weber River 228
Weddell Sea *401*, **429**, 502,
503
Weddell Seal *429*, *503*
Wegener, Alfred 86
Wellman, Harold 133
Welwitschia Mirabilis *292*
Werner, Abraham Gottlob 45,
52, 74
West Antarctic Ice Sheet (WAIS)
270, 273
West Eifel Field **163**
West Indies 412
West Svartisen Ice-cap **266**

Western Cordillera (North
America) 137, 139
Western Hemlock 303, 305
Wet Tropics World Heritage Site
317
wetlands **319–27**
animal life 321–27
Biebrza Marshes **325**
Camargue 213, **325**
Coorong **327**
environmental issues 319,
323
Everglades 228, **323**
Flow Country **324**
freshwater 320, 322, 323,
324, 325, 326, 327
Great Dismal Swamp **322**
Hortobágy **325**
Llanos 209, **324**, 333
Okavango Delta *321*, **326**
Okefenokee Swamp **322**
Pantanal 208, **324**, 333
plant life 321–27
saltwater 320, 323, 325
seasonal *318*, 321, 326
Sudd 217, **327**
Waituna Lagoon **327**
Awarua Wetland 327
Whale Shark *424*
whales 13, 26, *422*
Beluga (White Whale) *408*
Blue Whale 383
Bowhead Whale *402*
military sonar and 385
wheat 332, *350*
Whin Sill **184**
whirlpools *395*, 413
White Dome Geyser *185*
White Fir 305
White Island (New Zealand) *499*
White Nile 217, 234, 327
White Rock Canyon 208
White Sands (New Mexico, USA)
55
White Sea **404**
White Spruce 304
White-clawed Bear 149
White-tailed Deer 322
Whitehaven Beach (Australia) *94*
Wicken Fen *320*
Widmanstätten structure, in
meteorites 83
Wieliczka salt mines 53
Wild Boar 143, 311
wild vines 316
Wildebeest 292, *328*, 335
wildlife, and agriculture 343
Wilkins Ice-shelf *270*
Willis, Bailey 124
willows 31
Arctic Willow *339*
Dwarf Willow *339*
Wilson, John Tuzo 89
wind **482–87**; *see also* storms
anabatic 483
atmospheric circulation
450–51, 452
Beaufort Wind Force
Scale 482
blizzards *474*
Coriolis effect 394, 409, 450,
452, 466, 469, 482
daily winds 483
desert landforms and 279,
289, 293, *296*
dustdevils 484
erosion 93
harmattan 289
high and low pressure
weather 468
hurricanes 98, 407, *412*, *424*,
452, 469, 470, *471*
Intertropical Convergence
Zone (ITCZ) *451*

wind cont.
jet streams 450, 451, 453,
467, 468
katabatic 483
land-sea breezes 483
ocean currents and 394, 452
oceanic winds 483
pressure and wind speeds
482
prevailing winds 451
Saffir-Simpson scale 470
sandstorms 95, 484
seasonal winds 483
Southern Ocean 428, 452
squall lines 484
systems 451
thunderstorms 484
tornadoes *409*, 452, 479,
482, **485**
tree growth and *302*
tropical high winds *482*
typhoons 175, *354*, 424,
452, 469, 470
wave generation 432
wind power and wind farms
375, 413
wind-borne deposition 95
Windsor Caves **243**
Windward Passage 412
wine production *see* grapevines
Winsford, Cheshire, salt
mines 53
Wire Rush 327
Witwatersrand Basin *104*
Wizard Island (Crater Lake) *158*
Wolfe Creek Crater *100*
wolframite **55**
Woolly Monkey *114*
wolverines *312*
wolves 139
Grey Wolf 311
Maned Wolf *333*
wombats *114*
Wood Anemone *302*
Wood Duck 322
Woodleigh Crater **105**
woodpeckers 311
Great Spotted
Woodpecker *303*
Red-cockaded
Woodpecker 322
wool 378
World Heritage Site 236, 260,
313, 325
worms *see* nematodes
Worthington Glacier **260**
Wrangell Mountains 256
Wrangell St. Elias National Park
256, *270*
Wren, Sir Christopher 368
Wright, Frank Lloyd 360
Wright Valley (Antarctica) *503*
Wuha Lake *313*
Wülfen, F. X. 55
wulfenite 44, **55**

XY

xenoliths *38*, 66, 67, 68,
182, 183
Yamuna 368
Yangsan Fault 132
Yangshuo plain, tower karst *241*
Yangtze 214, **220**, 423,
424, 494
Yangtze River Dolphin 220
yardangs 278, 289, 292, 293
Yellow Coneflower 297
Yellow Perch 322
Yellow River 92, **221**, 423
Yellow Sea **423**
Yellowstone *45*, 69, 139, *185*,
186–87

Yellowstone Falls *199*
Yen Mountains 369
Yenisey 404
Yokohama 369
Yongding 369
Yosemite National Park 182,
305, *490*
Younghusband Peninsula 327
Yucatán Channel 409, 412
Yucatán Peninsula 24, *25*, 103
Yucca Mountain, nuclear storage
283
yuccas 283
Joshua Tree *282*

Z

Zagros Mountains **148**
Zambezi 128, **219**, 235,
416, 419
Zanskar River 269
zebras 335
Burchell's Zebra *113*, *321*
zeolite **60**, 68
Zheng He 417
zinc 50, 71, 82, 142, 376
zincblende 50
Zion National Park *199*
zircon 16, 52, **56**
zircon dating 56
zirconium 56
Zhugqu (China) landslide 98–99

ACKNOWLEDGMENTS

Dorling Kindersley would like to thank the following people for their help in the preparation of this book: Douglas Palmer for suggestions for photo sources, comments on the Glossary, and general advice; Vanessa Hamilton for original design work and Rebecca Milner for design assistance; Cathy Meeus for editorial help; Gemma Casajuana and Mark Bracey for DTP support; Erin Richards for administrative support; Julia Lunn and Rob Stokes for additional cartography; Robyn Bissette and Ellen Nanney of the Smithsonian Institution; Dr. A.G. Smith and Lawrence Rush of Cambridge University for plotting maps of the Earth's past tectonic plates; Philip Eales, Andrew Wayne, and Kevin Tildsley of Planetary Visions Ltd; and Sean Mulshaw and Geoffrey Gilbert for advice on the selection of features.

Revised editions Dorling Kindersley would like to thank: Carron Brown, Lili Bryant, Ruth O'Rourke Jones, Rohini Deb, Susmita Dey, Nandini Gupta, Neha Pande, Rupa Rao, Aashline R. Avarachan, Ekta Chadha, Dharini Ganesh, and Sarah Mathew for editorial assistance; Michael Yeowell, Alison Gardner, Adam Brackenbury, Thomas Morse, Anna Hall, Upasana Sharma, Shaarang Bhanot, and Debjyoti Mukherjee for design assistance; Simon Mumford for help with cartography and artwork; Deepak Negi for picture research assistance; Rohan Sinha and Sudakshina Basu for project management at DK Delhi; and Sachin Gupta, Bimlesh Tiwary, Anita Yadav, Satish Gaur, and Rajdeep Singh for technical assistance. Data for sizes of urban areas has been reproduced from the Demographia World Urban Areas survey.

Smithsonian Enterprises Carol LeBlanc, President; Brigid Ferraro, Vice President, New Business and Licensing; Jill Corcoran, Senior Director, Licensed Publishing; Avery Naughton, Licensing Coordinator; Paige Towler, Editorial Lead.

This book was produced in collaboration with the **Smithsonian** in Washington DC, USA, the world's largest museum and research complex. This renowned research centre is dedicated to public education, national service, and scholarship in the arts, sciences, and history.

PICTURE CREDITS
Dorling Kindersley would like to thank the following for their help with images: Romaine Werblow and Rob Nunn in the DK Picture Library; Giovanni Cafagna at Corbis; Tony Waltham at Geophotos; Caroline Thomas; and Mariana Sonnenberg.

Key: L = Left, R = Right, C = Centre, T = Top, B = Bottom, a = Above, b = Below.

Abbreviations
B&C: Bryan and Cherry Alexander Photography; **DK:** DK Picture Library www.dkimages.com; **FLPA:** FLPA – Images of Nature; **Geo:** Tony Waltham Geophotos; **Getty:** Getty Images; **RHPL:** Robert Harding Picture Library; **NG:** National Geographic Image Collection; **NHM:** The Natural History Museum, London; **OSF:** Oxford Scientific Films; **SI:** Smithsonian Institution; **SPL:** Science Photo Library; **Still:** Still Pictures; **WWI:** Woodfall Wild Images.

Sidebar Images (from top to bottom)
Earth - NASA: Hubble Space Telescope Center (PR96-27B); **Corbis:** NASA, L Clarke, Ken Straiton, Michael T Sedam; **DK**.
Land – Corbis: Darrell Gulin, Walter Hodges, Wolfgang Kaehler, Gary Braasch, José F Poblete, David Muench.**Oceans – Corbis**: FLPA / Winifred Wisniewski, David Pu'u, Darrell Gulin, Stuart Westmorland, Jeffrey L Rotman, Robert Yin.
Atmosphere – Corbis: Stuart Westmorland, Paul A Souders, William James Warren, Matthias Kulka, Craig Tuttle, William A Bake.

1 Still: UNEP / D Stanfill; **Dreamstime.com:** Egor Baliasov C. **2/3 Getty Images / iStock:** unikatdesign. **4 Corbis:** Darrell Gulin TC. **4/5 Getty Images / iStock:** John Morrison / E+ B. **5 alamy.com**: ImageState / Adam Jones TL; **Corbis**: William James Warren C; **Galaxy Contact**: Johnson Space Center (STS066-208-205) TCRb; **Getty**: Image Bank / Peter Cade TC. **8/9 Corbis:** Darrell Gulin. **10/11 Dreamstime.com:** Dudlajzov. **12 Geo:** CLb; **University of Edinburgh, Grant Institute:** TRb. **12/13 Getty Images:** Michele Falzone CB. **13 Corbis:** George HH Huey CT; James L Amos TCL; **NG:** Louis Mazzatenta C; **SPL:** James King-Holmes CRb. **14 SPL:** NASA BR; Patrice Loiex, Cern BL. **14/15 SPL:** Roger Harris. **15 NASA:** Chandra X-Ray Observatory (MSFC-9905485) C; Hubble Space Telescope Center (PR96-38B) CRa. **16 NHM:** C. **16/17 Corbis:** Ric Ergenbright. **17 Courtesy of the Archives, California Institute of Technology:** BR. NASA: Saturn Apollo Program (MSFC-690096) CL. **SPL:** Dr Kari Lounatmaa Cb. **18 Roger Buick:** Department of Earth and Space Sciences & Astrobiology Program, University of Washington Cb, BCR; **Corbis:** Roger Garwood BL. **18/19 Corbis:** Ric Ergenbright. **19 Dr Peter Crimes:** BR; **Dr Bruce Runnegar:** Centre for Astrobiology, Institute of Geophysics, University of California. **20 alamy.com:** Stephen Frink Collection / Masa Ushioda BL; **Corbis:** Bettmann CR; Ric Ergenbright LC.; **DK:** Colin Keates BC. **20/21 FLPA:** L West. **21 DK:** Royal Museum of Scotland / Harry Taylor Cb; NHM / Harry Taylor CRa. **22 Science Photo Library:** British Antarctic Survey Ca; **Corbis:** Jonthan Blair CbR; **FLPA:** L West (main); **Geo:** CLb. **23 Corbis:** Douglas Peebles BC; Kevin Schafer BL; Layne Kennedy R (main); **DK:** Hunterian Museum C; Jon Hughes BR. **24 Corbis:** Layne Kennedy (main); Tom Bean BL; **DK:** Senekenberg Nature Museum / Andy Crawford CaR; **NHM:** BC. **25 Corbis:** Richard Cummins CR; Roger Ressmeyer CLb; Royalty-Free R (main). **26 alamy.com:** Ukraft C; **Nature Picture Library:** Jurgen Freund CR; **Corbis:** Jonathan Blair BCL; **DK:** Harry Taylor BCR. **26/27 Corbis:** Royalty-Free. **27 Corbis:** Gallo Images BR; Torleif Svensson BL; **SPL:** Geospace C. **28 Alamy Stock Photo:** Francis Specker C; **DK:** BLR; Museum of London / Dave King CRb; Pitt Rivers Museum / Dave King CRa; **NHM:** BL; **SPL:** John Reader CLb, BC. **28/29 Corbis:** Royalty Free. **29 DK:** Gables CL **Dr Sandy Tudhope:** Institute of Geology and Geophysics, Edinburgh University CLa. **30 NHM:** BL; **Geo:** BR. **30/31 Corbis:** Royalty-Free. **31 Corbis:** William Findlay BL; **DK:** Royal British Columbia Museum / Andrew Nelmerm TL. **32/33 OSF:** Olivier Grunewald. **34 Hutchison Library:** Patricia Goycolea CLb; **NASA:** NSSDC / Apollo 8 BLa. **35 Corbis:** Stuart Westmorland CbR; **SPL:** Alfred Pasieka BR. **37 DK:** Clive Streeter CLa; **DK:** CRb. **38** Science Photo Library: (TR); **Corbis:** Raymond Gehman CLb; **Dr Sally Gibson:** Department of Earth Sciences, University of Cambridge CL. **39 Corbis:** Roger Ressmeyer Ca. Computer tomography scan reproduced from Schmandt, B. and E. Humphreys. 2010. "Complex subduction and small-scale convection revealed by body-wave tomography of the western United States mantle." Earth and Planetary Science Letters, 297, 435-445, doi:10.1016/j.epsl.2010.06.047 and Trabant, C., A. R. Hutko, M. Bahavar, R. Karstens, T. Ahern and R. Aster (2012), Data products at the IRIS DMC: stepping-stones for research and other application, Seismological Research Letters, 83(6), 846:854. doi: 10.1785/0220 120032: BR. **40 Geo:** CLb; Stephen Bond BCRa; Philip James Corwin TC; **Corbis: Guy** Motil CLa; **Geo:** CbL, CbR. **41 alamy.com:** David Noton Photography / David Noton Ca; **Geo:** BCL; Worldwide Picture Library / John Fowler CLb; **Still:** Staffan Andersson TL; Tiziana and Gianni Baldizzone Cb; **SPL:** George Bernard BR; **Geo:** TLb. **42 Philip E Batson:** CLa; **Corbis:** Kevin Schafer BR. **42/43 Dreamstime.com:** KPixMining. **43 DK:** *diamond*; Colin Keates *hematite*, *rock crystal*; Harry Taylor *azurite*, *calcite*, *galena*, *halite*, *kaolinite*, *magnetite*, *milky quartz*, *muscovite*, *smokey quartz*, *talc*; NHM / Colin Keates *opal*; NHM / Harry Taylor *gold*; **SI:** BL. **44 DK:** *stibnite*; Colin Keates *gold*, *olivine*; Harry Taylor *anhydrite*, *apatite*, *borax*, *calcite*, *chromite*, *green fluorite*, *wulfenite*. **44/45 Dreamstime.com:** Kenneth Keifer. **45 Corbis:** Robert Garvey TC. **DK:** Colin Keates *burmese ruby crystal*, *cusioned mix-cut*; Harry Taylor *amethyst quartz*, *fluorite*, *sphalerite*; **NHM:** CL. **Dreamstime.com:** Michael Turner CRa. **SPL:** TCL; Dr Jeremy Burgess LC. **46 Corbis:** Bettmann CRa. **DK:** CLb; Harry Taylor CLa, TCL; NHM / Harry Taylor CbR; **SI:** BR; **Geo:** BCL. **47 DK:** Colin Keates TC; Harry Taylor CRa, CbL, BL; NHM / Harry Taylor BCL; **Getty Images / iStock:** andersen_oystein CLa; **Getty Images:** Patrick Robert - Corbis CRb; **P A Photos:** Peter Jordan Bla; **SI:** BR. **48/49** Getty Images. **50 DK:** Harry Taylor CRa, CLb, CRb, BR, CaL; **NHM:** BCL. **51 AISA - Archivo Iconográfico S. A., Barcelona:** TCR; **Corbis:** Fulvio Roiter CRb; **DK:** Harry Taylor TR, CLa, Cb, BL, BCR; **Geo:** Ca. **52 Corbis:** Elio Ciol Ca; José Manuel Sanchis Calvete BCL; **DK:** Colin Keates CR; Harry Taylor TL, TC, CLb; NHM / Harry Taylor CRa (inset); **NHM:** Cb; **SPL:** Astrid & Hanns-Freider Michler BR. **53 Corbis:** Tim Thompson BL; **Paul Deakin:** TRb; **DK:** Harry Taylor CLa, C, BR, CaR, CbR, TCL. **54 Tony Dickson:** Department of Earth Sciences, University of Cambridge CRa; **DK:** Harry Taylor TR, CLa, CLb, BR, BCL, BCLa, CaL, CbR. **55 Corbis:** Scott T Smith CL; **DK:** Colin Keates BC; Harry Taylor CRa, BL, BRa, CbR, TCL, TCRb; NHM / Harry Taylor BCL. **56 Corbis:** Jim Sugar Photography BR; **DK:** Colin Keates Ca; Harry Taylor CRa, Cb, CRb, BL, BC (*andradite*), BC (*pyrope*), BC (*spessartine*), TCL; **NHM:** TCb; **SPL:** Andrew Syred CaL; **Geo:** TCb. **57 DK:** Colin Keates CRa, CaR; Harry Taylor TL, CL, CRb, BL, BCR, CaL, TCLb; NHM / Colin Keates BCLa; **SI:** TR. **58 Corbis:** Mike Zend TCR; **DK:** Harry Taylor Ca, CL, CR, CRb, BL, BR, BCR, CaR; **SI:** BCL, CbR. **59 Corbis:** Ecoscene / Ian Harwood TR; Lowell Georgia BR; **DK:** Harry Taylor CLa, BL, CaR, CbR, TCL; NHM / Colin Keates C; **SI:** CaL. **60 DK:** Harry Taylor TR, CLa, CR, BL, BC, CbL (*labradorite*), CbL (*microline*), CbL (*yellow*), TCLb; NHM / Harry Taylor CbL (*moonstone*); **SPL:** Philippe Gontier BRa. **61 Corbis:** David Muench B; George HH Huey CRb; Gianni Dagli Orti TR. **DK:** Harry Taylor CLa, CbR, TCL; NHM / Harry Taylor Ca. **62 SI:** BCLa; **Dr AG Tindle:** Open University BCLb. **63 Geo:** BC. **64 Corbis:** Hubert Stadler Ca; **Landform Slides:** Ken Gardner Cb; **Alamy Stock Photo:** Minden Pictures Cb; **SPL:** Tony Craddock TL; **Geo:** BCR. **64/65 Corbis:** Lester Lefkowitz B. **65 Science Photo Library:** Edward Kinsman BC; **SPL:** CR; Martin Dohrn Cb; Mike McNamee C; **SI:** CRb. **66 DK:** Clive Streeter CR; Harry Taylor CRa, CaR; **SI:** CL; **Geo:** TR, B, CaL. **67 Alamy Stock Photo:** Jens Andersen BL; **DK:** Harry Taylor CRa, CL, CRb, BCR, TCL; **SPL:** CbL; **Geo:** TR. **68 Corbis:** Raymond Gehman BL; **DK:** Harry Taylor Cb, BR; **SI:** PD & L CaR, TCLb; **Geo:** BC, TRb. **69 DK:** Harry Taylor TL, BL, CaL; **Getty Images / iStock:** Mh Photoz BR; **SI:** CRa; **Geo:** TCRb. **70 Corbis:** Douglas Peebles BR; **DK:** Colin Keates CL, BCa, TCL; **SPL:** CRb; **Geo:** CRa, BL. **71 Corbis:** Jason Hawkes TR; **DK:** Dinesh Khama BR; Harry Taylor C, CL, BL; **Alamy Stock Photo:** Johann Hinrichs BC. **72 Corbis:** Tiziana and Gianni Baldizzone BR; **DK:** Harry Taylor TCL, TR, CLa, Ca, BL, BC; **FLPA:** Maurice Nimmo Cb. **73 DK:** Harry Taylor TR, C; **FLPA:** E & D Hosking CRa; **Alamy Stock Photo:** Harvey Wood BL; **SI:** CR; **Geo:** BR. **74 DA Carswell:** University of Sheffield BC; **DK:** Andreas Einsiedel BL; Clive Streeter CbR; Harry Taylor Cla; **NHM:** CaL; **SI:** BR; **Geo:** TR, Ca. **75 Getty Images:** Federica Grassi / Moment TL; **DK:** Harry Taylor CLb, BR; **SI:** CRa, BC, CbR. **76 alamy.com:** Worldwide Picture Library / Ivan J Belcher BC; **DK:** Harry Taylor Ca, CR, BL, BR, TCb. **77 DK:** Andreas Einsiedel BL; Harry Taylor CRa, CRb, BCR, CbL; Peter Wilson Cla; **Geo:** TR. **78 Corbis:** Richard Cummins BL; **DK:** Harry Taylor CLa, BC, CbL; NHM TC; **Shutterstock.com:** Dianne Wickenden BR; **Getty Images:** Melvyn Longhurst / Corbis Documentary CR. **79 Corbis:** TCL; Michael St Maur Sheil CR; **DK:** Andreas Einseidel RC; Colin Keates CL; **SPL:** TR; **Science Photo Library:** Sinclair Stammers B. **80 Still:** UNEP / S Compoint. **81 Still:** Chris James BCL. **82 Corbis:** Wolfgang Kaehler Cra; **DK:** Harry Taylor TC, BR, BCL; Steve Gorton Cb; **SI:** CLb; **Geo:** CLa. **83 DK:** Colin Keates CRb; Harry Taylor BL; NHM: CL; **SI:** BCR; CaR. **84/85 Corbis:** Yann Arthus-Bertrand. **86 Corbis:** L Clarke BC; Pat O'Hara BL; **SPL:** TR. **86/87 Visible Earth:** Image Courtesy Jacques Descloitres, MODIS Land Rapid Response Team, NASA / GSFC B. **87 Alamy Stock Photo:** NB / ROD / NASA / Jerry Svarc, USGS BR; **DK:** Peter Anderson CRb; **SPL:** Mehau Kulyk Cb. **88 Corbis:** Dewitt Jones CR; Galen Rowell CRa; **SPL:** Dr Ken MacDonald BR. **88/89 Corbis:** Yann Arthur-Bertrand C. **89 Corbis:** Hubert Stadler CRa; **Ontario Science Centre:** BR. **90 Corbis:** James Randklev TR; **Geo:** B. **91 alamy.com:** Jon Arnold Images / James Montgomery CaL; **Corbis:** David Muench BR; Paul A Souders TCR; Royalty-Free BC; Wolfganf Kaehler CRb; **SPL:** Adam Hart-Davis CRa. **92 Association of American Geographers:** TR; **Corbis:** David Muench CL; Michael S Yamashita BC; Paul A Souders C. **92/93 Corbis:** Royalty-Free C. **93 Corbis:** Galen Rowell TR; Jon Sparks BRa; Wolfgang Kaehler CRa; **Geo:** BCR. **94 Corbis:** Charles Krebs Photography LC; Lowell Georgia BC; Travel Ink / Derek M Allan BCR; **SPL:** William Ervin Cb. **94/95 Corbis:** Yann Arthus-Bertrand T. **95 NASA:** Johnson Space Center (STS026-43-98) TR; Johnson Space Center / Image courtesy of Earth Science and Image Analysis Laboratory (STS043-22-23) BR; Johnson Space Center / Image courtesy of Earth Science and Image Analysis Laboratory (STS41D-32-14) BCR; **SPL:** Dr Juerg Alean BL; **Geo:** CR. **96 Corbis:** Hans Strand Ca; Lloyd Cluff BC; **Geo:** CLb. **96/97 Corbis:** SP Gillette C. **97 Corbis:** Jon Sparks BC; ML Sinibaldi BRa; **DisasterMan Ltd:** Bill McGuire CR; **Geo:** TR, Ca. **98 Corbis:** David Lees C; Natalie Fobes BC. **99 Rex Features:** Sipa Press. **98/99** Getty Images: AFP / STR. **100 Corbis:** Roger Ressmeyer CbL; **DK:** Harry Taylor BR; **Getty:** Photodisc CL; **Images Of Africa Photobank:** Vanessa Burger TR; **SPL:** Dr David King BC. **100/101 Corbis:** Yann Arthur-Bertrand C. **101 Corbis:** Roger Ressmeyer TC; Sygma / Duthei / Didier BCRa; **Department of Geology, University of Witwatersrand:** BCL. **102 Aurora & Quanta Productions Inc.:** Peter Essick CL; **Corbis:** Paul A Souders BL; Royalty-Free BR; **NASA:** Haughton-Mars Project 2002 TCL; Marshall Space Flight Center (MSFC-0201920) CR. **103 AirPhoto:** Jim Wark Ca; **The Barringer Crater Company:** TRb; **Corbis:** NASA TC; Image courtesy of **Virgil L Sharpton, University of Alaska – Fairbanks / Lunar and Planetary Institute and NASA:** Cb; **SPL:** Prof Walter Alvarez BR. **104 Getty Images / iStock:** holgs TL; **Corbis:** Michael S Lewis BCR; **Landsat:** CR; **NASA:** Johnson Space Center (STS51I-33-56A) Cb; **Stadt Noerdlingen:** TL. **105 Getty Images / iStock:** Purva Joshi CRa; **Geological Survey of Western Australia:** BL; **Global Impact Studies Project:** TR; **NASA:** Johnson Space Center (STS41D-14-41-028) CRb; **Novosti (London):** CLA; **Skyscan:** photo by skycam.com.au BR; Geo: CRa. **106 Corbis:** Lowell Georgia BC; Natalie Fobes Cb; Ricki Rosen BL; Science Pictures Limited CLa; **106/107 Getty:** Image Bank / Harald Sund C. **107 Alamy Stock Photo:** Krys Bailey CRb; **Corbis:** David Muench TR; Galen Rowell CRa; Papilio / Bob Marsh BC; **SPL:** David Parker C. **108 Corbis:** Tom Bean TC; **DK:** Brian Cosgrove Collection Ca, CRa; **Getty:** Taxi / Andrew Mounter B. **109 Corbis:** Doug Wilson TCR; Sygma / Aim Patrice BRa; **Galaxy Contact:** CLa; **Getty:** Stone / David Muench CLb. **110 alamy.com:** Peter Bowater C; **Corbis:** Tony Roberts CLb. **111 Still:** UNEP / A Ishokon. **112 Alamy Stock Photo:** Tom Stack Cb; **Science Photo Library:** Dr Gary Gaugler CRb; **Corbis:** Michael & Patricia Fogden BL; **Corbis:** Visuals Unlimited / Dr. Terry Beveridge BC; **DK:** BC (*vertebrate*); Christine M Douglas BC (*non flowering*) pl; Dave King TR; Kim Taylor BC (*invertebrate*); Neil Fletcher BL (*fungi*); Steve Wooster BC (*flowering plant*); **OSF:** BLa; Gary Gaugler / OKAPIA BC (*monerans*); SPL: CDC CaL. **112/113 Alamy Stock Photo:** R.M. Nunes TC. **113 Gerald D Carr:** Department of Botany, University of Hawaii BR; **DK:** Colin Keates BL, BLa. **OSF:** Richard Packwood B. **114 alamy.com:** Jan Baks CLb; **Corbis:** Agliolo / Sanford CLa; Eric and David Hosking BCR; **DK:** Geoff Dann BL; Ken Findlay BR; Peter Anderson BCL; **FLPA:** David Hosking CL. **115 alamy.com:** David Boag CRb. **116 Corbis:** James L Amos BCa; Tom Bean BL; Wolfgang Kaehler TCb. **117 FLPA:** Minden Pictures/Frans Lanting. **118/119 alamy.com:** ImageState / Adam Jones. **120/121 Corbis:** Roger Ressmeyer. **122 Corbis:** Bettmann CRa; Roger Ressmeyer BL; **DK:** Joe Cornish C; **James Jackson:** Department of Earth Sciences, University of Cambridge Cb. **122/123 Corbis:** Yann Arthur-Bertrand B. **123 Martin Miller:** Department of Geological Sciences, University of Oregon TL; **SPL:** Fred McConnaughey TR; Massonnet et al / CNES BRa. **124 Corbis:** CL; Bettmann BL; Lloyd Cluff B; **School of Earth Sciences, Stanford University:** TRb. **125 Corbis:** BR; Progressive Image / Bob Rowan TR **Martin Miller:** Department of Geological Sciences, University of Oregon TCL, BCR, TCR. **126 Dreamstime.com:** Sophie Mcaulay TR; Reproduced by permission of the **British Geological Survey** © NERC All Rights Reserved IPR/38-31c CL, BL; **SPL:** Robert Brook CR; **Art Directors & TRIP:** C Sanders BR. **127 Shutterstock.com:** Uellue BL; **Agence France Presse:** Sergei Chirikov Ca; **FLPA:** Walter Rohdich BRa; **NASA:** Johnson Space Center (STS41G-34-6) TCR. **128 alamy.com:** ImageState / Georgette Douwma TRb; **Corbis:** Bettmann BL; Stephen Frink Cb; **DK:** NHM / Harry Taylor CLb; **Images Of Africa Photobank:** David Keith Jones TC; **NG:** Chris Johns BCR. **128-129 Alamy Stock Photo**: RGB Ventures / SuperStock. **129 OSF:** NASA / Johnson Space Center (STS40-152-180). **130/131 Corbis:** Epa / David Wethey. **130 Corbis:** Michael S Yamashita. **131 Corbis:** Ocean TR; **NG:** Reza BCL; **SPL:** James Stevenson TRb. **132 Corbis:** Lloyd Cluff CLb; **Art Directors & TRIP:** TCR; M Fairman BCL; **Visible Earth:** Image courtesy Jacques Descloitres, MODIS Rapid Response Team, NASA/GSFC TR. **133 Institute of Geological and Nuclear Sciences, New Zealand:** TC, TR, B. **134/135** Press Association Images: AP Photo / Saitama Shimbun via kyodo News. **136 Corbis:** Galen Rowell TR, Yann Arthus-Bertrand TRb. **136/137 Dreamstime.com:** Petergarcia87 B. **137 Corbis:** CR; NASA BR; Yann Arthus-Bertrand TR. **138 Corbis:** Hulton-Deutsch Collection C; **Getty:** Stone / Joe Cornish BL; **Magnum:** Stuart Franklin TC; **SI:** CR. **139 Dreamstime.com:** Funniefarm5 C; **Getty Images / iStock:** John Morrison / E+ B; **Corbis:** Galen Rowell CLa; **RHPL:** Patrick Endres CR; Scott Darsney TCL; **Still:** Peter Arnold Inc / Alan Maichrowicz TCR. **140 Corbis:** Ecoscene / Graham Neden CR; Galen Rowell C; Hubert Stadler CRa, CRb; Jonathan Blair BCL; Lynda Richardson TCL; NASA CLb; William Manning CLa. **141 AKG London:** Alte Nationalgalerie, Berlin TRb; **Corbis:** Brian Vikander BCR; **DK:** TL; **South American Pictures:** Kathy Jarvis B (main); **Still:** Alan Watson TCR. **142 123RF.com:** gbs097 Cb; **naturepl.com:** Konrad Wothe C; **Corbis:** James L Amos CRa; John Noble BR. **143 Alamy Stock Photo:** imageBROKER.com GmbH & Co. KG B; **Corbis:** Jon Sparks CRa; Michael Busselle Cb; Roger Ressmeyer CL; J&C Sohns TR; R Wilmshurst TCL. **144 Corbis:** Galen Rowell BLa; Patrick Johns CR; Randy Wells CRa; Sandro Vannini BL, BC; **P A Photos:** European Press Agency CbR; **SPL:** TCR. **144-145 Getty Images:** Alessandro Bellani / Moment. **145 Getty:** Taxi / Pascal Tournaire. **146 Corbis:** A Alamary & E Vicens CLa; Bojan Brecelij TR; Natalie Fobes CbL; **NG:** John Eastcott and Yva Momatiuk BL; Randy Olsen CRa; **Konstantin Mikhailov:** BR. **147 Corbis:** Christine Osborne CaL; Richard Bickel CRa; Yann Arthus-Bertrand CLa, CbL; **FLPA:** Panda Photo TC; **NG:** Steve McCurry B. **148 Alamy Stock Photo:** Jon Arnold Images Ltd BC **Corbis:** CLa; Earl & Nazima Kowall CaL; Gallo Images / Roger de la Harpe TCR; **DK:** Frank Greenaway CbR; Ken Findlay CR; **FLPA:** Fritz Polking BR; **NPA Group:** CbL. **149 alamy.com:** Michael Grant BL; **Corbis:** NASA BCL; **DK:** Gables CRa; **Getty:** Taxi / Walter Bibikow T; **RHPL:** Gavin Hellier Cb. **150 Corbis:** Bettmann BL; David Keaton TRb; Galen Rowell Ca; Howard Davies CRa. **150/151 Mountain Images:** Ian Evans B; **151 Dreamstime.com:** Arpita Ray Ca; **Corbis:** Ecoscene / Robert Weight CRa; **NASA:** Johnson Space Center (STS41G-120-22) TCR. **152 Shutterstock.com:** Junn Kae CR; **Corbis:** Eric & David Hosking CRb; **DK:** Rob Reichenfeld CLa; **Getty:** Image Bank / John William Banagan TR; **NG:** Paul Chesley B; **WWI:** Joe Cornish Cb. **153 Alamy Stock Photo:** Colin Harris / era-images C; **Science Photo Library:** British Antarctic Survey CR; **British Antarctic Survey:** David Cantrill CRa; **Corbis:** Galen Rowell B, TCLb; **FLPA:** Eric & David Hosking TCR; **OSF:** Kim Westerskov Ca. **154 Corbis:** NASA CL. **154/155 Getty Images:** bomboman C. **155 Corbis:** Gary Braasch CbR; Roger Ressmeyer Cb; **DK:** Rob Reichenfeld CbL; **SI:** CRb; CR; **Geo:** TR, C. **156 Agence France Presse:** TC; **Corbis:** TR; Yann Arthur-Bertrand Ca; **US Geological Survey:** JD Griggs B; **Geo:** CR. **157 Getty Images:** Agoes

Rudianto / Anadolu Agency CR; **Corbis:** Sygma / Patrick Robert C; **Geo:** TCL, TCR. **158 Corbis:** David Muench BL; Terry W Eggers CRa; **NG:** Bates Littlehales BR; **Geo:** Ca, CRb. **159 Corbis:** Gary Braasch B (main); **RHPL:** Calvin W Hall Ca; **NG:** James C Richardson TRb; **Gary Rosenquist:** B (inset); **SI:** TCR; **US Geological Survey:** TCL. **160 Corbis:** BL; Roger Ressmeyer Ca, CL, BCL, BR; **Getty Images:** James L. Amos / Corbis Documentary CRa. **161 Corbis:** Nik Wheeler BLa; Sygma / Chloe Harford BR; **NG:** Sarah Leen Ca, CL; **Rex Features:** Sipa Press CRb; **SI:** Olger Aragon BC; William I Rose, Michigan Tech University, Department of Geological and Mining Engineering and Science CRa. **162 Corbis:** Galen Rowell BL; Jeremy Horner C; Roger Ressmeyer CRa; **Rex Features:** Sipa Press TCb; **SI:** Oscar Gonzalez-Ferran, Departmento de Geolgia y Geofisica, Universidad de Chile CRb; **US Geological Survey:** CbL. **163 Corbis:** Ric Ergenbright TRb; Roger Ressmeyer CRa; **Magnum:** Patrick Zachmann BL; **Geo:** BR, TCL; **Mats Wibe-Lund:** CL. **164/165 Press Association Images:** AP Photo / Francisco Negroni. **166 Agence France Press:** BC; **Corbis:** Roger Ressmeyer CL; **Getty:** Image Bank / Guido Alberto Rossi TCL; **Tom Pfeiffer:** Ca (inset). **167 Still:** Otto Hahn. **168/169 Dreamstime.com:** Absente. **169 Corbis:** Michael S Yamashita CRa; Ric Ergenbright CbL; Roger Ressmeyer CRb. **170 Alamy Stock Photo:** Avalon.red CLb; **Corbis:** Bettmann Ca; Roger Ressmeyer CR, BR, TR; **NG:** Sisse Brimberg CRb; **N.H.P.A.:** Kevin Schafer CLb. **171 Corbis:** Ecoscene / Chinch Gryniewicz CRb; Torleif Svensson B; Yann Arthus-Bertrand CLa; **Magnum:** Bruno Barbey TR, CRa; **NG:** George F Mobley CRb. **172 alamy.com:** RHPL BL; **Corbis:** Peter Turnley TC; **Katz/FSP:** Gamma CLa; **NG:** Chris Johns CRb; Medford Taylor BR; **SI:** Krafft Collection CRa. **173 Shutterstock.com:** Emre Akkoyun BR; **Corbis:** Yann Arthur-Bertrand CLa; **NG:** Steve Raymer CRa; **OSF:** NASA BR; **SI:** Alexander Belousov BC. **174 Shutterstock.com:** Shuttertong CR; **Corbis:** Michael S Yamashita BC; **Magnum:** Chris Steele-Perkins CL; **SI:** Liu Xiang, Changchun University of Science and Technology, Department of Geology BL; **US Geological Survey:** Jack Lockwood BR. **175 Corbis:** Roger Ressmeyer CLa; Saba / Alberto Garcia C; **Katz/FSP:** Van Cappellen / Rea TCR; **SI:** Chris Newhall, Department of Earth and Space Sciences, University of Washington BR; Rizal Dasoeki / Volcanological Survey of Indonesia Cb. **176 Corbis:** Dean Conger TC; Roger Ressmeyer CLa; **Getty Images:** DigitalGlobe TC; **NG:** Paul Chesley CR. **Popperfoto:** CRa; **SI:** Krafft Collection BL; **Still:** Alain Compost BR. **177 Getty Images:** Doug Allan / Photodisc B; **Getty Images / iStock:** Tammy616 / E+ TC; **naturepl.com:** Tui De Roy TR. **Corbis:** Galen Rowell CRb; **Hulton Archive/Getty:** Cb; **OSF:** Doug Allan B; Tammy Peluso TCL; Tui de Roy TR. **178 Geo:** Cb, CLb; **Martin Miller:** Department of Geological Sciences, University of Oregon Ca; **SPL:** Alfred Pasieka BC. **178/179 PJ Fleisher:** C. **179 Martin Miller:** Department of Geological Sciences, University of Oregon CbR; **PJ Fleisher:** TR; **Dr Parvinder S Sethi:** Cb; **Geo:** BCL. **180 Jens C Andersen:** Camborne School of Mines CLa; **Corbis:** Charles E Rotkin CRb; Kelly Mooney Photography BL; **Kurt Hollocher:** Union College TCR; **Geological Survey of Canada:** Dr Robert H Rainbird CRa. **181 Corbis:** Bill Ross BC; Galen Rowell TCR; Phil Shermeister CLa; Robert Holmes TR. **182 AirPhoto:** Jim Wark TL; **Corbis:** Charles O'Rear BRa; David Muench CRa; Jonathan Blair C; **Still:** Peter Arnold Inc / Robert MacKinlay BL. **183 alamy.com:** Neil Cameron CRa; RHPL CLa; **Art Directors & TRIP:** T Bognar BC; **Geo:** Tim Fogg BRa; **Dreamstime.com:** Stephen Young BL. **184 Corbis:** Yann Arthus Bertrand BC, BL; **DK:** Joe Cornish CLa; **Art Directors & TRIP:** B Gadsby TCL; **METI / ERSDAC:** TR. **185 Getty:** Photodisc Green CL, C; **Dreamstime.com:** James Mattil TR; **Geo:** BRa; **National Park Service:** BC. **186 Getty:** Image Bank / Michael Melford BR; **Getty Images / iStock:** MichaelJust Ca; **Corbis:** Jeff Vanuga CR; Kevin R Morris BLa; **Still:** Peter Arnold Inc / John Keiffer TC; **WWI:** Mark Hamblin Ca. **186/187 Corbis:** Raymond Gehman C. **Still:** Peter Arnold Inc / John Keiffer TC; **WWI:** Mark Hamblin Ca. **188 Corbis:** Charles O'Rear BC; Dave G Houser BL; Nik Wheeler BR; **Getty:** Stone / Jack Dykinga Ca; **NG:** E.C. Kolb CLa. **189 Alamy Stock Photo:** Compagnon Bruno / Sagaphoto.com CR; **Corbis:** Macduff Everton TR; **SPL:** Klaus Guldbrandsen BC; **Geo:** BR. **190/191 Getty Images:** National Geographic / Carsten Peter. **192 Corbis:** Eye Ubiquitous / Mike Powles CaL; Michael Freeman CRa; Michael S Yamashita BL; Robert Holmes Cb, BR. **193 Getty:** Stone / John Lamb C; Taxi / Travel Pix BL; **Geo:** CRb, BRa; **Waimangu Volcanic Valley Ltd, New Zealand:** TR, TLb. **194 Corbis:** Roger Ressmeyer. **195 Corbis:** Bob Krist Cb. **196/197 Corbis:** Yann Arthus-Bertrand. **198 Getty:** Stone / Hans Strand BL. **198/199 Dreamstime.com:** Scosens C. **199 Getty:** Image Bank / China Tourism Press BL; **RHPL:** Simon Harris C; **Still:** Peter Arnold Inc / Aldo Brando CRb; **Louise Thomas:** CL; **Geo:** BR. **200 © 2002 Environment Agency, National Centre for Environmental Data & Surveillance:** CR; **NG:** Paul Nicklen TC; Philip Schermeister TR. **200/201 Getty:** Image Bank / China Tourism Press B. **201 Corbis:** Nathan Benn CR; **Getty:** Image Bank / NASA CRb; **Still:** Peter Arnold Inc / Jim Wark CL; **Visible Earth:** NASA / GSFC / JPL, MISR Team BC. **202/203 Corbis:** NASA. **204 Corbis:** Charles E Rotkin BL; Lowell Georgia TR; Paul A Souders CLa; **RHPL:** Maurice Joseph CRb; **FLPA:** Steve McCutcheon TCL. **205 AirPhoto:** Jim Wark TR; **Corbis:** Richard Hamilton Smith Ca; **Still:** Peter Arnold Inc / Alex S MacLean B; **Telegraph Herald:** TCL. **206 AirPhoto:** Jim Wark TC. **Corbis:** Annie Griffiths Belt C; Buddy Mays BL; **Russ Finley Stock Photography:** BLa. **207 Aurora & Quanta Productions Inc.:** Peter Essick. **208 Getty:** Image Bank / Jack Dykinga TCL; Image Bank / Michael Melford TR; Stone / Donald Nausbaum B; **Landsat:** Cb. **209 Corbis:** Bettmann BL; James Marshall CR; **DK:** Cyril Laubscher TC; **Getty Images:** Elizabeth Fernandez / Moment TC. **210 Corbis:** Yann Arthus-Bertrand BL, TCL; **Getty:** Stone / Will & Deni McIntyre TCLb; **South American Pictures:** Tony Morrison BLa, BCR. **210/211 Getty Images:** Mark Fox / Moment. **212 Corbis:** Adam Woolfitt CRb; Eye Ubiquitous / Tim Hawkins CaL; **Getty:** Image Bank / Daniel Barbier CRa; Taxi / David Noton BL. **213 Corbis:** Papilio / Pat Jerrold CbL; **Getty:** Image Bank / Werner Dieterich BL; **RHPL:** Michael Busselle Ca; **Magnum:** Thomas Hoepker C; **P A Photos:** European Press Agency CL. **214 Corbis:** Joe McDonald BC; Lester Lefkowitz CL. **215 Still:** Jim Wark. **216 Agence France Presse:** Pavel Neubauer CLa; **Getty:** Stone / Simeone Huber TR; **Landsat:** CRb; **Russia and Eastern Images:** Mark Wadlow BL. **217 DK:** Peter Hiscock C; **Getty:** Image Bank / NASA B; **Images Of Africa Photobank:** David Keith Jones TR; **NG:** David S Boyer TCL; **Still:** Chris Caldicott CR. **218 Aurora & Quanta Productions Inc.:** Robert Caputo / IPN B, TCL; **Corbis:** Dave G Houser TR. **219 Corbis:** CL; Ed Kashi BR; Galen Rowell CR; Nik Wheeler BL; **NG:** Chris Johns TR. **220 Corbis:** Dean Conger B; How-Man Wong TC; Tom Nebbia TRb; **Still:** Dera CL; Mark Carwardine C. **221 Corbis:** Liu Liqun BL; Michael Freeman CR; Michael S Yamashita BR; Photowood Inc TCL; **RHPL:** Gina Corrigan CL. **222 Getty:** Taxi / Gavin Hellier B; **RHPL:** CLa; **NASA:** Johnson Space Center (STS087-707-092) CR; **Still:** Andre Maslennikov C; K McCullough TR. **223 Australian Portraits:** David Hancock BR; **Corbis:** Rik Ergenbright CL; **DK:** Cyril Laubscher BCL; The British Museum, London / Peter Hayman TRb; **RHPL:** TC. **224 Getty:** Image Bank / Art Wolfe B; **NG:** Emory Kristof C; **Geo:** RC; **SC Porter:** University of Washington CR. **225 DK:** Dave King TR; **Mike Grandmaison:** BR; **RHPL:** Paolo Koch Ca; **FLPA:** GT Andlewartna BL. **226 Corbis:** Visions of America / Joseph Sohm CbL; **Getty:** Stone / Vito Palmisano BL; **NG:** Medford Taylor TC; Raymond K Gehman CbR. **227 NG:** Medford Taylor. **228 alamy.com:** Masrdis CaL; **Corbis:** Scott T Smith BR; **RHPL:** S Grandadam CaR; **Leland Howard:** BL. **229 Corbis:** Craig Lovell CR; Kevin Schafer TR; **Alamy Stock Photo:** carlos sanchez pereyra B; **South American Pictures:** Tony Morrison CLa. **230 Still:** Glen Christian. **231 Corbis:** Ian Harwood / Ecoscene CbL; **Greenpeace:** Shailendra Yashwant Cb; **Still:** Gilles Corniere TCbR. **232 DK:** Paul Harris BC; Steve Shott CLa; **FLPA:** F Ardito / Panda TR; **Art Directors & TRIP:** O Semenenko BL; **Dreamstime.com:** Wirestock CRb. **233 Corbis:** Blaine Harrington III TRb; Bo Zaunders B; **Getty:** Image Bank / Macduff Everton CR; Image Bank / Werner Dieterich TCL. **234 Corbis:** Bernard and Catherine Desjeux TC; Jonathan Blair BR; **Images Of Africa Photobank:** David Keith Jones CR, BL; **Getty Images:** Orjan F. Ellingvag / Dagens Naringsliv / Corbis TR. **235 Corbis:** Caroline Penn Ca; **Getty:** Image Bank / Guido Alberto Rossi BL; **Images Of Africa Photobank:** David Keith Jones CL; Friedrich von Hörsten BR. **236 Corbis:** Galen Rowell BC; Hulton-Deutsch Collection CL; **DK:** Jerry Young TC; **NG:** Barry Tessman BR; Sarah Leen BL; **Alamy Stock Photo:** Image Professionals GmbH C. **237 Corbis:** David Turnley TCL, TRb; Yann Arthur-Bertrand B. **238 Aurora & Quanta Productions Inc.:** Evan Roberts BR; **Corbis:** Diego Lezama Orezzoli TR; Mark Garanger TCLb; NASA CLb; **Getty Images / iStock:** Valerii Evlakhov C. **239 Corbis:** Galen Rowell BL; Ted Spiegel BR; **RHPL:** CCD Tokeley Cb.; **Getty Images / iStock:** Mlenny / E+ CRa; **NG:** Reza TR. **240 Geo:** CLb, TRb. **240/241 Tom Till Photography:** B; **241 Paul Deakin:** TCb; **Geo:** CRa, CR, CbR; **Jenolan Caves Reserve Trust:** TCL. **242 Corbis:** Craig Lovell TR; David Muench CR; **Russ Finley Stock Photography:** TCL; **Alamy Stock Photo:** Avalon.red CLa; **Terra Galleria Photography:** T Luong B. **243 Aurora & Quanta Productions Inc.:** Stephen Alvarez TC; **Corbis:** Bojan Brecelj CR; **NG:** Wes Skiles BL; **Arthur Palmer:** CaL; **Geo:** BR. **244/245 Getty Images:** National Geographic / Speleoresearch & Films / Carsten Peter. **246 Caving Club of Touraine:** Jean-Luc Roch BRa; **RHPL:** CaL. **NG:** Sisse Brimberg CRa; **Geo:** CLa, BL, BCR. **247 Cango Caves:** Steve Mouton BL, CbL; **Geo:** TL, BR, CLa; **Alamy Stock Photo:** Eric Nathan BL. **248 Corbis:** Robert Holmes CR; **Getty Images / iStock:** Kaszojad Ca, B; **OSF:** Clive Bromhall TCL; **Geo:** TR, B. **249 Geo:** CLa, CRa; **Jenolan Caves Reserve Trust:** CRb, BR; **Koonalda Caves:** Peter Bell BL. **250/251 SPL:** Bernhard Edmaier. **252 Corbis:** David Muench BCL; Marc Muench Ca; **RHPL:** Andrew Sanders BL. **252/253 Corbis:** Tom Bean C. **253 Corbis:** Phil Schermeister BR; **Getty Images:** Fabrice Coffrini / AFP CB; Image Bank / Darrell Gulin TR; **RHPL:** Tony Waltham CR. **254 AirPhoto:** Jim Wark B; **Getty Images:** Michele D'Amico supersky77 / Moment CRa; **Geo:** CL, CaR; **Tom Lowell:** University of Cincinnati CaL. **255 AirPhoto:** Jim Wark CRa; **Geo:** TR, CLb, Cb, TCL. **256 Corbis:** Tom Bean CRa, BR; **RE Johnson Photography:** CRb; **Dan Stone:** CLa; **US Geological Survey:** Rod March BL; **Trapridge Glacier photo archive:** CLa. **257 Anthony Arendt:** University of Alaska CRa; **Jamie Buscher:** Department of Geological Sciences, Virginia Tech TC; **Corbis:** Neil Rabinowitz CLa; **Still:** Peter Arnold Inc / Jim Wark B. **258/259 Getty Images / iStock:** Matt Palmer. **260 Alamy Images:** Niebrugge Images TR; **NASA:** Earth Observatory TR; **Corbis:** Danny Lehman TR; David Muench CLa; David Samuel Robbins CRa; Jim Zuckerman BL; The Purcell Team BR. **261 Corbis:** Tom Bean CLa; **NASA:** Goddard Space Flight Center / USGS EDC B; **Still:** Peter Arnold Inc / Jim Wark TR. **262 B&C. 263 Jason E Box:** Polar Research Center, Ohio State University CRa; **Corbis:** Staffan Widstrand BC; **Geo:** CRb. **264 Josh Beck:** BL; **Peter Burden:** CLa; Simon Heyes, **Last Frontier Ltd:** BCL; **NASA:** Goddard Space Flight Center BR; **Still:** Peter Arnold Inc / SJ Krasemann CRa; **Nozomu Takeuchi:** Research Institute for Humanity and Nature (RIHN), Kyoto TCbR. **265 Corbis:** Eye Ubiquitous / James Davies CLb; Yann Arthus-Bertrand TR; **NG:** Steve Winter BL; **Freysteinn Sigmundsson:** Nordic Volcanological Institute, Reykjavik BR; **Still:** Alan Watson TL. **266 Kim Holmén:** Department of Meteology, Stockholm University Ca; **Dr Miriam Jackson:** Norweigan Water Resources & Energy Directorate Hydrology Department BR; **Dr John Wood:** Department of Geography and Earth Sciences, Brunel University TR; **Jøran Zahl Marken:** CbR. **267 Corbis:** Marc Garanger CaR; Julia K Davidson, www.flickr.com/photos/juliakathleendavidson: BL; **Still:** Peter Arnold Inc / Bill O'Connor BL; Peter Arnold Inc / Gordon Wiltsie CaL; Peter Arnold Inc / Helmut Gritscher BR. **268 Corbis:** NASA BC; **Jamie McPherson:** BL; **Klaus E Schwartz:** BR; **Still:** Theresa de Salis TCR, CLa; **Dreamstime.com:** Miroslav Liska TC; **Getty Images / iStock:** unikatdesign BL. **269 Getty Images / iStock:** Ghulam Hussain TL; **Corbis:** Galen Rowell BL. **Thomas E Dietz:** International Society for Mountain Medicine CR; **NG:** Bobby Model TL. **270 Science Photo Library:** British Antarctic Survey BC; **U.S. Geological Survey:** Courtesy of GNP Archives TC; Karen Holzer CRa; **Alamy Stock Photo:** DGP_travel BC; **Benjamin Stefanko:** CA. **271 Corbis:** Frans Lanting. **272 Alamy Stock Photo:** Ashley Cooper pics B; **Corbis:** Yann Arthus-Bertrand TCL. **272/273 Getty:** Image Bank / Joseph van Os B. **273 Corbis:** Galen Rowell TCR; Hulton-Deutsch Collection TCL; **SPL:** JG Paren CRa. **274 alamy.com:** Stock Connection Inc / James Kay BC; **Getty Images / iStock:** Ernest Kung / E+ BL; **Robert Dinwiddie. Corbis:** Paul A Souders TR. Tom Lowell: University of Cincinnati BR; **Still:** Peter Arnold Inc / Walter H Hodge CLb. **275 David Summers. 276/277 Corbis:** Yann Arthus-Bertrand. **278 Corbis:** Kevin Schafer CRa; Scott T Smith CLb; **DK:** Linda Whitwarm Cb; **Getty:** PhotoDisc BR. **278/279 alamy.com:** Geoffrey Morgan C. **279 Corbis:** Ric Ergenbright CRa; **Getty:** Taxi / Josef Beck CRb; **Geo:** CLa, TR. **280 Corbis:** CLa; David Muench TC; Gallo Images / Hein von Horsten C; Jeremy Horner CRa; **DK:** Alistair Duncan BC; Frank Greenaway BL; **NG:** Des and Jen Bartlett Cb. **280/281 Science Photo Library:** C.K. Lorenz BC. **281 Corbis:** Ecoscene / Andrew Brown TR; George HH Huey CLa; **DK:** Geoff Dann TC. **282 DK:** Dave King CRb; **Getty:** Stone / RGK Photography BR; Taxi / Alan Kearney BL; Jim Harding: CLa. **282/283 Scott T Smith:** T; **283 Agence France Presse:** John Gurzinski / STR CRa; **Corbis:** B.S.P.I. BL; Marko Modic BR; **DK:** Frank Greenaway Cb. **284 D Donne Bryant Stock Picture Agency:** LLC / David Ryan B; **Corbis:** Charles O'Rear CLa, CRa. **SPL:** CNES, Distribution Spot Image TCL. **285 Getty Images / iStock:** reptiles4all BC; **Royal Geographical Society Picture Library:** BR; **Still:** Michael Gunther T. **286/287 Getty Images:** Daniel J Bryant. **288/289 Alamy Stock Photo:** imageBROKER.com GmbH & Co. KG. **289 DK:** Andy Crawford CRb; Frank Greenaway Ca; **Getty:** Image Bank / Jean du Boisberranger BR; **NG:** Michael S Lewis CLb; **Still:** Cyril Ruoso CRa; **© M.P.F.T.:** TR. **290 Magnum:** Steve McCurry BLA. **290-291 Alamy Stock Photo:** Eric Nathan. **292 Corbis:** Gallo Images / Peter Lillie Ca; Michael & Patricia Fogden CbL; Papilio / Robert Gill BL; **Gerald Cubitt:** BR; **DK:** Frank Greenaway TR. **293 DK:** Colin Walton TCL; **RHPL:** Gina Corrigan BR; **Russia and Eastern Images:** Mark Wadlow CRa; **Wax Visual Stock Photography:** BL. **294 Nature Picture Library:** Gertrud Neumann-Denzau BC; **Corbis:** Dean Conger CRa; Steve Kaufman CRb; **Geo:** CR. **294/295 Alamy Stock Photo:** Imauritius images GmbH. **296 Corbis:** Australian Picture Library BCL; Eric & David Hosking CLa; Gallo Images / 318 TR; **296 Getty Images:** TTed Mead / Stone CR, BR CRa, CRb. **297 Corbis:** Gallo Images / 318 Cb; Michael & Patricia Fogden BL; **DK:** Cyril Laubscher CRa; **Getty Images:** Ted Mead / Stone TL, BR. **298/299 Magnum:** Stuart Franklin. **300 DK:** Matthew Ward CR; Neil Fletcher & Matthew Ward TR, CRa; Steve Gorton C; **FLPA:** David Hosking CL. **Still:** Roland Seitre BCR. **300/301 Natural Image:** Bob Gibbons C; **Alamy Stock Photo:** Herv Lenain B. **301 DK:** Peter Chadwick CLa; **FLPA:** B Borrel Casals Ca; Chris Mattison CaL; Fritz Polking CRa; **Still:** Frank Vidal TR. **302 Corbis:** Ecoscence / Andrew Brown CL; **Getty:** Stone / Jane Gifford CR; **Alamy Stock Photo:** Helen Dixon B; **Rex Features:** Sipa Press LC; **World Wildlife Fund:** MedPo / Pedro Regato RC. **303 Corbis:** Jonathan Blair BCL; Raymond Gehman BL, BC; **DK:** Eric Crichton Ca; Neil Fletcher TCb; Peter Anderson CRa; **FLPA:** M Moffett / Minden Pictures CLa; Martin B Withers CaR; **NASA:** CRa; **Art Directors & TRIP:** W Jacobs Cb. **304 Corbis:** Gunter Marx Photography CL, TCR; **FLPA:** Mark Newman B; **OSF:** Lon E Lauber TCL; **Getty Images:** Henrik Karlsson / Moment CRa. **305 Corbis:** Bob Rowan / Progressive Image CRb; Galen Rowell TRb; Marc Muench BL; **Still:** Art Wolfe Ca; David Drain TR; **Art Directors & TRIP:** M Barlow CR; **Alamy Stock Photo:** Lee Rentz TR. **306 Corbis:** David Muench TCL; Mark Muench CLa; **DK:** Jerry Young BL; **Getty Images / iStock:** Oxford Scientific / The Image Bank CR, BR; **OSF:** BR; **Alamy Stock Photo:** Bert de Ruiter BC. **307 Corbis:** Brian Vikander CL; **DK:** Jerry Young TR; Pau D'Arco CR; **South American Pictures:** Tony Morrison C; **Still:** Juan Carlos Munoz TCL; **Alamy Stock Photo:** Sue Cunningham Photographic B. **308 Corbis:** Macduff Everton CLb; Wayne Bennett BC. **308/309 Getty Images:** Matt Hoover Photo / Image Source. **310 FLPA:** Roger Tidman BCRa; T de Roy / Minden Pictures Ca; **Getty Images / iStock:** Alberto Carrera BR; Sue Cunningham Photographic BL; **South American Pictures:** Tony Morrison TL; **Still:** Alan Watson CRa. **311 Corbis:** Adrian Arbib T; Michael Busselle BL; **DK:** Neil Fletcher BC; **Getty:** Stone / Stephen Studd CL; **Alamy Stock Photo:** Angela Hampton Picture Library BR; **Rex Features:** Isopress CR. **312 NG:** Maria Stenzel BL; **Dorling Kindersley:** Cotswold Wildlife Park & Gardens, Oxfordshire, UK Cb; **Russia and Eastern Images:** Mark Wadlow Ca; **Still:** Patrick Bertrand TC; Roland Seitre R. **313 Corbis:** Keren Su CR; **Gerald Cubitt:** BL; **DK:** Kim Taylor CL; **Nigel Hicks:** BR; **Alamy Stock Photo:** imageBROKER.com GmbH & Co. KG CRa; **Still:** Michael Gunther CLa. **314 FLPA:** Frans Lanting. **315 Associated Press AP:** Yenni Kwok, Stringer BR; **Getty:** Stone / Manoj Shah TR; **RHPL:** Louise Murray CbL; **SPL:** Geoff Tompkinson CbR. **316 Corbis:** Hulton-Deutsch Collection BC; Paul A Souders CRb; Peter Johnson BR; **DK:** NHM / Frank Greenaway TCR; **FLPA:** Terry Whittaker CRa; Tim Rushforth BL. **Still:** Mark Edwards TR; **World Wildlife Fund:** Steve Nelson & Zovtaigi c/o World Wise Ecotourism Network CLa. **317 Corbis:** FLPA / Pam Gardner C; Paul A Souders L; Yann Arthus-Bertrand Cb; **N.H.P.A.:** Dave Watts TR; **Skyrail, Cairns, Australia:** BR; **Alamy Stock Photo:** Uwe Bergwitz TR. **318/319 WWI:** Ted Mead, **Getty Images:** Ted Mead / Photodisc. **320 Corbis:** David Muench BR; Ecoscene / Ian Harwood CbL; Nik Wheeler CL; **NG:** courtesy Moesgård Museum, Arhus, Denmark / Ira Block TRb. **321 alamy.com:** Steve Bloom Images / Steve Bloom B; **Corbis:** FLPA / Fritz Polking CaL; **DK:** Frank Greenaway TLb; Geoff Dann TC; Kim Taylor & Jane Burton TR; Philip Dowell CaR. **322 Corbis:** Patrick Johns Cb; Raymond Gehman CaR, BL; **DK:** Karl Stone BR; **Dave Liebman:** TR; **Virginia Museum of Natural History:** Susan B Felkner TL. **323 AirPhoto:** Jim Wark CbR; **Corbis:** David Muench CLb; Kevin Fleming BRa; Tony Arruza BCL; **Shutterstock.com:** Jason Heid T. **324 Corbis:** Tom Brakefield CLa; **FLPA:** Jurgen & Christine Johns CRb; **OSF:** Chris Catton CRa; **Still:** Gunter Ziesler TCL; Roland Seitre TR; **Getty Images / iStock:** JSabel BR; **Getty Images:** Nat Photos / Photodisc BL. **325 Getty:** Image Bank / Art Wolfe BL; **FLPA:** Terry Andrewartha TCL; **Still:** Nigel Dickinson BRa; Peter Arnold Inc / Gunter Ziesler CRa; **Dreamstime.com:** Artushfoto TR; Danolsen CB; **Getty Images / iStock:** Bernard Bialorucki CL. **326 Corbis:** Wolfgang Kaehler TCR; Yann Arthus-Bertrand Ca; **FLPA:** Minden Pictures / Frans Lanting B; **Still:** Paul Springett CRa. **327 Aurora & Quanta Productions Inc.:** Robert Caputo / IPN CL; **Australian Portraits:** David Hancock BR; **Corbis:** Gallo Images / 318 TC; **DK:** Andrew Butler CR; Exmoor Zoo / Peter Cross Cb; **NewsPhotos:** Chris Crerar BL. **328/329 Still:** Fritz Polking. **330 Corbis:** Michael S Yamashita BL; Peter Johnson CbR; Tom Bean Cb; **OSF:** Mills Tandy CbL; **Getty Images / iStock:** AlesHostnik BC; **South American Pictures:** Tony Morrison BR, CLb. **330/331 Corbis:** Tom Bean B. **331 Corbis:** D Robert & Lorri Franz Cb; Staffan Widstrand CLb; W Perry Conway TR; **DK:** Frank Greenaway CbR; Jerry Young BRa; NHM / Frank Greenaway CRb. **332 Corbis:** David Muench T; Joe McDonald Cb; Tom Bean BR; **FLPA:** Daphne Kinzler BCL; **OSF:** Scott Camazine BL; **Still:** Roland Seitre CRb. **333 Corbis:** CRb; Macduff Everton CbR; **DK:** Jerry Young CLa; **South American Pictures:** Mike Harding B; **Still:** Roland Seitre Ca. **334 Corbis:** Sygma TRb; Wolfgang Kaehler TC; **Art Directors & TRIP:** J Farmar CLa. **334/335 Corbis:** Brian A Vikander B. **335 DK:** Mike Dunning TCR; **Images Of Africa Photobank:** David Keith Jones CRa; **FLPA:** Terry Andrewartha Ca. **336 DK:** Jerry Young TRb; **Russia and Eastern Images:** Mark Wadlow CaR; **Still:** Adrian Arbib B; Stephen Pern CLa. **337 Corbis:** O Alamany & E Vicens BR; **DK:** Geoff Dann BCL; **FLPA:** E&D Hosking C; **Alamy Stock Photo:** Papilio TC; **NG:** Thad Samuels Abell II Cb; **Still:** Mark Edwards TR. **338 AirPhoto:** Jim Wark B; **Geo:** CR, CaR, C. **339 Alamy Stock Photo:** Arterra Picture Library TR; **Getty Images:** Iri_sha BR; **B&C:** CbL; **Corbis:** Tom Brakefield CL; **Hutchison Library:** Andrey Zvoznikov CRb; **NG:** Rich Reid C. **340/341 Corbis:** Keren Su. **342 Corbis:** Sylvain Saustier CRa; W Wayne Lockwood M.D. BL; **NG:** Maria Stenzell CL. **342/343 Corbis:** Charles. **343 Corbis:** Kennan Ward CL; Pallava Bagla TRb; Yann Arthus-Bertrand BL, Ca; **Getty:** Stone / Bruce Forster Cb. **344**

528 ACKNOWLEDGMENTS

ACKNOWLEDGMENTS

Corbis: Joe McDonald CL; Peter Johnson TL; **NG:** Jodi Cobb BCL; Maria Stenzel TR; **Still:** Nigel Dickinson BR; **Geo:** CR. **345 Corbis:** Fulvio Roiter CLa; Macduff Everton CRa; Michael Freeman TR; Richard Hamilton Smith TCLb; **Images Of Africa Photobank:** Ivor Migdoll BCR; **NG:** Michael Nicols Cb; **Still:** UNEP / C.K. Au BR. **346/347 Corbis:** NASA. **348 Science Photo Library:** Jim West CL; **NG:** William Albert Allard CRa; **SPL:** Francois Sauze CRb; Gordon Garradd TR; **Geo:** BR. **349 Corbis:** David Muench BL; Dean Conger Cb; Jonathan Blair TCL; Morton Beebe, SF TR; Pablo Corral BR; **Images Of Africa Photobank:** Friedrich von Hörsten Ca; **SPL:** Kaj R Svensson CRb. **350 Corbis:** Lanz von Horsten / Gallo Images TC; Photo courtesy of **Deere & Company, Moline, Illinois, USA:** CLb; **DK:** Andrew McRobb *wheat*; Frank Greenaway *rye*; Neil Fletcher & Watthew Ward *rice*; Steve Gorton *barley*, *maize*; **Getty:** Taxi / Walter Bibikow B; **NG:** Kenneth Garrett CRb; **Rex Features:** Ray Tang TRb. **351 alamy.com:** Cephas Picture Library / Mick Rock BL; **Corbis:** Dennis Degnan Cb; Raymond Gehman CLa; Tony Arruza TR; **NG:** James L Stanfield BR. **352/353 NG:** Steve Raymer **354 Corbis:** Michel Setboun BL; Tom Nebbia TR; **NASA:** Goddard Flight Center / C Mayhew & R Simmon CLb. **354/355 Getty Images:** Siegfried Layda / The Image Bank BC; **355 Corbis:** Sygma / Kontos Yannis TC; **Still:** Mark Edwards TR; **Art Directors & TRIP:** ASK Images CRb. **356 Corbis:** Bob Krist TR; Michael S Yamashita CRa. **356/357 Magnum:** Hiroji Kubota B. **357 Corbis:** Owen Franken CRa; **RHPL:** Ca; **Magnum:** Dennis Stock TL. **358/359 Getty Images:** National Geographic / Jim Richardson. **360 Corbis:** Alan Schein Photography Inc CLa; Bettmann TRb; Kelley Mooney Photography TCR; Lloyd Cluff CRb; Sandy Felsenthal Ca; **DK:** Neil Setchfield BR; **Getty:** Taxi / Walter Bibikow BL. **361 Corbis:** Danny Lehman TC; Paul Almasy CRb; Stephanie Maze CbR; Yann Arthus-Bertrand B; **Art Directors & TRIP:** ASK Images CRa. **362 Rex Features:** John Sutcliffe BL. **363 Getty Images:** China Photos / Stringer. **364 Corbis:** Bojan Brecelj CaR; David Turnley BR; John & Dallas Heaton BL; Yann Arthus-Bertrand CLa; **Getty:** Image Bank / Jorg Greuel TR; Stone / Chad Ehlers CaL; Stone / John Lamb CRb. **365 Corbis:** Craig Aurness BR; Owen Franken BL; WildCountry CLa, Ca; **Getty:** Image Bank / Pete Seaward Photography Cb; **Getty Images:** Greg Fonne CLa; **Art Directors & TRIP:** H Rogers TR. **366 Corbis:** Archivo Iconográfico S. A TCR; Nik Wheeler CRa; Yann Arthus-Bertrand BL; **Magnum:** Steve McCurry CLa; **Panos Pictures:** Mark Henley Cb; **Art Directors & TRIP:** A Tovy Ca; J Highet BR. **367 alamy.com:** Peter Bowater BL; **Alamy Stock Photo:** Joy Saha / Pacific Press / Sipa USA CRb; **RHPL:** Robert Harding CRa; **Images Of Africa Photobank:** David Keith Jones TCR; **Dreamstime.com:** Quazi Md Hasibul Hasan BR; **Art Directors & TRIP:** M Barlow CLa. **368 Corbis:** Bettmann TRb; Lindsay Hebberd Ca; Ted Spiegel TCR; **Getty:** Stone / Markus Amon CRa; **NG:** Steve McCurry BL; **Rex Features:** Sipa Press BRa, CRb. **369 Corbis:** Jose Fuste Raga CLa; Liu Liqun BL; Roger Ressmeyer CRa; Tibor Bognár TC; **RHPL:** P Scholey BR; R McLeod TR; **Still:** Andy Crump CRb. **370 Getty Images / iStock:** journey2008 Ca; **Corbis:** Bob Krist CRa; **Corbis:** Keren Su BCL; Travel Ink / Derek M Allan CLa; **Panos Pictures:** Jeremy Horner CbR; **Art Directors & TRIP:** P Treanor BR; **Dreamstime.com:** Darren Patterson TR. **371** Ed Wheeler TL; **Corbis:** Penny Tweedie BR; Sergio Dorantes BL; **RHPL:** Gavin Hellier TR. **372/373 Getty Images:** VCG. **374 Corbis:** Bohemian Nomad Picturemakers / Kevin R Morris BCR; Manfred Vollmer CRb; Michael S Yamashita CRa; **Getty:** Taxi / Benelux Press BL. **375 Corbis:** David H Wells CLb; Derek Croucher CLa; Kevin Schafer TR; Roger Ressmeyer CRa; Stock Photos TCL; **Magnum:** Gueorgui Pinkhassov BL; **Still:** Thomas Rampack BR. **376 Corbis:** Galen Rowell CR; Jim Richardson B; **Still:** Harmut Schwarzbach CL; Thierry Montford TC; **Art Directors & TRIP:** A Kuznetsov TR. **377 Corbis:** Gunter Marx CLa; Paul Hardy BL; Yann Arthus-Bertrand Ca; **DK:** Harry Taylor TC; **Getty:** Image Bank / Andy Caulfield CRb; Stone / Paul Chesley BR; **Still:** Nigel Dickinson TR. **378 Corbis:** Christopher Morris BL; Michael S Yamashita CL; Natalie Fobes CR; Owen Franken BC; Richard Hamilton Smith TCR; Saba / Keith Dannemiller Cb; Sygma / Bisson Bernard BR; **Magnum:** Patrick Zachmann CLa. **379 Corbis:** London Aerial Photo Library TL; Staffan Widstrand CRb; Tim McKenna BR; **Magnum:** David Hurn TRb; **Getty Images:** Nathan Laine / Bloomberg BL. **380/381 Getty:** Image Bank / Peter Cade. **382/383 Getty:** Stone / Arnulf Husmo. **384 Robert Dinwiddie:** CaR, Ca; **DK:** Andreas Einseidel Cb; Frank Greenaway CRaR; **© 1999 MBARI:** CRa, C. **385 Corbis:** Brian A Vikander TRb. **386 Getty Images:** Ralph White / The Image Bank Unreleased BC; **FLPA:** S Jonasson Ca; **Richard Lutz:** CLb; **Corbis:** Arctic-Images CR. **386/387 NG:** Image from Volcanoes of the Deep, a giant screen motion picture, produced for IMAX theaters by the Stephen Low Company in association with Rutgers University. Major funding for the project is provided by the National Science Foundation C. **387 Corbis:** Reuters / Handout / Japan Coast Guard CR; **NG:** Paul Chesley CL; **Visible Earth:** Provided by the SeaWiFS Project, NASA/Goddard Space Flight Center, and ORBIMAGE CL; NOAA: NSF TR. **388 Robert Dinwiddie:** CL; **SPL:** ER Degginger CaL. **389 Getty Images:** Spencer Platt TCL; **Press Association Images:** AP / Nathan Becker – Pacific Tsunami Warning Center TR. **390/391 Corbis:** Nippon News / Aflo / Mainichi. **392 Alamy Stock Photo:** R. Hughes CRa; **Shutterstock.com:** Neil Bromhall Cb; **Getty Images:** Stephen Frink / Photodisc Ca; **Getty Images / iStock:** KGrif C. **392/393 Corbis:** Jonathan Blair B. **393 SPL:** Eric Grave CaR; Institute of Oceanographic Sciences / NERC CbR; Jan Hinsch CRa; Manfred Cage TCR; Sinclair Stammers TR. **394 Corbis:** Paul A Souders CaR; **SPL:** Geospace CRb. **395 Japanese National Tourist Organization:** T. **396** Jeffrey L Rotman BC; **FLPA:** Minden Pictures / Fred Bavendam BR; Robert Yin Cb; **Southampton Oceanography Centre:** Dr Alex Mustard BL. **396/397 Corbis:** Jose Fuste Raga T; **397 Corbis:** Yann Arthus-Bertrand CbR; **NASA:** Image courtesy of Earth Sciences and Image Analysis Laboratory, Johnson Space Center (STS046-77-31) Ca; **Southampton Oceanography Centre:** Dr Alex Mustard TC; **Corbis:** Bettmann BL. **398 123RF.com:** tasfoto CRb; Paul A Souders BCL; Wolfgang Kaehler BR; **FLPA:** Fritz Polking BCR; **OSF:** Ben Osborne BR; **Alamy Stock Photo:** Rosemary Calvert Ca. **398/399 NG:** Maria Stenzel T. **399 Corbis:** Rick Price BL; Wolfgang Kaehler BR, BC. **400 Corbis:** Jon Hicks TR; Onne van der Wal CLa; Stuart Westmorland Ca; **DK:** CLb; **FLPA:** Silvestris TLa. **401 Depositphotos Inc:** Panther Media Seller TL; **Corbis:** Gianni Dagli Orti TRb; Peter Johnson BRA; Stephen Frink CRa; **DK:** Frank Greenaway TCb; **Getty Images / iStock:** Aiden Conners CR. **402 naturepl.com:** Bryan and Cherry Alexander CLb; **Corbis:** Bettmann TC; **NG:** Emory Kristof BR; Paul Nicklen CRb; **NOAA:** CaR. **403 Getty:** Stone / Hans Strand. **404 alamy.com:** F.R. CLa; **naturepl.com:** Bryan and Cherry Alexander BR; **Corbis:** Staffan Widstrand CRa; **NOAA:** V Juterzenka, Piepenburg, Schmid TCL; **Visible Earth:** Provided by the SeaWiFS Project, NASA/Goddard Space Flight Center, and ORBIMAGE BCL. **405 Biosphoto:** Blue Planet Archive / Bryan & Cherry Alexander CRa; **Corbis:** Bruce Burkhardt CR; Raymond Gehman BL; Wolfgang Kaehler BR; **Visible Earth:** Image courtesy Jacques Descloitres, MODIS Rapid Response Team, NASA/GSFC TCL. **406 Getty Images:** MathieuRivrin / Moment. **407 Corbis:** Neil Rabinowitz Cb; Richard Cummins BR; **DK:** Frank Greenaway CRa; **NG:** Nick Caloyianis CRb; **David Robinson:** TR. **408 Corbis:** Kevin Schafer CLa; Raymond Gehman BL; The Purcell Team CRb; **Hulton Archive/Getty:** Stock Montage / Archive Photos TR; **NG:** James P Blair BR; Norbert Rosing Ca. **409 Corbis:** BR; Darren Gulin CRb; Kevin R Morris BL; Lynda Richardson TR; Stephen Frink TCL; Tony Arruzal Cb; **Henry Genthe:** Ca. **410 Alamy Stock Photo:** Nature Picture Library. **411 Corbis:** Phil Schermeister CRb; Stephen Frink Cb; **Getty:** Taxi / Ron Whitby CRa. **412 Corbis:** TRb; Gavriel Jecan TCL; Macduff Everton B; Stephen Frink CR; **NG:** Wes Skiles CL. **413 Gunnar Britse:** windpowerphotos.com CRb; **Corbis:** Chris Hellier BL; Nik Wheeler BR; Patrick Ward CRa; **Geological Survey of Denmark and Greenland:** Peter K Warna-Moors Cb; **Geo:** CLa, Ca; **Visible Earth:** Provided by the SeaWiFS Project, NASA/Goddard Space Flight Center, and ORBIMAGE TCR. **414 Agence France Presse:** TRb; **Corbis:** Jeffrey L Rotman TCL. **414/415 Getty:** Image Bank / Macduff Everton B. **415 Corbis:** Jonathan Blair TRb; **Magnum:** Stuart Franklin TC; **NASA:** Johnson Space Center (STS41G-17-34-081) TL. **416 Corbis:** Brandon D Cole TR; Charles & Josette Lenars TC. **416/417 Getty Images:** fmajor B. **417 Dreamstime.com:** Theodor Bunica CLa; **42 Degrees South:** Rob Walls TR. **418 Corbis:** Dean Conger BL; Lawson Wood R; **Robert Dinwiddie:** CLb; **Geo:** TC. **419 Corbis:** Ca; Bojan Brecelj CRa; Kevin R Morris CRb; Yann Arthus-Bertrand BR, TCL; **Karen Hissman:** CbL; **Images Of Africa Photobank:** Brian Charlesworth BL. **420 Corbis:** CLb; Sygma / Haruyoshi Yamaguchi BR; Tim McKenna TC; **Robert Dinwiddie:** Cb; **Professor Peter Franks:** University of California, San Diego CRa; . **421 Corbis:** Amos Nachoum. **422 alamy.com:** RHPL CRb; **Corbis:** Annie Griffiths Belt BL; Natalie Fobes Ca, TRb; Ralph A Clevenger CL. **423 Corbis:** BL; Joel W Rogers CRa; Michael S Yamashita CLa; Rick Price TCL; Hosik Kim: TRb; **Visible Earth:** Provided by the SeaWiFS Project, NASA/Goddard Space Flight Center, and ORBIMAGE BR. **424 Corbis:** Catherine Karnow Ca; Jeffrey L Rotman Cb; Nik Wheeler BL; **P A Photos:** European Press Agency BR; **Visible Earth:** Provided by the SeaWiFS Project, NASA/Goddard Space Flight Center, and ORBIMAGE TR; **© 2003 Norbert Wu:** www.norbertwu.com CLa. **425 Corbis:** FLPA / Pam Gardner CRa; Stuart Westmorland BL; Yann Arthus-Bertrand BR; **Jezohare.com:** Ca; **NG:** Paul Chesley CLa. **426 naturepl.com:** Espen Bergersen. **427 Corbis:** Hans Georg Roth CRa; Margaret Courtney-Clarke Cb. **428 Corbis:** Peter Johnson CLa; Yann Arthus-Bertrand TC; **Getty:** Stone / Kim Westerskov B; **SPL:** Simon Fraser CRa. **429 Alamy Stock Photo:** Nature Picture Library Cb. **Corbis:** Galen Rowell TR, CLa; Rick Price TCL; Wolfgang Kaehler CRb, BR; **© 2003 Norbert Wu:** www.norbertwu.com BL. **430/431 Geo. 432 Getty Images:** STR / AFP TR; **Corbis:** Brandon D Cole CRa; **Getty:** Stone / Natalie Fobes Cb; Stone / Philip Long CLb. **432/433 Getty:** Stone / H Richard Johnston C. **433 Corbis:** B.S.P.I. CLa; Barry Davies CaL; Sygma / Frederick Astier Ca; **Shutterstock.com:** PX Media CRa; **Rex Features:** Timepix / G Davies BR. **434** Paul A Souders CaR. **434/435 Getty:** Image Bank / Peter Hendrie B. **435 Corbis:** Catherine Karnow TC; NASA BR; Yann Arthus-Bertrand TR, Ca; **Geo:** BRa. **436 Corbis:** John Heseltine BC; Micahel S Yamashita CbR; **DJ Sauchyn:** Prairie Research Collaborative Ca. **436/437 Corbis:** Dallas and John Heaton C. **437 Corbis:** Carol Havens BC; Ecoscene / John Farmer TRb; Joel W Rogers CRb. **438 Shutterstock.com:** Alec Vorobiov. **439 Corbis:** Gallo Images / Martin Harvey BC; Joe McDonald Cb; **Getty:** Stone / Jeremy Walker CRb. **440/441 Corbis:** William James Warren. **442/443 alamy.com:** IMAGINA / Atsushi Tsunoda. **444 Corbis:** Eye Ubiquitous / Bennett Dean BCL; **Getty:** Image Bank / Kevin Kelley CLa; **NG:** Thad Samuels Abell II CLb. **445 Corbis:** Ron Watts CL; **Tom Ekland:** CRa; **Getty:** Image Bank / Guido Alberto Rossi CR; **P A Photos:** European Press Agency BL. **SPL:** David Parker TCLb; **US Geological Survey:** AM Sarna-Wojcicki BR; **Visible Earth:** Image courtesy Jacques Descloitres, MODIS Rapid Response Team, NASA/GSFC CRb. **446 Corbis:** Grafton Marshall Smith CRa; **Transition Region and Coronal Explorer (TRACE)** – a mission of the Stanford-Lockheed Institute for Space Research, part of the NASA Small Explorer Program CRb. **447 Corbis:** Sygma / Herve Collart CaR; **Getty:** Image Bank / Stephen Frink BCa; Stone / Kim Blaxland TLb; Stone / Kim Westerskov CR; Stone / Mark Muench BR. **448 SPL:** David Hay Jones. **449 DK:** NASA BCa; **Rex Features:** Brian Harris CRb. **450 Bridgeman Art Library, London / New York:** Giraudon. *Portrait of Gustave Gaspard Coriolis* (1792–1843) engraved by Zephirin Felix Jean Marius Belliard (b.1798) (litho), Academie des Sciences, Paris, France BL; **Corbis:** Stocktrek CLb. **451 alamy.com:** Geoffry Morgan CRb; **Corbis:** NASA TR; **National Trust Photographic Library:** Mike Howarth CRa; **Visible Earth:** Image courtesy GOES Project Science Office BCL. **452 Corbis:** Paul A Souders CLb; **FLPA:** H Hoflinger TR; **Visible Earth:** Provided by the SeaWiFS Project, NASA/Goddard Space Flight Center, and ORBIMAGE BRa; Image courtesy Liam Gurnley, MODIS Atmosphere Team, University of Wisconsin – Madison CLa; Image courtesy Jacques Descloitres, MODIS Rapid Response Team, NASA/GSFC BCR. **453 Corbis:** Lowell Georgia TR; Sygma / Telegram Tribune / Jason Mellom CLb; **P A Photos:** European Press Agency BCL. **454 Corbis:** Frank Lane Picture Agency / Ron Boardman CL; **SPL:** BR; Eye of Science CRa. **454/455 Magnum:** Steve McCurry T. **455 Corbis:** BR; Jonathan Blair Cb; **Royal Swedish Academy of Sciences:** TR. **456 Corbis:** In Pictures / Richard Baker Ca; **NASA:** TC, TR; **Press Association Images:** AP TL. **456/457 Corbis:** Paul Souders B. **458 Corbis:** Hans Strand. **457 Alamy Images:** David Lock CRb; **Corbis:** Jim Reed CRa; **Getty Images:** AFP / HOANG DINH NAM TR; LatinContent / Rodrigo Baleia TC; **Rough Guides:** Karen Trist BCR. **458/459 Getty Images:** Mlenny / E+. **459 Corbis:** Douglas Faulkner Cb; **NG:** Michael S Lewis CRa; **Corbis:** Sygma / Kapoor Baldev CR; **Science Photo Library:** Nature's Images BC. **460 Corbis:** ChromoSohm Inc / Joseph Sohm B; Saba / James Leynse CR; **Magnum:** Bruno Barbey CRa. **461 Corbis:** Adrian Arbib CLa; Brian Vikander BR; Eye Ubiquitous / Ben Spencer BL; The Purcell Team Ca; **DK:** Marwell Zoological Park / Frank Greenaway TL. **462 Corbis:** Maurizio Lanini Cb; Michael Freeman CLa; **DK:** Joe Cornish CRa; Max Alexander BL; **Getty:** Stone / Oliver Strewe BR. **463 Corbis:** Bob Krist CbL; James L Amos CaR; Richard Hamilton Smith CLa; Yann Arthus-Bertrand BR; **Visible Earth:** Provided by the SeaWiFS Project, NASA/Goddard Space Flight Center, and ORBIMAGE BL. **464/465 Corbis:** Tom Bean. **466 SPL:** BR. **466/467 Corbis:** A&J Verkaik T. **467 Corbis:** Richard Hamilton Smith Ca; **Dreamstime.com:** Nickwigram CR. **468 Corbis:** Paul A Souders TCR; Roger Tidman CLa; **Getty:** Image Bank / Pascal Perret TR; **Ferdinand Valk:** CRb; **Visible Earth:** Image courtesy Jacques Descloitres, MODIS Rapid Response Team, NASA/GSFC CRa. **469 Agence France Presse:** STR BR; **Corbis:** Stocktrek TCL; **NASA:** Johnson Space Center (STS51I-37-83) TR. **470 Corbis:** Sygma / Diaro Listen CLb; **NASA:** Goddard Space Flight Center (GPN-2000-001331) Cb. **471 Popperfoto:** Reuters / Andrew Winning. **470/471 Corbis:** Reuters / Lucas Jackson (c). **472 Getty:** Stone / David Olsen BL; **Getty Images / iStock:** Zoonar RF TR. **472/473 Getty:** Taxi / Ron Chapple B. **473 Getty:** Photodisc Green Cb; **Getty Images:** Michael Leach / Photodisc CR; **SPL:** Simon Fraser TCR; **Visible Earth:** Provided by the SeaWiFS Project, NASA/Goddard Space Flight Center, and ORBIMAGE TR; Provided by the SeaWiFS Project, NASA/Goddard Space Flight Center, and ORBIMAGE BR. **474 Corbis:** Galen Rowell B; Wolfgang Kaehler TC; **Magnum:** Steve McCurry CRa; **NG:** George Grall Ca; Medford Taylor C; Patricia Rasmussen: Cb; **SPL:** Astrid & Hanns-Frieder Michler CRb; National Center for Atmospheric Research CR. **475 Corbis:** Raymond Gehman CLa; The Purcell Team TCL; William A Bake Ca; **Royal Meteorological Society:** CLb; **Visible Earth:** NASA Goddard Space Flight Center Image by Reto Stöckli. Enhancements by Robert Simmon BL. **476/477 Getty Images:** Barcroft Media. **478 Corbis:** Charles O'Rear BR; Ted Spiegel CLa; Wolfgang Kaehler CRa; **DK:** Brian Cosgrove Cloud Collection Ca, TCR, BL; **Getty:** Stone / Donovan Reese CRb. **479 Corbis:** Scott T Smith TCRb; **Getty:** Image Bank / Alan R Moller CLa; **North American Weather Consultants Inc, Utah:** CR; **Still:** Francois Suchel B; Keith Kent TR. **480 Corbis:** Galen Rowell CLa; Gallo Images Cb; John McAnulty BCL; Richard Hamilton Smith CRb; **DK:** Brian Cosgrove Cloud Collection CRa, BR; **Getty:** Stone / Paul A Souders TCR. **481 Corbis:** Maurice Nimmo CRa; Royalty-Free CLa; Tom Bean Cb; **DK:** Brian Cosgrove Cloud Collection TR, BL, BR. **482 Corbis:** Bettmann TR. **482/483** Darrell Gulin B; **NG:** Priitt Esilind CL. **483 Corbis:** Ecoscene / Nick Hawkes TC; Kit Kittle TR; **Alamy Stock Photo:** Wallace Garrison Ca; **SPL:** Picture Researchers Inc BC. **484 AirPhoto:** Jim Wark CL; **Corbis:** Inge Ypseert BL; NASA CRb; **NASA:** Johnson Space Center (STS41B-41-2347) TC; Johnson Space Center (STS51G-46-5) CR; **NG:** Chris Johns BR; **SPL:** Peter Menzel CRa. **485 Corbis:** Jim Zuckerman B; W Perry Conway TCR; **Gene E. Moore:** CLa, CRa, CaL, CaR. **486/487 NASA. 488 Corbis:** James L Amos Cb; Tom Bean BL; Yann Arthus-Bertrand CLb. **490 alamy.com:** Peter Haigh C; **Corbis:** David Muench Cb, BC; **Getty:** Stone / James Balog Ca; **Hutchison Library:** Nigel Smith BL. **491 alamy.com:** Winston Fraser TCL; **Corbis:** David Muench Ca, CR; Galen Rowell CRa; James L Amos TR; **RHPL:** Roy Rainford TCb; **Landsat:** BCL; **Visible Earth:** Image courtesy Jacques Descloitres, MODIS Rapid Response Team, NASA/GSFC C. **492 Corbis:** Wolfgang Kaehler CL; **Hutchison Library:** T Moling Cb; **Magnum:** Burt Glinn CLb; **Visible Earth:** NASA / GSFC / MITI / ERSDAC / JAROS, and US / Japan ASTER Science Team BCR, C. **493 Corbis:** FLPA Ca; Larry Lee BCL; Richard List TCL; **Getty:** Stone / Glen Allison Cb; **Still:** Michel Roggo TCb; UNEP / S Rocha TR. **494 alamy.com:** B&C BL; **Corbis:** Michael S Yamashita BCR; Yann Arthus-Bertrand Cb; **Getty:** Image Bank / Weinberg / Clark C; **Hutchison Library:** Trevor Page BCLa; **Geo:** CLb; **Visible Earth:** MISR Instrument Team, MISR Project CL. **495 Landsat:** Ca; **Magnum:** Hiroji Kubota C, Cb; **Rex Features:** Richard Gardner TR; **Geo:** BCL; **Visible Earth:** Jeff Schmaltz, MODIS Rapid Response Team, NASA / GSFC TCL. **496 Nature Picture Library:** Bruce Davidson BRa; **Corbis:** Charles O'Rear BCR; Owen Franken Cb; Wolfgang Kaehler C; **Getty:** Taxi / Hans Christian Heap CbL; **SPL:** Dr Ken MacDonald BCL; **Visible Earth:** Image Courtesy Luca Pietranera Telespazio, Rome, Italy BCa. **497 Corbis:** FLPA / Winifred Wisniewski Cb; Jon Hicks TR; Stephen Frink TC; **Getty:** Image Bank / AEF BC; **NASA:** Goddard Space Flight Center Scientific Visualization Studio and USGS Ca; **Visible Earth:** NASA / JPL TCb. **498 Getty:** Image Bank / Theo Allofs Cb; Image Bank / Thomas Schmitt BCR; **Landsat:** C. **499 Corbis:** Ecoscene / Wayne Lawler TCb; Hubert Stadler CRb; O Alamany & E Vicens TR; Yann Arthus-Bertrand TCL; **Getty:** Image Bank / Guido Alberto Rossi C; Stone / Art Wolfe Cb; Stone / Kim Westerkov BC; **Dreamstime.com:** Nyker1 BR; **Visible Earth:** Image courtesy Jacques Descloitres, MODIS Rapid Response Team, NASA/GSFC BCa; Image courtesy NASA / GSFC / LaRC / JPL / MISR Team CRa. **500 Corbis:** James L Amos BCR; Michael S Yamashita C; Neil Rabinowitz BCa; **NG:** Emory Kristof Cb; **Visible Earth:** NASA JPL BL. **501 alamy.com:** David Noton Photography / David Noton C; **Corbis:** Douglas Peebles Ca; Eye Ubiquitous / David Batterbury CR; Jim Zuckerman BCL; Kevin Schafer TRb; **Visible Earth:** Image courtesy Jacques Descloitres, MODIS Rapid Response Team, NASA/GSFC TCL. **502 Corbis:** Anne Hawthorne BCR; Galen Rowell Cb; **Landsat:** BLa; **Visible Earth:** Image courtesy Jacques Descloitres, MODIS Rapid Response Team, NASA/GSFC CL, C; Image courtesy Greg Shirah, GSFC Scientific Visualization Studio, based on data from the TOMS science team BL. **503 Corbis:** Galen Rowell Ca; Kevin Schafer BC; Sygma TC; Wolfgang Kaehler TR; Yann Arthus-Bertrand C, Cb.

Additional illustrations Centerpiece artworks on the following pages by **Planetary Visions Ltd:** 34–35, 36–37, 40–41, 384–385, 444–445, 446–447, 450–451, 490–491, 492–493, 494–495, 496–497, 498–499, 500–501, 502–503. Population density maps in the Tectonic Atlas use data sourced from **Center for International Earth Science Information Network** – CIESIN – Columbia University. 2016. Gridded Population of the World, Version 4 (GPWv4): Population Density. Palisades, NY: NASA Socioeconomic Data and Applications Center (SEDAC).

Data for the Keeling Curve on p.459 from https://keelingcurve.ucsd.edu/

All other images © Dorling Kindersley

DEFINITIVE VISUAL GUIDES

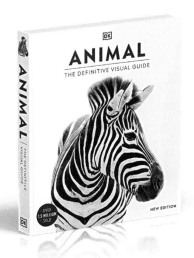

ANIMAL
THE DEFINITIVE VISUAL GUIDE

BIRD
THE DEFINITIVE VISUAL GUIDE

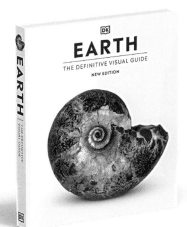

EARTH
THE DEFINITIVE VISUAL GUIDE

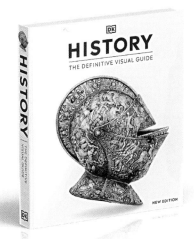

HISTORY
THE DEFINITIVE VISUAL GUIDE

OCEAN
THE DEFINITIVE VISUAL GUIDE

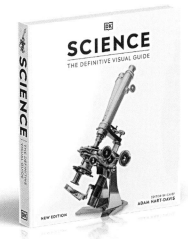

SCIENCE
THE DEFINITIVE VISUAL GUIDE

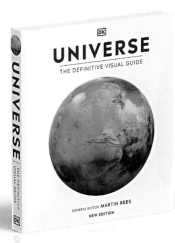

UNIVERSE
THE DEFINITIVE VISUAL GUIDE